Universitext

Universitext

Universitext is a series of textbooks that presents material from a wide variety of mathematical disciplines at master's level and beyond. The books, often well class-tested by their author, may have an informal, personal even experimental approach to their subject matter. Some of the most successful and established books in the series have evolved through several editions, always following the evolution of teaching curricula, to very polished texts.

Thus as research topics trickle down into graduate-level teaching, first textbooks written for new, cutting-edge courses may make their way into *Universitext.*

More information about this series at http://www.springer.com/series/223

Karl Hinderer • Ulrich Rieder • Michael Stieglitz

Dynamic Optimization

Deterministic and Stochastic Models

Springer

Karl Hinderer
Karlsruher Institut für Technologie (KIT)
Karlsruhe, Germany

Ulrich Rieder
University of Ulm
Ulm, Germany

Michael Stieglitz
Karlsruher Institut für Technologie (KIT)
Karlsruhe, Germany

ISSN 0172-5939 ISSN 2191-6675 (electronic)
Universitext
ISBN 978-3-319-48813-4 ISBN 978-3-319-48814-1 (eBook)
DOI 10.1007/978-3-319-48814-1

Library of Congress Control Number: 2016960732

Mathematics Subject Classification (2010): 90C39, 90C40, 90B10, 93E20, 60J20, 90-01

Printed on acid-free paper

This Springer imprint is published by Springer Nature
The registered company is Springer International Publishing AG
The registered company address is: Gewerbestrasse 11, 6330 Cham, Switzerland

Preface

Around 1980, Karl Hinderer started the book project *Dynamic Optimization*.

From then until 2003, he collected almost all scientific results in the area of dynamic programming, including applications. During those years, he presented several courses on this topic at the Karlsruher Institut für Technologie (KIT). Earlier versions of most parts of the book have also been tested by Dieter Kadelka, Alfred Müller, and Ulrich Rieder and particularly by Michael Stieglitz who worked intensively with Karl Hinderer in his later years. On April 17, 2010, Karl Hinderer died.

In the last four years, we have revised some versions of the book and put them together in the present volume. We believe that this textbook offers some advantages over other books. The most interesting aspects are the following:

- A detailed introduction of deterministic models.
- A firm theoretical basis without advanced measure theory.
- Precise modeling of a variety of applications in different areas.
- A large number of examples and exercises.

We would like to thank Wolfram Hinderer for his tireless help and for his excellent preparation of the TEX-manuscript.

We also thank anonymous referees for their valuable comments and suggestions.

Ulm and Sindelfingen, Ulrich Rieder
December 2016 Michael Stieglitz

Contents

List of Symbols for the Deterministic Models of Part I

A	action space, 17
a	action, 17
$A^N(s_0)$	set of N-stage action sequences admissible for s_0, 19
A_0	essential action space, 95
a_t	action at time t, 17
$ADP(s)$	absorbing DP depending on state s, 70
$b(\cdot)$	bounding function, 42
$B_b = B_b(S)$	S as the Banach space under norm $\|\cdot\|_b$, 43
$B_b = B_b(S)$	Banach space of functions on S with norm $\|\cdot\|_b$, 173
$\beta \in \mathbb{R}^+$	discount factor, 18
$c(s, a)$	one-stage cost for state s and action a, 35
$C_0(\cdot)$	terminal cost function, 35
$C_N(\cdot)$	N-stage cost function, 35
$C_{Ny}(s_0)$	N-stage objective function for action sequence y and initial state s_0, 35
CPA	set of convex and piecewise affine functions, 116
$D(s)$	set of admissible actions for state s, 17
D_a	a-section of D, 95
$D_n^*(s)$	set of optimal actions for stage n and state s, 24
DP_f	DP with $D(s) := D_f(s) := \{f(s)\}$, $s \in S, f \in \mathbb{F}$, 176
$DP_N(s_0)$	DP with horizon N and initial state s_0, 19
f	decision rule, 22
$\mathbb{F}$	set of all decision rules, 22
f_n	decision rule at stage n, 22
ID	set of functions with increasing differences, 129
$ILIP(\lambda)$	set of decision rules which are increasing and Lipschitz continuous with constant λ, 137
L	operator on $\mathfrak{V}_0$, 87
LIP	set of Lipschitz functions, 159
$MCP(i_0)$	shortest-path problem for non-sink i_0, 80
$N \in \mathbb{N}$	horizon, 17

$R(s) = R_1(s)$	set of reachable states in one step from state s, 37
$r(s, a)$	one-stage reward for state s and action a, 18
$r_f(s)$	one-stage reward for decision rule f and state s, 87
$R_n(s)$	set of reachable states in $n \geq 1$ steps starting from s, 37
$R_n(S_0)$	set of reachable states in $n \geq 1$ steps from a subset $S_0 \subset S$, 69
S	state space, 17
s	state, 17
s_0	initial state, 17
s_N	final state, 19
s_t	state at time t, 17
t	period, time point, 18
$T(s, a)$	transition function for state s and action a, 17
$T_f(s)$	transition function for decision rule f and state s, 87
U	optimal reward operator on $\mathfrak{V}_0$, 87
U_f	operator for the decision rule f on $\mathfrak{V}_0$, 87
$\mathfrak{V}_0$	set of real-valued functions on S, 87
$V(s)$	$\lim_{n\to\infty} V_n(s)$, 170
$V_0(\cdot)$	terminal reward function, 18
$V_N(\cdot)$	N-stage value function, 21
$V_N(s_0)$	maximal N-stage reward for initial state s_0, 20
$V_{1a}(\cdot)$, $V_{n,(a,x)}(\cdot)$	cf. (2.5), 21
$V_{N\pi}(s_0)$	cf. (3.9), 45
$V_{Ny}(s_0)$	cf. (2.3), 20
$W_n(s, a)$	cf. (2.6), 22
$y = (a_t)_0^{N-1}$	action sequence, 45

List of Symbols for the Stochastic Models of Part II

B_b^+	set of functions on S whose positive part belongs to B_b, 319
B_b $[MB_b]$	Banach space of [measurable] functions on S with finite b-norm, 303
$CM_N(s_0)$	control model with horizon N and initial state s_0, 191
$\mathbb{E}_{\pi s_0}$	expectation for policy π and state s_0, 202
$\eta^N := (\eta_t)_1^N$	vector of i.i.d. disturbances on Z^N, 191
f^∞	stationary policy $(f)_0^\infty$, 214
J_0	absorbing state space, 225
J	essential state space, 225
L	operator on $\mathfrak{V}_0$, 202
L_t	operators on $\mathfrak{V}_{0,t+1}$ for an MDP, 307
MDP_f	MDP with $D(s) := D_f(s) := \{f(s)\}$, $s \in S, f \in \mathbb{F}$, 212
MDP_∞	MDP with infinite horizon, 214
$MDP_N(p_0)$	MDP_N with initial distribution p_0, 228
$MDP_N(s_0)$	MDP with horizon N and initial state s_0, 202
MDP_N	$MDP_N(s_0)$ jointly for all s_0, 202
MDP'	uniformization of an MDPvar, 272
MRP'	Markov Renewal Program with infinite time horizon model, 385
OC_N	optimality criterion for horizon N, 299
$p_f(s, s')$	$p(s, f(s), s')$, 202
π	$(\pi_t)_0^\infty \in \Pi := \mathbb{F}^{\mathbb{N}_0}$, 214
$Pv(s, a)$	$\int P(s, a, s') \cdot v(s, a, s')$, $v : D \times S \to \overline{\mathbb{R}}$, $(s, a) \in D$, 303
$pv(s, a)$	$\sum_{s' \in S} p(s, a, s') \cdot v(s')$, 261
$r_f(s)$	one-stage reward for decision rule f and state s, 192
$r(s, a)$	one-stage reward for state s and action a, 202
$T_f(s, \eta_1)$	$T(s, f(s), \eta_1)$, 192
U	operator on $\mathfrak{V}_0$, 202

U_f	operator on $\mathfrak{V}_0$, 202
U_t	operators on $\mathfrak{V}_{0,t+1}$ for an MDP, 307
$\mathfrak{V}_{0t}$	set of measurable functions v on S_{t+1} for which $P_t v$ exists, 306
$\mathfrak{V}_0$	$\{v: S \to \overline{\mathbb{R}} : \mathbb{E}\, v \circ T(s, a, \eta_1) \in \overline{\mathbb{R}}\}$ for a CM, 193
$\mathfrak{V}_0$	$\{v: S \to \overline{\mathbb{R}} : \sum_{s'} p(s, a, s') \cdot v(s') \in \overline{\mathbb{R}}\}$ for an MDP, 202
$V(s_0)$	maximal expected N-stage reward for initial state s_0, 202
$V_{N\pi}(s_0)$	expected N-stage reward for policy π and initial state s_0, 202
V_{nf}	$V_{n\pi}$ for $\pi := (f)_0^{n-1}$, 211
$W_n(s, a)$	$r(s, a) + \beta \cdot \mathbb{E}\, V_{n-1}(T(s, a, \eta_1))$, 192
$\zeta_{t\pi}^{s_0}$	decision process up to time t generated by s_0 and π, 190

List of Symbols for Generalizations of Markovian Decision Processes of Part III

$(BMDPD, \hat{\mu}, \varphi)$	Bayesian model, 439
$BCM_N(\mu_0)$	N-stage Bayesian control model with prior μ_0, 392
$\beta(s, a)$	expected discount function, 356
$BMDPD_N(\mu_0)$	cf. Definition 25.1.3, 438
$CVPL$	set of convex and piecewise linear functions, 471
Δ_N	set of history-dependent policies, 459
Δ_{NZ}	set of disturbance-dependent policies, 391
$Kv(s, a)$	$\int K(s, a, d(z, s')) \cdot v(s, a, z, s')$, 358
$K_f(s, d(z, s'))$	$K(s, f(s), d(z, s'))$, 356
L	operator on $\mathfrak{V}_0$, 362
MRP_∞	Markov Renewal Program with infinite horizon, 375
Φ	Bayes operator, 393
φ	sufficient statistic, 394
$(POM, \hat{\mu})$	POM with transition probability $\hat{\mu}$, 461
$POM_N(i)$	$POM_N(\hat{\mu}(i))$, $i \in I$, 461
$POM_N(\mu_0)$	POM with prior μ_0, 460
$(POMCM, \hat{\mu}, \varphi)$	POM with sufficient statistic φ for prior $\hat{\mu}$, 467
$\mathbb{Q}_N^\mu$	cf. (23.3), 391
$*$-σ-algebra	Definition 23.1.5, 395
U	operator on $\mathfrak{V}_0$, 361
$\mathfrak{V}_0$	$\{v : S \to \overline{\mathbb{R}} : \int K(s, a, d(z, s')) \cdot \tilde{\beta}(z) \cdot v(s') \in \overline{\mathbb{R}}\}$, 360
V_N	N-stage value function, 360
$v_N(\mu, s)$	maximal N-stage Bayes reward for prior μ and initial state s, 392
$v_N(i, s)$	expected N-stage Bayes reward for index i and state s, 398
$v_{N\pi}(i, s)$	expected N-stage Bayes reward for policy π, index i and initial state s, 398
$v_{N\pi}(\mu, s)$	expected N-stage Bayes reward for policy π, prior μ and initial state s, 391

$V_{N\pi}(s_0)$	expected N-stage reward for policy π and initial state s_0, 360
MDP′	adjoined process to an MDPD, 362
MRP^{t_0}	cf. Definition 22.3.1, 385

List of Special Symbols

$\mathrm{Be}(\alpha_1, \alpha_2)$	Beta distribution
$\mathrm{Bi}(n, p)$	Binomial distribution
♦	end of example
χ_n^2	Chi-squared distribution with n degrees of freedom
$\diamond$	end of remark
$\mathrm{Exp}(\alpha)$	Exponential distribution
$\Gamma^-(\alpha, b)$	Inverse Gamma distribution
$\Gamma_{\alpha,b}$	Gamma distribution
$\mathrm{Geo}(p)$	Geometric distribution
$\mathrm{NBi}(n, \vartheta)$	Negative Binomial distribution
$\mathrm{Poi}(\alpha)$	Poisson distribution
$\square$	qed
$N(m, \sigma^2)$	Normal distribution
$U(B)$	Uniform distribution on $B \in \mathfrak{B}_n$
$p * q$	convolution of the probability measures p and q, 335
∂M	boundary of a set M
$\mathbb{E}$	expectation
$\preceq_M$	relation on a set M, 89
$\mathbb{N}$	set of positive integers
$\mathbb{N}_0$	$\{0\} + \mathbb{N}$
$\mathbb{N}_k$	$\{1, 2, \ldots, k\}$, $k \in \mathbb{N}$
$\|\cdot\|$	maximum norm, 211
$\mathfrak{P}(A)$	power set of A
$\propto$	cf. (23.1), 404
$\overline{\mathbb{R}}$	$[-\infty, \infty]$
$\mathbb{R}$	set of reals
$\mathbb{R}^+$	$(0, \infty)$
$\mathbb{R}_+$	$[0, \infty)$
$\sigma_n(x)$	cf. (2.14), 31

$\mathcal{T}$	system of open subsets of a metric space, 498
$\vee$	or
$\wedge$	and
$\lfloor x \rfloor$	largest integer $\leq x \in \mathbb{R}$
$\mathbb{Z}$	set of integers

List of Abbreviations

ADP	absorbing DP, 70
BCM	Bayesian control model, 390
BMCM	Bayesian Markovian control model, 443
BMDP	Bayesian Markovian decision process, 443
BMDPD	Bayesian Markovian decision process with disturbances, 436
CM	control model, 300
CM-RE	CM with random environment, 309
DP	deterministic program, 18
DP_∞	DP with infinite horizon, 179
(DR)	drive assumption, 375
e.g.	example given, 52
(EN)	essentially negative case, 263
(EP)	essentially positive case, 263
i.i.d.	independent and identically distributed, 63
lsc	lower semicontinuous, 149
(LUBF)	lower or upper bounding function, 360
$(MA1)_\infty$	first minimal assumption in an $MDPD_\infty$, 364
(MA1)	first minimal assumption, 263
(MA2)	second minimal assumption, 264
MCM	Markovian control model, 358
MDP	Markovian decision process, 200
MDP-RE	MDP with random environment, 307
MDPad	adjoined MDP, 210
MDPD	Markovian Decision Process with disturbances, 355
$MDPD_\infty$	Markovian Decision Process with disturbances and infinite horizon, 364
MDPvar	MDP with variable discount factor, 262
NE	north-east, 128
OC	optimality criterion, 22

OE	optimality equation, 74
POCM	partially observable control model, 457
POM	partially observable model, 456
POMCM	partially observable Markovian control model, 457
POMDP	partially observable Markovian decision process, 457
RCM	Renewal Control Model, 374
RI	reward iteration, 21
(RI)	reward iteration, 361
RMDP	Renewal Markovian Decision Process, 373
usc	upper semicontinuous, 149
[Var]	indicates that the result also holds for MDPvar, 263
VI	value iteration, 22
VI_N	the value iteration up to stage N, 265
VIF	value iteration with finite V_n and W_n, 341

Chapter 1
Introduction and Organization of the Book

In this treatise we deal with optimization problems whose objective functions show a sequential structure and hence are amenable to sequential methods. The corresponding field mostly goes by the name *Discrete-time Dynamic Programming*. Other names are *Discrete-time Stochastic Control* and *Multi-stage Optimization*. In order to avoid possible confusion with programming in computer science we speak of *Dynamic Optimization*.

We now start with a rough description of Part I *Deterministic Models* (DPs for short). In this part we discuss the control of a general discrete-time *deterministic* system which has *horizon* $N \geq 1$, i.e. operates for N time units. Many applications in Dynamic Optimization arise just in this way.

A certain system starts in an *initial state* s_0 and visits states s_1, s_2, ..., s_N at times $1 \leq t \leq N$ under the control of a sequence of *actions* $y = (a_0, a_1, \ldots, a_{N-1})$, respectively. Neither the states nor the actions need be real. More precisely, the sequence $(s_t)_1^N$ is determined as $s_{t+1} = T(s_t, a_t)$, $0 \leq t \leq N-1$, for a given *transition function* T. The states and actions belong to a *state space* S and an *action space* A, respectively. The choice of actions is subject to a restriction determined by the momentary state. The *objective function* $y \mapsto V_{Ny}(s_0)$ is in general the sum of discounted rewards earned in the N *periods*. It includes a *terminal reward* $V_0(s_N)$ when the evolution of the system ends at time N in state s_N. We call the set of all data (except for N and s_0) just introduced, a *deterministic model DP*, which is the topic of this part. Such a DP, the number N of stages and the initial state s_0 determine the following *optimization problem* $DP_N(s_0)$:

(i) find the supremum $V_N(s_0)$ of the objective function $y \mapsto V_{Ny}(s_0)$,
(ii) find, if possible, an s_0*-optimal action sequence* y^*, i.e. a maximum point of $y \mapsto V_{Ny}(s_0)$.

Then $V_N(s_0)$ and y^* constitute a *solution* of problem $DP_N(s_0)$.

K. Hinderer et al., *Dynamic Optimization*, Universitext,
DOI 10.1007/978-3-319-48814-1_1

In Part II *Markovian decision processes* (MDPs for short) we study stochastic models. These models also go by the names *Stochastic Dynamic Programs* and *Discrete-time Stochastic Control Models.*

In an MDP with horizon N the control of the system starting in s_0 cannot be exercised as in a DP, since now the states visited are random variables ζ_t, $1 \leq t \leq N$, whose realizations s_t are not known in advance with certainty. Thus one must decide in advance for each state $s_t \in S$ which action $\pi_t(s_t)$ to take, $0 \leq t \leq N-1$. The mapping π_t is called a *decision rule* at time t, and the sequence $\pi = (\pi_0, \pi_1, \ldots, \pi_{N-1})$ is called an *N-stage policy*. The evolution of the random states is no longer given by a transition function but by a *random transition law*, usually of a Markovian nature. DPs turn out to be special cases of MDPs.

An N-stage policy π, an initial state s_0 and a transition law jointly determine the random evolution of the system, the random variable *total reward* and its expectation $V_{N\pi}(s_0)$. Usually the latter is the expected sum of discounted rewards. A model MDP, the number N of stages and the initial state s_0 determine the following *optimization problem* $MDP_N(s_0)$:

(i) find the supremum $V_N(s_0)$ of the objective function $\pi \mapsto V_{N\pi}(s_0)$,
(ii) find, if possible, an s_0-optimal policy π^*, i.e. a maximum point of $\pi \mapsto V_{N\pi}(s_0)$. (Usually there exist policies which are s_0-optimal for *all* states $s_0 \in S$.)

Then $V_N(s_0)$ and π^* constitute a *solution* of problem $MDP_N(s_0)$.

The solution of a $DP_N(s_0)$ and similarly of an $MDP_N(s_0)$ usually proceeds in two steps:

Step I: For its mathematical content we use the term *Basic Theorem*, which avoids the controversial term *principle of optimality* (see Supplement 3.6.1). The Basic Theorem consists of two parts as follows:

(a) To find $V_N(s_0)$ one derives a recursion for computing successively the so-called n-stage *value function* $s \mapsto V_n(s) :=$ *supremum of the objective function of problem* $DP_n(s)$ from V_{n-1}, $1 \leq n \leq N$, $s \in S$. This recursion, which starts with the terminal reward function V_0, is called the *value iteration* (VI for short). In general, the VI is a difference equation of a rather complicated nature: it is a partial difference equation except for the trivial case where $|S| = 1$, and it is non-linear because it includes a maximization with respect to actions. The derivation of the VI is the same for problems with any kind of variables, whether real, vector, set or otherwise.
(b) In many examples one can find in principle from V_{n-1}, $1 \leq n \leq N$, a decision rule f_n, which we call a *maximizer at stage n*, such that the following *Optimality Criterion* (OC for short) holds: for each $N \geq 1$ and $s_0 \in S$ the policy $(f_N, f_{N-1}, \ldots, f_1)$ is s_0-optimal for problem $DP_N(s_0)$, and it determines in a simple way an s_0-optimal action sequence for problem $DP_N(s_0)$. Consequently, in DPs we rarely deal explicitly with s_0-optimal action sequences, but rather with maximizers.

Step II: The one-stage problems must be solved somehow, for which no help from Dynamic Optimization theory can be expected, but other methods from optimization theory must be brought in.

Fortunately in many examples the action variable of the one-stage problems varies in an interval or in a discrete interval. For both cases many methods are known. Nevertheless, only rarely does the VI have an explicit solution so that the OC leads to an explicit s_0-optimal action sequence, but the Basic Theorem serves the following important purposes:

(i) The value functions V_n can be found numerically if S and A are finite and *not too large*. If both spaces are large, we have to deal with the *curse of dimensionality*.
(ii) Sometimes the VI shows that V_n has the form $d_n v$, $n \geq 0$, for a known function v on S and numbers d_n which satisfy a relatively simple recursion, solvable at least numerically.
(iii) From the VI one often can derive useful structural properties both of the value functions and of the maximizers f_n at stage n. Examples are the monotone or concave dependence of $V_N(s_0)$ on the initial state s_0 and/or N and monotone dependence of $f_n(s)$ on s and/or n.

Although of mathematical interest in themselves, DPs and MDPs alone would not have brought about such a wealth of research since their origin in about 1960. Responsible is the abundance of more or less realistic applications in such fields as: agriculture; allocation of resources such as water and lumber; cash balancing; computer science and communication theory; engineering; extraction of geological resources; gambling; harvesting; insurance; inventory control; medicine; growth problems in economic theory; financial theory; queues; reliability; revenue problems such as air plane management; risk theory; search; statistics, stopping; and even some applications within mathematics.

The application of Dynamic Optimization to real situations encounters several difficulties: practitioners may not be aware of the potentials of the theory; real situations mostly require rather complicated models; often there is a lack of data (in particular for MDPs) and not seldom a considerable computational burden. However, the following facts should be stressed:

(i) In cases where no numerical results can be found, the precise modeling, possibly combined with structural properties of the solution, may be of considerable utility for understanding the problem.
(ii) The sensitivity of the solution on the data can often be tested numerically even on a PC, using small-scaled and simplified models.

These arguments should be taken into account when making the criticism that—compared e.g. with Linear Programming—only in a moderate number of real situations has Dynamic Optimization found an implementation.

MDPs and ideas for their solution emerged in the fields of statistical sequential analysis (Wald 1947) and of inventory theory Dvoretzky et al. (1952b). To Richard Bellman (1920–1984) we owe the creation of a unifying approach for a host of different problems. Since the appearance of his pioneering book *Dynamic Programming* (cf. Bellman 1957) the subject has experienced a vigorous development, both in theory and in applications, from subtle measure-theoretic and topological problems to numerical challenges for complex models. Accordingly there exist many excellent texts for different purposes and for a variety of readerships. The present book aims at both the applied mathematician and quantitatively oriented non-mathematicians. Scientists from Operations Research, Mathematics, Engineering Science, Biology, Agriculture, Computer Science and other areas have contributed to the progress. Here we mention the big influence of the book by Howard (1960) on finite-state and finite-action MDPs and of the paper by Blackwell (1965) about MDPs with very general state and action spaces. More information about the development of the field—from subtle measure-theoretic and topological aspects to numerical problems for complex applications—is given in the supplements of the sections of the present treatise. For several areas there exists so much material that special monographs have been written. This applies in particular to inventory problems, optimal stopping, statistical sequential problems and so-called bandit problems.

In Part III *Generalizations of Markovian Decision Processes* we consider important ramifications of MDPs, e.g. Bayesian models.

The goals of our treatise are:

(i) To present a firm theoretical basis without advanced measure theory.
(ii) To exhibit in a meticulous manner a large number of examples, not too involved, but hopefully still interesting and representative for their area. A list of the most important examples will be provided at the end of the book.
(iii) To exhibit the mathematical structures underlying the theory.
(iv) To provide a basis for future research.

Behind these goals lies the endeavor to build a bridge between the more applied sciences and those more theory oriented parts of mathematics which are relevant for our topic. We made efforts to provide complete and simple proofs; often they are succinct and the reader should always have paper and pencil within reach. For most theoretical results we give an example. The book should not only serve as a graduate textbook, but should also be suitable for private study. It is rather self-contained by the inclusion of four appendices (see the table of Contents). The book contains a substantial number of results which seem to be new, e.g. Proposition 4.1.4 and Theorem 5.1.7.

The book centers around the following three aspects:

(i) Students often have difficulties in finding the correct *model* for a verbally given application. Therefore we pay considerable attention to a careful introduction of a variety of models. For didactic reasons and because of lack of space we mostly restrict ourselves to simplified models. Much more involved and realistic models can be found in the literature.
(ii) We give due attention to the main (analytic and numerical) *solution methods* and testify their usefulness by many examples. As far as possible, results are stated constructively as algorithms suitable for computer programs. Of course,

numerical methods are very important for dealing with the proverbial *curse of dimensionality* in Dynamic Optimization.

(iii) The third main theme of our treatise is the study of the *structure of solutions*. This area is particularly fascinating with respect to maximizers; their structure can sometimes be conjectured from the problem. On the one hand we deal with properties of the value functions like monotonicity in a general sense, convexity, concavity, upper [lower] semicontinuity, Lipschitz-continuity and computability by recursion in the state space. On the other hand we study maximizers which have a structural property such as being monotone, of bang-bang type, of control-limit type or of being increasing and at the same time Lipschitz-continuous.

A subtitle of this book could be *Theory for Applications*. In fact, despite the stress we lay on the mathematical structure of the optimization problems, only a few results are included for their mathematical beauty alone. Therefore the present book differs very much from the first author's early foundational treatise Hinderer (1970).

In order to get quick access to stochastic models many textbooks either contain no deterministic models or only a short introduction to them. We choose another approach by starting with a detailed presentation of deterministic models, i.e. of DPs, for the following reasons:

(i) There are many interesting applications, e.g. in computer science, which do not require the more involved probabilistic setting. In particular, one need not care about measurability questions and the validity of the VI since for DPs the latter always holds.

(ii) Already deterministic models require a lot of concepts unfamiliar to most readers. In addition, the separation of a large portion of optimization methods from probabilistic modeling can smoothen the path to the stochastic problems.

(iii) Our definition of a DP, which is in the spirit of general optimization theory, is not based on the concept of policies (which becomes indispensable for stochastic models) but uses action sequences. Therefore, readers only interested in one of the many deterministic problems need not dip into the more demanding stochastic theory.

(iv) Many results for DPs serve as a basis for corresponding results for MDPs.

(v) In a course devoted to stochastic models only, the lecturer should not have serious difficulties in inserting auxiliary results from Part I whenever needed.

Our exposition of examples puts some emphasis on *discrete* problems (where the state space S and the action space A are finite or at least countable) since in general only these are accessible to a numerical computation. If one or both of the two spaces are uncountable, one should firstly consider whether or not a realistic discrete version exists; if not, a discretization of S and/or A is required for computation.

From an empirical point of view all real problems have a *horizon* N which is finite, though often large and/or not exactly known. From the historical beginning of the subject one was dealing with the *Approximation of the solution of problem*

$MDP_N(s_0)$ (and similarly for problems $DP_N(s_0)$) for large N. There are available the following two logical completely different approaches:

Approximation approach A: *Approximation by an infinite-stage model* MDP_∞. Such a problem has an expected infinite-stage reward $V_{\infty\pi}(s_0)$ for each infinite-stage policy π. Typically there exists a *stationary* optimal policy $\pi = (g, g, \ldots)$ where g is some decision rule determined by $V_\infty := \sup_\pi V_{\infty\pi}$. One hopes that for large N the solution of $MDP_\infty(s_0)$ is a *good* approximation to the solution of $MDP_N(s_0)$, i.e. that $V_\infty(s_0)$ is *near* to $V_N(s_0)$ and that the N-stage stationary policy $\pi_N = (g, g, \ldots, g)$ is in some sense *good* for $MDP_N(s_0)$.

Approximation approach B: *Approximation by the limit of the solution of problem* $MDP_N(s_0)$ *for* $N \to \infty$. If the pointwise limit $V(s_0)$ of $V_N(s_0)$ for $N \to \infty$ exists as finite number, it is an approximation of $V_N(s_0)$ by definition. Moreover, in general a *good* stationary N-stage policy for $MDP_N(s_0)$ is given by $\pi_N = (f, f, \ldots, f)$, where f, called an *asymptotically optimal* decision rule, is obtained in the same way from V as g is obtained from V_∞ in the approximation approach A.

In the early days of the theory the two approaches were sometimes mixed up, probably because in many (but not all) cases, the number $V_\infty(s_0)$ coincides with $V(s_0)$.

In contrast to most textbooks we decidedly prefer the second approach for the following reasons:

(i) In cases where $V_\infty(s_0)$ differs from $V(s_0)$ the approximation approach A cannot be a realistic model, since $V(s_0)$ is by definition the unique approximation.
(ii) In the stochastic case it is a natural requirement for a model that the system behavior should be reproducible by simulation, a requirement which does not hold for the approximation approach A since an infinite number of trials never can be realized in the strict sense.

Another point where we deviate in the stochastic case from most parts of the literature is the *treatment of the value iteration.* Its validity—which looks so obvious to the applied scientist—poses serious problems for stochastic models if, for example, both state space and action space are uncountable. In most textbooks one either proves the VI only for countable S or one uses the rather sophisticated machinery of Borel spaces. Here we utilize another idea, which goes back to Porteus (1971/1972), Dynkin and Yushkevich (1979) and Bertsekas (1975), but whose usefulness for applications seems to have not been realized for a long time. We carry through this program—maybe for the first time in the textbook literature—consistently and in detail under the name *Structure Theorem* since its application is usually connected with a structural property of the value functions. (This approach has been used by the author since about 1982 in his courses, but published only in Hinderer and Stieglitz 1996).

We now turn to a survey of the contents of Part I (DPs) and of Part II (MDPs).

Part I opens with a leisurely introduction of DPs, centering around the Basic Theorem. There follows a large number of examples, some of these even with an explicit solution, such as in the classical one-dimensional linear-quadratic system. Then we turn to DP networks. In Chap. 6 a first structural property is the monotonicity of value functions. Here we work on a very broad basis by using arbitrary relations rather than orderings on the state space. Moreover we explain a general guideline for obtaining structural results about value functions. Chapter 7 deals with concavity and convexity of the value functions, including the case where S and A are discrete intervals. Monotonicity of maximizers, a sometimes computationally attractive property, is studied in Chap. 8 where the action space is totally ordered. One also finds conditions under which there even exist maximizers which are both monotone and Lipschitz. This can result in a further acceleration of the value iteration. In Chap. 10 we treat the approximation of the solution of $DP_N(s_0)$ by its limit.

The design of Part II is guided by our intention to proceed step by step from models with modest mathematical prerequisites to more complicated ones and to treat interesting examples as soon as the necessary tools are available. In particular we defer measure theoretical probability as far as possible in order not to obscure the understanding of the optimization aspect.

We start with the simplest MDP, the *Control Model* (CM for short), having i.i.d. disturbances and a finite set Z of disturbances. Already here the proof of the VI requires considerably more work than in the case of DPs. The reason for the validity of the VI, i.e. of the possibility to compute $C_N(s)$ sequentially, is the fact that not only the objective function but also the joint probability distribution of the random states has a sequential structure, namely that of a Markov chain. From Chap. 12 on we turn to the more general models of MDPs with a finite state space or, more generally, with finite transition law. In each of these models finite probability spaces suffice since here we only consider finite horizon models. (For infinite horizon models, however, measure theory is unavoidable.) A number of numerically solvable examples are dealt with. Then in Chap. 14 we study the asymptotic behavior for $N \to \infty$ of the solution of $MDP_N(s_0)$ when the discount factor equals one. In particular we present the basics for models with the average reward criterion. The next step leads in Chap. 15 to MDPs with discrete transition law. Then examples treatable within this framework are presented. From Chap. 17 onwards we use basic measure theoretic probability; no topology is required. Then the problem of the existence of maximizers within the metric space setting follows. Here we encounter the so-called *measurable selection problem.* We present a few elementary results which suffice for some examples. Finally, we treat in Chap. 20 the approximation of the solution of $MDP_N(s_0)$ with *large N* by an infinite-stage model and also by the limit of the solution. In the latter method the models having a so-called bounding function prevail.

For topics treated in Part III see the table of Contents. This part is devoted to an important generalization of MDPs which contains as special cases Markov renewal programs, Bayesian models and models with partial observations. The treatment of Bayesian models is markedly influenced by lecture notes of U. Rieder.

In Table 1.1 on page 8, 13 stationary models and their mutual relations are given. Moreover, for each of these models we point out in Table 1.2 on page 9 the dependency of the data and the transitions on the relevant variables.

Table 1.1 Dependencies of model data and their transition functions on variables

Model	Data / Transition	Chapter
DP	$(S, A, D, T(s,a), r(s,a), V_0(s), \beta)$	Chap. 2, p. 18
	$(s,a) \overset{\text{det.}}{\mapsto} s'$	
CM	$(S, A, D, Z, Q(dz), T(s,a,z), r_Z(s,a,z), V_0(s), \beta)$	Chap. 16, p. 300
	prob. z, $(s,a,z) \overset{\text{det.}}{\mapsto} s'$	
MDP	$(S, A, D, P(s,a,ds'), r_S(s,a,s'), V_0(s), \beta)$	Chap. 16, p. 293
	$(s,a) \overset{\text{prob.}}{\mapsto} s'$	
MCM	$(S, A, D, Z, Q(s,a,dz), T(s,a,z), r_Z(s,a,z), V_0(s), \beta)$	Chap. 21, p. 358
	$(s,a) \overset{\text{prob.}}{\mapsto} z$, $(s,a,z) \overset{\text{det.}}{\mapsto} s'$	
MDPD	$(S, A, D, Z, K(s,a,d(z,s')), \tilde{r}(s,a,z,s'), V_0(s), \tilde{\beta}(z))$	Chap. 21, p. 355
	$(s,a) \overset{\text{prob.}}{\mapsto} (z,s')$	
BCM	$(\Theta, S, A, D, Z, Q(\vartheta,dz), T(s,a,z), r_Z(\vartheta,s,a,z), V_0(\vartheta,s), \beta)$	
	$\vartheta \overset{\text{prob.}}{\mapsto} z$, $(s,a,z) \overset{\text{det.}}{\mapsto} s'$	Chap. 23, p. 390
BMCM	$(\Theta, S, A, D, Z, Q(\vartheta,s,a,dz), T(s,a,z), r_Z(\vartheta,s,a,z), V_0(\vartheta,s), \beta)$	
	$(\vartheta,s,a) \overset{\text{prob.}}{\mapsto} z$, $(s,a,z) \overset{\text{det.}}{\mapsto} s'$	Chap. 25, p. 443
BMDP	$(\Theta, S, A, D, P(\vartheta,s,a,ds'), r_S(\vartheta,s,a,s'), V_0(\vartheta,s), \beta)$	
	$(\vartheta,s,a) \overset{\text{prob.}}{\mapsto} s'$	Chap. 25, p. 443
BMDPD	$(\Theta, S, A, D, Z, Q(\vartheta,s,a,dz), K_S(s,a,z,ds'), \tilde{r}(\vartheta,s,a,z,s'), V_0(\vartheta,s), \tilde{\beta}(z))$	
	$(\vartheta,s,a) \overset{\text{prob.}}{\mapsto} z$, $(s,a,z) \overset{\text{prob.}}{\mapsto} s'$	Chap. 25, p. 436
POM	$(\Theta \times S, A, D_S, Z, P(\vartheta,s,a,d(\vartheta',z)), K_S(s,a,z,ds'), \tilde{r}(\vartheta,s,a,z,\vartheta',s'), V_0(\vartheta,s), \tilde{\beta}(z))$	
	$(\vartheta,s,a) \overset{\text{prob.}}{\mapsto} (\vartheta',z)$, $(s,a,z) \overset{\text{prob.}}{\mapsto} s'$	Chap. 26, p. 456
POMCM	$(\Theta \times S, A, D_S, Z, P(\vartheta,s,a,d(\vartheta',z)), T(s,a,z), r_{\vartheta\times Z}(\vartheta,s,a,\vartheta',z), V_0(\vartheta,s), \tilde{\beta}(z))$	
	$(\vartheta,s,a) \overset{\text{prob.}}{\mapsto} (\vartheta',z)$, $(s,a,z) \overset{\text{det.}}{\mapsto} s'$	Chap. 26, p. 457
POCM	$(\Theta \times S, A, D_S, P(\vartheta,d(\vartheta',z)), T(s,a,z), r_{\Theta\times Z}(\vartheta,s,a,\vartheta',z), V_0(\vartheta,s), \tilde{\beta}(z))$	
	$\vartheta \overset{\text{prob.}}{\mapsto} (\vartheta',z)$, $(s,a,z) \overset{\text{det.}}{\mapsto} s'$	Chap. 26, p. 457
POMDP	$(\Theta \times S, A, D_S, P(\vartheta,s,a,d(\vartheta',s')), r_{\Theta\times S}(\vartheta,s,a,\vartheta',s'), V_0(\vartheta,s), \tilde{\beta}(z))$	
	$(\vartheta,s,a) \overset{\text{prob.}}{\mapsto} (\vartheta',s')$	Chap. 26

Almost all of the chapters end with a set of problems and with a supplement. The problem sections contain (i) exercises for the modeling and analysis of applications, (ii) examples suitable for numerical and graphical experiments with a PC, (iii) complements to results in the main text and (iv) exercises for a better understanding of concepts. Needless to say that a firm understanding of dynamic optimization cannot be gained without solving a substantial number of problems.

The supplements are primarily aimed at the advanced reader. They contain additional comments on related work and bibliographic notes.

Table 1.2 Relations between models

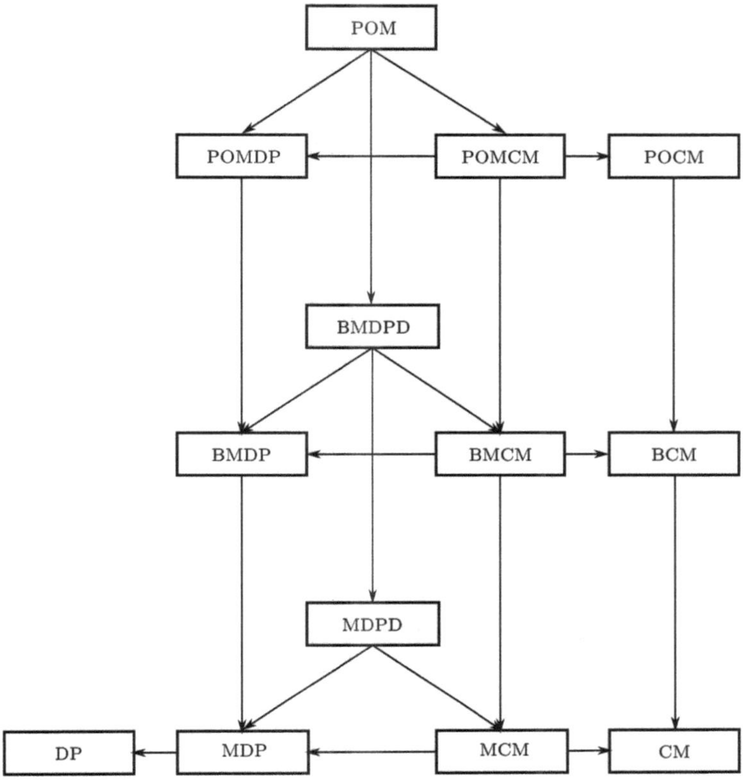

Our method of citation is as follows: Formula n of Chapter m is cited as (m.n). Formula n of Appendix X is cited as Appendix (X.n).

For the difficult task of a balanced selection out of the large number of publications we used the following principles. Firstly, we tried to give due credit for the main results presented, although it is sometimes hard to trace them back to their origin. Secondly, we included most review papers known to us. Thirdly, when citing literature for some specific topic, we refer to papers published until 2003 which initiated the topic or set new directions of research in it, and finally to some of the latest papers, from which the interested reader can trace back to earlier papers.

The book should be useful for a variety of courses. Each one should start with Chaps. 2, 3 and some examples from Chap. 4.

Course 1, a one-semester introduction, requiring only elementary mathematics, in particular no measure theory, could consist of Chaps. 5, 9, 11 and 12, some examples from Chap. 13, then Chap. 16 with some examples from Chap. 17.

Course 2, a two-semester graduate course, partly requiring basic measure theory, could comprise course 1 and in addition parts of Chaps. 6, 8, 18, 20, 21 and 22.

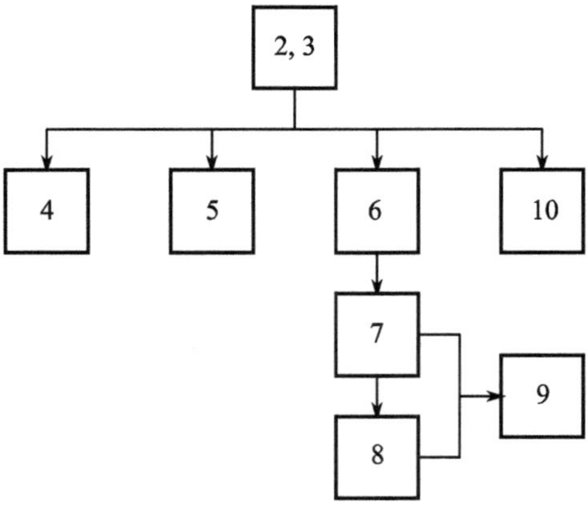

Fig. 1.1 Dependencies of the chapters of Part I

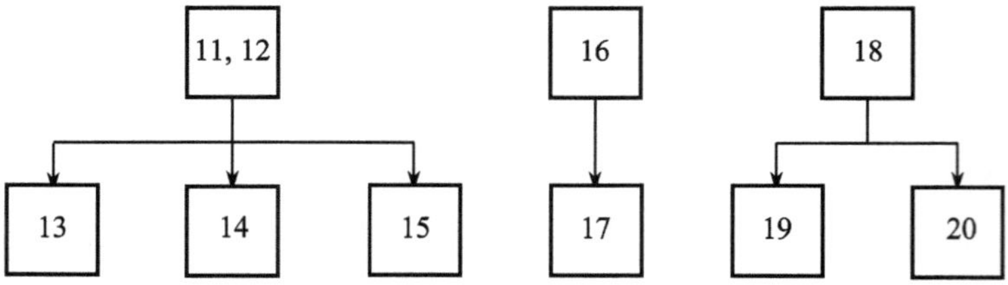

Fig. 1.2 Dependencies of the chapters of Part II

Course 3, a two-semester advanced course, could start after Chaps. 2, 3 and some examples from Chap. 4 directly with Chap. 18, treating the main results from Chaps. 11, 12, 13 and 16 as special cases, and then could proceed to Chaps. 19, 20 and Chaps. 21, 22, then Chap. 23 and parts of Chaps. 24, 25 and 26.

The dependencies of the chapters of Parts I, II and III, respectively, are shown in following figures (Figs. 1.1, 1.2 and 1.3).

In Appendices A, B, C and D we provide standard facts from optimization, convexity, metric spaces, measure and probability in order to keep the treatise self-contained.

There are excellent books which contain further mathematically interesting topics and more information on subjects we treated briefly, e.g. the average reward problems, control of queueing systems, bandit problems and linear-quadratic systems. References to such literature are given in the supplements. Here we mention but a few of them, with their main topics in parentheses: Bertsekas (2005, 2001) (theory and applications), Bertsekas and Shreve (1979) (foundations), Schäl (1990) (theory), Hernández-Lerma and Lasserre (1996, 1999) (theory), Puterman (1994) (theory, applications and numerical methods), Sennott (1999) (control of queueing systems), Altman (1999) (constrained MDPs), and Bäuerle and Rieder (2011) (MDPs with applications to finance).

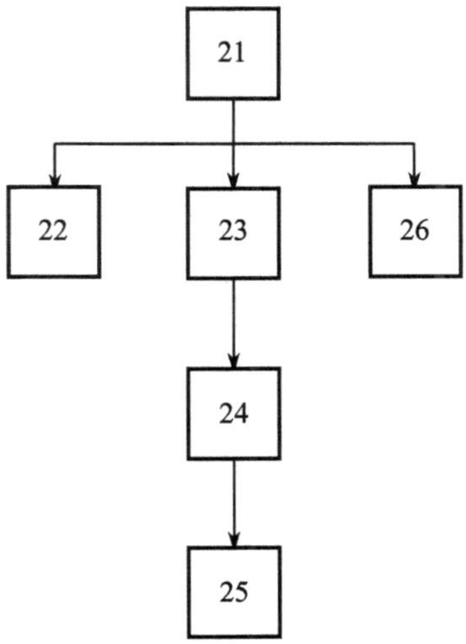

Fig. 1.3 Dependencies of the chapters of Part III

Numerical methods for breaking the *curse of dimensionaltity* are presented in Powell (2007) and Chang et al. (2007).

The topics of adaptive problems, gambling houses (see Maitra and Sudderth 1996), and of Markov games (see Filar and Vrieze 1997) are omitted.

Part I
Deterministic Models

Chapter 2
The Stationary Deterministic Model and the Basic Solution Procedure

We introduce as a prototype of deterministic dynamic optimization problems a simple allocation problem, give firstly an intuitive and then a formal description of the general problem, and derive the basic solution technique: value iteration and optimality criterion. This allows us to derive structural properties of the solution of the allocation problem.

2.1 A Motivating Example

Example 2.1.1 (Discrete allocation problem) Consider the process of allocating to a single project some parts a_0, a_1, a_2 and a_3 of a resource (such as units of material) of total amount $K := 10$ sequentially at the times $t = 0, 1, 2, 3$. (A simultaneous single allocation to four different projects leads to the same mathematical problem.) The allocation $a_t \in A := \mathbb{N}_{0,10}$ at time t is often called the *consumption.* Obviously the allocations must obey the restrictions $a_0 \leq K$ and $a_t \leq K - \sum_{i=0}^{t-1} a_i$ for $1 \leq t \leq 3$. We assume that a_t yields a reward $u(a_t)$ for some function u on A. The resource still available at time $1 \leq t \leq 4$ is $s_t := 10 - \sum_{i=0}^{t-1} a_i$, and the resource $s_t - a_t$ *not consumed at time t* is often called the investment. (Therefore allocation models often run under the heading ***consumption and investment.***) We denote the sum of allocations in the sequence $y = (a_0, a_1, a_2, a_3)$ by $c(y) := \sum_{t=0}^{3} a_t$. Then $10 - c(y)$ is the terminal resource left over at time 4. We assume that the terminal resource s_4 at the end of the allocation process yields the terminal reward $d \cdot u(s_4)$ for some constant $d \in \mathbb{R}_+$. Thus the case where the terminal resource is worthless can be modeled by the choice $d = 0$.

K. Hinderer et al., *Dynamic Optimization*, Universitext,
DOI 10.1007/978-3-319-48814-1_2

How should the sequence $y = (a_0, a_1, a_2, a_3)$ of allocations be made in order to maximize on the set $A^4(10) := \{y \in A^4 : c(y) \leq 10\}$ the sum of rewards

$$V_{4y}(10) := \sum_{t=0}^{3} u(a_t) + d \cdot u\left(10 - \sum_{t=0}^{3} a_t\right)?$$

We want to find $V_4(10) := \sup_{y \in A^4(10)} V_{4y}(10)$ and a maximum point $y^* = (a_t^*)_0^3$ of $y \mapsto V_{4y}(10)$. This problem is denoted by $DP_4(10)$, and the problem $DP_n(s)$ for $n \geq 1$ and $s \in \{0, 1, \ldots, 10\}$ is defined analogously. ♦

If in the preceding allocation problem we allowed non-negative *real* actions a_t, some cases where u is concave and differentiable would be solvable by the classical optimization method based on Fermat's criterion in Appendix A.4.16. No counterpart is available in the discrete case of Example 2.1.1, but ***Dynamic Optimization*** will help, as shown in Example 2.4.1 below. Crucial for its applicability is the following property: If we knew an optimal first input a_0^* for $DP_4(10)$, then the vector $(a_t^*)_1^3$ could be found as a maximum point of the problem $DP_3(10 - a_0^*)$.

For arbitrary DPs the basic solution method (explained in detail in Sect. 2.3), runs as follows: If $V_n(s_0)$, $1 \leq n \leq N$, denotes the maximum value of the objective function $y \mapsto V_{ny}(s_0)$, $y \in A^n(s_0)$, one can obtain $V_n(s_0)$ from the function V_{n-1} by maximizing certain functions $a \mapsto W_n(s, a)$, $s \in S$, determined by V_{n-1}, over a certain set $D(s)$. By iterating this step, we see that the problem $DP_N(s_0)$ of maximizing $y \mapsto V_{Ny}(s_0)$ with respect to the N variables $a_0, a_1, \ldots, a_{N-1}$ reduces to a sequence of N maximization problems, each parametrized by s, with respect to a single variable a. We call this approach for the moment the ***DP method***, in contrast to other standard *static* methods. The DP method has several favorable features, as will become apparent later on in many examples:

(a) The sets $D(s)$ and the functions $a \mapsto W_n(s, a)$, $1 \leq n \leq N$, are much simpler than $A^N(s_0)$ and $y \mapsto V_{Ny}(s_0)$, respectively.
(b) The problems of the existence and of uniqueness of a maximum and of maxima on the boundary of $A^N(s_0)$ reduce to the corresponding problems for the function $a \mapsto W_n(s, a)$ of one variable only.
(c) The approach is particularly suited to the many examples where the objective function $y \mapsto V_{Ny}(s_0)$ is recursively defined and where the explicit representation, needed in general for static methods, is cumbersome; see Equation (2.4) below.
(d) The approach provides a general method for studying the important dependence of the maximal n-stage reward $V_n(s_0)$ on the initial state s_0. As an example, if S and all sets $D(s)$ are convex, a direct checking of convexity of the value functions $s \mapsto V_N(s)$ may be impossible, while the Dynamic Optimization method may work; see Chap. 8 below.
(e) In many concrete applications the number N is not known exactly, but only bounds $1 \leq N_1 \leq N \leq N_2 \leq \infty$ are known. Then it is desirable to know V_N for all N within the bounds. In static methods this requires us in general to solve

an N-stage problem for *each* of these N's. On the other hand, as we shall see below, the DP solution for N_2 automatically also contains also the solution for all $N < N_2$.

(f) The DP method provides an exact numerical algorithm for many important problems where S and all sets $D(s)$ are finite.

2.2 The Model

We now give a detailed intuitive background and basic concepts for the general ***problem*** $\boldsymbol{DP_N(s_0)}$; see Fig. 2.1. The object of investigation is some system which starts at ***time*** $t = 0$ in an ***initial state*** s_0, belonging to a set S, called the ***state space.*** The system moves at times $t = 0, 1, \ldots, N-1$ successively to ***states*** s_t, i.e. $s_1, s_2, \ldots, s_N$. This movement is controlled by ***actions*** a_t, i.e. $a_0, a_1, \ldots, a_{N-1}$, respectively, taken by a decision maker at the times $t = 0, 1, \ldots, N-1$, respectively, from a set A, the ***action space.*** When discussing facts which concern states and actions at all times t we often write s and a rather than s_t and a_t, respectively, and we call s the *momentary state* and a the *momentary action.*

In examples, often the state s_t has one of the following meanings: (i) it is a summary of the *history* of the process up to time $t-1$, (ii) it represents information, necessary for the choice of an optimal action a_t, (iii) it depicts the environment in which the process is running at time t.

The number $N \in \mathbb{N}$ is called the ***horizon***, and the time interval $[t, t+1)$ is called the t-th ***period***; at ***stage*** n means *at time* $t := N-n$, $1 \leq n \leq N$. Each time an action is taken the momentary state of the system is assumed to be known to the decision maker. In general, when the system is in state s, not all actions from the action space A, but only those in a certain non-empty set $D(s) \subset A$ will be admissible. We call $D(s)$ the set of ***admissible actions for state s*** and

$$D := \{(s, a) \in S \times A : a \in D(s)\}$$

the ***constraint set.*** The influence of the decision maker on the transition of the system is described by a mapping $T: D \to S$, the so-called ***transition function***: If at time t the system is in state s_t and if action $a_t \in D(s_t)$ is taken, then the system moves to the new state $s_{t+1} := T(s_t, a_t)$. At time t, i.e. at the beginning of ***period*** t,

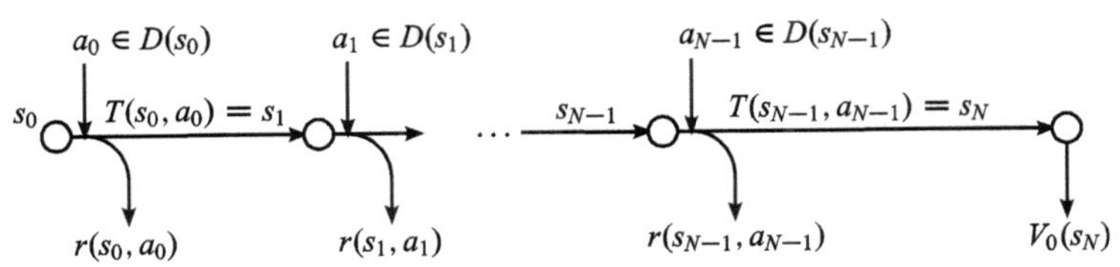

Fig. 2.1 Development of states s_t

a ***one-stage reward*** $r(s_t, a_t) \in \mathbb{R}$, given by the so-called ***one-stage reward function*** r is obtained. In addition, if the movement of the system ends at time N in state s_N, then a ***terminal reward*** $V_0(s_N) \in \mathbb{R}$ is obtained. The same monetary units, obtained at different time points, will have different cash value due to interest. This fact is taken into consideration by a so-called ***discount factor*** $\beta \in \mathbb{R}^+$; this means that the reward $r(s_t, a_t)$ obtained at time t and the terminal reward $V_0(s_N)$ at time N enter the account (2.3) below relative to time $t = 0$ as $\beta^t r(s_t, a_t)$ and $\beta^N V_0(s_N)$, respectively. In most applications early gains are more profitable than later ones, which means that $\beta < 1$.

Summing up we arrive at the following definition.

Definition 2.2.1 A ***(stationary) deterministic (dynamic) program*** (**DP** for short) is a tuple $(S, A, D, T, r, V_0, \beta)$ of the following kind:

- S is the state space.
- A is the action space.
- $D \subset S \times A$ such that all s-sections $D(s) := \{a \in A : (s, a) \in D\} \neq \emptyset$, $s \in S$. D is called the constraint set and $D(s)$ is called the set of admissible actions for state s.
- $T: D \to S$ is the transition function.
- $r: D \to \mathbb{R}$ is the one-stage reward function.
- $V_0: S \to \mathbb{R}$ is the *terminal reward function.*
- $\beta \in \mathbb{R}^+$ is the discount factor.

General assumption Throughout this book we require that both S and A are non-empty sets.

We also call the tuple $(S, A, D, T, r, V_0, \beta)$ the *data* of the DP. The data for the allocation problem from Example 2.1.1 with arbitrary $K \in \mathbb{N}$ are as follows: $S := A := \mathbb{N}_{0,K}$, $D(s) := \mathbb{N}_{0,s}$ for all s, $T(s, a) = s - a$, $r(s, a) := u(a)$ for $(s, a) \in D$; $V_0 = d \cdot u$, and β is arbitrary. One also could model the problem by using for a_t the investment at time t. Then S, A, D, V_0 and β would remain unchanged, while $T(s, a) = a$ and $r(s, a) := u(s - a)$ for $(s, a) \in D$.

Remark 2.2.2 In applications the states and/or actions are often integers or reals, but sometimes they are elements of $\mathbb{Z}^d$ or of $\mathbb{R}^d$ for $d \geq 2$ or they are sets. ◇

Remark 2.2.3 Sometimes the one-stage reward $r(s, a, s')$ also depends on the next state $s' = T(s, a)$. That case is covered by simply replacing s' by $T(s, a)$.

In a few cases, r also depends on β; in particular, if $r(s, a)$ consists of a reward $g(s, a)$ obtained at the end of the momentary period then $r(s, a) = \beta g(s, a)$. A dependence of r on β requires changes only for those few results which deal with the dependence of the solution on β. ◇

Remark 2.2.4 (The discount factor) One must not distinguish in the theory between $\beta < 1$, $\beta = 1$ and $\beta > 1$. On the other hand, for many models the pointwise limit of the value functions V_n for $n \to \infty$, dealt with in Chap. 10, is not defined unless $\beta < 1$. Also in economical applications β is usually smaller than one. If the N

periods have nothing to do with time but mean that a certain activity is executed N times, then only $\beta = 1$ is meaningful. ◇

Any concrete problem $DP_N(s_0)$ must be modeled by an appropriate choice of S, A, D, T, r, V_0 and β as done above for the allocation problem from Example 2.1.1. Particularly important is the choice of the state space S. Sometimes several choices are possible; then it is up to the decision maker's skill to find a formulation that easily admits theoretical analysis and computation. This skill can be acquired only by experience. Dreyfus and Law (1977) speak in the title of their book of the *Art of Dynamic Programming*, i.e. the art of finding an appropriate model. The same authors also suggest (loc. cit., p. 17) a useful mental device, called the *consultant question*, for a skillful choice of the state s_t. Essentially it reads as follows: *The momentary state should consist of the minimal information about the momentary situation you would have to acquire from a firm in case you would be hired to take over the problem and do things optimally from now on.*

We call a ***set continuous [discrete]*** if it is an interval or a product of intervals [an interval in $\mathbb{Z}$ or a product of such intervals]. We call a DP ***continuous [discrete]*** if either S or $D(s)$, $s \in S$, are continuous [if both S and $D(s)$, $s \in S$, are discrete]. The modeling procedure should also include a reflection about the question of whether to use a discrete or a continuous DP. More information on this feature is given before Example 2.4.3 below.

Now we define for a given model DP the ***maximization problem*** $DP_N(s_0)$, determined by an arbitrary horizon N and an arbitrary initial state $s_0 \in S$. Firstly, we say that a ***sequence*** $y := (a_t)_0^{N-1}$ is a ***sequence of admissible actions*** for s_0, if y obeys the restrictions

$$
\begin{aligned}
a_0 &\in D(s_0), \\
a_1 &\in D(s_1), \text{ where } s_1 := T(s_0, a_0), \\
a_2 &\in D(s_2), \text{ where } s_2 := T(s_1, a_1), \\
&\vdots \\
a_{N-1} &\in D(s_{N-1}), \text{ where } s_{N-1} := T(s_{N-2}, a_{N-2}).
\end{aligned}
\tag{2.1}
$$

(As an example, in the allocation problem from Example 2.1.1 with $N = 4$ and $K = 8$ the action sequence $(4, 3, 1, 0)$ is admissible for s_0 if and only if $s_0 \geq 8$.) In the final state $s_N := T(s_{N-1}, a_{N-1})$ no action is taken. The set of action sequences admissible for s_0 will be denoted by $A^N(s_0)$; it is non-empty because $D(s) \neq \emptyset$ for each s; we have $A^N(s_0) = A^N$ for all s_0 if $D(s) = A$ for all s; $A^N(s_0)$ is finite if all sets $D(s)$ are finite. Even for simple sets $D(s)$ the sets $A^N(s_0)$ can be complicated as seen, for instance, from the allocation problem from Example 2.1.1.

An initial state s_0 and a sequence $(s_t)_1^N$ of states as introduced in (2.1) above describes the evolution of the system under an admissible action sequence $y \in A^N(s_0)$. We call $(s_t)_1^N$ the ***decision process*** generated by (s_0, y). Notice that $s_t = s_t(s_0, (a_i)_0^{t-1})$ is a function of s_0 and of y, since

$$s_1 = T(s_0, a_0), \; s_2 = T(s_1, a_1) = T(T(s_0, a_0), a_1),$$
$$s_3 = T(T(T(s_0, a_0), a_1), a_2), \; \ldots .$$

It follows from (2.1) that the sets $A^n(s)$, $n \geq 1$, $s \in S$, have the following sequential structure: $A^1(s) = D(s)$ and

$$A^n(s) = \left\{(a, x) \in D(s) \times A^{n-1} : x \in A^{n-1}(T(s, a))\right\}, \; n \geq 2. \tag{2.2}$$

For initial state $s_0 \in S$, admissible action sequence $y = (a_t)_0^{N-1} \in A^N(s_0)$ and for $(s_t)_1^N$ generated by (s_0, y) the ***N-stage reward*** is the real number

$$V_{Ny}(s_0) := \sum_{t=0}^{N-1} \beta^t r(s_t, a_t) + \beta^N V_0(s_N) = r(s_0, a_0) \tag{2.3}$$
$$+ \sum_{t=1}^{N-1} \beta^t r(s_t(s_0, (a_i)_0^{t-1}), a_t) + \beta^N V_0(s_N(s_0, (a_i)_0^{N-1})).$$

Thus $y \mapsto V_{Ny}(s_0)$ is the ***objective function*** of the problem $DP_N(s_0)$. Notice that $V_{Ny}(s_0)$ means the total reward *discounted back* to time $t = 0$; the total reward accumulated at time N is $V_{Ny}(s_0)/\beta^N$. The complicated explicit representation (check it for $N = 3$)

$$V_{Ny}(s_0) = r(s_0, a_0) + \beta r(T(s_0, a_0), a_1) + \beta^2 r(T(T(s_0, a_0), a_1), a_2)$$
$$+ \cdots + \beta^N V_0(T(T(\ldots), a_{N-1})) \tag{2.4}$$

is rarely needed, but for some applications the explicit expression may be useful for checking the correct choice of the data. Also keep in mind that $y \mapsto V_{Ny}(s_0)$ is a function of the form $\sum_{t=0}^{N-1} g_t(s_0, a_0, a_1, \ldots, a_t)$.

Now the N-stage maximization problem $DP_N(s_0)$ for $N \geq 1$ and $s_0 \in S$ reads as follows:

(i) Compute the ***maximal N-stage reward*** *for initial state* s_0

$$V_N(s_0) := \sup\{V_{Ny}(s_0) : y \in A^N(s_0)\}.$$

(ii) Find, if possible, an ***s_0-optimal action sequence***, i.e. a maximum point of the objective function $y \mapsto V_{Ny}(s_0)$.

The function $V_N: S \to (-\infty, \infty]$ is called the ***N*-stage value function**. The set of problems $DP_N(s_0)$, $s_0 \in S$, is called the ***problem* DP_N**. Notice that $V_N(s_0)$ is finite if there exists an s_0-optimal action sequence. The sequence $(V_n)_1^N$ of value functions plays a central role for solving DP_N. From now on we shall mostly write $V_N(s)$ instead of $V_N(s_0)$. Only in rare cases will the value functions have an explicit form. Of course, s_0-optimal action sequences need not exist, and if they exist they need not be unique.

2.3 The Basic Solution Procedure

In the following generalization of Appendix A.4.5 the set $M(b)$ is the b-section of M (cf. Appendix A.3.8).

Lemma 2.3.1 (The joint supremum equals the iterated supremum) *Let B and C be non-empty sets and let v be a function on a set $M \subset B \times C$ for which $M(b) \neq \emptyset$ for all $b \in B$. Then*

$$\sup_{(b,c)\in M} v(b,c) = \sup_{b\in B} \sup_{c\in M(b)} v(b,c).$$

Proof Put $h(b) := \sup\{v(b,c) : c \in M(b)\}$. We have to show that $\sup v = \sup h$. From

$$v(b,c) \le h(b) \le \sup h$$

we get $\sup v \le \sup h$. On the other hand, from $v(b,c) \le \sup v$ we firstly obtain $h(b) \le \sup v$ and then $\sup h \le \sup v$. □

The first step towards the value iteration (2.7) is the next result. For $a \in D(s)$ and $x \in A^{n-1}(T(s,a))$ we denote by (a,x) the n-stage action sequence which first uses a and then x.

Lemma 2.3.2 (The reward iteration, RI for short) *The following holds*

$$\begin{aligned} V_{1a}(s) &= r(s,a) + \beta V_0(T(s,a)), (s,a) \in D, \\ V_{n,(a,x)}(s) &= r(s,a) + \beta V_{n-1,x}(T(s,a)), \\ &\qquad n \ge 2,\ (s,a) \in D,\ x \in A^{n-1}(T(s,a)). \end{aligned} \tag{2.5}$$

Proof The form of V_{1a} follows from (2.3) with $N := 1$ since $a_1 = T(s,a)$. Now assume $n \ge 2$. For $0 \le t \le n-2$ put $s'_t = s_{t+1}$ and $a'_t = a_{t+1}$, and put $s'_{n-1} := s_n$. Then $s'_t = T(s'_{t-1}, a'_{t-1})$ and $x = (a'_t)_0^{n-2}$. It follows easily that x and $(s'_t)_0^{n-1}$ satisfy (2.1) with $N := n-1$ and with $(s_t)_1^N$ and $(a_t)_0^{N-1}$ replaced by $(s'_t)_1^{n-1}$ and $(a'_t)_0^{n-2}$, respectively. This means that $(a'_t)_0^{n-2} \in A^{n-1}(s'_0)$ and that $(s'_t)_1^{n-1}$ is the decision process generated by (s_0', x). Now (2.3) yields $V_{n,(a,x)}(s_0) = r(s_0, a_0) + \beta \cdot B$ where

$$B := \sum_{t=1}^{n-1} \beta^{t-1} r(s_t, a_t) + \beta^{n-1} V_0(s_n) = \sum_{t=0}^{n-2} \beta^t r(s_t', a_t') + \beta^{n-1} V_0(s_{n-1}')$$
$$= V_{n-1,x}(s_0') = V_{n-1,x}(T(s_0, a_0)).$$

Inserting B into $V_{n,(a,x)}(s_0) = r(s_0, a_0) + \beta \cdot B$ completes the proof. □

The RI expresses the following fact: The n-stage discounted reward for the initial state s under the action sequence (a, x) equals the sum of the reward in the first period and the discounted $(n-1)$-stage reward for the initial state $T(s, a)$ under the action sequence x. Thus the recursion (2.5) exhibits the sequential structure of the objective functions $y \mapsto V_{ny}(s)$. Moreover, in case of finite S and A the RI is a convenient recursive algorithm for evaluating $V_{Ny}(s)$ on a computer.

We often use the functions $W_n: D \to (-\infty, \infty]$, defined by

$$W_n(s, a) := r(s, a) + \beta V_{n-1}(T(s, a)), \; n \geq 1. \tag{2.6}$$

A mapping f from S into A is called a ***decision rule*** if $f(s) \in D(s)$ for all s. Denote the ***set of all decision rules*** by $\mathbb{F}$. A ***decision rule*** $\boldsymbol{f_n}$ ***at stage*** $\boldsymbol{n}$ such that $f_n(s)$ is a maximum point of $a \mapsto W_n(s, a)$ for all s is called a ***maximizer at stage*** $\boldsymbol{n}$. Intuitively, $f_n(s)$ is an *optimal action* at state s when n periods are still ahead. A sequence $(f_n)_N^1 := (f_N, f_{N-1}, \ldots, f_1) \in \mathbb{F}^N$ of decision rules f_n at stage n is called an ***N-stage policy*** and it is called an ***N-stage maximizing policy*** if f_n is a maximizer at stage n for $1 \leq n \leq N$.

Of course, a maximizing policy need not exist, and if it exists, it need not be unique. Sometimes we need the following generalization of the concept of a maximizer at stage n: If w is a function on D, we call a decision rule f a ***maximizer of*** w if $f(s)$ is a maximum point of $a \mapsto w(s, a)$ for all s.

Theorem 2.3.3 (Basic Theorem for stationary DPs)

(a) The value functions V_n satisfy the following recursion, called ***value iteration (VI*** *for short):*

$$V_n(s) = \sup\{r(s, a) + \beta V_{n-1}(T(s, a)) : a \in D(s)\},$$
$$= \sup\{W_n(s, a) : a \in D(s)\}, \quad n \geq 1, s \in S. \tag{2.7}$$

(b) Let $N \geq 1$, let s_0 be an arbitrary initial state, let the action sequence $y^* = (a_t^*)_0^{N-1}$ be admissible for s_0 and let $(s_t)_1^N$ be the decision process generated by s_0 and y^*. Then y^* is s_0-optimal if and only if a_t^* is a maximum point of $a \mapsto W_{N-t}(s_t, a)$, $0 \leq t \leq N-1$.

(c) The ***Optimality Criterion (OC*** *for short): Let $N \geq 1$, let s_0 be an arbitrary initial state. If there exists a maximizing policy $(f_n)_N^1$ then:*

(c1) An s_0-optimal action sequence $(a_t^*)_0^{N-1}$ is given by the following ***forward procedure***

$$a_t^* := f_{N-t}(s_t), \quad s_{t+1} := T(s_t, a_t^*), \quad 0 \le t \le N-1. \tag{2.8}$$

If the maximizing sequence $(f_n)_N^1$ is unique, then $(a_t^*)_0^{N-1}$ is the unique s_0-optimal action sequence.

(c2) V_n, $n \ge 1$, is determined by f_n and W_n, since

$$V_n(s) = W_n(s, f_n(s)), \; s \in S.$$

Proof

(a) Fix s. Equation (2.7) follows for $n = 1$ immediately from (2.5). For $n \ge 2$ we use Lemma 2.3.1 for $b := a$, $B := D(s)$, $c := x$, $C := A^{n-1}$, $M := A^n(s)$ and $v(a, x) := V_{n,(a,x)}(s)$. From the recursive property (2.2) of $A^n(s)$ we see that the a-section $M(a)$ of M equals $A^{n-1}(T(s,a))$. Using the RI (2.5) and noting that $\beta > 0$, we obtain

$$\begin{aligned} V_n(s) &= \sup\{V_{n,(a,x)}(s) : (a,x) \in M\} \\ &= \sup_{a \in D(s)} \sup_{x \in A^{n-1}(T(s,a))} [r(s,a) + \beta V_{n-1,x}(T(s,a))] \\ &= \sup_a [r(s,a) + \beta \sup_x V_{n-1,x}(T(s,a))] \\ &= \sup_a [r(s,a) + \beta V_{n-1}(T(s,a))]. \end{aligned}$$

(b) From the VI we infer for $0 \le t \le N-1$, since $s_{t+1} = T(s_t, a_t^*)$, that

$$V_{N-t}(s_t) \ge r(s_t, a_t^*) + \beta V_{N-t-1}(s_{t+1}),$$

with equality if and only if a_t^* is a maximum point of $W_{N-t}(s_t, \cdot)$. Thus

$$\begin{aligned} V_N(s_0) &\ge r(s_0, a_0^*) + \beta V_{N-1}(s_1) \\ &\ge r(s_0, a_0^*) + \beta r(s_1, a_1^*) + \beta^2 V_{N-2}(s_2) \\ &\ge \ldots \ge \sum_0^{N-1} \beta^t r(s_t, a_t^*) + \beta^N V_0(s_N) = V_{Ny^*}(s_0), \end{aligned}$$

and equality holds throughout if and only if for $0 \le t \le N-1$ the action a_t^* is a maximum point of $W_{N-t}(s_t, \cdot)$.

(c) (c1) follows from (b) since $(a_t^*)_0^{N-1}$ satisfies the condition in (b); the assertion about uniqueness is obviously true. (c2) is obvious from (2.6). □

Remark 2.3.4 The VI says that the maximal reward for n periods and initial state s equals the maximum—over the initially admissible actions a—of the sum of the reward earned in the first period and the discounted maximal reward for the last $n-1$ periods and the next state $T(s,a)$ as initial state. This fact is intuitively plausible to such an extent that part of the literature refrains from a proof. ◇

Remark 2.3.5 The computation of $V_N(s_0)$ for some $N \geq 2$ by the VI also yields $V_n(s)$, $1 \leq n \leq N-1$, $s \in S$. Yet the solution of $DP_n(s)$ requires in addition the computation of an s-optimal $y \in A^n(s)$ by means of (2.8). ◇

Remark 2.3.6 The OC yields an s_0-optimal action sequence y^* for *each* s_0. If only an s_0-optimal action sequence for *a single* s_0 is required, it follows from (2.8) that it suffices to compute a maximum point $f_N(s_0)$ of $W_N(s_0,\cdot)$ instead of a whole maximizer f_N at stage N.

The s_0-optimal action sequence obtained by (2.8) is called the ***action sequence*** *generated by* s_0 and the ***maximizing policy*** $(f_n)_N^1$. ◇

Remark 2.3.7 We call Theorem 2.3.3 the *Basic Theorem* as it will play a dominant role throughout Part I and in modified form in the other chapters. Other names used in the literature include ***DP algorithm***, ***method of backward induction*** and above all, ***Bellman's principle of optimality.*** We reserve the latter name for another result, see Supplement 3.6.1. ◇

Remark 2.3.8 While the Basic Theorem 2.3.3 reduces the global N-stage problem to a sequence of N interconnected parametric one-stage optimization problems it does not tell us anything about how to solve the latter problems. For these one depends on methods of non-dynamic optimization. ◇

Remark 2.3.9 The VI holds whether or not there exist s_0-optimal action sequences. Since r is finite, the right-hand side of (2.7) is also defined in the case $V_{n-1}(T(s,a)) = \infty$ by our convention $x + \infty := \infty$ for real x. ◇

The essence of the proof of the VI may be phrased in the simple equation

$$\sup_{(a,x)}[g(a) + \beta h(a,x)] = \sup_a[g(a) + \beta \sup_x h(a,x)].$$

Unfortunately this simple method is not applicable for stochastic DPs.

For $s \in S$ and $n \geq 1$ we call the (possibly empty) set $\boldsymbol{D_n^*(s)}$ of maximum points of $a \mapsto W_n(s,a)$ the ***set of optimal actions*** for stage n and at state s. Thus $(f_n)_N^1$ is maximizing if and only if $f_n(s) \in D_n^*(s)$ for $1 \leq n \leq N$ and all s. The sequence $(D_n)_N^1$ determines for each s_0 *all* solutions of $DP_N(s_0)$ since by Theorem 2.3.3(b) $(a_t^*)_0^{N-1}$ is s_0-optimal if and only if $a_t^* \in D_{N-t}^*(s_t)$ for $0 \leq t \leq N-1$.

Only in rare cases will the value functions and s_0-optimal action sequences have an explicit form. Therefore, the computer-aided numerical solution, possibly after suitable discretization of the state and action space, is important. Assume for simplicity that both S and A are finite. We call the method provided by the Basic Theorem for solving problem $DP_N(s_0)$ the ***VI algorithm.*** It runs as follows.

One computes, starting with V_0, recursively by means of the ***backward procedure*** (2.7) (i.e. backward in stages) the value functions $V_1, V_2, \ldots, V_{N-1}$ by maximizing the (possibly infinite) functions $a \mapsto W_n(s,a)$ for all $s \in S$ and $1 \le n \le N-1$. After having computed V_n, the function V_{n-1} can be deleted from the memory. In the final step one computes $V_N(s_0)$ by maximizing $a \mapsto W_N(s_0,a)$. In practice one often computes $V_N(s_0)$ for *all* $s_0 \in S$.

The computation of an s_0-optimal action sequence $(a_t^*)_0^{N-1}$ for given initial state s_0 according to Theorem 2.3.3(b) cannot be done simultaneously with the recursive computation of the value functions, as one does not know in advance which sequence of states $s_1, s_2, \ldots$ is generated by s_0 and the optimal action sequence to be constructed. However, while maximizing $W_n(s,\cdot)$, $s \in S$, for $1 \le n \le N-1$ and $W_N(s_0,\cdot)$ one can compute and store a maximum point $f_n(s)$ and $f_N(s_0)$, respectively. Then one obtains an s_0-optimal action sequence by the forward procedure in the OC.

2.4 First Examples

Example 2.4.1 (Solution of Example 2.1.1) We treat this problem for arbitrary $K \in \mathbb{N}$ and N rather than only $K = 10$ and $N = 4$.

(a) After the definition of a DP we have seen that $S = A = \mathbb{N}_{0,K}$, $D(s) = \mathbb{N}_{0,s}$, $T(s,a) = s-a$, $r(s,a) = u(a)$ and $V_0 = d\cdot u$ for some $d \in \mathbb{R}_+$ and an arbitrary function u on $\mathbb{N}_{0,K}$. Note that $A^N(s_0) = \{(a_t)_0^{N-1} \in A^N : \sum_0^{N-1} a_t \le s_0\}$ since $\sum_0^{N-1} a_t \le s_0$ implies $a_t \le s_0 - \sum_{i=0}^{t-1} a_i = s_t$ for $0 \le t \le N-1$. Because of $T(s,a) \le s$, $(s,a) \in D$, the solution of $DP_N(s_0)$ is the same for each DP with $S = A = \mathbb{N}_{0,K}$ whenever $K \ge s_0$; even $S = A = \mathbb{N}_0$ could be used. As a consequence, it suffices to solve $DP_N(K)$ with the choice $S = A = \mathbb{N}_{0,K}$.

From now on assume that u is increasing and that $u(0) = 0 < u(K)$. (The case $u(K) = 0$, i.e. $u \equiv 0$, is trivial.) By Theorem 2.3.3(a) the VI has the form

$$V_n(s) = \max\{u(a) + \beta V_{n-1}(s-a) : a \in \mathbb{N}_{0,s}\}, \quad n \ge 1, s \in \mathbb{N}_{0,K}. \tag{2.9}$$

This implies by induction on $n \ge 0$ that $V_n(0) = 0$. Due to the discreteness of S and A, even for simple utility functions u one can expect only in very rare cases an explicit solution. However, for arbitrary u we can find a numerical solution by means of (2.9). As we show below, for u with *sufficient structure* we also can find structural properties of the solution, i.e. of V_n, of the smallest maximizer f_n at stage n and of those s_0-optimal action sequences $y^* = y^N(s_0) \in A^N(s_0)$ which are generated by s_0 and $(f_n)_N^1$.

Table 2.1 $V_n(s)$, $s \leq K := 8$, for $u(a) = \sqrt{a}$, $d = 1.5$, $\beta = 0.8$

	s								
n	0	1	2	3	4	5	6	7	8
1	0.000	1.200	2.200	2.697	3.111	3.493	3.814	4.132	4.415
2	0.000	1.000	1.960	2.760	3.174	3.572	3.903	4.221	4.526
3	0.000	1.000	1.800	2.568	3.208	3.622	3.954	4.272	4.590
4	0.000	1.000	1.800	2.440	3.054	3.566	3.981	4.312	4.630
5	0.000	1.000	1.800	2.440	2.952	3.444	3.858	4.267	4.599
6	0.000	1.000	1.800	2.440	2.952	3.366	3.776	4.169	4.500
7	0.000	1.000	1.800	2.440	2.952	3.366	3.776	4.107	4.435
8	0.000	1.000	1.800	2.440	2.952	3.366	3.776	4.107	4.435
9	0.000	1.000	1.800	2.440	2.952	3.366	3.776	4.107	4.435

Table 2.2 $f_n(s)$, $s \leq K := 12$, for $u(a) = \sqrt{a}$, $d = 1.5$, $\beta = 0.8$

	s												
n	0	1	2	3	4	5	6	7	8	9	10	11	12
1	0	0	1	1	2	2	2	3	3	4	4	5	5
2	0	1	1	1	2	2	2	3	3	4	4	4	5
3	0	1	1	1	1	2	2	2	3	4	4	4	4
4	0	1	1	1	1	1	2	2	3	4	4	4	4
5	0	1	1	1	1	1	2	2	2	3	4	4	4
6	0	1	1	1	1	2	2	2	2	2	3	4	4
7	0	1	1	1	1	2	2	2	2	3	3	4	4
8	0	1	1	1	1	2	2	2	2	3	4	4	4
9	0	1	1	1	1	2	2	2	2	3	4	4	4

(b) The Tables 2.1 and 2.2 and Fig. 2.2 show the result of computations for $u(a) = \sqrt{a}$, $d = 1.5$ and $\beta = 0.8$. We denote by f_n the smallest maximizer at stage n. One quickly obtains by the forward procedure (2.8) the subsequent K-optimal action sequences y^*; the resulting terminal state s_N can be used as control, since the sum of the actions and of s_N equals K.

$$\begin{aligned}
&N = 4,\ K = 8,\ \ y^* = (3, 2, 1, 1), && s_4 = 1,\ V_4(8) = 4.630\\
&N = 6,\ K = 8,\ \ y^* = (2, 2, 1, 1, 1, 0), && s_6 = 1,\ V_6(8) = 4.500\\
&N = 9,\ K = 12,\ y^* = (4, 2, 2, 1, 1, 1, 1, 0, 0), && s_9 = 0,\ V_9(12) = 5.548.
\end{aligned}$$

From the Tables 2.1 and 2.2 below one will conjecture that $V_n = V_7$ and that $f_n = f_7$ for $n \geq 8$. This rare property can be confirmed by Proposition 4.1.4.

(c) In later sections we systematically study structural properties of the solution of general DPs and apply these to our allocation problem. Here we give several results (c1)–(c7) which were suggested either by intuition or by numerical computations. Some of these results can be proved already here, and some

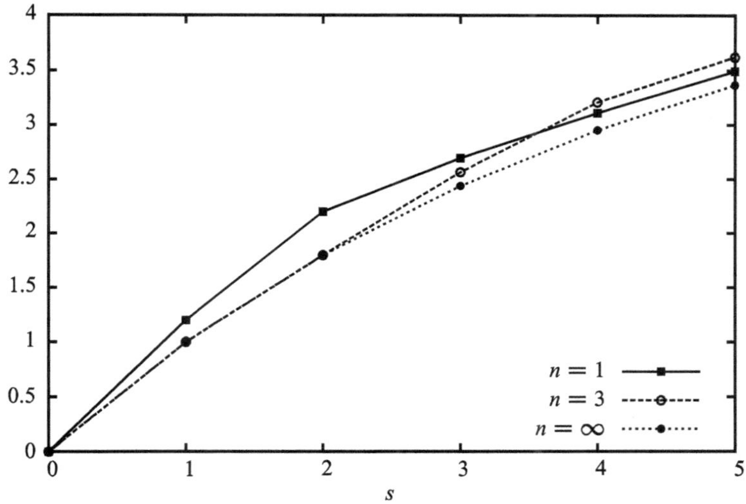

Fig. 2.2 Functions V_n for $n = 1$, $n = 3$ and $n = \infty$ (limit function V) for $K = 8$, $u(a) = \sqrt{a}$, $d = 1.5$ and $\beta = 0.8$

use an ad hoc method rather than the Basic Theorem. Discrete concavity and discrete convexity are defined in Appendix (D.4).

(c1) $V_n(s)$ is increasing in s. This is plausible since we expect from a larger initial resource a larger maximal reward. A proof can be given by Theorem 6.3.5, by Example 6.4.1(c1) or directly as in Problem 2.5.1.

(c2) The number $\overline{s} := \max\{0 \leq x < K : u(x) = 0\}$ can be interpreted as follows: If one allocates energy to a technical system, then $\overline{s} + 1$ is the minimal allocation which causes the system to work with profit. For $0 \leq s \leq \overline{s}$ and $n \geq 0$ we have $V_n(s) = 0 = f_{n+1}(s)$. Moreover, each $y \in A^N(s)$ is s-optimal. These statements are simple consequences of Problem 2.5.1(d) and of the VI (2.7).

(c3) $V_n(s)$ is increasing in n for *small* d [decreasing in n for *large* d], e.g. if $d \leq 1$ [if $\beta < 1, d \geq 1/(1-\beta)$]. (As seen from Table 2.1, $V_n(3)$ is in general neither increasing nor decreasing in n.)

For the *proof* one easily derives from the VI (2.7) by induction on $n \geq 0$ that $V_n(s)$ is increasing [decreasing] in n if $V_1 \geq V_0 = d \cdot u$ [$V_1 \leq V_0 = d \cdot u$].

(i) If $d \leq 1$ then the VI yields for $s \in \mathbb{N}_{0,K}$:

$$V_1(s) = \max_{0 \leq a \leq s} [u(a) + \beta d u(s-a)] \geq u(s) \geq du(s) = V_0(s).$$

(ii) Recall that u is increasing on $\mathbb{N}_{0,K}$ (cf. Definition 6.2.1(vii)). Then if $\beta < 1, d \geq 1/(1-\beta)$ we have:

$$V_1(s) \leq \max_{0 \leq a \leq s} u(a) + \beta d \max_{0 \leq a \leq s} u(s-a) = u(s) \cdot (1 + \beta d) \leq du(s).$$

(c4) Let u be discretely concave (cf. Appendix (D.1)). Then all value functions V_n are discretely concave, f_n is increasing and its upward jumps have size one; see also Table 2.2. Moreover, if g_n denotes the largest maximizer at stage $n \geq 1$, then a mapping $f: \mathbb{N}_{0,K} \to \mathbb{N}_{0,K}$ is a maximizer at stage n if and only if $f_n \leq f \leq g_n$. All results follow from Theorem 7.1.2 below by using as actions the investments.

(c5) Let u be discretely convex. In Theorem 7.3.3 below we compute the value functions explicitly and show that either $(0, 0, \ldots, 0)$ or $(s_0, 0, \ldots, 0)$ are s_0-optimal. These two action sequences are *extreme* in the sense that they prescribe to consume nothing at all times $0 \leq 1 \leq N - 1$ or to consume everything at time $t = 0$, respectively. Moreover, Example 7.3.5 below shows in case $\beta < 1$ and $s > s_0$ that $V_n = u$ for all $n \geq m$ and some $m \in \mathbb{N}$, and that for each s_0 the action sequence $(s_0, 0, \ldots, 0) \in A^N(s_0)$ is s_0-optimal for all $N \geq m$. In particular, if $\beta < 1$, then for some $m \in \mathbb{N}$ we have $V_n = u$ for $n \geq m$, and $(s_0, 0, \ldots, 0) \in A^N(s_0)$ is s_0-optimal for $N \geq m$.

(c6) The value functions V_n, $n \geq 1$, are Lipschitz continuous in d and also in β, both uniformly in s. In fact, for each s and for d, $d' \in \mathbb{R}_+$ we get, using Appendix A.4.4

$$
\begin{aligned}
|V_n(s, d) - V_n(s, d')| &= |\max_y V_{ny}(s, d) - \max_y V_{ny}(s, d')| \\
&\leq \max_y |V_{ny}(s, d) - V_{ny}(s, d')| = \beta^n |d - d'| \cdot \max u.
\end{aligned}
$$

Moreover, for each s and for β, $\beta' \in (0, 1]$ we get

$$
\begin{aligned}
|V_n(s, \beta) - V_n(s, \beta')| &\leq \max_y |V_{ny}(s, \beta) - V_{ny}(s, \beta')| \\
&\leq \left(\sum_0^{n-1} |\beta^t - \beta'^t| + d|\beta^n - \beta'^n| \right) \max u \\
&\leq n(n - 1 + 2d) \cdot \max u \cdot |\beta - \beta'|/2.
\end{aligned}
$$

Here we used that $|\beta^t - \beta'^t| = |\beta - \beta'| \cdot \sum_{i=0}^{t-1} \beta^i \beta'^{t-i} \leq t|\beta - \beta'|, t \geq 1$.

(c7) If u is discretely concave and if $\beta = d = 1$ one expects that it is optimal to allocate the resources among the N stages as evenly as possible in the sense that there exists an s_0-optimal action sequence $y^* \in A^N(s_0)$ whose actions differ from each other by not more than one unit. This is true as one can show, using $b := \lfloor s_0/(N+1) \rfloor$, that y^* is s_0-optimal if $(N+1)(b+1) - s_0$ of the components of y^* equal b, and if the remaining ones equal $b + 1$. ♦

We conclude our investigation of the allocation problem from Example 2.4.1 by studying the asymptotic behavior of the solution for $n \to \infty$. Such problems are studied in detail for general DPs in Chap. 10; here we can solve it by an ad hoc approach.

Proposition 2.4.2 (Asymptotic properties of Example 2.4.1) *Assume that $\beta < 1$ and $V(0) := 0$ and define $V(s)$, $1 \leq s \leq K$, by induction on s according to*

$$V(s) = \max_{1 \leq a \leq s} \{u(a) + \beta V(s-a)\}, \quad 1 \leq s \leq K. \tag{2.10}$$

Then:

(a) $V_n(s)$ converges for $n \to \infty$ to $V(s)$, $s \in \mathbb{N}_{0,K}$.
(b) V is increasing and, if u is discretely concave, also discretely concave.
(c) Let $f(s)$ be a maximum point of $a \mapsto u(a) + \beta V(s-a)$, $0 \leq s \leq K$. For $n \geq 1$ and $1 \leq s \leq K$ put $V_{nf}(s) := V_{ny}(s)$, where $y \in A^n(s)$ is generated by s and the policy $(f)_0^{n-1}$. Then the decision rule f is **asymptotically optimal** *in the sense that for $1 \leq s \leq K$*

$$|V_n(s) - V_{nf}(s)| \to 0 \text{ for } n \to \infty. \tag{2.11}$$

Proof

(a1) Firstly, assertion (a) holds for $s = 0$ since $V_n(0) = 0 \to 0 = V(0)$ for $n \to \infty$. Next, V is real-valued on the finite set $S = \mathbb{N}_{0,K}$, hence bounded. The same holds for the value functions since $0 \leq V_n \leq [1/(1-\beta)+d] \cdot \max u$. In fact, the lower bound holds trivially since $u \geq 0$ and $V_n(0) = 0$, and the upper bound follows from (2.10).

(a2) Fix $1 \leq s \leq K$. Since $\beta < 1$ and $V \geq 0$ we see, using $W(s,a) := u(a) + \beta V(s-a)$ for $0 \leq a \leq s$, that

$$V(s) = \max_{1 \leq a \leq s} W(s,a) \leq \max_{0 \leq a \leq s} W(s,a) = \max\{\beta V(s), V(s)\} = V(s),$$

hence $V(s) = \max_{0 \leq a \leq s} W(s,a)$. Let $\|\cdot\|$ be the maximum-norm on S. Now we get, using Appendix A.1.3(b)

$$\begin{aligned} |V_n(s) - V(s)| &= |\max_a [u(a) + \beta V_{n-1}(s-a)] - \max_a [u(a) + \beta V(s-a)]| \\ &\leq \beta \max_a |V_{n-1}(s-a) - V(s-a)| \leq \beta \|V_{n-1} - V\|. \end{aligned}$$

Here a runs over $\mathbb{N}_{0,s}$. Now induction on $n \geq 0$ implies $\|V_n - V\| \leq \beta^n \|V_0 - V\|$, which proves (a).

(b) This follows from (a) and properties (c1) and (c4) from Example 2.4.1 above since isotonicity and discrete concavity of V_n are easily seen to be preserved when n tends to ∞.

(c) Firstly, from the definition of f we know that $V(s) = u(f(s)) + \beta V(s - f(s))$. The RI (2.5) shows that $V_{nf}(s) = u(f(s)) + \beta V_{n-1,f}(s - f(s))$, $n \geq 1$, where $V_{0f} := V_0$. Now one easily obtains, using induction on $n \geq 0$, that $\|V - V_{nf}\| \leq$

$\beta^n \|V - V_0\|$. Finally the assertion follows, using (a2), from

$$\|V_n - V_{nf}\| = \|V_n - V + (V - V_{nf})\| \leq \|V_n - V\| + \|V - V_{nf}\| \leq 2\beta^n \|V - V_0\|. \quad \square$$

Relation (2.11) means that for each s the performance of the ***stationary policy*** consisting of n copies of the decision rule f becomes arbitrarily close to the performance of each s-optimal action sequence in $A^n(s)$ when n tends to ∞.

Now we turn to *continuous* DPs. There are problems such as the freighter problem from Example 4.1.1 below where only a discrete model makes sense. On the other hand, for many problems both a discrete and a continuous version may be formulated; see the allocation problems Example 2.4.1 and Example 2.4.3 below or the linear-quadratic problems in Example 3.1.2 and Example 4.1.7 below.

Here are a few comments on the appropriateness of discrete or continuous versions and on their solutions.

(i) From a rigorous point of view continuous DPs cannot be completely realistic models for applications since they assume infinite divisibility of states and/or of actions. This does not hold in reality; e.g. in the allocation problem arbitrary small investments do not make sense.
(ii) Continuous versions are often considered as *good* approximate descriptions of a discrete model in the sense that the solution of the continuous version is a *good* approximation to the solution of the latter model. In fact this seems to be true in many cases where the actions are measured in small units, e.g. in micro seconds when the resource means time. However, the discrete version often describes the problem equally well or even better than the continuous version.
(iii) We mention some difficulties when using continuous versions according to (ii) as approximations:

    (a) The solution of the continuous version requires, except for a few cases where an explicit solution exists, a discretization of the state and/or action space. Examples of this approach in the literature often include an analysis of the *discretization error*.
    (b) However, in the literature one rarely cares about the *continuation error* made when approximating the discrete version by the continuous version. Moreover, the continuation error and the discretization error should be added.
    (c) In continuous versions one must care about the existence of s_0-optimal action sequences or maximizers, a question often connected with the question of continuity of the value functions.

In view of the preceding comments we emphasize discrete DPs, and keep the treatment of the continuous versions brief. We now treat a continuous counterpart of the discrete Example 2.4.1. The only essential difference in the assumptions is the inclusion of a deterioration/expansion factor $z \in \mathbb{R}^+$.

Example 2.4.3 (Continuous allocation with utility function u)

(a) Consider the following DP: (i) the momentary resource s, consumption a and investment $s - a$ are non-negative reals; $D(s) = [0, s]$; (ii) the resource at time $t + 1$ equals $T(s_t, a_t) := z \cdot (s_t - a_t)$ for some $z \in \mathbb{R}^+$. Denote by s_0 the initial resource. We distinguish case 1: $z \leq 1$ and case 2: $z > 1$. As an example, if the resource consists of a perishable good, $z < 1$ may be a factor for the deterioration of the investment $s_t - a_t$. On the other hand, $z > 1$ occurs as interest factor for the investment when the resource is a capital. The maximal resource after N stages equals s_0 in case 1 and $z^N s_0$ in case 2. Therefore in case 1 we use $S = A = [0, K]$ where $K \geq s_0$, and $S = A = \mathbb{R}_+$ in case 2.

Again $r(s, a) := u(a)$ with increasing and non-negative utility function u and $V_0 = d \cdot u$ for some $d \in \mathbb{R}_+$. By Theorem 2.3.3 the VI holds and it reads

$$V_n(s) = \sup \{u(a) + \beta V_{n-1}(z \cdot (s - a)) : a \in [0, s]\}, s \in S. \tag{2.12}$$

Again f_n denotes the smallest maximizer at stage n, provided it exists. Sometimes another choice of actions a' (not applicable for the discrete version) is useful: If $s > 0$, then $a' :=$ the proportion a/s of the momentarily available amount s of resource, which is consumed; if $s = 0$, then a' may be chosen arbitrarily in $[0, K]$. The resulting DP$'$ differs from the original DP in the following data: $A' = [0, 1] = D'(s)$, $T'(s, a') = s \cdot (1 - a')$, $r'(s, a') = u(sa')$. Then DP$'$ has the value functions, starting with $V_0' := du$,

$$s \mapsto V_n'(s) = \sup \left\{u(sa') + \beta V_{n-1}(zs \cdot (1 - a')) : a' \in [0, 1]\right\}, \ n \geq 1, \tag{2.13}$$

which intuitively equals V_n. A formal proof uses induction on $n \geq 0$, and for fixed $s > 0$ the bijective substitution $a' := a/s$ in $W_n'(s, a')$.

As seen from (2.12) and (2.13), a function h_n from S into $[0, 1]$ is a maximizer at stage n in DP$'$ if and only if $s \cdot h_n(s)$ is a maximizer at stage n in DP.

We often use the following *abbreviation*: for $x \in \mathbb{R}$ and $n \in \mathbb{N}_0$ put

$$\sigma_n(x) := \sum_{t=0}^{n-1} x^t = \begin{cases} (1 - x^n)/(1 - x), & \text{if } x \neq 1, \\ n, & \text{if } x = 1; \end{cases} \tag{2.14}$$

in particular, $\sigma_0(x) = 0$ and $\sigma_1(x) = \sigma_n(0) = 1$ for $n \geq 1$.

(b) Explicit solutions exist rarely, e.g. if $u(a) = \sqrt{a}$ (cf. Example 4.1.6(a)) or if $\beta z = 1$ (cf. Problem 4.3.1). However, as we now indicate, for relatively general utility u the subsequent structural properties (b1)–(b7) of the solution are valid.

(b1) $V_n(s)$, $n \geq 1$, is increasing in s, non-negative and finite. Moreover,

$$V_n(s) \leq \sum_{\nu=0}^{n} \beta^\nu u(z^\nu s) + \beta^n d u(z^n s), \ s \in S, \text{ in case 1 and 2,}$$

$$V_n \leq [\sigma_n(\beta) + \beta^n d]u \text{ in case 1 },$$

$$V_n \leq [1/(1-\beta) + \beta^n d]u \text{ in case 1 and if } \beta < 1.$$

(b2) In case 1, if $0 \leq s \leq \bar{s} := \max\{0 \leq x \leq K : u(x) = 0\}$ and $n \geq 0$, then $V_n(s) = 0$. Moreover, each $y \in A^N(s)$ is s-optimal. These statements may be proved as in Example 2.4.1(c2), observing that $u(\bar{s}) = 0$ by continuity of u.

(b3) $V_n(s)$ is increasing in n [decreasing in n] if $d \leq 1$ [if $\beta < 1, d \geq 1/(1-\beta)$]. This may be proved as in Example 2.4.1(c3).

(b4) Let u be concave. Then all value functions V_n are concave, the smallest maximizers f_n, $n \geq 1$, exist and are increasing and $f_n(s') - f_n(s) \leq s' - s$ for $s \leq s'$. This follows from Example 8.2.14 below with $\eta_1 = u_2 \equiv 0$, $\eta_2(x) = zx$, and $u_1 = u$.

(b5) Let u be convex and $\beta < 1$. Then for some $m \in \mathbb{N}$ we have $V_n = u$ for all $n \geq m$, and for each s_0 the action sequence $(s_0, 0, \ldots, 0) \in A^N(s_0)$ is s_0-optimal for all $N \geq m$. This is shown in Example 7.3.5.

(b6) The value functions V_n, $n \geq 1$, are Lipschitz continuous in d and also in β, both uniformly in s. This may be proved as Example 2.4.1(c6).

(b7) As in the discrete version (see Proposition 2.4.2), in case $\beta < 1$ the sequence of value functions converges for $n \to \infty$ uniformly to some function V. However, the proof given in Theorem 10.1.10 differs from the proof of Proposition 2.4.2, and V cannot be defined recursively. ♦

2.5 Problems

Problem 2.5.1 Consider a DP where $S = A = \mathbb{N}_{0,K}$ for some $K \in \mathbb{N}$, $D(s) = \mathbb{N}_{0,s}$, $T(s,a) = s - a$, $r(s,a) = u(a)$ for some function u on A, $V_0(s) = d_0 \cdot u(s)$ for some $d_0 \in \mathbb{R}_+$, $\beta \in [0,1]$. Then for $n \geq 1$ and $s \in S$:

(a) $y = (a_t)_0^{n-1} \in A^n$ belongs to $A^n(s)$ if and only if $\sum_0^{n-1} a_t \leq s$.
(b) if $s \leq s'$ then $A^n(s) \leq A^n(s')$.
(c) if $s \leq s'$ and if u is increasing then $V_{ny}(s) \leq V_{ny}(s')$.
(d) if u is increasing, then $s \mapsto V_n(s)$ is increasing.

Problem 2.5.2 Consider a DP where $S = \mathbb{R}_+$, $A = [0,1]$, $D(s) = [0, \min\{1, s\}]$ and $T(s,a) = s - a$. Then for $n \geq 1$ and $s \in S$:

(a) $y = (a_t)_0^{n-1} \in A^n$ belongs to $A^n(s)$ if and only if $\sum_0^{n-1} a_t \leq \min\{1 + \sum_0^{n-2} a_t, s_0\}$;
(b) the properties (b)–(d) from Problem 2.5.1 remain true.

Problem 2.5.3 (Existence of an optimal action sequence without existence of maximizer) Consider the DP with $S = A = [0, 1]$; $D(s) := [0, s]$; $r(s, a) = 0$ for $s = a = 1$ and $= a/2$ else; $T(s, a) = (1 - a) \cdot s$; $V_0 \equiv 0$ and $\beta = 1$. Show that $(s, (1 - s) \cdot s)$ is the (unique) s-optimal action sequence for $DP_2(s)$, $s \in S$, and there exists no maximizer at stage 1.

2.6 Supplements

Supplement 2.6.1 (The discount factor) In economical applications β is usually smaller than one. In particular, if the length l of each period equals the k-th part of a year, if the annual interest rate equals i percent and if compound interest per period is assumed, then, since discounted rewards correspond to cash values at time zero, we have $1+i = 1/\beta^k = 1/\beta^{1/l}$, hence $\beta = \frac{1}{(1+i)^l} < 1$. Thus the larger l and/or i, the smaller β, and β approaches 1 when l tends to zero. If e.g. $i = 8\,\%$ then $\beta = 0.981$ if l is a quarter of a year and $\beta = 0.9936$ if l is one month. Moreover, if e.g. $l :=$ one hour and if N is not too large, let's say $N = 40$, then β can be practically taken equal to one.

If the N periods have nothing to do with time but mean that a certain activity is executed N times, then only $\beta = 1$ is meaningful.

The case $\beta > 1$ models the situation where the genuine discount factor equals some $\gamma < 1$ and where the one-stage reward increases from period to period (and similarly for the terminal reward) by the factor β/γ.

Supplement 2.6.2 (Changing the definition of an action) By another definition of the action in the continuous allocation problem from Example 2.4.3 one obtains three other formulations as follows, where S, V_0 and β remain unchanged.

(a) If a denotes the amount of the resource *not allocated* momentarily then $A = \mathbb{R}_+$, $D(s) = [0, s]$ for all s, $T(s, a) = z \cdot a$ and $r(s, a) := u(s - a)$ for $(s, a) \in D$.
(b) If a denotes the momentarily allocated *proportion* of the resource then $A = [0, 1] = D(s)$ for all s, $T(s, a) = zs(1 - a)$, $r(s, a) := u(s(1 - a))$ for $(s, a) \in D$.
(c) A further formulation is obtained if a denotes the momentarily not allocated *proportion* of the resource.

For some investigations the above formulations (a)–(c) have slight advantages over the formulation in Example 2.4.3.

Chapter 3
Additional General Issues

We present the basic theorems for cost minimization and for DPs with an absorbing set of states. We also prove the basic theorem using reachable states. The important notion of a bounding function is introduced.

3.1 The Basic Theorem for Cost Minimization

Our model for problems of ***cost minimization*** is a tuple $(S, A, D, T, c, C_0, \beta)$, where c denotes the ***one-stage cost function***, and C_0 the ***terminal cost function.*** The N-stage objective function for initial state s_0 is

$$y := (a_t)_0^{N-1} \mapsto C_{Ny}(s_0) := \sum_{t=0}^{N-1} \beta^t c(a_t, s_t) + \beta^N C_0(s_N), \quad y \in A^N(s_0).$$

Here $(s_t)_1^N$ is the decision process generated by (s_0, y). A minimum point of this function is called an ***s_0-optimal action sequence.*** The ***minimal N-stage cost for initial state s_0*** is

$$C_N(s_0) := \inf\{C_{Ny}(s_0) : y \in A^N(s_0)\}.$$

We call $C_N : S \to [-\infty, \infty)$ the N-stage ***minimal cost function.*** Moreover, we call a decision rule f_n a ***minimizer at stage $n \geq 1$*** if $f_n(s)$ is a minimum point of

$$a \mapsto W_n(s, a) := c(s, a) + \beta \cdot C_{n-1}(T(s, a)), \quad a \in D(s),$$

K. Hinderer et al., *Dynamic Optimization*, Universitext,
DOI 10.1007/978-3-319-48814-1_3

for all $s \in S$. Moreover, if f_n is a minimizer at stage $1 \leq n \leq N$, then $(f_n)_N^1$ is called an ***N*-stage minimizing policy**. Since minimizing $y \mapsto C_{Ny}(s)$ is the same as maximizing $y \mapsto -C_{Ny}(s)$, the Basic Theorem 2.3.3 yields immediately the following minimization counterpart.

Theorem 3.1.1 (Basic Theorem for cost minimization)

(a) The minimal cost functions C_n satisfy the VI in the form

$$C_n(s) = \inf\{c(s,a) + \beta \cdot C_{n-1}(T(s,a)) : a \in D(s)\},\ n \geq 1, s \in S. \tag{3.1}$$

*(b) The **Optimality Criterion** (**OC** for short): Let s_0 be an arbitrary initial state. If there exists a minimizing policy $(f_n)_N^1$ then an s_0-optimal action sequence $(a_t^*)_0^{N-1}$ is given by the following **forward procedure***

$$a_t^* := f_{N-t}(s_t), \quad s_{t+1} := T(s_t, a_t^*), \quad 0 \leq t \leq N-1.$$

Moreover, if the minimizing policy $(f_n)_N^1$ is unique, then $(a_t^*)_0^{N-1}$ is the unique s_0-optimal action sequence.

Example 3.1.2 (A linear-quadratic system with discrete control) One of the best known problems is the cost-minimal linear control of the position of a system under quadratic state and control costs (i.e. action costs). In the standard model one makes the somewhat unrealistic assumption that the controls are reals whose absolute value may be arbitrary large and also arbitrary small. The popularity of this model stems mainly from its explicit solvability; see Example 4.1.7. We now consider a model where the controls (i.e. the actions) only assume values in the finite set $A = D(s) = \mathbb{Z}(-m, m)$ for an integer $m \geq 1$; here $\mathbb{Z}(x, y) = \{x, x+1, \ldots, y\}$ for $x \leq y$ in $\mathbb{Z}$. Sometimes we also admit the less realistic case where $m = \infty$ with $\mathbb{Z}(-\infty, \infty) := \mathbb{Z}$. Moreover we use $S = \mathbb{Z}$, $T(s,a) = s - a$, the costs $c(s,a) := s^2 + a^2$ and $C_0(s) = ds^2$ for some $d \in \mathbb{R}_+$. No explicit solution seems to be possible, but we have the subsequent *structural properties*:

(a) *The minimal cost functions are finite, even and discretely convex, increasing on $\mathbb{N}_0$ and $C_n(0) = 0 \leq C_n(s)$ for all $n \geq 0$, $s \in S$.*
(b) *Denote the value functions by C_{nm}, $n \geq 0$, $1 \leq m \leq \infty$. It is intuitively clear that $C_{nm}(s)$ is decreasing in m and that $C_{n,\infty} \leq C_{nm}$ for all $m < \infty$.*

Proof of (a) If $m < \infty$ the finiteness of C_n results from the finiteness of $D(s) = A$, while in case $m = \infty$ finiteness follows from part (b). Next, both $C_n(0) = 0$ and $C_n(-s) = C_n(s)$, $s \in \mathbb{Z}$, follow easily by induction on $n \geq 0$ by observing that $C_n \geq 0$ due to $c \geq 0$ and $C_0 \geq 0$. Convexity is shown in Example 8.3.7. Then we see for $s \in \mathbb{N}_0$ that $C_n(s+1) - C_n(s) \geq C_n(1) - C_n(0) \geq C_n(0) - C_n(-1)$, and the last inequality yields $C_n(1) - C_n(0) \geq 0$, which completes the proof of isotonicity of C_n on $\mathbb{N}_0$.

Proof of (b) The latter property as well as $C_{n,m+1} \leq C_{nm}$ for $m < \infty$ follows by induction on $n \geq 0$. ♦

3.2 The Basic Theorem Using Reachable States

In general the numerical solution of $DP_N(s_0)$ via the VI algorithm requires finiteness of S, since otherwise the functions V_n, $1 \leq n \leq N-1$, can neither be completely computed nor stored. This finiteness restriction excludes, for instance, the linear-quadratic system with discrete and bounded control (i.e. $m < \infty$) in Example 3.1.2 above. The restriction can be dropped if for each $s \in S$ the set of ***states reachable from s in one step***, i.e. the set

$$R(s) := \cup_{a \in D(s)} \{T(s,a)\}$$

is finite. This is due to the fact that for the computation of $V_N(s_0)$ and an s_0-optimal action sequence by the VI algorithm one often does not need $V_n(s)$ and $f_n(s)$, $1 \leq n \leq N-1$, *for all s*. For example, if $R(s_0) \neq S$, the VI algorithm does not need $V_{N-1}(s)$ and $f_{N-1}(s)$ for those states s which are not reachable in one step from s_0.

If $(s_t)_1^n$ is generated by (s_0, y) we put $dp_n(s_0, y) := s_n$, $n \geq 1$. We say that the state $s \in S$ is ***reachable from*** $\boldsymbol{s_0 \in S}$ ***in*** $\boldsymbol{n \geq 1}$ ***steps*** if $s = dp_n(s_0, y)$ for some $y \in A^n(s_0)$. We denote by $\boldsymbol{R_n(s_0)}$, $\boldsymbol{n \geq 1}$, the ***set of states reachable in n steps from*** $\boldsymbol{s_0 \in S}$, i.e. $R_n(s) = \{dp_n(s,y) : y \in A^n(s)\}$. We have $R_1(s) = R(s)$. We also use $R_0(s) := \{s\}$. The sets $R_n(s)$ are finite if all sets $R(s)$, $s \in S$, are finite. The latter holds in particular if all sets $D(s)$ are finite.

It is intuitively clear that a state is reachable from s in $m+n$ steps if and only if it is reachable in n steps from some state which is reachable from s in m steps. In fact, the mapping $s \mapsto R(s)$ has the so-called *semigroup property*

$$R_{m+n}(s) = \bigcup_{s' \in R_m(s)} R_n(s'), \quad m, n \geq 0, s \in S. \tag{3.2}$$

In particular, the choices $n = 1$, $m = n$ and $m = 1$ yield recursions for the computation of $R_n(s)$:

$$R_{n+1}(s) = \bigcup_{s' \in R_n(s)} R(s') = \bigcup_{s' \in R(s)} R_n(s'), \quad n \geq 0, s \in S. \tag{3.3}$$

Proof of (3.2) The state x belongs to the right-hand side of (3.3) if and only if there exists $s' \in R_m(s)$ such that $x \in R_n(s')$ if and only if there exists $y \in A^m(s)$ such that $x \in R_n(dp_m(s,y))$ if and only if there exist $y \in A^m(s)$ and, using $s_m := dp_m(s,y)$, $y' \in A^n(s_m)$ such that $x = dp_n(s_m, y')$ if and only if there exists $(y, y') \in A^{m+n}(s)$ such that $x = dp_{m+n}(s, (y, y'))$ if and only if x belongs to $R_{m+n}(s)$. □

We give some examples for $R(s)$.

Example 3.2.1

(a) If $T(s,a) = a$ for all (s,a), then $R(s) = D(s)$, hence $R_t(s) = A^t(s)$.
(b) In the discrete allocation problem from Example 2.4.1 we have $R(s) = \mathbb{N}_{0,s}$, $1 \le s \le m$.
(c) In the freighter problem from Example 4.1.1 below we have $R(s) = \mathbb{N}_3 - \{s\}$, $s \in \mathbb{N}_3$. ♦

The special case $S_0 := \{s_0\}$ of Proposition 3.2.2 below shows that the computation of $V_N(s_0)$ and an s_0-optimal action sequence by the VI algorithm requires $V_n(s)$ and $f_n(s)$, $1 \le n \le N-1$, only for $s \in R_{N-n}(s_0)$. Thus we can solve $DP_N(s_0)$ for s_0 in a finite set S_0 of states by applying Proposition 3.2.2 separately for the singletons $\{s_0\}$, $s_0 \in S_0$. In Proposition 3.2.2 we present a more general approach. It is based on the observation that often the sets $R_{N-n}(s_0)$ and $R_{N-n}(s_0')$ overlap considerably for s_0' *near* s_0, and that V_n on $R_{N-n}(s_0) \cup R_{N-n}(s_0')$ may be used for solving $DP_N(s_0)$ as well as $DP_N(s_0')$.

For each finite subset S_0 of S and for $t \ge 0$ we consider the set $R_t(S_0) := \cup_{s \in S_0} R_t(s)$ which is the set of states reachable from S_0 in t steps. The next result holds without finiteness of S, although in most of its applications S is finite.

Proposition 3.2.2 (The Basic Theorem using reachable sets) *Fix a finite set S_0 of states and $N \ge 1$. Then the following hold:*

(a) The Modified VI: The restriction $\tilde{V}_n$ of V_n to $R_{N-n}(S_0)$, $1 \le n \le N$, satisfies

$$\begin{aligned} \tilde{V}_n(s) &= \sup_{a \in D(s)} [r(s,a) + \beta \tilde{V}_{n-1}(T(s,a))] \\ &=: \sup_{a \in D(s)} \tilde{W}_n(s,a), \quad s \in R_{N-n}(S_0). \end{aligned} \tag{3.4}$$

where $\tilde{V}_0 := V_0 | R_N(S_0)$. Thus $V_N = \tilde{V}_N$ on S_0,

(b) Modified OC: If there exists a maximum point $f_n(s)$ of $a \mapsto \tilde{W}_n(s,a)$ for $1 \le n \le N$, $s \in R_{N-n}(S_0)$, then for each $s_0 \in S_0$ the action sequence $(a_t^*)_0^{N-1}$ from (2.8) is defined and it is s_0-optimal.

Proof

(a) Fix $1 \le n \le N$ and $s \in R_{N-n}(S_0)$. Then $s \in R_{N-n}(s')$ for some $s' \in S_0$. Consequently $T(s,a) \in R_{N-n+1}(s')$, $a \in D(s)$, by (3.4), hence $V_{n-1}(T(s,a)) = \tilde{V}_{n-1}(T(s,a))$. Now (3.4) follows from the VI.
(b) It suffices to verify that $(a_t^*)_0^{N-1}$ satisfies the condition in (b). Firstly, obviously $(a_t^*)_0^{N-1}$ is admissible. Next, $s_t \in R_t(S_0)$ as stated above. As a consequence, $W_{N-t}(s_t, \cdot) = \tilde{W}_{N-t}(s_t, \cdot)$ by (a). Finally the desired condition is fulfilled by the construction of the f_n's. □

Remark 3.2.3

(a) The application of (3.4) requires the computation of the sets $R_t(s_0)$ for $1 \le t \le N-1$ by (3.3). Unfortunately, if S is not finite, the cardinality of the sets $R_t(s_0)$ often increases for $t \to \infty$ rapidly to ∞. Then Proposition 3.2.2 is applicable only for small values of N.
(b) If S and A are finite, both the VI algorithm and Proposition 3.2.2 may be applied for the solution of $DP_N(s_0)$ for $s_0 \in S_0$, and the latter requires fewer operations. However, the gain in number of operations must be contrasted with the time to compute the reachable sets.
(c) The solution of $DP_N(s_0)$ via Proposition 3.2.2 also yields without further computations the solution of the problems $DP_n(s)$ for $1 \le n \le N-1$ and $s \in R_{N-n}(s_0)$. ◇

Example 3.2.4 (Continuation of the linear-quadratic system of Example 3.1.2) A numerical solution by the VI fails due to $|S| = \infty$. However, for given $N \ge 1$ and $k \ge 1$ a numerical solution of $DP_N(s_0)$ for $|s_0| \le k$ can be found via Proposition 3.2.2. Since $C_n(0) = 0$, $n \ge 0$, and since the value functions are even, it suffices to compute $C_N(s_0)$ for $s_0 \in S_0 := \mathbb{Z}(1,k)$. The VI may require $V_n(s) = V_n(|s|)$ for some $0 \le n \le N-1$ and some $s < 0$. All sets $R_t(S_0) = \mathbb{Z}(-tm, k+tm)$ are finite. Now we get from (3.4) for the restriction $\tilde{C}_n$ of C_n to $R_{N-n}(S_0)$, $1 \le n \le N$, using $\tilde{C}_0(s) = ds^2$ for $s \in R_N(S_0)$,

$$\tilde{C}_n(-s) = \tilde{C}_n(s) = s^2 + \min_{|a|\le m} [a^2 + \beta \tilde{C}_{n-1}(|s-a|)],$$

$$=: \min_{|a|\le m} \tilde{W}_n(s,a), \quad 1 \le s \le k + (N-n)m.$$

If $1 \le n \le N$ and $0 \le s \le k + (N-n)m$ and if $f_n(s)$ is a minimum point of $\tilde{W}_{N-n}(s,\cdot)$, it is also a minimum point of $\tilde{W}_{N-n}(-s,\cdot)$. And then an s_0-optimal action sequence is defined as in (2.8).

As an example, for $m = d = 2$, $k = 10$, $\beta = 1$ and $N = 8$ we get $C_8|\mathbb{N}_{0,10} = (0, 2, 7, 15, 27, 44, 67, 97, 135, 182, 239)$. For *small* s these values are close to the corresponding values (0.00, 1.62, 6.47, 14.56, 25.89, 40.45, 58.25, 79.28, 103.55, 131.06, 161.80) in the continuous version Example 4.1.7; see also Fig. 3.1. In the present discrete case $y = (2,2,2,2,1,1,0,0) \in A^8(10)$ is 10- optimal. ◆

3.3 DPs with an Absorbing Set of States

Consider the allocation Example 2.4.1 with $\bar{s} := \max\{0 \le x \le K : u(x) = 0\}$. In Example 2.4.1(c2) we saw that $V_n(s) = 0 = f_{n+1}(s)$ for $s \in J_0 := \mathbb{N}_{0,\bar{s}}$ and $n \ge 0$, and that each $y \in A^N(s)$ is s-optimal. This was due to the fact that the system which starts at some state $s_0 \in J_0$ stays there and earns neither rewards nor a terminal reward. There are many DPs with the same property, which is formalized as follows.

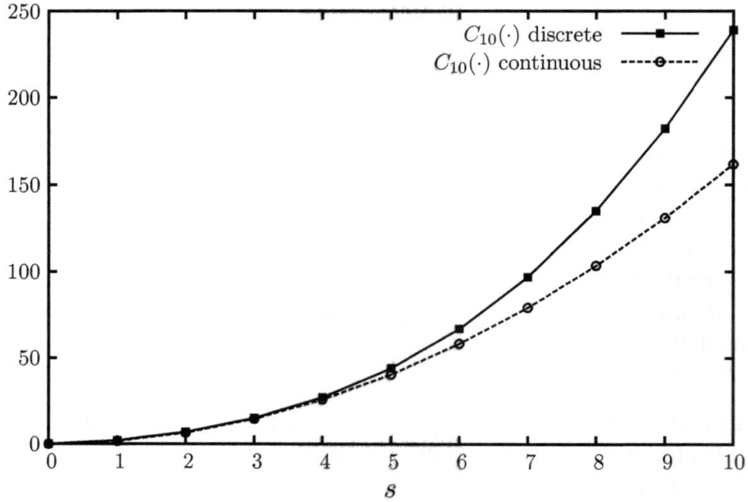

Fig. 3.1 Minimal cost function for the discrete linear-quadratic system in Example 3.2.4 and the corresponding continuous system in Example 4.1.7

Definition 3.3.1 A nonempty proper subset J_0 of the state space S is called an ***absorbing set*** for the DP under consideration if we have

- $T(s,a) \in J_0$ for all $s \in J_0$, $a \in D(s)$.
- $r(s,a) = 0$ for all $s \in J_0$, $a \in D(s)$.
- $V_0(s) = 0$ for all $s \in J_0$.

The set $J := S - J_0$ is called the ***essential state space*** (with respect to J_0). We do not call $\emptyset$ and S absorbing (although they trivially have the property of an absorbing set) since these two cases do not contain useful information.

From the VI Theorem 2.3.3 we obtain by induction on $n \geq 0$:

Lemma 3.3.2 (The value function on an absorbing set) *In a DP with absorbing set J_0 we have*

$$V_n = 0 \textit{ on } J_0 \textit{ for all } n \geq 0.$$

Thus one must perform the VI only for s in the essential state space. Moreover, for states s in an absorbing set we have $W_n(s,\cdot) = 0$, hence each action sequence admissible for such an s is s-optimal. In applications of a DP with an absorbing set J_0 we often omit $V_n(s)$ and $f_n(s)$ for $s \in J_0$. Note also that $R(J_0) = J_0$.

Remark 3.3.3

(a) Of course, a DP need not have an absorbing set of states; this holds, for example, if $r \equiv 1$.

(b) The union of absorbing sets is absorbing, unless the union equals S. Thus the union of all absorbing sets is the largest absorbing set, unless the union equals S.
(c) A subset of an absorbing set J_0 need not be absorbing. In particular, the singletons formed from the points in J_0 need not be absorbing, hence will not be called absorbing states. ◇

3.4 Finiteness of the Value Functions and Bounding Functions

Obviously $-\infty < V_n \leq \infty$ and $-\infty \leq C_n < \infty$, $n \geq 1$. For several purposes, e.g. for checking concavity or Lipschitz continuity of the value functions [minimal cost functions, one needs conditions under which $V_n[C_n]$ is finite or bounded. Already in Example 2.4.3 we have seen that the value functions are finite in the following continuous allocation problems I and II, defined as follows:

In model I we have $S = A = [0, K]$ where $K \geq s_0$; $D(s) = [0, s]$; $T(s, a) = z \cdot (s - a)$ for some $z \in (0, 1]$; $r(s, a) = u(a)$ for some increasing function u on A; $u(0) = 0$. Model II differs from model I only by $S = A = \mathbb{R}_+$; $z \in (1, \infty)$.

More models with finite value functions are treated in the next result.

Lemma 3.4.1

(a) All value functions are finite if there exists a maximizer at each stage, in particular if $D(s)$ is finite for all s. If $D(s)$ is a finite subset of $\mathbb{R}$ then there exists for $n \geq 1$ a smallest maximizer at stage n.
(b) All value functions are upper bounded (hence also finite) if r and V_0 are upper bounded. All value functions are bounded if r and V_0 are bounded.
(c) All minimal cost functions are lower bounded (hence also finite) if c and C_0 are lower bounded, in particular if $c \geq 0$ and $C_0 \geq 0$. All minimal cost functions are bounded if c and C_0 are bounded.

Proof

(a) If there exists a maximizer at stage n then there exists an s-optimal $y \in A^n(s)$ by Theorem 2.3.3(c). Thus $V_n(s) = V_{ny}(s) \in \mathbb{R}$. The second assertion is obviously true.
(b) From (2.3) we see that $V_{Ny}(s_0) \leq \sigma_N(\beta) \sup r + \beta^N \sup V_0$ and that $|V_{Ny}|(s_0) \leq \sigma_N(\beta) \sup |r| + \beta^N \sup |V_0|$. Now the assertions follow since $V_N(s_0) = \sup_y V_{Ny}(s_0)$ and $|V_N(s_0)| \leq \sup_y |V_{Ny}(s_0)|$.
(c) follows easily from (b). □

Boundedness of the value functions also holds if the DP has a bounding function (see Definition 3.4.2 below).

If both r and V_0 are bounded, then one easily derives from the VI by induction on $n \geq 0$, using $\sigma_n(\beta)$ from (2.14):

$$|V_n| \leq \sup |r| \cdot \sigma_n(\beta) + \beta^n \cdot \sup |V_0|, \quad n \geq 1. \tag{3.5}$$

In general, this bound is numerically rather conservative. We now generalize this result to the case where there exists a non-negative function b on S and constants $d_n \in \mathbb{R}_+$ such that $|V_n(s)| \leq d_n \cdot b(s) < \infty$, $n \geq 0$, $s \in S$. It will turn out that DPs with such value functions have properties very similar to properties of the DPs with bounded value functions.

Definition 3.4.2 A function $b \geq 0$ on S is called a ***bounding function*** for the DP if there exists a non-negative constant δ such that for all $(s, a) \in D$

$$|r(s,a)| \leq \delta b(s), \; |V_0(s)| \leq \delta b(s) \text{ and } b(T(s,a)) \leq \delta b(s). \tag{3.6}$$

This definition is also used for cost minimization problems with r and V_0 replaced by c and C_0, respectively.

Remark 3.4.3

(a) Often one obtains a hint for a possible bounding function b from $\underline{b}(s) := \sup_{a \in D(s)} |r(s,a)|$ as we must have $\underline{b} \leq b$.
(b) One may replace (3.6) by the formally weaker condition that $|r(s,a)| \leq \delta_1 b(s)$, $|V_0(s)| \leq \delta_2 b(s)$ and $b(T(s,a)) \leq \delta_3 b(s)$ for non-negative δ_1, δ_2 and δ_3, since then (3.6) holds with $\delta := \max\{\delta_1, \delta_2, \delta_3\}$.
(c) In our examples most bounding functions b satisfy $b \geq 1$. This condition could be included in our definition (thus avoiding division by zero in the definition of $\|v\|_b$ below). It is no genuine restriction since $b' := 1 + b$ is a bounding function whenever $b \geq 0$ is a bounding function. However, b may yield in (3.7) below a better bound than b'.
(d) Not every DP has a bounding function as this requires that $a \mapsto r(s,a)$ is bounded; this does not hold, for instance, in the linear-quadratic model in Example 3.1.2 if $m := \infty$ is also admitted here.
(e) If both r and V_0 are bounded then $b \equiv 1$ is a bounding function.
(f) More information about bounding functions is given in Chap. 10, where their role for DPs with large horizons is expounded. ◇

When dealing with bounding functions b it is advantageous to use the weighted supremum norm with weight function b, as follows. For arbitrary functions $b \geq 0$ and functions v on S put

$$\|v\|_b := \sup_{s \in S} |v(s)|/b(s), \quad \text{where } 0/0 := 0.$$

Thus $\|v\|_b$ is the smallest constant $\delta \in [0, \infty]$ for which $|v(s)| \leq \delta \cdot b(s)$ for all s. It is easily seen that $\|\cdot\|_b$ is a norm, called the ***b-norm***, on the linear space $\boldsymbol{B_b} = \boldsymbol{B_b(S)}$

of all functions v on S for which $\|v\|_b$ is finite. For functions w on D we put

$$\|w\|_b := \sup_{(s,a)\in D} |w(s,a)|/b(s).$$

Now we see: A function $b \geq 0$ is a bounding function if and only if

$$\|r\|_b,\ \|V_0\|_b \text{ and } \beta_b := \beta\|b \circ T\|_b \text{ are finite.}$$

Proposition 3.4.4 (Bounds for value functions) *If the DP has a bounding function b then*

$$|V_n(s)| \leq d_n b(s) := \Big[\|r\|_b \cdot \sigma_n(\beta_b) + \beta_b^n \cdot \|V_0\|_b\Big] \cdot b(s),\ n \geq 0,\ s \in S. \tag{3.7}$$

In particular, the functions V_n/b, $n \geq 0$, are bounded, hence all value functions are finite. Moreover, (3.7) *holds for cost minimization problems when V_n, r and V_0 are replaced by C_n, c and C_0, respectively.*

Proof We use induction on $n \geq 0$ for the assertion (I_n) that (3.7) is true. (I_0) holds by definition of $\|V_0\|_b$. Assume (I_n) for some $n \geq 0$. Then we obtain from the VI and from Appendix A.4.3 for $s \in S$

$$\begin{aligned} |V_{n+1}(s)| &= |\sup_{a\in D(s)} W_{n+1}(s,a)| \leq \sup_{a\in D(s)} |W_{n+1}(s,a)| \\ &\leq \sup_a |r(s,a)| + \beta\, d_n \cdot \sup_a b(T(s,a)) \\ &\leq (\|r\|_b + \beta_b d_n)\, b(s) = d_{n+1} b(s). \end{aligned}$$

The other assertions are obvious. □

Example 3.4.5

(a) *The discrete linear-quadratic model in Example* 3.1.2 *with $m < \infty$*. We already know from Example 3.1.2 that $0 \leq C_n < \infty$. We now show that $s \mapsto b(s) := 1 + s^2$ is a bounding function. In fact, we have $|c(s,a)| = s^2 + a^2 \leq s^2 + m^2 \leq m^2 b(s)$, $|C_0(s)| = ds^2 \leq db(s)$ and $b(T(s,a)) = 1 + (s-a)^2 \leq 1 + (s+m)^2 \leq 1 + (s+m)^2 + (m-s)^2 \leq 2(1+m^2)b(s)$.

   It follows from Proposition 3.4.4 that $C_n(s) \leq d_n(1+s^2)$, $s \in \mathbb{Z}$, for the constants d_n from (3.7) with $\|c\|_b \leq m^2$, $\|C_0\|_b \leq d$ and $\|b \circ T\|_b \leq 2(1+m^2)$.

(b) *Affine upper bounds for the value functions of an allocation problem.* Consider that variant of the continuous allocation problem from Example 2.4.3(a) where $S = \mathbb{R}_+$, $z \in \mathbb{R}^+$, u is non-negative and concave, and $V_0 = du$. Thus $V_n \geq 0$. We assert that for some $\delta \in \mathbb{R}_+$

$$V_n(s) \leq \delta \cdot [\sigma_n(\beta\delta) + d(\beta\delta)^n] \cdot (1+s), \quad n \geq 1, s \in \mathbb{R}_+. \tag{3.8}$$

As u is concave, there exists by Appendix D.2.4 some $\alpha, \gamma \in \mathbb{R}_+$ such that $u(s) \leq \alpha + \gamma s$, $V_0(s) = du(s) \leq d(\alpha + \gamma s)$, $s \in \mathbb{R}_+$. Put $\delta := \max\{\alpha, z\}$. It follows that $s \mapsto b(s) := 1+s$ is a bounding function as we obtain for $(s, a) \in D$ with $\delta := \max\{\alpha, \gamma, d\alpha, d\gamma, 1, z\}$,

$$\begin{aligned}
&0 \leq r(s,a) = u(a) \leq \alpha + \gamma s \leq \delta b(s), \\
&0 \leq V_0(s) \leq \delta b(s), \\
&b(T(s,a)) = 1 + z(s-a) \leq 1 + zs \leq \max\{1,z\}(1+s) \leq \delta b(s).
\end{aligned}$$

Now the upper bound in (3.8) follows from (3.7). ♦

3.5 Problems

Problem 3.5.1 (The number of policies and action sequences in $A^N(s_0)$) For finite S and A put $\underline{D} := \min_{s \in S} |D(s)|$, $\overline{D} := \max_{s \in S} |D(s)|$. Let $N \in \mathbb{N}$ and $s_0 \in S$. Then

(a)

$$\underline{D}^{N-1} \cdot |D(s_0)| \leq |A^N(s_0)| \leq \overline{D}^{N-1} \cdot |D(s_0)|,$$
$$|A^N(s_0)| \leq (\overline{D}/\underline{D})^N \cdot \underline{D}^{N \cdot |S|} \leq |F^N| = (\prod_{s \in S} |D(s)|)^N \leq (\underline{D}/\overline{D})^N \cdot \overline{D}^{N \cdot |S|}.$$

In particular, if $d := |D(s)|$ is independent of s, then

$$|A^N(s_0)| = d^N \text{ and } |F^N| = d^{N \cdot |S|}.$$

(b) We have $q := |F^N|/|A^N(s_0)| \geq \underline{D}^{N(|S|-1)}$. Equality holds if $|D(s)|$ is independent of s.

Problem 3.5.2 (The number of s_0-optimal policies) Let S be finite. Put $\alpha(s) := |D(s)|$. Then $\alpha := \prod_{s \in S} \alpha(s)$ equals the number of decision rules. Fix s_0 and assume that there is a single s_0-optimal action sequence for DP_N and let (s_ν) be the generated state sequence. Prove that then there are $\alpha^N / \prod_{\nu=0}^{N-1} \alpha(s_\nu)$ s_0-optimal N-stage policies.

Problem 3.5.3 (Numerical effort for solving $DP_N(s_0)$) Assume that S and A are finite and that $D(s) = A$ for all s.

(a) The VI requires in the last step only $|A|$ additions, the same number of multiplications and $|A| - 1$ comparisons. Simple enumeration requires (independent of $|S|!$) $N \cdot |A|^N$ multiplications, the same number of additions and $|A|^N - 1$ comparisons.

(b) The VI is superior to enumeration in numbers of multiplications, additions and comparisons if $|S| \le |A|^{N-1}/(N-1)$ but worse if $|S| \ge 2 \cdot |A|^{N-1}$.

Problem 3.5.4 Extend Remark 3.2.3(b) as follows. Assume that $D(s)$ is finite for all s (while S and A may be arbitrary). For a finite set $\emptyset \neq B \subset S$ put

$$A_B := \cup_{s \in B} D(s), \; D_B := \{(s,a) \in B \times A : a \in D(s)\}.$$

(a) The computation of $V_n|B$ and $f_n|B$ from V_{n-1} by the VI requires $|D_B| \le |A|$ multiplications, the same number of additions and $|D_B| - |B| \le |B| \cdot (|A| - 1)$ comparisons. If the DP is invariant, one needs only $|A_B| \le |A|$ multiplications.
(b) Use (a) to compute in terms of $|R_\nu(s_0)|, 0 \le \nu \le N-1$, the number of operations required for the solution of $DP_N(s_0)$ when combining the VI with the method of reachable state spaces $R_\nu(s_0)$.

Problem 3.5.5 (Equivalence of optimization by action sequences and by policies) Fix $s_0 \in S$. Let $\pi = (\pi_t)_0^{N-1} \in \mathbb{F}^N$. Here we use forward indexing of the decision rules, i.e. π_t is the decision rule applied in period t, $0 \le t \le N-1$. Each policy π generates an action sequence $y = g(s_0, \pi) = (a_t)_0^{N-1}$ (which is admissible for s_0, but in general not s_0-optimal) by means of

$$a_t := \pi_t(s_t), \quad s_{t+1} := T(s_t, a_t), \; 0 \le t \le N-1.$$

Thus $(s_t)_1^N$ is the decision process generated by s_0 and $y = g(s_0, \pi)$. Denote by $V_{N\pi}(s_0)$ the total N-stage reward $V_{Ny}(s_0)$ for $y := g(\pi, s_0)$, i.e.

$$V_{N\pi}(s_0) = \sum_{t=0}^{N-1} \beta^t \cdot r(s_t, \pi_t(s_t)) + \beta^N \cdot V_0(s_N). \tag{3.9}$$

We call $\pi^* \in \mathbb{F}^N$ an ***s_0-optimal policy***, if it is a maximum point of $\pi \mapsto V_{N\pi}(s_0)$. Moreover, π^* is called an ***optimal policy***, if it is s_0-optimal for all $s_0 \in S$. As seen from Problem 3.5.1, in general there are many more (s_0-optimal) policies than (s_0-optimal) action sequences. Then prove the following results:

(a) The problem of maximizing $y \mapsto V_{Ny}(s_0)$ on $A^N(s_0)$ is equivalent via $\pi \mapsto \Phi(\pi) := g(s_0, \pi)$ to the problem of maximizing $\pi \mapsto V_{N\pi}(s_0)$ on $\mathbb{F}^N$ in the following sense:

$$V_N(s_0) := \sup\{V_{Ny}(s_0) : y \in A^N(s_0)\} = \sup\{V_{N\pi}(s_0) : \pi \in \mathbb{F}^N\},$$

and a policy π^* is s_0-optimal if and only if the action sequence y^*, generated by (s_0, π^*), is s_0-optimal.
(b) Each s_0-optimal action sequence is generated by s_0 and some s_0-optimal policy.
(c) If f_n is a maximizer at stage n, $1 \le n \le N$, then $\pi^* := (f_n)_N^1$ is optimal.

Problem 3.5.6 (Non-stationary models) The theory developed for stationary DPs can be extended to N-stage non-stationary models, where the data may vary from period to period. From two possible frameworks, forward and backward indexing of data, we choose the former.

The model is determined by giving a terminal reward function and for each of the N periods a state space, an action space, a restriction set, a reward function and a discount factor. Most theoretical results carry over from Chap. 2 in a straightforward manner, though the formal apparatus becomes more complicated.

The formal definition of non-stationary models is as follows.

Definition A non-stationary N-stage DP_N (with forward indexing of data) is a tuple $((S_t)_0^N, (A_t)_0^{N-1}, (D_t)_0^{N-1}, (T_t)_0^{N-1}, (r_t)_0^{N-1}, V_N, (\beta_t)_0^{N-1})$ with the following meaning:

- S_t, A_t, $D_t \subset S_t \times A_t$, $\tilde{T}_t: \tilde{D}_t \to \tilde{S}_{t+1}$, $r_t: D_t \to \mathbb{R}$, and $\beta_t \in \mathbb{R}^+$ are the state space, action space, constraint set, transition function, reward function, and discount factor, respectively, for period t.
- $V_N: S_N \to \mathbb{R}$ is the terminal reward function.

The meaning of s_t and a_t remain unchanged, but now $y = (a_t)_0^{N-1}$ is admissible for s_0 if $a_t \in D_t(s_t)$ for $0 \le t \le N-1$. The N-stage reward for initial state s_0 and action sequence y is

$$V_{0y}(s_0) := \sum_{t=0}^{N-1} \left(\prod_{i=0}^{t-1} \beta_i \right) \cdot r_t(s_t, a_t) + \left(\prod_{i=0}^{N-1} \beta_i \right) \cdot V_N(s_N).$$

Of course, $V_0(s_0) := \sup\{V_{0y}(s_0) : y \text{ admissible for } s_0\}$ is the maximal N-stage reward for initial state s_0. Each non-stationary DP_N defines in an obvious way a family DP_{N-t}, $0 \le t \le N-1$, having as data that part of the data for DP_N which corresponds to the last $N-t$ stages of DP_N. Denote by V_t the value function for DP_{N-t}. A mapping $\pi_t: S_t \to A_t$ is called a *maximizer in period* t if for all $s \in S_t$ the action $\pi_t(s)$ is a maximum point of

$$a \mapsto W_t(s,a) := r_t(s,a) + \beta_t \cdot V_{t+1}(T_t(s,a)),\ 0 \le t \le N-1.$$

Then prove the

Basic Theorem for Non-stationary Deterministic DPs:

(a) V_0 may be computed by the value iteration

$$V_t(s) = \sup\{r_t(s,a) + \beta_{t+1} \cdot V_{t+1}(T_t(s,a)) : a \in D_t(s)\},$$
$$0 \le t \le N-1,\ s \in S_t.$$

(b) If π_t is a maximizer for period t, $0 \le t \le N-1$, then for arbitrary $s_0 \in S_0$ the sequence of actions

$$a_t^* := \pi_t(s_t^*),\ 0 \le t \le N-1,$$

is s_0-optimal, where $s_{t+1}^* := T_t(s_t^*, a_t^*)$ and $s_0^* := s_0$.

3.6 Supplements

Supplement 3.6.1 The so-called ***Principle of Optimality*** has played a central role in Dynamic Optimization Theory from the beginning. It is due to Bellman (1957, p. 83) who formulated it in his famous book *Dynamic Programming* as follows:

> … In each process, the functional equation governing the process was obtained by an application of the following intuitive:
> Principle of Optimality. *An optimal policy has the property that whatever the initial state and initial decisions are, the remaining decisions must constitute an optimal policy with regard to the state resulting from the first decision.*
> The mathematical transliteration of this simple principle will yield all the functional equations we shall encounter throughout the remainder of the book. A proof by contradiction is immediate …

There are two mathematical transliterations of the Principle, one in terms of action sequences and one in terms of policies:

(a) *If the action sequence* $(a_t)_0^{N-1}$ *is* s_0*-optimal for the* N*-stage problem and if* $s_1 := T(s_0, a_0)$*, then* $(a_t)_1^{N-1}$ *is* s_1*-optimal for the* $(N-1)$*-stage problem.*
(b) *If the policy* $(f, \sigma) \in F \times F^{N-1}$ *is* s_0*-optimal for the* N*-stage problem and if* $s_1 := T(s_0, f(s_0))$*, then* σ *is* s_1*-optimal for the* $(N-1)$*-stage problem.*

Discussion of the Principle of Optimality

(a) Obviously the Principle is only a necessary criterion for optimality of an action sequence or policy. However, sufficient criteria such as the OC Theorem 3.1.1(b) are much more useful. Therefore we cannot see much use for the Principle as formulated above. In fact, we do not know of any application. At most it could serve for the construction of counterexamples. As stated above, Bellman regarded the Principle as a basic tool for proving the VI. However, as far as we know, he did not use it in a formal proof of the VI. We will not use the Principle in the present book.
(b) The misinterpretations of the Principle in the literature seem to be due to a mixing up of the Principle with what we call the Basic Theorem (though these are definitely two different things). Another explanation for the misinterpretation may be that simply *necessity* and *sufficiency* was mixed up. In fact, the OC may be derived from the following result, which is in a sense the converse of the Principle and hence we call it the ***Inverse Principle of Optimality***: *If* a_0^* *is an*

optimal initial action for DP_N at state s_0 and if $(a_t^*)_1^{N-1}$ is an s_1^*-optimal action sequence for DP_{N-1} and $s_1^* := T(s_0, a_0)$, then $(a_t^*)_0^{N-1}$ is s_0-optimal for DP_N.

(c) Discussions of the Principle are given in Yakowitz (1969, p. 42), Hinderer (1970, p. 14), Porteus (1975) and in detailed form in Sniedovich (1992). The first paper which points out the need for a proof of the Basic Theorem (in its stochastic version) rather than relying on the Principle is Karlin (1955).

Supplement 3.6.2 (Uncertainty of the data) A problem in modeling is the fact that the data of the DP may not be known exactly. In particular, the following problems may arise:

(a) Often the data r and/or V_0 and/or β are not exactly known,
(b) often also N is not exactly known,
(c) often the data in (a) change with time.

These difficulties can be partially met (for fixed initial state s_0) as follows, assuming that optimal action sequences exist:

(a) One decides about a model with fixed horizon N which seems to be plausible and computes for it an optimal action sequence y^* and the maximal N-stage reward $V_N(s_0)$. Then, if we know bounds within which the data in (a) are likely to vary, one can try to find lower and upper bounds for the performance of y^*, i.e. for the N-stage reward obtained under y^* for any admissible choice of the data in (a). A cruder method for obtaining such bounds consists in computing the performance of y^* for some scenarios with respect to the data in (a) which are likely to occur. In particular, an analysis of the dependence of the solution on values of β in a (possibly two-sided) neighborhood of $\beta = 1$ may be desirable.

(b1) Assume we know that the exact horizon cannot be *large* and that we have decided to use an optimal N-stage action sequence $y^* = (a_t^*)_0^{N-1}$ for a plausible horizon N. If it turns out that we want or must run the decision process for N' more periods, then—whether or not the data has changed meanwhile—the best to do is to start a new N'-stage DP where the final state s_N^* of the first N-stage DP under y^* is used as the initial state of the new DP. This method is known under the heading *rolling horizon*.

(b2) Assume one knows that the exact horizon N must be very large. Then it is common to approximate problem DP_N by an infinite-stage problem DP_∞, or, as we prefer, to use in each period in state s an action $f(s)$ which is asymptotically optimal; cf. Chap. 10. Both methods may be questionable when β is near one and if the length of a period is *large*, as then typically r and/or V_0 change over time.

(c) If one knows at least bounds within which the data change with time one can try to apply a similar approach as in (a) above.

Supplement 3.6.3 (Historical remarks) Information about the impressive scientific work of R. Bellman (1920–1984) (e.g. Bellman 1957), the pioneer in the field of dynamic programming, can be found in Adomian and Lee (1986) and in the obituary of Gani (1984). A survey of his papers on applications of DP to computer science is given by Lew (1986). Maybe the first formal statement of the simple origin of the VI, namely Lemma 2.3.1, is due to Hinderer (1970, p. 17).

First computational efforts can be found in Dreyfus and Law (1977) and Morin (1978). Recent work on approximate dynamic programming includes Powell (2007) and Almudevar (2014). Simulation-based algorithms are contained in Chang et al. (2007). Of interest is also the neuro-dynamic programming approach by Bertsekas and Tsitsiklis (1996). For a recent survey on applications see van Roy (2002). The book of Sutton and Barto (1998) synthesized the relationship between dynamic programming and reinforcement learning.

Chapter 4
Examples of Deterministic Dynamic Programs

In this chapter we explicitly solve the following examples: optimal routing of a freighter, a production-inventory problem with linear costs, allocation and linear-quadratic problems and a scheduling problem. Then we discuss some further models: DPs with random length of periods and with random termination.

4.1 Examples with an Explicit Solution

Example 4.1.1 (Optimal routing of a freighter) Assume you own a freighter which sails between $l \geq 3$ different ports. It starts in a given port s_0, each trip lasts one week and after N trips the ship must sail into dry dock for overhaul in some of the ports. A trip from s to $a \neq s$ yields the freight revenue $r(s, a) \in \mathbb{R}_+$, and the overhaul in port s costs you $-V_0(s) \in \mathbb{R}_+$, $s \in S$. We use the following DP: $s_t \in S = A := \mathbb{N}_l$, $t \geq 1$, is the port after the t-th trip; the next port is $a_t := s_{t+1} \in D(s_t) := A - \{s_t\}$; $T(s, a) = a$; r and V_0 have the meaning stated above; $\beta = 1$. We want to find a route $(s_0, a_0, a_1, \ldots, a_{N-1})$ with maximal total net revenue. Note that $s_N = a_{N-1}$ is the port where the ship is overhauled. Obviously the total reward for the action sequence $y = (a_t)_0^{N-1}$ equals $V_{Ny}(s_0) = \sum_0^{N-1} r(s_t, a_t) + V_0(s_N)$. Assume e.g. $l = 3$, $s_0 = 1$, $V_0 = (-50, -40, -35)$, $N = 7$ and that r is given as in Table 4.1.

Firstly consider the naive ***myopic behavior*** where one sails from s to that port $g(s)$ for which $a \mapsto r(s, a)$ is maximal. Obviously $g(1) = 3$, $g(2) = 1$ and $g(3) = 2$. This determines in case $N = 7$ the route $(1, 3, 2, 1, 3, 2, 1, 3)$, yielding the total reward $25 + 10 + 15 + 25 + 10 + 15 + 25 - 35 = 90$. Why needn't this route be s_0-optimal for $s_0 = 1$? The answer is simple: While the trip from 1 to 3 yields maximal momentary reward, from port 3 only relatively unprofitable further trips are possible. In fact, the route $(s_0, y^*) := (1, 2, 1, 2, 1, 2, 1, 3)$ yields more, namely $20 + 15 + 20 + 15 + 20 + 15 + 25 - 35 = 95$. Let us see, whether y^* is 1-optimal.

K. Hinderer et al., *Dynamic Optimization*, Universitext,
DOI 10.1007/978-3-319-48814-1_4

Table 4.1 Freight revenues $r(s,a)$ for Example 4.1.1 with $l = 3$

	a		
s	1	2	3
1	–	20	25
2	15	–	10
3	0	10	–

Table 4.2 Solution of Example 4.1.1 for $N = 7$

$V_n(s)$			
	s		
n	1	2	3
0	−50	−40	−35
1	−10	−25	−30
2	−5	5	−10
3	25	10	15
4	40	40	25
5	60	55	50
6	75	75	65
7	95	90	85

$f_n(s)$			
	s		
n	1	2	3
0			
1	**3**	3	2
2	3	**1**	1
3	**2**	1	2
4	3	**1**	1
5	**2**	1	2
6	3	**1**	2
7	**2**	1	2

Using the VI

$$V_n(s) = \max_{a \neq s}\{r(s,a) + V_{n-1}(a)\}, \quad s \in \{1,2,3\},$$

both the value functions and the largest maximizers f_n can be found for a few n's by hand, and the reader is advised to do so. (When using a computer program for solving DP_N it is convenient to replace $D(s)$ in the VI by A and to put $r(s,s)$, $s \in S$, equal to a number λ which is so small that for each s and $1 \leq n \leq N$ the action s cannot belong to $D_n^*(s)$. As an example, for $N = 100$ one may take $\lambda = -6000$. For details see the remark below.)

We obtain e.g.

$$V_1(1) = \max_{a \neq 1}\{r(1,a) + V_0(a)\} = \max\{20 - 40, 25 - 35\} = -10, f_1(1) = 3.$$

Table 4.2 shows V_n and f_n for $1 \leq n \leq 7$. The bold-face numbers in the table for f_n mark a 1-optimal route in $A^7(1)$, which indeed equals $(1, y^*)$.

We claim that now V_N for arbitrary $N \geq 7$ can be easily found without further maximizations: as an example, for $N = 100$, we have $V_{100} = (1720, 1720, 1710)$. With additional work one sees that the route consisting of 49 cycles $2 \to 1 \to 2$, followed by $(1, 3)$ is s_0-optimal for $s_0 = 2$ and $N = 100$. If you cannot verify these two results, go to Proposition 4.1.4 below. Note that Table 4.2 also solves the problems $DP_n(s_0)$, $1 \leq n \leq 6$.

A variant of our problem treats the case where the ship must eventually be in port e (possibly with an $(N+1)$-st trip to port e) where overhaul costs $h \geq 0$. Then one uses $V_0(s) := r(s,e) + h$ for $s \neq e$ and $V_0(e) := h$. ♦

Remark 4.1.2 (Replacement of $D(s)$ by A) Consider a DP with finite state and action space and where $D(s) \neq A$ for at least one s. Consider the DP′ which differs from DP only by the following data: $D'(s) = A$; for $a \in D(s) : T'(s,a) = T(s,a)$ and $r'(s,a) = r(s,a)$; for $a \notin D(s)$: $T'(s,a) = s$ and $r'(s,a) = \lambda$, where $\lambda \in \mathbb{R}$. Then DP_N has the same solution as DP'_N, $N \geq 1$, provided λ is small enough, e.g. if

$$\lambda = -1 - 2\Big[\|r\| \cdot \sigma_N(\beta) + \max\{1, \beta^N\} \cdot \|V_0\|\Big].$$

Here $\|\cdot\|$ is the maximum norm. For the proof we firstly note that for $1 \leq n \leq N$ we have $V_n = V'_n$ and $D_n^* = (D')_n^*$ if we have both $(I_n) : V'_{n-1} = V_{n-1}$ and $(J_n) : W'_n(s,b) < W'_n(s,a)$ for $b \notin D(s)$ and $a \in D(s)$.

In fact, (I_n) implies $W_n(s,a) = W'_n(s,a)$, $a \in D(s)$, and then (J_n) ensures that $V_n = V'_n$ and $D_n^* = (D')_n^*$. We now prove $(I_n) \wedge (J_n)$ by induction on $1 \leq n \leq N$ and by observing that $\lambda \leq \|r\| - 1 - 2[\|r\| \cdot \sigma_n(\beta) + \beta^n \|V_0\|]$. Firstly (I_1) holds trivially. Next we get (J_1) from

$$W'_1(s,b) = \lambda + \beta V_0(s) < -\|r\| - \beta\|V_0\| \leq r(s,a) + \beta V_0(T(s,a)) = W'_1(s,a).$$

Now assume $(I_n) \wedge (J_n)$ for some $1 \leq n < N$. Then (I_{n+1}) holds as shown above. Finally (J_{n+1}) holds since

$$\begin{aligned}
W'_{n+1}(s,b) &= \lambda + \beta V_n(s) < \|r\| - 2\Big[\|r\|\sigma_{n+1}(\beta) + \beta^{n+1}\|V_0\|\Big] \\
&\quad + \beta\Big[\|r\|\sigma_n(\beta) + \beta^n\|V_0\|\Big] \leq -\|r\| - \beta\Big[\|r\|\sigma_n(\beta) \\
&\quad + \beta^n\|V_0\|\Big] \leq r(s,a) + \beta V_n(T(s,a)) = W'_{n+1}(s,a).
\end{aligned}$$

For the proof of the last inequality we used (3.5). ◇

Example 4.1.3 (A production-inventory problem with linear costs) A firm can produce at the beginning of each of N time periods at most $b \in \mathbb{N}$ pieces of a certain item and it can store at most $B \in \mathbb{N}$ pieces, $B \geq b$, of the items. During each period a known deterministic demand of $z \leq b$ pieces arises. The production of $a > 0$ pieces costs $e_1 + e_2 a$ monetary units, an inventory of y pieces at the end of any period results in $e_3 y$ units of holding costs for that period, and a final inventory of s_N pieces has a *scrap* value of $e_4 s_N$ units. Of course, e_1 through e_4 are assumed to be non-negative. How can one find, given an initial inventory of amount $s_0 \in \mathbb{N}_{0,B}$, a production schedule which minimizes the sum of all discounted production and inventory costs minus the scrap value of the final inventory? This problem can be modeled as follows. As state s_t we take the inventory at time t. Then $S = \mathbb{N}_{0,B}$. As action a_t we take the number of pieces produced at time t, hence $A = \mathbb{N}_{0,b}$. The new inventory

$$s_{t+1} = T(s_t, a_t) := s_t + a_t - z, \quad 0 \le t \le N-1,$$

must belong to S. Thus a_t must be chosen such that both $0 \le a_t \le b$ and $0 \le s_{t+1} \le B$, i.e. $z - s_t \le a_t \le B + z - s_t$. Therefore the set of admissible actions at state $s \in \mathbb{N}_{0,B}$ equals

$$D(s) = \{a \in \mathbb{N}_{0,b} : z - s \le a \le B + z - s\}.$$

$D(s)$ is non-empty for all s since $z \in D(s)$; see Fig. 4.1 for the numerical data given below; obviously $C_0(s) = -e_4 s$ and

$$c(s,a) = \begin{cases} e_1 + e_2 a + e_3(s + a - z), & \text{if } a > 0, \\ e_3(s - z), & \text{if } a = 0. \end{cases}$$

It is easy to set up the VI, to program it and to compute the numerical solution for real-world data. For the understanding of the VI algorithm it is useful to do some computation by hand for very small data, e.g. $z = 2, b = 3, B = 4, e_1 = 10, e_2 = 2, e_3 = e_4 = \beta = 1, N = 6$. Then (3.1) reads

$$C_n(s) = \min\{c(s,a) + C_{n-1}(s + a - 2) : a \in D(s)\}, \quad n \ge 1, 0 \le s \le 4, \qquad (4.1)$$

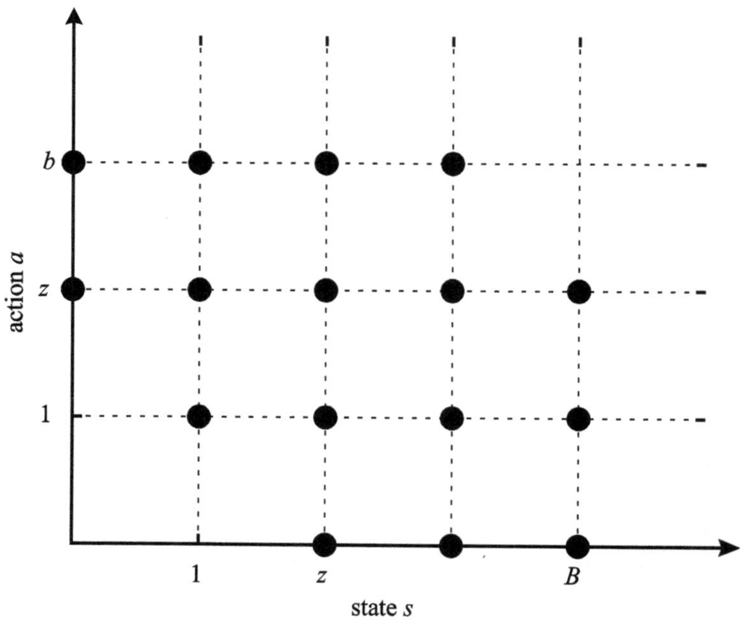

Fig. 4.1 Constraint set D for Example 4.1.3

Table 4.3 $C_n(s)$ and $D_n^*(s)$ for the production-inventory problem from Example 4.1.3

$C_n(s)$						$D_n^*(s)$					
	s						s				
n	0	1	2	3	4	n	0	1	2	3	4
0	0	−1	−2	−3	−4	0					
1	14	12	0	0	0	1	**2**	1	0	0	0
2	28	18	14	13	2	2	2	3	**0**	0	0
3	35	32	28	19	16	3	3	**3**	0	0	0
4	49	46	35	33	30	4	2, **3**	3	0	0	0
5	63	53	49	47	37	5	**2,3**	3	0	0	0
6	70	67	63	54	51	6	3	3	**0**	0	0

with $C_0(s) := -s$ and

$$D(s) = \{a \in \mathbb{N}_{0,3} : 2 - s \leq a \leq 6 - s\},$$

$$c(s,a) = \begin{cases} 8 + s + 3a, & \text{if } a > 0, \\ s - 2, & \text{if } a = 0. \end{cases}$$

From (4.1) one obtains the minimal cost functions C_n and the sets $D_n^*(s)$ of minimum points of $W_n(s, \cdot)$ for $n \leq 6$ as shown in Table 4.3. (It is useful first to tabulate the values of $T(s,a)$ and $c(s,a)$ for all $(s,a) \in D$). For $s_0 = 2$ we obtain the minimal 6-stage cost $C_6(2) = 63$ and the three s_0-optimal action sequences (0,2,2,3,3,0), (0,3,3,0,2,2) and (0,2,3,3,0,2). The s-position of the bold-face numbers a_t^* in the table for $D_n^*(s)$ marks the inventories

$$s_{t+1} := s_t + a_t^* - 2, \quad 0 \leq t \leq 5,$$

under the 2-optimal action sequence $(0, 2, 3, 3, 0, 2)$.

From $C_1 \geq C_0$ and the VI one easily derives (cf. Theorem 6.1.1) the intuitively obvious fact that $C_n(s)$ is increasing in $n \geq 0$. In particular the minimal n-stage cost functions are non-negative since $C_n \geq C_1 \geq 0$.

At first sight it seems hopeless to obtain an explicit solution for arbitrary horizon, and a numerical solution for large horizon, say $N = 600$, seems to require the aid of a computer. However, C_3 and C_6 differ only by the constant value 35 and hence the sets $D_3^*(s)$ and $D_6^*(s)$ coincide. This fact permits an *explicit* solution because of the next result, whose simple proof is omitted. ♦

Proposition 4.1.4 (Periodicity of value functions and of maximizers with respect to the stage number) *Assume that $\beta = 1$ and that there exist constants $m \geq 0$, $\tau \geq 1$ and $\alpha \in \mathbb{R}$ such that $V_{m+\tau} = V_m + \alpha$. Then for all $\nu \geq 1$ and $1 \leq i \leq \tau$ we have:*

(a) $V_{m+\nu\tau+i} = V_{m+i} + \nu \cdot \alpha$, hence $\frac{V_n}{n} \to \frac{\alpha}{\tau}$ for $n \to \infty$.
(b) If g_i is a maximizer at stage $m+i$, then g_i is also a maximizer at stage $m+\nu\tau+i$.

If V_n has been computed for $n \leq m+\tau$ and if the assumptions of Proposition 4.1.4 are fulfilled then V_N is known for all $N \geq 1$ by Proposition 4.1.4(a). Moreover, one obtains from maximizers f_n at stage $1 \leq n \leq m + \tau$ (if these maximizers exist) for each $N \geq m + \tau + 1$ a maximizing N-stage policy π as follows: Let $N = m + \nu\tau + i$ for appropriate $\nu \geq 1$ and $1 \leq i \leq \tau$; then π begins with $(f_{m+i}, f_{m+i-1}, \ldots, f_{m+1})$, followed by ν repetitions of $(f_{m+\tau}, f_{m+\tau-1}, \ldots, f_{m+1})$ and ends with $(f_m, f_{m-1}, \ldots, f_1)$.

Unfortunately the simple periodicity of the sequence $(f_n)_{m+1}^{\infty}$ expressed in Proposition 4.1.4(b) does not carry over to s_0-optimal action sequences.

As seen from Table 4.3 our production-inventory problem satisfies $C_6 = C_3 + 35$ and hence the first assumption of the cost minimization version of Proposition 4.1.4 for $m = \tau = 3$ and $\alpha = 35$. Therefore $C_{3(\nu+1)+i} = C_{3+i} + 35\nu$ for $\nu \geq 1$ and $i = 1$, 2, 3. As an example, $C_{600} = C_3 + 6965$. Moreover, since $f^* := (3, 3, 0, 0, 0)$ is a minimizer at stages $n = 3$, 4 and 5, f^* is a minimizer at all stages $n \geq 3$. ($N^* = 3$ is called a ***turnpike horizon*** for f^*.) Verify that for $s_0 = 0$ the (unique) s_0-optimal action sequence for horizon $N = 3k$ consists of k repetitions of the string $(3, 3, 0)$.

We will sometimes use

Lemma 4.1.5 (Linear first order difference equation) *The recursion for real numbers b_n, $n \geq 0$ of the form*

$$b_n = \gamma + \alpha b_{n-1}, \quad n \geq 1,$$

where γ and α are reals, has the solution

$$b_n = \gamma \cdot \sigma_n(\alpha) + \alpha^n \cdot b_0, \quad n \geq 1. \tag{4.2}$$

Here $\sigma_n(\alpha)$ is defined in (2.14). The easy proof by induction on $n \geq 0$ is omitted.

Example 4.1.6 (Continuous allocation with square root utility) The single-product problem. For the continuous allocation problem from Example 2.4.3 with *square root utility* $u(a) = \sqrt{a}$ of the allocation a we now derive an explicit solution. This problem and a few others can also be solved by the classical static method, but for reasons of illustration we present the DP method here; it requires about the same effort as the static method. Set $S = A = \mathbb{R}_+$ and assume $V_0(s) = d_0\sqrt{s}$ for some $d_0 \in \mathbb{R}_+$. Put

$$d_n^2 = \sigma_n(\beta^2 z) + (\beta^2 z)^n \cdot d_0^2, \quad n \geq 1. \tag{4.3}$$

Then the following assertion (I_n) holds:

(i) $V_n(s) = d_n\sqrt{s}$, $n \geq 1$, $s \in \mathbb{R}_+$.
(ii) For each $n \geq 1$ the decision rule $s \mapsto f_n(s) := s/d_n^2$ is the unique maximizer at stage n.

Proof We verify the assertion (I_n) by induction on $n \geq 1$. The VI has the form

$$V_n(s) = \sup_{0 \leq a \leq s} W_n(s,a), \quad n \geq 1, s \in [0,K],$$

where

$$W_n(s,a) := \sqrt{a} + \beta V_{n-1}(z \cdot (s-a)).$$

For the proof of (I_1) fix $s \geq 0$ and put

$$w(a) := W_1(s,a) = \sqrt{a} + \beta d_0 \sqrt{z \cdot (s-a)}, \quad 0 \leq a \leq s.$$

Obviously $V_1(0) = f_1(0) = 0$. Thus assume $s > 0$. Since $d_0 \geq 0$ and since u is concave, w is concave by Appendix D.2.9. In addition w' exists on $(0,s)$. Put $\gamma_0 := (\beta d_0)^2 z$. Now Appendix D.3.3(a) tells us that a point $a^* \in (0,s)$ is a maximum point of w if and only if $w'(a^*) = 0$. This is equivalent, as a simple calculation shows, to $a^* = s/(1+\gamma_0)$. Since in addition w is strictly concave, $s \mapsto f_1(s) := s/d_1^2$ is the unique maximizer at stage 1. Moreover, $V_1(s) = W(s, f_1(s)) = d_1\sqrt{s}$ where $d_1^2 = 1 + \beta^2 d_0^2 z$. Thus (I_1) is true.

Now assume that (I_n) holds for some $n \geq 1$. Then a computation similar to the proof of (I_1) with d_0 replaced by d_n verifies (I_{n+1}). The numbers $d_n \in \mathbb{R}^+$, $n \geq 1$, are given by the recursion $d_n^2 = 1 + \beta^2 d_{n-1}^2 z$, $n \geq 1$, which by (4.2) with $b_n := d_n^2$, $\gamma := 1$ and $\alpha := \beta^2 z$ has the solution (4.3). This proves (I_{n+1}). □

Note also that the forward procedure (2.8) implies for $N \geq 1$ by induction on t: for $s_0 \geq 0$ the unique s_0-optimal action sequence $(a_t^*)_0^{N-1}$ is given by

$$a_t^* := (\beta z)^{2t} s_0 / d_N^2.$$

The multi-project problem with pooled allocations. This problem has a similar explicit solution as the single-product model; in particular the constants d_n are the same as in (a). However, the proof is more demanding. Only for simplicity of notation we do restrict ourselves to the case $z := 1$. The problem concerns the managing of $m \geq 2$ projects. For project $1 \leq j \leq m$ we have an upper bound K_j for the available resource. At time $t \geq 0$ there is available for project j a resource of amount $s^j \in [0, K_j]$; we put $s := (s^j)_1^m \in S = \times_1^j [0, K_j]$. At time $t \geq 0$ the allocations of amount $a^j \in [0, s^j]$ for all projects $1 \leq j \leq m$ are pooled together with square root utility.

We use $A = \times_1^m [0, K_j]$. Thus the vector $a := (a^j)_1^m \in D(s) := \times_1^m [0, s^j]$ of allocations yields the utility $u(\sum_1^m a^j)$ where $u(x) := x^{1/2}$, $x \in \mathbb{R}_+$. The vector of resources at time $t+1$ equals $T(s,a) := s - a$. There is a discount factor $\beta \in \mathbb{R}^+$ and a terminal reward $V_0 = d_0 u$ for some $d_0 \in \mathbb{R}_+$.

Using d_n^2 from (4.3) with $z := 1$, the solution consists of the following assertions (I_n), $n \geq 1$:

(i) $V_n(s) = d_n(\sum_1^m s^j)^{1/2}$, $s \in S$,
(ii) *the decision rule* $s \mapsto f_n(s) := s/d_n^2$ *is a maximizer at stage* n.

Proof We verify (I_n) by induction on $n \geq 1$. For the proof of (I_1) fix $s \in S$. If $s = \mathbf{0}$, then $V_1(s) = 0$ since $\{\mathbf{0}\}$ is absorbing. Moreover, $f_1(s) = \mathbf{0}$ is obvious. If $s \neq \mathbf{0}$ the set $M := M(s) := \{1 \leq j \leq m : s^j > 0\}$ is non-empty. For $a \in D(s)$ put $a_M := (a^j, j \in M)$. We abbreviate $\sum_{j \in M}$ by $\sum_M$. Now we have for $a \in D(s)$

$$w(a) := W_1(s, a) = \left(\sum_M a^j\right)^{1/2} + \beta d_0 \left(\sum_M (s^j - a^j)\right)^{1/2}$$
$$=: w_M(a_M).$$

Put $D_M(s) := \times_{j \in M}[0, s^j]$. We consider the mapping h from $D(s)$ into $D_M(s)$, defined by $h(a) := a_M$. This mapping is bijective with inverse $h^{-1}(b) := a$, where $a^j := b^j$ if $j \in M(s)$, and $a^j := 0$, else. If w_M has the maximum point $b^* \in D_M(s)$ then bijectivity of h ensures that $h^{-1}(b^*)$ is a maximum point of w; cf. Appendix A.4.11.

It remains to find b^*. From concavity of u and of $a_M \mapsto \sum_M a^j$ we conclude by Appendix D.2.9 that w_M is concave. In addition $\operatorname{grad} w_M$ exists on the interior of $D_M(s)$. Put $\gamma_0 := (\beta d_0)^2$. From Appendix D.3.1(c) we know that an interior point b^* of $D_M(s)$ is a maximum point of w_M if $\operatorname{grad} w_M(b^*) = \mathbf{0}$. This holds, as a simple computation shows, if and only if $\sum_M b^{*j} = \sum_M s^j/(1 + \gamma_0)$, in particular if $b^* := (s^j, j \in M)/(1 + \gamma_0)$. Since $h^{-1}(b^*) = s/(1 + \gamma_0)$, a maximizer at stage 1 is $f_1(s) := s/(1 + (\beta d_0)^2)$. Moreover, $V_1(s) = W_1(s, f_1(s)) = d_1(\sum_1^m s^j)^{1/2}$. Thus (I_1) is true.

Now assume that (I_n) holds for some $n \geq 1$. Then a computation similar to the proof of (I_1) with d_0 replaced by d_n verifies (I_{n+1}). □

Note that the numbers d_n, $n \geq 0$, do not depend on the number m of projects. Moreover, the result above can be extended to the case where $u(x) := x^\alpha$ for some $\alpha \in (0, 1)$.

Dependence of the solution on parameters. We now study the dependence of the sequence $(d_n)_0^\infty$, which completely determines the solution, on the parameters n, d_0, β, z and $\alpha := \beta^2 z$. From (4.3) we see that d_n depends on β and z only via α. Figure 4.2 shows the following behavior of $n \mapsto d_n$ for several choices of α and d_0:

(i) For fixed n, the sequence d_n is increasing in d_0, β and z. This means, for example, that the optimal investment proportion $(s_t^* - a_t^*)/s_t^* = 1 - 1/d_{N-t}^2$ in period t is increasing in d_0, β and z, a fact which is to be expected.
(ii) One might expect that d_n always increases in n since more decisions allow a better distribution of the initial resource to the several periods. However, if

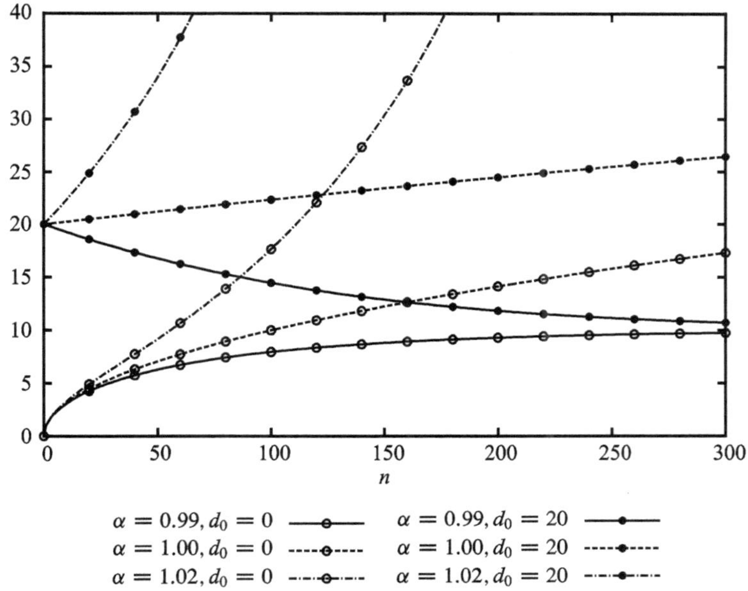

Fig. 4.2 The numbers d_n of Example 4.1.6 for $\alpha := \beta^2 z = 0.99$, 1 and 1.02 and $d_0 = 0$ and $d_0 = 20$

$\beta < 1$ the terminal reward $d_0 \cdot \beta^N \sqrt{s_N}$ may be decreasing in N which may destroy isotonicity of d_n. More precisely, the following holds:

If $\alpha \geq 1$, then d_n increases in n due to (4.3) strictly towards ∞.

If $\alpha < 1$, then α^n converges towards zero, hence d_n converges by (4.3) for $n \to \infty$ towards $d_\infty := 1/\sqrt{1-\alpha}$. Moreover, as

$$d_{n+1}^2 - d_n^2 = \alpha^n(1 - d_0^2/d_\infty^2) \stackrel{\geq}{\underset{<}{=}} 0 \quad \text{if and only if} \quad d_0 \stackrel{\leq}{\underset{>}{=}} d_\infty,$$

we see that

$$(d_n) \text{ is } \begin{cases} \text{strictly increasing,} & \text{if } d_0 < d_\infty, \\ \text{constant} = d_0, & \text{if } d_0 = d_\infty, \\ \text{strictly decreasing,} & \text{if } d_0 > d_\infty. \end{cases}$$

Example 4.1.7 (The one-dimensional linear-quadratic problem with quadratic costs) We consider the one-dimensional version of one of the most important and best studied problems in engineering and production smoothing; for the latter application see e.g. Heyman and Sobel (1984, p. 407) or Schneeweiss (1974, p. 84 and 114). Assume that the state $s_t \in \mathbb{R}$ of some system (e.g. the location of an object moving on the real line or an inventory with backlogging of a production process) can be controlled by arbitrary translations of amount $-a_t \in \mathbb{R}$. The goal consists in keeping the states $s_1, s_2, \ldots, s_N$, given an initial state s_0, in a cost-minimal manner close to the origin. Closeness to the origin of s_t, $1 \le t \le N-1$, and of s_N is measured by $\lambda \cdot s_t^2$ and $d_0 \cdot s_N^2$, respectively, and the costs for the control a are $\gamma \cdot a^2$ for some constants $\lambda, d_0 \in \mathbb{R}_+$ and $\gamma \in \mathbb{R}^+$. We want to minimize

$$y := (a_t)_0^{N-1} \mapsto \tilde{C}_{Ny}(s_0) := \gamma a_0^2 + \sum_{t=1}^{N-1} \beta^t \cdot (\lambda s_t^2 + \gamma a_t^2) + d_0 \beta^N s_N^2,$$

where $a_t \in \mathbb{R}$ and $s_{t+1} := s_t - a_t$, $0 \le t \le N-1$. More generally we treat the case of the transition law $s_{t+1} := g \cdot s_t - h \cdot a_t$ for real constants g and h.

It is convenient to solve the equivalent problem of minimizing

$$C_{Ny}(s_0) := \tilde{C}_{Ny}(s_0) + \lambda s_0^2 = \sum_{t=0}^{N-1} \beta^t \cdot (\lambda s_t^2 + \gamma a_t^2) + d_0 \beta^N s_N^2.$$

This problem is a DP_N with $S = A = D(s) = \mathbb{R}$, $T(s,a) = gs - ha$, $c(s,a) = \lambda s^2 + \gamma a^2$ and $C_0(s) = d_0 s^2$. (The name of the problem is due to the linear form of T and the quadratic form of c.) The VI reads

$$C_n(s) = \lambda s^2 + \inf_{a \in \mathbb{R}} \left[\gamma a^2 + \beta C_{n-1}(gs - ha)\right], \quad n \ge 1, s \in \mathbb{R}.$$

It is easy to compute $C_1(s)$ by minimizing the function

$$\mathbb{R} \ni a \mapsto w(a) := \gamma a^2 + \beta d_0 \cdot (gs - ha)^2,$$

which is strictly convex by Appendix D.2.5 since $a \mapsto w'(a) = 2 \cdot (\gamma + \beta d_0 h^2) a - 2\beta d_0 ghs$ is strictly increasing. By Appendix D.3.1(c) the unique zero of w' is the unique minimum point $f_1(s)$ of w. A simple computation shows that $f_1(s) := \beta d_0 gh \cdot s/(\gamma + \beta h^2 d_0)$, and that $C_1(s) = \lambda s^2 + w(f_1(s)) = d_1 s^2$, where $d_1 := \lambda + \beta\gamma g^2 d_0/(\gamma + \beta h^2 d_0)$.

As C_1 has the same functional form as C_0, we obtain by induction on n the following result:

Let $d_0 \in \mathbb{R}_+$, $\lambda \in \mathbb{R}_+$, $\gamma \in \mathbb{R}^+$, g and h real. Define the sequence $(d_n)_1^\infty$ of non-negative numbers recursively by

$$d_n = \lambda + \frac{\beta\gamma g^2 d_{n-1}}{\gamma + \beta h^2 d_{n-1}}, \quad n \ge 1. \tag{4.4}$$

Then for $n \ge 1$:

(a) $C_n(s) = d_n \cdot s^2$, $s \in \mathbb{R}$.
(b) The unique minimizer at stage n is

$$s \mapsto f_n(s) = \beta d_{n-1} gh \cdot s/(\gamma + \beta h^2 d_{n-1}). \tag{4.5}$$

The numerical computation of d_n by means of (4.4) is at least as easy as with the explicit formula. The function C_{10} for $\lambda = \beta = \gamma = g = h = 1$, $d_0 = 2$ is shown in Fig. 3.1. Moreover, much information about the s_0-optimal N-stage action sequence $y_N(s_0) = (a_t)_0^{N-1}$ and the state sequence $(s_t)_1^N$, generated by s_0 and $y_N(s_0)$, and about the asymptotic behavior of the sequence (d_n) can be obtained from (4.4) without use of the explicit solution, cf. Example 10.3.8.

Concerning the monotonicity of $C_n(s) = d_n s^2$ in n we have the following result:

$$\begin{aligned} d_1 > d_0 &\Longrightarrow n \mapsto C_n(s) \text{ is strictly increasing for } s \neq 0, \\ d_1 = d_0 &\Longrightarrow C_n = C_0 \text{ for } n \geq 1, \\ d_1 < d_0 &\Longrightarrow n \mapsto C_n(s) \text{ is strictly decreasing for } s \neq 0. \end{aligned} \tag{4.6}$$

We prove the first of these statements. Firstly we note that by (a) it suffices to verify that $(d_n)_0^\infty$ is strictly increasing, i.e. that $d_{n+1} > d_n$, $n \geq 0$. This follows by induction on $n \geq 0$ since applying (4.4) twice we see that $d_{n+1} > d_n$ if and only if $d_n > d_{n-1}$, $n \geq 1$.

By means of (4.4) the three conditions above can be written as conditions for the parameters $\lambda, \gamma, g, h, \beta$ and d_0. In particular, for the special case $\gamma = g = h = \beta = 1$ one sees that $d_1 > d_0$ if and only if $0 \leq d_0 < d^* := (\lambda + \sqrt{\lambda^2 + 4\lambda})/2$; $d_1 = d_0$ if and only if $d_0 = d^*$ and $d_1 < d_0$ if and only if $d_0 > d^*$. ♦

Example 4.1.8 (Minimizing weighted flow-time) We must process $m \geq 2$ jobs on a single machine without interruption (so-called *preemptive scheduling*). Job i has the processing time $\tau_i \in \mathbb{R}^+$. For each job i there accumulate costs $c_i \in \mathbb{R}^+$ per time unit as long as the processing of job i is not finished (maybe not even begun). In which order should one process the jobs in order to minimize the sum of costs, called the *weighted flow-time?* We describe the problem by the following problem $DP_m(\mathbb{N}_m)$. The state s_t is the set of jobs which wait for processing after t jobs have been processed. Thus S is the power set $\mathfrak{P}(\mathbb{N}_m)$ of $\mathbb{N}_m$. The action $a_t \in A := \mathbb{N}_m$ is the index of the $(t+1)$-st processed job, $0 \leq t \leq m-1$. If $s \neq \emptyset$ the action $a \in D(s) := s$ leads to the new state $T(s,a) := s - \{a\} =: s - a$. Moreover, we use $D(\emptyset) = A$ and $T(\emptyset, a) = \emptyset$. The cost due to processing job a in state s equals τ_a times the sum of c_a plus the costs c_i for the jobs $i \in s-a$ still waiting for processing. Thus $c(s,a) := \tau_a \cdot \sum_{i \in s} c_i$, in particular $c(\emptyset, a) = 0$. There is no terminal cost and $\beta = 1$. Thus $\{\emptyset\}$ is absorbing, hence $C_n(\emptyset) = 0$, $n \geq 1$, by Sect. 3.3. The weighted flow-time $C_m(\mathbb{N}_m)$ can be found by the Basic Theorem 3.1.1, starting with $C_0(\cdot) \equiv 0$, from

$$\begin{aligned} C_n(s) &= \min_{a \in s} \left[\tau_a \cdot \sum_{i \in s} c_i + C_{n-1}(s-a) \right] \\ &=: \min_{a \in s} W_n(s,a), \; s \subset \mathbb{N}_m, \; s \neq \emptyset. \end{aligned} \tag{4.7}$$

This procedure leads to an explicit solution, as follows.

Proposition 4.1.9 (Explicit solution of the scheduling problem)

(a) The minimal weighted flow-time $C_m(\mathbb{N}_m)$ *equals*

$$\sum_{i=1}^{m} \tau_i \cdot \left(\sum_{j=i}^{m} c_j \right).$$

(b) It is optimal to process the jobs in increasing order of $i \mapsto \tau_i/c_i$.

In the special case where all jobs have the same processing costs the optimal rule in (b) prescribes to schedule the jobs according to increasing processing times. This is called the ***SPT-rule*** (***S***hortest ***P***rocessing ***T***ime first).

Proof

(a) Without loss of generality we assume that $i \mapsto \tau_i/c_i$ is increasing. It suffices to show that for $1 \le n \le m$ the following assertion (I_n) holds: If $s \in S$ and $|s| = n > 0$ then

$$C_n(s) = \sum_{i \in s} \tau_i \sum_{j \in s: j \ge i} c_j =: \sum_{i \in s} \tau_i K(s, i).$$

Firstly, (I_1) follows trivially from (4.7). Assume that (I_n) holds for some $1 \le n \le m-1$. Select $s \in S$ with $|s| = n+1$. Then for $a \in s$, $i \in s - a$ we have

$$K(s-a, i) = \begin{cases} K(s,i) - c_a, & \text{if } i < a, \\ K(s,i), & \text{if } i > a. \end{cases}$$

Now we obtain from (4.7) and (I_n), since $|s - a| = n$,

$$\begin{aligned} W_{n+1}(s,a) &= c(s,a) + C_n(s-a) = \tau_a \cdot \sum_{i \in s} c_i + \sum_{i \in s-a} \tau_i K(s-a, i) \\ &= \tau_a \cdot \sum_{i \in s} c_i + \sum_{i<a} \tau_i \cdot K(s,i) - \sum_{i<a} \tau_i c_a + \sum_{i>a} \tau_i \cdot K(s,i). \end{aligned}$$

Here and in the next formula the index i in the sums belongs to s. Now a simple calculation shows that

$$W_{n+1}(s,a) = \sum_{i \in s} \tau_i K(s,i) + c_a \cdot \sum_{i<a} (\tau_a/c_a - \tau_i/c_i) c_i.$$

As the latter sum is non-negative and vanishes (as an empty sum) for the smallest a in s, the action $\min s$ is a minimum point of $W_{n+1}(s, \cdot)$, hence (I_{n+1}) is true.

(b) Since $D(s)$ is finite for all s, there exists a smallest minimizer f_n at stage $1 \leq n \leq m$. In the proof of (a) we have seen that $f_n(s) = \min s$ if $|s| = n$. Now (b) follows from Theorem 2.3.3(c). □♦

4.2 Further Examples

We discuss some further examples which can be modeled as DPs.

Example 4.2.1 (DPs with random length of periods) Consider an arbitrary DP, but assume that the length of period $t \in \mathbb{N}_0$ is a random variable $\eta_{t+1} \geq 0$. Then period t begins at the random time $\eta^t := \sum_{i=1}^{t} \eta_i$. We also allow that the reward r_Z in that period depends on η_{t+1}. We assume that $\eta_1, \eta_2, \ldots, \eta_N$ are i.i.d. and that $r_Z(s, a, \eta_1)$ is integrable for all (s, a). We follow the custom for DPs with random length of periods to write the discount factor, assumed to be ≤ 1, as $e^{-\alpha}$; $\alpha \geq 0$ is called the *discount rate.* The N-stage reward under initial state s_0 and action sequence $y \in A^N(s_0)$ is the random variable

$$R_{Ny}(s_0, \eta^N) := \sum_{t=0}^{N-1} e^{-\alpha\eta^t} \cdot r_Z(s_t, a_t, \eta_{t+1}) + e^{-\alpha\eta^N} \cdot V_0(s_N).$$

It is integrable since the expectation of its absolute value, i.e.

$$\mathbb{E}\,|R_{Ny}(s_0, \eta^N)| \leq \sum_{t=0}^{N-1} \mathbb{E}\,|r_Z(s_t, a_t, \eta_1)| + |V_0(s_N)|$$

is finite. Our goal consists in maximizing the expected N-stage reward

$$y \mapsto v_{Ny}(s_0) := \mathbb{E}\,R_{Ny}(s_0, \eta^N)$$

over the set $A^N(s_0)$. Observing that η^t and η_{t+1} are independent and that the expectation of a product of independent integrable random variables equals the product of the expectations, we obtain

$$v_{Ny}(s_0) := \sum_{t=0}^{N-1} \beta^t \cdot \mathbb{E}\,r_Z(s_t, a_t, \eta_1) + \beta^N \cdot V_0(s_N),$$

where $\beta := \mathbb{E}\,e^{-\alpha\eta_1}$. Therefore our problem reduces to the DP which has discount factor β and $r(s,a) := \mathbb{E}\,r_Z(s, a, \eta_1)$. In the important case where the times η^t define a Poisson process with intensity λ, we obtain the discount factor $\beta = \lambda \cdot \int e^{-(\alpha+\lambda)z}\,dz = \lambda/(\alpha + \lambda)$. ♦

Example 4.2.2 (DPs with random termination) Assume that in a problem $DP_N(s_0)$ the decision process terminates (i.e. discontinues earning rewards) randomly at one of the times $t = 0, 1, \ldots, N$ according to the following scheme: There is given an i.i.d. sequence of binary random variables X_t, $0 \le t \le N-1$, where X_0 has the binomial probability distribution $\mathrm{Bi}(1,p)$ for some $p \in [0,1)$. We assume that the decision process terminates at time $0 \le t \le N-1$ if and only if $X_t = 1$, provided it does not terminate earlier. More precisely, we assume that termination occurs at the time $\tau := \min\{0 \le t \le N-1 : X_t = 1\}$ with $\min\emptyset := N$. Thus $1_{[\tau>t]} = \prod_{i=0}^{t}(1-X_i)$ and $\delta_{\tau,t} = X_t\prod_{i=0}^{t-1}(1-X_i)$ for $0 \le t \le N-1$. In particular, $\delta_{\tau,0} = X_0$. Up to one period before termination the process earns rewards as in a DP; in addition we assume that termination in state s at time $0 \le t \le N-1$ earns $\beta^t\rho(s)$ for some function ρ on S, while termination in state s at time $\tau = N$ earns $\beta^N V_0(s)$. Then the N-stage total reward under an action sequence $y \in A^N(s_0)$ is the random variable

$$
\begin{aligned}
R_{Ny}(s_0, (X_i)_0^{N-1}) &= \sum_{t=0}^{\tau-1} \beta^t r(s_t, a_t) + \beta^\tau \rho(s_\tau)\cdot 1_{[\tau<N]} + \beta^N V_0(s_N)\cdot\delta_{\tau,N} \\
&= \sum_{t=0}^{N-1} \beta^t \left[r(s_t,a_t)\cdot 1_{[\tau>t]} + \rho(s_t)\cdot\delta_{\tau,t}\right] + \beta^N V_0(s_N)\cdot\delta_{\tau,N} \\
&= \sum_{t=0}^{N-1} \beta^t \left[r(s_t,a_t)\cdot\prod_{i=0}^{t}(1-X_i) + \rho(s_t)\cdot X_t\prod_{0}^{t-1}(1-X_i)\right] \\
&\quad + \beta^N V_0(s_N)\prod_{t=0}^{N-1}(1-X_t).
\end{aligned}
$$

We want to maximize the expectation $V'_{Ny}(s_0)$ of $R_{Ny}(s_0, (X_i)_0^{N-1})$. Now we get, using $q := 1-p$ and the independence of the X_i's,

$$
V'_{Ny}(s_0) = \sum_{t=0}^{N-1} (\beta q)^t \left[q\cdot r(s_t,a_t) + p\cdot\rho(s_t)\right] + (\beta q)^N V_0(s_N).
$$

It follows that $V'_{Ny}(s_0)$ is the N-stage reward in that DP′ which differs from the original DP only by $r'(s,a) := q\cdot r(s,a) + p\cdot\rho(s)$ and $\beta' := \beta q$. Now we can apply the theory of Chap. 2 to DP′. In particular, the VI for DP′ has the form

$$
V'_n(s) = p\cdot\rho(s) + q\cdot\sup_{a\in D(s)}\left[r(s,a) + \beta V'_{n-1}(T(s,a)\right], \; n\ge 1, \; s\in S,
$$

starting with $V'_0 := V_0$. ♦

Example 4.2.3 (Stopping a finite-stage DP) Consider a DP with an absorbing one-point set $\{\hat{s}\}$. (One-point sets are also called ***singletons.***) We call the DP a ***stopped DP*** if it has the following properties:

(i) There exists an action $\overline{a}$, called the ***stopping action***, which belongs to each set $D(s)$, $s \in J := S - \{\hat{s}\}$, and which leads from any $s \in J$ to $\hat{s}$, i.e. $T(s, \overline{a}) = \hat{s}$.
(ii) Each set $D(s)$, $s \in J$, contains besides $\overline{a}$ at least one more action, and no action $a \neq \overline{a}$ leads to $\hat{s}$.

Because of (i) and (ii) in each state in J stopping is neither enforced nor forbidden. The name stopped DP is justified by the fact that for each $N \geq 1$ the decision processes starting in some $s_0 \neq \hat{s}$ stop earning rewards as soon as the stopping action is taken. Intuitively one expects the following: If the decision process has not yet stopped before some fixed stage $1 \leq n \leq N$ and if s is the state at stage n, then one should stop if and only if the reward $r(s, \overline{a})$ for stopping is at least as large as $V_n(s)$, the maximal reward for the remaining n periods. This is easily verified as follows. From Sect. 3.3 we know that $V_n(\hat{s}) = 0$ for all $n \geq 0$ and that

$$V_n(s) = \max\{g(s), \overline{V}_n(s)\} \text{ for } s \neq \hat{s}, \tag{4.8}$$

where $g(s) := r(s, \overline{a})$ and

$$\overline{V}_n(s) := \sup\{W_n(s, a) : a \in D(s), a \neq \overline{a}\}.$$

If a maximizer f_n at stage n exists, it prescribes, if one has not stopped at times $0 \leq t \leq N - n$ and if the momentary state is s,

(i) whether to stop, i.e. whether $f_n(s) = \overline{a}$,
(ii) which action $f_n(s) \neq \overline{a}$ to take if one should not stop.

The set S_n of states $s \in S$ where $\overline{a}$ is optimal at stage $n \geq 1$ is called the ***stopping set*** at stage n. Obviously

$$S_n = \{s \in S : g(s) \geq \overline{V}_n(s)\} = \{s \in S : g(s) = V_n(s)\}, \quad n \geq 1.$$

(Note that $V_n(\hat{s}) = 0$ and (4.8) imply $\overline{V}_n(\hat{s}) = 0$, and that $\hat{s}$ belongs to each stopping set.) Thus one should stop at stage n in state $s \in S$ if and only if s belongs to S_n. (We use the convention to stop in case of indifference. Moreover, if $g(s) = \overline{V}_n(s)$, then $s \in S_n$ and this implies $f_n(s) = \overline{a}$ for some but not necessarily for all maximizers f_n at stage n.)

By far the best studied cases of the preceding model are stopped DPs where $D(s) = \{0, 1\}$ for each s, i.e. where only either stopping ($\overline{a} := 1$) or continuation ($a = 0$) is allowed. One may also generalize the preceding problem by admitting that stopping is forbidden at some states and/or enforced at some states in J. ♦

4.3 Problems

Problem 4.3.1 (A certain class of continuous allocation problems with closed solutions) Assume that $S = A = R_+$, $D(s) = [0, s]$, $T(s, a) = z \cdot (s - a)$ for some interest factor $z \in \mathbb{R}^+$, $r(s, a) = u(a)$, $V_0(s) = e_0 \cdot u(s/e_0)$ for some $e_0 \in R^+$. Put $e_n := \sigma_n(\beta) + \beta^n \cdot e_0$, $n \geq 0$; e_n coincides with d_n^2 in Example 4.1.6.

If u is concave and if $\beta z = 1$, then: $V_n(s) = e_n \cdot u(s/e_n)$, $s \mapsto f_n(s) := s/e_n$ is a maximizer at stage $n \geq 1$, and $y^* := (s_0/e_N)_0^{N-1}$, $N \geq 1$, is an s_0-optimal action sequence.

Note that y^* does not depend on u and that all components of y^* are the same, i.e. that y^* distributes the initial resource s_0 evenly among the N stages.

Problem 4.3.2 (A production-inventory problem with a single production mode) Modify Example 4.1.3 with the numerical data given there, by assuming that in each period only production in batches of size 3 (or no production) is allowed. In this DP denote the minimal costs by $C'_n(s)$.

(a) $C'_{3k}(0) = C_{3k}(0)$, $k \geq 1$.
(b) Infer from Table 4.3 for $DP_6(2)$ that $C'_6(2)$ exceeds 63.

Problem 4.3.3 We consider the continuous allocation Problem 4.3.1 with square root utility.

(i) Verify by means of Example 3.2.4 that the s_0-optimal N-stage action sequence y^* is given by (4.4), and that the state sequence $(s^*_{\nu N})_{\nu=1}^N$, generated by s_0 and y^* is given by (4.5).
(ii) $N \mapsto s_{\nu N}$ is increasing [decreasing] if and only if the same holds for (d_n). What is the limit of $s_{\nu N}$ for $N \to \infty$?
(iii) If $z = 1$ then $\nu \mapsto s_{\nu N}$ is decreasing.
(iv) Assume $z\beta^2 < 1$. Show that $f := \lim_n f_n$ exists and is asymptotically optimal (see Definition 10.1.1).

Problem 4.3.4 Solve Problem 4.3.1 by using

(i) $a_\nu :=$ the capital reinvested at time ν,
(ii) $a_\nu :=$ the proportion of capital consumed at time ν.

Compare with the solution given in Problem 4.3.1. (In case (ii) the maximizer $s \mapsto f_n(s)$ at stage n does not depend on s. This is a very rare situation.)

Problem 4.3.5 (i) Generalize Problem 4.3.1 by using $u(a) := a^\alpha$ and $V_0(s) := d_0 s^\alpha$ for $\alpha \in (0, 1)$.
(ii) Solve the corresponding minimization problem for $\alpha \in (-\infty, 0)$ and for $\alpha \in (1, \infty)$.

Problem 4.3.6 Consider the allocation Problem 4.3.1 with $u(a) = a$ and $V_0(s) = d_0 s$. Maximize and minimize.

Problem 4.3.7 If $u \geq 0$ and $V_0 = d \cdot u$ in Problem 4.3.1 with $0 \leq d \leq 1$, then $V_n(s)$ is increasing in n for all s. The same holds for Proposition 4.1.4(a).

Problem 4.3.8 Solve the extension of Problem 4.3.1 where the reward function has the form

$$r(s,a) = c_1\sqrt{a} + c_2\sqrt{s-a} + c_3\sqrt{s}$$

for $c_1, c_2, c_3 \geq 0$, $c_1 + c_2 + d_0 > 0$. Show that $V_n(s) = d_n\sqrt{s}$ for some $d_n \in \mathbb{R}_+$, and study the convergence of (d_n).

Problem 4.3.9 (Maximal reliability of a circuit) A circuit consists of N units, arranged in series. Each unit consists of identical components, arranged in parallel. Each component costs $c \in \mathbb{R}^+$ and has a reliability $p \in (0,1)$. There is a budget of amount $B \geq N \cdot c$. How many components should each of the N units contain in order to maximize the reliability of the circuit, provided that the components act stochastically independent? Find the numerical solution for $N = 8, p = 0.9, c = 1, B = 12$.

Problem 4.3.10 Maximize the function

$$x \mapsto Q_N(x) := \sum_{n=0}^{N-1} (s_{n+1} - s_n)^2/(e^{s_{n+1}} - e^{s_n}),$$

where $x = (s_1, \ldots, s_N)$ ranges over that part of $\mathbb{R}_+^N$ where $s_0 < s_1 < \cdots < s_N$.

In the last two problems we consider the linear-quadratic model in Example 4.1.7.

Problem 4.3.11 Assume $\lambda > 0$. Then:

(i) $\lambda < d < \lambda + \gamma(g/h)^2$.
(ii) (d_n) is strictly increasing [strictly decreasing] towards d if and only if $d_0 < d$ [$d_0 > d$], and $d_n = d$ for all n if $d_0 = d$.
(iii) If $|g| < 1 + \beta h^2\lambda$, then $s_N(s_0) \to 0$ for $N \to \infty$.

Problem 4.3.12 Assume $\lambda > 0$ and $\beta g^2 < 1$. Write (4.4) as $d_{n+1} = H(d_n)$ and show that $H: \mathbb{R}_+ \to \mathbb{R}_+$ satisfies

$$|H(x) - H(y)| \leq \beta g^2|x - y|.$$

For given $K, \varepsilon \in \mathbb{R}^+$ there exists an n_0 such that

$$|C_n(s) - C(s)| \leq \varepsilon \text{ for } n \geq n_0,\ 0 \leq s \leq K,$$

where $C(s) := \lim C_n(s)$. Compute such an n_0.

4.4 Supplement

The first example of an allocation problem (with $u(a) := \sqrt{a}$) seems to be due to Hotelling (1931). Later treatments include Bellman (1957, p. 4–9) and Hakansson (1970/1971). More detailed presentations, including specific algorithms, are given in Dreyfus and Law (1977, Chaps. 3 and 8), and Denardo (2003). A comprehensive account is given in the book by Ibaraki and Katoh (1988).

There is a huge literature mostly in the engineering and inventory area, on generalizations to multidimensional states and/or random disturbances, possibly with not completely known distributions and/or incomplete observations. We only mention Aoki (1989), Bertsekas (1987) and Dreyfus and Law (1977).

Chapter 5
Absorbing Dynamic Programs and Acyclic Networks

We study the problem of maximizing the sum of discounted rewards, earned not over a fixed number of periods, but until the decision process enters a given absorbing set. The basic theorem for absorbing DPs is derived. Moreover, we show how absorbing DPs can be used to find cost-minimal subpaths in acyclic networks.

5.1 Absorbing Dynamic Programs

In this section we consider a DP with a *fixed* absorbing set J_0. Thus we have (see Sect. 2.2) $\emptyset \neq J_0 \neq S$, and $T(s, a) \in J_0$, $r(s, a) = V_0(s) = 0$ for $s \in J_0$, $a \in D(s)$. Several concepts, such as $N(s)$, to be introduced below, depend on J_0, but in general this dependence will not be indicated.

For $s_0 \in S$, $n \geq 1$ and action sequence $y \in A^n(s_0)$ we denote by $dp(s_0, y)$ ***the decision process*** $(s_t)_1^n$ generated by (s_0, y). Moreover, we put $dp_t(s_0, y) := s_t$, $0 \leq t \leq n$. Recall from Sect. 2.2 the definitions $J := S - J_0 =$ the essential state space; $R_n(s) := \{dp_n(s, y) : y \in A^n(s)\} =$ the set of states reachable from $s \in S$ in $n \geq 1$ steps; $R_0(s) := \{s\}$ and $R_n(S_0) :=$ the set of states reachable from a subset $S_0 \subset S$ in $n \geq 1$ steps. Note that $R(s) = R_1(s)$ and $R(J_0) \subset J_0$.

As explained below in detail, we study in this section the problem of maximizing in a DP the sum of discounted rewards, earned not over a fixed number of periods but until the decision process enters a given absorbing set. The problem and results are easily understood, but it seems that an exact formal treatment requires considerable effort.

Definition 5.1.1 For each DP with absorbing set J_0 we call

$$S \ni s \mapsto N(s) := N_{J_0}(s) := \inf\{n \in \mathbb{N}_0 : R_n(s) \subset J_0\}$$

(with $\inf \emptyset := \infty$) the ***absorption time*** of the DP (with respect to J_0).

K. Hinderer et al., *Dynamic Optimization*, Universitext,
DOI 10.1007/978-3-319-48814-1_5

$N(s)$ belongs to $\overline{\mathbb{N}}_0 := \mathbb{N}_0 + \{\infty\}$. If $N(s)$ is finite, it is the earliest time $n \geq 0$ such that all states which are reachable from s in n steps belong to J_0. Note that $N(s) = 0$ if and only if $s \in J_0$.

Lemma 5.1.2 *Consider a DP with absorbing set J_0. For each $s \in J$ with $N(s) < \infty$:*

(a) $R_n(s) \subset J_0$ if and only if $n \geq N(s)$.
(b) If $s' \in R(s)$, then $N(s') \leq N(s) - 1$.

Proof

(a1) We prove that $n \geq N(s)$ implies $R_n(s) \subset J_0$. It suffices to show that $R_n(s) \subset J_0$ for some $n \geq 0$ implies $R_{n+1}(s) \subset J_0$. This holds since $R_{n+1}(s) = \cup_{s' \in R_n(s)} R(s')$ by (3.3) and since $s' \in R_n(s) \subset J_0$ implies $R(s') \subset J_0$.
(a2) If $R_n(s) \subset J_0$ then $n \geq N(s)$ by definition of $N(s)$.
(b) Put $n := N(s)$. Then $J_0 \supset R_n(s) = \cup_{x \in R(s)} R_{n-1}(x)$ by (3.3), hence $R_{n-1}(s') \subset J_0$. Thus $N(s') \leq n - 1$. □

Definition 5.1.3 We call a DP with an absorbing set J_0 an ***absorbing DP*** (**ADP** for short) if

- $N(s)$ is finite for all $s \in S$.
- $V_0 \equiv 0$.

Consider an ADP and fix $s \in J$, hence $1 \leq N(s) < \infty$. For $n \geq 1$ we call the action sequence $y \in A^n(s)$ an ***s-target sequence*** if $dp_{n-1}(s, y) \in J$ and $dp_n(s, y) \in J_0$, i.e. if the decision process generated by (s, y), enters J_0 at its last state. Denote by $K_n(s)$ the set of s-target sequences in $A^n(s)$. If $K_n(s)$ is non-empty then $n \leq N(s)$ by definition of $N(s)$. Thus, if $y \in A^n(s)$ belongs to K_n, the time where $dp(s, y)$ enters J_0 is not larger than $N(s)$, but it may well be smaller. For each $y = (a_t)_0^{m-1} \in A^m(s)$ with $m := N(s)$ we have $dp_m(s, y) \in J_0$ by Lemma 5.1.2(a). Therefore there exists a unique $1 \leq k = k(s) \leq m$ such that $y^k := (a_t)_0^{k-1}$ is an s-target sequence, called the ***target part*** of y. As a consequence, the set $K(s) := \cup_{n=1}^{\infty} K_n(s)$ ***of all s-target sequences*** is non-empty. As in Sect. 2.1 we denote by $V_{ny}(s)$ the sum of discounted rewards obtained by the action sequence $y \in A^n(s)$. Then $y \in K_n(s)$ earns, using $s_0 := s$,

$$V_y(s) := V_{ny}(s) = \sum_{t=0}^{n-1} \beta^t r(s_t, a_t) \quad \text{for } s =: s_0 \in J \text{ and if } y \in K_n(s). \tag{5.1}$$

Here $(s_t)_1^n$ is generated by (s, y).

Our goal is to solve the ***problem ADP*** **(*s*)**, $s \in J$, of maximizing the sum of discounted rewards earned by the s-target sequences. More precisely, problem $ADP(s)$ is the task of finding

$$V(s) := \sup_{y \in K(s)} V_y(s), \tag{5.2}$$

the maximal sum of discounted rewards earned by the s-target sequences, and an s-optimal target sequence, i.e. a maximum point of $y \mapsto V_y(s)$ on $K(s)$.

We put $V|J_0 :\equiv 0$ and call $V: S \to (-\infty, \infty]$ the ***value function*** of the ADP. From (5.1) we see that V does not depend on $D(s)$ and $T(s,a)$ for $s \in J_0$. Therefore we often do not specify these data in examples; one possible choice is $D(s) := A$ and $T(s,a) := s$.

We now relate problem $ADP(s_0)$ to problem $DP_n(s_0)$ when n is large enough. For a numerical exploitation of the next result one requires in general finiteness of S and of A.

Proposition 5.1.4 *For $s_0 \in J$ and $n \geq N(s_0)$:*

(a) $V(s_0) = V_n(s_0)$. Thus $V(s_0)$ can in principle be found by the VI method: Choose $n \geq N(s_0)$ and compute $V_\nu(s) = \sup_{a \in D(s)} W_\nu(s,a) =: UV_{\nu-1}(s)$, $1 \leq \nu \leq n-1$, $s \in S$, starting with $V_0 :\equiv 0$. Then $V(s_0) = UV_{n-1}(s_0)$.
(b) The target part of each s_0-optimal action sequence for DP_n is s_0-optimal for problem $ADP(s_0)$.

Proof Put $s := s_0$ and $m := N(s)$.

(a1) For each $z \in A^n(s)$ let $y \in A^\nu(s)$ be its target part.
It follows from (5.2) that $V_{nz}(s) = V_{\nu y}(s) \leq V(s)$. Thus $V_n(s) \leq V(s)$.
(a2) Select $y = (b_t)_0^{\nu-1} \in K(s)$ for some $1 \leq \nu \leq m$. Then there exists a $z \in A^n(s)$ which has y as target part: one only has to choose an arbitrary $b_\nu \in D(s_\nu)$, put $s_{\nu+1} := T(s_\nu, b_\nu)$, choose an arbitrary $b_{\nu+1} \in D(s_{\nu+1})$ and so on. It follows from (5.2) that $V_{\nu y}(s) = V_{nz}(s) \leq V_n(s)$, hence $V(s) \leq V_n(s)$.
(b) Let y^* be s-optimal for DP_n and let y^k be its target part. Then we get, using (5.2) and (a), that $V_{y^k}(s) = V_{ny^*}(s) = V_n(s) = V(s)$. □

The VI method requires the knowledge of $N(s)$—which often is difficult to compute exactly—or at least an upper bound n for $N(s)$. $N(\cdot)$ even has a finite upper bound M, then due to Proposition 5.1.4(a) one knows $V(s) = V_M(s)$ *for all* $s \in J$ as soon as V_M has been found by the VI. Sometimes the VI yields $V_{n+1} = V_n$ for some $n < M$, in which case Proposition 5.1.4(a) tells us that already V_n equals V. (If one only knows that $V_{n+1}(s) = V_n(s)$ one cannot conclude that $V(s) = V_n(s)$!)

Sometimes, e.g. in Example 5.1.9 below, the VI method is computationally inferior to the method of recursion in state space, which does not require $N(s_0)$ or an upper bound of $N(s_0)$. The latter method is stated in Theorem 5.1.7(b) below. For any function g from S into $\mathbb{N}_0$ put $[g = i] := \{s \in S : g(s) = i\}$, $i \geq 0$. In the special case $h := N(\cdot)$ the idea in Theorem 5.1.7(b) below is the following: Firstly, using that $V = 0$ on $J_0 = [N(\cdot) = 0]$, one computes V on the set $J_1 := [N(\cdot) = 1]$. The latter has the property that all states reachable in one step from a point in J_1 belong to J_0. Next, using V on $J_0 + J_1$, one computes V on the set $J_2 := [N(\cdot) = 2]$, which has the property that all states reachable in one step from a point in J_2 belong to $J_0 + J_1$. This procedure is repeated until one arrives for $k := N(s_0)$ at the set $J_k := [N(\cdot) = k]$, which contains s_0. The existence of such a k is ensured since $\sum_{i=0}^{\infty} J_i = S$ due to finiteness of $N(\cdot)$.

For the application of the preceding method the absorption function $N(\cdot)$ can be found by the following recursive computation of the sets $J_i := [N(\cdot) = i]$, $i \geq 1$:

$$J_i = \{s \in S - \sum_{n=0}^{i-1} J_n : R(s) \subset \sum_{n=0}^{i-1} J_n\}, \quad i \geq 1.$$

For the proof of this equality denote its right-hand side by $\tilde{J}_i$. We firstly assume that $s \in J_i$. Then $N(s) = i$, hence $s \in S - \sum_{n=0}^{i-1} J_n$. In addition, by Lemma 5.1.2(b) we have $N(s') \leq i - 1$ for $s' \in R(s)$, hence $R(s) \subset [N(\cdot) \leq i - 1]$. Thus $s \in \tilde{J}_i$. On the other hand, let $s \in \tilde{J}_i$. Then $N(s) \geq i$. In addition, $N(s') \leq i - 1$, hence $R_{i-1}(s') \subset J_0$ for $s' \in R(s)$. Now (3.3) implies $R_i(s) = \cup_{s' \in R(s)} R_{i-1}(s') \subset J_0$. It follows that $N(s) \leq i$. Altogether we know that $s \in J_i$.

Part (b) of the proof of Theorem 5.1.7 below for the special case $h := N(\cdot)$ shows that the applicability of the method of recursion in state space depends crucially on the Lemma 5.1.2(b) of the absorption time. Therefore we now study functions h with the corresponding property, as follows.

Definition 5.1.5 Consider a DP with absorbing set J_0. We call a function $h: S \to \mathbb{N}_0$ a ***drift function*** (with respect to J_0) if

- $J_0 = [h = 0]$.
- $$s \in J, s' \in R(s) \text{ implies } h(s') \leq h(s) - 1. \tag{5.3}$$

Concerning (5.3) note that $1 \leq h(s) < \infty$. The relation (5.3) implies that for each $s \in J$ the sets $R_t(s)$, $t \geq 1$, of states reachable from s in t steps *drift* with increasing t towards J_0 in the sense that for $s \in J$ and $t \geq 0$

$$R_t(s) \subset [h \leq (h(s) - t)^+], \quad \text{hence } R_t(s) \subset J_0 \text{ for } t \geq h(s). \tag{5.4}$$

The proof by induction on $t \geq 0$ uses that $R_{t+1}(s) = \cup_{s' \in R_t(s)} R(s')$ by (3.3).

Since $h(S) \subset \mathbb{N}_0$, S is the disjoint union of the sets $[h = k]$, $k \geq 0$. By Example 5.1.6(a) below we could use $N(\cdot)$ in each ADP as drift function, but often other drift functions are easier to find (see Example 5.1.6(c) below) and easier to apply in Theorem 5.1.7 below. In the search for h one may be guided by the idea that $h(s)$ should be a *measure of the distance* of the state s from J_0.

Example 5.1.6 (Drift functions)

(a) A DP with absorbing set J_0 and with $V_0 \equiv 0$ is absorbing if and only if it has a drift function, and then $N(\cdot)$ is the pointwise smallest drift function, i.e. $0 \leq N(s) \leq h(s) < \infty$, $s \in S$, for each drift function h of the DP. Moreover, $V(s) = V_n(s)$ for $s \in J$ and $n \geq h(s)$.

Proof

(a1) If the DP is absorbing, then $J_0 = [N(\cdot) = 0]$ and (5.3) for $h := N(\cdot)$ follows from Lemma 5.1.2(b). Thus $N(\cdot)$ is a drift function.

(a2) Let h be a drift function of the DP. Then $R_t(s) \subset J_0$ for $s \in S$ and $t := h(s)$ by (5.4), hence $N(s) \leq h(s) < \infty$. Thus the DP is absorbing, and $N(\cdot)$ is the pointwise smallest drift function. □

(b) *Consequence of (a):* In an ADP we have $V(s_0) = V_{h(s_0)}(s_0)$, $s_0 \in J$, for each drift function h, due to Proposition 5.1.4(a) for $n := h(s)$. Moreover, by Proposition 5.1.4(b) the target part of each s_0-optimal action sequence for problem $DP_{h(s_0)}(s_0)$ is s_0-optimal for problem $ADP(s_0)$.

(c) Consider the following DP: $S = \mathbb{N}_{0,m}$ for some $1 \leq m \leq \infty$ with $\mathbb{N}_{0,\infty} := \mathbb{N}_0$; assume that $J_0 := \mathbb{N}_{0,k}$ for some fixed $0 \leq k \leq m-1$ is absorbing; $D(\cdot)$ arbitrary; $T(s,a) \leq s-1$ for $s \in J$ and $a \in D(s)$. Then $s \mapsto h(s) := (s-k)^+$ is a drift function. The easy proof uses isotonicity of $\mathbb{R} \ni x \mapsto x^+$, and is left to the reader. On the other hand, depending on the explicit form of $D(\cdot)$ and $T(\cdot,\cdot)$, the absorption time may have a complicated form. As an example, if $m = 7, k = 3$, $D(4) = \{1,2,3,4\}$, $D(5) = \{3\}$, $D(6) = \{2\}$, $D(7) = \{1\}$ and $T(s,a) = s-a$ then $N(4) = h(4) = 1$, $N(5) = 1 < h(5) = 2$, $N(6) = 2 < h(6) = 3$, $N(7) = 3 < h(7) = 4$.

(d) Consider a DP with absorbing set J_0 and an ordering (i.e. a transitive, reflexive and antisymmetric relation) $\leq_S$ on S. Then $s \mapsto h(s) := |\{j \in J : j \leq_S s\}|$ is a drift function if

(i) $h(s) < \infty$ for $s \in S$,
(ii) $s' <_S s$ (i.e. $(s' \leq_S s) \wedge (s' \neq s)$) for all $s \in J$, $s' \in J_0$,
(iii) $s' <_S s$ for all $s \in J$, $s' \in R(s)$.

Proof For $s \in S$ put $M(s) := \{j \in J : j \leq_S s\}$, hence $h(s) = |M(s)|$, $s \in S$.

(d1) If $s \in J$ then reflexivity of $\leq_S$ ensures that $s \in M(s)$, hence $h(s) \geq 1$. Moreover, if $h(s') \geq 1$ for some $s' \in J_0$, then there exists some $s \in J$ with $s \leq_S s'$. In addition we get from (ii) that $s' \leq_S s$. Now antisymmetry of $\leq_S$ implies $s' = j \in J$, which is a contradiction to $s \in J_0$. Altogether we obtain $J_0 = [h = 0]$.

(d2) It remains to verify (5.3). Thus select $s \in J$ and $s' \in R(s)$. Because of (iii) we have $s' \leq_S s$. Case 1: $s' \in J_0$. Then (5.3) holds since (d1) implies $h(s') = 0 < 1 \leq h(s)$. Case 2: $s' \in J$. Then $M(s') \subset M(s)$: in fact, if $j \in M(s')$ we have $j \leq_S s'$, and $s' \leq_S s$ yields $j \leq_S s$. Now we complete the proof of (5.3) in case 2 by showing that $M(s') \neq M(s)$. In fact, $s \in M(s)$ by reflexivity of $\leq_S$, but $s \notin M(s')$; otherwise we obtain, using (iii), that $s \leq_S s' \leq_S s$, hence $s' = s$, in contrast to $s' <_S s$ by (iii). □

(e) Note that (c) is a special case of (d). As another example for (d) consider a DP with an arbitrary $S \subset \mathbb{N}_0^2$ with $(0,0) \in S$, $J_0 = \{(0,0)\}$ and $\leq_S$ the usual ordering. Then conditions (i) and (ii) hold trivially, while (iii) holds if $(T(s,a) \leq_S s) \wedge (T(s,a) \neq s)$ for all $s \in S - \{(0,0)\}$ and $a \in D(s)$. ♦

Theorem 5.1.7 (Basic Theorem for absorbing DPs) *For each ADP with absorbing set J_0 the following holds:*

(a) The value function V is the unique solution $v: S \to (-\infty, \infty]$ of the ***optimality equation (OE)***

$$v(s) = \sup_{a \in D(s)} [r(s,a) + \beta v(T(s,a))], \; s \in J,$$

for which $v = 0$ on J_0.

(b) Let h be a drift function. Then $V(s_0)$, $s_0 \in J$, can be found in $h(s_0)$ steps by the following method of ***recursion in the state space****: compute from the OE, starting with $V = 0$ on $[h = 0] = J_0$, successively V on $[h = 1]$, on $[h = 2]$, ..., on $[h = h(s_0)]$.*

(c) Let h be a drift function. There holds the following ***optimality criterion (OC)*** *for $s_0 \in J$. Assume that for all $s \in S(s_0) := [1 \le h \le h(s_0)]$ there exists a maximum point $f(s)$ of*

$$D(s) \ni a \mapsto W(s,a) := r(s,a) + \beta V(T(s,a)).$$

Then we have:

(c1) There exists a unique $1 \le n \le N(s_0)$ such that the algorithm

$$a_t^* := f(s_t), \; s_{t+1} := T(s_t, a_t^*), \quad 0 \le t \le n-1, \tag{5.5}$$

is defined (i.e. such that $s_t \in S(s_0)$ for $0 \le t \le n-1$), and such that $s_n \in J_0$.

(c2) The s_0-target sequence $y^* := (a_t^*)_0^{n-1}$ is s_0-optimal for ADP(s_0).

Proof of (a) $\wedge$ (b) =: (α)

(α1) Firstly, $V|J_0 \equiv 0$ holds by definition.

(α2) We show that V is a solution of the OE. Fix $s \in J$, hence $n := N(s) \ge 1$. Now $s' := T(s,a) \in R(s)$ for $a \in D(s)$, hence $N(s') \le n - 1$ by Lemma 5.1.2(b) and then $V_{n-1}(s') = V(s')$ by Proposition 5.1.4(a). Thus Proposition 5.1.4(a) and the VI for DP yield

$$\begin{aligned} V(s) = V_n(s) &= \sup_{a \in D(s)} [r(s,a) + \beta V_{n-1}(T(s,a))] \\ &= \sup_{a \in D(s)} [r(s,a) + \beta V(T(s,a))]. \end{aligned} \tag{5.6}$$

(α3) We show by induction on $n \ge 0$ the assertion (I_n) that V is uniquely defined on $M_n := [h \le n]$. Firstly, (I_0) is clear since $V = 0$ on $M_0 = [h = 0] = J_0$. Assume that for some $n \ge 0$ there holds (I_t) for $0 \le t \le n$. Select $s \in [h = n + 1]$ and $a \in D(s)$. For $s' := T(s,a) \in R(s)$ we have $h(s') \le h(s) - 1 = n$ by (5.3), hence $T(s,a) \in M_n$. Thus we can compute V on $[h = n+1]$ by (5.6). Now the proof of (a) and (b) is complete.

(c1) Firstly we note that $S(s_0)$ is the domain of f and that $S(s_0) \subset J$. Since $s_0 \in [h \leq h(s_0)]$ holds trivially and since $h(s_0) \geq 1$ due to $s_0 \in J$, we have $s_0 \in S(s_0)$. Thus a_0^* and s_1 are defined, and then (5.3) shows that $s_1 \in J_0 + S(s_0)$. In case $s_1 \in J_0$, the algorithm stops with $n = 1$, while in case $s_1 \in S(s_0)$ the algorithm continues for at least one more step, and then $n \geq 2$. In the latter case we have two more cases.

Case (i): $s_t \in S(s_0)$ for $2 \leq t \leq N(s_0) - 1$. Then $n = N(s_0)$ since $s_{N(s_0)} \in J_0$ by Lemma 5.1.2(a).

Case (ii): There is some $2 \leq k \leq N(s_0) - 2$ such that $s_t \in S(s_0)$ for $2 \leq k$ and $s_{k+1} \in J_0$. Then $n = k + 1$.

(c2) Put $m := N(s_0)$. We extend the algorithm (5.5) by selecting for $n \leq t \leq m - 1$ an arbitrary action $a_t^* \in D(s_t)$, where $s_{t+1} := T(s_t, a_t^*)$. Then y^* is the target part of $y' := (a_t^*)_0^{m-1}$. From Proposition 5.1.4(b) with $n := m$ we know that y^* is s_0-optimal for $ADP(s_0)$ if y' is s_0-optimal for DP_m. The latter is true by Theorem 2.3.3(b) if there holds for $0 \leq t \leq m - 1$ the following property (I_t): a_t^* is a maximum point of $a \mapsto W_{m-t}(s_t, a) = r(s_t, a) + \beta V_{m-t-1}(T(s_t, a))$.
We now prove $(I_t)_0^{m-1}$.

Case (i): $n \leq t \leq m - 1$. Then $s_t \in J_0$, and hence for $a \in D(s_t)$ we have $r(s_t, a) = 0$ and $s' := T(s_t, a) \in J_0$. Then we infer from (3.5) that $V_{m-t-1}(s') = 0$, hence $W_{m-t}(s_t, a) = 0$. This proves (I_t).

Case (ii): $0 \leq t \leq n - 1$. Fix $a \in D(s_t)$ and put $s' := T(s_t, a)$. Then $s' \in R(s_t)$. Moreover $R(s_t) \subset R_{t+1}(s_0)$ by (3.3) with $n := t$, $s := s_0$ and $s' := s_t$. Thus $s' \in R_{t+1}(s_0)$. Since $N(\cdot)$ is a drift function by Example 5.1.6(a) we get from (5.4) with s and t replaced by s_0 and $t+1$, respectively, that $N(s') \leq m-t-1$. Hence Proposition 5.1.4(a) implies $V(s') = V_{m-t-1}(s')$. This proves (I_t).

□

Note that the numerical exploitation of Theorem 5.1.7(b) for the computation of $V(s_0)$ requires finiteness of the set $S(s_0)$. As an example, if the assumptions of Example 5.1.6(c) hold then $[h = i] = \{k + i\}$. Then Theorem 5.1.7(b) yields successively $V(k + 1), V(k + 2), \ldots, V(m)$.

Remark 5.1.8 In the situation of Theorem 5.1.7(c) the following fact may be useful for finding the target part of y^*: If J_0 is a singleton and if $a_{t+1}^* \neq a_t^*$ for some $0 \leq t \leq h(s_0) - 1$, then $s_t \in J$. In fact, from $f(s_{t+1}) = a_{t+1}^* \neq a_t^* = f(s_t)$ we conclude that $s_{t+1} \neq s_t$. Now $s_t \in J_0$ would imply $s_{t+1} \in J_0$ in contrast to $|J_0| = 1$.

◇

Example 5.1.9 (Cost-minimal transportation from a point in $\mathbb{N}$ to the origin under quadratic costs) (This example fits into the problem of cost-minimal paths and hence can also be solved by Theorem 5.2.6.) A system is moving in the set $S := \mathbb{N}_0$. It shall be controlled under quadratic costs in a cost-minimal manner until it is absorbed in $J_0 := \{0\}$. If the system is not yet absorbed in J_0, i.e. if $s \in J := \mathbb{N}$, it is transported from s to the next state $s' := a \in D(s) := \mathbb{N}_{0,s-1}$ with cost

$c(s,a) = (s-2a)^2$. We use the DP with $S = A = \mathbb{N}_0$, J_0, J and c as above and with the following data: $D(0) := \{0\}$, $T(s,a) = a$, $C_0 \equiv 0$, and $\beta = 1$. Obviously J_0 is absorbing. The computation of $C(s_0)$, $s_0 \geq 1$, i.e. of the minimal cost until absorption in 0 when the systems starts in s_0, can be done via the method Theorem 5.1.7(b) of recursion in state space, as follows. From Example 5.1.6(c) with $k := 0$ we see since $R(s) = D(s) = \mathbb{N}_{0,s-1}$ for $s \in J$, that the DP has the drift function $s \mapsto h(s) := s$, hence the DP is absorbing by Example 5.1.6(a). It follows from the minimization version of Theorem 5.1.7(a) that $C(0) = 0$ and that

$$\begin{aligned} C(s) &= \min_{0 \leq a \leq s-1} \{(s-2a)^2 + C(a)\} \\ &= \min_{0 \leq a \leq s-1} W(s,a), \quad s \geq 1. \end{aligned} \tag{5.7}$$

As an example, for $s_0 = 5$ the computation can be done by hand, and yields $C(i)$, $1 \leq i \leq 5$, and the smallest minimizer f of W from (5.7) as follows:

$$C(1) = 1, f(1) = 0;\ C(2) = \min\{4, 1\} = 1, f(2) = 1;$$

$$C(3) = \min\{9, 2, 2\} = 2, f(3) = 1;\ C(4) = \min\{16, 5, 1, 6\} = 1, f(4) = 2;$$

$$C(5) = \min\{25, 10, 2, 3, 10\} = 2, f(5) = 2.$$

Thus $C(5) = 2$. Moreover, since $a_0^* = f(5) = 2$; $s_1 = a_0^* = 2$; $a_1^* = f(s_1) = 1$; $s_2 = a_1^* = 1$; $a_2^* = f(s_2) = 0$; $s_3 = a_2^* = 0$, we get from Theorem 5.1.7(c) that $(a_t^*)_0^2 = (2, 1, 0)$ is a 5-optimal target sequence for $ADP(5)$. ♦

Example 5.1.10 (A knapsack problem) You want to pack a knapsack with units of m different items. Item i has weight $g(i) \in \mathbb{N}$ and utility $u(i) \in \mathbb{R}^+$, $1 \leq i \leq m$. How many units of each of the items should be packed in order to maximize the sum of utilities under the restriction that the total weight does not exceed $K \in \mathbb{N}$? How large is the corresponding maximal value $v(K)$ of utilities? In order that at least one unit of at least one item can be packed we assume $K \geq \underline{g} := \min_i g(i)$.

(a) Denoting by x_i the number of packed units of item i, one can formulate the problem in a static way as the following integer linear program:

$$\text{maximize } (x_i)_1^m \mapsto \sum_1^m u(i)x_i$$

$$\text{under the constraint } \sum_1^m g(i)x_i \leq K,\ x_i \in \mathbb{N}_0.$$

(b) Alternatively we now solve the problem by selecting and packing at each time t one of those items a_t which are feasible i.e. for which $g(a_t) \leq s_t$. We use the following ADP: $a_t \in A = \mathbb{N}_m$ denotes the item chosen at time t; $s_t \in S = \mathbb{N}_{0,K}$ is the weight still remaining before the decision at time t; thus $s_0 = K$. The

Table 5.1 Data for Example 5.1.10

	a				
	1	2	3	4	5
$g(a)$	3	4	6	7	8
$u(a)$	3	5	6	8	9

packing process (s_t) stops as soon as it enters the set $J_0 := \{0, 1, \ldots, \underline{g} - 1\}$, as then no further packing is possible. J_0 is absorbing; hence $J = S - J_0 = \{\underline{g}, \underline{g} + 1, \ldots, K\}$; for $s \in J$ we put $D(s) = \{a \in \mathbb{N}_m : g(a) \leq s\}$, $T(s, a) = s - g(a)$, and $r(s, a) = u(a)$; $V_0 \equiv 0$ and $\beta = 1$. Example 5.1.6(c) with $k := \underline{g} - 1$ shows that the DP is indeed absorbing with the drift function $s \mapsto (s - \underline{g} + 1)^+$. Thus $V(K)$ can be found by solving by recursion in state space the OE

$$V(s) = \max\{u(a) + V(s - g(a)) : 1 \leq a \leq m,\ g(a) \leq s\},\ s \in J,$$

(c) As a numerical example we consider the case $m = 5$, $g(a)$ and $u(a)$ given in Table 5.1.

One obtains the following results:

(i) For the weight limit $K = 11$ we obtain the maximal total utility $V(11) = 13$. Moreover, there are five 11-optimal action sequences, yielding two different solutions: either pack one unit of item 1 and two units of item 2 or one unit of item 2 and one unit of item 4. It follows that both $(1, 2, 2)$ and $(2, 4)$ are 11-optimal for $ADP(11)$. In each of the two cases the weight limit 11 is exhausted.

(ii) For $K = 9$ we have $V(9) = 10$ and there exists a unique 9-optimal action sequence which tells to pack two units of item 2. Then $(2, 2)$ is 9-optimal for $ADP(9)$ and one weight unit is unused. ◆

Example 5.1.11 (Optimal routing of a freighter when trips have unequal duration) Consider that variant of Example 4.1.1 with finite set I of ports where a trip from any port i to any port $a \neq i$ lasts $\lambda(i, a) \in \mathbb{R}^+$ time units and where the number of trips is not fixed in advance but where all trips must be made within $L \in \mathbb{R}^+$ time units. We want to maximize the sum of rewards obtainable when the initial port is i_0. In order that there exists at least one sequence of trips from any initial port i we assume that $\mu := \min_{i \in I} \min_{a \neq i} \lambda(i, a) \leq L$. The two computational algorithms we present below are rather time consuming, but it is instructive at least to model the problem as follows by an ADP.

Since we must keep track of both the momentary port i and the momentarily remaining time l, the state is the pair $s = (i, l) \in S := I \times [0, L]$; put $\mu(i) := \min_{a \neq i} \lambda(i, a)$; the state $s = (i, l) \in J = \{(i, l) \in S : \mu(i) \leq l\}$ means that from i at least one more trip can be made within the remaining time l; the state $s = (i, l) \in J_0 := S - J = \{(i, l) \in S : \mu(i) > l\}$ means that from i no trip can be made within the remaining time l; $A = I$; for $(i, l) \in J$ we use $D(i, l) = \{a \in I : a \neq i, \lambda(i, a) \leq l\} \neq \emptyset$, $T(i, l, a) = (a, l - \lambda(i, a))$, $a \in D(i, l)$; for $s \in J_0$ we use

$D(s) = I$ and $T(i,l,a) = (i,l)$; for $(i,l) \in J$ the reward $r(i,l,a)$ is independent of l, while $r(i,l,a) = 0$ for $(i,l) \in J_0$; $V_0 \equiv 0$; $\beta = 1$. Obviously J_0 is absorbing.

One easily sees that each decision process starting in some $s \in S$ is absorbed in J_0 after at most $M := L/\mu$ steps. Thus $DP_M(i_0, L)$ solves our problem.

Also the method of recursion in state space from Theorem 5.1.7(b) is applicable since our model has the following drift function: $h(i,l) := \lfloor (l - \mu(i))/\mu \rfloor + 1$ if $(i,l) \in J$ and $h(i,l) := 0$, if $(i,l) \in J_0$. The proof runs as follows. Firstly $h(s) \geq 1$ if and only if $s \in J$ is obvious. In order to verify (5.3) select $s = (i,l) \in J$. Each $s' \in R(s)$ has the form $s' = T(i,l,a) = (a, l - \lambda(i,a))$ for some $a \in D(i,l)$. We have to show that $h(s') \leq h(s) - 1$, which holds trivially if $s' \in J_0$. Thus assume that $s' \in J$. Then $\mu \leq \mu(a)$ and $\mu(i) \leq \lambda(i,a)$ implies

$$h(s') = \lfloor (l - \lambda(i,a) - \mu(a) + \mu)/\mu \rfloor \leq \lfloor (l - \mu(i))/\mu \rfloor = h(s) - 1. \qquad \blacklozenge$$

5.2 Cost-minimal Subpaths in Acyclic Networks

We now show how absorbing DPs can be used for finding cost-minimal subpaths in a certain type of graphs.

Definition 5.2.1 A (directed) ***graph*** is a pair (I, Γ), where I is a set with at least two elements, and Γ is a subset of $I \times I$. The elements i in I are called ***nodes*** and the elements (i,j) in Γ are called the (directed) ***arcs*** from i to j. The nodes in the (possibly empty) i-section

$$\Gamma(i) := \{j \in I : (i,j) \in \Gamma\}$$

of Γ are called the ***successors*** of i. Note that $\Gamma = \{(i,j) \in I^2 : j \in \Gamma(i)\}$.

A graph (I, Γ) is geometrically visualized by depicting the nodes as points in the plane and the arcs (i,j) as arrows from i to j. Figure 5.1 shows a graph with 9 nodes ($I = \mathbb{N}_9$) and 15 arcs, where e.g. $\Gamma(3) = \{4, 6, 8\}$ and $\Gamma(9) = \emptyset$.

Definition 5.2.2 Let (I, Γ) be a directed graph.

- A node i is called a ***sink***, if it has no successor, i.e. if $\Gamma(i)$ is empty; otherwise i is called a ***non-sink.*** The set of sinks and of non-sinks is denoted by J_0 and $J := I - J_0$, respectively. In many applications J_0 is a singleton.
- A node is called ***isolated***, if i is a sink and if no arc ends in i.
- A sequence $(i_t)_0^n$, $n \geq 1$, of at least 2 nodes (not necessarily different from each other) is called a ***path*** from i_0 to i_n if $i_t \in \Gamma(i_{t-1})$ for $1 \leq t \leq n$. Note that for each path $(i_t)_0^n$ there holds $i_t \in J$ for $0 \leq t < n$, hence at most i_n is a sink.
- If $i_0 \in J$ we call the sequence $(i_t)_1^n$ a ***subpath from i_0 to i_n*** if $(i_t)_0^n$ is a path from i_0 to i_n. Moreover, we call a subpath $(i_t)_1^n$ from i_0 to i_n a ***subpath from i_0 into J_0*** if $i_n \in J_0$.

Fig. 5.1 An acyclic net

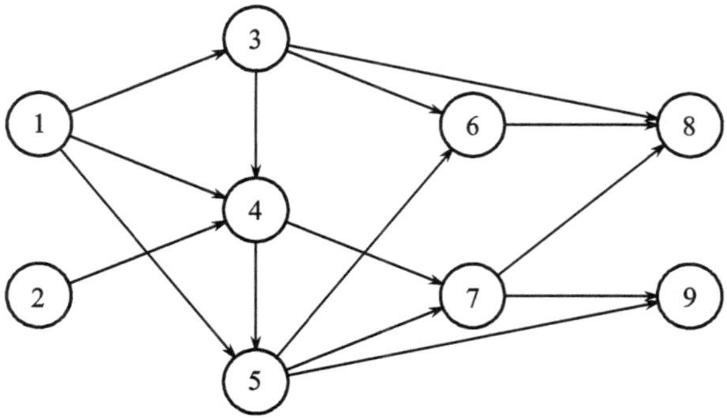

- A path $(i_t)_0^n$ from $i_0 \in J$ to $i_n \in J$ is called a ***cycle*** if $i_0 = i_n$.
- We call a graph an ***acyclic net***, if it has only finitely many nodes, no isolated nodes and no cycles. Then $2 \leq |I| < \infty$.

The graph in Fig. 5.1 is an acyclic net with the sinks 8 and 9. There are 18 different subpaths from the non-sink 1 into the set $J_0 = \{8, 9\}$ of sinks.

Lemma 5.2.3 *Each acyclic net (I, Γ) has the following properties:*

(a) There exists at least one path.
(b) Each path consists of at most $|J| + 1$ nodes, and all these nodes are different from each other.
(c) There is at least one sink and at least one non-sink.
(d) From each non-sink i there exists at least one subpath from i into J_0.

Proof

(a) Assume that there exists no path. Then $\Gamma(i) = \emptyset$ for all $i \in I$, hence all nodes are isolated, which is excluded for acyclic nets.
(b1) Let $(i_t)_0^n$ be a path with at least $|J| + 2$ nodes, i.e. $n + 1 \geq |J| + 2$. Since $i_t \in J$ for $0 \leq t \leq n-1$ the path $(i_t)_0^{n-1}$ consists of $n \geq |J| + 1$ non-sinks. Then there exists a pair of equal nodes, say $i_\mu = i_\nu$, for some $0 \leq \mu < \nu \leq n$. Thus $(i_t)_\mu^\nu$ is a cycle, which is excluded for acyclic nets. This proves the first assertion in (b).
(b2) Let $(i_t)_0^n$ be a path with $i_\mu = i_\nu$ for some $0 \leq \mu < \nu \leq n$. Then $(i_t)_\mu^\nu$ is a cycle, which is excluded for acyclic nets. This proves the second assertion in (b).
(c1) Assume that J_0 is empty. Then each node i has a successor, say $g(i) \in J$. Let $i_0 \in J$ be arbitrary. Then we obtain a path $(i_t)_0^n$, where $n = |J| + 1$, by $i_{t+1} := g(i_t)$, $0 \leq t \leq n-1$. This contradicts (a).
(c2) Assume that J is empty. Select some $i \in J_0$. Since i is not isolated, there exists some arc (j, i). Then j belongs to J, which is a contradiction.
(d) As $i_0 := i$ belongs to J, there exists some $i_1 \in \Gamma(i_0)$. If i_1 is a sink, we are ready. Otherwise there exists some $i_2 \in \Gamma(i_1)$, etc. As there are no cycles and as by (c) there are at most $|I| - 1$ non-sinks we arrive after at most $|I| - 1$ steps at some sink. □

It follows from Lemma 5.2.3(d) that for each $i \in J$ the *set* $P(i)$ *of subpaths from* i *into* J_0 is non-empty and finite.

Definition 5.2.4

- An ***acyclic network*** (for cost minimization) is a triple (I, Γ, c) consisting of an acyclic net (I, Γ) and a function c on Γ. Here $c(i,j)$ is interpreted as the cost (or reward, if $c(i,j) < 0$) for traveling along the arc (i,j) from i to j.
- The cost of a subpath $z := (i_t)_1^n$ from $i_0 \in J$ into J_0 (i.e. for which $i_n \in J_0$) is defined as the real number

$$C_z(i_0) := \sum_{t=0}^{n-1} c(i_t, i_{t+1}).$$

The case where there is also a cost $\tau(j)$ for subpaths ending in a sink j is reduced to our model by replacing $c(i,j)$ for $j \in J_0$ by $c(i,j) + \tau(j)$. Our aim is to study for a given non-sink i_0 the ***shortest-path problem MCP*** (i_0) of finding

(i) the minimal cost

$$C(i_0) = \min_{z \in P(i_0)} C_z(i_0)$$

within the set $P(i_0)$ of subpaths from $i_0 \in J$ into J_0, and

(ii) the cost-minimal subpaths within $P(i_0)$, i.e. the minimum points of $P(i_0) \ni z \mapsto C_z(i_0)$.

However, this name (with *path* replaced by *subpath*) is justified only if the nodes are points in a Euclidean space and if the reals $c(i,j)$ are the (non-negative) distances between the points i and j. Note that our setting includes by the case $c \leq 0$ also longest path problems.

Example 5.2.5 (Cost-minimal subpaths in an acyclic network) Figure 5.2 shows an acyclic network with the same net (I, Γ) as in Fig. 5.1. The nodes are represented by circles, the left half of which contains the node number. The numbers at the arcs are the costs for traveling along the arc. The minimal cost $C(i)$ of subpaths from any node $i \in J = \mathbb{N}_7$ into $J_0 = \{8, 9\}$ and all cost-minimal subpaths from node 1 into J_0 are computed in Example 5.2.9 below. The numbers $C(i)$ are written in the right half of the circle for node i, while the arcs of the cost-minimal subpaths are depicted in bold face. In this simple example one realizes at once that the arcs $(1, 4)$, $(3, 8)$ and $(5, 9)$ could be deleted as there exists for each of these arcs (i,j) a cheaper subpath from i to j. For large networks such a reduction is time-consuming and in general not advisable.

The cost-minimal subpath problem *MCP*(i_0) has a number of applications outside the area of transportation. Examples are:

(i) *cost-minimal production schedules*. Here $m \geq 2$ different jobs have to be processed on a single machine. Let $c(i,j) \geq 0$ be the set-up time when job

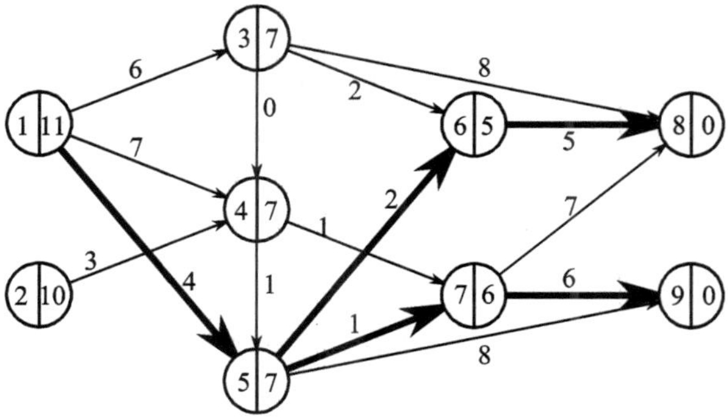

Fig. 5.2 The acyclic network for Example 5.2.5

j is processed after job i. In which order should one schedule the jobs in order to minimize the total set-up time? We leave it to the reader to set up the corresponding acyclic network.

(ii) A message must be routed in a net (I, Γ) of acyclic communication channels with $I := \mathbb{N}_m$ from node 1 to the single sink m. With given probability $p(i,j) \in (0, 1]$ the message is transmitted from node i to node $j \in \Gamma(i)$. Find a route which maximizes the probability that the message arrives in node m. It is left to the reader to model the problem under a *Markovian* assumption for the random transmission as an acyclic network with $c(i,j) := -\log p(i,j)$. ♦

We now solve problem $MCP(i_0)$ by proving its equality with problem $ADP(s_0)$ with $s_0 := i_0$ for the following special DP $(S, A, D, T, c, C_0, \beta)$ for cost minimization: The states and actions are the nodes, i.e. $S = A = I$;

$$D(s) = \begin{cases} \Gamma(s), & \text{if } s \in J, \\ \{s\}, & \text{if } s \in J_0; \end{cases}$$

$T(s,a) = a$; c is extended from Γ to D by $c(s,s) := 0$ for $s \in J_0$; $C_0 \equiv 0$; $\beta := 1$. Obviously $(S, A, D, T, c, C_0, \beta)$ is an ADP with the absorbing set J_0 of sinks.

Note that $a_t = i_{t+1} = s_{t+1}$, $t \geq 1$. Now the equality of $MCP(i_0)$ with $ADP(i_0)$ follows from the first two of the following three facts:

(i) The s_0-target sequences are the subpaths from $i_0 := s_0$ to J_0 since $y = (a_t)_0^{n-1} = (i_t)_1^n$ belongs to $K(i_0)$ if and only if $i_{n-1} \in J$, $i_n \in J_0$ if and only if $z := (i_t)_1^n \in P(i_0)$. Thus $K(i_0) = P(i_0)$.

(ii) The cost of an s_0-target sequence equals the cost for the corresponding subpath from $i_0 := s_0$ to J_0 since

$$C_y(s_0) = \sum_{t=0}^{n-1} c(s_t, a_t) = \sum_{t=0}^{n-1} c(i_t, i_{t+1}) = C_z(i_0), \; i_0 := s_0.$$

(iii) Due to Lemma 5.2.3(b) there holds $N(s_0) \leq |J|$.

Now Theorem 5.1.7 and the equality $MCP(i_0) = ADP(i_0)$ yield the next result.

Theorem 5.2.6 (Basic Theorem for acyclic networks) *Let (I, Γ, c) be an acyclic network with the set J_0 of sinks and the set J of non-sinks. Let $C(i)$, $i \in J$, be the minimal cost for subpaths from i into J_0.*

(a) There holds $C|J_0 = 0$ and

$$C(i) = \min_{a \in \Gamma(i)} \{c(i,a) + C(a)\}, \quad i \in J. \tag{5.8}$$

(b) Let h be a drift function. Then $C(i_0)$, $i_0 \in J$, can be found numerically from (5.8) in $h(i_0)$ steps by the following method of ***recursion in the set I of nodes:*** *starting with $C = 0$ on $[h = 0] = J_0$, compute successively C on $[h = 1]$, on $[h = 2]$, ..., on $[h = h(i_0)]$.*

(c) Let h be a drift function. There holds the following ***optimality criterion (OC)*** *for $i_0 \in J$. For $i \in I_0 := [1 \leq h \leq h(i_0)]$ let $f(i)$ be a (certainly existing) minimum point $f(i)$ of*

$$\Gamma(i) \ni a \mapsto c(i,a) + C(a).$$

Then we have:

(c1) There exists a unique $1 \leq n \leq |J|$ such that the algorithm

$$i_{t+1} := f(i_t), \quad 0 \leq t \leq n-1.$$

is defined (i.e. such that $i_t \in I_0$ for $0 \leq t \leq n-1$), and such that $i_n \in J_0$.

(c2) $z^* := (i_t^*)_1^n$ is a cost-minimal subpath from i_0 into J_0.

Remark 5.2.7 Let C_n be the n-stage minimal cost function in an ADP. It follows from Proposition 5.1.4(a) and Lemma 5.2.3(b), since $N(i_0) \leq |J|$ that $C = C_{|J|}$. Thus $C(i_0)$, $i_0 \in J$, can also be found, using $C_n := 0$ on J_0, from the VI

$$C_n(i) = \min_{a \in \Gamma(i)} \{c(i,a) + C_{n-1}(a)\}, \quad 1 \leq n \leq |J| - 1, i \in J,$$

$$C(i_0) = \min_{a \in \Gamma(i_0)} \{c(i_0,a) + C_{|J|-1}(a)\}.$$

This method requires $(|J| - 1) \cdot |J| + 1$ minimizations and should often be inferior to the method in Theorem 5.2.6, which requires only $|J|$ minimizations. ◇

Remark 5.2.8 (Infinite many of nodes) Most of the results of Sect. 5.2 remain true if finiteness of I in the definition of an acyclic network is replaced by the following conditions: (i) The set J_0 of sinks is nonempty, (ii) from each non-sink there is a path to J_0; cf. Lemma 5.2.3. ◇

Table 5.2 Solution of Example 5.2.9

i	$\Gamma(i)$	$C(i)$	$f(i)$
1	3,4,5	11	5
2	4	10	4
3	4,6, 8	7	4, 6
4	5,7	7	7
5	6,7,9	7	6, 7
6	8	5	8
7	8, 9	6	9

Example 5.2.9 (Example 5.2.5 continued) We apply Theorem 5.2.6 with $h := N(\cdot)$ to the acyclic network in Fig. 5.2. As $J_0 = \{8, 9\}$, we firstly have $J_1 = \{6, 7\}$, and then we compute $C(i)$ and $f(i)$ for $i \in J_1$. Then we compute $C(i)$ and $f(i)$ for $i \in J_2 = \{5\}$, then in $J_3 = \{4\}$, then in $J_4 = \{2, 3\}$ and finally in $J_5 = \{1\}$. The results are shown in Table 5.2. As example, the minimal cost $C(1)$ from node 1 to $\{8, 9\}$ is 11 which results from each of the two cost-minimal subpaths $(1, 5, 6, 8)$ and $(1, 5, 7, 9)$; the latter are indicated in Fig. 5.2 by bold lines. The number $C(i)$, $i \in \mathbb{N}_7$, is also shown in Fig. 5.2 in the right half of the circle around node i. ♦

5.3 Problems

Problem 5.3.1 Derive the VI for that variant of the knapsack problem from Example 5.1.10 where also the sum of volumes $d(i) \in \mathbb{N}$ of packed items of type i may not exceed some given constant $L \geq d := \min_i d(i)$.

Problem 5.3.2 (Traveling salesman problem) Show that the path (3,2,5,4,1), starting in $i := 1$, is a solution of the problem with the five cities $1 :=$ Bonn, $2 :=$ Dresden, $3 :=$ Hamburg, $4 :=$ Karlsruhe and $5 :=$ Leipzig, and the railway distances $d(i,j)$ as given in Table 5.3. Note that even this small problem is time consuming, while an inspection by eye of a map of Germany will immediately suggest the optimal path stated above.

Problem 5.3.3 (For (i) and (ii) cf. Beckmann (1968, p. 11)) An airline has available m types of planes (in unlimited number) for flights from A to B. Each plane of type i, $1 \leq i \leq m$, has $g(i)$ seats and causes costs $u(i) \in \mathbb{N}$ per flight. Unoccupied seats are worthless. One wants to find a cost-minimal combination of planes in order to transport K passengers.

(i) Formulate the problem as an absorbing DP. (Notice the similarity with the knapsack problem from Example 5.1.10.)
(ii) Compute the solution for $m = 3$, $K = 160$ and the data in Table 5.4. How many seats remain unoccupied?
(iii) Verify that (ii) has the explicit solution given in Table 5.5, for $n \in \mathbb{N}_0$.

Table 5.3 Data for Problem 5.3.2

	j				
i	1	2	3	4	5
1	–	730	500	290	550
2	730	–	490	640	120
3	500	490	–	680	400
4	290	640	680	–	520
5	550	120	400	520	–

Table 5.4 Data for Problem 5.3.3

	i		
	1	2	3
$g(i)$	30	50	80
$u(i)$	3	4	5

Table 5.5 Solution for Problem 5.3.3

s	$C(s)$	$f(s)$
$80n + 1 \leq s \leq 80n + 30$	$5n + 3$	1
$80n + 31 \leq s \leq 80n + 50$	$5n + 4$	2
$80n + 51 \leq s \leq 80n + 80$	$5n + 5$	3

Problem 5.3.4 Consider an acyclic network in which $I := \mathbb{N}_m$ and where there is an arc from i to j for $i, j \in I$ if and only if $i < j$. Show that the sets J_k for the recursion in state space in Theorem 5.2.6, are the singletons $\{m - k\}$ for $0 \leq k \leq m - 1$ and $J_k = \emptyset$ for $k \geq m$.

Problem 5.3.5 Consider a DP with $S = \mathbb{N}_{0,m}$ and absorbing set $J_0 = \mathbb{N}_{0,b}$ for some $0 \leq b < m$. If $\max_{a \in D(s)} T(s, a) \leq s - k$ for $s \in J = \{b + 1, b + 2, \ldots, m\}$ and some $k \in \mathbb{N}_{b+1}$, then $N(m) = \lfloor \frac{m-b}{k} \rfloor$.

Problem 5.3.6 (The postman problem) For most optimization purposes *undirected graphs* can be defined as graphs (I, Γ) with symmetric Γ, so that necessarily cycles occur. A prominent example of a problem of shortest path in an undirected network is the so-called *postman problem*: a postman starts in some node i of a undirected network, whose nodes represent junctions and whose arcs represent streets. The postman wants to traverse all streets at least once with minimal time before returning to node i. Unfortunately this problem does not fit into Theorem 5.2.6, but it can be solved by reaching methods; cf. Supplement 5.4.6.

5.4 Supplements

Supplement 5.4.1 (Computational efficiency) Consider a DP network with $I := \mathbb{N}_n$ and $\Gamma(i) := \mathbb{N}_{i+1,n}$ for $1 \leq i \leq n - 1$ and $\Gamma(n) := \emptyset$. One easily sees that the computation of the minimal cost $C(1)$ for paths from 1 to n by means of the OE requires

$$\sum_{i\in J}|\Gamma(i)| = \sum_{i=1}^{n-1}(n-i) = n(n-1)/2 \text{ additions and}$$

$$\sum_{i\in J}(|\Gamma(i)|-1) = (n-1)(n-2)/2 \text{ comparisons,}$$

hence altogether $\sim n^2$ elementary operations := additions or comparisons. On the other hand, as any path $(i_t)_1^k$ of length $k \leq n-1$ from 1 to n is determined by the $k-1$ different nodes $1 < i_1 < i_2 < \cdots < i_{k-1} < i_k = n$, there are

$$\sum_{k=1}^{n-1}\binom{n-2}{k-1} = 2^{n-2}$$

different paths from 1 to n. This shows clearly the computational superiority of the DP approach over simple enumeration.

Supplement 5.4.2 (Adding forbidden arcs) Sometimes the structure of Γ in a DP network can be simplified by adding *forbidden* arcs (i.e. arcs not in Γ) and assigning to them the cost ∞. (For an implementation on a computer, it suffices to replace ∞ by $\sum|c(i,j)|+1$, where the sum runs over all (i,j) in Γ). No change of the theory is necessary, and an optimal path from a non-sink to J_0 does not contain a forbidden arc.

Supplement 5.4.3 (Parametrizing an acyclic network)

(a) In an acyclic network (I,Γ) put $I_k := \{k\} + RI(k)$ and $\Gamma_k := \Gamma \cap I_k$, $k \in I$. It is easy to show that (I_k,Γ_k) is again an acyclic network with $J_0 \cap RI(k)$ as the set of sinks.

(b) The computation of the minimal cost for paths from some non-sink k to nodes i in the set $RI(k)$ can be done according to the *backward OE* in Theorem 5.2.6(a), but also by the *forward OE*

$$K(i) = \min\{c(a,i) + K(a) : a \in \Gamma^{-1}(i)\},\ i \in J, \tag{5.9}$$

where the inverse Γ^{-1} of Γ is defined by

$$\Gamma^{-1}(i) =: \{j \in I : i \in \Gamma(j)\}.$$

(The elements in $\Gamma^{-1}(i)$ are called the *precursors* of the node i.) Note that (5.9) can be solved by recursion in state space starting with $K(k) := 0$. It follows from Theorem 5.2.6 with I replaced by I_k and Γ replaced by Γ^{-1} that $K(i)$ is the minimal cost for paths from node k to node i.

Supplement 5.4.4 (Literature on knapsack problems) There is an extensive literature on knapsack problems. Special efficient algorithms have been devised within the field of combinatorial optimization (cf. e.g. Nemhauser and Wolsey 1988 and Neumann and Morlock 1993). The book by Martello and Toth (1990) is exclusively devoted to knapsack problems.

Supplement 5.4.5 (Literature on the theory of graphs) Introductions to the theory of graphs and optimization problems connected with them are given in many books on operations research such as Gondran and Minoux (1984), Jungnickel (1987), Nemhauser et al. (1989) and Neumann and Morlock (1993).

Supplement 5.4.6 (Literature on the reaching method) Sometimes the solution of a DP network can be done more efficiently than with the forward or backward procedure in Theorem 5.2.6 by a method, due to Dijkstra (1959), which is sometimes called *reaching*. It can also be interpreted in a somewhat complicated way as a DP, cf. Dreyfus and Law (1977, p. 56). Another advantage of reaching is that it still works – in contrast to Theorem 5.2.6 – for networks which have some cycles, however none with negative cost.

Chapter 6
Monotonicity of the Value Functions

Only a small number of DPs possess an explicit solution. Therefore structural (i.e. qualitative) properties of the value functions and maximizers are of considerable importance. Here we study the monotonicity of $V_n(s)$ in n and, with arbitrary relations on S, also in s. Examples of structural properties considered in later sections are concavity and Lipschitz continuity of the value function. As a first benefit of each structural result it can serve to check the correctness of numerical computations. Further benefits, specific to the structure considered, are discussed at the corresponding place.

The tool for most structural investigations is the VI. For its efficient handling we now introduce a short-hand notation using operators. No knowledge of operator theory is required; in the few cases where we need notions and facts from functional analysis, we state them in detail.

For the set $\mathbb{F}$ of all decision rules we use the abbreviations

$$T_f(s) := T(s,f(s)), \quad r_f(s) := r(s,f(s)), \quad s \in S, f \in \mathbb{F}.$$

The operators L, U_f for $f \in \mathbb{F}$, and U are defined on $\mathfrak{V}_0 := \{v: S \to \overline{\mathbb{R}}\}$ as follows:

$$\begin{aligned}(Lv)(s,a) &:= r(s,a) + \beta v(T(s,a)),\ (s,a) \in D, \\ (U_f v)(s) &:= Lv(s,f(s)) = r_f(s) + \beta v(T_f(s)),\ s \in S, \qquad (6.1) \\ (Uv)(s) &:= \sup_{a \in D(s)} Lv(s,a) = \sup_{f \in \mathbb{F}} U_f v(s),\ s \in S.\end{aligned}$$

The operator U is called the ***optimal reward operator.***

The last equality in (6.1) follows from Appendix A.4.11 as for fixed s the mapping $f \mapsto f(s)$ from $\mathbb{F}$ into $D(s)$ is surjective.

K. Hinderer et al., *Dynamic Optimization*, Universitext,
DOI 10.1007/978-3-319-48814-1_6

Using U the VI may be written as

$$V_n = UV_{n-1}, \quad n \geq 1, \tag{6.2}$$

and hence $V_n = U^n V_0$, $n \geq 1$. Moreover, f is a maximizer at stage n if and only if $U_f V_{n-1} = UV_{n-1}$, which by (6.2) is equivalent to $U_f V_{n-1} = V_n$.

In order not to overburden the terminology; in minimization problems we again use the notations L, U_f and U, but with the following different, yet obvious meaning: for $v \in \mathfrak{V}_0$ and $(s,a) \in D$

$$(Lv)(s,a) := c(s,a) + \beta v(T(s,a)),$$
$$(U_f v)(s) := Lv(s,f(s)) = c_f(s) + \beta v(T_f(s)),$$
$$(Uv)(s) := \inf_{a \in D(s)} Lv(s,a) = \inf_{f \in \mathbb{F}} U_f v(s).$$

6.1 Monotone Dependence on the Horizon

The usual ordering on $\mathbb{R}$ and also on $\overline{\mathbb{R}}$ is denoted by $\leq$. For sequences $(x_n)_0^\infty$ of reals we agree: Instead of *non-decreasing* we say *increasing* or *isotone*, and instead of *non-increasing* we say decreasing or *antitone*. If v_1, v_2 and v_n, $n \geq 1$, are functions on a set M, then $v_1 \leq v_2$ and monotonicity of $n \mapsto v_n$ is to be understood pointwise.

Theorem 6.1.1 (Monotone dependence on the horizon)

(a) *Each of the operators L, U_f, $f \in \mathbb{F}$, and U is isotone, i.e. $v_1 \leq v_2$ implies $Lv_1 \leq Lv_2$, $U_f v_1 \leq U_f v_2$ and $Uv_1 \leq Uv_2$.*

(b) *If $V_0 \leq V_1$ [$V_0 \geq V_1$], then $V_n(s)$ is increasing [decreasing] in n. In particular, if $V_0 \equiv 0$ and if $r \geq 0$ [$r \leq 0$], then $V_n(s)$ is increasing [decreasing] in n.*

Proof (a) is obvious from the definition of the operators involved, while (b) uses (a) and induction on $n \geq 1$. □

Remark 6.1.2

(i) Obviously Theorem 6.1.1(b) also holds when the value functions V_0, V_1 and V_n are replaced by the minimal cost functions C_0, C_1 and C_n, respectively.
(ii) It is *not* true that $V_0(s) \leq V_1(s)$ for *some* $s \in S$ implies that $n \mapsto V_n(s)$ is increasing; see Problem 6.5.4 below.
(iii) Obviously Theorem 6.1.1(b) generalizes as follows: If $V_k \leq V_{k+1}$ ($V_k \geq V_{k+1}$) for some $k \geq 0$, then $n \mapsto V_n$ is increasing (decreasing) for $n \geq k$. ◇

Example 6.1.3 (Monotone dependence on the horizon)

(a) In Example 2.4.1(c3) we gave conditions under which $V_n(s)$ is increasing or decreasing in n in the discrete allocation problem from Example 2.4.1. Similar conditions are given in Example 2.4.3(b3) for the continuous allocation problem from Example 2.4.3. By the way, the assumption made for Example 2.4.1(c3) and Example 2.4.3(b3), that the utility function u is increasing, can be dropped.

(b) Already in Example 4.1.7 (4.6) we studied the monotonicity of $n \mapsto C_n(s)$, the linear system with quadratic costs and with $C_0(s) := d_0 s^2$. Consider that variant where $0 < \beta < 1$, $\gamma = g = 1$ and where instead of $C_0(s) := d_0 s^2$ the terminal cost function C_0 satisfies $C_0(s) \geq \lambda s^2/(1-\beta)$, $s \in \mathbb{R}$.

Then $C_n(s)$ is decreasing in n. ♦

6.2 Relations and Orderings

In most important cases both S and A are subsets of $\mathbb{R}$ with the usual total ordering $\leq$. Therefore one can concentrate for first reading on this case by skipping this section and going immediately to Sect. 6.3. We decided to work (on arbitrary sets S and A) within the framework of relations rather than orderings since the restriction to orderings would not result in an essential simplification.

A (binary) ***relation*** on a non-empty set M is defined as an arbitrary subset R of M^2. We then call (M, R) (or only M, if the relation R is clear from the context) a ***structured set.*** (Note that (M, R) is nothing but a directed graph. See Sect. 5.2 with M as the set of nodes and R as the set of arcs.) In applications relations often are defined verbally, and then R must be chosen as the set of all pairs $(x, y) \in M^2$ which share the relation. Instead of $(x, y) \in R$ one writes $x \preceq_M y$ or simply $x \preceq y$. By $x \not\preceq y$ we mean $(x, y) \notin R$, and $y \succeq x$ is the same as $x \preceq y$.

Unless stated otherwise, we endow $\mathbb{R}^d$ with the *argumentwise ordering* which we denote by $\leq^d$ or just by $\leq$ (cf. Definition 6.2.2(iv)). When dealing with general relations $\preceq$ one must beware of fallacies which may occur by unconsciously assuming that $\preceq$ has the same properties as the usual total ordering $\leq$ on $\mathbb{R}$. As an example, a constant mapping from the structured set $(M, \preceq_M)$ into $(B, \preceq)$ need not be increasing if $\preceq_B$ is not reflexive.

We now summarize some basic definitions.

Definition 6.2.1

(i) The relation $\preceq$ on M is called

- ***reflexive,*** if $x \preceq x$ for all $x \in M$,
- ***transitive,*** if $x \preceq y$ and $y \preceq z$ implies $x \preceq z$ for all $x, y, z \in M$,
- ***antisymmetric,*** if $x \preceq y$ and $y \preceq x$ implies $x = y$ for all $x, y \in M$,
- ***total,*** if any two elements $x \neq y \in M$ are ***comparable,*** i.e. if at least one of the two statements $x \preceq y$, $y \preceq x$ holds.

(ii) A relation is called

- a ***preordering***, if it is reflexive and transitive,
- an ***ordering***, (also called a ***partial ordering***) if it is reflexive, transitive and antisymmetric,
- a ***total ordering,*** if it is reflexive, transitive, antisymmetric and total.

As a rule, arbitrary relations will be denoted by $\preceq$, $\preceq_M$ etc., while orderings will be denoted by $\leq$, $\leq_M$ etc. A set endowed with a (pre-)ordering is called a ***(pre-)ordered set.***

(iii) Let $\preceq$ be a relation on M, defined by a set $R \subset M^2$. Unless stated otherwise, we use on subsets $B \neq \emptyset$ of M the *restriction* of $\preceq$ on B, defined by the set $R \cap B^2$.

(iv) Each relation $\preceq$ on M defines the new relation ***strictly smaller:***

$$x \prec y :\Longleftrightarrow (x \preceq y) \wedge (x \neq y).$$

This relation is not reflexive. If $\preceq$ is an ordering, then $\prec$ is transitive.

(v) An element b of a structured set $(M, \preceq)$ is called

- a ***smallest [largest] element*** of M if $b \preceq x$ $[b \succeq x]$ for all $x \in M$,
- a ***minimal [maximal] element*** of M if $b \preceq x$ $[b \succeq x]$ for all $x \in M$ which are comparable with b.

Obviously each smallest [largest] element of M is also a minimal [maximal] element, and the converse holds if $\preceq$ is a total relation. Minimal [maximal] elements (and hence also smallest [largest] elements) need not exist. If $\preceq$ is antisymmetric, M has at most one smallest [largest] element. As an example, consider the triangle $M \subset M' := \mathbb{R}^2$ with vertices $(0,0)$, $(1,0)$ *and* $(0,1)$. *M has the unique smallest (and hence unique minimal) element* $(0,0)$, *it has no largest element, and the points on the segment joining* $(0,1)$ *and* $(1,0)$ *are the maximal elements.*

(vi) A function v on a structured set $(M, \preceq_M)$ is called ***$\preceq_M$-increasing*** [***$\preceq_M$-decreasing***] if $x, y \in M$, $x \preceq_M y$ implies that $v(x) \leq v(y)$ $[v(y) \leq v(x)]$.

(vii) As a special case of (vi) a function v on $M := \mathbb{N}_{0,K}$ with $K \in \overline{\mathbb{N}}_0$ and $\preceq_M := \leq^1$ is increasing if and only if $v(n) \leq v(n+1)$ for $0 \leq n < K$. This property is also called isotonicity of v.

(viii) As an extension of (vi), needed, for example, in Theorem 6.3.5 below for $s \mapsto v(s) := T(s, a)$, let v be a mapping from the structured set $(M, \preceq_M)$ into the structured set $(B, \preceq_B)$. Then v is called ***$(\preceq_M, \preceq_B)$-increasing*** [***$(\preceq_M, \preceq_B)$-decreasing***] if $x, y \in M, x \preceq_M y$ implies that $v(x) \preceq_B v(y)$ $[v(y) \preceq_B v(x)]$. We often simply say *increasing* or *decreasing*, if it is clear from the context which relations are used on M and on B. Instead of *increasing* [*decreasing*] we also say ***isotone [antitone].*** A mapping v is called ***monotone*** if it is increasing or decreasing. All constant mappings from M into B are both increasing and decreasing if $\preceq_B$ is reflexive. In particular, all constant functions on M are both increasing and decreasing.

(ix) A subset B of a structured set $(M, \preceq)$ with a reflexive relation $\preceq$ is called ***increasing*** (or an upper set) if $x \in B$ and $y \in M$ with $y \succeq x$ implies $y \in B$. As an example, the set $\{(x, y) \in (\mathbb{R}_+)^2 : x^2 + y^2 \geq 1\}$ is an increasing subset of $\mathbb{R}^2$, while $\{(x, y) \in (\mathbb{R}_+)^2 : x^2 + y^2 \leq 1\}$ is not. Moreover, B is called ***decreasing*** (or a lower set) if $x \in B$, $y \in M$ and $y \preceq x$ implies $y \in B$. Due to reflexivity of $\preceq$ the set B is increasing if and only if B^c is decreasing if and only if 1_B is increasing with respect to $\preceq$. A subset of $(\mathbb{R}, \leq)$ is increasing if and only if it is a right unbounded interval.

In the sequel we summarize the most important relations and orderings. In order to visualize the subsequent relations $\preceq_M$ on M the reader is advised to draw figures, whenever possible, for an appropriate subset $M \neq \emptyset$ of $\mathbb{R}$ or of $\mathbb{R}^2$ and, moreover, for several $x \in M$ the set of lower bounds $\{a \in M : a \preceq_M x\}$ of x, and of the set of upper bounds $\{a \in M : a \succeq_M x\}$ of x. With a view towards later applications we indicate in brackets for some of the relations the meaning of $\preceq_M$-isotonicity for a function g on M. This is applied for $M := S$ and $g := V_n$ in the context of Theorem 6.3.5 below.

Definition 6.2.2 The following relations are given on an arbitrary set M.

- The ***inverse relation*** $\preceq_{inv}$ of the relation $\preceq$ on M is defined on M by: $x \preceq_{inv} y :\Longleftrightarrow x \succeq y$.

  A mapping v from $(M, \preceq_M)$ into $(B, \preceq_B)$ is decreasing with respect to $\preceq_M$ and $\preceq_B$ if and only if it is increasing with respect to the inverse of $\preceq_M$ and with respect to $\preceq_B$ (or with respect to $\preceq_M$ and the inverse of $\preceq_B$). Thus any statement about increasing mappings yields corresponding statements about decreasing mappings. Similarly, from properties of smallest [minimal] elements of a structured set one obtains corresponding properties for largest [maximal] elements.
- The ***diagonal ordering*** $\leq_{\text{diag}}$ on M is defined by the *diagonal* $R := \{(x, x) : x \in M\}$, i.e. $x \leq_{\text{diag}} y$ if and only if $x = y$. [g is arbitrary.] This relation is useful for a unified formulation of convexity results, see Theorem 7.1.2 below.
- The ***trivial preordering*** $\preceq_{tr}$ on M is defined by the set $R := M^2$, i.e. $x \preceq_{tr} y$ holds for all x, y.
- Let $(M_i, \preceq_i)$, $1 \leq i \leq m$, be structured sets. The ***product relation*** $\preceq = \times_1^m \preceq_i$ on $M := \times_{i=1}^m M_i$ is defined by

$$(x_i)_1^m \preceq (y_i)_1^m :\Longleftrightarrow x_i \preceq_i y_i \text{ for all } i.$$

  The product relation is reflexive or a preordering or an ordering if and only if all relations $\preceq_i$ have the same properties. Note that the argumentwise ordering $\leq^d$ on $\mathbb{R}^d$ is the d-fold product of $\leq^1$ on $\mathbb{R}$.
- From two relations $\preceq_1$ and $\preceq_2$ on M_1 and M_2, respectively, and from their inverse relations, one obtains four different product relations on $M_1 \times M_2$ which will be called the ***quadrant relations*** (the ***NE-, NW-, SW-,*** and ***SE-*** relations) generated by $\preceq_1$ and $\preceq_2$. For example, in $\mathbb{R}^2$ the NE-relation is the usual ordering, and for general M_1 and M_2 we have

$$(x_1, x_2) \preceq_{\text{SE}} (y_1, y_2) \Longleftrightarrow (x_1 \preceq_1 y_1) \wedge (x_2 \succeq_2 y_2).$$

- For fixed $b \in \mathbb{R}$ consider on $\mathbb{R}$ the ***mode-ordering***

$$x \preceq y \Longleftrightarrow (x \leq y \leq b) \vee (b \leq y \leq x).$$

Then g is increasing with respect to $\preceq$ if and only if it is unimodal with mode b, i.e. increasing on $(-\infty, b]$ and decreasing on $[b, \infty)$. Then b is a maximum point of g, and $g(b) \geq g(b-) \vee g(b+)$.

- The ***induced relation*** $\preceq_h$ induced on a set by a mapping h from M into a structured set $(B, \preceq_B)$ is defined by $x \preceq_h y :\Longleftrightarrow h(x) \preceq_B h(y)$; [$g = v \circ h$ for a function $v: h(M) \to \mathbb{R}$, increasing with respect to $\preceq_B$]. If $(B, \preceq_B) = (\mathbb{R}, \leq)$ then $\preceq_h$ is a preordering, and even a total ordering if h is injective. If in addition $M = \mathbb{R}$ and $h(x) := -|x|$, then $g: \mathbb{R} \to \mathbb{R}$ is increasing with respect to $\preceq_h$ if and only if g is unimodal and even.
- The ***set inclusion ordering*** $\leq$ is defined on the power set $\mathcal{K}$ of a set K by $B \leq C :\Longleftrightarrow B \subset C$.
- Let $\preceq_l$, $1 \leq l \leq k$, be relations on a set M. The ***intersection relation*** $\preceq := \cap_{l=1}^{k} \preceq_l$ of these relations is defined by $x \preceq y :\Longleftrightarrow x \preceq_l y$ for $1 \leq l \leq k$. In practice we will consider each of the relations $\preceq_l$ separately.

An important term is ***isotonicity in arguments.*** This is a special case of isotonicity of a mapping v from a structured set $(M, \preceq_M)$ into a structured set $(B, \preceq_B)$ which arises when M is a subset of a Cartesian product $X := \times_1^m X_i$ for some $m \geq 2$. For $\emptyset \neq I \subset \mathbb{N}_m$ we call $x_I := (x_i, i \in I)$ the x_I-argument of v.

Definition 6.2.3 Let $\preceq_i$ be a relation on X_i, $i \in I$, and endow $X_I := \times_{i \in I} X_i$ with the product relation $\preceq_I := \times_{i \in I} \preceq_i$. We say that the mapping v is ***increasing in*** $\boldsymbol{x_I}$ with respect to $\preceq_B$ if for each $z \in proj(M \to \times_{i \notin I} X_i)$ the mapping $x_I \mapsto v_z(x_I) := v(x_I, z)$ is $(\preceq_I, \preceq_B)$-increasing on the non-empty z-section $M_z \subset X_I$ of M. (Note that $M_z = X_I$ if $M = X$.) Then v is clearly increasing in x_I with respect to $\preceq_B$ if and only if it is $(\preceq_I^{\arg}, \preceq_B)$-increasing where

$$(x_i)_1^m \preceq_I^{\arg} (y_i)_1^m :\Longleftrightarrow (x_i \preceq_i y_i, i \in I) \wedge (x_i = y_i, i \notin I).$$

Some special cases of isotonicity are

(a) $I = \mathbb{N}_m$. This kind of isotonicity could be named ***joint isotonicity*** in analogy to joint continuity of v.
(b) $I = \{j\}$ for some $1 \leq j \leq m$. This defines ***isotonicity of v in the single argument*** $\boldsymbol{x_j}$, i.e. isotonicity of $x_j \mapsto v((x_i)_1^m)$. (This is analogous to continuity of v in the argument x_j.) If v is isotone in x_j for *all* $1 \leq j \leq m$, we say that v is ***argumentwise isotone.***

As a rule, argumentwise isotonicity is easier to deal with than joint isotonicity. The properties of the following example are easily seen.

Example 6.2.4

(a) If all relations $\preceq_i$, $1 \leq i \leq m$, are reflexive, joint isotonicity of v on $M \subset X$ implies argumentwise isotonicity of v.
(b) If $M = X$ and if $\preceq_B$ is transitive, argumentwise isotonicity implies joint isotonicity of v. This becomes clear by considering for each x, $y \in M$ with $x \preceq_X y$ the values of v along a path parallel to the axes and connecting x with

y. The assumption $M = X$ cannot be dropped: If $m = 2$, $X_1 = X_2 = \mathbb{R}_+$, $M := \{(x, y) \in \mathbb{R}_+^2 : x \leq y \leq 2x\}$, $v(0, 0) := 1$ and $v(x, y) := 0$ else, then v is argumentwise increasing but not jointly increasing.

(c) The assumption $M = X$ in (b) can be weakened to the requirement that for each $x, y \in M$ there exists a path from x to y with line segments parallel to the axes. As an example, if $S \subset \mathbb{R}_+ = A$ and $D(s) = [0, d(s)]$, $s \in S$, with an increasing function $d \geq 0$ on S, then a function w on D is jointly increasing if (and only if) it is argumentwise increasing.

(d) As a consequence of (a) and (b), for mappings v from the Cartesian product of structured sets $(X_i, \preceq_i)$ into a structured set $(B, \preceq_B)$ joint isotonicity and argumentwise isotonicity with respect to $\preceq_B$ are equivalent if all relations $\preceq_i$ are reflexive and if $\preceq_B$ is transitive. ♦

6.3 Monotone Dependence on the Initial State

In the rare cases where the value functions have an explicit form, this can be used for a direct study of structural properties. As an example we mention the multi-product allocation problem from Example 4.1.6(b) with square root utility and $d_0 > 0$. There we have $V_n(s) = d_n(\sum_{j=1}^m s^j)^{1/2}$ for $s = (s^j)_1^m \in S := \times_1^m[0, K_j]$ and for constants $d_n > 0$. Then V_n is obviously jointly increasing, and it follows from Appendix D.6.2 that it is also concave.

In some cases where the value functions do not have an explicit form, monotonicity of $s \mapsto V_n(s)$ is intuitively clear from the problem formulation. As an example, in the discrete allocation problem from Example 2.4.1 one expects that the value functions are increasing functions of the initial capital s; indeed this was shown in Example 2.4.1(c1) by an ad hoc method.

One of the main themes of the present treatise is the search for assumptions such that: (a) All value functions belong to a certain set $\mathfrak{V}$ of functions (e.g. the set of continuous functions or the set of concave functions), (b) there exist maximizers (possibly of a special form) at each stage. The typical proof proceeds by induction on n and applies the VI. It is useful to formalize this approach in the next result whose proof is omitted since it is very simple.

Theorem 6.3.1 (Structure Theorem for DPs) *Let $\mathfrak{V}$ be a set of functions on S which contains V_0. Then:*

(a) Assume (S1): $Uv \in \mathfrak{V}$ for all $v \in \mathfrak{V}$. Then $V_n \in \mathfrak{V}$ for all $n \geq 1$.

(b) Assume besides (S1) also (S2): For each $v \in \mathfrak{V}$ there exists a maximizer f_v of Lv. Then for each $n \geq 1$ the decision rule $f_n := f_{V_{n-1}}$ is a maximizer at stage n and $(f_n)_N^1$ is an optimal policy for each $N \geq 1$ by the Basic Theorem 2.3.3.

The preceding method entails the following guideline: The value functions can be expected to possess a given structural property if D, T, r and V_0 share a *similar* property. Thus our main task consists in detecting these structural properties for D, T, r and V_0.

Firstly we collect without proof simple *closure* properties of the set of increasing mappings on a structured set, useful for checking monotonicity of the value functions $s \mapsto V_n(s)$. (In most cases where S is an interval in $\mathbb{R}$, the usual differential criteria for monotonicity cannot be used since $V_n(\cdot)$ as a supremum of functions is typically not differentiable everywhere.)

Lemma 6.3.2 (Closure properties of increasing mappings on structured sets)

(a) The set of increasing functions on a structured set is a convex cone, i.e. (a1) the sum of two increasing functions is increasing, and (a2) the product λv of some $\lambda \in \mathbb{R}_+$ and an increasing function v is increasing.
(b) The composition $v \circ g$ of an $\preceq_M$-increasing [$\preceq_M$-decreasing] function v on $(M, \preceq_M)$ with a mapping g from $(M_1, \preceq_1)$ into $(M, \preceq_M)$ is $\preceq_1$-increasing [$\preceq_1$-decreasing] if g is $(\preceq_1, \preceq_M)$-increasing.
(c) The product of two non-negative increasing functions is increasing.
(d) The positive function v is increasing if and only if $1/v$ is decreasing.

Concerning isotonicity of $V_n(s)$ in s we obviously should assume isotonicity of $T(s,a)$, $r(s,a)$ and of $V_0(s)$ in s. A bit more thought is required for the restriction set D, as follows. We formulate monotonicity properties of D by means of the set-valued mapping $s \mapsto D(s)$ which assigns to each state $s \in S$ the set of actions admissible at s. This mapping will be denoted by $D(\cdot)$ in order to distinguish it from the set D. Note that $D(\cdot)$ is a mapping from S into the system $\mathcal{P}'(A)$ of non-empty subsets of A. The mapping $D(\cdot)$ is called a ***correspondence from S into A***. More generally, each mapping $s \mapsto G(s)$ from S into the system of non-empty subsets of A is called a ***correspondence with graph*** $G := \{(s,a) \in S \times A : a \in G(s)\}$.

If S is structured the correspondence $D(\cdot)$ is called ***increasing [decreasing]*** (with respect to $\preceq_S$) if $s \preceq_S s'$ implies $D(s) \subset [\supset] D(s')$, i.e. if $D(\cdot)$ is increasing with respect to $\preceq_S$ and the set inclusion ordering on $\mathcal{P}'(A)$.

If $\preceq_A$ is reflexive we say that $D(\cdot)$ has the ***interval form*** $D(\cdot) = [c,d]$ if

$$D(s) = [c(s), d(s)] := \{a \in A : c(s) \preceq_A a \preceq_A d(s)\}, \quad s \in S, \tag{6.3}$$

for mappings c and $d \succeq_A c$ from S into A. Reflexivity of $\preceq_A$ ensures that $D(s) \neq \emptyset$ for all s. If A is an interval in $\mathbb{R}$ [in $\mathbb{Z}$], then the sets $[c(s), d(s)]$ from (6.3) are intervals in $\mathbb{R}$ [in $\mathbb{Z}$] in the usual sense. If $A = \times_{i=1}^m A_i$ for intervals A_i and if $c(s) = (c_i(s))_1^m$ and $d(s) = (d_i(s))_1^m$ for functions $c_i \leq d_i$ from S into A_i, then $[c(s), d(s)]$ is the m-dimensional interval $\times_{i=1}^m [c_i(s), d_i(s)]$.

Remark 6.3.3 If S is structured and if $\preceq_A$ is a preordering, the correspondence $D(\cdot)$ of the interval form $D(\cdot) = [c,d]$ is increasing if and only if (c is decreasing) $\wedge$ (d is increasing), both with respect to $\preceq_S$ and $\preceq_A$. Figure for $S = A := \mathbb{R}_+$! $\diamond$

In applications the state space is uniquely determined, but one may replace without change of the solution the given action space A by any superset of the ***essential action space*** $A_0 := \cup_{s \in S} D(s)$; A_0 is also the projection $\mathrm{pr}(D \to A)$

of D into A. In particular, A may be replaced by A_0. E.g. for $S = A = \mathbb{R}_+$, $D(s) := [0, 1 \wedge s]$, hence $A_0 = [0, 1]$! The set

$$D_a := \{s \in S : (s, a) \in D\}, \quad a \in A,$$

is called the *a-section of D*. As an example, if $S = A = \mathbb{R}_+$ and $D(s) = [0, s]$ for all s, then $D_a = [a, \infty)$; figure! Note that $D_a \neq \emptyset$ if and only if $a \in A_0$, and that $s \in D_a$ if and only if $a \in D(s)$ if and only if $(s, a) \in D$. A mapping w from D into some set defines the mapping $w(\cdot, a)$ only for $a \in A_0$. Therefore the statement that some property of $w(\cdot, a)$ holds *for all a* is to be understood as *for all* $a \in A_0$.

Lemma 6.3.4 (Supremum of a family of increasing functions) *Let S be structured and let w be a function on D. If D(s) and w(s, a) are increasing in s, then the function*

$$s \mapsto w^*(s) := \sup\{w(s, a) : a \in D(s)\}$$

is increasing on S.

Proof Fix s and s' with $s \preceq s'$. If $a \in D(s)$ then also $a \in D(s')$, hence $w(s', a)$ is defined and $\geq w(s, a)$. Thus

$$\begin{aligned} w^*(s) &= \sup\{w(s, a) : a \in D(s)\} \leq \sup\{w(s', a) : a \in D(s)\} \\ &\leq \sup\{w(s', a') : a' \in D(s')\} = w^*(s'). \end{aligned}$$

□

The next result is central for the present section. It is most useful in the predominant case where $S \subset \mathbb{R}$ and $\leq_S = \leq^1$. Consistent with our guideline stated above, the result says roughly that each value function V_n is increasing if D, T, r and $V_0 \circ T$ are increasing in s. We do not require finiteness of the value functions. A DP is called ***invariant*** if $T(s, a) =: T_1(a)$ is independent of s. Sometimes this property simplifies the study of the structure of the solution, see e.g. Example 6.4.1 with choice I of actions, and the VI (6.5).

Theorem 6.3.5 (Isotonicity of the value functions) *Let* $(S, \preceq_S)$ *be structured. Then* $V_n(\cdot)$, $n \geq 1$, *is* $\preceq_S$*-increasing under the following conditions:*

(i) $D(\cdot)$ *is* $\preceq_S$*-increasing,*
(ii) $s \mapsto T(s, a)$ *is* $(\preceq_S, \preceq_S)$*-increasing,*
(iii) $s \mapsto r(s, a)$ *is* $\preceq_S$*-increasing,*
(iv) $s \mapsto V_0(T(s, a))$ *is* $\preceq_S$*-increasing.*

Moreover, (iv) holds under (ii) if V_0 *is* $\preceq_S$*-increasing. In addition, (ii) and (iv) hold if the DP is invariant.*

Proof We prove by induction on $n \geq 1$ the assertion (I_n) that V_n is $\preceq_S$-increasing. Firstly, $W_1(s,a)$ is $\preceq_S$-increasing in s by Lemma 6.3.2. Then Lemma 6.3.4 implies (I_1). Assume (I_n) for some $n \geq 1$. Then $V_n(T(s,a))$ and hence $W_{n+1}(s,a)$ are $\preceq_S$-increasing in s by Lemma 6.3.2. Finally (I_{n+1}) follows from Lemma 6.3.4. □

Remark 6.3.6

(i) It follows from Theorem 6.3.5 that in the minimization problem the minimal cost $C_n(s)$, $n \geq 1$, is $\preceq_S$-decreasing in s if $D(s)$ and $T(s,a)$ are $\preceq_S$-increasing in s and if both $c(s,a)$ and $C_0(s)$ are $\preceq_S$-decreasing in s.
(ii) From Theorem 6.3.5 we obtain by inverting $\preceq_S$ that $V_n(s)$ is *decreasing* in s, if $D(\cdot)$ is decreasing (i.e. if $s \preceq_S s'$ implies $D(s) \supset D(s')$), if $T(s,a)$ is increasing (not decreasing!) in s and if $r(s,a)$ and $V_0(s)$ are decreasing in s.
(iii) From (ii) we see that $C_n(s)$ is increasing in s, if $D(\cdot)$ is decreasing, if $T(s,a)$ is increasing in s and if both $c(s,a)$ and $C_0(s)$ are increasing in s. ◇

If $S \subset \mathbb{R}^m$ for some $m \geq 2$ often $\preceq_S$ is the product of orderings $\leq_i$, $1 \leq i \leq m$, where for each i the ordering $\leq_i$ either equals the usual ordering $\leq$ or its inverse or the diagonal ordering. This case and more general ones are treated in the next result which is obviously a special case of Theorem 6.3.5.

Proposition 6.3.7 (Isotonicity of the value functions in an argument s_I) *Let $S \subset \times_1^m X_i$, $\emptyset \neq I \subset \mathbb{N}_m$, and for $i \in I$ let $\preceq_i$ be a relation on X_i. Then $(s_i)_1^m \mapsto V_n((s_i)_1^m)$, $n \geq 1$, is increasing in $s_I := (s_i, i \in I)$ if:*

(i) $D(\cdot)$ is increasing in s_I,
(ii) $s \mapsto T(s,a)$ is increasing in s_I with respect to $\preceq_I^{\text{arg}}$,
(iii) $s \mapsto r(s,a)$ is increasing in s_I,
(iv) $s \mapsto V_0(T(s,a))$ is increasing in s_I.

Moreover, (iv) holds under (ii) if V_0 is increasing in s_I. In addition, (ii) and (iv) hold if the DP is invariant.

From Example 6.2.4(b) above we see that sufficient conditions for argumentwise isotonicity and for (joint) isotonicity of the value functions are obtained from Proposition 6.3.7 by the choice $I := \{j\}$, $1 \leq j \leq m$, and by the choice $I := \mathbb{N}_m$, respectively.

We consider a simple example for $m = 2$: X_1, X_2 are subsets of $\mathbb{R}$, $S = X_1 \times X_2$, $s = (s_1, s_2)$, $\leq_1 := \leq$, $\leq_2 :=$ the inverse of $\leq$, $D = S \times A$ and $T = (T_1, T_2)$. We are looking for conditions under which V_n is (jointly) increasing with respect to $(\leq_1, \leq_2)$, i.e. for which $s_1 \leq s_1'$, $s_2' \leq s_2$ implies $V(s_1, s_2) \leq V(s_1', s_2')$. We could apply Proposition 6.3.7. However, since in our example argumentwise isotonicity equals (joint) isotonicity by Example 6.2.4(b), one obtains quickly from Proposition 6.3.7 the following sufficient conditions: $T_1(x_1, x_2, a)$, $T_2(x_1, x_2, a)$, $r(x_1, x_2, a)$ and $V_0(x_1, x_2)$ are increasing in x_1 and decreasing in x_2.

6.4 Splitting Models

Example 6.4.1 (Splitting a resource into two parts – the management of resource models)

(a) Some aspects of several classical problems can be studied in a unified manner within the following model. It concerns the ***optimal sequential splitting*** of some resource of amount s_t into a first part, retained for future use, and a second part for immediate use; see, for example, the reservoir problem from Table 6.1. We allow the resource to be measured either on the continuous scale (the continuous case) or on the discrete scale (the discrete case) and the problem may concern maximization or minimization. Thus S is an interval or a discrete interval, both in general non-degenerate; see Appendix D.5.1 for the notion of a discrete interval. Mostly bounded sets S are more realistic than unbounded ones, but unboundedness is required in some models for reasons of mathematical tractability.

We can model the same problem by two DPs which differ only by the following choices I and II of actions:

In *choice I* the action a is the first part of the resource, e.g. the dam content after release in the reservoir problem from Table 6.1. From the examples in Table 6.1 we see that in general the next state s' does not simply equal a, but a number $T_1(a)$ for a certain function T_1 from A into S. Therefore we use $s' = T(s,a) = T_1(a)$, and thus the DP is invariant; this property is often an advantage of choice I.

In *choice II* the action a is the second part of the resource, e.g. the amount of water released in the reservoir problem. Now we must assume that the next state s' equals $T_1(s-a)$ with the same function T_1 as in choice I.

For both choices I and II we assume that A and $D(s)$, $s \in S$, are intervals or discrete intervals and without loss of generality that $A = A_0 := \cup_{s \in S} D(s)$.

Characteristic for the splitting model is the form of the one-stage reward (respectively the one-stage cost) which for choice I is the sum

$$r(s,a) = u_1(a) + u_2(s-a) \text{ respectively } c(s,a) = c_1(a) + c_2(s-a) \quad (6.4)$$

for $(s,a) \in D$ and for functions u_1 (respectively c_1) on A and u_2 (respectively c_2) on $\tilde{A} := \{s - a : (s,a) \in D\}$.

The VI for maximization and choice I reads: For $s \in S$ and $n \geq 1$

$$V_n(s) = \sup_{a \in D(s)} \Big[u_1(a) + u_2(s-a) + \beta V_{n-1}(T_1(a)) \Big]. \quad (6.5)$$

The VI for maximization and choice II is obtained from (6.5) by interchanging a with $s - a$ for s fixed. Then $\tilde{D}(s) := \{a \in A : s - a \in D(s)\}$ is an interval or a discrete interval, and we get for $s \in S$ and $n \geq 1$

$$V_n(s) = \sup_{a \in \tilde{D}(s)} \Big[u_1(s-a) + u_2(a) + \beta V_{n-1}(T_1(s-a)) \Big]. \tag{6.6}$$

The value functions remain the same for both choices I and II, but the maximizers change: f_n is a maximizer at stage n for choice I if and only if $s \mapsto s - f_n(s)$ is a maximizer at stage n for choice II. We use (6.6) besides (6.5) since sometimes the derivation of some properties of the value function by the two recursions leads to different sufficient conditions; see (c1) below.

(b) We now turn to important special cases of the splitting model (some with maximization, some with minimization), which for obvious reasons (see Table 6.1) we name ***ARCHI models.*** Their continuous versions with choice I of actions are collected in Table 6.1; note that $A = S$ by our assumption that

Table 6.1 ARCHI models with choice I: r and c as in (6.4)

Model	$S = A$	$D(s)$	$s' = T_1(a)$
1. **A**_llocation_, maximization $s =$ capital before investment $a =$ capital after investment $z \in [1, \infty)$ interest factor $s - a =$ consumption	$\mathbb{R}_+$	$[0, s]$	za
2. **R**_eservoir_, maximization $s =$ dam content before release $a =$ dam content after release $z \in \mathbb{R}^+$ inflow $s - a =$ release $B \in \mathbb{R}^+$ dam capacity	$[0, B]$	$[0, s]$	$(a + z) \wedge B$
3. **C**_ash balance_, minimization $s =$ cash before transfer $a =$ cash after transfer $z \in \mathbb{R}$ cash flow $-(s - a) =$ transfer	$\mathbb{R}$	$\mathbb{R}$	$a - z$
4. **H**_arvesting_, maximization $s =$ asset before harvest $a =$ asset after harvest $\alpha, z \in \mathbb{R}^+$ growth parameters $s - a =$ harvest	$\mathbb{R}_+$	$[0, s]$	$\alpha \cdot (e^{za} - 1)$
5. **I**_nventory_, minimization			
5a. _backlogging_	$(-\infty, B]$	$[s, B]$	$a - z$
5b. _lost sales_ $s =$ stock before ordering $a =$ stock after ordering $z \in \mathbb{R}^+$ demand $-(s - a) =$ order $B \in \mathbb{R}^+$ inventory capacity	$[0, B]$	$[s, B]$	$(a - z)^+$

$A = A_0$, and that $T(s, a) = T_1(a)$. For the discrete versions only the intervals S and $D(s)$ must be replaced by the corresponding discrete intervals.

We add the following remarks on the models with choice I.

(b1) In reality there are often restrictions (e.g. of technological or ecological nature) on the amount a transferred. As an example, in the allocation problem we may have an upper bound $\eta \in \mathbb{R}^+$, which results in the new set $D(s) = [0, \eta \wedge s]$. Many results without such an additional restriction easily carry over to the case of an additional restriction.

(b2) The models are often simplifications. In particular, the constant z may be only a crude approximation as the expectation of a more realistic random quantity, e.g. the random inflow in the reservoir model. Moreover, data such as the random inflow may change with time. Both generalizations are treated in Theorem 17.12.

(b3) In the allocation problem with $z = 1$ one may use $S = [0, B]$, where B is the maximal initial capital.

(b4) In the reservoir problem u_1 may be an ecological reward for the water retained and u_2 may be the reward for the production of electrical power and/or for irrigation from the water released. In the cash balance problem $c_2(s - a)$ are the costs for the transfer of $a - s$. In the harvesting problem u_2 may be the gain from the harvest.

(b5) In the inventory models ***backlogging*** means that in each period unsatisfied demand is postponed for later service, which entails the possibility of negative stocks. ***Lost sales inventory*** means that in each period unsatisfied demand is lost. Usually $c_1(a)$ means holding costs (plus penalty costs for unsatisfied demand in the lost sales model), while $c_2(s - a)$ denotes the cost for the order $a - s$.

(b6) Allocation models are also called ***consumption-investment models*** or ***optimal growth models.***

(b7) In the continuous splitting models with $D(s) = [0, s]$ (but not in its discrete analogue) one may use as action a the *proportion* of the resource retained. This may lead to new conditions for isotonicity of the value functions.

(c) From Theorem 6.3.5 we obtain the following monotonicity results both for the discrete and the continuous versions:

(c1) In the general splitting model all value functions V_n, $n \geq 1$, are increasing if either $D(\cdot)$ and u_2 are increasing (cf. (6.5)) or if $\tilde{D}(\cdot)$, T_1, u_1 and V_0 are increasing (cf. (6.6)).

We now comment on this result when applied to the ARCHI models.

(c2) The allocation model, the reservoir model and the harvesting model have the plausible property that the value functions are increasing as soon as u_2 is increasing. This can be seen quickly from the VI (6.5).

(c3) In the cash balance model with $c_1 :\equiv 0$ the function c_2 will not be monotone, but rather $-c_2$ will be unimodal, and the same will hold under natural assumptions for the functions C_n.

(c4) In both inventory models isotonicity does not hold for the minimal cost functions C_n under plausible conditions, since these exclude isotonicity of u_2 or of C_0. ♦

Remark 6.4.2 (Lost sales inventory) In the literature the lost sales inventory model is considered to be *equivalent* to the reservoir model, as follows: Making the two bijective transformations $\tilde{s} := B - s$ = unused dam content (the so-called freeboard) before release and $\tilde{a} := B - s + a = \tilde{s} + B$ = freeboard after release, we see that the new state $\tilde{s}' :=$ freeboard one period later (before release) equals $B - s' = B - (s - a + z) \wedge B = (\tilde{a} - z)^+$. Thus the reservoir data in terms of $\tilde{s}$ and $\tilde{a}$ coincide with the lost sales inventory data with states $\tilde{s}$ and actions $\tilde{a}$. However, while the reservoir model concerns maximization, the lost sales model concerns minimization, and hence results from one model typically seem not to be very useful for the other one. ◇

Example 6.4.3 (Splitting a resource into three parts) (This is one of the simplest examples where the action space is a subset of $\mathbb{R}^2$.) Here $S = [0, B]$, $B \in \mathbb{R}^+$, in the continuous case, and $S = \mathbb{N}_{0,B}$, $B \in \mathbb{N}$, in the discrete case. That part of the momentary resource which is not retained for later use is allocated to two projects; this implies, as seen below, that $A = S^2$. As in Example 6.4.1 of splitting a resource into two parts, there are the following two choices of actions:

In *choice* I we use the actions $a = (a^3, a^1)$, where a^3 is the part of the resource retained for later use and a^1 is the part of the resource allocated to project 1. Thus $s - a^3 - a^1$ is allocated to project 2. Then

$$D(s) = \{(a^3, a^1) \in S^2 : a^3 + a^1 \le s\}$$

is a triangle in the continuous case, and in the discrete case the set of grid points in a triangle; figure! It is natural to assume that the resource available at the next time and the one-stage reward are given by

$$T(s, a^3, a^1) = T_3(a^3),$$
$$r(s, a^3, a^1) = u_1(a^1) + u_2(s - a^3 - a^1) + u_3(a^3).$$

In particular, the DP is invariant. As $D(\cdot)$ is increasing, it follows from Theorem 6.3.5 that *V_n is increasing for all $n \ge 1$ if u_2 is increasing*.

In *choice II* the resource of momentary amount $s \in S$ is split up into a part a^1 allocated to project 1, a part a^2 allocated to project 2 and a part $s - a^1 - a^2$ retained for later use. It follows from the DP with choice I of actions that we now have, using the actions $a = (a^1, a^2)$,

$$D(s) = \{(a^1, a^2) \in S^2 : a^1 + a^2 \le s\},$$
$$T(s, a^1, a^2) = T_3(s - a^1 - a^2),$$
$$r(s, a^1, a^2) = u_1(a^1) + u_2(a^2) + u_3(s - a^1 - a^2).$$

Finally, it follows from Theorem 6.3.5 that V_n is increasing for all $n \geq 1$ if T_3, u_3 and V_0 are increasing. The model can easily be extended to the case of splitting the resource into an arbitrary number of parts; see Example 7.1.6. ♦

Example 6.4.4 (An example with argumentwise monotone value functions) We consider the following problem of allocating capital during $N \geq 1$ periods in an optimal way for the reduction of toxic waste. At time $t = 0$ we have an amount $x_0 \in \mathbb{N}_K$ of waste and an amount $z_0 \in \mathbb{N}_L$ of capital; both K and L are positive integers. If at time $1 \leq t \leq N$ an amount $x \in \mathbb{N}_{0,K}$ of waste and $z \in \mathbb{N}_{0,L}$ of capital are still present, we may use an amount $0 \leq a \leq z$ of capital for reducing x to $(x - g(a))^+$ for some increasing function g on $\mathbb{N}_{0,L}$ with $g(0) = 0$. There is no one-stage reward, and the terminal reward has the form $V_0(x, z) = z - h(x)$ for some increasing function $h \geq 0$ on $\mathbb{N}_{0,K}$ with $h(0) = 0$. Our goal consists in maximizing our discounted terminal reward.

We use the following DP: $s = (x, z) \in S = \mathbb{N}_{0,K} \times \mathbb{N}_{0,L}$; $A = \mathbb{N}_{0,L}$; $D(x, z) = \mathbb{N}_{0,z}$; $T(x, z, a) = ((x - g(a))^+, z - a)$; $r :\equiv 0$; V_0 as above and $\beta \in \mathbb{R}^+$.

Of course, one will conjecture that $V_n(x, z)$ is decreasing in x and increasing in z. In fact this follows easily by applying Proposition 6.3.7 for $m = 2$, $X_1 = \mathbb{N}_{0,K}$, $X_2 = \mathbb{N}_{0,L}$, once for $j = 1$ with $\preceq_1$ = the inverse of $\leq^1$, and once for $j = 2$ with $\preceq_2 = \leq^1$. Then the assumptions (i)–(iii) in Proposition 6.3.7 are fulfilled: All terms $D(x, z)$, $V_0(T(x, z, a)) = z - a - h((x - g(a))^+$ and $r(x, z, a)$ are decreasing in x and increasing in z. ♦

Example 6.4.5 (Isotonicity of $s \mapsto V_n(s)$ with respect to non-standard relations)

(a) Assume $S = A = \mathbb{R}_+$. We want to find conditions under which each function V_n is unimodal with given mode $b \in \mathbb{R}^+$. This holds according to Definition 6.2.2(vi) if and only if V_n is increasing with respect to the ordering

$$s \preceq s' :\Longleftrightarrow (s \leq s' \leq b) \vee (b \leq s' \leq s).$$

Now Theorem 6.3.5 yields: If $D(s) = [0, s \wedge b]$, $T(s, a) = (b - a - |s - b|)^+$, $r(s, a) = (s - a) \wedge (b - a)$ and $V_0(s) = s \wedge b$, then V_n is unimodal with node b.

For the proof note that $A_0 = [0, b]$, $D_a = [a, \infty)$ for $0 \leq a \leq b$ and that $T(\cdot, a)$, $r(\cdot, a)$ and V_0 are unimodal with mode b.

(b) We want to find conditions under which $s \mapsto V_n(s)$ is increasing on some given proper subset S' of S, given an arbitrary relation $\preceq_S$ on S. This holds if and only if V_n is increasing with respect to the relation

$$s \preceq s' :\Longleftrightarrow (s, s' \in S') \wedge (s \preceq_S s').$$

From Theorem 6.3.5 we obtain the following conditions of isotonicity with respect to $\preceq_S$:

(i) $D(\cdot)|S'$ is increasing,
(ii) $T(\cdot, a)$ is increasing on $D_a \cap S'$ for $a \in A' := \{a \in A : D_a \cap S' \neq \emptyset\}$,
(iii) $T(s, a) \in S'$ for $s \in S'$ and $a \in D(s)$,

(iv) $r(\cdot, a)$ is increasing on $D_a \cap S'$ for $a \in A'$,
(v) V_0 is increasing on S'.

For the proof note that (ii) and (iii) imply isotonicity of $T(\cdot, a): D_a \to S$ with respect to $\preceq$ for $a \in A_0$. In fact, this property means that

$$s, s' \in D_a, s \preceq s' \Longrightarrow T(s,a) \preceq T(s',a), \text{ i.e.}$$
$$s, s' \in D_a \cap S', s \preceq_S s' \Longrightarrow (T(s,a), T(s',a) \in S') \wedge (T(s,a) \preceq_S T(s',a)),$$

and the latter holds by (ii) $\wedge$ (iii). ♦

6.5 Problems

Problem 6.5.1 In the inventory problem from Example 4.1.3 the minimal costs $C_n(s)$ are increasing in n if the discounted scrap value does not exceed the holding cost per unit, i.e. if $\beta e_4 \leq e_3$.

Problem 6.5.2

(i) If $v < w$ then $U_f v < U_f w$ for all f. Does $v < w$ imply $Uv < Uw$?
(ii) Assume that there exists a maximizer at each stage. If $V_1 > V_0$ [$V_1 < V_0$] (e.g. if $V_0 \equiv 0$ and $\inf r > 0$ [$\sup r < 0$]) then $V_n(s)$ is strictly increasing [strictly decreasing] in n.

Problem 6.5.3 Assume that V_0 is bounded below and put for $s \in S$

$$B(s) := \sup\{r(s,a) : a \in D(s)\} + \beta \cdot \inf\{V_0(T(s,a)) : a \in D(s)\}.$$

If $V_0 \leq B$, then $V_n(s)$ is increasing in n. Find an analogous sufficient condition for antitonicity of $V_n(s)$ in n.

Problem 6.5.4 Consider the discrete allocation problem from Example 2.1.1 with VI (2.9), where $\beta = 1$, $u(0) = -1$, $u(1) = V_0(0) = 1$ and $V_0(1) = 0$. Then $V_0(1) < V_1(1)$, but $V_1(1) > V_2(1)$.

Problem 6.5.5 Let $(M_i, \leq_i)$ be structured sets, $i = 0, 1, 2, 3$, and $f_j: M_0 \mapsto M_j$ be mappings, $j = 1, 2$.

(i) The function $f = (f_1, f_2)$ from M_0 to $M_1 \times M_2$ is increasing with respect to $\leq_0$ and $\leq_1 \times \leq_2$ if and only if f_j is increasing with respect to $\leq_0$ and $\leq_j$, $j = 1, 2$.
(ii) If f_1, f_2 and the mapping v from $M_1 \times M_2$ into M_3 are increasing then $x \mapsto v(f_1(x), f_2(x))$ is increasing.

Problem 6.5.6 If A is structured, if the assumptions of Theorem 6.3.5 hold and if $a \mapsto T(s_0, a)$ and $a \mapsto r(s_0, a)$ are increasing for some s_0, then $a \mapsto W_n(s_0, a)$ is increasing for all n.

Problem 6.5.7 If $\preceq_A$ is a preordering, the correspondence $D(\cdot)$ of the interval form $D(\cdot) = [c, d]$ is increasing if c is decreasing and d is increasing.

Problem 6.5.8 In an invariant DP with structured state space the value functions V_n, $n \geq 1$, are increasing if $\preceq_S$ is reflexive, if $D(\cdot)$ is increasing and if $r(s, a)$ is increasing in s. (V_0 need not be increasing.)

Problem 6.5.9 (Strictly increasing mappings) A mapping v from a structured set $(M, \preceq_M)$ into a structured set $(B, \preceq_B)$ is called *strictly increasing* if x, $y \in M$ and $x \prec_M y$ implies that $v(x)$ and $v(y)$ are comparable and that $v(x) \prec_B v(y)$. Prove the following properties:

(i) If v is strictly increasing and if $\prec_B$ is reflexive then v is increasing.
(ii) The composition of two strictly increasing mappings is strictly increasing.
(iii) The sum of finitely many increasing functions (if defined) is strictly increasing if at least one of them is strictly increasing.
(iv) If v is a strictly increasing function and $c \in \mathbb{R}^+$ then $c \cdot v$ is strictly increasing.
(v) The product of two positive increasing functions is strictly increasing if at least one of the functions has this property.
(vi) Prove without differentiation that

$$x \mapsto \log(x^4 + x^2 + 1) + 2x \cdot \exp(\sqrt{x})$$

is strictly increasing on $\mathbb{R}^+$, using the fact that $x \mapsto x^\alpha$ for $\alpha \in \mathbb{R}^+$, $x \mapsto \log(x)$ and $x \mapsto \exp(x)$ are strictly increasing on $\mathbb{R}^+$.

Problem 6.5.10 Assume that the isotonicity assumptions of Theorem 6.3.5 hold, that there exists for each $n \geq 1$ a maximizer at stage n and that either $r(s, a)$ is strictly increasing in s or that both $V_0(s)$ and $T(s, a)$ are strictly increasing in s. Then $V_n(s)$ is finite and strictly increasing in s for all $n \in \mathbb{N}$.

Problem 6.5.11 (Alfred Müller) Example for a DP where $s \mapsto V_n(s)$ is increasing on $[0, 1]$ and on $(1, \infty)$ for all $n \geq 1$, but where V_2 is not increasing on $\mathbb{R}_+$. The data are as follows: $S = \mathbb{R}_+$; $A = D(s) = [0, 1]$ for all s; $r(s, a) = s \cdot \sqrt{a}$;

$$T(s, a) = \begin{cases} 2 + s \cdot (1 - a), & \text{if } 0 \leq s \leq 1, \\ 1 - a, & \text{if } s > 1; \end{cases}$$

$V_0(s) = 2$ for $0 \leq s \leq 1$ and $V_0(s) = 1$ for $s > 1$.

Problem 6.5.12 Consider the discrete allocation problem from Example 2.1.1 and prove the following properties:

(a) If $\beta < 1$ and $d \geq 1/\sqrt{1 - \beta^2}$, then $V_n(s)$ is decreasing in n.
(b) If $d \leq 1$, then $V_n(s)$ is increasing in n.
(c) $V_n(s, \beta, d)$ is increasing in β and in d.

6.6 Supplement

Literature on monotonicity properties Monotonicity properties of the value functions were studied early on, mostly for the usual ordering on $S \subset \mathbb{R}^d$. The crucial Lemma 6.3.4 can be found in Beckmann (1968, p. 91). A survey of results is given in Hinderer (1985).

Chapter 7
Concavity and Convexity of the Value Functions

Here we deal with the following questions, assuming that the functions under consideration are defined on convex sets or on a non-degenerate discrete interval.

- When is V_n concave or convex or affine? The validity of such a property may be used to check the correctness of computations, and it is also crucial for the existence of monotone maximizers; see Chap. 8. If S is a discrete interval, discrete concavity, discrete convexity and discrete affinity are also of interest.
- When is $W_n(s,\cdot)$ convex or discretely convex on $D(s)$? If this is the case then under some assumptions on D one can restrict the maximization of $W_n(s,\cdot)$ to the extreme points of the set $D(s)$. In particular, if $D(s)$ is a compact interval or a finite discrete interval, a so-called *bang-bang maximizer at stage n* (see Sect. 7.2 below) can be found from the values of $W_n(s,\cdot)$ at the two endpoints of $D(s)$.
- When is $W_n(s,\cdot)$ concave or discretely concave? If this is the case then special methods of maximizations can be applied; see Proposition 7.2.4.
- If S is a subset of $\mathbb{R}^m$ for some $m \geq 2$, and if the one-dimensional sections of S are intervals (in particular if S is convex) or discrete intervals, then argumentwise concavity and argumentwise convexity are also of interest. We only touch upon these two problems by treating simple DPs in Propositions 7.1.7 and 7.2.11.

7.1 Concave Value Functions

When speaking of convexity of sets such as S, A and D we tacitly assume that these are sets in a Euclidean space.

Elementary properties of convex sets and convex or concave functions, discussed in Appendix D, are often used without mention. If D is convex, then also S, the essential action space A_0 and the sets $D(s)$, $s \in S$, are convex. In particular, if $D \subset \mathbb{R}^2$ is convex, S, A_0 and $D(s)$, $s \in S$, are intervals. Euclidean sets are endowed with the usual ordering, unless stated otherwise.

K. Hinderer et al., *Dynamic Optimization*, Universitext,
DOI 10.1007/978-3-319-48814-1_7

Convex constraint sets are often of the form $S \times A$ for convex S and convex A or of the interval form $D(s) = [c(s), d(s)]$, where S is convex, $A \subset \mathbb{R}$ is an interval and where c is convex and d is concave. The proof of convexity of D is left to the reader. We also use the following generalization: Let S and $A \subset \mathbb{R}^m$ be convex, $\preceq_A := \leq^m$, and let $D(s) := \{a \in A : c(s) \preceq_A a \preceq_A d(s)\}$, $s \in S$, for mappings $c \preceq_A d$ from S into A. Then D is convex if and only if (c is $\preceq_A$-convex) $\wedge$ (d is $\preceq_A$-concave); see the definition after Lemma 7.1.1.

When studying concavity of V_n on a convex state space S one might try to use assumptions on $T(\cdot, a)$, $r(\cdot, a)$ and V_0 such that at least $W_1(\cdot, a)$, $a \in A_0$, becomes concave. However, concavity of $W_1(\cdot, a)$, $a \in A_0$, does not guarantee that $s \mapsto V_1(s) = \sup_a W_1(s, a)$ is concave, even if $D = S \times A$, since the supremum over a of a family of concave functions $w(\cdot, a)$, $a \in A$, is in general not concave. (An example of this fact is as follows: $S = A = [0, 2]$, $w(s, a) = -s^2$ if $0 \leq a \leq 1$ and $w(s, a) = -(s-2)^2$ if $1 < a \leq 2$.) This difficulty disappears, as shown by the next result, if D is convex and if we can ensure that $W_1(s, a)$ is concave not only in s but jointly in (s, a).

Lemma 7.1.1 (The supremum of a family of concave functions) *Let w be a [strictly] concave function on the convex set D. Then*

$$s \mapsto w^*(s) := \sup\{w(s, a) : a \in D(s)\}$$

is [strictly] concave, provided $w^* < \infty$, in particular, if the sup *is attained for all s.*

Proof

(a) We prove concavity of w^* under concavity of w. Select $s, s' \in S$ and $\alpha \in (0, 1)$. Since D is convex, we get for $a \in D(s)$ and $a' \in D(s')$ that $\alpha(s, a) + (1 - \alpha)(s', a') = (\alpha s + (1-\alpha)s', \alpha a + (1-\alpha)a')$ belongs to D, hence $\alpha a + (1-\alpha)a' \in D(\alpha s + (1-\alpha)s')$. Now concavity of w implies

$$\begin{aligned} w^*(\alpha s + (1-\alpha)s') &\geq w(\alpha s + (1-\alpha)s', \alpha a + (1-\alpha)a') \\ &= w(\alpha \cdot (s, a) + (1-\alpha) \cdot (s', a')) \\ &\geq \alpha w(s, a) + (1-\alpha) w(s', a'). \end{aligned}$$

As $a \in D(s)$ and $a' \in D(s')$ were arbitrary, we obtain

$$w^*(\alpha s + (1-\alpha)s') \geq \alpha w^*(s) + (1-\alpha) w^*(s').$$

(b) Strict concavity of w^* follows from strict concavity of w, if the sup of $w(s, \cdot)$ is attained at $f(s)$, $s \in S$, by the same reasoning as in (a) for $a := f(s)$ and $a' := f(s')$. □

In case $S \subset \mathbb{R}^k$ there exist a variety of conditions for the transition function $T = (T_i)_1^k$ which together with conditions on the other data of the DP ensure concavity of the value functions. We cover the whole variety using the following setup. Let $\preceq_K$

be a relation on the convex set $K \subset \mathbb{R}^k$. A mapping g from the convex set $M \subset \mathbb{R}^m$ into K is called ***$\preceq_K$-concave*** if $x \neq y \in M$ and $\alpha \in (0, 1)$ imply

$$g(\alpha x + (1-\alpha)y) \succeq_K \alpha g(x) + (1-\alpha)g(y).$$

Most important is the case where $\preceq_K$ is the product of orderings $\le_i$ on $\mathbb{R}$, $1 \le i \le k$, some of which equal the usual ordering $\le^1$, some equal the inverse of $\le^1$ and the rest equals $\le_{\text{diag}}$. Then $g = (g_i)_1^k$ is $\preceq_K$-convex if and only if the corresponding component functions $g_i: M \to \mathbb{R}$ are concave or convex or affine, respectively. Thus, when $M = D$, $K = S \subset \mathbb{R}^k$ and $g = T = (T_i)_1^k$, the next result covers, for example, the three cases where either (T_i, $1 \le i \le k$, is concave) $\wedge$ (V_n is $\le^k$-increasing for $n \ge 0$) or where (T_i, $1 \le i \le k$, is convex) $\wedge$ (V_n is $\le^k$-decreasing for $n \ge 0$) or where T_i, $1 \le i \le k$, is affine.

Theorem 7.1.2 (Concavity of value functions) *Assume*

(i) S is structured and D is convex,
(ii) T is $\preceq_S$-concave,
(iii) r and V_0 are concave,
(iv) all value functions V_n, $n \ge 0$, are $\preceq_S$-increasing.

Then:

(a) If V_n is finite for all $n \ge 1$ then V_n and W_n are concave, and the (possibly empty) set $D_n^*(s)$, $s \in S$, of maximum points of $W_n(s, \cdot)$ is convex.
(b) If $D(s)$ is bounded for each s, then V_n is finite for $n \ge 1$.

Proof

(a1) We prove by induction on $n \ge 1$ the assertion (I_n) that V_{n-1} is concave. (I_1) holds by (iii). Assume (I_n) for some $n \ge 1$. Then $V_{n-1} \circ T$ is finite and concave by (ii) and (iv), due to Appendix D.6.2(a) with $M := D$, $K := S$, $g := T$ and $v := V_{n-1}$. Moreover, W_n is finite and concave by Appendix D.2.9. Finally, V_n is concave by Lemma 7.1.1.
(a2) From (a1) we know that $W_n(s, \cdot)$ is concave for all n and s, which implies convexity of $D_n^*(s)$ by Appendix D.3.1.
(b) Firstly V_0 is finite as in each DP. Assume that V_{n-1} is finite for some $n \ge 1$. Then W_n and hence $v(a) := W_n(s, a)$ is concave in a for fixed s by (a). There exists by Appendix D.2.4(a) an affine upper bound $a \mapsto h(a)$ of v. Thus, as $D(s)$ is bounded, h and hence v is upper bounded on $D(s)$. This ensures finiteness of V_n. □

Remark 7.1.3

(a) By Theorem 6.3.5 all value functions are $\preceq_S$-increasing if $D(\cdot)$ is $\preceq_S$-increasing and if $r(s, a)$, $T(s, a)$ and $V_0(s)$ are $\preceq_S$-increasing in s.
(b) Concavity of V_0 in Theorem 7.1.2(iii) can be replaced by concavity of $V_0 \circ T$. ◇

Remark 7.1.4 A minimization version of Theorem 7.1.2 reads as follows. The minimal cost functions C_n, the functions W_n, $n \geq 1$, and the set $D_n^*(s)$, $s \in S$, of minimum points of $W_n(s, \cdot)$ are convex under the following conditions:

(i) S is structured and D is convex,
(ii) T is $\preceq_S$-concave,
(iii) c and C_0 are non-negative and convex,
(iv) all functions C_n, $n \geq 0$, are $\preceq_S$-decreasing. ◇

Remark 7.1.5 We consider strict concavity of value functions, which is of computational interest, since there exist efficient numerical methods for maximizing strictly concave functions. Strict concavity can be checked by the following result.

Under the assumptions (i)–(iv) of Theorem 7.1.2 and under finiteness of V_n for $n \geq 1$:

(a) If $r(s, a)$ is strictly concave in a then at each stage there exists at most one maximizer.
(b) Assume that V_0 is strictly concave, that T is injective and that there exists at each stage $n \geq 1$ a maximizer f_n. Then V_n and W_n (hence also $a \mapsto W_n(s, a)$, $n \geq 1$, $s \in S$) are strictly concave, and the f_n's are unique.

For the proof of part (a) we notice that we saw in part (a) of the proof of Theorem 7.1.2 that $u := \beta V_n \circ T$, $n \geq 0$, is concave, hence $u(s, a)$ is concave in a. Now the assertion follows since $W_n(s, \cdot)$, $s \in S$, is the sum of the strictly concave function $r(s, \cdot)$ and the concave function $u(s, \cdot)$.

For the proof of part (b) we show by induction on $n \geq 0$ the assertion (I_n) that V_n is strictly concave. (I_0) holds by assumption. Assume (I_n) for some $n \geq 0$. Then W_{n+1} is strictly concave if $V_n \circ T$ has the same property. This follows from Appendix D.6.2(b2) with $M := D$, $K := S$, $v := V_n$ and $g := T$. Now uniqueness of $f_{n+1}(s)$ is obvious. Moreover, V_{n+1} is strictly concave by Lemma 7.1.1 since $V_{n+1}(s) = \sup_{a \in D(s)} W_{n+1}(s, a) = W_{n+1}(s, f_{n+1}(s)) < \infty$. ◇

Example 7.1.6 (Concave value functions in a continuous multi-allocation problem) We generalize the continuous allocation problem from Example 6.4.3 to the following case: in each period some part a_i of the momentary resource s is allocated to project i, $1 \leq i \leq m$, and the remainder is kept for the next period. We use the following DP: $S = [0, \bar{s}]$ for some $\bar{s} \in \mathbb{R}^+$, $A = S^m$; a is the vector $(a_i)_1^m$; (no confusion should occur since a_i usually denotes the action at time i); thus $D(s)$ is the simplex $\{(a_i)_1^m \in A : \sum_{i=1}^m a_i \leq s\}$; $T(s, a) = s - \sum_{i=1}^m a_i$; the one-stage reward is the sum of utilities of the m allocations plus the utility of the remaining resource, i.e.

$$r(s, a) = \sum_{i=1}^{m} u_i(a_i) + u_{m+1}\left(s - \sum_{i=1}^{m} a_i\right)$$

for functions u_i, $1 \leq i \leq m+1$, on S. If u_i, $1 \leq i \leq m+1$, and V_0 are concave then:

For all $n \geq 1$ the value functions V_n and $W_n(s, \cdot)$, $s \in S$, are finite and concave, and the (possibly empty) set $D_n^*(s)$, $s \in S$, of maximum points of $W_n(s, \cdot)$ is convex.

The proof is a simple application of Theorem 7.1.2 with $\preceq_S := \preceq_{\text{diag}}$; observe that affinity of T implies concavity of $(s,a) \mapsto u_{m+1}(s - \sum_{i=1}^{m} a^i)$ by Appendix D.2.10(b). ♦

We now treat in a simple DP the problem of argumentwise concavity.

Proposition 7.1.7 (Value functions which are concave in one argument) *Consider a DP with*

(i) $S = S_1 \times S_2$ *and* $D = S \times A$ *with intervals* S_1, S_2 *and* A,
(ii) $T = (T_1, T_2)$ *where* $T_1(s_1, s_2, a)$ *is independent of* (s_2, a) *and where* $T_2(s_1, s_2, a)$ *is concave in* (s_2, a) *on* $B := S_2 \times A$,
(iii) $r(s,a)$ *is concave in* (s_2, a),
(iv) $V_n(s)$ *is increasing in* s_2 *for* $n \geq 0$ *and finite for* $n \geq 1$,
(v) $V_0(s)$ *is concave in* s_2.

Then all value functions $s \mapsto V_n(s)$, $n \geq 1$, *are concave in* s_2.

Proof We prove by induction on $n \geq 1$ the assertion (I_n) that $v := V_{n-1}$ is concave in s_2. (I_1) holds by (v). Assume that (I_n) holds for some $n \geq 1$.

(a) Fix $s_1 \in S_1$ and put $z := T_1(s_1)$. We assert that $v(z, T_2(s,a))$ is concave in $b := (s_2, a)$. For the proof select $b, b' \in B$ and $\alpha \in (0,1)$. Firstly (ii) implies $T_2(s_1, \alpha b + (1-\alpha)b') \geq \alpha T_2(s_1, b) + (1-\alpha)T_2(s_1, b')$. Now the assertion follows since (iv) and concavity of $v(s_1, s_2)$ in s_2 yield

$$v(z, T_2(s_1, \alpha b + (1-\alpha)b')) \geq \alpha v(z, T_2(s_1, b)) + (1-\alpha)v(z, T_2(s_1, b')).$$

(b) From (a) we see that $W_n(s,a)$ is concave in (s_2, a). Since V_n is finite, we get concavity in s_2 of V_n by applying Lemma 7.1.1 with s replaced by s_2 and $w(s,a)$ replaced by $W_n(s_1, s_2, a)$ with s_1 fixed. □

In examples like Example 7.1.6 one wants to know conditions under which the sets $D_n^*(s)$ of maximum points (or minimum points) of $W_n(s, \cdot)$ are not only convex, but also non-empty. In addition, if $A \subset \mathbb{R}$ we are interested in the existence of a *smallest [largest] maximizer at stage n*, which is to be understood pointwise. In general such problems are connected with the problem of existence of maximizing policies. We start by solving this problem for Example 7.1.8 in an elementary way. Then we study continuity of the value functions, since often this property is related to the existence of maximizing policies. The same problem is studied in more depth in Chap. 9 for the case when S and A are metric spaces.

Example 7.1.8 (Existence of minimizers in continuous linear-convex systems)

(a) We generalize the continuous linear-quadratic system of Example 4.1.7 with $S = A = \mathbb{R}$, $D = S \times A$ and $T(s,a) = gs - ha$ with $g \in \mathbb{R}_+$, $h \in \mathbb{R}^+$, by assuming

 (i) $c \geq 0$ is convex,
 (ii) $c(s,a) \to \infty$ for $|a| \to \infty$, $s \in \mathbb{R}$,

(iii) $C_0 \geq 0$ is convex.

Then:
For all $n \geq 1$ the minimal cost functions C_n and $a \mapsto W_n(s,a)$, $s \in \mathbb{R}$, are convex, and the set $D_n^*(s)$, $s \in \mathbb{R}$, of minimum points of $W_n(s,\cdot)$ is a compact interval $[f_n(s), g_n(s)]$. Thus f_n $[g_n]$ is the smallest [largest] minimizer at stage n.

(b) We consider the multi-dimensional generalization of the model in (a). Here $S = \mathbb{R}^k$, $A = \mathbb{R}^m$, $D = S \times A$; the state is controlled by means of a $k \times m$-matrix H, i.e. $T(s,a) = Gs + Ha + z$, where G is a $k \times k$-matrix and $z \in S$. We assume:

(i) $c \geq 0$ is convex,
(ii) $c(s,a) \to \infty$ for $|a| \to \infty$, $s \in S$,
(iii) $C_0 \geq 0$ is convex.

Then:
For all $n \geq 1$ the minimal cost functions C_n and $a \mapsto W_n(s,a)$, $s \in S$, are convex, and the set $D_n^*(s)$, $s \in S$, of minimum points of $W_n(s,\cdot)$ is non-empty and compact. Thus there exists a minimizer at stage n.

For the proof of part (a) we fix $n \geq 1$.

(a1) We use the minimization version from Remark 7.1.4 above with $\preceq_S := \preceq_{\text{diag}}$. There the assumptions (i) and (iii) hold obviously, and (iv) is true since each function on S is $\preceq_{\text{diag}}$-decreasing. In addition, (ii) is true, i.e. T is affine, since a simple computation shows that for (s,a), $(s',a') \in \mathbb{R}^2$ and $\alpha \in (0,1)$

$$T(\alpha s + (1-\alpha)s', \alpha a + (1-\alpha)a') = \alpha T(s,a) + (1-\alpha)T(s',a').$$

Now we obtain from Remark 7.1.4 that C_n and $a \mapsto W_n(s,a)$, $s \in \mathbb{R}$, are convex, and that $D_n^*(s)$, $s \in \mathbb{R}$, is convex, hence empty or an interval.

(a2) It remains to show that $D_n^*(s)$, $s \in \mathbb{R}$, is non-empty and compact. Fix s. As $v := W_n(s,\cdot)$ is convex on the open set $\mathbb{R}$ by (a), it is continuous by Appendix D.2.2. Now the assertion follows from Appendix A.4.15 since $v(a) \to \infty = \sup v$ for $|a| \to \infty$, hence v is not constant.

(b) The proof is nearly the same as for the model in (a). ♦

Remark 7.1.9 Conditions (i) and (ii) in (a) are satisfied, for example, if

$$c(s,a) = \sum_{i=1}^{m} u_i(\alpha_i s - \varepsilon_i a)$$

for convex functions $u_i \geq 0$, α_i, $\varepsilon_i \in \mathbb{R}$ for all i and if without loss of generality in addition $b \mapsto u_1(b) \to \infty$ for $|b| \to \infty$ and $\varepsilon_1 \neq 0$. Note that convexity of c follows from Appendix D.2.10(a) since $(s,a) \mapsto \alpha_i s - \varepsilon_i a$ is affine. A special case for u_i is

$$u_i(b) := \gamma_i |b|^{\delta_i}, \; b \in \mathbb{R},$$

for $\gamma_i > 0$ and $\delta_i \in [1, \infty)$. These special assumptions include (with a slightly different assumption on g and h in the transition function) the linear-quadratic model Example 4.1.7 by choosing the special case $m = 2$, $\gamma_1 = \lambda$, $\gamma_2 = \gamma$, $\delta_1 = \delta_2 = 2$, $\alpha_1 = \varepsilon_2 = 1, \alpha_2 = \varepsilon_1 = 0$. ◇

We now turn briefly and with elementary tools to the problem of continuity of the value functions.

Often the maximal reward operator U does not preserve continuity but upper semicontinuity. Therefore we basically deal with the latter notion rather than with continuity. A function v on a set $M \in \mathbb{R}^d$ is called ***upper semicontinuous*** (***usc*** for short), if for all $x \in M$ and all sequences (x_n) in M which converge to x we have $\limsup_n v(x_n) \leq v(x)$. (It suffices to consider only sequences (x_n) for which the sequence $(v(x_n))$ is also convergent in $\overline{\mathbb{R}}$.) Moreover, v is called ***lower semicontinuous*** (***lsc*** for short) if $-v$ is usc. Obviously a function on M is continuous if and only if it is both usc and lsc. Properties of usc and lsc functions can be found in Proposition 9.1.1 and Lemma 9.1.4 and in Appendix C. In particular: (a) A concave function on an interval is usc if and only if it is continuous; (b) The set of usc [lsc] functions on M is a convex cone.

If in the rest of this section S and A are intervals, we use on $S \times A$ the maximum metric, which is a product metric; see Appendix C.1.

Proposition 7.1.10 (Existence of smallest and largest maximizers when value functions are upper semicontinuous) *Assume that*

(i) S and A are intervals,
(ii) for each compact subinterval I of S the set $M := (I \times A) \cap D$ is compact,
(iii) T is continuous,
(iv) r and V_0 are usc.

Then for all $n \geq 1$ the value function V_n and $W_n(s, \cdot)$, $s \in S$, are finite and usc and there exists a smallest and a largest maximizer at stage n.

Proof We use the Structure Theorem 6.3.1 with $\mathfrak{V}$ as the set of usc functions on S. We verify (S1) and (S2).

(a) Proof of (S1): Select $v \in \mathfrak{V}$. It easily follows from (iii) and (iv) that the function $w := Lv$ is usc on D; see Proposition 9.1.1 and Lemma 9.1.4. Then for $s \in S$ the function $w(s, \cdot)$ is usc on the set $D(s)$, which is compact by the following Remark 7.1.11(c). Next, the set $D^*(s)$ of maximum points of $w(s, \cdot)$ is non-empty and compact by Proposition 9.1.1(e). It follows that $D^*(s)$, $s \in S$, has a smallest [and a largest] element $f_v(s)$ $[g_v(s)]$, hence f_v $[g_v]$ is the smallest [the largest] maximizer of Lv.
(b) Proof of (S2): We show that for $v \in \mathfrak{V}$ the function $s \mapsto w^*(s) := Uv(s)$ is usc. Select $s_0 \in S$ and a sequence $(s_n)_1^\infty$ in S converging to s_0 and such that $(w^*(s_n))$ also converges in $\overline{\mathbb{R}}$. Then $(s_n)_0^\infty$ is contained in a compact interval $I \subset S$. Put $a_n := f_v(s_n)$. As M is compact by (ii) and as $(s_n, a_n) \in M$, there exists a sequence $(n_k) \subset \mathbb{N}$ such that (s_{n_k}, a_{n_k}) converges to some point (s, a_0) in M. Obviously $s = s_0$ and $a_0 \in D(s_0)$. Now $w^* \in \mathfrak{V}$ since (S1) and upper

semicontinuity of Lv, proven in (a), imply

$$\lim_n w^*(s_n) = \lim_k w^*(s_{n_k}) = \lim_k Lv(s_{n_k}, a_{n_k}) \leq Lv(s_0, a_0) \leq w^*(s_0).$$

□

Remark 7.1.11

(a) Condition (ii) of Proposition 7.1.10 holds under (i) if A is closed and if D is compact.
For the proof notice that M is bounded as a subset of the compact, hence bounded set D, and M is closed as the intersection of the closed sets $I \times A$ and D.

(b) Condition (ii) of Proposition 7.1.10 holds under (i) if $D(\cdot) = [c(\cdot), d(\cdot)]$ for continuous functions c and $d \geq c$ from S into A.
For the proof we firstly note that on the compact set I the functions c and d assume their infimum and supremum, respectively, hence M is bounded as a subset of the bounded set $I \times [\min c|I, \max d|I]$. Next, M is closed as the intersection of the sets $\{(s,a) \in S \times A : d(s) - a \geq 0\}$ and $\{(s,a) \in S \times A : a - c(s) \geq 0\}$, which are closed by Proposition 9.1.1(b), since $(s,a) \mapsto d(s) - a$ and $(s,a) \mapsto a - c(s)$ are continuous.

(c) Under (i) the condition (ii) of Proposition 7.1.10 implies compactness of all sets $D(s)$, $s \in S$. For the proof select $s \in S$. Then $B := \{s\} \times D(s) = (I \times A) \cap D$ for $I := [s, s]$ is compact by (ii). Let (a_n) be a sequence in $D(s)$. Then the sequence $((s_n, a_n))$ with $s_n := s$ for all n belongs to B, hence has a subsequence $((s_{n_k}, a_{n_k}))$ which converges to some point (s', a') in B. It follows that $s' = s$ and that (a_{n_k}) converges to $a' \in D(s)$. Thus $D(s)$ is compact. ◇

Proposition 7.1.12 (Existence of smallest and largest maximizers for special restriction sets) *Assume that*

(i) S and A are intervals,
(ii) $D(\cdot) = [\underline{d}(\cdot), d(\cdot)]$ for a decreasing and continuous function $\underline{d}$ and an increasing and continuous function $d \geq \underline{d}$ on S,
(iii) T is continuous and $T(s,a)$ is increasing in s,
(iv) r is usc and $r(s,a)$ is increasing in s,
(v) V_0 is usc and increasing.

(a) Then for all $n \geq 1$ the functions V_n and $W_n(s, \cdot)$, $s \in S$, are finite and usc, V_n is increasing, and there exists a smallest and a largest maximizer at stage n.
(b) Assume in addition that $\underline{d}$ is convex and that d, T, r and V_0 are concave. Then for all $n \geq 1$ the functions V_n and $W_n(s, \cdot)$, $s \in S$, are concave and continuous.

Proof

(a) Firstly, $D(\cdot)$ is increasing by Remark 6.3.3. Now the assumptions in Theorem 6.3.5 are fulfilled, hence V_n, $n \geq 1$, is increasing. The validity of the assumptions in Proposition 7.1.10 is obvious, except for condition (ii), which holds by Remarks 7.1.11(b) above. Now Proposition 7.1.10 ensures the finiteness and upper semicontinuity of V_n and $W_n(s,\cdot)$, $n \geq 1$, and the existence of a smallest and a largest maximizer at each stage.

(b) Firstly, D is convex due to (ii). Also the assumptions in Theorem 7.1.2(a) hold. Thus V_n and $W_n(s,\cdot)$, $n \geq 1$, are concave (hence lsc by Appendix D.2.2(a)) and by (a) also usc, hence continuous. □

Example 7.1.13 (Existence of maximizers) We jointly treat the continuous versions of the allocation model and of the reservoir model in Example 6.4.1. We use choice I of the actions; see Example 6.4.1. From Table 6.1 we see: $S = A \subset \mathbb{R}_+$ is an interval; $D(s) = [0, s]$, $T(s,a) = T_1(a)$ and $r(s,a) = u_1(a) + u_2(s-a)$ for functions T_1 and u_1 on A and u_2 on $\tilde{A} := \{s - a : (s,a) \in D\} = A$.

Assume: T_1, u_1, u_2 and V_0 are concave and continuous, and u_2 and V_0 are increasing. Then for all $n \geq 1$ the value function V_n is finite, concave, continuous and increasing, and there exists a smallest and a largest maximizer at stage n.

The proof is an application of Proposition 7.1.12(a) and (b). It suffices to verify the assumptions (i)–(v) of Proposition 7.1.12 and property (vi), that T and r are concave. Firstly, (i)–(iii) and (v) are obviously true. (iv) r is usc since u_1 and u_2 are continuous; isotonicity of $r(s,a)$ in s follows from isotonicity of u_2. (vi) Concavity of T and of $(s,a) \mapsto u_1(a)$ follows from concavity of T_1 and of u_1, respectively, by Appendix D.2.12. Concavity of $(s,a) \mapsto u_2(s-a)$ follows from concavity of u_2 and from Appendix D.2.10(b). Now concavity of r follows by Appendix D.2.9(a). ♦

7.2 Convex Value Functions and Bang-Bang Maximizers

From the conditions for convexity of V_n, derived in this section, one easily obtains conditions for *concavity* of the minimal cost functions C_n by replacing V_n by $-C_n$. For checking finiteness and $\preceq_S$-isotonicity of V_n see Lemma 3.4.1 and Theorem 6.3.5, respectively; in particular, V_n is finite for all n if $D = S \times A$ with finite A. The next result is a certain counterpart to Theorem 7.1.2.

Proposition 7.2.1 (Convex value functions when $D(s)$ is independent of s) *Assume:*

(i) S is structured and convex, A is an arbitrary set and $D = S \times A$,
(ii) $T(s,a)$ is $\preceq_S$-convex in s,
(iii) $r(s,a)$ is convex in s,
(iv) V_n is $\preceq_S$-increasing for $n \geq 0$,
(v) V_0 is convex,
(vi) all value functions V_n, $n \geq 1$, are finite.

Then all value functions V_n and $W_n(\cdot, a)$, $a \in A$, $n \geq 1$, are convex.

Proof We prove by induction on $n \geq 1$ the assertion (I_n) that V_{n-1} is convex. (I_1) holds by (v). Assume that (I_n) holds for some $n \geq 1$. For fixed $a \in A$ we get from (ii), (iii), from Appendix D.6.2 with $v := V_{n-1}$, $K := S$, $\preceq_K := \preceq_S$, $M := S$ and $g := T(\cdot, a)$, and from Appendix D.2.9(a) and (b) that $W_n(s,a)$ is convex in s. Now Appendix D.2.9(c) together with the finiteness of V_n implies the convexity of $s \mapsto V_n(s) = \sup_{a \in A} W_n(s,a)$. □

Remark 7.2.2

(a) In Proposition 7.2.1 with $S \subset \mathbb{R}^k$ most important are the special cases where $\preceq_S = \times_1^k \leq_i$ and where each ordering $\leq_i$ either equals the usual ordering $\leq$ on $\mathbb{R}$ or its inverse or the diagonal ordering.
(b) In most results of the present subsection we assume that $D(s)$ does not depend on s. If this does not hold, one may try to use the obvious continuous version of Proposition 7.3.4 below. If $D(s) = [0, s]$ one may try a transformation $b = as$, $b \in [0, 1]$, of the actions. ◇

Assume that $A \subset \mathbb{R}^k$ and that $D(s)$ is convex for each s. We call a maximizer f_n at stage n a ***bang-bang maximizer*** if $f_n(s)$ is an extreme point of $D(s)$ for all s; see Appendix D.4.1. If each set $D(s)$ is a polytope $\text{co}(B(s))$ for some finite set $B(s)$ (see Appendix D.1.1), the set $E(s)$ of extreme points of $D(s)$ equals the finite set of vertices of $D(s)$ and it is contained in $B(s)$. For a DP with $S = \mathbb{R}^+$ and $A = \mathbb{R}_+^n$ for some $n \geq 1$ we get an example from Appendix D.4.2(a) with $K := s \in S$. Then $D(s) = \Delta_n(s)$ and $B(s) = E(s) = \{\mathbf{0}, se_1, se_2, \ldots, se_n\}$.

The next result requires that $r(s,a)$ is convex in s and in a. This is a property weaker than convexity of r as seen by $D := \mathbb{R}^2$ and $r(s,a) := sa$.—A similar remark concerns the assumptions on $T(s,a)$.

Theorem 7.2.3 (Existence of bang-bang maximizers when $D(s)$ is independent of s) *Besides (i)–(v) from Proposition* 7.2.1 *assume:*

(vii) $A \subset \mathbb{R}^k$ *equals the polytope* $\text{co}(B)$ *for some finite set* $B \subset \mathbb{R}^k$*,*
(viii) $T(s,a)$ *is* $\preceq_S$*-convex in* a*,*
(ix) $r(s,a)$ *is convex in* a*.*

Then:

(a) V_n*,* $n \geq 1$*, is finite and convex. Moreover, if* E *denotes the set of extreme points of* A*, the value functions satisfy*

$$V_n(s) = \max_{a \in B} W_n(s,a) = \max_{a \in E} W_n(s,a), \quad n \geq 1, s \in S. \tag{7.1}$$

(b) If $f_n(s)$ *is chosen from the non-empty set of maximum points of* $W_n(s,\cdot)|E$ *for* $s \in S$*, then* f_n *is a bang-bang maximizer at stage* n*.*

Proof

(a) We prove by induction on $n \geq 1$ the assertion (I_n) that V_{n-1} is finite and convex. (I_1) holds by (v). Assume that (I_n) holds for some $n \geq 1$. From Proposition 7.2.1 we know that $W_n(s,a)$ is convex in s. Now fix $s \in S$. Then we get from (iv), (viii), (ix), from Appendix D.6.2 with $v := V_{n-1}$, $K := S$, $\preceq_K := \preceq_S$, $M := A$

and $g := T(s, \cdot)$, and from Appendix D.2.9(a) and (b) that $W_n(s, a)$ is convex in a. Now (Equation 7.1) follows from Appendix D.4.3(d), observing that $E \subset B \subset A$ by Appendix D.4.1 and that B is finite. Finally finiteness and convexity of V_n follow from (7.1).

(b) follows obviously from (a). □

If the action space A is given as co(B), the last term in (7.1) requires the computation of the set E. Then the evaluation of $V_n(s)$ by the second term will in general be more efficient than the evaluation by the last term.

We mention without proof the following two extensions.

Proposition 7.2.4 (Convexity of w^* when $D(s)$ is not independent of s) *Assume that*

(a) S and A are convex,
(b) for all s the set $D(s)$ is the convex hull of the set $\bigcup_{i=1}^m \{d_i(s)\}$ for certain mappings d_i from S into A,
(c) w is a function on D such that $w(s, \cdot)$ is convex on $D(s)$, $s \in S$,
(d) the functions $s \mapsto w(s, d_i(s))$, $1 \le i \le m$, are convex on S.

Then $s \mapsto w^*(s) := \sup w(s, \cdot)$ equals $\max_{1 \le i \le m} w(s, d_i(s))$, w^* is finite and convex, and w has a bang-bang maximizer.

Proposition 7.2.5 (Convexity of value functions and existence of bang-bang maximizers) *Assume that*

(i) S and A are convex,
(ii) for all s the set $D(s)$ is the convex hull of the set $\bigcup_{i=1}^m \{d_i(s)\}$ for mappings d_i from S into A,
(iii) $T(s, \cdot)$, $s \in S$, and $s \mapsto T(s, d_i(s))$ are affine or (convex and V_n, $n \ge 0$, is increasing) or (concave and V_n, $n \ge 0$, is decreasing),
(iv) $r(s, \cdot)$, $s \in S$, $s \mapsto r(s, d_i(s))$ and V_0 are convex.

Then V_n is finite and convex for $n \ge 1$, there exists at each stage a bang-bang maximizer, and for $n \ge 1$

$$V_n(s) = \max_{1 \le i \le m} \left[r(s, d_i(s)) + \beta V_{n-1} \circ T(s, d_i(s)) \right], \quad s \in S.$$

Example 7.2.6 (A multi-location harvesting problem) A farmer has available at each year $0 \le t \le N-1$ momentarily $s \in (0, \infty)$ tons of seed of a certain agricultural product. It is used for planting at year t the proportion $a^i \in [0, 1]$ with $\sum_1^m a^i \le 1$ at some location $1 \le i \le m$; the seed not planted, i.e. $s(1 - \sum_1^m a^i)$ is kept for the next year. (The less experienced reader may choose $m = 1$.) Planting $a^i s$ at year t yields next year $b_i a^i s$ tons of new seed; here $b_i \in [1, \infty)$. Thus the seed available at year $t+1$ equals $T(s, a) := s(1 + \sum_1^m a^i(b_i - 1))$, where $a = (a^i)_1^m$. At year $0 \le t \le N-1$ there arise planting costs $g_i(a^i s) \in \mathbb{R}_+$, $1 \le i \le m$, and storage costs $h(s(1 - \sum_1^m a^i)) \in \mathbb{R}_+$ for concave functions g_i and h on $\mathbb{R}_+$ with $g_i(0) = h(0) = 0$. After N years the finally available seed is sold for $d_0 \in \mathbb{R}^+$

dollars per ton. We want to find a planting policy $(f_n)_N^1$ which maximizes the terminal reward minus the discounted sum of planting and storage costs.

We use the following DP: $S = \mathbb{R}^+$; $D(s) = A$ is the polytope $\{(a^i)_1^m \in [0,1]^m : \sum_{i=1}^m a^i \leq 1\}$; $T(s,a)$ as above; $r(s,a) = -\sum_1^m g_i(a^i s) - h(s(1 - \sum_1^m a^i))$ and $V_0(s) = d_0 s$. Using $\preceq_S := \leq_{\text{diag}}$, all assumptions (i)–(v) and (vii)–(ix) in Theorem 7.2.3 are fulfilled. This is easily verified, except for the following assumptions:

(ii) and (viii): As $T(s,a)$ is $\leq_{\text{diag}}$-convex in s and in a if $T(s,a)$ is affine in s and in a, respectively; both properties are verified by elementary computations. (iv) holds, since each function on S is $\leq_{\text{diag}}$-increasing. (iii) and (ix): One easily shows that for $1 \leq i \leq m$ the function $a \mapsto v(a) := a^i$ is affine. Then $a \mapsto a^i$ and $a \mapsto s(1 - \sum_1^m a^i)$ are also affine. Now Appendix D.2.10 ensures concavity of $g(a^i s)$ and of $h(s(1 - \sum_1^m a^i))$, both in s and in a. This proves convexity of $r(s,a)$ both in s and in a. (vii): $A = \text{co}(e_0, e_1, \ldots, e_m)$, where $e_0 := \mathbf{0}$ and where e_i, $1 \leq i \leq m$, is the i-th unit vector in $\mathbb{R}^m$.

If g_i, $1 \leq i \leq m$, and h are concave, then we obtain from Theorem 7.2.3:

(a) The value functions are finite and convex, and

$$V_n(s) = \max\left\{-h(s) + \beta V_{n-1}(s), \max_{1 \leq i \leq m} [-g_i(s) + \beta V_{n-1}(b_i s)]\right\}, \; s \in \mathbb{R}^+.$$

(b) A bang-bang maximizer f_n is given as follows:
Put $v_{n0}(s) := -h(s) + \beta V_{n-1}(s)$ and $v_{ni}(s) := -g_i(s) + \beta V_{n-1}(b_i s)$; then $f_n(s) = \mathbf{0}$ if $\max_{1 \leq i \leq m} v_{ni}(s) < v_{n0}(s)$; else $f_n(s)$ equals the μ-th unit vector, where μ is the smallest $1 \leq i \leq m$ with $v_{ni}(s) \geq v_{n0}(s)$. ♦

Consider the DP where $S = \mathbb{R}_+$, $D(s) = A = \mathbb{N}_5$, $r(s,a) = |s - a| = \max\{s - a, a - s\}$, $T(s,a) = 2s + 3a$ and $V_0(s) = \max\{1 - s, 0, s - 2\}$. Then V_n, $n \geq 1$, is finite by finiteness of A and it is convex by Proposition 7.2.1, using $\preceq_S := \leq_{\text{diag}}$. However, the special form of both $r(\cdot, a)$ and of V_0 as maxima of finitely many affine functions implies by Proposition 7.2.8 below that V_n is not only convex, but the maximum of finitely many affine functions. We need some preparations.

We call a function v on the convex set $S \subset \mathbb{R}^m$ ***convex and piecewise affine*** ($v \in CPA$ for short) if v is the maximum of finitely many affine functions. Since affine functions on S are convex, the functions in CPA are indeed convex by Appendix D.2.9(c). It is easily seen that for each finite set $K \neq \emptyset$, $p_k \in \mathbb{R}^m$ and $b_k \in \mathbb{R}$ the function

$$s \mapsto v(s) = \max_{k \in K}[\langle p_k, s \rangle + b_k], \quad s \in S \subset \mathbb{R}^m$$

belongs to CPA.

Lemma 7.2.7 (The function class *CPA*)
The following hold:
(a) CPA is a convex cone.
(b) CPA is closed under forming the maximum of finitely many functions.

Proof

(a) Let $v = \max_{i\in I} v_i$, $v' = \max_{j\in J} v'_j$ for finite I and J, affine v_i and v'_j, and $\gamma \in \mathbb{R}_+$. Now $\gamma v = \max_i(\gamma v_i) \in CPA$ since γv_i is affine. For $s \in S$ we have $v(s) + v'(s) = \max_{(i,j)\in I\times J}[v_i(s) + v'_j(s)]$ by Lemma 2.3.1 with $B := I$, $C := J$ and $M := I \times J$. Thus $v + v' \in CPA$ since $v_i + v'_j$ is affine.
(b) Let $v = \max_{i\in I} v_i$ for finite I and $v_i \in CPA$. Then $v_i = \max_{j\in M(i)} v_{ij}$ for some finite sets $M(i)$, $i \in I$, and affine functions v_{ij}. Let $M := \{(i,j) \in I \times \cup_{i\in I} M(i) : j \in M(i)\}$. Then for $s \in S$ we have $v(s) = \max_{(i,j)\in M} v_{ij}(s)$ by Lemma 2.3.1 with $B := I$, $C := \cup_{i\in I} M(i)$. Thus $v \in CPA$. □

Proposition 7.2.8 (Convex and piecewise affine value functions) *V_n, $n \geq 1$, is convex and piecewise affine if:*

(i) S is convex,
(ii) A is finite and $D = S \times A$,
(iii) $T(\cdot, a)$ is affine for all a,
(iv) $r(\cdot, a) \in CPA$ for all a,
(v) $V_0 \in CPA$.

Proof Firstly, V_n, $n \geq 1$ is finite due to finiteness of A. We now use induction on $n \geq 0$ of the assertion (I_n) that $V_n \in CPA$. (I_0) holds by (v). Assume (I_n) for some $n \geq 0$. Then $V_n = \max_{i\in I} v_i$ for a finite set I and affine functions v_i. Fix $a \in A$. Then $v_i(T(\cdot, a))$ is affine, hence $V_n(T(\cdot, a)) \in CPA$. By (iv) and Lemma 7.2.7(a) $W_{n+1}(\cdot, a) \in CPA$ and then $V_{n+1} \in CPA$ by Lemma 7.2.7(b). □

We now turn to DPs with affine value functions. Combining Theorem 7.1.2, Proposition 7.2.1 and Theorem 7.2.3 we obtain

Proposition 7.2.9 (Affine value functions when $D(s)$ is independent of s) *Assume:*

(i) S is structured, S and A are convex and $D = S \times A$,
(ii) T is $\preceq_S$-affine,
(iii) r and V_0 are affine,
(iv) V_n is $\preceq_S$-increasing for $n \geq 0$ and finite for $n \geq 1$.

Then V_n is affine. Moreover, (7.1) *holds if $A \subset \mathbb{R}^k$ is a polytope* co(B) *for some finite set $B \subset \mathbb{R}^k$ and with the set E of vertices.*

Example 7.2.10 (Affine value functions) We consider the following DP: $S = A = [0, 1]$; $D = S \times A$; $\preceq_S = \leq^1$, $T(s, a) = e_1 s + e_2 a$, $r(s, a) = e_3 s + ca$; $V_0(s) = e_0 s$ with $e_0, e_1, e_2, e_3 \in \mathbb{R}_+$, $c \in \mathbb{R}$; $\beta \in \mathbb{R}^+$. We also assume $e_1 + e_2 \leq 1$ in order that $T(s, a) \in S$ for all $(s, a) \in D$. The model looks very simple, but its solution is non-trivial. The computation of the solution via VI is rather tedious. More efficient is the use of Proposition 7.2.9, whose assumptions are fulfilled: T, r and V_0 are affine by Appendix D.2.3(a); V_n, $n \geq 1$, is increasing for $n \geq 0$ by Theorem 6.3.5 and upper bounded for $n \geq 1$ by Lemma 3.4.1(b); A is a polytope with the set $E = \{0, 1\}$. We obtain for $n \geq 1$ and $s \in [0, 1]$ from (7.1):

$$\begin{aligned} V_n(s) &= \max\{W_n(s,0), W_n(s,1)\} \\ &= \max\{e_3 s + \beta V_{n-1}(e_1 s), e_3 s + c + \beta V_{n-1}(e_1 s + e_2)\}. \end{aligned} \tag{7.2}$$

From Proposition 7.2.9 and from Appendix D.2.3(a) we know that $V_n(s) = b_n s + d_n$, $n \geq 0$, $s \in S$, for appropriate reals b_n and d_n with $b_0 = e_0$ and $d_0 = 0$. We are going to compute b_n and to derive a recursion for d_n. Assume that $V_{n-1}(s) = b_{n-1}s + d_{n-1}$, $s \in S$, for some $n \geq 1$. Now (7.2) yields

$$V_n(s) = (\beta e_1 b_{n-1} + e_3)s + \beta d_{n-1} + (c + \beta e_2 b_{n-1})^+ =: b_n s + d_n, s \in S.$$

Thus $b_n = \beta e_1 b_{n-1} + e_3$, and we get from Lemma 4.1.5 for $n \geq 1$

$$b_n = e_3 \sigma_n(\beta e_1) + (\beta e_1)^n e_0, \qquad d_n = (c + \beta e_2 b_{n-1})^+.$$

Finally, for $N \geq 1$ a maximizing sequence $(f_n)_N^1$ is given by the constant (!) maximizers $f_n \equiv 0$ if $c + \beta e_2 b_{n-1} \leq 0$ and $f_n \equiv 1$, else. ♦

We now treat in a simple DP the problem of argumentwise convexity. Note that isotonicity of $V_n(s_1, s_2)$ in s_2, required in assumption (iv) of the DP, can be checked by Theorem 6.3.5 with $\preceq_S := \leq_{\text{diag}} \times \leq^1$.

Proposition 7.2.11 (Value functions which are convex in one argument) *Consider a DP with*

(i) $S = S_1 \times S_2$ *with intervals* S_1 *and* S_2, A *arbitrary and* $D = S \times A$,
(ii) $T = (T_1, T_2)$ *where* $T_1(s_1, s_2, a)$ *is independent of* s_2 *and where* $T_2(s, a)$ *is convex in* s_2,
(iii) $r(s, a)$ *is convex in* s_2,
(iv) $V_n(s)$ *is increasing in* s_2 *for* $n \geq 0$ *and finite for* $n \geq 1$,
(iv) $V_0(s)$ *is convex in* s_2.

Then all value functions $s \mapsto V_n(s)$, $n \geq 1$, *are convex in* s_2.

Proof

(a) Fix $n \geq 0$ and $a \in A_0$. Drop in this part of the proof the action a in T and s_2 in T_1. Assume that $v := V_n$ is convex in s_2. Then $v(T_1(s_1), T_2(s_1, s_2))$ is convex in s_2. For the proof fix s_1, s_2, s_2' and $\alpha \in (0, 1)$ and put $z := T_1(s_1)$, $y := T_2(s_1, s_2)$ and $y' := T_2(s_1, s_2')$. Firstly (ii) implies $T_2(s_1, \alpha s_2 + (1-\alpha)s_2') \leq \alpha y + (1-\alpha)y'$. Now (iv) and convexity of V_n in s_2 yield

$$v(z, T_2(s_1, \alpha s_2 + (1-\alpha)s_2')) \leq v(z, \alpha y + (1-\alpha)y') \leq \alpha v(z, y) + (1-\alpha)v(z, y').$$

(b) We now prove the assertion (I_n) that $V_n(s)$ is convex in s_2 by induction on $n \geq 0$. Firstly, (I_0) holds by (v). Assume (I_n) for some $n \geq 0$. From (a), (iii) and Appendix D.2.9(a) and (b) we see that $W_{n+1}(s,a)$ is convex in s_2. Since V_{n+1} is finite, we get convexity in s_2 of V_{n+1} from Appendix D.2.9(c). □

7.3 Discretely Convex Value Functions

A function v on a discrete non-degenerate interval M is called ***discretely convex*** if

$$x \mapsto v(x+1) - v(x),\ x < \sup M, \tag{7.3}$$

is increasing, or equivalently in case $|M| \geq 3$, if

$$v(x) \leq [v(x+1) + v(x-1)]/2, \text{ for } \inf M < x < \sup M$$

(cf. also Appendix D.5.1).

Moreover, v is called ***discretely concave*** if $-v$ is discretely convex, and v is called ***discretely affine*** if it is both discretely convex and discretely concave. Finally, v is called ***strictly discretely convex [concave]*** if isotonicity in (7.3) is strict. We do not treat convexity or concavity of functions on sets $M \subset \mathbb{Z}^d$, $d \geq 2$, since this concept seems not to be a fruitful one.

Remark 7.3.1 We firstly note the following simple result: If S and A are discrete intervals and if the sets $D(s)$, $s \in S$, are bounded (hence finite), it follows from Lemma 3.4.1(a) for each $n \geq 1$ that a smallest and largest maximizer at stage n exists, and hence V_n and W_n are finite. ◇

Below we show that several of our results in Sect. 7.2 about convex value functions on intervals carry over to results about discretely convex value functions on discrete non-degenerate intervals. The situation is quite different for discrete concavity since there appears to be no discrete analogue of Lemma 7.1.1, which was crucial for the proof of Theorem 7.1.2. An example with discretely concave value functions is treated in Example 8.3.6 below.

The next two results are discrete counterparts to Proposition 7.2.1 and Theorem 7.2.3, respectively. We restrict ourselves to the three most important relations on the discrete non-degenerate-interval S, namely $\leq^1$, the inverse of $\leq^1$ and the diagonal ordering, and we state the resulting sufficient conditions explicitly.

Proposition 7.3.2 (Discretely convex value functions when $D(s)$ is independent of s) *Assume:*

(i) S is a discrete non-degenerate interval, A is arbitrary and $D = S \times A$,
(ii) $T(\cdot, a)$, $a \in A$, is either discretely affine or (discretely convex and V_n, $n \geq 0$, is increasing) or (discretely concave and V_n, $n \geq 0$, is decreasing),
(iii) $r(s, a)$ is discretely convex in s,

(iv) *V_0 is discretely convex,*
(v) *V_n is finite for $n \geq 1$.*

Then V_n is discretely convex for $n \geq 1$.

Proof for the case where $T(\cdot, a)$ is discretely convex and where V_n, $n \geq 0$, is increasing. We show by induction on $n \geq 0$ the assertion (I_n) that V_n is discretely convex. (I_0) holds by (iv). Assume (I_n) for some $n \geq 0$ and fix $a \in A$. Now Appendix D.5.3(a) with $v := V_n$, $I := M := S$ and $g := T(\cdot, a)$ implies discrete convexity of $V_n(T(\cdot, a))$. Next, (iii) ensures that $W_{n+1}(\cdot, a)$ is discretely convex. Finally (I_{n+1}) follows since the supremum of a family of discretely convex functions, if finite, is discretely convex; see Appendix D.5.2(e). □

Theorem 7.3.3 (Discretely convex value functions and existence of bang-bang maximizers when $D(s)$ is independent of s) *Assume (i)–(iv) from Proposition 7.3.2 and in addition*

(vi) *A is a finite discrete non-degenerate interval,*
(vii) *$T(s, \cdot)$, $s \in S$, either is discretely affine or (discretely convex and V_n, $n \geq 0$, is increasing) or (discretely concave and V_n, $n \geq 0$, is decreasing),*
(viii) *$r(s, a)$ is discretely convex in a.*

Then V_n, $n \geq 1$, is finite and discretely convex and

$$V_n(s) = \max\{W_n(s, \min A), W_n(s, \max A)\} \quad s \in S. \tag{7.4}$$

Thus, the smallest maximizer f_n at stage $n \geq 1$ is given by $f_n(s) := \min A$ if $W_n(s, \min A) \geq W_n(s, \max A)$ and $f_n(s) := \max A$, else, and f_n is a bang-bang maximizer.

Proof We treat the case where $T(s, \cdot)$ is discretely concave and where V_n, $n \geq 0$, is decreasing. Fix $n \geq 1$. Firstly, finiteness of $D(s)$ ensures finiteness of V_{n-1} by Lemma 3.4.1(a). Thus V_{n-1} is discretely convex by Proposition 7.3.2 and (for $n = 1$) by (v). Next, we verify that $W_n(s, \cdot)$ is discretely convex. In fact, from Appendix D.5.3(a) we see that that $V_{n-1}(T(s, \cdot))$ is discretely convex. Now $W_n(s, \cdot)$ is discretely convex by Appendix D.5.2(c). This verifies (7.4), which implies the stated properties of f_n. □

In the discrete allocation problem from Example 2.1.1 (see Example 7.3.5 below) we cannot use as action the proportion, but only the amount of momentary resource consumed. This leads to $D(s) = \mathbb{N}_{0,s}$, hence Theorem 7.3.3 cannot be used. Fortunately this case is covered by the next result.

Proposition 7.3.4 (Discretely convex value functions and existence of bang-bang maximizers) *Assume that*

(i) *S and A are discrete non-degenerate intervals,*
(ii) *$D(s)$ is the finite discrete interval with endpoints $d_1(s)$ and $d_2(s) \geq d_1(s)$, $s \in S$, where d_1 and d_2 are functions from S into A,*

(iii) $a \mapsto T(s, a)$ *and* $s \mapsto T(s, d_i(s))$, $i = 1, 2$, *are discretely affine,*
(iv) $a \mapsto r(s, a)$, $s \in S$, $s \mapsto r(s, d_i(s))$, $i = 1, 2$, *are discretely convex,*
(v) V_0 *is discretely convex.*

Then V_n *is finite and discretely convex for* $n \geq 1$ *and*

$$V_n(s) = \max\{W_n(s, d_1(s)), W_n(s, d_2(s))\}, \ s \in S. \tag{7.5}$$

Moreover, the smallest maximizer at stage n *is given by* $f_n(s) := d_1(s)$ *if* $W_n(s, d_1(s)) \geq W_n(s, d_2(s))$ *and* $f_n(s) := d_2(s)$, *else, hence is a bang-bang maximizer.*

Proof Firstly, finiteness of $D(s)$ ensures finiteness of V_n for $n \geq 1$ by Lemma 3.4.1. Next, we show by induction on $n \geq 0$ the assertion (I_n) that V_n is discretely convex. (I_0) holds by (v). Assume (I_{n-1}) for some $n \geq 1$. Then $a \mapsto V_{n-1}(T(s, a))$ and $s \mapsto V_{n-1}(T(s, d_i(s)))$ are discretely convex by Appendix D.5.3(a). It follows from Appendix D.5.2(c) that $a \mapsto W_n(s, a)$ and $s \mapsto W_n(T(s, d_i(s)))$ are discretely convex. Now we get (7.5), and then (I_n) by Appendix D.5.2(e). Finally (7.5) implies the properties of f_n. □

Example 7.3.5 (A discrete allocation problem with convex utility) Consider the discrete allocation problem from Example 2.1.1. Recall that $S = A = \mathbb{N}_{0,K}$ for some $K \in \mathbb{N}$, $D(s) = \mathbb{N}_{0,s}$, $T(s, a) = s - a$, $r(s, a) = u(a)$, that $V_0 = du$ for some $d \in \mathbb{R}_+$ and an increasing discretely convex function u on $\mathbb{N}_{0,K}$ with $u(0) = 0 < u(K)$ and that $\beta \leq 1$. The VI has the form

$$V_n(s) = \max\{u(a) + \beta V_{n-1}(s - a) : a \in \mathbb{N}_{0,s}\}, \ n \geq 1, s \in \mathbb{N}_{0,K}.$$

Put $\overline{s} := \max\{0 \leq x \leq K - 1 : u(x) = 0\}$. Then isotonicity of u implies that $u(s) = 0$ for $0 \leq s \leq \overline{s}$ and that $u(s) > 0$ for $\overline{s} < s \leq K$. Put $\overline{S} := \{\overline{s}, \overline{s} + 1, \ldots, K\}$. Already in Example 2.4.1(c) we have seen for $0 \leq s \leq \overline{s}$ that $V_n(s) = 0$, $n \geq 1$, and that for each $N \geq 1$ the action sequence $\mathbf{0} \in A^N(s)$ is s_0-optimal.

We now prove that for $s \in \overline{S}$:

$$V_n(s) = \max\{\beta^n d, 1\} \cdot u(s), \ n \geq 1. \tag{7.6}$$

Moreover, for $N \geq 1$ the action sequence $\mathbf{0} \in A^N(s)$ belongs to the set $B^N(s)$ of s-optimal action sequences if $\beta^N d > 1$ and $se_0 := (s, 0, \ldots, 0)$ belongs to $B^N(s)$ if $\beta^N d \leq 1$. (Note that $\mathbf{0}$ and se_0 are extreme in the sense that they prescribe either to consume nothing at all times $0 \leq t \leq N - 1$ or to consume everything at time $t = 0$, respectively.)

For the proof fix $n \geq 1$.

(a) We infer from Proposition 7.3.4 that the smallest maximizer f_n at stage $n \geq 1$ is bang-bang, i.e. that $f_n(s) \in \{0, s\}$ for all $s \in S$. In fact, the assumptions of Proposition 7.3.4 hold: (i) is obvious, (ii) holds with $d_1(s) = 0$ and $d_2(s) = s$, (iii) $a \mapsto T(s, a) = s - a$, $s \mapsto T(s, d_1(s)) = s$ and $s \mapsto T(s, d_2(s)) = 0$

are discretely affine, (iv) $a \mapsto r(s,a) = u(a)$, $s \mapsto r(s,d_1(s)) = 0$, and $s \mapsto r(s,d_2(s)) = u(s)$ are discretely convex.

(b) If $s \in \overline{S}$, then we get, observing $u(0) = V_{n-1}(0) = 0$, that $W_n(s,0) = \beta V_{n-1}(s)$ and $W_n(s,s) = u(s)$, hence $V_n(s) = \max\{\beta V_{n-1}(s), u(s)\}$.

(c) Fix $s \in \overline{S}$. We prove by induction on $n \geq 1$ the assertion (I_n) that (7.6) holds. Firstly (I_1) follows from (b) since $V_0(s) = du(s)$ and since $u(s) \geq 0$. Assume (I_n) for some $n \geq 1$. Then, since $\beta \leq 1$, we get from (b)

$$\begin{aligned} V_{n+1}(s) &= \max\{\beta V_n(s), u(s)\} = \max\{\max\{\beta^{n+1}d, \beta\} \cdot u(s), u(s)\} \\ &= \max\{\beta^{n+1}d, \beta, 1\} \cdot u(s) = \max\{\beta^{n+1}d, 1\} \cdot u(s). \end{aligned}$$

(d) For $y := \mathbf{0} \in A^N(s)$ we get $V_{Ny}(s) = \sum_{t=0}^{N-1} \beta^t u(0) + \beta^N du(s) = \beta^N du(s)$ which equals $V_N(s)$ if $\beta^N d > 1$. Then $y \in B^N(s)$. For $y' := se_0 \in A^N(s)$ we get $V_{Ny'}(s) = u(s)$ which equals $V_N(s)$ if $\beta^N d \leq 1$. Then $y \in B^N(s)$. ♦

7.4 Problems

Problem 7.4.1 Assume that $S \subset \mathbb{R}^m$ is convex, that $A = \mathbb{R}^d$ and that D has the interval form $D(\cdot) = [c(\cdot), d(\cdot)]$. Then D is convex if and only if (c is convex) $\wedge$ (d is concave).

Problem 7.4.2 Consider the allocation problem from Example 6.4.1 where $S = A = \mathbb{R}_+$, $t_1 \equiv 0$, $t_2(s-a) = \sqrt{s-a}$, $u_1(a) = a$ for $a > 0$ and $u_1(0) = -1$, $u_2 \equiv 0$, $V_0 = u_1$ and $\beta = 1$. Show that V_1 is concave but not continuous, and that there does not exist a maximizer at stage one.

Problem 7.4.3 Consider the linear-quadratic model of Example 4.1.7 with arbitrary convex $C_0 \geq 0$. Then C_n is convex for all n.

Problem 7.4.4 Consider the allocation problem from Example 7.1.6. If $s > 0$ we say that an N-stage action sequence $y = (a_t)_0^{N-1} \in A^N(s)$ is a *zero-all sequence* if it allocates at some time t^* the whole momentarily available resource $s_{t^*} > 0$ to one of the m activities, say to activity i^*, and nothing to the other activities and also nothing to each activity at all other times. Prove the following result: Assume that there exists a bang-bang maximizer f_n at each stage n and that for each $s > 0$ we have $T(s, \mathbf{0}) > 0$ and $T(s, s \cdot e_i) = T(0, \mathbf{0}) = 0$ for $1 \leq i \leq k$. Then there exists for each $s > 0$ an s-optimal zero-all action sequence. Moreover, if $(a_t^*)_0^{N-1}$ is generated by s and $(f_n)_N^1$, then $t^* = \min\{0 \leq t \leq N-1 : a_t^* \neq \mathbf{0}\}$.

Problem 7.4.5 If in the continuous allocation problem from Example 2.4.3 $r(s,a) = u(sa)$ for convex u and $u(0) = V_0(0) = 0$, then we have for $N \geq 1$ the following explicit solution:

(a) $V_N(s) = \max\{\beta^N V_0(z^N s), \max_{1 \leq t \leq N}[\beta^t u(z^t s)]\}$, $s \in \mathbb{R}_+$.

(b) For $s_0 \in S$ there exists an s_0-optimal action sequence for problem $DP_N(s_0)$ of the following form: If $\beta^N V_0(z^N s_0) > \max_{1 \le t \le N} \beta^t u(z^t s_0)$ then consume nothing at all times $0 \le t \le N-1$; otherwise there exists some $0 \le t_0 \le N-1$ such that one consumes the whole available resource at time t_0 and nothing at the other times. In the latter case we have $V_N(s_0) = \beta^t u(z^t s_0)$.

7.5 Supplements

Supplement 7.5.1 (Discussion of Theorem 7.2.3)

(a) Theorem 7.2.3 is not valid without the assumption (*) that $D(s)$ is independent of s, as shown by the following counterexample: $S = [-1, +1]$, $A = [0, 1]$, $D(s) = [s^2, 1]$, $T \equiv V_0 \equiv 0$, $r(s, a) = -a$, β arbitrary. Then $V_n(s) = -s^2$, which is not convex. On the other hand, the DP satisfies all assumptions in Theorem 7.2.3 except for (*).

(b) If Theorem 7.2.3 is not applicable because $D(s)$ depends on s, a transformation of actions to a new DP′ for which the assumptions of Theorem 7.2.3 hold, may help. This approach can be tried in particular when $S = A = \mathbb{R}_+$ and $D(s) = [0, s]$ with the transformation $a' := a/s$ (with $0/0 := 0$), which yields $D'(s) := [0, 1]$ for all s.

Supplement 7.5.2 (Non-finite convex and concave functions) For theoretical purposes it is sometimes advantageous to admit convex and concave functions which assume the values $+\infty$ and $-\infty$; cf. e.g. Rockafellar (1970) and Stoer and Witzgall (1970).

Chapter 8
Monotone and ILIP Maximizers

We present conditions under which the pointwise smallest maximizer $f_n(s)$ depends monotonely on the current state s, is increasing and Lipschitz in s, is monotone in n and is monotone both in n and s. We also introduce several algorithms for computing the smallest maximizer.

8.1 Several Algorithms for Computing the Smallest Maximizer

Throughout this chapter the arbitrary state space S carries a relation $\preceq_S$ and the arbitrary action space A carries a total ordering $\leq_A$, unless stated otherwise. Often the subscripts S and A in $\preceq_S$ and $\leq_A$ are dropped if no confusion is possible. In most applications A is a subset of $\mathbb{R}$ and endowed with the usual ordering. If at each stage n a pointwise smallest maximizer f_n exists, we want to find conditions under which f_n has one of the following structural properties:

- $f_n(s)$ depends monotonely on the current state s.
- $f_n(s)$ is (increasing) $\wedge$ (Lipschitz) in s.
- $f_n(s)$ is monotone in n.
- $f_n(s)$ is monotone both in n and in s.

Note that a solution of problem $DP_N(s_0)$, $s_0 \in S$, by means of Theorem 2.3.3(c), using any maximizing sequence $(f_n)_N^1$, does not require that f_n, $1 \leq n \leq N$, is the *smallest* maximizer at stage n. However, the latter property appears in Lemma 8.2.7, which we need for obtaining increasing maximizers f_n, $1 \leq n \leq N$.

An important benefit of isotonicity of $f_n(\cdot)$ is the *accelerated numerical computation* of f_n (and hence of V_n), $1 \leq n \leq N$, for some $N \geq 1$. For instance,

K. Hinderer et al., *Dynamic Optimization*, Universitext,
DOI 10.1007/978-3-319-48814-1_8

Example 8.2.14 with $N = 1000$, $K = 10$, $u_1(a) = |1-a|$, $u_2(a) = \sqrt{a}$, $V_0(a) = a^2$ and $\beta = 1$ Algorithm 2 below requires only 17 % of the time needed by the Algorithm 1 below.

We now introduce several algorithms. Denote by $I(i,j)$ the discrete interval with integer endpoints $i \leq j$. Throughout this part of the section we use without further mention the following assumption: S is a finite subset of $\mathbb{R}$, $\leq_S = \leq^1$, $A = I(\underline{a}, \overline{a})$ for some $\underline{a} <^1 \overline{a}$ and $\leq_A = \leq^1$. Then we know from Lemma 3.4.1(a) that V_n is finite, and that there exists a smallest maximizer f_n, $n \geq 1$, since $D(s) \subset \mathbb{R}$ is finite for $s \in S$.

Now the OC in Theorem 2.3.3(c) yields

Algorithm 1 If V_{n-1} (and hence also W_n) is known for some fixed n, $1 \leq n \leq N$, then the (unique) smallest minimizer f_n is obtained by

$$f_n(s) := \text{ the smallest maximum point of } W_n(s,\cdot) \text{ on } D(s). \tag{8.1}$$

Note that f_n need not be increasing. The crucial idea for an acceleration of Algorithm 1 is as follows: If f_n is known to be $\leq_S$-increasing then one can sometimes find $f_n(s)$ by replacing $D(s)$ in (8.1) by a smaller set $\Delta_{nx}(s)$. This can be achieved by different algorithms, some of which are now explained.

Algorithm 2 Select an arbitrary state $s_0 \in S$, compute $f_n(s_0)$ by (8.1) and put

$$\Delta_{n2}(s) := \begin{cases} D(s) \cap I(f_n(s_0), \overline{a}), & \text{if } s > s_0, \\ D(s) \cap I(\underline{a}, f_n(s_0)), & \text{if } s < s_0. \end{cases}$$

Then (8.1) holds with $D(s)$ replaced by $\Delta_{n2}(s)$, $s \in S$. This is obvious since isotonicity of f_n implies $f_n(s) \in \Delta_{n2}(s)$ for $s \neq s_0$. Note also that in case $f_n(s_0) = \underline{a}$ we have $\Delta_{n2}(s) = D(s)$ for $s > s_0$ and $\Delta_{n2}(s) = \{\underline{a}\}$ for $s < s_0$. Similar properties hold in case $f_n(s_0) = \overline{a}$.

Algorithm 3 Assume that $S = I(\underline{s}, \overline{s})$ for some $\underline{s} < \overline{s}$. Compute $f_n(\underline{s})$ by (8.1) and put for $\underline{s} \leq s < \overline{s}$

$$\Delta_{n3}(s+1) := D(s+1) \cap I(f_n(s), \overline{a}).$$

Then (8.1) holds with $D(s)$ replaced by $\Delta_{n3}(s)$, $s \in S$.

Algorithm 4 A modification of Algorithm 3 is the following backward-algorithm: Let S be as in Algorithm 3. Compute $f_n(\overline{s})$ by (8.1) and put for $\underline{s} < s \leq \overline{s}$

$$\Delta_{n4a}(s-1) := D(s-1) \cap I(\underline{a}, f_n(s)).$$

Then (8.1) holds with $D(s)$ replaced by $\Delta_{n4a}(s)$, $s \in S$.

Another modification of Algorithm 3 is the following

Algorithm 5 Let S be as in Algorithm 3. Compute $f_n(s_0)$ for some $\underline{s} < s_0 \leq \overline{s}$ by (8.1). If one has found $f_n(s)$ for $s \in I(x,y) \subset S$ then put, if $x-1 \in S$ or if $x+1 \in S$,

$$\Delta_{n5b-}(x-1) := D(x-1) \cap I(\underline{a}, f_n(x)),$$

$$\Delta_{n5b+}(x+1) := D(x+1) \cap I(f_n(x), \overline{a}),$$

respectively. Then (8.1) holds with $D(s)$ replaced by $\Delta_{n5b-}(s)$ and $\Delta_{n5b+}(s)$, $s \in S$, respectively.

The computational advantages of monotonicity properties other than isotonicity of $f_n(s)$ in s are expounded below at the appropriate places.

8.2 Monotone Dependence on the Current State

For first reading one should concentrate on the most important case where both S and A are intervals with the usual ordering; figures are helpful.

We start with the following elementary example.

Example 8.2.1 We call a function w on D ***separable*** if $w(s,a) = u_1(s) + u_2(a)$ for functions u_1 on S and u_2 on A_0. Consider an invariant DP with $S = A = \mathbb{N}_0$, $D(s) = \mathbb{N}_{0,s}$, $s \in S$, and with separable r. Since $D(s) \subset \mathbb{R}$ is finite for all s, there exists a smallest maximizer f_n at stage $n \geq 1$ by Lemma 3.4.1(a). Then f_n is increasing for $n \geq 1$.

For the proof select $s \leq s'$. Note that $f(s) := f_n(s)$ is the smallest maximum point of $u := u_2 + \beta V_{n-1} \circ T_1$ on $\mathbb{N}_{0,s}$, where $T(s,a) = T_1(a)$. Assume that $f(s') < f(s)$. Then, since $u(f(s')) \geq u(a)$ for all $a \in D(s') \supset D(s)$ we have $u(f(s')) \geq u(f(s))$. On the other hand, since $f(s)$ is the smallest maximum point of u on $\mathbb{N}_{0,s}$, we have $u(f(s')) < u(f(s))$. This is a contradiction.

We split our general problem into two parts:

(a) Find conditions for a function w on the constraint set D under which a smallest (or largest) maximizer of w is increasing; such a condition is given in Lemma 8.2.7 below.
(b) Find assumptions on D, r and V_0 such that all functions $w := W_n$, satisfy the conditions of (a); such assumptions are given in Theorem 8.2.9 below.

Recall from Sect. 6.3 that a correspondence from S into A is a mapping $\Gamma(\cdot)$ from S into the system of non-empty subsets of A. We then write $\Gamma \in \mathrm{gr}(S,A)$.

Definition 8.2.2 A correspondence $\Gamma \in \mathrm{gr}(S, A)$ is called ***NE-complete*** (also called **monotonely complete**) if

$$(s, a'), (s', a) \in D \text{ with } s \preceq s', a \leq a' \text{ implies that } (s, a), (s', a') \in D.$$

The importance of this concept becomes apparent from

Lemma 8.2.3 (Smallest and largest elements of NE-complete correspondences)

Assume that $\Gamma \in \mathrm{gr}(S, A)$ is NE-complete and that $\Gamma(s)$ has a smallest element $\underline{f}(s)$ [a largest element $\overline{f}(s)$] for $s \in S$. Then $\underline{f}$ [$\overline{f}$] is increasing.

Proof Firstly we note that $\underline{f}$ [and also $\overline{f}$] is unique, if it exists; this holds since the ordering on A is antisymmetric. We now prove that $\underline{f}$ is increasing; the proof for $\overline{f}$ is similar. Assume that $\underline{f}$ is not increasing. Then, as A is totally ordered, there exist s and $s' \succeq s$ such that $a' := \underline{f}(s) > a := \underline{f}(s')$. Then $a \in \Gamma(s)$ since $\Gamma(\cdot)$ is NE-complete. As a' is the smallest element in $\overline{\Gamma}(s)$, we obtain $a' \leq a$, which is a contradiction. □

Example 8.2.4 (NE-complete constraint sets)

(a) If $D = S \times A$, then D is NE-complete.
(b) Assume that D has the interval form $D(s) = [c(s), d(s)] := \{a \in A : c(s) \leq a \leq d(s)\}$, $s \in S$. Then D is NE-complete if and only if both c and d are increasing.
 In fact, if c and d are increasing, select $s \preceq s'$ and $a \leq a'$ such that $a' \in D(s)$, $a \in D(s')$. Then $c(s) \leq c(s') \leq a \leq a' \leq d(s) \leq d(s')$, hence $a \in D(s)$ and $a' \in D(s')$. On the other hand, if D is NE-complete, we reason as follows. Let $s \preceq s'$ be arbitrary. Then $a := c(s) \in D(s)$ and $a' := d(s) \in D(s')$. Assume that $a > a'$. Then $(s, a'), (s, a) \in D$ as D is NE-complete. It follows that $a \leq a'$, which is a contradiction, hence c is increasing. In the same way one shows that d is increasing. Special cases are:
(b1) $D(s) = \{0, 1, \ldots, s\}$ in case $S \subset \mathbb{N}_0$.
(b2) $D(s) = \{s, s+1, \ldots, K\}$ for some $K \in \mathbb{N}$ in case $S = \{a \in \mathbb{Z} : a \leq K\}$. ♦

If there exists a smallest maximizer f_n at stage $n \geq 1$, then $f_n(s)$ is the smallest element of $D_n^*(s)$, i.e. of the set of maximum points of $W_n(s, \cdot)$, $s \in S$. Thus f_n is increasing by Lemma 8.2.3 provided D_n^* is NE-complete. This fact is useless unless the latter property can be checked without computing the sets $D_n^*(s)$. Fortunately, as we show in Lemma 8.2.7 below, NE-completeness of D_n^* is implied by NE-completeness of D, provided W_n has increasing differences in the sense of the following definition.

Recall the notation $D_a := \{s \in S : (s, a) \in D\}$ for the a-section of D, the fact that $s \in D_a$ if and only if $a \in D(s)$, and that D_a is non-empty if and only if $a \in A_0$. For $s, s' \in S$ put $I(s, s') := D(s) \cap D(s')$.

Definition 8.2.5 If D is NE-complete then a function w on D is said to have ***increasing differences*** (***w*** $\in$ ***ID*** for short) if

$$w(s', a) - w(s, a) \leq w(s', a') - w(s, a') \tag{8.2}$$

whenever $s \preceq_S s'$ and $a \leq_A a'$. The function w is said to have ***decreasing differences*** if $-w$ has increasing differences.

Remark 8.2.6

(a) $w \in ID$ if and only if $a \mapsto w(s', a) - w(s, a)$ is $\leq_A$-increasing on $I(s, s')$ whenever $(s, s' \in S) \wedge (s \preceq_S s') \wedge (|I(s, s')| \geq 2)$.
(b) Obviously $w \in ID$ if and only if $s \mapsto w(s, a') - w(s, a)$ is $\preceq_S$-increasing on $D_a \cap D_{a'}$, whenever $(a, a' \in A_0) \wedge (a \leq_A a') \wedge (|D_a \cap D_{a'}| \geq 2)$.
(c) If w is the restriction of a function w' on a restriction set $D' \supset D$ and if w' has increasing differences, then w also has increasing differences. ◇

Lemma 8.2.7 (Increasing maximizers of functions on D**)** *If $w \in ID$, then the smallest [the largest] maximizer of w, if it exists, is increasing.*

Proof According to (8.2) it suffices to show that $s \mapsto B(s) :=$ set of maximum points of $w(s, \cdot)$ is NE-complete. Assume that $s \preceq s'$, $a \leq a'$ and $a' \in B(s)$, $a \in B(s')$. Then $a' \in D(s)$, $a \in D(s')$, and as D is NE-complete we know that $a \in D(s)$ and $a' \in D(s')$. Since a' and a are maximum points of $w(s, \cdot)$ and $w(s', \cdot)$, respectively, and as $w \in ID$ we obtain from (8.2)

$$0 \leq w(s, a') - w(s, a) \leq w(s', a') - w(s', a) \leq 0.$$

This implies $a \in B(s)$ and $a' \in B(s')$, hence $B(\cdot)$ is NE-complete. □

We call two functions v_1 on a structured set X_1 and v_2 on a structured set X_2, respectively, *iso-monotone* if they are monotone in the same sense. Recall that discrete interval, discrete convexity and discrete concavity are defined in Appendix D.5.1.

Lemma 8.2.8 (Closure properties of the set *ID***)**

(a) ID is a convex cone.
(b) If the function w on D is separable, in particular if w does not depend on s or on a, then $w \in ID$.
(c) Let I be an interval and let g_S and g_A be functions on S and A, respectively, such that $g_S(s) - g_A(a) \in I$ for all $(s, a) \in D$. If v is concave [convex] on I and if g_S and g_A [and $-g_A$] are iso-monotone then ID contains

$$(s, a) \mapsto v(g_S(s) - g_A(a)).$$

In particular, if $S, A \subset \mathbb{R}$, if v is concave, $\alpha_1, \alpha_2, \alpha_3 \in \mathbb{R}$ with $\alpha_1 \cdot \alpha_2 \geq 0$, and if $\alpha_1 s - \alpha_2 a + \alpha_3 \in I$ for all $(s, a) \in D$, then ID contains

$$(s, a) \mapsto v(\alpha_1 s - \alpha_2 a + \alpha_3).$$

(d) *Let v be an increasing function on an interval I and let $w: D \to I$ have increasing differences. If (v is concave) $\wedge$ (w is SE-monotone) or if (v is convex) $\wedge$ (w is NE-monotone), then $v \circ w \in ID$. (Recall from Definition 6.2.2(v) the definition of monotonicity with respect to the quadrant relations.)*

(e1) *Let I be a discrete non-degenerate interval and let g_S and g_A be functions on S and A, respectively, such that $g_S(s) - g_A(a) \in I$ for all $(s, a) \in D$. If v is discretely concave [discretely convex] on I and if g_S and g_A [and $-g_A$] are iso-monotone then ID contains $(s, a) \mapsto v(g_S(s) - g_A(a))$.*

(e2) *Let v be an increasing function on a discrete non-degenerate interval I and let $w: D \to I$ have increasing differences. If (v is discretely concave) $\wedge$ (w is SE-monotone) or if (v is discretely convex) $\wedge$ (w is NE-monotone), then $v \circ w \in ID$.*

(f) *If $f \geq 0$ and $g \geq 0$ are increasing functions on S and on A, respectively, then $(s, a) \mapsto f(s) \cdot g(a)$ has increasing differences on $D := S \times A$.*

Proof The parts (a) and (b) are obvious from (8.2).

(c) We consider only the case where v is concave and where both g_S and g_A are increasing; the other cases are similar. Fix $a \leq a'$. Then $\delta := g_A(a') - g_A(a) \geq 0$, and isotonicity of g_S and concavity of v implies by Appendix D.2.6(c) with v replaced by $-v$, isotonicity of

$$\begin{aligned} s \mapsto\ & v(g_S(s) - g_A(a')) - v(g_S(s) - g_A(a)) \\ &= v(g_S(s) - g_A(a) - \delta) - v(g_S(s) - g_A(a)). \end{aligned}$$

(d1) We give a proof for the cases when v is increasing and concave and when w is SE-increasing. Let $\leq_D$ be the restriction of the product relation $\leq_S \times \leq_A$ to D. Select $(s, a) \leq_D (s', a')$ in D such that $(s, a'), (s', a) \in D$. Then $h := w(s, a) - w(s', a) \leq 0$ since w is SE-increasing. As $w \in ID$, $w(s', a') \geq w(s', a') + h \geq w(s, a')$, hence $w(s', a') + h \in I$. Since $w(s', a) + h = w(s, a)$ we now obtain

$$\begin{aligned} & v \circ w(s, a') + v \circ w(s', a) - v \circ w(s, a) - v \circ w(s', a') \\ &= [v \circ w(s, a') - v(w(s', a') + h)] \\ &+ [v(w(s', a') + h) - v \circ w(s', a') - v(w(s', a) + h) + v \circ w(s', a)]. \end{aligned}$$

The first bracket is ≤ 0 as $w(s', a') + h \geq w(s, a')$ and as v is increasing. SE-isotonicity of w implies that $x := w(s', a') \leq y := w(s', a)$. Now concavity of v and $h \leq 0$ imply by Appendix D.2.6(c) that $v(x+h) - v(x) \leq v(y+h) - v(y)$. Therefore the second bracket is also ≤ 0, hence $v \circ w \in ID$.

(d2) In the first of the two cases and when w is SE-decreasing one can copy the reasoning in (d1) with s and a interchanged with s' and a', respectively.

(d3) The proof for the two cases where v is increasing and convex and where w is NE-increasing [NE-decreasing] is similar to (d1) and (d2), respectively.

(e) The proof is nearly the same as for (c) and (d), using Appendix D.5.2(b) instead of Appendix D.2.6(c).

(f) The elementary proof is left to the reader. □

Combining Lemma 8.2.7 with Lemma 8.2.8 and observing that all value functions are finite if there exists a maximizer at each stage yields the next result.

Theorem 8.2.9 (Increasing maximizers) *Assume that*

(i) D is NE-complete,
(ii) r has increasing differences,
(iii) there exists a smallest [a largest] maximizer $\underline{f}_n$ $[\overline{f}_n]$ *at each stage* $n \geq 1$.

Then:

(a) The maximizers $\underline{f}_n$ $[\overline{f}_n]$, $n \geq 1$, *are increasing under each of the following conditions:*

(a1) the DP is invariant, i.e. $T(s,a)$ *is independent of s.*

(a2) S is an interval, $T(s,a) = g_S(s) - g_A(a)$ $[T(s,a) = g_S(s) + g_A(a)]$ *for isomonotone functions* g_S *and* g_A *on S and A, respectively, such that* $g_S(s) - g_A(a) \in S$ $[g_S(s) + g_A(a) \in S]$ *for all* $(s,a) \in D$, *and* V_n, $n \geq 0$, *is concave [convex].*

(a3) S is an interval, T has increasing differences and for $n \geq 0$: V_n, $n \geq 1$, *is increasing, and* (V_n *is concave*) $\wedge$ (*T is SE-monotone*) *or* (V_n *is convex*) $\wedge$ (*T is NE-monotone*).

(b) The assertion in (a) remains true, when S is a discrete non-degenerate interval and when convexity [concavity] is replaced by discrete convexity [discrete concavity].

Proof Use induction on $n \geq 1$. Then one uses Lemma 8.2.8(b) for (a1), Lemma 8.2.8(c) for (a2), Lemma 8.2.8(d) for (a3) and (8.2.8(b)) $\wedge$ (8.2.8(e)) for (b). □

Remark 8.2.10 (Increasing maximizers need not entail increasing optimal action sequences) Assume $N = 3$, $S = \mathbb{N}_0$, $A = D(0) = \{0,1\}$, $D(s) = \{1\}$ for $s \geq 1$; $T(s,a) = 1 - a + h(s)$, where $h(s) = s + 1$ if s is even and $h(s) = s - 1$ if s is odd; $r(s,a) = 1 - a$, $\beta = 1$, $V_0 \equiv 0$. Moreover $a_1(0) = 1$ and $a_1(1) = 0$. Thus $s \mapsto a_1(s)$ is not increasing. ◇

Remark 8.2.11 It can happen that some but not all maximizers of w are monotone. As an extreme example of this kind consider the constant function $w \equiv 1$ on $D := \{(s,a) \in \mathbb{R}_+ \times \mathbb{R} : -s \leq a \leq s\}$. Then $s \mapsto s$, the (pointwise) largest maximizer of w is strictly increasing while $s \mapsto -s$, the smallest maximizer is strictly decreasing, and there are many other maximizers which neither increase nor decrease. ◇

Remark 8.2.12

(a) Inverting one of the relations $\preceq_S$ and $\leq_A$ in Theorem 8.2.9 yields sufficient conditions for a smallest [largest] maximizer at stage n to be *decreasing* with respect to $\preceq_S$ and $\leq_A$.
(b) One obtains sufficient conditions for a smallest [a largest] *minimizer* at stage n to be increasing by replacing in Theorem 8.2.9 r, V_n, *concave* and *convex* by $-r$, $-V_n$, *convex* and *concave*, respectively. Then (ii) means that $c := -r$ has decreasing differences.
(c) If there exists at stage $n \geq 1$ a smallest increasing [a largest increasing] maximizer, many more increasing (and also many non-increasing) maximizers may exist. This is illustrated by the following example: $S = A = D(s) = [0, 1]$, T is arbitrary, $r \equiv 1 = \beta$, $V_0 \equiv 0$. Then $V_n \equiv n$ and $D_n^*(s) = [0, 1]$ for $n \geq 1$, $s \in S$. ◇

Example 8.2.13 (Isotonicity of minimizers in the continuous linear-convex system of Example 7.1.8*)* The underlying DP is given by $S = A = \mathbb{R}$, $D = S \times A$ and $T(s, a) = gs - ha$, with $g \in \mathbb{R}_+$, $h \in \mathbb{R}^+$, and with

(i) $c \geq 0$ is convex,
(ii) $c(s, a) \to \infty$ for $|a| \to \infty$, $s \in \mathbb{R}$,
(iii) $C_0 \geq 0$ is convex.

In the mentioned example we have shown for $n \geq 1$ that C_n is convex and that there exists a smallest minimizer f_n. In addition we get from the minimization version of Theorem 8.2.9(a2) that f_n *is increasing* if we also have

(iv) c has decreasing differences on $D = \mathbb{R}^2$.

It also follows from Lemma 8.2.8(a) and (c) that condition (iv) holds in the special case where $c(s, a) := \sum_{i=1}^m u_i(\alpha_i s - \varepsilon_i a)$ and where u_i, α_i and ε_i satisfy the conditions stated in Remark 7.1.9 and also the condition $\alpha_i \varepsilon_i \geq 0$ for all i. ◆

Example 8.2.14 (The discrete splitting problem from Example 6.4.1*)* We consider the problem with choice I of actions (i.e. action a denotes the amount of resource retained for further use) and by specifying $S = A = \mathbb{N}_{0,K}$ for some $K \in \mathbb{N}$ and $D(s) = \mathbb{N}_{0,s}$. T and r are as in Example 6.4.1, i.e. $T(s, a) = a$, $r(s, a) = u_1(a) + u_2(s - a)$ for functions u_1 and u_2 on $\mathbb{N}_{0,K}$; V_0 and $\beta \in \mathbb{R}^+$ are arbitrary. By (6.5) the VI has the form

$$V_n(s) = \max_{a \in \mathbb{N}_{0,s}} [u_1(a) + u_2(s - a) + \beta V_{n-1}(a)], \quad n \geq 1, s \in \mathbb{N}_{0,K}.$$

Finiteness of $D(s) \subset \mathbb{R}$ for all s implies that there exists a smallest maximizer f_n at stage $n \geq 1$, and then V_n is finite for $n \geq 1$ by Lemma 3.4.1. If in addition u_2 is discretely concave, then f_n *is increasing for each* $n \geq 1$.

The proof is a direct application of Theorem 8.2.9(a1), observing that D is NE-complete by Example 8.2.4(b1) and that $(s, a) \mapsto r(s, a) = u_1(a) + u_2(s - a)$ belongs to ID by Lemma 8.2.8(a), (b) and (e1) with $I := \mathbb{N}_{0,K}$, $g_S(s) := s$ and $g_A(a) := a$. ◆

Example 8.2.15 (A deterministic inventory problem with linear ordering cost and backlogging)

(α) *The discrete version.* In each of N periods a firm has a demand of $z \in \mathbb{N}$ units for some good which can be stored up to $K \in \mathbb{N}$ units. There is an ordering cost of $\lambda \in \mathbb{R}^+$ per unit. Moreover, in period t there is a holding cost $H(a_t) \geq 0$ if $a_t \geq 0$ and a shortage cost $H(a_t) \geq 0$ if $a_t < 0$; here H is a function on A. In order to meet the demand non-negative orders are made in each period. Denote by $s_t \in S = \{K, K-1, \ldots\}$ and $a_t \in A := S$ the inventory in period t before and after ordering, respectively; thus $D(s) = \{a \in \mathbb{Z} : s \leq a \leq K\}$; $T(s, a) = a - z$, hence the DP is invariant; $a_t - s_t$ is the order in period t.

We allow backlogging, i.e. demand unsatisfied in some period is noted and, if possible, delivered in a later period.

Therefore $c(s, a) = \lambda \cdot (a - s) + H(a)$. The terminal cost $C_0(s)$ is a penalty cost for unsatisfied demand if $s < 0$, and $-C_0(s)$ is a (non-negative) scrap value, if $s > 0$. It seems reasonable to assume that $H(0) = \varepsilon$ for some $\varepsilon \in \mathbb{R}^+$, that H is decreasing and discretely concave on $\{0, -1, -2, \ldots\}$ and increasing and discretely concave on $\mathbb{N}_{0,K}$. Moreover, it seems reasonable to assume that $C_0(0) = 0$, that C_0 is decreasing and discretely concave on $\{0, -1, -2, \ldots\}$ and decreasing and discretely convex on $\mathbb{N}_{0,K}$.

As an example, one might choose

$$H(a) := \varepsilon + \delta_h \cdot (a^+)^{\alpha_h} + \delta_{sh} \cdot (a^-)^{\alpha_{sh}},$$
$$C_0(s) \;=\; -\delta_{sc} \cdot (s^+)^{\alpha_{sc}} + \delta_p \cdot (s^-)^{\alpha_p}.$$

Here δ_h, δ_{sh}, δ_{sc} and δ_p are positive reals belonging to the holding costs, to the shortage costs, to the scrap value and to the penalty costs, respectively. Moreover, α_h, α_{sh}, α_{sc} and α_p are reals in $(0, 1]$, attached to the holding costs, to the shortage costs, to the scrap value and to the penalty costs, respectively.

For all $n \geq 1$ we assert the following:

(a) C_n is finite.
(b) There exists a smallest and a largest minimizer f_n $[g_n]$ at stage n.
(c) f_n and g_n are increasing.

Note that isotonicity of f_n means that the optimal ordering $\tilde{f}_n(s) := f_n(s) - s$, which is prescribed at stage n in state s, satisfies

$$\tilde{f}_n(s') - \tilde{f}_n(s) \geq -(s' - s) \text{ for } s < s'.$$

This property could be phrased as $\tilde{f}_n$ *is lower Lipschitz with constant 1*.

The proof of the assertions (a) and (b) follows from Lemma 3.4.1(a), since the sets $D(s)$ are finite. Moreover, (c) follows from the minimization version of Theorem 8.2.9(a) as D is NE-complete by Example 8.2.4(b2) and $-c \in ID$ by Lemma 8.2.8(a) and (b).

(β) *The continuous version* differs from the discrete version by using $z, K \in \mathbb{R}^+$, $S = A = (-\infty, K]$; $D(s) = [s, K]$. No change is necessary for T, r and C_0.
If H and C_0 are lsc (see Sect. 7.1), then for all $n \geq 1$ the following holds:

(a) C_n is finite and lsc.
(b) There exists a smallest and a largest minimizer f_n [g_n] at stage n.
(c) f_n and g_n are increasing.

The assertions (a) and (b) follow easily from the minimization version of Proposition 7.1.10, using Remark 7.1.11(b). Moreover, (c) follows from the minimization version of Theorem 8.2.9(a1) as D is NE-complete by Example 8.2.4(b) and $-c \in ID$ by Lemma 8.2.8(a) and (b). ♦

Example 8.2.16 (Optimal control-limit policies in binary DPs) Later on we shall encounter so-called binary models (replacement, admittance in queues, bandit problems, etc.) which are characterized by the assumption that $A = D(s) = \{0, 1\}$, $s \in S$. (As an example, in replacement models the actions $a = 1$ and $a = 0$ mean that one should replace or not replace, respectively.) Then a smallest maximizer f_n exists at each stage n and it is obviously increasing (for an arbitrary relation in S) by Lemma 8.2.7 if $s \mapsto W_n(s, 1) - W_n(s, 0)$ is increasing. Note that f_n is increasing if and only if it has the following property: if it prescribes action 1 at some state it also prescribes action 1 at all larger states, or equivalently, if and only if the set $\{s \in S : f_n(s) = 1\}$ is increasing as defined in Chap. 6. In case $S \subset \mathbb{R}$ with the usual ordering the decision rule f_n is increasing if and only if there is some state $s^* \in \mathbb{R}$, called a ***control-limit***, such that

$$f_n(s) = \begin{cases} 0, & \text{if } s < s^*, \\ 1, & \text{if } s > s^*. \end{cases}$$

Therefore in binary DPs with arbitrary structured state space a policy consisting of increasing maximizers will be called a ***control-limit policy.*** ♦

Now we turn to further criteria for a function w on D to have increasing differences; in examples the criteria are mainly applied to $w := r$.
If $A \subset \mathbb{R}$ and if $D(s)$, $s \in S$, is an interval then:

(a) $D(s)^\circ$, $s \in S$, either is empty or an open interval.
(b) $D(s)^\circ \cap D(s')^\circ = (D(s) \cap D(s'))^\circ$, $s' \in S$. A simple proof may be given, using the fact that a set $X \subset \mathbb{R}$ is open if and only if for each $x \in X$ there exists some $\varepsilon \in \mathbb{R}^+$ such that $(x - \varepsilon, x + \varepsilon) \subset X$.
(c) For $s, s' \in S$ recall that $I(s, s') := D(s) \cap D(s')$. Thus

$$\begin{aligned} |I(s, s')| \geq 2 &\iff D(s) \cap D(s') \text{ is a non-degenerate interval} \\ &\iff D(s)^\circ \cap D(s')^\circ \text{ is an open interval} \\ &\iff D(s)^\circ \cap D(s')^\circ \text{ is non-empty.} \end{aligned}$$

Proposition 8.2.17 (Functions with increasing differences) *Let S be a structured set, $A \subset \mathbb{R}$, $\preceq_A := \leq^1$ and let w be a function on D.*

Then the following assertions hold:

(a) Assume for all $s \in S$ that $D(s)$ is an interval in $\mathbb{Z}$.

(a1) $w \in ID$ if and only if

$$w(s', a) - w(s, a) \leq w(s', a+1) - w(s, a+1) \tag{8.3}$$

whenever $s, s' \in S$ with $s \preceq_S s'$, and $a, a+1 \in I(s, s')$.

(a2) If in addition $S \subset \mathbb{Z}$ with $\preceq_S := \leq^1$ and if for all $a \in A_0$ the set D_a is a discrete interval then $w \in ID$ if and only if

$$w(s+1, a) - w(s, a) \leq w(s+1, a+1) - w(s, a+1) \tag{8.4}$$

whenever $s, s+1 \in S$ and $a, a+1 \in I(s, s+1)$.

(b) Assume:

(i) For $s \in S$ the constraint set $D(s)$ is a (possibly degenerate) interval,
(ii) for $s \in S$ with non-empty $D(s)^\circ$ the function $w(s, \cdot)$ is continuous on $D(s)$,
(iii) the partial derivative $\frac{\partial}{\partial a} w$ exists on $\hat{D} := \{(s, a) \in S \times A : a \in D(s)^\circ\}$.

Then $w \in ID$ if and only if $s \mapsto \frac{\partial}{\partial a} w(s, a)$ is $\preceq_S$-increasing on the a-section $\hat{D}_a := \{s \in S : a \in D(s)^\circ\}$ of $\hat{D}$ for all $a \in \cup_{s \in S} D(s)^\circ$ for which $\hat{D}_a \neq \emptyset$.

(c) Let S and A be non-degenerate intervals and $D = S \times A$. Assume:

(i) $w(s, \cdot)$ is continuous on A for $s \in S$,
(ii) $\frac{\partial}{\partial a} w$ exists on $S \times A^\circ$ and $s \mapsto \frac{\partial}{\partial a} w(s, a)$ is continuous on S for $a \in A^\circ$,
(iii) there exists $\frac{\partial^2}{\partial s \partial a} w$ on $S^\circ \times A^\circ$.

Then w belongs to ID if and only if $\frac{\partial^2}{\partial s \partial a} w \geq 0$ on $S^\circ \times A^\circ$.

(d) If w is defined on $D := \mathbb{R} \times \mathbb{R}$, and if there exists $\frac{\partial^2}{\partial s \partial a} w$ then $w \in ID$ if and only if $\frac{\partial^2}{\partial s \partial a} w \geq 0$.

Proof

(a11) The only-if part of (a1) is obvious from Remark 8.2.6(a).

(a12) The if-part means that (8.3) implies (8.2) if $D(s)$, $s \in S$, is an interval in $\mathbb{Z}$. Thus select $s, s' \in S$ and $a, a' \in I(s, s')$ such that $s \preceq s'$ and $a \leq a'$. Since (8.2) holds trivially if $a = a'$, we assume $a' \geq a + 1$. As $D(s)$ is a discrete interval containing a and $a - 1$, we have $b, b+1 \in D(s)$ for $a \leq b+1 \leq a'-1$ and also for $a \leq b \leq a' - 1$. Thus there is defined $\sum_{b=a}^{a'-1} [w(s, b+1) - w(s, b)]$, and this sum equals $x := w(s, a') - w(s, a)$.

With the same reasoning we see that there is defined $\sum_{b=a}^{a'-1} [w(s', b+1) - w(s', b)]$, and this sum equals $y := w(s', a') - w(s', a)$. From (8.3) we know that $w(s, b+1) - w(s, b) \leq w(s', b+1) - w(s', b)$ for $a \leq b \leq a' - 1$. Thus $x \leq y$ which implies (8.2).

(a21) The necessity of (8.4) follows from (a1) with $s' := s + 1$.

(a22) It remains to show that (8.4) implies (8.3). From (8.4) we obtain

$$\Delta(s) := w(s, a+1) - w(s, a) \leq \Delta(s+1) \leq \cdots \leq \Delta(s')$$

(which proves (8.3)), provided that $s, s+1, \ldots, s' \in S$ and that $a, a+1 \in D(s) \cap D(s+1) \cap \cdots \cap D(s')$. Since $S = \cup_{a \in A_0} D_a$ and since the sets $D(s)$, $s \in A_0$, are discrete intervals, S is also a discrete interval. Now $s, s' \in S$ implies that $s, s+1, \ldots, s' \in S$. Moreover, $a, a+1 \in D(s) \cap D(s+1) \cap \cdots \cap D(s')$ means that $s, s+1, \ldots s' \in D_a \cap D_{a+1}$, hence $a, a+1 \in \cap D(s) \cap D(s+1) \cap \cdots \cap D(s')$. This completes the proof of (a2).

(b1) Put $\Delta(s, s', a) := w(s', a) - w(s, a)$ for $s, s' \in S$ and $a \in I(s, s')$.

We assert: If $s, s' \in S$ with $s \preceq_S s'$ and $|I(s, s')| \geq 2$ then $a \mapsto \Delta(s, s', a) := w(s', a) - w(s, a)$ is continuous on $I(s, s')$, $\frac{\partial}{\partial a}\Delta(s, s', \cdot)$ exists on $I(s, s')^\circ$, and $\Delta(s, s', \cdot)$ is increasing on $I(s, s')$ if and only if $\frac{\partial}{\partial a}\Delta(s, s', \cdot) \geq 0$ on $I(s, s')^\circ$.

For the proof fix s, s' and drop s, s' in $I(s, s')$ and $\Delta(s, s', a)$. Firstly, (iii) ensures the existence of $\frac{\partial}{\partial a}\Delta$ on $I^\circ = D(s)^\circ \cap D(s')^\circ$. Next, Δ is continuous on I by (ii). Finally Appendix A.4.12(a) implies that Δ is increasing on I if and only if $\frac{\partial}{\partial a}\Delta \geq 0$ on I°.

(b2) Now we prove (b) by a sequence of four equivalences. The validity of the first, second and third equivalence is ensured by the definition of *ID*, by (b1) and by a trivial rewriting, respectively. For the last equivalence observe that $a \in I(s, s')^\circ$ if and only if $s, s' \in \hat{D}_a$. Now we obtain, using the property $P(s, s') := s, s' \in S, s \preceq_S s'$ and $|I(s, s')| \geq 2$,

$$\begin{aligned} & w \in ID \\ \Longleftrightarrow\ & \Delta(s, s', \cdot) \text{ is increasing on } I(s, s') \text{ whenever } P(s, s') \text{ holds} \\ \Longleftrightarrow\ & \frac{\partial}{\partial a}\Delta(s, s', \cdot) \geq 0 \text{ on } I(s, s')^\circ \text{ whenever } P(s, s') \text{ holds} \\ \Longleftrightarrow\ & \frac{\partial}{\partial a}w(s, a) \leq \frac{\partial}{\partial a}w(s', a) \text{ for } a \in I(s, s')^\circ \text{ whenever } P(s, s') \text{ holds} \\ \Longleftrightarrow\ & s \mapsto \frac{\partial}{\partial a}w(s, a) \text{ is } \preceq_S\text{-increasing on } \hat{D}_a \text{ for all } a \in \cup_{s \in S} D(s)^\circ \\ & \text{for which } \hat{D}_a \neq \emptyset. \end{aligned}$$

(c) We have $D(s) = A = A_0 = I(s, s')$, hence $D(s)^\circ = A^\circ \neq \emptyset$. By assumption $w(s, \cdot)$ is continuous on A for $s \in S$, and $\frac{\partial}{\partial a}w$ exists on $S \times A^\circ$. Now we obtain from Proposition 8.2.17(b) that $w \in ID$ if and only if $v := \frac{\partial}{\partial a}w(\cdot, a)$ is increasing on $\hat{D}_a = S$ for all $a \in A^\circ$. Finally Appendix A.4.12(a) implies that $w \in ID$ if and only if $\frac{\partial^2}{\partial s \partial a}w = \frac{\partial}{\partial s}v \geq 0$ on $S^\circ \times A^\circ$.

(d) This follows from (c) with $S = A := \mathbb{R}$. Note that continuity of $w(s, \cdot)$ on $A = \mathbb{R}$ for $s \in \mathbb{R}$ follows from the existence of $\frac{\partial}{\partial a} w$ on $S \times A = \mathbb{R}^2$. □

From Lemma 8.2.8 and Proposition 8.2.17 one obtains many examples for functions in *ID*, two of which are as follows:

(a) Assume $S = A = \mathbb{R}_+$ and $D := \{(s, a) \in \mathbb{R}_+^2 : 0 \leq a \leq s\}$ and let g and h be increasing functions on $\mathbb{R}_+$. Then *ID* contains the function $\big(g(s) + h(a)\big)^2$, e.g. $(s^3 + e^a)^2$ by Lemma 8.2.8(c).
(b) From Remark 8.2.6(c) and from Proposition 8.2.17(d) we see: If w on an arbitrary restriction set $D \subset \mathbb{R} \times \mathbb{R}$ is the restriction of a function w', defined on $\mathbb{R} \times \mathbb{R}$, and if there exists $\frac{\partial^2}{\partial s \partial a} w'$ with $\frac{\partial^2}{\partial s \partial a} w' \geq 0$ on $\mathbb{R} \times \mathbb{R}$, then $w \in ID$. As an example satisfying the assumptions we mention $w(s, a) := sa$ and more generally

$$(s, a) \mapsto w(s, a) := \sum_{(i,j) \in B} \alpha_{ij} \cdot s^i a^j, B \text{ a finite subset of } \mathbb{N}^2, \alpha_{ij} \in \mathbb{R}_+.$$

8.3 Increasing and Lipschitz Continuous Dependence on the State

The basic model in this subsection is a DP with $S, A \subset \mathbb{R}$, hence $D \subset \mathbb{R}^2$. In this DP the decision rule f_n denotes the smallest maximizer at stage $n \geq 1$, if it exists. We use the following subsets of the set $\mathbb{F}$ of decision rules:

(i) $\mathbb{F}_i$ is the set of increasing decision rules.
(ii) $\mathbb{F}_i(\lambda)$, $\lambda \in \mathbb{R}_+$, is the set of decision rules f for which $s \mapsto \lambda s - f(s)$ is increasing, i.e. for which

$$f(s') \leq f(s) + \lambda \cdot (s' - s) \text{ for } s \leq s'.$$

$f \in \mathbb{F}_i(0)$ means that f is decreasing. If S is a discrete non-discrete interval, $f \in \mathbb{F}_i(\lambda)$ means that the size of the upward jumps of f has amount $\leq \lambda$.
(iii) ***ILIP*** $(\boldsymbol{\lambda}) := \mathbb{F}_i \cap \mathbb{F}_i(\lambda)$, $\lambda \in \mathbb{R}_+$, which equals the ***set of decision rules f which are both increasing and Lipschitz continuous with constant*** $\boldsymbol{\lambda}$.

Note that $f \in ILIP(\lambda)$ if and only if

$$0 \leq f(s') - f(s) \leq \lambda \cdot (s' - s) \text{ for } s \leq s'.$$

Recall that Lipschitz continuity of f with constant λ means that that

$$|f(s) - f(s')| \leq \lambda \cdot |s - s'|, \quad s, s' \in S.$$

If S is an interval and $f \in ILIP(\lambda)$, then (α) f is continuous by Lipschitz continuity of f, and (β) by a theorem of Lebesgue, f has almost everywhere a derivative f', and $0 \leq f' \leq \lambda$. Note that $f \in ILIP(0)$ if and only if $s \mapsto f(s)$ is constant.

The method of Sect. 8.2 for verifying that f_n, $n \geq 1$, is increasing can also be used for showing that f_n belongs to $ILIP(\lambda)$ for given $\lambda \in \mathbb{R}_+$; see Proposition 8.3.2(a) below.

The two-step algorithm at the beginning of the present section for the accelerated computation of $(f_n)_N^1$ and of V_N in certain DPs with increasing smallest maximizers f_n can be further accelerated in case that $f_n \in ILIP(\lambda)$, $1 \leq n \leq N$, for some $\lambda \in \mathbb{N}_0$. In fact, $f_n \in ILIP(\lambda)$ if and only if for $s, s+1 \in S$

$$f_n(s+1) \in A_\lambda(f_n(s)) := \{f_n(s), f_n(s)+1, \ldots, f_n(s)+\lambda\}.$$

Thus in the two-step algorithm one can replace $A(f_n(s))$ by the *smaller* set $A(f_n(s)) \cap A_\lambda(f_n(s))$.

Lemma 8.3.1 (ILIP maximizers of functions on $D \subset \mathbb{R}^2$) *Consider our basic model with S, $A \subset \mathbb{R}$. For $f \in \mathbb{F}$ and $\lambda \in \mathbb{R}_+$ define the decision rule $s \mapsto f_\lambda(s) := \lambda s - f(s)$, and for $w: D \to \mathbb{R}$ define the function*

$$w_\lambda(s, a') := w(s, \lambda s - a'),\ (s, a') \in D_\lambda := \{(s, \lambda s - a) \in S \times \mathbb{R} : a \in D(s)\}.$$

Then f is the smallest [largest] maximizer of w if and only if f_λ is the largest [smallest] maximizer of w_λ.

Proof For the first assertion it suffices to show for fixed $s \in S$ that an action a^* is the smallest maximum point of $w(s, \cdot)$ if and only if $\lambda s - a^*$ is the largest maximum point of $w_\lambda(s, \cdot)$. (The second assertion is proved in a similar way.)

(a) For fixed $s \in S$ the function $a \mapsto \lambda s - a$ from $D(s)$ onto $D_\lambda(s)$ is bijective. Therefore $a \in D(s)$ if and only if $\lambda s - a \in D_\lambda(s)$.
(b) Keep s fixed and put $b := \lambda s - a$, $b^* := \lambda s - a^*$ for $a, a^* \in D(s)$. Drop s in $w(s, a)$ and $w_\lambda(s, a)$. Then we obtain
   a^* is the smallest maximum point of $w(\cdot)$

$\Longleftrightarrow$ $w(a^*) \geq w(a)$ for all $a \in D(s)$, $w(a^*) > w(a)$ for all $a \in D(s)$, $a < a^*$
$\Longleftrightarrow$ $w(\lambda s - b^*) \geq w(\lambda s - b)$ for all $b \in D_\lambda(s)$, $w(\lambda s - b^*) > w(\lambda s - b)$ for all $b \in D_\lambda(s)$, $b > b^*$
$\Longleftrightarrow$ $w_\lambda(b^*) \geq w_\lambda(b)$ for all $b \in D_\lambda(s)$, $w_\lambda(b^*) > w_\lambda(b)$ for all $b \in D_\lambda(s)$, $b > b^*$
$\Longleftrightarrow$ $\lambda s - a^*$ is the largest maximum point of $w_\lambda(\cdot)$. □

For $\lambda \in \mathbb{R}_+$ consider the model DP$_\lambda$ which differs from our basic DP by $D_\lambda := \{(s, \lambda s - a) \in S \times \mathbb{R} : a \in D(s)\}$, $T_\lambda(s, a) := T(s, \lambda s - a)$ and $r_\lambda(s, a) := r(s, \lambda s - a)$, $(s, a) \in D_\lambda$. From Lemma 8.3.1 we obtain the next result Proposition 8.3.2, which shows how Theorem 8.2.9 may be used to verify that $f := f_n$ belongs to $ILIP(\lambda)$. For the proof of part (b) of Proposition 8.3.2 use induction on $n \geq 0$.

We omit the proof of

Proposition 8.3.2 (ILIP maximizers) *Assume $S, A \subset \mathbb{R}$, that V_n, $n \geq 1$, is finite, and that there exists the smallest maximizer f_n at stage $n \geq 1$. Then:*

(a) $f_n \in ILIP(\lambda)$ if both DP and DP$_\lambda$ satisfy the assumptions in Theorem 8.2.9.
(b) $V_{n\lambda} = V_n$, $n \geq 0$.

Example 8.3.3 (ILIP maximizers in a continuous splitting model) As in Example 7.1.13 we jointly treat the continuous versions of the allocation model and of the reservoir model in Example 6.4.1. We use choice I of the actions. From Table 6.1 we see: $S = A \in \{[0,K], \mathbb{R}_+\}$ for some $K \in \mathbb{R}^+$; $D(s) = [0,s]$, $T(s,a) = T_1(a)$ and $r(s,a) = u_1(a) + u_2(s-a)$ for functions T_1 and u_1 on A and u_2 on $\tilde{A} = A$.

Assume that (i) T_1 is continuous, (ii) u_2 is increasing, (iii) u_1, u_2 and V_0 are concave and continuous. Then by Example 7.1.13 for all $n \geq 1$ the value function V_n is finite, concave and continuous, and there exists a smallest maximizer f_n at stage n. We now show that $f_n \in ILIP(1)$.

For the proof we firstly note that f_n is increasing by Theorem 8.2.9(a) since D is NE-complete and $r(s,a) = u_1(a) + u_2(s-a)$ belongs to ID by Lemma 8.2.8(a) and (c) with $I := A$ since u_2 is concave. Because of Proposition 8.3.2(a) with $\lambda = 1$ it remains to show that $s \mapsto f_{n1}(s) := s - f(s)$ is also increasing. Now f_{n1} is a maximizer of $(s,a) \mapsto W_{n1}(s,a) := r(s, s-a) + \beta V_{n-1}(s-a)$, $(s,a) \in D$. Thus the assertion follows from Theorem 8.2.9(b), since concavity of u_1 implies by Lemma 8.2.8(a) and (c) with $I := A$ that $(s,a) \mapsto r(s, s-a)$ belongs to ID and since V_{n-1} is concave. ◆

Example 8.3.4 (ILIP minimizers in a continuous linear-convex system) Consider the continuous linear-convex model of Example 7.1.8 with one-stage cost function $(s,a) \mapsto c(s,a)$. In Example 8.2.13 we have seen that the smallest minimizer f_n at stage $n \geq 1$ exists and is increasing under the following conditions:

(ii) $c \geq 0$ is convex,
(ii) $c(s,a) \to \infty$ for $|a| \to \infty$, $s \in \mathbb{R}$,
(iii) c has decreasing differences,
(iv) $C_0 \geq 0$ is convex.

We assert that f_n even belongs to $ILIP(g/h)$, provided that in addition to conditions (i)–(iv) we have

(v) $(s,a) \mapsto c_\lambda(s,a) := c(s, \lambda s - a)$ has decreasing differences; here $\lambda := g/h$.

For the proof it suffices by Proposition 8.3.2, observing that $T_\lambda(s,a) = T(ha)$ and that $D_\lambda = \mathbb{R}^2$, to show that DP$_\lambda$ satisfies Theorem 8.2.9(a). This is indeed true since c_λ has decreasing differences.

As a special example where condition (v) holds, consider as in Example 8.2.13

$$c(s,a) := \sum_{i=1}^{m} u_i(\alpha_i s - \varepsilon_i a)$$

for convex $u_i \geq 0$, $\alpha_i \varepsilon_i \geq 0$ for all i, $\varepsilon_1 \neq 0$ and $b \mapsto u_1(b) \to \infty$ for $|b| \to \infty$. (From Remark 7.1.9 and from Example 8.2.13 we know that then also (i), (ii) and (iii) hold.)

Now we see from Lemma 8.2.8(a) and (c), since

$$c_\lambda(s,a) = \sum_i u_i((\alpha_i - \varepsilon_i \lambda)s + \varepsilon_i a),$$

that condition (v) holds if for all i we have $\alpha_i \varepsilon_i \leq \varepsilon_i^2 \lambda$. ♦

The results in the next Lemma 8.3.5–8.3.7 deal with special problems; the reader may skip them in a first reading.

When trying to treat the discrete version of the splitting problem by means of Theorem 8.2.9(b), (c) and (d) we encounter the difficulty that convexity of D, when $D \subset \mathbb{Z}^m$ for some $m \geq 2$, is not defined and hence there does not exist discrete counterparts to Lemma 7.1.1 and Theorem 7.1.2 for proving concavity of V_n. Fortunately the following result will help in a few cases. Recall from Appendix D.5.1 that each function on a discrete interval M with $|M| = 2$ is discretely concave.

Lemma 8.3.5 (*ILIP*(λ) **maximizers for functions on** $D \subset \mathbb{Z}^2$) *For* $\lambda \in \mathbb{N}_0$ *assume that*

(i) *S, A are discrete non-degenerate intervals and $D(s)$, $s \in S$, are discrete intervals,*
(ii) *if $|S| \geq 3$ and s, $s+2 \in S$ then:*

$$a \in D(s), b \in D(s+2) \text{ and } a < b \Rightarrow \{a+\lambda, b-\lambda\} \subset D(s+1),$$

(iii) *D and $D_\lambda := \{(s, \lambda s - a) \in S \times \mathbb{Z} : a \in D(s)\}$ are NE-complete,*
(iv) *$A_0 := \cup_{s \in S} D(s)$ and $A_{0,\lambda} := \cup_{s \in S} D_\lambda(s)$ are discrete non-degenerate intervals, and K and u are discretely concave functions on A_0 and on $A_{0,\lambda}$, respectively,*
(v) *there exists the smallest maximizer f of*

$$D \ni (s,a) \mapsto w(s,a) := K(a) + u(\lambda s - a).$$

Then f belongs to ILIP(λ) and

$$s \mapsto w^*(s) := \sup\{w(s,a) : a \in D(s)\}$$

is discretely concave.

Proof

(a) By definition we have $f \in \mathit{ILIP}(\lambda)$ if both f and $s \mapsto f_\lambda(s) := \lambda s - f(s)$ are increasing. Now f is increasing by Lemma 8.2.7 if w belongs to ID. This holds by the following reasoning: $(s,a) \mapsto K(a)$ belongs to ID by Lemma 8.2.8(b).

Next, $(s, a) \mapsto u(\lambda s - a)$ belongs to *ID* by Lemma 8.2.8(e1) with $g_S(s) := \lambda s$, $g_A(a) := a$, and $I := A_{0,\lambda}$. Now $w \in ID$ by Lemma 8.2.8(a). In a similar way one shows that f_λ is increasing: one uses that by Lemma 8.3.1 f_λ is the largest maximizer of $(s, a) \mapsto K(\lambda s - a) + u(a)$.

(b) We show that w^* is discretely concave. This is trivially true if $|S| = 2$. Thus assume $|S| \geq 3$. Fix $s \in S$ such that $s + 2 \in S$; then also $s + 1 \in S$. By (D.5) it suffices to show that $\Delta := w^*(s+2) - 2w^*(s+1) + w^*(s) \leq 0$. For the proof put $a := f(s) \in D(s)$ and $b := f(s+2) \in D(s+2)$, hence $a \leq b$ since f is increasing.

Case 1: $a = b$, hence $f(s+1) = a \in D(s+1)$ since f is increasing by (a). Put $x := \lambda s - a$. Then $\Delta \leq 0$ as u is discretely concave and as

$$\Delta = [u(x + 2\lambda) - u(x + \lambda)] - [u(x + \lambda) - u(x)].$$

Note that the latter term is defined since $x \in D(s)$ implies that $x+\lambda = \lambda(s+1) - f(s+1) \in D_\lambda(s)$ and $x+2\lambda = \lambda(s+2) - f(s+2) \in D_{2\lambda}(s)$, hence x, $x + \lambda$, $x + 2\lambda \in A_{0,\lambda}$.

Case 2: $a < b$. Then, observing that $\{a + \lambda, b - \lambda\} \subset D(s+1) \subset A_0$, and discrete concavity of K, we obtain

$$\begin{aligned}\Delta &\leq w(s+2, b) - w(s+1, a+\lambda) - w(s+1, b-\lambda) + w(s, a)\\ &= [K(b) - K(b-\lambda)] - [K(a+\lambda) - K(a)] \leq 0.\end{aligned}$$

□

Example 8.3.6 (ILIP(1) maximizers in a discrete splitting problem) We consider once more Example 8.2.14. There we have $S = A = \mathbb{N}_{0,K}$ for some $K \in \mathbb{N}$; $D(s) = \mathbb{N}_{0,s}$; $T(s, a) = a$ since a denotes the resource retained for further use, and $r(s, a) = u_1(a) + u_2(s - a)$ for some functions u_1 and u_2 on $\mathbb{N}_{0,K}$. In Example 8.2.14 we have seen for $n \geq 1$ that V_n is finite and that f_n exists, and that f_n is increasing if u_2 is discretely concave. We now prove:

If u_1, u_2 and V_0 are discretely concave, then for $n \geq 1$:

(a) V_n is discretely concave.
(b) $f_n \in ILIP(1)$.

Proof We apply Lemma 8.3.5 with $\lambda = 1$, $K(a) := u_1(a) + \beta V_n(a)$, $a \in A = \mathbb{N}_{0,K}$, and $u := u_2$. Note that $A_0 = A_{0,\lambda} = \mathbb{N}_{0,K}$ since $D(s) = D_\lambda(s) = \mathbb{N}_{0,s}$.

(a) We use induction on $n \geq 0$. Firstly, V_0 is discretely concave by assumption. Assume that V_n is discretely concave for some $n \geq 0$. From Appendix D.5.2(c) we know that K is discretely concave. Next, it is easily seen that s, $s + 2 \in S$, $a \in D(s)$ and $a < b \in D(s+2)$ implies $\{a + 1, b - 1\} \subset D(s+1)$. Now the assertion follows from Lemma 8.3.5, since $A_0 = A_{0,\lambda} = \mathbb{N}_{0,K}$ is a discrete non-degenerate interval, and since $W_{n+1}(s, a) = K(a) + u(s - a) = w(s, a)$.

(b) This is contained in Lemma 8.3.5. ♦

The minimization version of Lemma 8.3.5 has the same assumptions as Lemma 8.3.5 except that K and u are discretely convex and that there exists a smallest minimizer f of w. Then $f \in ILIP(\lambda)$ and w^* is discretely convex.

Example 8.3.7 (ILIP(λ) minimizers in a discrete linear-convex system) Due to technical requirements in linear-convex (in particular in linear-quadratic) systems the states s_t and the controls a_t are often multiples of some unit, taken here to be one. Moreover, it is realistic to assume that the set of admissible actions is bounded. This model also deserves attention since most numerical computations require discrete states and actions.

Let us consider the discrete counterpart of Example 8.2.13 where $S = \mathbb{Z}$, $A = D(s) = \mathbb{Z}(-m, m)$ for some $m \in \mathbb{N} + \{\infty\}$, where $\mathbb{Z}(-m, m) := \{-m, -m+1, \ldots, m\}$ for $m \in \mathbb{N}$, and $\mathbb{Z}(-\infty, \infty) := \mathbb{Z}$;

$$T(s,a) = h \cdot (\lambda s - a) \text{ for some } h \in \mathbb{N}, \lambda \in \mathbb{N}_0,$$

$$c(s,a) = u_0(s) + u_1(a) + u_2(h \cdot (\lambda s - a)).$$

We assume: (i) in case $m < \infty$ we have $\lambda \in \{0, 1\}$; (ii) the four functions u_0 and C_0 on $S = \mathbb{Z}$, u_1 on $A_0 = \cup_{s \in S} D(s) = A$ and u_2 on $A_{0,\lambda} = \cup_{s \in S} D_\lambda(s) = \mathbb{Z}$ are non-negative and discretely convex, (iii) in case $m = \infty$ we have $u_1(b) \to \infty$ or $u_2(b) \to \infty$ for $|b| \to \infty$.

We firstly note two elementary properties:

(a) $0 \leq C_n < \infty$ for all $n \geq 1$.
(b) There exists a smallest minimizer f_n at stage $n \geq 1$.

Both assertions hold trivially if m is finite. Otherwise (a) holds since $C_n \geq 0$ due to $c \geq 0$ and $C_0 \geq 0$, and $C_n(s) \leq C_{ny}(s) < \infty$ for each $y \in A^N(s)$, while (b) is obvious.

Now we show: For all $n \geq 1$ the minimal cost function C_n is discretely convex and f_n belongs to $ILIP(\lambda)$. We firstly prove by induction on $n \geq 0$ that C_n is discretely convex. C_0 is discretely convex by assumption. Assume that C_n is discretely convex for some $n \geq 0$. We have $D = S \times A$, $D_\lambda(s) := \{\lambda s - a : a \in D(s)\} = \mathbb{Z}$ if $m = \infty$ and $D_\lambda(s) = \mathbb{Z}(\lambda s - m, \lambda s + m)$, otherwise. Now we apply the minimization version of Lemma 8.3.5 with $K := u_1$ and $u(b) := u_2(hb) + \beta C_n(hb)$. There the assumptions (i)–(v) are fulfilled: for (ii) observe that $\lambda \in \{0, 1\}$ in case $m < \infty$; for (iii) apply Example 8.2.4, and for (iv) note that $A_0 = A$, $A_{0,\lambda} = \mathbb{Z}$ and that u is discretely convex on $\mathbb{Z}$ by Appendix D.5.2(c) and (d). Now Lemma 8.3.5 with

$$(s,a) \mapsto w(s,a) := W_{n+1}(s,a) - u_0(s) = K(a) + u(\lambda s - a)$$

shows that $C_{n+1} - u_0$ is discretely convex, and Appendix D.5.2(c) ensures that also C_{n+1} is discretely convex. Finally, $f_n \in ILIP(\lambda)$ follows for fixed $n \geq 1$ from Lemma 8.3.5 with $w := W_n - u_0$. ♦

8.4 Monotone Dependence on the Stage Number

Since no general theory for our topic seems to exist, we restrict ourselves to a few examples.

Example 8.4.1 (Continuous splitting problem where f_n is increasing in n) Consider the following special case of the splitting problem from Example 8.3.3 with choice I of the actions: $S = A = [0, K]$ for some $K \in \mathbb{R}^+$; $D(s) = [0, s]$, $T(s, a) = a$ and $r(s, a) = u_1(a) + u_2(s - a)$ for functions u_1 and u_2 on A.

Then:

$f_n(s)$ exists and is increasing in n on $\mathbb{N}$ if

(i) u_1, u_2 and V_0 are concave and continuous,
(ii) u_1 is strictly increasing and u_2 and V_0 are increasing,
(iii) $\beta = 1$.

Proof

(a) By Example 7.1.13 at each stage $n \geq 1$ the decision rule f_n exists, and V_n is concave and continuous. By Example 8.3.3 we have $f_n \in ILIP(1)$, which implies that $h_n(s) := s - f_n(s)$ is increasing in s.
(b) We now fix $s \in S$ and prove that $f_n(s)$ is increasing in n, i.e. that $h_n(s)$ is decreasing in n, i.e. that for fixed $n \geq 1$

$$h_{n+1}(s) \leq h_n(s) =: b. \tag{8.5}$$

For the proof we consider for $(z, a) \in D$ the number

$$G_n(z, a) := W_n(z, z - a) = u_1(z - a) + u_2(a) + V_{n-1}(z - a).$$

From Lemma 8.3.1 with $\lambda = 1$ we know that $h_n(z)$ is the largest maximum point of $a \mapsto G_n(z, a)$, and from Proposition 8.3.2(b) with $\lambda = 1$ we know that $\sup_{a \in D(z)} W_n(z, z - a) = V_n(z)$. Next, (8.5) holds if

$$G_{n+1}(s, a) < G_{n+1}(s, b) \text{ for all } a \in D(s) \text{ for which } a > b. \tag{8.6}$$

A proof of (8.6) $\Longrightarrow$ (8.5) by contradiction can be given, using that $G_{n+1}(s, h_{n+1}(s)) = \sup_{a \in D(s)} G_{n+1}(s, a)$. Thus we have to verify $G_{n+1}(s, a) < G_{n+1}(s, b)$ for each fixed $a \in D(s)$ with $a > b$. Then $a^* + a - b > a^*$ for $a^* := h_n(s - a)$. Now concavity of u_2 and the law of decreasing marginal returns in Appendix D.2.6(d) show that

$$u_2(a) - u_2(b) \leq u_2(a^* + a - b) - u_2(a^*). \tag{8.7}$$

As u_1 is strictly increasing and as $s-a < s-b$, we have $u_1(s-a)-u_1(s-b) < 0$. Combining this with (8.7) and with $V_n(s) = G_n(s, b)$ (shown above) we obtain, using $b^* := h_n(s-b)$,

$$\begin{aligned}
&G_{n+1}(s,a) - G_{n+1}(s,b)\\
&= [u_1(s-a) + u_2(a) + V_n(s-a)] - [u_1(s-b) + u_2(b) + V_n(s-b)]\\
&< u_2(a^* + a - b) - u_2(a^*) + G_n(s-a, a^*) - G_n(s-b, b^*)\\
&= u_2(a^* + a - b) - u_2(a^*) + u_1(s-a-a^*) + u_2(a^*)\\
&- u_1(s-b-b^*) - u_2(b^*) + V_{n-1}(s-a-a^*) - V_{n-1}(s-b-b^*)\\
&= G_n(s-b, a^* + a - b) - G_n(s-b, b^*) \le 0.
\end{aligned}$$

The last inequality holds as $b^* = h_n(s-b)$ is a maximum point of $G_n(s-b, \cdot)$. Now the proof of (8.6) is complete. □♦

Proposition 8.4.2 (Isotonicity of $f_n(s)$ in n in invariant DPs) *Let S be structured and A totally ordered. Assume that*

(i) *$D(\cdot)$ is NE-complete,*
(ii) *$T(s, a) = T_1(a)$ for some increasing mapping $T_1: A \to S$ (hence the DP is invariant),*
(iii) *r has increasing differences,*
(iv) *there exists the smallest maximizer f_n at each stage $n \ge 1$ (hence V_n, $n \ge 1$, is finite),*
(v) *$V_0(s') - V_0(s) \le V_1(s') - V_1(s)$ for $s \preceq s'$.*

Then $n \mapsto V_n(s') - V_n(s)$ is increasing on $\mathbb{N}$ for $s \preceq s'$, and $f_n(s)$ is increasing in n.

Proof

(a) We prove the first assertion which, including (v), means that

$$(B_n): \quad V_{n-1}(s') - V_{n-1}(s) \le V_n(s') - V_n(s) \text{ for } n \ge 1, s \preceq s'.$$

(a1) We prove $(B_n)_1^\infty$ by induction on $n \ge 1$. (B_1) holds by (v). Now assume that (B_n) holds for some $n \ge 1$. We want to verify (B_{n+1}), i.e.

$$V_n(s') + V_{n+1}(s) \le V_n(s) + V_{n+1}(s') \text{ for } s \preceq s'.$$

Define on $\tilde{D} := \mathbb{N} \times D$ the function $(k, s, a) \mapsto w(k, s, a) := W_k(s, a)$. We assert that for all $s \preceq s'$ and (s, a'), $(s', a) \in D$ we have

$$w(n, s', a') + w(n+1, s, a) \le w(n, s, a \wedge a') + w(n+1, s', a \vee a'). \qquad (8.8)$$

Considering the two cases $a \le a'$ and $a \ge a'$ and noting that D is NE-complete, we see that (s, a), (s', a'), $(s, a \wedge a')$ and $(s', a \vee a')$ belong to D, hence (8.8) is defined. For the proof of (8.8) we firstly note that

$$r(s',a') + r(s,a) \le r(s, a \wedge a') + r(s', a \vee a')$$

holds if $a > a'$ by (iii), and it holds trivially if $a \le a'$. It follows that (8.8) is true if

$$V_{n-1}(T_1(a')) + V_n(T_1(a)) \le V_{n-1}(T_1(a \wedge a')) + V_n(T_1(a \vee a')).$$

This is trivially true if $a > a'$, and it is true by (B_n) if $a \le a'$ since $T_1(a) \le T_1(a')$ by (ii). Thus (8.8) is true.

(a2) Now we obtain from (8.8) for $s \preceq s'$, $a' := f_n(s')$ and $a := f_{n+1}(s)$, observing that $a \vee a' \in D(s)$ and $a \wedge a' \in D(s')$ since $(s, a \wedge a')$ and $(s', a \vee a')$ belong to D,

$$\begin{aligned} V_n(s') + V_{n+1}(s) &= w(n+1, s, a) + w(n, s', a') \\ &\le w(n, s, a \wedge a') + w(n+1, s', a \vee a') \\ &\le \sup_{b \in D(s)} W_n(s,b) + \sup_{b \in D(s')} W_{n+1}(s', b) \\ &= V_n(s) + V_{n+1}(s'). \end{aligned}$$

Thus (B_{n+1}) is true.

(b) We now prove the second assertion. Fix $s \in S$. Now we apply Lemma 8.2.7 with S replaced by $\tilde{S} := \mathbb{N}$ and for the function $(n,a) \mapsto w(n,a) := W_n(s,a)$ on $\tilde{D} := \mathbb{N} \times D(s)$. Obviously $g(n) := f_n(s)$ is the smallest maximum point of $w(n, \cdot)$ on $D(s)$. $\tilde{D}$ is NE-complete by Example 8.2.4(a). Isotonicity of $f_n(s)$ in n follows from Lemma 8.2.7 if w has increasing differences, which means that

$$W_{n'}(s,a) - W_n(s,a) \le W_{n'}(s,a') - W_n(s,a') \text{ if } n \le n', a \le a',$$

and which is obviously true if

$$W_{n+1}(s,a) - W_n(s,a) \le W_{n+1}(s,a') - W_n(s,a') \text{ if } a \le a'.$$

This follows from $W_n(s,a) = r(s,a) + \beta V_{n-1}(T_1(a))$ and from (a), observing that T_1 is increasing. □

Remark 8.4.3

(a) Consider once more the preceding model. From Theorem 8.2.9(a1) we know that (even without assumption (v)) $f_n(s)$ is also increasing in s. Thus $f_n(s)$ is argumentwise isotone. By Example 6.2.4(b) this is equivalent to isotonicity in (n,s) on $\mathbb{N} \times S$, i.e. to joint isotonicity.

(b) We indicate the usefulness of isotonicity of $f_k(s)$ in k for the numerical computation of f_n and of V_n for $2 \leq n \leq N$ when S and A are finite: Assume that $A = \{x, x+1, \ldots, z\}$ for some integers $x < z$. For $f \in \mathbb{F}$ and $s \in S$ put $A(f(s)) := \{f(s), f(s)+1, \ldots, z\}$. Then use the following algorithm:

Step 1: Compute f_1 and V_1 by Theorem 2.3.3.
Step 2: If f_n and V_n (and hence also W_{n+1}) is computed for some fixed n, $1 \leq n < N$, then $f_{n+1}(s)$ is the smallest maximum point of $W_{n+1}(s, \cdot)$ on $D(s) \cap A(f_n(s))$, and $V_{n+1}(s) = W_{n+1}(s, f_{n+1}(s))$, $s \in S$.

(c) The algorithm in (c) can be combined, if $f_k(s)$ is increasing in s or if even $f_k \in ILIP(\lambda)$, $2 \leq k \leq N$, with the corresponding algorithm described above. $\diamond$

8.5 Problems

Unless stated otherwise, S is structured and A is totally ordered.

Problem 8.5.1 Prove the following two properties:

(a) Assume that for some function $d: S \to A$

$$D(s) = (\cdot, d(s)], \ s \in S.$$

Then $D(\cdot)$ is NE-complete if and only if d is increasing. The same holds if

$$D(s) = [d(s), \cdot), \ s \in S.$$

(b) If $D_i \subseteq S \times A$ is monotonely complete for $i \in I$ and if for all $s \in S$

$$D(s) := \bigcap_{i \in I} D_i(s)$$

is nonempty, then D is monotonely complete.

(c) The graph of a mapping d from S into A is monotonely complete if and only if d is increasing.

Problem 8.5.2 (Sufficient conditions for monotone completeness) Let $\preceq_S$ be reflexive and consider the following conditions:

(M1) The constraint set D is SE-increasing.
(M2) The sets $D(s)$, $s \in S$ are decreasing and $D(\cdot)$ is expanding.
(M3) D is monotonely complete and $D(\cdot)$ is expanding.

Show that (M1) $\Leftrightarrow$ (M2) $\Rightarrow$ (M3), and that (M3) $\Rightarrow$ (M2) if (S is totally ordered) $\wedge$ ($A = A_0$).

Problem 8.5.3 Assume $A = \mathbb{R}_+$ and $D(\cdot) = [c(\cdot), d(\cdot)]$. Then D satisfies condition (M2) of Problem 8.5.2 if and only if $(c \equiv 0) \wedge (d$ is decreasing).

Problem 8.5.4 Assume that $A \subseteq \mathbb{R}$. Then:

(a) If D is monotonely complete then $s \mapsto \inf D(s)$ and $s \mapsto \sup D(s)$ are increasing.
(b) If A is an interval in $\mathbb{R}$ or in $\mathbb{Z}$, then D is monotonely complete if and only if $s \mapsto \inf D(s)$ and $s \mapsto \sup D(s)$ are increasing.

Problem 8.5.5 Let w_1 and w_2 be finite functions on $A \subseteq \mathbb{R}$. If the set of maximum points of w_i has an infimum α_i [supremum α_i], then $\alpha_1 \leq \alpha_2$ provided that $d \cdot w_2 - w_1$ is increasing for some constant $d \in \mathbb{R}^+$. Give an example with $A = \mathbb{R}$ where $d \cdot w_2 - w_1$ is increasing for some $d \in \mathbb{R}^+$, but not for $d = 1$.

Problem 8.5.6 Assume that S and A_0 are intervals in $\mathbb{N}_0$. If $s, s+1 \in S$ and $(s, a+1)$, $(s+1, a) \in D$ implies (s, a), $(s+1, a+1) \in D$, then D need not be NE-complete.

Problem 8.5.7 Assume that $S = A = \mathbb{Z}$ and $D = \mathbb{Z}^2$. If $u \in ID$ and if $u(a, \cdot)$ is concave for all $a \in \mathbb{Z}$, then ID contains

$$(s, a) \mapsto u(a, s - a).$$

Problem 8.5.8 (Increasing maximizers when the value functions are convex) Assume

(i) S is an interval in $\mathbb{R}$,
(ii) $A \subset \mathbb{R}$ and $D(s) = A$ for all s,
(iii) $r(\cdot, a)$, $a \in A$, and V_0 are convex,
(iv) r has increasing differences,
(v) V_n is finite for all n,
(via) $T(s, a) = \alpha_1 s + \alpha_2 a$ for constants $\alpha_1 \cdot \alpha_2 \geq 0$ such that $T(s, a) \in S$ for $(s, a) \in D$, or
(vib) S is right unbounded, $T(\cdot, a)$ is convex, T, $r(\cdot, a)$, and V_0 are increasing for $a \in A$, and T has increasing differences.

Then V_n is convex and the smallest [largest] maximizer at stage n, if it exists, is increasing.

Problem 8.5.9 (Increasing maximizers when the value functions are concave) Assume that

(i) S and A are intervals in $\mathbb{R}$,
(ii) $D(s) = A$ for all s or $D(s) = [0, d(s)]$ for some increasing and concave function $d \geq 0$,
(iii) r and V_0 are concave,
(iv) r has increasing differences,
(v) V_n is finite for all n,

(via) $T(s,a) = \alpha_1 s - \alpha_2 a$ for constants $\alpha_1 \cdot \alpha_2 \geq 0$ such that $T(s,a) \in S$ for $(s,a) \in D$, or

(vib) S is right unbounded, T is SE-increasing, concave, and has increasing differences, and $r(\cdot, a)$, $a \in A_0$, and V_0 are increasing.

Then V_n is concave and the largest [smallest] maximizer at each stage n, if it exists, is increasing.

Problem 8.5.10 Consider the discrete version of Example 6.4.3. Thus $S = \mathbb{N}_0$. Show that all minimal maximizers (at least one exists, as $D(s)$ is finite for all s) are increasing, if $t_3 \equiv 0$ and if u_3 is concave.

Problem 8.5.11 Assume that $D(s) = [c(s), d(s)]$ for all s, so that $[c(s), d(s)] := \{c(s), c(s)+1, \ldots, d(s)\}$ in case $A = \mathbb{Z}$. Then D is NE-complete if and only if both c and d are increasing.

Problem 8.5.12 (Closure properties of the set *ID***)** Prove the following properties:

(a) *ID* is a convex cone.

(b) If the function w on D does not depend on s or on a, then $w \in ID$.

(c) Let I be an interval in $\mathbb{R}$ or in $\mathbb{Z}$. Let g and h be functions on S and A, respectively, such that $g(s)-h(a) \in I$ for all $(s,a) \in D$. If v is concave [convex] on I and if g and h are iso-monotone [anti-monotone], then *ID* contains the function

$$(s,a) \mapsto v(g(s) - h(a)).$$

In particular, if v is concave, $\alpha_1 \cdot \alpha_2 \geq 0$, $\alpha_3 \in \mathbb{R}$ and if $\alpha_1 s - \alpha_2 a + \alpha_3 \in I$ for all $(s,a) \in D$, then *ID* contains the function

$$(s,a) \mapsto v(\alpha_1 s - \alpha_2 a + \alpha_3).$$

(d) The function $(s,a) \mapsto g(s) \cdot h(a)$ belongs to *ID*, if g and h are iso-monotone.

(e) Assume that $v \geq 0$ and $w \geq 0$ belong to *ID*. Then $v \cdot w \in ID$ under each of the following conditions:

(e1) $w(s,a) \geq 0$ is independent of s or of a.

(e2) $v(s,\cdot)$, $w(\cdot,a)$ are iso-monotone, and $v(\cdot,a)$, $w(s,\cdot)$ are iso-monotone.

(f) Let v be an increasing function on a right unbounded interval I in $\mathbb{R}$ or in $\mathbb{Z}$ and let $w: D \to I$ have increasing differences. If (v is concave) $\wedge$ ($w(s,a)$ is SE-monotone) or if (v is convex) $\wedge$ ($w(s,a)$ is NE-monotone), then $v \circ w \in ID$.

(g) Part (f) remains true if increasing differences, concave and convex are replaced by decreasing differences, convex and concave, respectively.

Chapter 9
Existence of Optimal Action Sequences

In many special models the existence of maximizers at all stages (and hence by the OC in Theorem 2.3.3(c) also of s-optimal action sequences for all initial states s) can be established by ad hoc methods. The existence of a maximizer at each stage is also obvious if the set $D(s)$ of admissible actions is finite for all s. In Proposition 7.1.10 we gave a result which covers many applications where S and A are intervals. The existence problem for maximizers under more general conditions is most easily dealt with under assumptions which are so strong that continuity (or at least upper semicontinuity) of W_n and also of V_n is implied.

9.1 Upper and Lower Semicontinuous Functions

A useful weakening of continuity of a function is upper (or lower) semicontinuity. A function v on a metric space (M, ρ) is called **upper semicontinuous**, *usc* for short, if for all $x \in M$

$$\limsup_{n\to\infty} v(x_n) \leq v(x) \tag{9.1}$$

for each sequence (x_n) in M that converges to x. The function v is called **lower semicontinuous**, *lsc* for short, if $-v$ is usc, i.e. if for all $x \in M$

$$v(x) \leq \liminf_{n\to\infty} v(x_n)$$

for each sequence (x_n) in M that converges to x. Upper and lower semicontinuity of a function v on M depends on the metric in M, but topologically equivalent metrics in M lead to the same notion since the property *limit of a sequence* is a topological property. In proofs we often tacitly use that v is usc if (and only if) (9.1) holds for

K. Hinderer et al., *Dynamic Optimization*, Universitext,
DOI 10.1007/978-3-319-48814-1_9

all $x \in M$ and all $(x_n) \subset M$ for which $x_n \to x$ and for which $(v(x_n))$ converges in $\overline{\mathbb{R}}$. For the proof note that $\limsup_{n\to\infty} v(x_n)$ is the largest cluster value of $(v(x_n))$.

Note that part (e) of the next proposition generalizes the classical result Appendix A.4.14 for $k = 1$ about the existence of maximum points of functions.

Proposition 9.1.1 (Properties of upper and lower semicontinuous functions) *Let v be a function on a metric space (M, ρ), and for $\alpha \in \overline{\mathbb{R}}$ put $[v \geq \alpha] := \{x \in M : v(x) \geq \alpha\}$. Then:*

(a) v is continuous if and only if it is both usc and lsc.
(b) v is usc if and only if the sets $[v \geq \alpha]$, $\alpha \in \mathbb{R}$, are closed.
(c) The indicator function of a set $B \subset M$ is usc [lsc] if and only if B is closed [open].
(d) If v is convex on an interval, then v is usc; moreover, v is continuous if and only if v is lsc.
(e) Let v be usc [lsc]. Then the set M^* of maximum points [minimum points] of v is closed. If in addition M is compact, then M^* is non-empty and compact.
(f) If v is usc and if $B_\alpha := \{x \in M : v(x) \geq \alpha\}$ is compact for some $\alpha \in (-\infty, \sup v)$, then the set M^* of maximum points of v is non-empty and compact.
(g) If v is lsc and if $B_\alpha := \{x \in M : v(x) \leq \alpha\}$ is compact for some $\alpha \in (\inf v, \infty)$, then the set M^* of minimum points of v is non-empty and compact.

Proof

(a) The proof is trivial.
(b1) Let v be usc. Select $\alpha \in \mathbb{R}$ and a sequence of points x_n in $[v \geq \alpha]$, converging to some $x \in M$. Then $v(x) \geq \limsup_n v(x_n) \geq \alpha$, i.e. $x \in [v \geq \alpha]$. Thus $[v \geq \alpha]$ is closed.
(b2) Assume that $[v \geq \alpha]$ is closed for all real α and that v is not usc. Then there exists some x in M and a sequence (x_n) in M, converging to x, such that $\limsup_n v(x_n) > v(x)$. Select $\alpha \in \mathbb{R}$ such that $\limsup_n v(x_n) > \alpha > v(x)$. Then $x_n \in [v \geq \alpha]$ for $n > n_0$. Since $[v \geq \alpha]$ is closed, we get $x \in [v \geq \alpha]$, which contradicts $v(x) < \alpha$.
(c) follows from (b) as $[1_B \geq \alpha] \in \{\emptyset, B, M\}$ for each real α.
(d) follows from Appendices D.2.8(f) and D.2.2.
(e) We give a proof when v is upper semicontinuous. M^* is closed by (b), since $M^* = [v \geq \sup v]$. Then M^* as a closed subset of the compact set M is compact. In order to show that M^* is non-empty, select $x_n \in M$ such that $v(x_n) \to \sup v$. Since M is compact, a subsequence (x_{n_k}) of (x_n) converges to some $x \in M$. Then $x \in M^*$, since upper semicontinuity of v implies $\sup v = \limsup_k v_{n_k}(x) \leq v(x)$.

(f) and (g) follow from (e) since v on M and on $[v \geq \alpha]$ have the same set M^*. □

Remark 9.1.2

(a) From Proposition 9.1.1 and from Lemma 9.1.4 below one easily obtains properties of lsc functions. As an example, Proposition 9.1.1(b) implies that v is lsc if and only if for each $\alpha \in \mathbb{R}$ the sets $[v \leq \alpha]$ are closed.
(b) We tacitly use the following simple facts: (b1) The restriction of an usc function on M to a subset of M is usc. (b2) Let (M, ρ) and (M', ρ') be metric spaces, and $\emptyset \neq B \subset M \times M'$. If $h: B \to \overline{\mathbb{R}}$ is usc with respect to a product metric, then each of the x-sections $h(x, \cdot): B_x \to \mathbb{R}$ is usc with with respect to ρ'. (b3) If $w: M' \to \overline{\mathbb{R}}$ is usc with respect to ρ', then $(x, x') \mapsto w(x')$ is usc on $M \times M'$ with respect to each product metric. (b4) If v is usc then $[v = \infty]$ is closed since $[v = \infty] = \cap_n [v \geq n]$.
(c) Be cautious in applying Proposition 9.1.1(b): If, for example, v is usc on $(0, 1)$ then $B_\alpha = \{x \in (0, 1) : v(x) \geq \alpha\}$ is closed in $(0, 1)$, but not necessarily in $\mathbb{R}$; this is seen, for instance, from $v :\equiv 1$, hence $B_1 = (0, 1)$. ◇

Remark 9.1.3 If $\|\cdot\|$ is any norm on $\mathbb{R}^k$, the distance (with respect to $\|\cdot\|$) of a point $b \in \mathbb{R}^k$ from a nonempty set $M \subset \mathbb{R}^k$ is defined by $d(b, M) := \inf\{\|b - x\| : x \in M\}$. We apply Proposition 9.1.1(g) in order to show that the infimum is attained if M is closed. The assertion holds trivially if the continuous function $x \mapsto v(x) := \|b - x\|$ is constant on M. Otherwise there exists some $x_0 \in M$ with $\alpha := v(x_0) \in (\inf v, \infty)$. Firstly, continuity of v implies by Proposition 9.1.1(b) that $B_\alpha := [v \leq \alpha]$ is closed. In addition, B_α is bounded. In fact, if (x_n) is a sequence in M such that $\|x_n\| \to \infty$ for $n \to \infty$, then $v(x_n) \to \infty$, hence $x_n \notin B_\alpha$ for some n. ◇

Lemma 9.1.4 (Closure properties of the set of upper semicontinuous functions on a metric space)

(a) If v_1 and v_2 are usc functions on M, then αv_1 is usc for $\alpha \in \mathbb{R}_+$, and $v_1 + v_2$, if defined, is usc. Thus the set of usc functions on M is a convex cone.
(b) A usc function of a continuous mapping is usc.
(c) If $(v_i, i \in I)$ is a family of usc functions on M, then $\inf\{v_i, i \in I\}$ is usc. In particular, the limit of a decreasing sequence of continuous (or only usc) functions is usc.
(d) The maximum of finitely many usc functions is usc.
(e) If v_1 and v_2 are non-negative and lsc [usc and finite], then $v_1 \cdot v_2$ is lsc [usc].
(f) The uniform limit of a sequence of usc functions is usc.
(g) (Theorem of Baire) A function is usc and upper bounded if and only if it is the decreasing limit of a sequence of bounded continuous functions.

Proof

(a) This follows from the fact, known from calculus, that

$$\limsup_{n\to\infty}(\alpha_n + \beta_n) \leq \limsup_{n\to\infty} \alpha_n + \limsup_{n\to\infty} \beta_n$$

for all sequences of numbers α_n and β_n in $(-\infty, \infty]$.

The simple proof of (b) is left to the reader.

(c) and (d) follow from Proposition 9.1.1(b) since for any $\alpha \in \mathbb{R}$ we have

$$[\inf v_i \geq \alpha] = \bigcap_i [v_i \geq \alpha], \quad [\max_i v_i \geq \alpha] = \bigcup_i [v_i \geq \alpha],$$

and since the intersection of any number of closed sets and the union of finitely many closed sets is closed.

(e) This follows from the following facts: If $\alpha_n, \beta_n \in [0, \infty]$ for $n \geq 1$, then

$$\liminf_{n\to\infty}(\alpha_n \beta_n) \geq (\liminf_{n\to\infty} \alpha_n) \cdot (\liminf_{n\to\infty} \beta_n).$$

If in addition the sequences (α_n) and (β_n) are bounded, then

$$\limsup_{n\to\infty}(\alpha_n \beta_n) \leq (\limsup_{n\to\infty} \alpha_n) \cdot (\limsup_{n\to\infty} \beta_n).$$

(f) Assume that v_k is usc on M for $k \in \mathbb{N}$, and that (v_k) converges uniformly to the function v. Then, for $\varepsilon > 0$, we have $\sup_{x \in M} |v_k(x) - v(x)| \leq \varepsilon$ for $k > k_0 = k_0(\varepsilon)$. Select some sequence (x_n) in M converging to some $x \in M$. Then, for $k > k_0$ and $n \geq 1$ we have $v(x_n) \leq v_k(x_n) + \varepsilon$ and $v_k(x) \leq v(x) + \varepsilon$. Since v_k is usc, we get

$$\limsup_{n\to\infty} v(x_n) \leq \limsup_{n\to\infty} v_k(x_n) + \varepsilon \leq v_k(x) + \varepsilon \leq v(x) + 2\varepsilon.$$

Since $\varepsilon > 0$ is arbitrary, the assertion follows.

(g) We refer to Bertsekas and Shreve (1978, p. 147). □

9.2 Existence of Maximizers

For the rest of this section we assume that the state space and action space are metric spaces (S, ρ_S) and (A, ρ_A), and D is endowed with a product metric ρ_D, e.g. the l_1-metric $\rho_S + \rho_A$.

It follows that each sequence $((s_n, a_n))_1^\infty \in D$ converges with respect to ρ_D to (s, a) if and only if (s_n converges with respect to ρ_S to s) $\wedge$ (a_n converges with respect to ρ_A to a). Note that the metric space (D, ρ_D) is always closed, but D need not be closed as a subset of $S \times A$, when the latter is endowed with a product metric.

Assume that $D(s)$, $s \in S$, is compact. Then the existence of a maximizer at stage n is ensured by Proposition 9.1.1(e) if $W_n(s, \cdot)$ is usc for $s \in S$. Thus, according to our scheme in Chap. 6 for proving structural results, we shall look for conditions which ensure that upper semicontinuity of $W_n(s, \cdot)$, $s \in S$, for some $n \geq 1$ implies upper semicontinuity of V_n and that this in turn implies upper semicontinuity of $W_{n+1}(s, \cdot)$, $s \in S$. As we show in Theorem 9.2.5 below, this method is successful, provided we replace compactness of $D(s)$, $s \in S$, by the stronger property of *quasi-continuity* of $D(\cdot)$, as defined below.

Besides $D(\cdot)$ we also encounter the correspondence $s \mapsto D_n^*(s) :=$ set of maximum points of $a \mapsto W_n(s,a)$, provided $D_n^*(s)$ is non-empty for all s. Therefore we now consider an arbitrary ***correspondence*** $\Gamma(\cdot)$ from S into A. $\Gamma(\cdot)$ is said to be ***closed-valued*** or ***compact-valued*** or ***convex-valued***, if $D(s)$ is closed or compact or convex, respectively, for all $s \in S$.

Definition 9.2.1 Let $\Gamma(\cdot)$ be a correspondence from S into A.

- $\Gamma(\cdot)$ is called ***quasi-continuous*** if it has the following property: If $s_n \to s$ and $a_n \in \Gamma(s_n)$ for $n \geq 1$, then $(a_n)_1^\infty$ has a cluster value in $\Gamma(s)$.
- $\Gamma(\cdot)$ is called ***lower semicontinuous*** (lsc for short) if it has the following property: If $s_n \to s$, then each point in $\Gamma(s)$ is a cluster value of a sequence of points $a_n \in \Gamma(s_n)$ for $n \geq 1$.
- $\Gamma(\cdot)$ is called ***continuous*** if it is both quasi-continuous and lower semicontinuous.

Remark 9.2.2

(a) If $\Gamma(\cdot)$ is quasi-continuous, then $\Gamma(\cdot)$ is compact-valued. This is seen by taking $s_n := s$ for all n.
(b) Quasi-continuity of $D(\cdot)$ lies between the two properties of D of being compact or closed. In fact, $D(\cdot)$ is quasi-continuous if and only if each sequence $((s_n, a_n))$ in D such that $(s_n)_1^\infty$ converges has a cluster value in D.
(c) Quasi-continuity of $\Gamma(\cdot)$ depends on ρ_S and ρ_A, but not on ρ_D. ◇

For specific applications one rarely needs to work with Definition 9.2.1, as several easily checkable sufficient criteria such as the next one are available. Part (a) of Lemma 9.2.3 also provides examples of quasi-continuous correspondences which are not continuous.

Lemma 9.2.3 (Quasi-continuous correspondences from S into A)

(a) Assume that A is an interval and that $\Gamma(\cdot)$ has the interval form $[c(\cdot), d(\cdot)]$. Then $\Gamma(\cdot)$ is quasi-continuous [continuous] if and only if c is lsc [continuous] and d is usc [continuous].
(b) Endow $S \times A$ with a product metric and let A be compact. Then $D(\cdot)$ is quasi-continuous if and only if D is closed in $S \times A$.
(c) If $\Gamma(s) = A$ for all s then $\Gamma(\cdot)$ is continuous if and only if A is compact.
(d) If S is an interval and if $\{(s,a) \in D : s \in I\}$ is compact for each compact subinterval I of S, then $D(\cdot)$ is quasi-continuous.
(e) A mapping g from S into A is continuous if and only if the one-point correspondence $s \mapsto \Gamma(s) := \{g(s)\}$ is quasi-continuous if and only if $\Gamma(\cdot)$ is continuous.
(f) If B is a closed subset of (D, ρ_D) with $\mathrm{pr}(B \to S) = S$ and if $D(\cdot)$ is quasi-continuous, then $B(\cdot)$ is quasi-continuous.

Proof The proofs of parts (b)–(f) and of the necessity part of (a) are left to the reader. Proof of the sufficiency part of (a):

Case (i): c is lsc and d is usc. Select $(s_n, a_n) \in D$ for $n \geq 1$ such that $s_n \to s$. For each $\varepsilon > 0$ there exists an $n_0 = n_0(\varepsilon)$ such that for all $n \geq n_0$

$$c(s) - \varepsilon \leq \liminf_k c(s_k) - \varepsilon \leq c(s_n) \leq a_n \leq d(s_n)$$
$$\leq \limsup_k d(s_k) + \varepsilon \leq d(s) + \varepsilon.$$

Thus a_n is contained for $n \geq n_0$ in the compact interval $I(\varepsilon) := [c(s)-\varepsilon, d(s)+\varepsilon]$. Therefore $(a_n)_1^\infty$ has a cluster value in $I(\varepsilon)$ for each $\varepsilon > 0$, hence in $[c(s), d(s)] = \Gamma(s)$. Thus $\Gamma(\cdot)$ is quasi-continuous.

Case (ii): c and d are continuous. Then $\Gamma(\cdot)$ is quasi-continuous by case (i). Select $s_n \to s$ and $a_0 \in \Gamma(s)$ and define $a_n \in \Gamma(s_n)$, $n \geq 1$ by

$$a_n := \begin{cases} c(s_n), & \text{if } a_0 < c(s_n), \\ a_0, & \text{if } c(s_n) \leq a_0 \leq d(s_n), \\ d(s_n), & \text{if } d(s_n) < a_0. \end{cases}$$

Since $c(s_n) \to c(s)$ and $d(s_n) \to d(s)$ by continuity of c and d, there exists for each $\varepsilon > 0$ some $n_0 = n_0(\varepsilon)$ such that $a_0 \in [c(s_n) - \varepsilon, d(s_n) + \varepsilon]$ for $n > n_0$. It follows from the construction of a_n that $|a_0 - a_n| \leq \varepsilon$ for $n \geq n_0$, hence $(a_n)_1^\infty$ converges towards a_0. Thus $\Gamma(\cdot)$ is also lsc. □

Lemma 9.2.4 (Existence of maximizers of functions on *D*) *Let w be a function on D, and put $w^*(s) := \sup_{a \in D(s)} w(s, a)$, $s \in S$.*

(a) Let $D(\cdot)$ be quasi-continuous and let w be usc.
(a1) w has a maximizer. If $A \subset \mathbb{R}$, then w has a smallest and a largest maximizer.
(a2) w^* is usc. If w is finite then w^* is finite.
(b) Let $D(\cdot)$ and w be continuous, hence w has a maximizer by (a1).
(b1) w^* is continuous.
(b2) If w has a unique maximizer f, then f is continuous. If $A \subset \mathbb{R}$, then the smallest [largest] maximizer of w is lsc [usc].

Proof

(a1) Fix $s \in S$. As $w(s, \cdot)$ is usc by Remark 9.1.2(b1) and as $D(s)$ is compact by Remark 9.1.2(b), $w(s, \cdot)$ assumes its supremum by Proposition 9.1.1(e). The assertion in case $A \subset \mathbb{R}$ follows from the compactness of the set of maximum points of $w(s, \cdot)$, see Proposition 9.1.1(e).

(a2) For $s_0 \in S$ select a sequence $(s_n)_1^\infty$ in S, converging to s_0, and such that $\lim w^*(s_n)$ exists in $\overline{\mathbb{R}}$. We have to show that $\lim w^*(s_n) \leq w^*(s_0)$. By (a1) there exists a maximum point a_n of $w(s_n, \cdot)$. By quasi-continuity of $D(\cdot)$ there is a subsequence (a_{n_k}) of (a_n) converging to some $a_0 \in D(s_0)$. Now the assertion follows since upper semicontinuity of w implies

$$\lim_n w^*(s_n) = \lim_k w^*(s_{n_k}) = \lim_k w(s_{n_k}, a_{n_k}) \le w(s_0, a_0) \le w^*(s_0).$$

Moreover, as $w(s, \cdot)$ assumes its supremum, $w^*(s)$ is finite if w is finite.

(b1) In view of (a2) and Proposition 9.1.1(a) we have to show that w^* is lsc, i.e. that $w^*(s_0) \le \lim w^*(s_n)$ for each sequence $(s_n)_1^\infty$ in S which converges to some $s_0 \in S$ and for which $\lim w^*(s_n)$ exists in $\overline{\mathbb{R}}$. We know from (a1) that $w^*(s_0) = w(s_0, a_0)$ for some $a_0 \in D(s_0)$. Since $D(\cdot)$ is lsc, there exists a subsequence (s_{n_k}) of (s_n) and a sequence of points $a_{n_k} \in D(s_{n_k})$, converging to a_0; hence $(s_{n_k}, a_{n_k}) \to (s_0, a_0)$. It follows by lower semicontinuity of w that

$$\begin{aligned} w^*(s_0) = w(s_0, a_0) &\le \lim_k w(s_{n_k}, a_{n_k}) \\ &\le \lim_k w^*(s_{n_k}) = \lim_n w^*(s_n). \end{aligned}$$

(b2) From (b1) we know that w^* is continuous. As $D(\cdot)$ is quasi-continuous, D is closed by Remark 9.2.2(b). It follows easily that

$$D^* := \{(s, a) \in D : w(s, a) = w^*(s)\}$$

is a closed subset of D. Then $D^*(\cdot)$ is quasi-continuous by Lemma 9.2.3(f) with $B := D^*$. Thus, if $D^*(s) = \{f(s)\}$ for all s, f must be continuous by Lemma 9.2.3(e). Finally assume that $A \subset \mathbb{R}$. As $D^*(\cdot)$ is quasi-continuous, $D^*(s)$ has a largest element $f(s)$, $s \in S$, hence f is the largest maximizer of w. Select s_0, $s_n \in S$ such that $s_n \to s_0$ and $\lim f(s_n)$ exists. As $a_n := f(s_n) \in D^*(s_n)$ and $D^*(\cdot)$ is quasi-continuous, (a_n) has a cluster value in $D^*(s_0)$, which must be $\lim f(s_n)$. It follows that $\lim f(s_n) \le \sup D^*(s_0) = f(s_0)$ which proves that f is usc. In the same way one shows that the smallest maximizer of w is lsc. □

Now we obtain the main result of this section.

Theorem 9.2.5 (Existence of maximizers and upper semicontinuous value functions) *Assume that:*

(i) $D(\cdot)$ is quasi-continuous,
(ii) T is continuous,
(iii) r and V_0 are upper semicontinuous.

Then for $n \ge 1$:

(a) V_n and W_n are finite and usc, and there exists a maximizer f_n at stage n. If in addition $A \subset \mathbb{R}$, then there exists a largest maximizer $\overline{f}_n$ and a smallest maximizer $\underline{f}_n$ at stage n.
(b) If $D(\cdot)$, r and V_0 are continuous, then:
(b1) V_n and W_n are continuous.

(b2) If f_n is unique, then f_n is continuous.
(b3) If $A \subset \mathbb{R}$, then $\overline{f}_n$ is usc and $\underline{f}_n$ is lsc.

Proof

(a) We prove by induction on $n \geq 1$ the assertion
(I_n): V_n and W_n are finite and usc, and there exists a maximizer f_n at stage n.
(I_1): Obviously $W_1 = r + \beta \cdot V_0 \circ T$ is finite, and it is usc by (ii), (iii) and Lemma 9.1.4(b). It follows from Lemma 9.2.4(a1) and (a2) that W_1 has a maximizer f_1 and that V_1 is finite and usc, respectively. Assume (I_n) for some $n \geq 1$. Then (I_{n+1}) follows by a reasoning as in the proof of (I_1) with V_0 replaced by V_n. If in addition $A \subset \mathbb{R}$, then the existence of $\overline{f}_n$ and of $\underline{f}_n$ follows from Lemma 9.2.4(a1).
(b) The proof is very similar to the proof of (a). Instead of Lemma 9.2.4(a1) one uses Lemma 9.2.4(b1); the details are left to the reader. □

Remark 9.2.6

(a) Uniqueness of f_n is guaranteed if $D(\cdot)$ is convex-valued and if $W_n(s, \cdot)$ is strictly concave for $s \in S$; see Appendix D.3.1. Sufficient conditions for the latter property are given in Theorem 7.1.2.
(b) Proposition 7.1.10 is contained in Theorem 9.2.5, as can be seen from Lemma 9.2.3(d).
(c) If $D(\cdot)$ is not compact-valued then $D(\cdot)$ cannot be quasi-continuous, so that Theorem 9.2.5 cannot be applied. Sometimes Corollary 9.3.11 below can help, see Example 9.3.12. ◇

Since V_n is usc [continuous] if and only if $C_n := -V_n$ is lsc [continuous], and since a decision rule f is a maximizer of $r+\beta\cdot V_{n-1}\circ T$ if and only if f is a minimizer of $-r + \beta \cdot C_{n-1} \circ T$, Theorem 9.2.5 immediately yields a minimization version, as follows.

Corollary 9.2.7 (Existence of minimizers) *Assume that:*

(i) $D(\cdot)$ is quasi-continuous,
(ii) T is continuous,
(iii) c and C_0 are lower semicontinuous.

Then for $n \geq 1$:

(a) C_n and W_n are finite and lsc, and there exists a minimizer f_n at stage n. If in addition $A \subset \mathbb{R}$, then there exists a largest minimizer $\overline{f}_n$ and a smallest minimizer $\underline{f}_n$ at stage n.
(b) If $D(\cdot)$, r and C_0 are continuous, then:
(b1) C_n and W_n are continuous.
(b2) If f_n is unique, then f_n is continuous.
(b3) If $A \subset \mathbb{R}$, then $\overline{f}_n$ is usc and $\underline{f}_n$ is lsc.

Example 9.2.8 (Splitting of a resource into $m+1$ parts) At each time t the available resource s is split into $m+1$ parts $a_1, a_2, \ldots, a_m$ and $s - \sum_{i=1}^m a_i$. Thus $A = \mathbb{R}_+^m$ and $D(s)$ is the simplex $\{(a_i)_1^m \in A : \sum_{i=1}^m a_i \leq s\}$ for all s. From Example 9.3.6(iv) and Lemma 9.3.7 below we infer that $D(\cdot)$ is continuous. Assume that $T(s,a)$ and $r(s,a)$ depend only on the parts into which s is split, i.e. that for $a = (a_i)_1^m \in D(s)$

$$T(s,a) = \eta(a, s - \sum_{i=1}^m a_i) \text{ and } r(s,a) = g(a, s - \sum_{i=1}^m a_i)$$

for a mapping η from $M := \{(a,b) \in A \times \mathbb{R}_+ : a \in D(s), b \leq s\}$ into S and a function g on M. Now we obtain from Theorem 9.2.5 the following result: If η is continuous and g and V_0 are usc [continuous], then there exists at each stage a maximizer and V_n is usc [continuous] for all n. ♦

9.3 Lipschitz Continuity

In many examples where S is an interval, the value functions not only possess the qualitative property of continuity but even the quantitative property of Lipschitz continuity with computable bounds for the Lipschitz module. Lipschitz continuity means boundedness of the difference quotients, a generalization of the property of boundedness of the derivative. This generalization is important since value functions are typically not everywhere differentiable. Moreover, Lipschitz continuity makes sense and is useful, in contrast to continuity, also for models with discrete state spaces like $\mathbb{N}$.

We derive in the sequel sufficient conditions for Lipschitz continuity of the value functions V_N. This property ensures by definition that V_N cannot oscillate wildly.

In our study Lipschitz continuity of the mapping $s \mapsto D(s)$ plays an important role. This requires us to deal with Lipschitz continuity of mappings from a metric space into another metric space. On subsets of $\mathbb{R}^d$ we use, unless stated otherwise, the maximum-metric. The next definition covers for the choice $M' = \mathbb{R}$ and $\rho' :=$ Euclidean metric the usual Lipschitz functions.

Definition 9.3.1 Let (M, ρ) and (M', ρ') be metric spaces. On the set of mappings $v: M \to M'$ define the so-called ***Lipschitz module*** $l(\cdot) = l_{\rho,\rho'}(\cdot)$ by

$$l(v) := \sup_{x \neq y} \frac{\rho'(v(x), v(y))}{\rho(x,y)} \in [0, \infty]. \tag{9.2}$$

The mapping $v: M \to M'$ is called ***Lipschitz continuous*** or shortly ***Lipschitz*** (with respect to ρ and ρ') if it has a finite Lipschitz module.

Obviously v is Lipschitz if and only if there exists some $K \in \mathbb{R}_+$, called a ***Lipschitz constant*** of v, such that

$$\rho'(v(x), v(y)) \leq K \cdot \rho(x, y) \text{ for all } x, y \in M. \tag{9.3}$$

Then $l(v)$ is the smallest of the Lipschitz constants K.

If v is a function, the term $\rho'(v(x), v(y))$ in (9.2) and (9.3) equals $|v(x) - v(y)|$. If v on M is Lipschitz, it is also Lipschitz on each non-empty subset B of M and $l(v|B) \leq l(v)$.

Note that $\rho'(v(x), v(y))/\rho(x, y)$ is the rate of change of the mapping v between x and y, relative to ρ. It generalizes the difference quotient of a function on an interval in $\mathbb{R}$. Thus $l(v)$ is the maximal rate of change over M. For a function v the Lipschitz module yields the upper bound $l(v) \cdot \rho(x, y)$ for the difference $|v(x - v(y)|$. Note that the bound $l(v) \cdot \rho(x, y)$ depends in a complicated way on the choice of ρ since a smaller ρ yields a larger $l(v)$. This dependence is one of the reasons why we admit in Theorem 9.3.9 several metrics on D.

For simplicity of formulations and of proofs we prefer to study Lipschitz continuity of mappings v via properties of the Lipschitz module $l(v)$, whether or not $l(v)$ is finite. As an example, Lemma 9.3.2(b) implies that the composition $v \circ g$ is Lipschitz if both v and g are Lipschitz.

Lemma 9.3.2 (Properties of the Lipschitz module and of Lipschitz continuous mappings) *The following properties hold:*

(a) *Lipschitz continuity implies (uniform) continuity.*

(b) *Let (M_i, ρ_i) be metric spaces for $i = 1, 2, 3$ and let $g: M_1 \to M_2$ and $v: M_2 \to M_3$. Then $l(v \circ g) \leq l(v) \cdot l(g)$.*

(c) *On the vector space $\mathbb{R}^M$ of functions on a metric space M the L-module $l(\cdot)$ has the following properties: $l(v) \geq 0$, $l(\alpha v) = |\alpha| \cdot l(v)$ and $l(v_1 + v_2) \leq l(v_1) + l(v_2)$ for $v, v_1, v_2 \in \mathbb{R}^M$ and $\alpha \in \mathbb{R}$. In addition, $l(v) = 0$ if and only if v is a constant.*

(d) *Let v be a continuous function v on an interval I with endpoints a, b, $-\infty \leq a < b \leq \infty$. If v has a right derivative v'_+ on (a, b), then $l(v) = \sup |v'_+|$. If in addition v is convex or concave then $l(v) = \max\{|v'_+(a+)|, |v'_+(b-)|\}$.*

(e) *For finitely many functions v_i on the same metric space we have $l(\max_i v_i) \leq \max_i l(v_i)$ and $l(\min_i v_i) \leq \max_i l(v_i)$.*

(f) *If v is the finite limit of a sequence of functions v_n, $n \geq 1$, on the same metric space, then $l(v) \leq \liminf_n l(v_n)$.*

(g) *For a function v on an interval I in $\mathbb{Z}$ we have*

$$l(v) = \sup\{|v(i+1) - v(i)| : i \in I, i < \sup I\}.$$

(h) *Let (M_1, ρ_1) and (M_2, ρ_2) be metric spaces and endow $M_1 \times M_2$ with the taxicab metric (cf. Appendix C.1). For each function v on M_1 we have $l(v) = l(w)$ for $w(x, y) := v(x)$, $(x, y) \in M_1 \times M_2$.*

Proof The proof of (a)–(c) and of (f)–(h) is easy and is left to the reader.

(d) Put $w := v'_+$.

(d1) We have $\sup|w| \leq l(v)$ (in particular, $\sup|w| = l(v)$ if $\sup|w| = \infty$) since we get for $x \in (a, b)$, using continuity of $x \mapsto |x|$ on $\mathbb{R}$,

$$|w(x)| = \left|\lim_{y \downarrow x} \frac{v(y) - v(x)}{y - x}\right| = \lim_{y \downarrow x} \frac{|v(y) - v(x)|}{|y - x|} \leq l(v).$$

(d2) Assume $l(v) < \infty$. If v is differentiable on (a, b) it follows easily from the first mean value theorem that $\sup|v'| \geq l(v)$. In the general case we use Scheeffer's theorem. This implies that $(v(y) - v(x))/(y - x)$, $x \neq y$, belongs to the compact interval $[\inf w, \sup w]$. Thus we get $l(v) \leq \max\{|\inf w|, |\sup w|\} = \sup|w|$.

(d3) Now the assertion for convex or concave functions follows, observing that then v'_+ is monotone.

(e) We know from Appendix A.1.4 that

$$|\max_i a_i - \max_i b_i| \leq \max_i |a_i - b_i|, \quad (a_i)_1^d, (b_i)_1^d \in \mathbb{R}^d.$$

Now we get for fixed $x, y \in M$, putting $a_i := v_i(x)$ and $b_i := v_i(y)$, that

$$|\max_i v_i(x) - \max_i v_i(y)| \leq \max_i |v_i(x) - v_i(y)| \leq \rho(x, y) \cdot \max_i l(v_i).$$

This proves the first assertion, from which the second one follows, using (c). □

By Lemma 9.3.2(c) the set ***LIP*** of Lipschitz functions on a metric space is a subspace of the vector space $\mathbb{R}^M$. Since the restriction $l(\cdot)|\,\mathit{LIP}$ of $l(\cdot)$ from $\mathbb{R}^M$ to *LIP* is finite and has there the three first properties of part (c), $l(\cdot)$ is a so-called seminorm on *LIP*.—Parts (d) and (e) yield many examples of Lipschitz continuous functions. As an example, $x \mapsto \sqrt{x}$ is Lipschitz on each interval $[a, \infty)$, $a > 0$, but not on $[0, \infty)$; and $x \mapsto \max_{1 \leq i \leq n}(c_i x + d_i)$ for reals c_i and d_i is Lipschitz on $\mathbb{R}$, but in general not everywhere differentiable.

We define Lipschitz-continuity of the correspondence $D(\cdot)$ only if all sets $D(s)$ are closed. Then we use the Hausdorff metric which we now introduce. Let (M, ρ) be a metric space and $\mathcal{F} := \mathcal{F}(M, \rho)$ the system of non-empty closed subsets of M. For $b \in M$ and $C \in \mathcal{F}$ the finite number

$$\rho(b, C) := \inf_{c \in C} \rho(b, c)$$

is called the *distance of the point b from the set C*. (As C is closed, the infimum is attained.) Thus

$$\sup_{b \in B} \rho(b, C) = \sup_{b \in B} \inf_{c \in C} \rho(b, c) \in [0, \infty]$$

is the maximal distance of points in B from the set C; draw a figure for the case $M = \mathbb{R}^2$.

Definition 9.3.3 (The Hausdorff distance) Let (M, ρ) be a metric space and $\mathcal{F}$ the system of non-empty closed subsets of M. For $B, C \in \mathcal{F}$ the number

$$h(B, C) = h_\rho(B, C) := \max\left\{\sup_{b\in B} \rho(b, C), \sup_{c\in C} \rho(c, B)\right\} \in [0, \infty]$$

is called the ***Hausdorff distance*** (induced by ρ) between B and C.

Lemma 9.3.4 (The Hausdorff metric) *The function $h = h_\rho$ on $\mathcal{F} \times \mathcal{F}$ is a distance, i.e. it has all properties of a metric except that it need not be finite. Thus h is a metric on each subsystem $\mathcal{H}$ of $\mathcal{F}$ for which h is finite on $\mathcal{H} \times \mathcal{H}$. It is called the **Hausdorff metric** induced by ρ on $\mathcal{H}$.*

Proof If $B \subset C$, then $\rho(b, C) = 0$ for all $b \in B$, hence $h(B, C) = \sup_{c\in C} \rho(c, B)$. Moreover, $h(\{x\}, \{y\}) = \rho(x, y)$ for $x, y \in M$. Note that in general $h(B, C)$ differs from the distance between B and C, defined as the finite number $\rho(B, C) := \inf\{\rho(b, c) : b \in B,\ c \in C\}$. □

Example 9.3.5 (Hausdorff metrics on systems $\mathcal{H}$) We consider the following systems $\mathcal{H}$:

(a) $\mathcal{H}$:= system of non-empty bounded sets in $\mathcal{F}$ ($= \mathcal{F}$, if ρ is bounded). Finiteness of h on $\mathcal{H}$ is shown as follows: The triangle inequality for ρ implies for bounded closed subsets B and C of M and arbitrary $x \in B$

$$\begin{aligned}\sup_{b\in B} \inf_{c\in C} \rho(b, c) &\leq \sup_{b\in B} \rho(b, x) + \inf_{c\in C} \rho(x, c)\\ &< \sup_{b,b'\in B} \rho(b, b') + \rho(x, C) < \infty,\end{aligned}$$

and similarly $\sup_{c\in C} \inf_{b\in B} \rho(b, c) < \infty$.

(b) $\mathcal{H}$:= system of all non-empty compact subsets of M. Finiteness of h on $\mathcal{H}$ follows from (a) since compact sets are bounded. Moreover, for compact intervals in $M := \mathbb{R}$ we have

$$h([\alpha, \beta], [\alpha', \beta']) = \max\{|\alpha - \alpha'|, |\beta - \beta'|\}.$$

This can be seen by assuming without loss of generality $\alpha \leq \alpha'$ and considering the three cases where $\beta < \alpha'$, $\alpha' \leq \beta < \beta'$ and $\beta \geq \beta'$.

(c) $\mathcal{H}$:= system of all (non-compact) closed intervals $[t, \infty)$, $t \in \mathbb{R}$. Then $h([\alpha, \infty), [\beta, \infty)) = |\alpha - \beta|$ for $\alpha, \beta \in \mathbb{R}$.

(d) $\mathcal{H}$:= system of simplexes $D(s)$, $s \in S := \mathbb{R}_+$, with vertices $\mathbf{0}$ and $s \cdot e_1$, $s \cdot e_2$, …, $s \cdot e_n$, i.e. $D(s) := \{(a_i)_1^m \in \mathbb{R}_+^m : \sum_{i=1}^m a_i \leq s\}$, provided S is endowed with the norm $\|s\| := |s|$. This system $\mathcal{H}$ arises in the problem of splitting a resource into $m + 1$ parts; cf. Example 9.2.8 above. Finiteness of h on $\mathcal{H}$ holds since $h(D(s), D(s')) = |s - s'|$, $s, s' \in S$.

(e) $\mathcal{H} := \{M\}$, whether or not M is compact, and $h(M, M) = 0$. ♦

For a DP with closed-valued $D(\cdot)$ the system $\mathcal{H}_D := \{D(s),\ s \in S\}$ and the Hausdorff distance, now denoted by h_A, which is induced by ρ_A on $\mathcal{F}(A)$ play an important role. If h_A is finite on $\mathcal{H}_D$ (hence h_A is the Hausdorff metric), we have three different notions of continuity of $D(\cdot)$, all of which depend on ρ_S and ρ_A, but not on a metric on D:

(i) continuity in the sense of Definition 9.2.1,

(ii) continuity of the mapping $D(\cdot)$ from (S, ρ_S) into $(\mathcal{H}_D, h_A)$, which we call *Hausdorff continuity*,

(iii) *Lipschitz continuity* of $D(\cdot)$ with respect to ρ_S and h_A. This means that the L-module of $D(\cdot)$, which we denote by $l(D)$, i.e. the number

$$l(D) := \sup_{s \neq s'} \frac{h_A(D(s), D(s'))}{\rho_S(s, s')}$$

is finite. This holds if and only if

$$h_A(D(s), D(s')) \leq K \cdot \rho_S(s, s') \text{ for } s, s' \in S,$$

for some $K \in \mathbb{R}_+$. Lipschitz continuity implies Hausdorff continuity.

Example 9.3.6 (Hausdorff continuity and Lipschitz continuity of $\mathbf{D(\cdot)}$*)*

(a) If $D(s) = A$ for all s, then $D(\cdot)$ is Lipschitz and $l(D) = 0$.

(b) Assume that $A := \mathbb{R}$ and $D(\cdot) = [c, d]$. Then we have:

(b1) $D(\cdot)$ is Hausdorff continuous if and only if both c and d are continuous. This holds since by Example 9.3.5(b)

$$h(D(s), D(s')) = \max\{|c(s) - c(s')|, |d(s) - d(s')|\}. \tag{9.4}$$

(b2) It follows from (9.4) that $l(D) = \max\{l(c), l(d)\}$. Thus $D(\cdot)$ is Lipschitz if and only if both c and d are Lipschitz.

(c) In some inventory problems, we have $S \subset= \mathbb{R} = A$ and $D(s) = [s, \infty)$ for all s. Then it follows from Example 9.3.5(c) that $l(D) = 1$.

(d) In Example 9.3.5(d) we have $l(D) = 1$.

(e) $l(D) = l(f)$ if $D(s) = \{f(s)\}$, $s \in S$, for some $f\colon S \to A$. ♦

Lemma 9.3.7 *Each compact-valued and Hausdorff continuous (in particular, each compact-valued and Lipschitz continuous) correspondence $D(\cdot)$ from S into A is continuous.*

Proof Put for the moment $\rho := \rho_A$.

(a) In order to show that $D(\cdot)$ is quasi-continuous, we select a sequence $(s_n)_1^\infty$ in S converging to some s in S. Select $a_n \in D(s_n)$ for $n \in \mathbb{N}$. Since $D(s)$ is closed there exists an $a_n' \in D(s)$ such that $\rho(a_n, a_n') = \rho(a_n, D(s))$. Moreover (a_n') has a cluster value a in $D(s)$ since $D(s)$ is compact. Now we obtain

$$\rho(a_n, a_n') = \rho(a_n, D(s)) \le \sup\{\rho(a, D(s)) : a \in D(s_n)\} \le h(D(s_n), D(s)),$$

which converges to zero as $D(\cdot)$ is Hausdorff continuous. It follows from

$$\rho(a_n, a) \le \rho(a_n, a_n') + \rho(a_n', a)$$

that a is also a cluster value of (a_n).

(b) In order to complete the proof for continuity of $D(\cdot)$ we have to show that for each $s \in S$, each $a \in D(s)$ and each sequence (s_n) in S, converging to s, there exist a sequence $(a_n) \subset D(s_n)$ having the cluster value a. In fact, as $D(s_n)$ is closed there is some a_n in $D(s_n)$ such that $\rho(a, a_n) = \rho(a, D(s_n))$. Now

$$\rho(a, a_n) \le \sup_{b \in D(s)} \rho(b, D(s_n)) = \rho(D(s); D(s_n)) \le h(D(s), D(s_n)),$$

which converges to zero by Hausdorff continuity of $D(\cdot)$. Thus (a_n) converges to a, hence a is a cluster value of (a_n). □

The assumption $A = A_0$ in the next result avoids unnecessary large Lipschitz modules for functions on A, where only the restriction to A_0 matters.

Lemma 9.3.8 (Preservation of Lipschitz continuity under maximization) *Consider a DP with $A = A_0$ and a function w on D of the form $w(s, a) = w_S(s) + w_A(a) + w_D(s, a)$ for functions w_S on S, w_A on A and w_D on D. Assume:*

(i) the product metric ρ_D satisfies $\rho_D \le \rho_S + \rho_A$,
(ii) $D(\cdot)$ is closed-valued and h_A is finite on $\mathcal{H}_D$,
(iii) $w^*(s) := \sup_{a \in D(s)} w(s, a)$ is finite for all s.

Then $l(w^*) \le K(w^*) := l(w_S) + l(w_A) \cdot l(D) + l(w_D) \cdot [1 + l(D)]$.

Proof Fix $s, s' \in S$. Firstly observe that

$$\sup_{a \in D(s)} \rho_A(a, D(s')) \le h(D(s), D(s')) \le l(D) \cdot \rho_S(s, s').$$

Now we obtain

$$w^*(s) - w^*(s') = \sup_{a \in D(s)} w(s, a) - \sup_{a' \in D(s')} w(s', a')$$

$$= \sup_a \inf_{a'} [w(s, a) - w(s', a')] \le \sup_a \inf_{a'} |w(s, a) - w(s', a')|$$

$$\le \sup_a \inf_{a'} [|w_S(s) - w_S(s')| + |w_A(a) - w_A(a')| + |w_D(s,a) - w_D(s',a')|]$$

$$\le \sup_a \inf_{a'} \big[l(w_S) \cdot \rho_S(s,s') + l(w_A) \cdot \rho_A(a,a') + l(w_D) \cdot \rho_D((s,a),(s',a'))\big]$$

$$\le \sup_a \inf_{a'} \big[l(w_S) \cdot \rho_S(s,s') + l(w_A) \cdot \rho_A(a,a') + l(w_D) \cdot [\rho_S(s,s') + \rho_A(a,a')]\big]$$

$$\le [l(w_S) + l(w_D)] \cdot \rho_S(s,s') + [l(w_A) + l(w_D)] \cdot \sup_a \inf_{a'} \rho_A(a,a')$$

$$\le K(w^*) \cdot \rho_S(s,s').$$

Exchanging $w^*(s)$ and $w^*(s')$ yields $-(w^*(s) - w^*(s')) \le K(w^*) \cdot \rho_S(s,s')$, which completes the proof. □

Now we are ready for the main result of this section. For its application note that $l(D) = 0$ if $D(s) = A$ for all s; sometimes this property can be achieved by a substitution as in Example 9.3.12(b). The representation of r in Theorem 9.3.9 is no restriction; it allows us to treat several special cases jointly. Condition (i) in the next result is fulfilled if ρ_D is one of the three standard metrics d_1, d_2 and d_3 (cf. Appendix C.1) since $d_1 \le d_2 \le d_3 = \rho_S + \rho_A$. Moreover, $l(w_S) = l_{\rho_S}(w_S)$, $l(w_A) = l_{\rho_A}(w_A)$ and $l(w_D) = l_{\rho_D}(w_D)$.

Theorem 9.3.9 (Lipschitz continuous value functions) *Consider a DP where (without loss of generality) $A = A_0$ and where the reward function r on D is written in the form $r(s,a) = r_S(s) + r_A(a) + r_D(s,a)$ for functions r_S on S, r_A on A and r_D on D. Assume:*

(i) $\rho_D \le \rho_S + \rho_A$,
(ii) *$D(\cdot)$ is closed-valued and h_A is finite on $\mathcal{H}_D$,*
(iii) *V_n is finite for all n.*

(a) We have

$$l(V_n) \le \alpha \cdot \sigma_n(\gamma) + \gamma^n \cdot l(V_0), \quad n \ge 1, \tag{9.5}$$

where

$$\alpha := l(r_S) + l(r_A) \cdot l(D) + l(r_D) \cdot [1 + l(D)], \quad \gamma := \beta \cdot l(T) \cdot [1 + l(D)].$$

Thus, all value functions V_n, $n \ge 1$, are Lipschitz if $D(\cdot)$, T, r_S, r_A, r_D and V_0 are Lipschitz.

(b) If $T(s,a) = \eta(a)$ does not depend on s, then (9.5) holds with $\gamma := \beta \cdot l(\eta) \cdot l(D)$.

(c) Assume that $V := \lim_{n\to\infty} V_n$ exists (pointwise) as finite limit and that γ in (a) and/or in (b) satisfies $\gamma < 1$. Then V is Lipschitz and $l(V) \le \alpha/(1-\gamma)$ for the corresponding γ.

Proof

(a) Fix $n \geq 1$ and put $w_D := r_D + \beta \cdot V_{n-1} \circ T$. It follows from Lemma 9.3.2(b) and (c) that

$$l(w_D) \leq l(r_D) + \beta \cdot l(V_{n-1} \circ T) \leq l(r_D) + \beta \cdot l(V_{n-1}) \cdot l(T).$$

Now $V_n = UV_{n-1}$ from the VI, and Lemma 9.3.8 with $w := r$, $w_S := r_S$ and $w_A := r_A$ yields $l(V_n) \leq \alpha + \gamma \cdot l(V_{n-1})$, from which (9.5) follows by induction on $n \geq 1$.

(b) This follows as in (a) by applying Lemma 9.3.8 with $w := r$, $w_S := r_S$, $w_A := r_A + \beta \cdot V_{n-1} \circ \eta$ and $w_D := r_D$.

(c) This follows from Lemma 9.3.2(f). □

Remark 9.3.10 Theorem 9.3.9 and the subsequent Proposition 9.3.13 remain true for the minimization problem if r_S, r_A, r_D, V_0 and V_n are replaced by c_1, c_2, c_3, C_0 and C_n, respectively. This is due to the fact that $l(-v) = l(v)$ for each function v on (M, ρ) by Lemma 9.3.2(c). ◇

Assumption (iii) in Theorem 9.3.9 is fulfilled if both r and V_0 are bounded or if a bounding function exists; cf. Sect. 3.4. Another case is treated in the next result.

Corollary 9.3.11 (Upper and lower semicontinuous maximizers) *Consider the DP from Theorem* 9.3.9. *If $D(\cdot)$ in Theorem* 9.3.9 *is compact-valued and if $D(\cdot)$, T, r_S, r_A, r_D and V_0 are Lipschitz, then V_n is finite for all $n \geq 1$ and the assertions (a)–(c) in Theorem* 9.3.9 *hold. In addition we have:*

(a) There exists a maximizer f_n at each stage n.
(b) If f_n is unique, it is continuous.
(c) In case $A \subset \mathbb{R}$ there exists a smallest [largest] maximizer f_n at each stage n, and f_n is lower [upper] semicontinuous.

Proof From Lemmas 9.3.7 and 9.3.2(a) we know that $D(\cdot)$, T, r and V_0 are continuous. Then V_n is finite and the result follows from Theorem 9.2.5. □

Example 9.3.12 (Allocation problem) Assume that $S = A = [0, s_0]$ for some $s_0 \in \mathbb{R}^+$, $D(s) = [0, s]$, $T(s, a) = s - a$, $r(s, a) = r_S(s) + r_A(a) + u(s - a)$. On D we use the taxicab metric. Assume that the functions r_S, r_A, u and V_0, which are defined on $[0, s_0]$, are Lipschitz.

(a) A direct application of Theorem 9.3.9 and of Corollary 9.3.11 shows that all value functions are finite and that by (9.5)

$$l(V_n) \leq \alpha \cdot \sigma_n(2\beta) + (2\beta)^n \cdot l(V_0), \tag{9.6}$$

where $\alpha := l(r_S) + l(r_A) + 2 \cdot l(u)$. This bound is very poor for the relevant values of β, and it tends to ∞ for $\beta \geq 1/2$.

(b) Transforming the DP by $b := s - a$ leads to a DP′ which again has the value functions V_n and for which the following better bound can be derived from

Theorem 9.3.9(b):

$$l(V_n) \leq \alpha' \cdot \sigma_n(\beta) + \beta^n \cdot l(V_0), \tag{9.7}$$

where $\alpha' := l(r_S) + 2 \cdot l(r_A) + l(u)$. (Note that for small n the bound in (9.7) need not be better than the one in (9.6).) For the proof we firstly note that DP$'$ differs from DP only by $T'(s,b) := b \in [0,s]$, $r'_A(a) := u(a)$ and $r'_D(s,b) := r_A(s-b)$. The conditions (i)–(iii) in Theorem 9.3.9 are fulfilled: Firstly, (i) holds by assumption. Next, (ii) holds since $D(\cdot)$ is compact-valued and since $h_A([0,s],[0,s']) = |s-s'|$ by Example 9.3.6(b). Finally (iii) holds by Corollary 9.3.11 due to the following facts: $l(D) = 1$; $l(T') \leq 1$ is trivial; $l(r_S) < \infty$ and $l(r'_A) < \infty$ hold by assumption; $r'_D = r_A \circ g$, where $g(s,b) = s-b$, hence $l(r'_D) \leq l(r_A) \cdot l(g) = l(r_A) < \infty$ by Lemma 9.3.2(b).

(c) On the compact sets D and S the functions r and V_0, respectively, are Lipschitz, hence continuous, hence bounded. It follows from Theorem 10.1.10 below that V exists as a finite limit if $\beta < 1$, and then Theorem 9.3.9(c) shows that V is Lipschitz with $l(V) \leq \alpha'/(1-\beta)$. ♦

For a mapping w from $S \times A$ into (M', ρ') put

$$l^S(w) := l^S_{\rho',\rho_S}(w) := \sup_{a \in A} l(w(\cdot,a)) \in [0,\infty].$$

(Neither on A nor on $S \times A$ a metric is required.) The mapping w is called ***uniformly Lipschitz*** in a if $l^S(w)$ is finite, i.e. if

$$\rho'(w(s,a), w(s',a)) \leq K \cdot \rho_S(s,s') \text{ for } s,s' \in S \text{ and } a \in A, \tag{9.8}$$

for some $K \in \mathbb{R}_+$. Then $l^S(w)$ is the smallest of the constants K in (9.8). If w does not depend on s, then $l^S(w) = 0$. The example $w(s,a) := s + a^2$, $s \in S := \mathbb{R}$, $a \in A := \mathbb{R}$, shows that w need not be Lipschitz with respect to the taxicab metric on $\mathbb{R}^2$ if it is uniformly Lipschitz in a. On the other hand, if A is endowed with a metric ρ_A then one easily sees that Lipschitz continuity of a function w on $D := S \times A$ with respect to one of the three standard product metrics implies uniform Lipschitz continuity of w in a, and $l^S(w) \leq l(w)$. Moreover, if $s \mapsto w^*(s) := \sup_{a \in A} w(s,a)$ is finite for all s, then $l(w^*) \leq l^S(w)$.

Proposition 9.3.13 (Lipschitz continuity of value functions under uniform Lipschitz conditions) *Consider a DP with $D(s) = A$ for all s and where without loss of generality the one-stage reward function r on D is written in the form $r(s,a) = r_S(s) + r_A(a) + r_D(s,a)$ for functions r_S on S, r_A on A and r_D on D. Assume that V_n is finite for all n.*

(a) We have

$$l(V_n) \leq \alpha \cdot \sigma_n(\gamma) + \gamma^n \cdot l(V_0), \quad n \geq 1, \tag{9.9}$$

where $\alpha := l(r_S) + l^S(r_D)$, $\gamma := \beta \cdot l^S(T)$. Thus all value functions are Lipschitz if r_S and V_0 are Lipschitz and if T and r_D are uniformly Lipschitz in a.

(b) *If $T(s, a)$ does not depend on s, then $l(V_n) \le l(r_S) + l^S(r_D)$ for $n \ge 1$.*
(c) *If $V := \lim_{n\to\infty} V_n$ exists (pointwise) as a finite limit and if $\gamma < 1$, then V is Lipschitz and $l(V) \le \alpha/(1-\gamma)$.*

Proof The bound (9.9) is easily derived by induction on $n \ge 0$ using that $|V_n(s) - V_n(s')| \le \sup_{a\in A} |W_n(s, a) - W_n(s', a)|$ by Appendix A.4.4. Now (a)–(c) follow, using $l(T) = 0$ for (b) and Lemma 9.3.2(f) for (c). □

Remark 9.3.14 If the assumptions in Proposition 9.3.13 hold and if both T and r_D are Lipschitz then Theorem 9.3.9 also provides a bound for $l(V_n)$; but it is inferior to (9.9) since $l^S(T) \le l(T)$ and $l^S(r_D) \le l(r_D)$.

Example 9.3.15 (Continuation of the allocation problem from Example 9.3.12*)* By the substitution $d := a/s \in A'' := [0, 1]$ one obtains from the original DP a DP″ to which Proposition 9.3.13 can be applied. Then one obtains the following improvement of (9.7):

$$l(V_n) \le \alpha'' \cdot \sigma_n(\beta) + \beta^n \cdot l(V_0),$$

where $\alpha'' := l(r_S) + l(r_A) + l(u)$. For the proof one uses that many properties of $l(\cdot)$, such as $l(v_1 + v_2) \le l(v_1) + l(v_2)$, easily imply the same property for $l^S(\cdot)$. The details are left to the reader. ♦

Example 9.3.16 (A linear system) Consider the minimization problem for the DP where $S = A = D(s) = \mathbb{R}$ for all s, $T(s, a) = gs - ha$ for $h \in \mathbb{R}^+$ and $g \in \mathbb{R}_+$, $c(s, a) = |s| + u(a)$ for a function $u \ge 0$ on $\mathbb{R}$, $C_0 \ge 0$. Then C_n is finite since $c \ge 0$ and $C_0 \ge 0$ implies $0 \le C_n \le C_{n\pi} < \infty$ for each policy π. Moreover:

(a) The direct application of Proposition 9.3.13 yields $l(C_n) \le \sigma_n(\beta g) + (\beta g)^n \cdot l(C_0)$, $n \ge 1$ since $l(c_S) = 1$ for $c_S(s) := |s|$ and $l^S(T) = g$.
(b) For most values of $g > 0$, h, u and C_0 the bound in (a) is improved by

$$l(C_n) \le 1 + g \cdot l(u)/h. \tag{9.10}$$

For the proof one makes the substitution $b := s - a$ which leads to a DP′ having again the value functions V_n. Now (9.10) follows easily from Theorem 9.3.9(b). ♦

9.4 Problems

For problems Problem 9.4.1 to Problem 9.4.7 S and A are endowed with metrics ρ_S and ρ_A, respectively, and D is endowed with a product metric. For the remaining problems let $(M, \rho,)$, (M', ρ') etc. denote metric spaces.

Problem 9.4.1 Let $D_i(\cdot)$, $i \in I$, be a family of correspondences from S into $A_i \subset A$ such that $D(s) := \cap_i D_i(s)$ is non-empty for all $s \in S$. If all D_i are closed in $S \times A$ and if $D_i(\cdot)$ is quasi-continuous for one i, then $D(\cdot)$ is quasi-continuous.

Problem 9.4.2 Let (A_i, ρ_i) be metric spaces, and endow $A := \times_1^m A_i$ with a product metric. If $D_i(\cdot)$ is a quasi-continuous [continuous] correspondence from S into A_i, $1 \leq i \leq m$, then $s \mapsto \times_1^m D_i(s)$ is quasi-continuous [continuous].

Problem 9.4.3 Assume that $A = \times_1^m A_i$ for intervals $A_i \subset \mathbb{R}$ and that $D(\cdot) = [c, d] \subset \mathbb{R}^m$, where $c = (c_i)_1^m$ and $d = (d_i)_1^m$ are mappings from S into A with component functions $c_i: S \mapsto A_i$ and $d_i: S \mapsto A_i$ with $c_i \leq d_i$. Derive conditions under which $D(\cdot)$ is quasi-continuous or continuous.

Problem 9.4.4 Assume that $S = \mathbb{R}_+, A \subset \mathbb{R}^m$, and that D is a cone (i.e. $(s, a) \in D$ and $\lambda \in \mathbb{R}_+$ imply $\lambda \cdot (s, a) \in D$)), that $D(0)$ is bounded and that $D(1)$ is compact. Then $D(\cdot)$ is continuous.

Problem 9.4.5 Assume that $S = A = [0, 2]$, $D(s) = [0, 1]$ for $0 \leq s \leq 1$, $D(s) = [0, 2]$ for $1 < s < \infty$. Then $(s, a) \mapsto w(s, a) := a$ is continuous on D, but w^* is not usc.

Problem 9.4.6 If $A \subset \mathbb{R}^m$, if $D(\cdot)$ is quasi-continuous and if $D(s)$ has a smallest [largest] element $f(s)$ with respect to $\leq^m$, then f is componentwise lsc [usc].

Problem 9.4.7 Assume that $D(\cdot)$, r and V_0 are lsc and that T is continuous. Then V_n and W_n are lsc for all n.

Problem 9.4.8 There following hold:

(a) If the mapping w from $S \times A$ into (M', ρ') is Lipschitz with respect to $\rho_S + \rho_A$, then w is uniformly Lipschitz in s and $l^S(w) \leq l(w)$.
(b) Endow $S \times A := \mathbb{R}^2$ with any of the three standard product metrics, denoted by ρ'. Then $(s, a) \mapsto w(s, a) := s + a^2$, $s, a \in S \times A$ is uniformly Lipschitz in s but not Lipschitz with respect to ρ'.

Problem 9.4.9 Consider a DP with bounding function $b > 0$; cf. Sect. 3.4. Put $B(s, a) := b \circ T(s, a)/b(s)$, $\tilde{r} := r/b$ and $\tilde{V}_n := V_n/b$. If $\rho_D \leq \rho_S + \rho_A$ then

$$l(\tilde{V}_n) \leq (1 + l(D)) \cdot [l(\tilde{r}) + \beta \cdot l(B) \cdot \|V_{n-1}\|_b + \beta_b \cdot l(\tilde{V}_{n-1}).$$

Problem 9.4.10 If $B(s) := \{f(s)\}$, $s \in S$, for a decision rule f, then $l(B) = l(f)$.

Problem 9.4.11 If $D(s) = A$ for all s and if w is a function on D such that w^* is finite, then $l(w^*) \leq l^S(w)$.

9.5 Supplement

Further versions of Lemma 9.2.4 may be found, for example, in Schäl (1975), Bertsekas and Shreve (1978, p. 148) (for the case $D(s) = A$ for all s), Dubins and Savage (1965, p. 38), Hinderer (1970, p. 34) and Dynkin and Yushkevich (1979, p. 54).

Chapter 10
Stationary Models with Large Horizon

In this book we prefer to place more emphasis, at least from the conceptual point of view, on $\lim DP_n$ than on DP_∞. There are a number of reasons for our approach. We present the general theory for $\lim DP_n$ and discuss the optimality equation (Bellman equation) for the limit value function. DPs with infinite horizon are also considered.

10.1 The General Theory

The allocation problem from Example 4.1.6 with square root utility had an explicit solution in case $z\beta^2 < 1$ such that both the sequence of value functions and the sequence of unique maximizers converge pointwise to limit functions V and f, respectively. Moreover, it is easy to see that f is a maximizer of LV. In Problem 4.3.3 we asked to show that f is *asymptotically optimal* in the following sense.

Definition 10.1.1 Assume that $V_N(s)$ is finite for all s and N. For $s \in S$ and a decision rule f denote by $V_{Nf}(s)$ the N-stage reward $V_{Ny}(s)$ for the action sequence $y = y(N,f) = (a_t)_0^{N-1} \in A^N(s)$, generated by s and f

$$a_t := f(s_t),\ s_{t+1} := T(s_t, a_t),\ s_0 := s,\ 0 \le t \le N-1.$$

Then f is called ***asymptotically optimal*** if for each $s \in S$

$$V_N(s) - V_{Nf}(s) \to 0 \text{ for } N \to \infty.$$

If $V(s) := \lim_{n\to\infty} V_n(s)$ exists and is finite for all s and if f is asymptotically optimal, then $V(s)$ and $y(N,f) \in A^N(s)$ may be used for large horizon N as approximate solution of $DP_N(s)$ for *large N*.

K. Hinderer et al., *Dynamic Optimization*, Universitext,
DOI 10.1007/978-3-319-48814-1_10

In the literature there is often used a similar yet conceptually different method (*infinite-stage models*) for the approximate solution of DP_N for large N. We give some information about this method in Sect. 10.5 below.

Now the following questions arise for general DPs.

Question (i): Under what conditions does there exist, as a limit in $\overline{\mathbb{R}}$,

$$V(s) := \lim_{n\to\infty} V_n(s) \text{ for all } s \in S?$$

Question (ii): Under what conditions are maximizers f of LV (assuming that V and f exist) asymptotically optimal?

In case questions (i) and (ii) can be answered affirmatively, one will search for methods for the approximate computation of V. In general the starting point is the fact that in general V is a solution v of the so-called **optimality equation (OE)** or *Bellman equation*

$$v(s) = \sup_{a\in D(s)} [r(s,a) + \beta \cdot v \circ T(s,a)]\,, \; s \in S, \tag{10.1}$$

or shortly

$$v = Uv.$$

The OE is a functional equation of a usually complicated type. We already encountered it in connection with absorbing DPs where convergence of $(V_n)_0^\infty$ is trivial and where the OE could be solved by recursion in state space; cf. Theorem 5.1.7. Each function v on S which satisfies (10.1) is called a *solution of the OE* or a *fixed point of the optimal reward operator* U. Note that $v \equiv \infty$ and $v \equiv -\infty$ are always solutions of the OE, the so-called two trivial solutions, of U. The OE for V, if V exists, is formally obtained by putting n equal to infinity in the VI

$$V_n(s) = \sup_{a\in D(s)} [r(s,a) + \beta \cdot V_{n-1}(T(s,a))],$$

but this of course is not a proof.

From the preceding discussion we see the importance of

Question (iii): Under which conditions is V, if it exists, a solution of the optimality equation and how can it be identified within the set of all solutions?

In general, questions (i)–(iii) do not have simple answers:

(a) V need not exist; cf. Example 10.1.2(a2).
(b) If V exists it need not satisfy the OE (cf. Example 10.1.2(b)).
(c) The OE may have several finite solutions (cf. Example 10.1.2(c)).
(d) The OE may have only the two trivial solutions (cf. Example 10.1.2(d1)).
(e) Maximizers of LV need not be asymptotically optimal (cf. Example 10.1.2(b)).

Concerning (b) we remark the following. If V exists, put

$$W := LV := r + \beta \cdot V \circ T.$$

Then $UV(s) = \sup\{W(s,a) : a \in D(s)\}$ for all s. Obviously the convergence of (V_n) towards V implies convergence of (W_n) towards W. Therefore, and as $V_n(s) = \sup\{W_n(s,a) : a \in D(s)\}$ for all s, the validity of the OE for V means that for all s

$$\lim_{n\to\infty} \sup_{a\in D(s)} W_n(s,a) = \sup_{a\in D(s)} \lim_{n\to\infty} W_n(s,a),$$

i.e. that one may interchange lim and sup in $(n,a) \to W_n(s,a)$. One cannot expect this to always hold.

Of course, the preceding questions also arise for cost minimization problems. There we ask (i) for the existence of

$$C(s) := \lim_{n\to\infty} C_n(s) \text{ for all } s \in S,$$

(ii) for minimizers f of LC and (iii) for solutions v of the OE, the latter now having the form

$$v(s) = \inf_{a\in D(s)} [c(s,a) + \beta \cdot v \circ T(s,a)], \; s \in S,$$

which we again write as $v = Uv$.

Example 10.1.2 (Solutions of the optimality equation)

(a) Assume $S = A = \mathbb{R}_+, D(s) = [0,s], T(s,a) = (s-a)/\beta, r(s,a) = 1-1/(1+a)$ for all (s,a) and $\beta < 1$. Then both $s \mapsto v_1(s) := s/[1 + (1-\beta)s]$ and $s \mapsto s$ are finite solutions of the OE.
(b) Assume $S = A = \mathbb{R}_+$, $s \in D(s) \subset \mathbb{R}_+$, $T(s,a) = a$, $r(s,a) \geq 1$ for all (s,a), and $\beta = 1$. Then V exists, equals ∞ and is a solution of the OE.
(c) If $D(s) = \mathbb{R}_+$ for all s, then the OE has only the two trivial solutions.
(d) If $D(s) = [s,\infty)$ for all s, then the OE has only the following non-trivial solutions v: for some $t \in \mathbb{R}_+$

$$v(s) = \begin{cases} \infty, & \text{if } s < t, \\ -\infty \text{ or } \infty, & \text{if } s = t, \\ -\infty, & \text{if } s > t. \end{cases} \quad \blacklozenge$$

Before answering questions (i)–(iii) affirmatively under appropriate assumptions we present the following two auxiliary results without proof.

Lemma 10.1.3 *If g and h are upper bounded functions on some set M, then*

$$\sup_x g(x) - \sup_x h(x) \le \sup_x [g(x) - h(x)],$$

$$|\sup_x g(x) - \sup_x h(x)| \le \sup_x |g(x) - h(x)|.$$

Lemma 10.1.4 (Convergence of suprema) *Let* $(g_n)_1^\infty$ *be a sequence of functions on a set M, converging pointwise towards the function g.*

Then:

(a) *Each of the following conditions is sufficient for the convergence of* $(\sup g_n)_1^\infty$ *towards* $\sup g$*, i.e. for the validity of*

$$\lim_n \sup_x g_n(x) = \sup_x \lim_n g_n(x).$$

(a1) $g_n(x)$ *is increasing in n.*
(a2) $(g_n)_1^\infty$ *converges uniformly towards g (which holds trivially if M is finite).*
(a3) *M is a compact metric space,* $g_n(x)$ *is decreasing in n and* g_n *is usc on M.*
(b) *Assume that the conditions of (a3) hold. Each cluster value of any sequence* $(x_n^*)_1^\infty$ *of maximum points of the functions* g_n *is a maximum point of g.*

Note that in Lemma 10.1.4(a1) *increasing* cannot be replaced by *decreasing*. This is illustrated by the example $g_n(x) = 1_{(0,1/n)}(x)$, $0 \le x \le 1$.

A first positive result is

Proposition 10.1.5 (Validity of the Optimality Equation for V)

(a) *If V exists then* $UV \le V$.
(b) *If* $V_1 \ge V_0$ *or if the value functions* V_n *are finite and converge uniformly to V, then* $V = UV$.
(c) *Assume that A is a metric space and that*

(i) (V_n) *is decreasing,*
(ii) $D(\cdot)$ *is compact-valued,*
(iii) *V is finite,*
(iv) $a \mapsto W_n(s, a)$ *is usc for all n and s.*

Then $V = UV$*. Moreover, if* f_n *is a maximizer at stage n,* $n \in \mathbb{N}$*, and if* $f(s)$ *is any cluster value of* $(f_n(s))_1^\infty$*,* $s \in S$*, then f is a maximizer of LV.*

Proof

(a) From the VI we obtain $W_n(s, a) \le V_n(s)$, and hence $W(s, a) \le V(s)$ for $(s, a) \in D$. Thus

$$UV(s) = \sup_a W(s, a) \le V(s) \text{ for } s \in S.$$

(b) follows from Lemma 10.1.4(a4) and (a2), applied to $g_n(a) := W_n(s,a)$, $s \in S$ fixed.
(c) Finiteness of V_0 and $V_n > -\infty$ implies inductively finiteness of all functions V_n as in case of a function V_{n-1} the finite function $W_n(s,\cdot)$ is usc on the compact set $D(s)$. Now (c) follows similarly as (b) from Lemma 10.1.4(a3) and (b). □

The simplest positive answer to all questions (i)–(iii) can be given in the case where both r and V_0 are bounded (in particular if both S and A are finite) and where $\beta < 1$. We immediately turn to the more general case where a bounding function, as defined in Sect. 3.4, exists. Recall that for functions v on S and w on D we denote the b-norm by $\|v\|_b$ and by $\|w\|_b$, respectively, and that $\beta_b := \beta \cdot \|b \circ T\|_b$. The set $B_b := B_b(S)$ of functions v on S for which $\|v\|_b$ is finite is a Banach space under the norm $\|\cdot\|_b$, hence a complete metric space with the metric $\rho(u,v) := \|u - v\|_b$. Convergence of functions in B_b in b-norm obviously implies pointwise convergence and, if $\inf b > 0$, even uniform convergence.

Remark 10.1.6 Let b be a bounding function for the DP. Then by (3.7)

$$|V_n| \le \left[\|r\|_b \cdot \sigma_n(\beta_b) + \|V_0\|_b \cdot \beta_b^n\right] \cdot b,\ n \in \mathbb{N}. \tag{10.2}$$

As a consequence, $V_n \in B_b$ for all n. ◇

Remark 10.1.7 Recall from Sect. 3.4:

(a) $b \equiv 1$ is a bounding function with $\beta_b = \beta$ if r and V_0 are bounded.
(b) For $(s,a) \in D$ and any function v on S the following inequalities

$$|v(s)| \le \|v\|_b \cdot b(s), \quad \beta \cdot |v \circ T(s,a)| \le \beta_b \cdot \|v\|_b \cdot b(s) \tag{10.3}$$

obviously hold. ◇

An important tool for our purpose is Banach's well-known fixed point theorem for contractions. We prove a generalization due to Weissinger (1952).

Proposition 10.1.8 (Banach's fixed point theorem and Weissinger's generalization) *Let K be a mapping from a complete metric space (X,ρ) into itself such that the following holds for some $\alpha_t \in \mathbb{R}_+$, $t \ge 1$,*

$$\rho(K^t x, K^t y) \le \alpha_t \cdot \rho(x,y), \quad t \ge 1, \quad x, y \in X. \tag{10.4}$$

If $\sum_1^\infty \alpha_t < \infty$, then the following holds:

(a) For each $x_0 \in X$ the sequence $(x_n)_0^\infty := (K^n x_0)_0^\infty$ converges for $n \to \infty$ to some $x^* \in X$.
(b) x^* is the unique fixed point of K, hence independent of $x_0 \in M$.

(c) Using $\alpha_0 := 1$, *the following estimate holds*

$$\rho(x^*, x_n) \le \alpha_n \cdot \rho(x^*, x_0) \le \alpha_n \cdot \sum_{t=0}^{\infty} \alpha_t \cdot \rho(x_1, x_0), \ n \ge 1. \tag{10.5}$$

(d) (Banach) If K is contracting, i.e. if

$$\rho(Kx, Ky) \le \gamma \cdot \rho(x, y), \quad x, y \in X,$$

for some $\gamma \in [0, 1)$, *then (a) and (b) hold, and*

$$\rho(x^*, x_n) \le \gamma^n \cdot \rho(x^*, x_0) \le \gamma^n \rho(x_1, x_0)/(1 - \gamma), \ n \ge 1.$$

Proof

(a) For $m, n \ge 0$ the triangle inequality yields

$$\begin{aligned}\rho(x_n, x_{n+m}) &\le \sum_{t=0}^{m-1} \rho(x_{n+t}, x_{n+t+1}) = \sum_{t=0}^{m-1} \rho(K^t x_n, K^t x_{n+1}) \\ &\le \rho(x_n, x_{n+1}) \cdot \sum_{t=0}^{m-1} \alpha_t \le \alpha_n \cdot \sum_{t=0}^{\infty} \alpha_t \cdot \rho(x_1, x_0).\end{aligned} \tag{10.6}$$

The latter term becomes arbitrarily small when n and m are large enough. Thus $(x_n)_0^\infty$ is a Cauchy sequence, which ensures the existence of x^* since (X, ρ) is complete.

(b) From $\rho(Kx_n, Kx^*) \le \alpha_1 \cdot \rho(x_n, x^*) \to 0$ for $n \to \infty$, from continuity of $\rho(\cdot, x^*)$ and from (a) we see that $Kx^* = \lim_n Kx_n = \lim_n x_{n+1} = x^*$. Thus x^* is a fixed point of K. Moreover, if $y \in X$ is any fixed point, then $y = x^*$. In fact, the following holds for $n \ge 1$, since $\alpha_n \to 0$ for $n \to \infty$,

$$\rho(x^*, y) = \rho(K^n x^*, K^n y) \le \alpha_n \cdot \rho(x^*, y) \to 0 \quad (n \to \infty).$$

(c) The first inequality follows from (b) and from (10.4), while the second one is obtained from the first one by letting $m \to \infty$ in (10.6) with $n := 0$ and from continuity of $\rho(x_0, \cdot)$.

(d) This is the special case of (a)–(c) where $\alpha_n = \gamma^n$. □

Lemma 10.1.9 (**U is Lipschitz with constant β_b**) *If b is a bounding function of the DP, then U and U_f, $f \in F$, map B_b into itself and for v and w in B_b*

$$\|Uv - Uw\|_b \le \sup_{f \in F} \|U_f v - U_f w\|_b \le \beta_b \cdot \|v - w\|_b. \tag{10.7}$$

Proof

(a) Put $\alpha := \|r\|_b + \beta_b \cdot \|v\|_b$. If $v \in B_b$ then by (10.3) for all (s, a)

$$|Lv(s,a)| = |r(s,a) + \beta \cdot v \circ T(s,a)| \leq |r(s,a)| + \beta \cdot |v \circ T(s,a)| \leq \alpha b(s).$$

Thus $|Lv(s, f(s))| \leq \alpha b(s)$, hence $U_f v \in B_b$. Moreover, $Uv \in B_b$ since

$$|Uv(s)| = |\sup_a Lv(s,a)| \leq \sup_a |Lv(s,a)| \leq \alpha b(s).$$

(b) We use Lemma 10.1.3 for fixed s, $M := D(s)$, $g := Lv(s,\cdot)$ and $h := Lw(s,\cdot)$. Because of (10.7), g and h are finite, and they are upper bounded as $\sup_a Lv(s,a) = Uv(s) < \infty$ by (a), and similarly for w. Therefore we obtain from (10.3)

$$\begin{aligned}|Uv(s) - Uw(s)| &= |\sup_a Lv(s,a) - \sup_a Lw(s,a)| \\ &\leq \sup_a |Lv(s,a) - Lw(s,a)| \\ &= \beta \cdot \sup_a |(v - w) \circ T(s,a)| \leq \beta_b \cdot \|v - w\|_b \cdot b(s).\end{aligned}$$

□

Now we obtain the main result of this section.

Theorem 10.1.10 (The optimality equation and asymptotically optimal decision rules) *Assume that b is a bounding function for the DP such that $\beta_b < 1$.*

(a) The value functions V_n belong to B_b and $(V_n)_0^\infty$ converges in b-norm towards a limit function $V \in B_b$ which is independent of $V_0 \in B_b$.
(b) V is the unique solution of the OE within B_b, and

$$\|V\|_b \leq \|r\|_b / (1 - \beta_b).$$

(c) The following estimate holds

$$\|V_N - V\|_b \leq \beta_b^N \cdot \|V_0 - V\|_b \quad N \geq 1. \tag{10.8}$$

(d) For any decision rule f the sequence $(V_{nf})_0^\infty$ converges in b-norm to some function $V_f \in B_b$, and V_f is the unique fixed point of U_f within B_b.
(e) The decision rule f is asymptotically optimal if and only if f is a maximizer of LV if and only if $V = V_f$, and then, using $\beta_{bf} := \beta \cdot \|b \circ T_f\|_b$,

$$0 \leq V_N - V_{Nf} \leq (\beta_b^N + \beta_{bf}^N) \cdot \|V_0 - V_f\|_b \cdot b(\cdot), \quad N \geq 1. \tag{10.9}$$

(f) If LV has a maximizer (in particular if D(s) is finite for all s) then

$$V = \max_{f \in F} V_f.$$

Proof

Assertion (a) follows from Banach's fixed point theorem Proposition 10.1.8 with $B := B_b$, $K := U$, ρ the metric induced by $\|\cdot\|$ and $\gamma := \beta_b$. In fact, we have $V_n = UV_{n-1}$ for $n \in \mathbb{N}$, U maps B_b into itself by Lemma 10.1.9, and U is a contraction on B_b since $\beta_b < 1$.
The bound in (b) is obtained from (10.2) for $n \to \infty$.
The estimate in (c) follows from (10.5).

(d) Consider the DP_f which differs from the given DP only by $D(s) := D_f(s) := \{f(s)\}$, $s \in S$. If b is a bounding function for the DP, then it is also a bounding function of DP_f, and $\beta_{bf} \leq \beta_b$. Therefore (a), (b) and (c) also hold for DP_f. In particular, (d) follows from (a) for DP_f.

(e1) We show that f is a maximizer of LV if and only if $V = V_f$. In fact, if $V = V_f$, then (b) and (d) imply $U_f V = U_f V_f = V_f = V = UV$. On the other hand, if f is a maximizer of LV, then $U_f V = UV = V$, which implies $V_f = V$ by (d).

(e2) If f is asymptotically optimal then $V = V_f$ since $V_N \to V$ and $V_{Nf} \to V_f$ for $N \to \infty$.

(e3) If $V = V_f$ then we get from (10.8) for DP_f

$$\|V_f - V_{nf}\|_b \leq \beta_{bf}^N \cdot \|V - V_0\|_b. \tag{10.10}$$

Now (10.9) and hence asymptotic optimality of f follows by inserting (10.8) and (10.10) into

$$V_N - V_{Nf} = (V_N - V) - (V_f - V_{Nf}) \leq \|V_N - V\| + \|V_f - V_{Nf}\|.$$

(f) This follows from (e) since $V_{nf} \leq V_n$ for all n and f implies $V_f \leq V$, hence $\sup_f V_f \leq V$. □

Remark 10.1.11 Assume that both S and A are finite. Obviously the finite fixed point V_f of U_f can be computed as the unique finite solution v of the system of linear equations

$$v_f - \beta \cdot v(T_f) = r_f. \tag{10.11}$$

Therefore and by Theorem 10.1.10(f) one can in principle compute V by solving the $|F|$ systems of linear equations (10.11) for all $f \in F$, and by comparing the resulting functions V_f. ◇

Example 10.1.12 (Allocation problem with square root utility and general terminal reward) Example 4.1.6 with $S = A = \mathbb{R}_+$, $T(s,a) = z(s-a)$ for some $z \in [1,\infty)$, $r(s,a) = \sqrt{a}$, $V_0(s) = d_0\sqrt{s}$ for some $d_0 \in \mathbb{R}_+$ and $\beta \in \mathbb{R}^+$ had the explicit solution

$$V_n(s) = d_n\sqrt{s}, f_n(s) = s/d_n^2, d_n^2 = \sigma_n(\gamma) + \gamma^n d_0^2, \gamma := \beta^2 z.$$

From this we derived that $V_n(s) \to \infty$ for $\gamma \geq 1$, $s \in S$ and $n \to \infty$, while for $\gamma < 1$

$$V_n(s) \to V(s) := \sqrt{s/(1-\gamma)}, f_n(s) \to f(s) := (1-\gamma) \cdot s.$$

Now assume that V_0 is arbitrary but such that $|V_0(s)| \leq h \cdot \sqrt{s}$ for $s \in S$ and some $h \in \mathbb{R}_+$. (As an example, $V_0(s) := \log(1+s)$ has this property for $h := 1$.) There is in general no hope of an explicit solution. However, for $\gamma < 1$, and large N, we may still use V_N and the action sequence generated by s and f as an approximate solution of $DP_N(s)$. This follows from Theorem 10.1.10 (observing that f is a maximizer of LV) as $s \mapsto b(s) := \sqrt{s}$ is a bounding function with constant $\delta := \max\{\sqrt{z}, 1\}$ and $\delta_b := \beta\sqrt{z} = \sqrt{\gamma}$. In fact, we have for $(s,a) \in D$

$$b \circ T(s,a) = \sqrt{z(s-a)} \leq \sqrt{zs}, \; |r(s,a)| = \sqrt{a} \leq \sqrt{s}, \; |V_0(s)| \leq \sqrt{s}.$$

Note that (10.7) and (10.8) provide the estimates

$$|V_n(s) - \sqrt{s/(1-\gamma)}| \leq \gamma^{n/2} \|V_0 - V\|_b \cdot \sqrt{s},$$
$$|V_{nf}(s) - V_n(s)| \leq 2\gamma^{n/2} \|V_0 - V\|_b \cdot \sqrt{s}.$$

◆

10.2 The Structure of the Limit Value Function

Most of the results in Chaps. 6–8 about the structure of the value functions V_n and of maximizers f_n of W_n easily carry over to structure results for the limit value function V of the sequence (V_n) and to maximizers f of W, respectively. (The latter are often asymptotically optimal, cf. Theorem 10.1.10(d).) The reason for this simple situation is that most of the properties of V_n involved are preserved under the limit for $n \to \infty$. For the investigation of upper [lower] semicontinuity of V we use the following notion. A function b from the metric space (S, ρ_S) into $\mathbb{R}_+$ is called *locally bounded* if b is bounded on each non-empty compact subset of S, or equivalently, if $\sup_n b(s_n) < \infty$ for each convergent sequence (s_n) in S. If $S \subset \mathbb{R}^d$ then b is locally bounded if and only if it is bounded on each ε-ball of any point in S. In particular, if $S = \mathbb{R}_+^d$, then b is locally bounded if and only if b is bounded on each interval $[0, x]$, $x \in \mathbb{R}_+^d$.

Now we collect the pertinent results for V and f. In brackets we cite results about the corresponding property for the value functions.

Theorem 10.2.1 (Preservation of the structure of value functions under the limit $n \to \infty$) *Assume that $V := \lim_n V_n$ exists. For (b)–(f) we assume that V is finite, and for (d) and (e) that S and A are metric spaces.*

(a) If all functions V_n are increasing with respect to an arbitrary relation $\le_S$ on S, then V is increasing with respect to $\le_S$. [Theorem 6.3.5]
(b) If all functions V_n are concave then V is concave. [Theorem 7.1.2]
(c) If all functions V_n are convex then V is convex. [Proposition 7.2.1]
(d) Let all functions V_n be finite and usc or continuous. If (V_n) converges to V in b-norm for a locally bounded function b from S into $\mathbb{R}_+$ (in particular if (V_n) converges uniformly) then V is usc or continuous, respectively. [Theorem 9.2.5]
(e) If all functions V_n are finite and Lipschitz with $\liminf_n l(V_n) < \infty$, then V is Lipschitz and

$$l(V) \le \liminf_n l(V_n).$$

Proof Everything is simple, except for (d), for which we reason as follows.

(d1) Assume that V_k is usc for all k. As (V_k) converges in b-norm to V, we have

$$|V_k(s) - V(s)| \le \varepsilon \cdot b(s) \text{ for } s \in S,\ \varepsilon > 0 \text{ and } k > k_0 = k_0(\varepsilon).$$

Select some sequence (s_n) in S converging to some $s \in S$. Then

$$V(s_n) \le V_k(s_n) + \varepsilon \cdot b(s_n) \text{ for } n \in \mathbb{N} \text{ and } V_k(s) \le V(s) + \varepsilon \cdot b(s) \text{ for } k > k_0.$$

The number $\alpha := \sup_n b(s_n)$ is finite as b is locally bounded. Now we get for $k > k_0$ that

$$\limsup_{n\to\infty} V(s_n) \le \limsup_{n\to\infty} V_k(s_n) + \varepsilon \cdot \alpha \le V_k(s) + \varepsilon \cdot \alpha \le V(s) + \varepsilon \cdot [\alpha + b(s)].$$

As $\varepsilon > 0$ is arbitrary and as $\alpha + b(s)$ is finite, upper semicontinuity of V is shown.

(d2) If all functions V_k are continuous, then both V and $-V$ are usc by (d1) as the sequence $(-V_n)$ also satisfies the assumptions needed for (d1). But then V is continuous. □

Remark 10.2.2 In the case of an absorbing DP (cf. Sect. 5.1) often the easiest way to prove structural properties of V and of maximizers of LV seems to be to use Theorem 10.2.1. Alternatively, one may try to use recursion in state space. ◇

Remark 10.2.3 There are examples where none of the functions V_n has a certain property (P) but nevertheless V has property (P). In such cases V_0 is often responsible for V_n not having property (P). As an example, if C_0 in Example 10.1.2(c) is neither convex nor concave then Example 10.1.2(a1) shows that the same holds for all functions C_n. However, if C_0 is bounded or if $0.5 \le \beta < 1$, then $s \mapsto C(s) = 2s$

is both convex and concave.—In general, when no explicit solution is available and when one does not succeed in showing that all V_n's have a certain property (P), one may be able to verify (P) for V by applying Theorem 10.1.10 with V_0 replaced by an appropriate $V_0' \in B_b(S)$; often $V_0' :\equiv 0$ will be useful. ◇

Concerning maximizers of W we have the following result whose simple proof is omitted.

Theorem 10.2.4 (Existence and structure of maximizers of W) *Assume that V and hence also W exists. For (c) and (d) we assume that S is structured.*

(a) Assume that S and A are metric spaces such that:
(i) $D(s)$ is compact for all s,
(ii) $a \mapsto T(s,a)$ is continuous for all s,
(iii) $a \mapsto r(s,a)$ is usc for all s,
(iv) V exists and is usc. Then W has a maximizer. If in addition $A \subset \mathbb{R}$, then there exist a smallest and a largest maximizer of W.
(b) If $D(s)$ is a polytope and if $W(s,\cdot)$ is convex for all s, then W has a bang-bang maximizer.
(c) If A is totally ordered and if all functions W_n have increasing differences, then W has increasing differences.
(d) Assume that W has a maximizer f which is the limit of a sequence (f_n) of decision rules.
(d1) If all f_n are increasing, then f is increasing.
(d2) If S and A are subsets of $\mathbb{R}$ and if $f_n \in ILIP(\lambda)$ for all n and some $\lambda \in \mathbb{R}_+$, then $f \in ILIP(\lambda)$.

10.3 DPs with Infinite Horizon

Now we define DPs with infinite horizon.

Definition 10.3.1 A **DP with infinite horizon** (***DP***$_\infty$ for short) is a tuple (S, A, D, T, r, β) of the following kind:

- S, A, D and T are defined as in a DP.
- r is a bounded function on D, the one-stage reward function.
- $\beta < 1$ is the discount factor.

Notice that no terminal reward function V_0 is required. For an initial state s_0 denote by $A^\infty(s_0)$ the set of infinite action-sequences $y := (a_t)_0^\infty$ which satisfy the restrictions $a_t \in D(s_t)$, $s_{t+1} := T(s_t, a_t)$, $t \in \mathbb{N}_0$. Because of the first two properties in the definition the infinite-stage reward

$$V_{\infty y}(s) := \sum_{t=0}^{\infty} \beta^t \cdot r(s_t, a_t),$$

for any initial state $s := s_0$ and action sequence $y \in A^\infty(s)$ exists and is finite, and the same is true for the ***maximal infinite-stage reward***

$$V_\infty(s) := \sup_{y \in A^\infty(s)} V_{\infty y}(s).$$

We call $s \mapsto V_\infty(s)$ the ***infinite-stage value function.*** The action sequence $y \in A^\infty(s)$ is called *s-optimal* if it maximizes $y \mapsto V_{\infty y}(s)$. Note that the latter is a function of the infinitely many variables $a_0, a_1, \ldots$. An action sequence $y \in A^\infty(s_0)$ is said to be generated by a decision rule f if $a_t = f(s_t)$, $s_{t+1} := T(s_t, a_t)$, $t \in \mathbb{N}_0$.

By a reasoning similar to the proof of Theorem 10.1.10 one obtains the next result.

Theorem 10.3.2 (The optimality equation and optimal action sequences) *Assume that r is bounded and that $\beta < 1$.*

(a) The infinite-stage value function V_∞ equals V, and hence is the unique bounded solution of the OE, i.e.

$$V_\infty(s) = \sup_{a \in D(s)} [r(s,a) + \beta \cdot V_\infty(T(s,a))],\ s \in S.$$

(b) If f is a maximizer of LV_∞, then for any $s_0 \in S$ the action sequence y^*, generated by s_0 and f, is s_0-optimal.

*Example 10.3.3 (**DP**$_\infty$s with random length of periods)* We consider the infinite-stage version of Example 4.2.1 with an i.i.d. sequence $\eta^\infty := (\eta_t)_1^\infty$ of random lengths of the periods with probability distribution $Q(dz)$. We exclude the trivial case where $\eta_1 = 0$ almost surely so that the sequence of decision epochs $T_t := \sum_{i=1}^t \eta_i$ converges almost surely to ∞ for $t \to \infty$. In probability theory the sequence $(T_t)_1^\infty$ is called a *renewal process* with interarrival distribution $Q(dz)$. We assume that r_Z is bounded and that $\alpha > 0$. Put $\beta := \mathbb{E}\, e^{-\alpha\eta_1}$, which is smaller than one, as $P(\eta_1 = 0) < 1$. Then for any $s_0 \in S$ and $y = (a_t)_0^\infty \in A^\infty(s_0)$ we have

$$\mathbb{E}\left(\sum_0^\infty e^{-\alpha \cdot T_t} \cdot |r_Z(s_t, a_t, \eta_{t+1})|\right) \tag{10.12}$$
$$\le \sup|r_Z| \cdot \sum_0^\infty \mathbb{E}\left(\prod_{i=1}^t e^{-\alpha\cdot\eta_i}\right) = \sup|r_Z|/(1-\beta) < \infty.$$

Here the interchange of expectation and summation is allowed since $e^{-\alpha \cdot T_t} \cdot |r_Z(s_\nu, a_t, \eta_{t+1})| \ge 0$. Because of (10.12) there exists almost surely the random variable

$$R_{\infty y}(s_0, \eta^\infty) := \sum_0^\infty e^{-\alpha\cdot T_t} \cdot r_Z(s_t, a_t, \eta_{t+1}),$$

it is integrable, and the expected infinite-stage reward $\mathbb{E}\, R_{\infty y}(s_0, \eta^\infty)$ for initial state s_0 and action sequence y equals $\sum_0^\infty \beta^t \cdot \mathbb{E}\, r_Z(s_t, a_t, \eta_1)$. This shows that our problem reduces to the DP$_\infty$ with $r(s,a) := \mathbb{E}\, r_Z(s,a,\eta_1)$ and discount factor β. ♦

Example 10.3.4 (A batch service problem with an explicit solution) The batch service interpretation of the model is as follows. Customers arrive at a service station with a finite waiting room of $m \geq 1$ places according to a renewal process with interarrival distribution $Q(dz)$ concentrated on $\mathbb{N}$. At each arrival the server decides whether to serve the arriving and the waiting customers as a batch (action $a = 1$) with constant service time 1 and with cost $K > 0$, or to wait for the next arrival (action $a = 0$). (Of course, if the waiting room is full at the arrival of a customer, the m waiting customers and the arriving customer must be served as a batch.) In the latter case there is a waiting cost $K_1(z) \geq 0$ for each customer waiting for the momentary period of length z. We want to minimize the expected infinite-stage costs with discount rate $\alpha > 0$. (In the notation of queueing theory we control a system of type $GI/D^B/1/m$.)

We use the DP$_\infty$ from Example 10.3.3 with the following data: s is the momentary number of customers in the system, i.e. those waiting plus the one just arriving, hence $S = \mathbb{N}_{m+1}$; $A = D(s) = \{0,1\}$ for $s \leq m$ and, as a full waiting room enforces a batch service, $D(m+1) = \{1\}$; $T(s,0) = s+1$, $T(s,1) = 1$; the one-stage cost has the form $c_Z(s,0,z) = s \cdot K_1(z)$ and $c_Z(s,1,z) = K$, hence $c(s,0) = \gamma s$, $c(s,1) = K$. Here we assume that $\gamma := \mathbb{E}\, K_1(\eta_1)$ is finite and positive. In the most important special case $K_1(z)$ accumulates with some cost rate $c_1 \geq 0$, i.e.

$$K_1(z) = c_1 \cdot \sum_{x=0}^{z-1} e^{-\alpha x} = c_1 \cdot \frac{1 - e^{-\alpha \cdot z}}{1 - e^{-\alpha}}. \tag{10.13}$$

In this case we have $\gamma = c_1 \cdot (1-\beta)/(1-e^{-\alpha})$, where $\beta := \mathbb{E}\, e^{-\alpha \cdot \eta_1} < 1$. The DP$_\infty$ can also be interpreted as a maintenance problem for an aging equipment which has maximal age $m+1$ and which is maintained at each time point $t \in \mathbb{N}_0$. If the equipment is not replaced at age $s \in \mathbb{N}_m$, one has to pay maintenance costs proportional to age, i.e. $c(s,0) = \gamma s$ (where $\gamma > 0$ is a given factor), and the equipment ages by one period. If the equipment is replaced by a new one at age $s \in \mathbb{N}_{m+1}$ (it must be replaced at age $m+1$), one has to pay replacement cost $K > 0$. The duration of replacement is one period. There is a constant discount rate $\beta \in (0,1)$.

It follows from the minimization version of Theorem 10.3.2 and Example 10.3.3 that the minimal expected infinite-stage cost function C_∞ equals C and is the unique solution of the OE. Therefore the OE reads

$$\begin{aligned} C(s) &= \min\{\gamma s + \beta \cdot C(s+1), C(m+1)\} && (10.14)\\ &=: \min\{W(s,0), W(s,1)\}, \quad 1 \leq s \leq m, \\ C(m+1) &= K + \beta \cdot C(1); && (10.15) \end{aligned}$$

here we already plugged (10.15) into (10.14). The functional equation consisting of (10.15) and (10.14) has a *circular* form. As none of the values $C(s)$, $1 \leq s \leq m+1$ is known a priori, it cannot be solved numerically by recursion in state space. Surprisingly, it has an explicit solution, as given in Proposition 10.3.5(b) below. Moreover, we confirm the plausible conjecture that it is optimal to serve at any time if and only if the number of customers in the system reaches some critical level $s^* = s^*(m, K, \gamma) \in \mathbb{N}_{m+1}$. Note that under the optimal policy the sequence of states, starting in $s_0 := 1$, runs cyclically through the set $\{1, 2, \ldots, s^*\}$. ♦

An explicit solution of the above example is provided by the following proposition

Proposition 10.3.5 (Explicit solution of the batch service problem)

(a) *It is optimal to serve as soon as s^* customers are in the system, where $s^* = s^*(m)$ is the largest $k \in \mathbb{N}_{m+1}$ such that*

$$t_k := \frac{1}{1-\beta} \cdot \left(k - \frac{1-\beta^k}{1-\beta}\right) \leq \frac{K}{\gamma}.$$

(b) *The minimal expected infinite-stage costs satisfy*

$$C(s) = C(s^*) = \frac{K + \gamma \cdot \sum_{i=1}^{s^*-1} i \cdot \beta^i}{1 - \beta^{s^*}}, \; s^* \leq s \leq m+1, \tag{10.16}$$

and, if $s^* \geq 2$,

$$C(s) = \gamma s + \beta \cdot C(s+1), \quad s = s^* - 1, s^* - 2, \ldots, 1. \tag{10.17}$$

Proof

(a1) We show by downward induction on s the intuitively obvious fact that C is increasing on $\mathbb{N}_{m+1}$. Firstly, (10.14) shows that $C(m) \leq C(m+1)$. Assume $C(s-1) \leq C(s)$ for some $3 \leq s \leq m+1$. Then (10.14) and $\gamma > 0$ imply that

$$C(s-2) \leq \min\{\gamma \cdot (s-1) + \beta \cdot C(s), C(m+1)\} = C(s-1).$$

(a2) Put $W(0,0) := \beta \cdot C(1)$. (Note that $W(s,0)$ is defined by the DP only for $1 \leq s \leq m$.) From (a1) we see that $s \mapsto W(s,0)$ is strictly increasing on $\mathbb{N}_{0,m}$. As $W(0,0) = \beta \cdot C(1) < C(m+1)$ by (10.15), there is a largest number k in $\mathbb{N}_{0,m}$ such that

$$W(k,0) \leq C(m+1).$$

Let j be the largest of these numbers k. Then it follows from the OC and from (10.14) that

$$s \mapsto f(s) = \begin{cases} 0, & \text{if } 1 \le s \le j, \\ 1, & \text{if } j+1 \le s \le m+1, \end{cases}$$

is the smallest minimizer of LC. Moreover, the OE yields

$$C(s) = \begin{cases} \gamma \cdot s + \beta \cdot C(s+1), & \text{if } 1 \le s \le j, \\ C(j+1) = K + \beta \cdot C(1), & \text{if } j+1 \le s \le m+1. \end{cases} \tag{10.18}$$

Next one obtains from (10.18) that

$$\begin{aligned} C(j+1) &= K + \beta \cdot C(1) = K + \beta \cdot \gamma + \beta^2 \cdot C(2) \\ &= \cdots = K + \gamma \cdot \sum_{i=1}^{j} i \cdot \beta^i + \beta^{j+1} \cdot C(j+1), \end{aligned}$$

hence

$$C(j+1) = B := \frac{K + \gamma \cdot \sum_{i=1}^{j} i \cdot \beta^i}{1 - \beta^{j+1}}. \tag{10.19}$$

(a3) Put $\sigma := \sum_{i=0}^{j} \beta^i$. We assert that for $0 \le s \le m$ such that $(s = j) \vee (s = j+1)$

$$\sigma \cdot (W(s,0) - B) = \gamma \cdot t_{s+1} - K.$$

In fact, we have by (10.18) and (10.19)

$$\begin{aligned} &\sigma \cdot (W(s,0) - B) = \sigma \cdot (\gamma s + \beta \cdot C(s+1) - B) \\ &= \sigma \cdot (\gamma s - (1-\beta) \cdot B) = \sigma \cdot \gamma s - K - \gamma \cdot \sum_{i=1}^{j} i \cdot \beta^i \\ &= \gamma s \cdot \sum_{i=0}^{j} \beta^i - \gamma \cdot \sum_{i=0}^{j} i \cdot \beta^i - K = \gamma \cdot \sum_{i=0}^{j} (s-i) \cdot \beta^i - K \\ &= \sum_{i=0}^{s-1} (s-i) \cdot \beta^i - K = \gamma \cdot t_{s+1} - K. \end{aligned}$$

(a4) We now prove that $j + 1 = s^*$, which verifies (a). In fact, if $j < m$, then by definition on j we have by (10.14) and (10.19)

$$W(j,0) \le B = W(j+1,1) < W(j+1,0),$$

Table 10.1 $C(s)$, $1 \le s \le s^* = 6$

$C(1)$	$C(2)$	$C(3)$	$C(4)$	$C(5)$	$C(6)$
91.13	96.30	100.35	103.19	104.74	104.91

which implies by (a3) that $t_{j+1} \le K/\gamma < t_{j+2}$. If $j = m$, then $W(m,0) \le B$ by (10.19), which implies by (a3) that $t_{m+1} \le K/\gamma$. Now strict isotonicity of $(t_k)_1^\infty$ and the definition of s^* verify that $s^* = j + 1$.

(b) This part follows from (10.18) and (10.19) since $s^* = j + 1$ by (a4). □

Example 10.3.6 As a special case of Example 10.3.4 assume that Q is the geometric distribution Geo(p) with parameter p and K_1 as in (10.13) with $c_1 = 1$. Then, using $q := 1 - p$, we have

$$\beta = p/(e^\alpha - q), \quad \gamma = (1 - q \cdot e^{-\alpha})^{-1}.$$

For $\alpha = 0.05, p = 0.7$ we obtain $\beta = 0.9318, \gamma = 1.3993$. The numerical solution for $m = 11$ and $K = 20$ (hence $K/\gamma = 14.2929$) yields $s^* = 6$ and Table 10.1. Here the waiting room of size 11 is not fully used, a waiting room of size 5 would suffice. This leads to the question to find for given γ, K the largest size $m = m^*$ of the waiting room such that no waiting room is wasted (assuming that initially only one customer is in the system). It is easy to see that m^* is the largest $k \in \mathbb{N}_0$ such that

$$t_{k+1} \le K/\gamma < t_{k+2}. \blacklozenge$$

Remark 10.3.7

(a) The number $s^*(m)$ exists since t_k is obviously strictly increasing in k towards ∞. Moreover, if $s^*(\infty)$ denotes the largest integer in $\mathbb{N}$ such that $t_k \le K/\gamma$, then $s^*(m) = \min\{m + 1, s^*(\infty)\}$.
(b) For the numerical computation of t_k one can use that $t_{k+1} = \beta \cdot t_k + k$, $k \ge 1$, $t_1 = 0$. This also proves that $t_k = \sum_{i=0}^{k-2} (k - 1 - i) \cdot \beta^i$, $k \ge 1$.
(c) The series in (10.16) can be summed up.
(d) From (10.17) one can derive by induction on s an explicit form of $C(s)$ for $1 \le s < s^*$, which, however, is less useful for computations than the recursion (10.17). ◇

Example 10.3.8 (Asymptotic behavior of the solution of the linear-quadratic problem from Example 4.1.7*)*

(a) For each $\beta > 0$ the sequence (d_n) converges monotonely for $n \to \infty$ towards the unique positive solution d, independent of d_0, of the quadratic equation

$$I(x) := \beta h^2 x^2 + (\gamma - \beta \gamma g^2 - \beta \lambda h^2) \cdot x - \gamma \lambda = 0, \quad x \in \mathbb{R}. \tag{10.20}$$

Hence $C_n(s)$ converges for $n \to \infty$ to $C(s) := ds^2$, $s \in S$.

(b) The limit of the decision rules f_n, i.e. the decision rule

$$s \mapsto f(s) := h(d - \lambda)s/(\gamma g)$$

is asymptotically optimal.

Proof

(a1) The sequence $(d_n)_0^\infty$ is bounded since (4.4) implies for $n \geq 0$

$$0 \leq d_{n+1} - \lambda = \frac{\beta\gamma g^2 \cdot d_n}{\gamma + \beta h^2 \cdot d_n} = \frac{\beta h^2 \cdot d_n}{\gamma + \beta h^2 \cdot d_n} \cdot \frac{\gamma g^2}{h^2} \leq \frac{\gamma g^2}{h^2}. \tag{10.21}$$

The sequence $(d_n)_0^\infty$ is also monotone (hence convergent to some $d \in [\lambda, \infty)$). In fact, firstly $C_n(s) = d_n \cdot s^2$, $n \geq 1$, $s \in \mathbb{R}$, by Example 4.1.7. Now Theorem 6.1.1 implies that $C_n(s)$ is monotone in n if $(C_0 \leq C_1) \vee (C_0 \geq C_1)$, i.e. if $(d_0 \leq d_1) \vee (d_0 \geq d_1)$, which always holds.

(a2) Letting n tend to ∞ in (10.21) yields

$$d = \lambda + \beta\gamma g^2 d/(\gamma + \beta h^2 d), \tag{10.22}$$

which is equivalent to $I(d) = 0$. Thus d is independent of d_0. Moreover, since $I(0) = -\gamma\lambda < 0$, $\beta h^2 > 0$ and since $d \geq \lambda > 0$, the equation $I(x) = 0$ has the unique positive root $x := d$.

(b1) Put $\delta := g - h^2 \cdot (d - \lambda)/(\gamma g)$. Then $gs - hf(s) = \delta s$. It is easy to see (cf. corresponding computations in Example 4.1.7 with d_0 replaced by d) that $U_f C = C$. Now induction on $n \geq 0$, using $C - C_{N+1,f} = U_f C - U_f C_{Nf}$, yields

$$C(s) - C_{Nf}(s) = (d - d_0)(\beta\delta^2)^N \cdot s^2.$$

(b2) We show that $\beta\delta^2 < 1$. By (10.22) the assertion is equivalent to $\beta g^2 < (1 + \beta h^2 d/\gamma)^2$. Using $I(d) = 0$ we get

$$0 < \lambda \cdot (\gamma + \beta h^2 d) = I(d) + \beta\lambda h^2 d + \gamma\lambda = d \cdot (\gamma + \beta h^2 d - \beta\gamma g^2).$$

Therefore $\beta g^2 < 1 + \beta h^2 d/\gamma \leq (1 + \beta h^2 d/\gamma)^2$.

(b3) From (b2) we see that

$$\begin{aligned} C_N(s) - C_{Nf}(s) &= [C_N(s) - C(s)] + [C(s) - C_{Nf}(s)] \\ &= (d_N - d) \cdot s^2 + (d - d_0)(\beta\delta^2)^N \cdot s^2, \end{aligned}$$

which converges to zero by (a) and (b2). □

10.4 Problems

Problem 10.4.1 (The allocation problem with an increasing utility function) A possible setting of the well-known allocation problem is defined by $S = \mathbb{R}_+$, $A = D(s) = [0,1]$, $r(s,a) = u(as)$ for an arbitrary concave and increasing utility function u on S with $u(0) = 0$, $T(s,a) = (1-\alpha)s/\beta$ and $V_0(s) := d \cdot u(cs)$ for some $c \in (0, 1/\beta]$ and $d \leq 0$. There seems to be no hope for an explicit solution. Assume $\beta < 1$. Then:

(a) $s \mapsto b(s) := u(s/\beta)$ is a bounding function with $\beta_b = 1$.
(b) V_0 belongs to $B_b(S)$.
(c) $V_n \in B_b(s)$ for all n and $(V_n)_0^\infty$ converges in b-norm towards

$$s \to V(s) := u((1-\beta)s)/(1-\beta).$$

(d) $f := 1 - \beta$ is asymptotically optimal.
(e) The function $n \mapsto V_n(s)$ is increasing.
(f) A lower bound for V_n is

$$V_{nf}(s) = V(s) - \beta^n(V(s) - V_0(s)).$$

Problem 10.4.2 Consider Proposition 10.3.5 of the batch service problem from Example 10.3.4.

(a) We have $s^* = 1$ if and only if $K/\gamma < 1$, $2 \leq s^* \leq m$ if and only if $1 \leq K/\gamma < t_{m+1}$ and $s^* = m + 1$ if and only if $K/\gamma \geq t_{m+1}$. Interpretation!
(b) Show that s^* is increasing in γ and decreasing in β.

10.5 Supplement

In this book we prefer to lay more emphasis, at least from the conceptual point of view, on $\lim DP_n$ than on DP_∞. Here are a number of reasons for our approach.

(i) In reality all time spans are finite, and thus the number of decisions, either by persons or by technical devices, must be finite though possibly very large. Thus the ultimate goal, whether $\lim DP_n$ or DP_∞ is used, is the construction of an approximate solution for DP_N. Now, V is already by definition an approximation for V_N while V_∞, if it exists, may be different from $\lim_N V_N$ and hence useless as an approximation.
(ii) The analysis of DP_∞ presents a number of problems, e.g. the existence of stationary optimal policies, which do not arise in $\lim DP_n$.
(iii) DP_∞ does not take into account the influence of V_0. This means in fact that V_0 is put equal to zero. Therefore it may happen that V_∞ exists while $\lim V_n$ does not. In such cases it would be erroneous to use V_∞ as an approximation of V_N for large N.

Part II
Markovian Decision Processes

Chapter 11
Control Models with Disturbances

In this chapter we introduce control models with finite and i.i.d. disturbances. We prove the reward iteration and derive the basic solution techniques: value iteration and optimality criterion.

In the simplest N-stage stochastic model with initial state s_0, the transition from state s_t, $0 \leq t \leq N-1$, to s_{t+1} is determined by a transition function T which is *disturbed* by some random variable η_{t+1}. These random variables, which are called the ***disturbances***, take values in a *finite* set Z, called the ***disturbance space***, and they are assumed to be i.i.d. (independent and identically distributed).

Thus, if at time t we are in state s_t, take action a_t and if the disturbance η_{t+1} takes the value z_{t+1}, then the system moves to the new state s_{t+1} according to the so-called ***state equation***

$$s_{t+1} := T(s_t, a_t, z_{t+1}), \quad 0 \leq t \leq N-1.$$

We can and shall assume, unless stated otherwise, that the η_t's are defined on some finite probability space $(\Omega, \mathbb{P})$. The canonical choice is $\Omega := Z^N$, η_t is the t-th coordinate random variable and $\mathbb{P}$ is the probability distribution on Ω with discrete density

$$z^N := (z_t)_1^N \mapsto \prod_1^N q(z_t), \quad z^N \in Z^N,$$

where $z \mapsto q(z) := \mathbb{P}(\eta_1 = z)$ is the discrete density of η_1 on Z. We allow that the one-stage reward $r_Z(s, a, z)$ may depend not only on the momentary state s and the momentary action a but also on the momentary disturbance z. In many applications of the present model the state space S is countable or even finite. However, as other cases occur as well we allow S to be arbitrary. Then all deterministic DPs from Part I are contained in the present model by the case $|Z| = 1$. In the general case the

K. Hinderer et al., *Dynamic Optimization*, Universitext,
DOI 10.1007/978-3-319-48814-1_11

random variables η_t and the probability space Ω do not belong to the data defining the model; they are only needed for the formulation of the optimization problem in this section (see (11.2) below).

Definition 11.1 A **control model with finite and i.i.d. disturbances** (*CM* for short) is a tuple $(S, A, D, Z, Q, T, r_Z, V_0, \beta)$ of the following kind:

- S, A, D, V_0 and β have the same meaning as in the DP (cf. Definition 2.2.1).
- $Z \neq \emptyset$ is a finite set of disturbances.
- Q is the discrete probability distribution of each of the i.i.d. disturbances; thus Q has on Z the discrete density $z \mapsto q(z) := Q(\{z\})$.
- $T: D \times Z \to S$ is the transition function.
- $r_Z: D \times Z \to \mathbb{R}$ is the one-stage reward function.

We obtain a lot of examples of CMs by adding a disturbance space Z and a probability distribution Q on Z to any of the many examples of deterministic DPs in Part I.

The notions *decision rule* and *N-stage policy* are used as in Chap. 2.

For given initial state $s_0 \in S$ and *horizon* $N \in \mathbb{N}$ we now introduce the *N-stage optimization problem* $CM_N(s_0)$. We firstly explain the evolution of the sequence of random states under the control of the decision maker. We distinguish between the random state at time t from its realizations by denoting the former by ζ_t and the latter by s_t. Assume that the first action a_0 has been selected from $D(s_0)$. In a deterministic DP it suffices for the next step to specify a single action a_1 from the uniquely determined set $D(s_1)$, where $s_1 := T(s_0, a_0)$. In the stochastic case we do not know after the choice of a_0 which realization of $\zeta_1 := T(s_0, a_0, \eta_1)$ will occur. Thus we must specify the action a_1 in dependence of any possible realization s_1 of ζ_1, or more simply, for all states $s_1 \in S$. This means to select a decision rule π_1 from $\mathbb{F}$ and to take action $a_1 := \pi_1(s_1)$ whenever the realization s_1 of ζ_1 occurs. After this has been done, we know the next random state $\zeta_2 := T(\zeta_1, \pi_1(\zeta_1), \eta_2)$. This procedure is repeated: at time t, $1 \leq t \leq N-1$, one selects some decision rule π_t and then the next state is determined as $\zeta_{t+1} := T(\zeta_t, \pi_t(\zeta_t), \eta_{t+1})$. Thus the role played in deterministic DPs by the action sequences $y = (a_t)_0^{N-1} \in A^N(s_0)$ is now taken over by policies $\pi = (\pi_t)_0^{N-1} \in \mathbb{F}^N$. More precisely, for any N-stage policy $\pi = (\pi_t)_0^{N-1}$ the state at time t of the process governed by π when starting in some initial state s_0 is an S-valued random variable $\zeta_{t\pi} = \zeta_{t\pi}^{s_0}(\eta_1, \eta_2, \ldots, \eta_t)$ on $(\Omega, \mathbb{P})$, defined recursively by $\zeta_{0\pi} :\equiv s_0$ and

$$\begin{aligned}\zeta_{t+1,\pi} &= \zeta_{t+1,\pi}(s_0, \eta_1, \ldots, \eta_{t+1}) \\ &= T(\zeta_{t\pi}, \pi_t(\zeta_{t\pi}), \eta_{t+1}), \quad 0 \leq t \leq N-1.\end{aligned}$$

We call $(\zeta_{t\pi}^{s_0})_{t=0}^N$ the ***decision process*** generated by s_0 and π. The dependence of $\zeta_{t\pi}^{s_0}$ on s_0 will in general be suppressed. Note that

(a) $\zeta_{t\pi}$ depends on $\pi = (\pi_t)_0^{N-1}$ only through $\pi_0, \pi_1, \ldots, \pi_{t-1}$.
(b) For s_0 and π fixed, $\zeta_{t\pi}$ assumes only finitely many values, as Z is finite.

(c) Since the η_t's are i.i.d., the joint distribution of $\zeta_{1\pi}$, ..., $\zeta_{N\pi}$ is already determined by π, the initial state s_0 and by the probability distribution Q. The $\zeta_{t\pi}$'s are in general not independent.

Put $\eta^N := (\eta_t)_1^N$. The ***N*-stage random reward** earned by the decision process $(\zeta_{t\pi})_{t=0}^N$, when starting in s_0, is the real random variable on $(\Omega, \mathbb{P})$

$$G_{N\pi}(s_0, \eta^N) := \sum_{t=0}^{N-1} \beta^t \cdot r_Z(\zeta_{t\pi}, \pi_t(\zeta_{t\pi}), \eta_{t+1}) + \beta^N \cdot V_0(\zeta_{N\pi}). \tag{11.1}$$

Remark 11.2

(a) By (11.1) the case where the one-stage reward also depends on the next state s' can be reduced to our setting by substituting s' by $T(s, a, z)$.
(b) For most applications one can apply the basic solution method of Theorem 11.6 below directly to the data of the CM without writing up $G_{N\pi}$ explicitly. ◇

Fix $s_0 \in S$. One cannot expect to find a policy which maximizes $\pi \mapsto G_{N\pi}(s_0, \eta^N(\omega))$ simultaneously for each $\omega \in \Omega$. Therefore one is content to solve the following ***problem*** $\boldsymbol{CM_N(s_0)}$: maximize the *expected N-stage reward for policy π and fixed initial state* s_0, defined by the real number

$$V_{N\pi}(s_0) := \mathbb{E}\, G_{N\pi}(s_0, \eta^N) = \sum_{z^N \in Z^N} G_{N\pi}(s_0, z^N) \cdot \prod_{t=0}^{N} q(z_t). \tag{11.2}$$

(The latter equality in (11.2) shows that $V_{N\pi}(s_0)$ does not depend on the choice of the probability space and the state random variables.)

Even if one wants to solve $CM_N(s_0)$ only for a single initial state s_0, the sequential solution method of Theorem 11.6 below requires us to solve all problems $CM_n(s)$ for all $1 \le n \le N-1$ and for all initial states $s \in S$. Moreover, as $V_{N\pi}(s_0)$ depends on π_0 only through the single action $a_0 := \pi_0(s_0)$, we see that a maximum point π^* of $\pi \mapsto V_{N\pi}(s_0)$, if it exists, can be chosen independently of s_0. *Therefore we maximize $\pi \mapsto V_{N\pi}(s_0)$ jointly for all initial states $s_0 \in S$.* We call this the ***control problem*** $\boldsymbol{CM_N}$.

In the sequel it is convenient to denote an arbitrary initial state by s rather than by s_0.

Definition 11.3 Given a CM_N we define:

- An N-stage policy π^* is called ***optimal*** for the N-stage problem if it maximizes $\pi \mapsto V_{N\pi}(s)$ for all $s \in S$.
- By a ***solution of the N-stage problem*** $\boldsymbol{CM_N}$ we understand an N-stage optimal policy (if one exists) and the ***N-stage value function*** $V_N: S \to (-\infty, \infty]$,

defined by

$$s \mapsto V_N(s) := \sup\{V_{N\pi}(s) : \pi \in \mathbb{F}^N\}, \quad N \geq 1.$$

- Put

$$r(s,a) := \mathbb{E}\, r_Z(s,a,\eta_1), \quad (s,a) \in D.$$

The decision rule f_n is called a ***maximizer at stage*** $n \geq 1$ if for all $s \in S$ the action $f_n(s)$ is a maximum point on $D(s)$ of

$$a \mapsto W_n(s,a) := r(s,a) + \beta \cdot \mathbb{E}\, V_{n-1}(T(s,a,\eta_1)).$$

As Z^N is finite, (11.1) and (11.2) imply that

$$V_{N\pi}(s_0) = \sum_{t=0}^{N-1} \beta^t \cdot \mathbb{E}\, r_Z(\zeta_{t\pi}, \pi_t(\zeta_{t\pi}), \eta_{t+1}) + \beta^N \cdot \mathbb{E}\, V_0(\zeta_{N\pi}). \tag{11.3}$$

In view of this equation a naive approach to solving problem CM_N consists in maximizing separately in each period, i.e. to select (if possible) the actions $\pi_t^*(s)$ for $0 \leq t \leq N-1$ and all s as a maximum point of $a \mapsto \mathbb{E}\, r_Z(s,a,\eta_1)$. However, using this so-called *myopic policy* neglects that the action $a^* := \pi_t^*(s)$, which is optimal in period t in state s, may influence via the next state $s_{t+1} = T(s,a^*,\eta_{t+1})$ the sets of available actions and the achievable rewards at later periods in a negative manner. What is required is an appropriate balance in each period between the reward obtainable immediately and in the future.

One will expect that in the present stochastic setting the sequence of value functions can be computed in principle recursively by the following generalization of the value iteration of Part I:

$$V_n(s) = \sup_{a \in D(s)} \left[r(s,a) + \beta \cdot \mathbb{E}\, V_{n-1}(T(s,a,\eta_1))\right], \; n \geq 1, s \in S. \tag{11.4}$$

This recursion is plausible as it says that $V_n(s)$ is the supremum of the sum of the reward in the first period and the discounted maximal reward for the remaining $n-1$ periods, starting in the new state $T(s,a,\eta_1)$ and averaged over the possible disturbances η_1. As shown in Theorem 11.6 below, this is true, but the proof is not as simple as in the deterministic case.

For deterministic DPs the reward iteration (2.5) was a first step towards the Basic Theorem 2.3.3. We now proceed to the stochastic counterpart. We use the following *notation*:

$$T_f(s,\eta_1) := T(s,f(s),\eta_1), \quad r_f(s) := r(s,f(s)).$$

Lemma 11.4 (Reward iteration) *The expected N-stage reward $V_{N\pi}$ may be computed recursively for horizon N and policy π by the* ***reward iteration (RI*** *for short)*

$$V_{1f}(s) = r_f(s) + \beta \cdot \mathbb{E}\, V_0(T_f(s,\eta_1)),$$
$$V_{n(f,\sigma)}(s) = r_f(s) + \beta \cdot \mathbb{E}\, V_{n-1,\sigma}(T_f(s,\eta_1)),$$
$$2 \le n \le N, f \in \mathbb{F}, \sigma \in \mathbb{F}^{n-1}.$$

Proof The assertion for $n = 1$ follows easily from (11.3) with $N = 1$, $\pi = (f)$, and $s = s_0$. Consider the case $n \ge 2$. Fix $s \in S$ and $\pi := (f,\sigma) \in \mathbb{F} \times \mathbb{F}^{n-1}$. Put $X := (\eta_2, \eta_3, \dots, \eta_n)$, $x := (z_2, z_3, \dots, z_n)$. Then $\mathbb{P}(\eta_1^N = (z_1, x)) = \mathbb{P}(\eta_1 = z_1) \cdot \mathbb{P}(X = x)$ since the disturbances η_t are independent. From (11.1) one obtains $G_{n\pi}(s, \eta_1^N) = r_Z(s, f(s), \eta_1) + \beta \cdot G_{n-1,\sigma}(T_f(s,\eta_1), X)$. It follows from (11.2) that $V_{n\pi}(s) = r_f(s) + \beta \cdot \mathbb{E}\, G_{n-1,\sigma}(T_f(s,\eta_1), X)$. The second term divided by β may be written as

$$\sum_z \left[\sum_x G_{n-1,\sigma}(T_f(s,\eta_1), x) \cdot \mathbb{P}(\eta_1 = z) \right] \cdot \mathbb{P}(\eta_1 = z)$$
$$= \sum_z V_{n-1,\sigma}(T_f(s,\eta_1)) \cdot \mathbb{P}(\eta_1 = z) = \mathbb{E}\, V_{n-1,\sigma}(T_f(s,\eta_1)). \qquad \square$$

The RI (which also can be used for the computation of the single functions $V_{n\pi}$ on a computer in case that S and A are finite), says the following: The expected n-stage reward when starting in s, using the decision rule f in the first period and the policy σ in the last $n-1$ periods equals the sum of the reward, obtained under f in the first period and the discounted expected $(n-1)$-stage reward when starting in $s_1 := T_f(s, \eta_1)$ and using σ.

It is useful as in Part I to write the RI and other formulae in a succinct manner with the aid of certain *operators*. Another great advantage of the operators is the fact that many results for different models depend only on certain properties (cf. Lemma 11.5 below) of the operators defined by these models, so that proofs carry over quickly.

Let $\mathfrak{V}_0 := \{v: S \to \overline{\mathbb{R}} : \mathbb{E}\, v \circ T(s,a,\eta_1) \in \overline{\mathbb{R}}$ for all $(s,a) \in D\}$. The operators L, U_f and U are defined on $\mathfrak{V}_0$ as follows:

$$Lv(s,a) := r(s,a) + \beta \cdot \mathbb{E}\, v(T(s,a,\eta_1)), \tag{11.5}$$
$$U_f v(s) := Lv(s, f(s)) = r_f(s) + \beta \cdot \mathbb{E}\, v(T_f(s,\eta_1)), \tag{11.6}$$
$$Uv(s) := \sup_{a \in D(s)} Lv(s,a) = \sup_{f \in \mathbb{F}} U_f v(s). \tag{11.7}$$

Note that U_f and U map $\mathfrak{V}_0$ into itself and that $U_f v$ is the expected reward in a one-stage CM with policy $\pi := f$ and terminal reward function v.

The simple proof of the equality in (11.7) uses surjectivity of $f \mapsto f(s)$, s fixed. As η_1 assumes only finitely many values the expectation in (11.5) exists and hence Lv, $U_f v$ and Uv are well-defined.

Let $V_0 \in \mathfrak{V}_0$. Now the RI for policy $\pi = (\pi_t)_0^{N-1} =: (\pi_0, \sigma)$ may be written as

$$V_{N\pi} = U_{\pi_0} V_{N-1,\sigma} = U_{\pi_0} U_{\pi_1} \cdots U_{\pi_{N-1}} V_0.$$

Moreover, *f is a maximizer at stage n if and only if*

$$U_f V_{n-1} = UV_{n-1}. \tag{11.8}$$

(Note that we do not yet know whether or not $UV_{n-1} = V_n$.)

If the VI was of the form (11.4) it could be written as

$$V_n(s) = \sup_{a \in D(s)} W_n(s,a) = UV_{n-1}(s),\ n \geq 1, \tag{11.9}$$

which is equivalent to $V_n = U^n V_0$ for all $n \geq 1$.

Let us try to verify (11.9) along the lines of the proof in the deterministic case. At first,

$$V_1 = UV_0, \tag{11.10}$$

as $V_{1f} = U_f V_0$ by the RI. Using the RI once more and also the fact (see Lemma 2.3.1), that the two-dimensional supremum equals the iterated supremum, we obtain for $n \geq 2$

$$V_n(s) = \sup_{(f,\sigma)} V_{n(f,\sigma)}(s) = \sup_f \sup_\sigma \big[r_f(s) + \beta \cdot \mathbb{E}\, V_{n-1,\sigma}(T_f(s,\eta_1))\big],$$

hence

$$V_n(s) = \sup_f \big[r_f(s) + \beta \cdot \sup_\sigma \mathbb{E}\, V_{n-1,\sigma}(T_f(s,\eta_1))\big]. \tag{11.11}$$

At this point the proof in the deterministic case was complete, as there $\mathbb{E}\, V_{n-1,\sigma}(T_f(s,\eta_1))$ reduced to $V_{n-1,\sigma}(T_f(s))$. In the stochastic case there arises the problem to show that $\sup_\sigma$ and $\mathbb{E}$ in (11.11) can be interchanged, which at first sight seems to be hardly achievable. As a first step observe that

$$\sup_\sigma \mathbb{E}\, V_{n-1,\sigma}(T_f(s,\eta_1)) \leq \mathbb{E}\, V_{n-1}(T_f(s,\eta_1)) \tag{11.12}$$

as $V_{n-1,\sigma} \leq V_{n-1}$ and as this inequality is preserved when taking $\mathbb{E}(\ldots)$ and $\sup_\sigma(\ldots)$. Moreover, if there exists an n-stage optimal policy τ then the required interchange holds as

$$\sup_{\sigma} \mathbb{E}\, V_{n-1,\sigma}(T_f(s,\eta_1)) \geq \mathbb{E}\, V_{n-1,\tau}(T_f(s,\eta_1)) = \mathbb{E}\, V_{n-1}(T_f(s,\eta_1)).$$

For a formal proof of the VI we need the following properties of the operators introduced above. Note that if $v_1 \in \mathfrak{V}_0$ and if $v_2 \geq v_1$ then also $v_2 \in \mathfrak{V}_0$.

Lemma 11.5 (Properties of the operators L, U_f and U) *Let B denote any of the operators L, U_f or U. Then:*

(a) B is isotone, i.e. $Bv_1 \leq Bv_2$ for $v_1,\ v_2 \in \mathfrak{V}_0$ with $v_1 \leq v_2$.
(b) $B(v + \alpha) = Bv + \beta\alpha$ for $v \in \mathfrak{V}_0$ and real α.
(c) $V_1 = UV_0$, and $V_n \leq UV_{n-1}$ for $n \in \mathbb{N}$.

Proof Part (a) and (b) are easy and left to the reader, while (c) follows from (11.10) and (11.12) above. □

Theorem 11.6 (Basic Theorem for CMs with i.i.d. disturbances and finite disturbance space) *Let $V_0 \in \mathfrak{V}_0$. Then the following holds:*

(a) ***Optimality Criterion*** *(**OC** for short). If f_n is a maximizer at stage n for $1 \leq n \leq N$, then $(f_n)_N^1 := (f_N, f_{N-1}, \ldots, f_1)$ is optimal for CM_N, $N \in \mathbb{N}$.*
(b) The ***value iteration*** *(**VI** for short) holds in the form*

$$\begin{aligned} V_n(s) &= \sup\{r(s,a) + \beta \cdot \mathbb{E}\, V_{n-1}(T(s,a,\eta_1)) : a \in D(s)\} \\ &= \sup\{W_n(s,a) : a \in D(s)\} \\ &= UV_{n-1}(s),\ n \in \mathbb{N},\ s \in S. \end{aligned}$$

Proof

(a) We prove the assertion, denoted by (I_N), by induction on N. From Lemma 11.5(c) and (11.8) for $n = 1$ we obtain $V_1 = UV_0 = U_{f_1} V_0 = V_{1,f_1}$. Thus (I_1) is true. Assume (I_N) to be true for some $N \in \mathbb{N}$. Then $\sigma := (f_n)_N^1$ is optimal for CM_N. Now Lemma 11.5(c) and (11.8) for $f := f_{N+1}$ imply (I_{N+1}) as

$$V_{N+1} \leq UV_N = U_f V_N = U_f V_{N\sigma} = V_{N+1,(f,\sigma)} \leq V_{N+1}. \tag{11.13}$$

(b) From (11.13) we see that the VI holds at least when there exist maximizers at all stages. This covers many applications. For the proof in the general case we refer to Hinderer (1970). □

Remark 11.7 (Solution of a CM_N) The recursive solution of problem CM_N by means of the Basic Theorem 11.6 runs as follows. Firstly one computes, using V_0, the numbers $V_1(s)$ for all s by maximizing $W_1(s,\cdot)$, and this also yields a maximizer f_1 at stage 1 (if one exists). Next, if we have computed V_{n-1} for some $1 \leq n \leq N-1$, then we compute $V_n(s)$ for all s by maximizing $W_n(s,\cdot)$, and this also yields a maximizer f_n at stage n (if one exists). Thus we finally obtain V_N and an N-stage policy $(f_N, f_{N-1}, \ldots, f_1) =: (f_n)_1^N$ (if one exists), which is optimal for CM_N by the OC. Note that this method requires for the solution of CM_N also the solution of the problems CM_n, $1 \le n < N$, or saying it positively, the method allows us to find the solution for CM_{N+1} from the solution of CM_N by a single step in the VI. ◇

Maximizers exist at all stages if all sets $D(s)$ are finite, in particular if A is finite. More general conditions are given in Chap. 17. A policy $(\pi_t)_0^{N-1}$ is called ***stationary*** if π_t does not depend on t. Except for a few models optimal policies for finite horizon N are not stationary.

Remark 11.8

(a) Assume that S, A and Z are finite. The solution of CM_N by the VI requires $N \cdot (|D| - |S|)$ comparisons, $N \cdot |D| \cdot |Z|$ multiplications, the same number of additions and, unless $\beta = 1$, in addition $|Z|$ multiplications (of the form $\beta \cdot q_z$). Unless $r_Z(s, a, z)$ is independent of z, the computation of r requires in addition $|D| \cdot |Z|$ multiplications and $|D| \cdot (|Z| - 1)$ additions. If the CM is invariant one needs only $N \cdot (|D| + (|Z| - 1) \cdot |A_0|) \le N \cdot (|D| + |Z| \cdot |A_0|$ additions and (independent of $|S|$!) only $N \cdot |Z| \cdot |A_0|$ multiplications. – As an example, for $|S| = 100$, $|D| = 500$, $|Z| = 10$, $\beta = 1$, $N = 20$ and if $r_Z(s, a, z)$ is independent of z we need 100 000 multiplications, 100 000 additions and 8 000 comparisons. The curse of dimensionality, mentioned in Chap. 1, is here even stronger.

(b) For the numerical solution of $CM_N(s_0)$ finiteness of Z and of $D(s)$ for all s suffices when the VI is combined with the concept of reachable state spaces, introduced in Chap. 13 below. ◇

In *minimization problems* we use as in Part I the following terminology: $c_Z := -r_Z$ is the ***one-stage cost function*** with the expected value $c(s, a) := \mathbb{E}\, c_Z(s, a, \eta_1)$, $(s, a) \in D$; $C_N(s) := -V_N(s)$ is the ***minimal expected N-stage costs*** for initial state s. For simplicity of notation we retain the symbols L, U_f, U and W_n in minimization problems, i.e. in (11.5), (11.6) and (11.7) one replaces r and sup by c and inf, respectively. Moreover, let $V_0 \in \mathfrak{V}_0$. A decision rule f is called a ***minimizer at stage*** n if $f(s)$ minimizes $a \mapsto W_n(s, a)$ for all s, i.e. if $U_f V_{n-1} = U V_{n-1}$.

It is obvious that the Basic Theorem 11.6 remains true for minimization problems if in the OC maximizer is replaced by minimizer and if the VI is written as

$$\begin{aligned} C_n(s) &= \inf\{c(s, a) + \beta \cdot \mathbb{E}\, C_{n-1}(T(s, a, \eta_1)) : a \in D(s)\} \\ &= \inf\{[W_n(s, a) : a \in D(s)\} \\ &=: UC_{n-1}(s), \; n \in \mathbb{N}, \; s \in S. \end{aligned}$$

Having established the Basic Theorem one can now investigate the structure of solutions using the tools developed in Chaps. 7, 8 and 9.

Example 11.9 (Discrete state replacement problem with a failure state) A system (e.g. a mechanical tool for an industrial process) undergoes random deterioration.

Its degree of deterioration, or age for short, at time t is classified as being in some state $s_t \in S := \mathbb{N}_{0m}$ for some $m \in \mathbb{N}$. The ordering of the states corresponds to increasing age, $s_t = 0$ denotes a new system and $s_t = m$ means a failure state which enforces replacement. The action a_t is 1 or 0, according to whether or not the system is replaced at the end of period t by a new one. In the first case the next state is $s_{t+1} = 0$, while in the latter case there is a random increase η_{t+1} of age. States larger than m are treated as failure. We assume that the η_t's are i.i.d. random variables with values in some finite set $Z \subset \mathbb{N}_0$. Thus we have a CM with the following data: $A = D(s) = \{0, 1\}$ for $s < m$ and $D(m) = \{1\}$;

$$T(s,a,z) = \begin{cases} (s+z) \wedge m, & \text{if } a = 0, \\ 0, & \text{if } a = 1; \end{cases}$$

the reward for production in state s is $g(s) \in \mathbb{R}$, and the replacement costs are $\gamma \in \mathbb{R}^+$, hence

$$r(s,a) = \begin{cases} g(s), & \text{if } a = 0, \\ g(s) - \gamma, & \text{if } a = 1. \end{cases}$$

It is natural to assume that both g and V_0 are decreasing. As A is finite, the value functions are finite. One will expect that

(i) $s \mapsto V_n(s)$ is decreasing,
(ii) there exist maximizers f_n at each stage n which prescribe to replace at time n in age s if s equals or exceeds some *critical age* $s_n^* \in \mathbb{N}_{0m}$, i.e.

$$f_n(s) = \begin{cases} 0, & \text{if } s < s_n^*, \\ 1, & \text{if } s \geq s_n^*. \end{cases} \tag{11.14}$$

(Notice that $s_n^* = 0$ [$s_n^* = m + 1$] means that one replaces at all states [only at state m].) ♦

A policy composed of decision rules of the form (11.14) is called a **control-limit policy** or *threshold-policy*. The existence of an optimal control-limit policy speeds up the numerical computation of the VI considerably, using the familiar bisection method for computing s_n^* in (11.15) below.

Proposition 11.10 (Structure of the solution of replacement Example 11.9) *Assume that the functions g and V_0 are decreasing. Then:*

(a) V_n is decreasing for all n.
(b) For $n \in \mathbb{N}$ put

$$s_n^* := \inf\{s \in \mathbb{N}_{0,m-1} : \beta \cdot \mathbb{E}\, V_{n-1}((s+\eta_1) \wedge m) < \beta \cdot V_{n-1}(0) - \gamma\} \tag{11.15}$$

(with $\inf \emptyset := m$*). Then* f_n *from (11.14) is the smallest maximizer at each stage* n *and* $(f_n)_N^1$ *is an optimal control-limit policy for* CM_N*,* $N \in \mathbb{N}$*.*

Proof

(a) As r_Z does not depend on z, the CM is equivalent to its adjoint MDP. The VI holds by Theorem 11.6(b) and it has the form

$$\begin{aligned} V_n(s) &= \max\{W_n(s,0), W_n(s,1)\} \\ &= g(s) + \max\{\beta \cdot \mathbb{E}\, V_{n-1}((s+\eta_1) \wedge m), -\gamma + \beta \cdot V_{n-1}(0)\}. \end{aligned}$$

We use induction on $n \geq 0$ for the assertion (I_n) that V_n is decreasing. (I_0) holds by assumption. If (I_n) holds then $s \mapsto W_{n+1}(s,a)$ is decreasing as $r(s,a)$ is decreasing in s and $T(s,a,z)$ is increasing in s. Now (I_{n+1}) follows from Remark 6.3.6(ii), observing that $D(\cdot)$ is decreasing.

(b) Fix $n \geq 1$. For the smallest maximum point $h_n(s)$ of $W_n(s,\cdot)$ we have $h_n(s) = 1$ if and only if $W_n(s,0) < W_n(s,1)$ if and only if

$$\beta \cdot \mathbb{E}\, V_{n-1}((s+\eta_1) \wedge m) < \beta \cdot V_{n-1}(0) - \gamma.$$

Moreover, $s \mapsto \mathbb{E}\, V_{n-1}((s+\eta_1) \wedge m)$ is decreasing as V_{n-1} is decreasing. This proves the assertion about f_n. Finally, optimality of $(f_n)_N^1$ follows from the OC Theorem 11.6(a). □

Chapter 12
Markovian Decision Processes with Finite Transition Law

Firstly we introduce MDPs with finite state spaces, prove the reward iteration and derive the basic solution techniques: value iteration and optimality criterion. Then MDPs with finite transition law are considered. There the set of reachable states is finite.

12.1 Finite Horizon

The control models CM with independent disturbances are not suitable for all applications, as is illustrated by the following example.

Example 12.1.1 (A marketing problem) A company sells a certain product. At the beginning of each period, say a month, the market situation for the product is evaluated and classified as being in some state s in a finite set S. Then the company decides on an appropriate sales-promoting action a such as an advertising campaign. We assume that from previous experience one has reasonably accurate estimates for the probabilities $p(s, a, s')$ that state s will be transformed into state s' under action a. Moreover we assume that (i) the reward r (promoting costs deducted) for period t is a function of state s_t and action a_t, and (ii) there is a terminal reward (e.g. the scrap value when selling the production equipment and the production rights to another company) $V_0(s_N)$ when the company discontinues the sale at time N in state s_N. Which policy maximizes the expected total reward for initial state s_0? ♦

In the sequel we often denote the sequence of states $(s_1, s_2, \ldots, s_N)$ by s^N. The deeper reason why we could solve CMs sequentially is the fact that for fixed initial state s_0 and policy π the sequence of states $X_t := \zeta_{t\pi}^{s_0}$, $1 \leq t \leq N$, forms a Markov chain in the following sense:

Let S be finite, $s_0 \in S$ and let $(p_t(s, s'),\ s, s' \in S)$, $0 \leq t \leq N-1$, be stochastic $S \times S$-matrices. A sequence $(X_t)_1^N$ of S-valued random variables on an arbitrary finite

K. Hinderer et al., *Dynamic Optimization*, Universitext,
DOI 10.1007/978-3-319-48814-1_12

probability space $(\Omega, \mathbb{P})$ is said to be an N-stage non-homogeneous **Markov chain** with state space S, initial state s_0 and transition matrices $(p_t)_0^{N-1}$ if the discrete density $s^N \mapsto \mathbb{P}((X_t)_1^N = s^N)$ of $(X_t)_1^N$ equals

$$p_0(s_0, s_1) \cdot p_1(s_1, s_2) \cdots p_{N-1}(s_{N-1}, s_N), \quad s^N \in S^N. \tag{12.1}$$

We build upon the following simple existence result: For given s_0 and $(p_t)_0^{N-1}$ there always exists a probability space $(\Omega, \mathbb{P})$ and on it a random vector $(X_t)_1^N$ which is a Markov chain with state space S, initial state s_0 and transition matrices p_t. In fact, the so-called *canonical construction* is as follows:

As sample space Ω we take S^N, the set of all state sequences s^N, for $\mathbb{P}$ the probability distribution with the discrete density from (12.1) and for X_t the t-th coordinate random variable on Ω, i.e. $X_t(s^N) := s_t$, $1 \le t \le N$.

Formula (12.1) describes the evolution of the process $(X_t)_1^N$ in a very intuitive way (which also tells us how to simulate a Markov chain): The process starts in s_0 and moves between time t and $t+1$, $0 \le t \le N-1$, from the momentary state s_t to the next state s_{t+1}, which occurs according to the discrete density $p_t(s_t, \cdot)$. The selection of s_{t+1} is made independent of s_0, s_1, ..., s_{t-1}, and this is the core of the Markovian nature of the process.

Now consider problem $CM_N(s_0)$ with a finite state space S. One proves easily that for any policy $\pi = (\pi_t)_0^{N-1}$ the discrete density of the sequence $(X_t)_1^N := (\zeta_{t\pi}^{s_0})_1^N$ of random states satisfies (12.1) with

$$p_{\pi_t}(s, s') := \mathbb{P}(T(s, \pi_t(s), \eta_1) = s'), \quad 0 \le t \le N-1, \quad s, s' \in S.$$

Therefore the sequence of states $\zeta_{t\pi}^{s_0}$, $1 \le t \le N$, is a Markov chain. The purpose of the present section is to show that the theory of Chap. 11 carries over to the case where the state random variables, now denoted by ζ_t, $1 \le t \le N$, form a Markov chain with *arbitrary* transition probabilities $p_{\pi_t}(s, s')$ of the form $p_{\pi_t}(s, s') := p(s, \pi_t(s), s')$ for a single given stochastic $D \times S$-matrix p. From (12.1) we obtain in (12.3) below a recursive structure of the joint distribution of the ζ_t's. Together with the recursive structure of the N-stage random reward (cf. (12.6)) we get a recursive structure of the expected N-stage reward, expressed in Lemma 12.1.4 below in the form of the reward iteration. From the latter everything follows very similar as in Chap. 11. Since our present model does not contain disturbances, the one-stage reward does not depend on some disturbance, but it may depend on the state at the end of the period.

Definition 12.1.2 A ***Markovian Decision Process*** (***MDP*** for short) ***with finite state space*** is a tuple $(S, A, D, p, r_S, V_0, \beta)$ of the following kind:

- S is finite.
- A, D, V_0 and β have the same meaning as in the CM. If $D(s) = A$ for all s, we drop D from the tuple defining the MDP.

- p is a transition matrix from D into S, i.e. a function $(s, a, s') \mapsto p(s, a, s')$ from $D \times S$ into $\mathbb{R}_+$ such that

$$\sum_{s' \in S} p(s, a, s') = 1 \text{ for } (s, a) \in D.$$

We call p the (stochastic) ***transition law*** of the MDP.
- $r_S : D \times S \to \mathbb{R}$ is the ***one-stage reward function.*** If $r_S(s, a, s')$ is independent of s', we replace r_S in the tuple by r.

For an MDP the notions of *decision rule* and *N-stage policy* are the same as for CMs. From what has been said about Markov chains in general we know that there exists, given an MDP, for each $s_0 \in S$ and each policy $(\pi_t)_0^{N-1}$ a Markov chain $\zeta^N = (\zeta_t)_1^N$ on some probability space $(\Omega, \mathbb{P}_{\pi s_0})$ with state space S, starting in s_0 and with transition matrices

$$p_{\pi_t} := (p(s, \pi_t(s), s'),\ s, s' \in S), \quad 0 \le t \le N - 1.$$

Thus the so-called *decision process* ζ^N has the discrete density

$$s^N \mapsto p_\pi(s_0, s^N) := p_{\pi_0}(s_0, s_1) \cdot p_{\pi_1}(s_1, s_2) \cdots p_{\pi_{N-1}}(s_{N-1}, s_N). \tag{12.2}$$

Obviously this discrete density has the following *recursive property:*

$$p_{(f,\sigma)}(s_0, s^N) = p_f(s_0, s_1) \cdot p_\sigma(s_1, (s_t)_2^N),\ (f, \sigma) \in \mathbb{F} \times (\mathbb{F}^{N-1}),\ N \ge 2. \tag{12.3}$$

We call $(\Omega, \mathbb{P}_{\pi s_0})$ the canonical probability space of the MDP. Note that Ω and $\mathbb{P}_{\pi s_0}$ and ζ_t may depend on N, but this dependence will in general be suppressed in the notation.

Remark 12.1.3

(a) Notice the difference between the stochastic framework for CMs as given in Chap. 11 and the one used for MDPs. In the latter the canonical probability space is constructed from the data, with the probability distribution $\mathbb{P}_{\pi s_0}$ (but not the state random variables ζ_t) depending on π and s_0. For CMs the state random variables $\zeta_{t\pi}^{s_0}$ (but not the canonical probability space) depend on π and on s_0.

(b) In the paragraph preceding Definition 12.1.10 below we show that MDPs comprise CMs in the sense that each CM can be reduced to some MDP. ◇

If the decision maker chooses $\pi = (\pi_t)_0^{N-1}$ he obtains, when starting in $s_0 =: \zeta_0$, the **N-stage random reward**

$$R_{N\pi}(s_0, \zeta^N) := \sum_{t=0}^{N-1} \beta^t \cdot r_S(\zeta_t, \pi_t(\zeta_t), \zeta_{t+1}) + \beta^N \cdot V_0(\zeta_N).$$

If $(\Omega, \mathbb{P}_{\pi s_0})$ is canonical, we get

$$R_{N\pi}(s_0, s^N) := \sum_{t=0}^{N-1} \beta^t \cdot r_S(s_t, \pi_t(s_t), s_{t+1}) + \beta^N \cdot V_0(s_N), \; s^N \in S^N. \tag{12.4}$$

By $\boldsymbol{MDP_N}$ we denote the family of maximum problems $MDP_N(s_0)$ jointly for all s_0: For each $s_0 \in S$ the ***maximum problem*** $\boldsymbol{MDP_N(s_0)}$ consists in maximizing the ***expected N-stage reward*** $\boldsymbol{V_{N\pi}(s_0)}$, defined by

$$V_{N\pi}(s_0) := \mathbb{E}_{\pi s_0} R_{N\pi}(s_0, \zeta^N) = \sum_{s^N \in S^N} R_{N\pi}(s_0, s^N) \cdot p_\pi(s_0, s^N), \tag{12.5}$$

over the set $\mathbb{F}^N$ of N-stage policies π. (Here expectations with respect to $\mathbb{P}_{\pi s_0}$ are denoted by $\mathbb{E}_{\pi s_0}$.) The latter equation shows that $V_{N\pi}(s_0)$ is independent of the choice of $(\Omega, \mathbb{P}_{\pi s_0})$ and of the ζ_t's; only the transition law p matters.

Note that $V_{N\pi}(s_0)$ exists and is finite for all N, π and s_0, as ζ^N assumes only finitely many values. Moreover, the ***maximal expected N-stage reward for initial state*** $\boldsymbol{s_0}$, i.e.

$$V_N(s_0) := \sup\{V_{N\pi}(s_0) : \pi \in \mathbb{F}^N\}$$

exists and $-\infty < V_N(s_0) \leq \infty$.

For simplicity we denote in the sequel the initial state by s instead of s_0. Notions such as value function, optimal policy, maximizer at stage n, etc. are retained literally from Chap. 11. We also use for $(s, a) \in D$, $s' \in S$ and $f \in \mathbb{F}$

$$r(s, a) := \sum_{s' \in S} p(s, a, s') \cdot r_S(s, a, s'),$$

$$r_f(s) := r(s, f(s)), \quad p_f(s, s') := p(s, f(s), s').$$

Let $\mathfrak{V}_0 := \{v : S \to \overline{\mathbb{R}} : \sum_{s'} p(s, a, s') \cdot v(s') \in \overline{\mathbb{R}} \text{ for all } (s, a) \in D\}$.

As in Chap. 11 we use the operators L, U_f, and U which are defined on $\mathfrak{V}_0$ as follows:

$$Lv(s, a) := r(s, a) + \beta \cdot \sum_{s' \in S} p(s, a, s') \cdot v(s'),$$

$$U_f v(s) := Lv(s, f(s)) = r_f(s) + \beta \cdot \sum_{s' \in S} p_f(s, s') \cdot v(s'),$$

$$Uv(s) := \sup_{a \in D(s)} Lv(s, a) = \sup_{f \in \mathbb{F}} U_f v(s).$$

The latter equality is verified as in Chap. 11. Note that U_f and U map $\mathfrak{V}_0$ into itself. In minimization problems we use the same notation L, U_f and U, but with r_S replaced by c_S and sup replaced by inf.

Lemma 12.1.4 (Reward iteration) *In each MDP with finite state space the expected N-stage reward under any policy may be computed by the* ***reward iteration*** *(**RI** for short)*

$$\begin{aligned} V_{1f} &= U_f V_0, \\ V_{n(f,\sigma)} &= U_f V_{n-1,\sigma},\ n \geq 2,\ (f,\sigma) \in \mathbb{F} \times \mathbb{F}^{n-1}. \end{aligned}$$

Proof The case $n = 1$ is trivial. For $n \geq 2$ the proof runs as the proof of Lemma 11.4, using now (12.3) for $N := n$ and the fact that (12.4) yields

$$R_{n(f,\sigma)}(s, (s_t)_1^n) = r_S(s, f(s), s_1) + \beta \cdot R_{n-1,\sigma}(s_1, (s_t)_2^n). \tag{12.6}$$

□

Lemma 11.5 on properties of L, U_f and U carries over with minor changes of the proof. As the proof of Theorem 11.6 required only the RI from Lemmas 11.4 and 11.5, Theorem 11.6 carries over to MDPs as follows.

Theorem 12.1.5 (Basic Theorem for MDPs with finite state space) *The following two results remain true:*

*(a) The optimality criterion (**OC** for short) of Theorem 11.6(a) remains true.*
*(b) The value iteration (**VI** for short) holds in the form*

$$\begin{aligned} V_n(s) &= \sup_{a \in D(s)} \left[r(s,a) + \beta \cdot \sum_{s' \in S} p(s,a,s') \cdot V_{n-1}(s') \right] \\ &= UV_{n-1}(s) =: \sup_{a \in D(s)} W_n(s,a),\ n \in \mathbb{N},\ s \in S. \end{aligned}$$

Example 12.1.6 (A special stochastic marketing problem) We return to Example 12.1.1 and want to get an idea about the numerical solution and its dependence on s, n, β and V_0. We assume that there are four states of the business: good ($s = 3$), medium ($s = 2$), bad ($s = 1$) and very bad ($s = 0$). There are two actions, *advertise* ($a = 1$) and *do nothing* ($a = 0$). We consider *two versions*. In the first one a transition from the very bad state under advertising to the bad state is possible with probability 0.1, but in the second one (whose data are given in Table 12.1 in parentheses) the process stops as soon as the very bad state is reached. Table 12.1 reflects for both versions that $a = 1$ increases the probability of a transition to a better state but decreases the reward due to advertising costs. (More precisely, $P(s, 1, \cdot)$ is stochastically larger than $P(s, 0, \cdot)$ for all s; cf. Chap. 18.) We make the natural assumption that V_0 is non-negative and increasing.

Table 12.1 Data for Example 12.1.6

	$p(s,a,s')$				$r(s,a)$
(s,a)	$s'=0$	$s'=1$	$s'=2$	$s'=3$	
(0, 0)	1	0	0	0	0
(0, 1)	0.9 (1)	0.1 (0)	0	0	−2 (0)
(1, 0)	0.4	0.5	0.1	0	2
(1, 1)	0.2	0.4	0.3	0.1	0
(2, 0)	0.2	0.3	0.4	0.1	6
(2, 1)	0.1	0.2	0.3	0.4	4
(3, 0)	0	0	0.4	0.6	8
(3, 1)	0	0	0.2	0.8	6

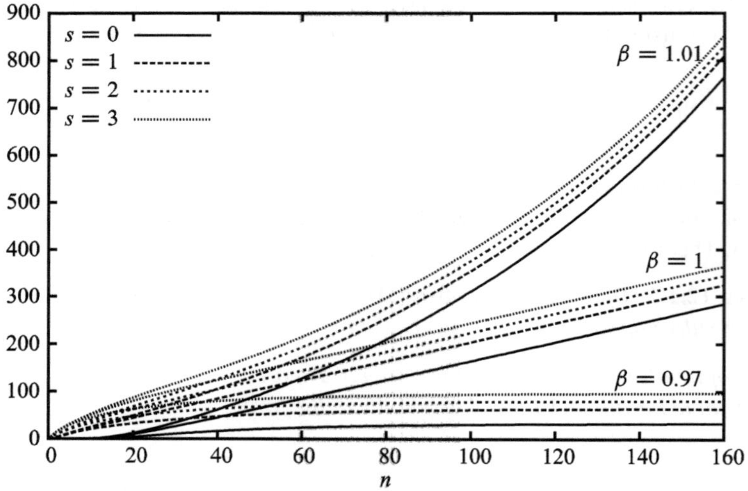

Fig. 12.1 $V_n(s)$ of Example 12.1.6 (first version) for $V_0 \equiv 0$ and $\beta = 0.97$, $\beta = 1$, and $\beta = 1.01$

In the sequel we consider numerical results for several choices of V_0 and β. The results for the first version are shown in Figs. 12.1 and 12.2 and for the second version in Fig. 12.3. We are going to comment on the findings which contain much information on the structure of the solution of the MDP.

(a) *Monotonicity of the value functions in s, n, β and V_0.*

(a1) In Figs. 12.1, 12.2 and 12.3 the value functions are increasing in s, which means that a better initial state implies a higher expected total reward. We shall verify in Chap. 18 that $V_n(s)$ is indeed increasing in s whenever $V_0(s)$ is increasing in s.

(a2) In Figs. 12.1 and 12.3 the value functions are increasing in n, $0 \le n \le 160$, which means that a longer horizon implies a higher expected total reward. In fact, $V_n(s)$ is increasing in n whenever $V_0 \equiv c \ge 0$ and either $\beta \ge 1$ or $c = 0$.

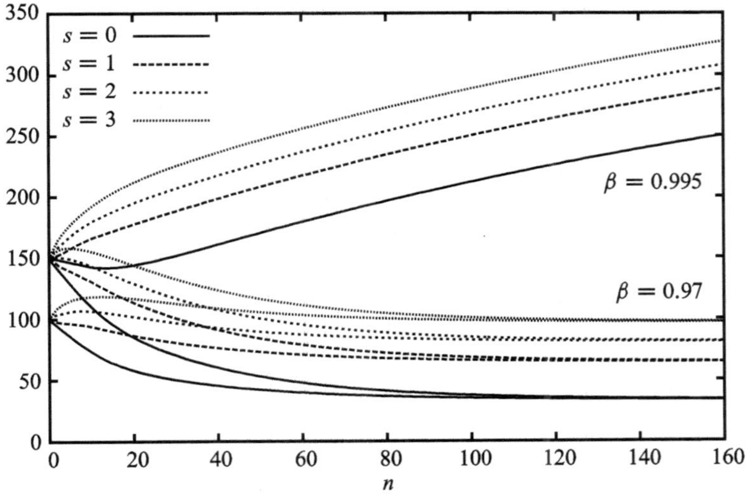

Fig. 12.2 $V_n(s)$ of Example 12.1.6 (first version) for $V_0 \equiv 100$, $\beta = 0.97$ and $V_0 \equiv 150$, $\beta = 0.97$, and $\beta = 0.995$

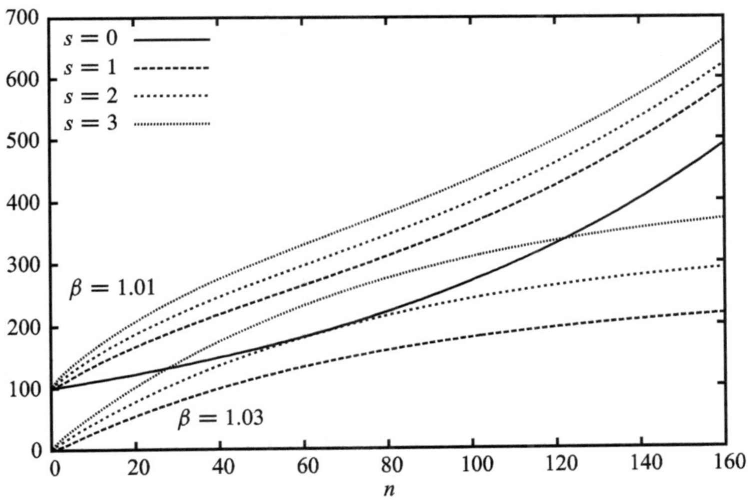

Fig. 12.3 $V_n(s)$ of Example 12.1.6 (second version) for $V_0 \equiv 0$, $\beta = 1.03$, and $V_0 \equiv 100$, $\beta = 1.01$

For then, as $\max_a r(s,a) \geq 0$,

$$V_1(s) = UV_0(s) = \max_{a \in A} r(s,a) + \beta c \geq c = V_0(s).$$

Now one easily derives from the VI by induction on n, using isotonicity of U, that $V_n \geq V_{n-1}$ for all n, cf. Chap. 6. On the other hand, Fig. 12.2 shows that in general $V_n(s)$ need not be increasing in n, and that moreover $n \mapsto V_n(s)$ may be increasing (decreasing) for some s and neither increasing nor decreasing for other states s.

(a3) In Fig. 12.1 the numbers $V_n(s)$ are increasing in β. Moreover, for fixed n, π and s, $\beta \mapsto V_{n\pi}(s,\beta)$ is a polynomial. Hence, as there are only finitely many policies, $\beta \mapsto V_n(s,\beta)$ is continuous and piecewise polynomial.

(b) *Pointwise convergence of the sequence of value functions.*

(b1) In Figs. 12.1 and 12.2 (first version) convergence seems to be present for $\beta < 1$ and divergence to infinity seems to occur for $\beta \geq 1$. On the other hand, Fig. 12.3 indicates that in the second version with $V_0 \equiv 0$ convergence is present even for some values of $\beta > 1$, e.g. for $\beta = 1.03$. This can be intuitively explained by the high probability that the process stops in state $s = 0$ after a moderate number of periods.

(b2) We know from (a3) that the value functions are increasing in β. Thus, if $(V_n(s))_0^\infty$ tends to infinity for some β_0 then it does so for all $\beta > \beta_0$. This is no contradiction to Fig. 12.3, where we seem to have convergence for $\beta = 1.03$, $V_0 \equiv 0$, and divergence to ∞ for $\beta = 1.01$, $V_0 \equiv 100$; for in the latter case we have

$$V_n(s) \geq V_n(0) = \beta^n V_0(0) = (1.01)^n \cdot 100 \to \infty \text{ for } n \to \infty.$$

We mention that convergence holds for all s and all $\beta < 1$ in the first version and for all s and all $\beta < \beta^* \approx 1.04$ in the second version, provided $V_0(0) = 0$.

(b3) Figure 12.2 indicates that the limit $V(s)$ of the sequence $(V_n(s))_{n=0}^\infty$, when it exists, is independent of V_0.

(c) *Dependence of the smallest maximizer f_n on β and on n.*
This problem is more complicated. Computational results for $1 \leq n \leq 200$ in the *second version* with $V_0 \equiv 0$ and several values of β are given in Fig. 12.4. The findings are as follows.

(c1) As $V_0 \equiv 0, f_1$ is independent of β and equal to $g_0 := (0,0,0,0)$, i.e. at the last stage one should not advertise, irrespective of the state. In the sequel f_1 will be excluded from the discussion.

(c2) It is easy to see that $f_n(0) = 0$ for all n and β. It seems that of the eight possible decision rules g for which $g(0) = 0$ only g_0 and the following four arise as smallest maximizers for some stage:

$$g_1 := (0,0,1,0),\ g_2 := (0,1,1,0),\ g_3 := (0,1,1,1),\ g_4 := (0,1,0,0).$$

Figure 12.4 shows the regions G_i in the (β, n)-plane where $f_n = g_i, 0 \leq i \leq 4$. Figure 12.4 reads as follows: If, for example, $\beta = 0.61$ then

$$f_n = \begin{cases} g_0, & \text{if } 1 \leq n < 5, \\ g_1, & \text{if } 5 \leq n < 7, \\ g_2, & \text{if } 7 \leq n \leq 200. \end{cases}$$

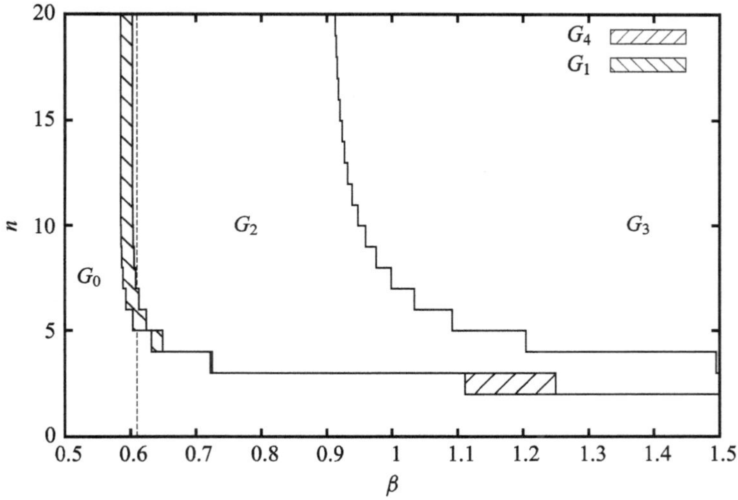

Fig. 12.4 Smallest maximizers for Example 12.1.6 (second version) for $V_0 \equiv 0$. A *dotted vertical line* indicates $\beta = 0.61$. Note that G_1 consists of three parts

It seems that large horizons and large discount factors are an incentive for advertising. ♦

In many MDPs with finite S and A one is led after a few steps in the VI to the conjecture that a certain decision rule f is a maximizer at all stages large enough, say for $n \geq n_0$. The smallest of these numbers n_0, i.e.

$$N^*(f) := \inf\{k \in \mathbb{N} : f \text{ is a maximizer at all stages } n \geq k\},$$

if finite, is called the ***turnpike horizon*** of f.

As an example, Fig. 12.4 suggests the following for the special marketing example: (i) For each β there seems to exist a turnpike horizon $N^*(g(\beta))$ with $g(\beta)$ being one of the decision rules g_0, g_1, g_2, g_3, g_4. (ii) There seem to be three critical values of β, namely $\beta_1 \approx 0.585\,385$, $\beta_2 \approx 0.602\,689$ and $\beta_3 \approx 0.906\,017$. At these points the decision rule $g(\beta)$ and the number $N^*(g(\beta))$ change abruptly. For example in a left neighborhood of β_2 we have $g(\beta) = g_1$ and $N^*(\beta) = 6$, while in a right neighborhood of β_2 we have $g(\beta) = g_2$, and $N^*(g(\beta))$ seems to tend to infinity when β approaches β_2 from the right.

If f has a turnpike horizon the VI simplifies significantly as no minimization is necessary for $n \geq N^*(f)$, i.e.

$$V_n = U_f V_{n-1},\ n \geq N^*(f).$$

In the rare cases where $N^*(f) = 1$ the MDP has for each N a *stationary* optimal N-stage policy, namely $(f)_N^1$.

MDPs with infinite state space usually require a more sophisticated theory (cf. Chaps. 16 and 17) and they are difficult to solve numerically. Yet some of these models (e.g. Example 13.2.2) can be treated essentially as simply as finite state MDPs, as follows.

Definition 12.1.7 An ***MDP with finite transition law*** is a tuple $(S, A, D, P, r_s, V_0, \beta)$ with

- S is arbitrary.
- A, D, r_S, V_0 and β have the same meaning as in the MDP with finite state space.
- The transition law p is a non-negative function on $D \times S$ such that for each (s, a) the set $R(s, a)$ of states with $p(s, a, s') > 0$ is finite and $\sum_{s' \in R(s,a)} p(s, a, s') = 1$.

The construction of the probability space $(\Omega, \mathbb{P}_{\pi s})$ and of the random states ζ_t with $\mathbb{P}_{\pi s}$ determined by its discrete density as in (12.3) carries over without difficulty, and the proof of the reward iteration of Lemma 12.1.4 and the Basic Theorem 12.1.5 carry over literally. Thus we obtain

Corollary 12.1.8 (The Basic Theorem for MDPs with finite transition law) *The Basic Theorem 12.1.5 remains true for MDPs with finite transition law.*

The method of reachable state spaces, introduced in Sect. 3.2 for the purpose of solving the problem $DP_N(s_0)$ for a single initial state s_0, easily carries over to an MDP with finite transition law. Thus we have $R_0(s_0) := \{s_0\}$, and $R_t(s_0)$, $t \geq 1$, is defined recursively by

$$R_1(s) := \cup_{a \in D(s)} R(s, a), \; s \in S,$$
$$R_{t+1}(s_0) := \cup_{s' \in R_t(s_0)} R_1(s'), \; t \geq 1.$$

One can show that $R_t(s_0)$, $1 \leq t \leq N$, is the set of those states which occur at time t with positive probability in at least one of the N-stage decision processes which start in s_0.

Proposition 12.1.9 (Solution of $MDP_N(s_0)$ with finite transition law using reachable state spaces) *Consider an MDP with finite transition law.*
Then:

(a) $V_N(s_0)$ can be found for fixed $N \geq 1$ and $s_0 \in S$ by the so-called partial VI

$$\begin{aligned} V_n(s) &= \sup_{a \in D(s)} \left[r(s,a) + \beta \cdot \sum_{s' \in R(s,a)} p(s,a,s') \cdot V_{n-1}(s') \right] \\ &=: \sup_{a \in D(s)} W_n(s,a), \quad 1 \leq n \leq N, \; s \in R_{N-n}(s_0). \end{aligned} \tag{12.7}$$

(b) Each policy $(f_n)_N^1$ such that $f_n(s)$ is a maximum point of $a \mapsto W_n(s, a)$ for $s \in R_{N-n}(s_0)$, $1 \leq n \leq N - 1$, is s_0-optimal.

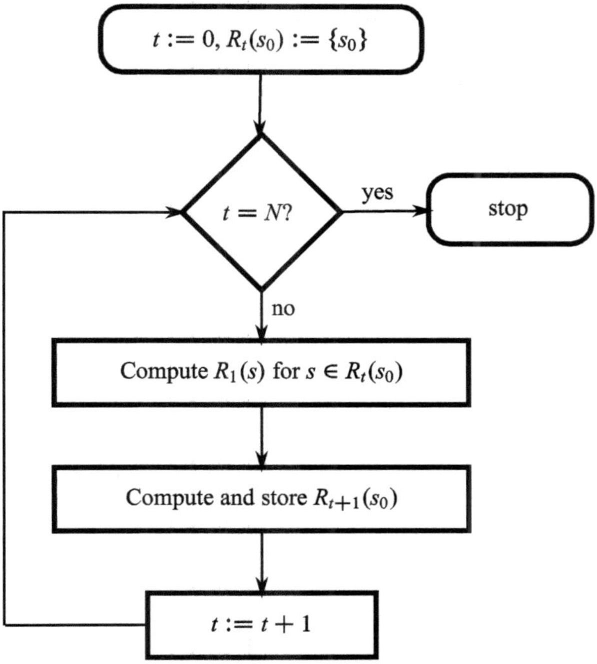

Fig. 12.5 Flow chart for computing the sets of reachable states

Proof For fixed $N \geq 1$ one obtains (12.7) by induction on $1 \leq n \leq N$, while (b) follows from (a) and the RI of Lemma 12.1.4. □

Note that the solution of $MDP_N(s_0)$ is determined by $V_N(s_0)$ and by $(f_n|R_{N-n}(s_0))_{n=N}^1$, and that in contrast to the usual VI the computations for $MDP_N(s_0)$ do not contain the solution for the problems $MDP_n(s_0)$, $1 \leq n < N$. The implementation of the algorithm for given s_0 and N runs as follows. In a *first* step one computes by a forward procedure the finite sets $R_t(s_0)$, $1 \leq t \leq N-1$, according to the flowchart Fig. 12.5.

From the first step we know $R_t(s_0)$, $0 \leq t \leq N-1$. Since $s \in R_{N-n}(s_0)$ implies $R(s,a) \subset R_{N-(n-1)}(s_0)$ for $a \in D(s)$, this allows us to compute in a second step by the backward procedure of the partial VI (12.7) the values $V_n(s)$ and $f_n(s)$ for $s \in R_{N-n}(s_0)$, $1 \leq n \leq N$.

The method of reachable state spaces solves numerically the problem $MDP_N(s_0)$, provided $D(s)$ is finite for all s. In case of a finite state space the method may accelerate the usual solution procedure; then, of course, the gain in time for computing the VI using Proposition 12.1.9 must be balanced against the time to compute the reachable state spaces.

So far it seems as if we must develop separate theories for MDPs and for CMs, since in these models the functions $V_{n\pi}$ are defined in different ways. Fortunately it mostly suffices to derive theoretical results for MDPs, because we can adjoin to each CM an MDPad (see below) which has the same solution (i.e. the same value functions and for each $n \geq 1$ the same set of maximizers at stage n) as the CM.

Definition 12.1.10 To each CM $(S, A, D, Z, Q, r_Z, V_0, \beta)$ we adjoin the MDP $(S, A, D, p, r_S, V_0, \beta)$ (**MDPad** for short) with

$$p(s, a, s') := Q(T(s, a, \eta_1) = s') = \sum_{\{z \in Z : T(s,a,z) = s'\}} q(z), \ (s, a) \in D, \ s' \in S,$$

and

$$r_S(s, a) := \mathbb{E}\, r_Z(s, a, \eta_1), \ (s, a) \in D.$$

Using the elementary fact from probability theory that

$$\mathbb{E}\, v(T(s, a, \eta_1)) = \sum_{s'} Q(T(s, a, \eta_1) = s') \cdot v(s') = \sum_{s'} p(s, a, s') \cdot v(s'),$$

we see that both of the definitions for $\mathfrak{V}_0$ coincide and therefore the CM and the MDPad have the same operator L, namely

$$\begin{aligned} Lv(s, a) &= r(s, a) + \beta \cdot \mathbb{E}\, v(T(s, a, \eta_1)) \\ &= r(s, a) + \beta \cdot \sum_{s'} p(s, a, s') \cdot v(s'). \end{aligned} \tag{12.8}$$

Then also both models have the same operators U_f. Now the RI of Lemma 11.4 for the CM and the RI of Lemma 12.1.4 for its adjoint MDPad prove

Proposition 12.1.11 (CM and MDPad) *The two models CM and MDPad have the same solution. In addition they have the same functions W_n and hence the same maximizers at each stage.*

Thus one can also utilize results derived for MDPs also for CMs. Since this situation also occurs for other models we formalize it in the following definition. By a ***model*** we understand a tuple of data.

Definition 12.1.12 (Equivalence of models) Consider two models M and $\hat{M}$, both of which contain the same set S. Assume that in M and $\hat{M}$ there are defined a set $\mathbb{F}$ and functions $V_{n\pi}$ and $\hat{V}_{n\pi}$ on S for each $n \geq 1$ and each parameter $\pi \in \mathbb{F}^n$. Then M and $\hat{M}$ are said to be ***equivalent*** if $V_{n\pi} = \hat{V}_{n\pi}$ for $n \geq 1$ and $\pi \in \mathbb{F}^n$.

Remark 12.1.13 As a rule, the equivalence of models in this book is intuitively clear. The proofs mostly use that in both models the operators L coincide and that the RI holds. Since the proofs are tedious and boring, we mostly omit them. ◇

12.2 Large Horizon

We now extend the most important results from Chap. 10 about DPs with large horizon to MDPs. We only treat the following case which is basic for computations:

Definition 12.2.1 The model $(S, A, D, p, r_S, V_0, \beta)$ is called an **MDP with large horizon** if the following conditions hold:

- D, p, r_s, V_0 have the same meaning as in the MDP with finite state space.
- $A \neq \emptyset$ is finite.
- $\beta < 1$.

The general problems arising in case of a large horizon are described in Sect. 20.1.

We repeat a few definitions and introduce some new ones

(i) $V := \lim_{n\to\infty} V_n$, in case the pointwise limit exists in $\overline{\mathbb{R}}$.
(ii) The decision rule $f \in \mathbb{F}$ is called ***asymptotically optimal*** if

$$V_{nf}(s) - V_n(s) \to 0 \text{ for } n \to \infty \text{ for all } s \in S.$$

Here V_{nf} means $V_{n\pi}$ for $\pi := (f)_0^{n-1}$.
(iii) For functions $v \in \mathfrak{V}_0$ the equation $v = Uv$ is called the ***optimality equation*** (**OE** for short) or ***Bellman equation***.

We now show that:

(a) V exists and is finite.
(b) V satisfies the optimality equation.
(c) A decision rule is asymptotically optimal if and only if it is a maximizer of LV.

For this goal we can use Banach's fixed point theorem from Proposition 10.1.8, which leads to symmetric upper and lower bounds for $V - V_N$. However, by exploiting the structure of U we obtain in (12.12) below better non-symmetric bounds for $V - V_N$ and in (12.13) also a better estimate for the quality of asymptotically optimal decision rules. For functions v on some finite set we denote by $\|v\|$ the maximum norm and by $\operatorname{sp} v := \max v - \min v$ the span. Note that $\operatorname{sp} v \leq 2\|v\|$.

Lemma 12.2.2 (Estimates for $U^t v - U^t w$) *For functions v and w on S and for $t \geq 0$ we have*

$$\beta^t \cdot \min(v - w) \leq U^t v - U^t w \leq \beta^t \cdot \max(v - w), \tag{12.9}$$

$$\|U^t v - U^t w\| \leq \beta^t \cdot \|v - w\|, \tag{12.10}$$

$$\operatorname{sp}(U^t v - U^t w) \leq \beta^t \cdot \operatorname{sp}(v - w). \tag{12.11}$$

Proof The upper bound in (12.9) follows by induction on t since

$$\begin{aligned} Uv - Uw &= \max_{f \in \mathbb{F}} U_f v - \max_{f \in \mathbb{F}} U_f w \\ &\leq \beta \cdot \max_{f \in \mathbb{F}} U_f(v - w) \leq \beta \cdot \max(v - w). \end{aligned}$$

The lower bound follows by exchanging v and w in the upper bound, and then (12.10) and (12.11) are obvious. □

Theorem 12.2.3 (The optimality equation and asymptotically optimal decision rules) *For an MDP with large horizon we have:*

(a) *The sequence of value functions V_n converges to a finite limit function V, and $\|V\| \leq \|r\|/(1-\beta)$.*
(b) *V is the unique finite solution of the optimality equation, hence independent of V_0.*
(c) *V_N and V satisfy for $N \geq 0$*

$$V + \beta^N \cdot \min(V_0 - V) \leq V_N \leq V + \beta^N \cdot \max(V_0 - V). \tag{12.12}$$

(d) *For each decision rule $f \in \mathbb{F}$ there exists $V_f := \lim_n V_{nf}$, and V_f is the unique finite fixed point of U_f.*
(e) *We have $V = \max_{f \in \mathbb{F}} V_f$, and $V = V_f$ if and only if f is a maximizer of LV.*
(f) *The decision rule f is asymptotically optimal if and only if f is a maximizer of LV, and then*

$$0 \leq V_N - V_{Nf} \leq \beta^N \cdot \mathrm{sp}(V - V_0), \quad N \geq 0. \tag{12.13}$$

Proof Part (a), except for the bound, and (b) follow from Banach's fixed point theorem Proposition 10.1.8. The bound holds since the definition of V_N shows that $\|V_N\| \leq \|r\| \cdot \sigma_N(\beta)$, $N \geq 1$.

(c) This follows from (12.9) since $V_N - V = U^N V_0 - U^N V$ by (b).
(d) This follows from (a) and (b), applied to the model MDP$_f$ which differs from the MDP only by $D(s) := D_f(s) := \{f(s)\}$, $s \in S$.

The proof of (e) and (f), except for (12.13), is the same as for the corresponding parts of Theorem 10.1.10 with $b :\equiv 1$. Moreover, if f is a maximizer of LV then $V = V_f$ by (e), and then the lower bound in (12.12) for MDP$_f$ yields $V_f - V_{Nf} \leq -\beta^N \cdot \min(V_0 - V_f)$. Together with the upper bound in (12.12) the inequality (12.13) and hence asymptotic optimality of f follows from

$$0 \leq V_N - V_{Nf} = (V_N - V) + (V_f - V_{Nf}) \leq \beta^N \cdot [\max(V_0 - V) - \min(V_0 - V)]\,. \square$$

Remark 12.2.4

(a) It follows from Definition 12.1.10 that Theorem 12.2.3 and also Theorem 12.3.2 below remain true for CMs.

(b) It is appropriate to use (12.12) instead of the inferior bounds

$$\beta^N \cdot \min(V_0 - V_1)/(1-\beta) \leq V_N - V \leq \beta^N \cdot \max(V_0 - V_1)/(1-\beta) \qquad (12.14)$$

for $N \geq 0$, since in the large horizon approach we assume that V is known. Note that (12.14) follows from (12.12) and by letting m go to infinity in

$$V_0 - U^k V_0 = \sum_{\nu=0}^{k-1} (U^\nu V_0 - U^\nu V_1) \leq \max(V_0 - V_1) \cdot \sigma_k(\beta). \qquad \diamond$$

Proposition 12.2.5 (Turnpike horizon and asymptotic optimality) *If f in an MDP with large horizon has a turnpike horizon then f is asymptotically optimal.*

Proof If $N := N_f^*$, it follows for $m \geq 0$ from (12.11), applied to MDP_f, that

$$\|V_{N+m} - V_{N+m,f}\| = \|U_f^m V_N - U_f^m V_{Nf}\| \leq \beta^m \cdot \|V_N - V_{Nf}\|,$$

which converges to zero for $k \to \infty$. □

Remark 12.2.6 (Computation of V and of an asymptotically optimal $\boldsymbol{f^*}$*)*

(a) Obviously the fixed point V_f of U_f can be computed as the unique solution v of the system of $|S|$ linear equations

$$(I - \beta \cdot p_f)v = r_f, \qquad (12.15)$$

where I denotes the $S \times S$-unit matrix. Therefore and by Theorem 12.2.3(e) one could compute V and f^* by solving the $|\mathbb{F}|$ systems of linear equations (12.15) for each $f \in \mathbb{F}$, and by comparing the resulting functions V_f.
(b) In special cases with small set D one may conjecture from the computation of V_n for a few small values of n that a certain decision rule f has a turnpike horizon, which would imply by Proposition 12.2.5 and Theorem 12.2.3(e) and (f) that $V = V_f$ and that f is asymptotically optimal. Then one can try the following: Either compute V_f from (12.15) or verify that $UV_f = V_f$.
(c) The converse assertion in Proposition 12.2.5 does not hold. $\diamond$

Denote by $\mathbb{F}(\beta)$, $0 \leq \beta < 1$, the set of asymptotically optimal decision rules. We call a decision rule f ***Blackwell-optimal*** if it is asymptotically optimal for all β in a left neighborhood of $\beta = 1$.

Proposition 12.2.7 (Dependence of the limit value function on the discount factor) *For an MDP with large horizon we have:*

(a) For each $s \in S$ the function $\beta \mapsto V^\beta(s)$ on $(0,1)$ is continuous and the quotient of a piecewise polynomial function and a polynomial.
(b) The mapping $\beta \mapsto \mathbb{F}(\beta)$ from (0,1) into $\mathbb{F}$ is piecewise constant. In particular, there exists a Blackwell-optimal policy.

Proof

(a) Fix s and $\beta < 1$. According to (12.15) and Cramer's rule $V_f(s) = d_f(s)/d_f$, where $d_f := \det(I - \beta \cdot p_f) \neq 0$ and where $d_f(s)$ is obtained from d_f by replacing the s-row by r_f. The elements of d_f and $d_f(s)$ are either constants or linear in β, hence both d and d_f are polynomials.

(b) This follows from Theorem 12.2.3(e) and (f). □

12.3 Infinite Horizon

Now we turn to MDPs with infinite horizon, characterized by the catchword *approximation of the model.* An MDP$_\infty$ has no terminal reward function V_0 and hence also no sequence of value functions $(V_n)_0^\infty$ exists.

Definition 12.3.1 An ***MDP with infinite horizon*** (**MDP**$_\infty$ for short) is a tuple (S, A, D, p, r_S, β) which is defined as an MDP with large horizon.

For the definition of the expected infinite-stage reward earned by an infinite-stage policy $\pi = (\pi_t)_0^\infty \in \Pi := \mathbb{F}^{\mathbb{N}_0}$ and initial state s we need an infinite sequence of state random variables, which we denote by ζ_t, $t \in \mathbb{N}$, on some probability space $(\Omega, \mathfrak{F}, \mathbb{P}_{\pi s})$ which describes the infinite-stage decision process. As in Markov chain theory, this cannot be done without measure theory. (This is another indication that the *limit of the solution* is a simpler notion than the *limit of the model.*) The canonical construction is as follows. Firstly we choose the (uncountable!) sample space $\Omega := S^{\mathbb{N}}$, endowed with the σ-algebra $\mathfrak{F} := \otimes_1^\infty \mathfrak{P}(S)$, and as ζ_t we take the t-th coordinate random variable. The crucial point, which follows from the Theorem of Ionescu Tulcea of Appendix B.3.5, is the existence of a probability distribution $\mathbb{P}_{\pi s}$ on $\mathfrak{F}$ such that for each $N \in \mathbb{N}$ the probability distribution of $\zeta^N := (\zeta_t)_1^N$ has the density (12.3) with $s_0 := s$. (This means that $\zeta := (\zeta_t)_1^\infty$ is a non-homogeneous Markov chain with state space S, initial state s and having the transition matrices p_{π_t}, $t \geq 0$.) Denote by $\mathbb{E}_{\pi s}(\cdot)$ the expectation with respect to $\mathbb{P}_{\pi s}$.

Since

$$\sum_{t=0}^{\infty} \beta^t \cdot |r_S(\zeta_t, \pi_t(\zeta_t), \zeta_{t+1})| \leq \|r_S\|/(1-\beta) < \infty$$

the following definitions make sense and moreover yield finite values:

- $R_{\infty\pi}(s, \zeta) := \sum_{t=0}^\infty \beta^t \cdot r_S(\zeta_t, \pi_t(\zeta), \zeta_{t+1})$.
- $V_{\infty\pi}(s) := \mathbb{E}_{\pi s} R_{\infty\pi}(s, \zeta)$.
- $V_\infty(s) := \sup\{V_{\infty\pi}(s) : \pi \in \Pi\}$.
- $\pi^* \in \Pi$ is called *optimal* if $V_{\infty\pi^*} = V_\infty$. Moreover, policies of the form $\pi := (f)_0^\infty =: f^\infty$ for some decision rule f are called ***stationary.***

We now point out another reason why we prefer the *limit of the solution* to the infinite-stage model. In the usual frequency interpretation of probabilities the expectation $V_{\infty\pi}(s)$ of the random variable $R_{\infty\pi}(s, \zeta)$ is considered as a prognostic value for the average of a large number of independent realizations of $R_{\infty\pi}(s, \zeta)$. However, the computation of each realization requires us to wait for infinitely many periods. As this is impossible, one will in practice stop observations after a large yet finite number of periods. But then we are back in the finite-stage case with large horizon!

Our goals for MDP_∞s are as follows.

(a) We show that V_∞ exists and equals V with arbitrary terminal reward function. As a consequence, V_∞ satisfies the optimality equation and methods for computing V are mostly also applicable for V_∞.
(b) For an approximate solution of MDP_∞ one will carry through the VI for a number n of steps, starting with a terminal reward function which we denote by v_0 rather than V_0 since it is not determined by the model but can be chosen freely, possibly near V_∞ in order to obtain good bounds. The resulting value functions are denoted by $v_1, v_2, \ldots, v_n$. From the information available after n steps of the VI we derive lower and upper bounds for V_∞. They are improving in n in the sense that the lower bounds are increasing in n and the upper bounds are decreasing. Moreover they converge to V_∞ for $n \to \infty$. In addition, we obtain a lower bound for the performance of the stationary policy f_n^∞ where f_n is an arbitrary maximizer at stage n.
(c) We show that for given $\varepsilon > 0$ we can find already after one step of the VI a number $N_0 \geq 1$ such that for all $n \geq N_0$ the stationary policy f_n^∞ is *ε-optimal*, i.e. $V_{f_n} \geq V_\infty - \varepsilon$.

Theorem 12.3.2 (Infinite-horizon MDPs) *Denote by v_n, $n \geq 0$, the value functions in the MDP, which has the same data as the MDP_∞ and in addition some terminal reward function v_0. (Thus $V = \lim_{n\to\infty} v_n$ by Theorem 12.2.3(a).) Then the following holds:*

(a) $V_\infty = V$.
(b) For each decision rule f the stationary policy f^∞ is optimal if and only if f is asymptotically optimal if and only if f is a maximizer of LV if and only if $V = V_f$.
(c) Let $n \geq 1$, let f_n be a maximizer at stage n and put

$$w_{n-} := v_n + \beta \cdot \frac{\min(v_n - v_{n-1})}{1-\beta}, \quad w_{n+} := v_n + \beta \cdot \frac{\max(v_n - v_{n-1})}{1-\beta}.$$

We have

$$w_{n-} \leq V_{f_n} \leq V_\infty \leq w_{n+}, \tag{12.16}$$

and the bounds for V_∞ are improving in n and converging to V_∞ for $n \to \infty$.

(d) *We have*

$$V_{f_n} \geq V_\infty - \frac{\beta}{(1-\beta)} \cdot \mathrm{sp}(v_n - v_{n-1}) \geq V_\infty - \frac{\beta^n}{(1-\beta)} \cdot \mathrm{sp}(v_1 - v_0). \quad (12.17)$$

Thus f_n is ε-optimal if

$$n \geq n_0 := \frac{\log[\varepsilon \cdot (1-\beta)/\,\mathrm{sp}(v_1 - v_0)]}{\log \varepsilon}.$$

Proof

(a1) For $N \geq 1$ and $\pi = (\pi_t)_0^\infty \in \Pi$ put $\pi^N := (\pi_t)_0^{N-1}$. As the probability distribution of ζ^N has the density (12.3) with $s_0 := s$, the expected reward $v_{N\pi^N}(s)$ from MDP_N can be represented as an expectation with respect to $\mathbb{E}_{\pi s}$ as follows:

$$v_{N\pi^N}(s) = \mathbb{E}_{\pi s} R_{N\pi^N}(s, \zeta^N). \quad (12.18)$$

As the mapping $\pi \mapsto \pi^N$ from Π into $\mathbb{F}^N$ is surjective, it follows that

$$v_N(s) = \sup_{\sigma \in \mathbb{F}^N} v_{N\sigma}(s) = \sup_{\pi \in \Pi} \mathbb{E}_{\pi s} R_{N\pi^N}(s, \zeta^N), \; s \in S. \quad (12.19)$$

(a2) Using (12.18) we obtain

$$\begin{aligned}
&\mathbb{E}_{\pi s} |R_{\infty\pi}(s, \zeta) - R_{N\pi^N}(s, \zeta^N)| \\
&\leq \mathbb{E}_{\pi s} \Big(\sum_N^\infty \beta^t |rs(\zeta_t, \pi_t(\zeta_t), \zeta_{t+1})| + \beta^N |v_0(\zeta_N)| \Big) \\
&\leq \|rs\| \cdot \sum_N^\infty \beta^t + \|v_0\| \beta^N = \Big[\|rs\|/(1-\beta) + \|v_0\| \Big] \cdot \beta^N =: d \cdot \beta^N.
\end{aligned}$$

(a3) Now we obtain from (12.19) and Lemma 10.1.3, as $v_{N\pi^N}$ and $V_{\infty\pi}$ are finite, that

$$\begin{aligned}
|V_\infty(s) - v_N(s)| &= |\sup_\pi V_{\infty\pi}(s) - \sup_{\pi \in \Pi} v_{N\pi^N}(s)| \\
&\leq \sup_\pi |\, \mathbb{E}_{\pi s} R_{\infty\pi}(s, \zeta) - \mathbb{E}_{\pi s} R_{N\pi^N}(s, \zeta^N)| \\
&\leq \sup_\pi \mathbb{E}_{\pi s} |R_{\infty\pi}(s, \zeta) - R_{N\pi^N}(s, \zeta^N)| \leq d \cdot \beta^N.
\end{aligned}$$

Letting N tend to ∞ we see that V and V_∞ coincide.

(b) This follows from Theorem 12.2.3(e) and (f) since $V = V_\infty$.

(c1) Put $f := f_n$. We verify the first inequality in (12.16). From the upper bound in (12.10), applied to MDP_f with $t := 1$ and $v := v_{n-1}$ we obtain $U_f v_{n-1} - V_f \leq \beta \cdot \max(v_{n-1} - U_f v_{n-1})/(1-\beta)$. Now the assertion follows since $U_f v_{n-1} = U v_{n-1} = v_n$.

(c2) The second inequality in (12.16) is trivial. Moreover, the third one follows from (a) and the lower bound in (12.10) with $t := 1$ and $v := v_{n-1}$.

(c3) The upper and the lower bounds converge to V_∞ since $v_n \to V_\infty$ for $n \to \infty$. Antitonicity of the upper bounds (and similarly isotonicity of the lower bounds) are obtained as follows. From (12.9) with $t := 1$, $v := v_n$ and $w := v_{n-1}$ we get

$$\begin{aligned} w_{n+1,+} &= v_n + (v_{n+1} - v_n) + \beta \cdot \max(v_{n+1} - v_n)/(1-\beta) \\ &\leq v_n + \beta \cdot \max(v_n - v_{n-1}) + \beta^2 \cdot \max(v_n - v_{n-1})/(1-\beta) = w_{n+}. \end{aligned}$$

(d) The first inequality in (12.17) is obvious from (12.16), while the second one follows from the first one and from (12.9) with $t := n-1$, $v := v_1$ and $w := v_0$. □

12.4 Problems

Problem 12.4.1 (Monotone dependence of the value functions on the horizon) Prove:

(a) If $V_{k+1} \geq V_k$ [$V_{k+1} \leq V_k$] for some $k \in \mathbb{N}_0$ then $n \mapsto V_n(s)$ is increasing [decreasing] for $n \geq k$ and all $s \in S$.

(b) Assume in addition that V_n is finite for all n and that there is a set $I \subset S$ such that for all $s \in I$: $D(s)$ is finite, $V_{k+1}(s) > V_k(s)$ and $\sum_{t \in I} p(s, a, t) > 0$ for $a \in D(s)$.

Then $n \mapsto V_n(s)$ is *strictly* increasing for $n \geq k$ and all $s \in I$.

Problem 12.4.2 Show that $V_n(s)$ in Example 12.1.6 is strictly increasing in $n \geq 0$ for $1 \leq s \leq 3$, if $(1-\beta) \cdot \min V_0 < 2$.

Problem 12.4.3 (Howard's toymaker problem) Many of the concepts and results for MDPs can be illustrated by the following simple example from the pioneering book by Howard (1960). It is a simplified version of the marketing Example 12.1.1 with only two states: $s = 0$ and $s = 1$ denotes a good and a bad business situation, respectively, and $a = 0$ and $a = 1$ means *do nothing* or *advertise*, respectively. The other data are as in Table 12.2.

(a) In order to get a firm grasp of the VI, compute V_1, V_2 and V_3 for $V_0 := (105, 100)$, $\beta = 0.98$ and $\beta = 1$.

(b) U^β has for all $\beta \in \mathbb{R}^+ - \{1, 10\}$ a unique fixed point v^β, and no fixed point for $\beta \in \{1, 10\}$. Compute the fixed points.

Table 12.2 Data for Howard's toymaker Problem 12.4.3

(s, a)	$p(s, a, 0)$	$p(s, a, 1)$	$r_S(s, a, 0)$	$r_S(s, a, 1)$	$r(s, a)$
$(0, 0)$	0.5	0.5	9	3	6
$(0, 1)$	0.8	0.2	4	4	4
$(1, 0)$	0.4	0.6	3	-7	-3
$(1, 1)$	0.7	0.3	1	-19	-5

(c) Put $\beta_0 := 20/29$. $L^\beta v^\beta$ has for $\beta \in (\beta_0, 10) - \{1\}$ the unique maximizer $f_1 :\equiv 1$, for $\beta \in (0, \beta_0) + (10, \infty)$ the unique maximizer $f_0 :\equiv 0$, and for $\beta = \beta_0$ all four decision rules as maximizer.

Problem 12.4.4 Compute the fixed points v^β and the maximizers f^β of $L^\beta v^\beta$ for the second version of Example 12.1.6.

12.5 Supplements

Supplement 12.5.1 (Howard's toymaker Problem 12.4.3) The numerical data in the second version of this example are taken from Hübner (1980). An example with only two states, used widely in the literature for didactic purposes, is the so-called toymaker problem of Howard (1960). An explicit solution of it is given in Hinderer and Hübner (1977) and Hinderer (1976).

Supplement 12.5.2 (Historical remarks) The first paper on an MDP problem seems to be a recreational problem posed by Caley in 1874. Next in time seems to be the application of MDPs to reservoir problems by Massé in 1944. Another widely unknown paper, which uses VI for a statistical problem is Hughes in 1949. However, the systematic investigation in the field, starting with deterministic DPs is the pioneering work of Bellman. It began around 1950 and had a first culmination in his famous book Bellman (1957). There also his *Principle of Optimality* is stated. Karlin (1955) was perhaps the first one to point out the need for a proof of the Basic Theorem in its stochastic version rather than relying on the Principle of Optimality.

In the basic papers Dvoretzky et al. (1952a,b) on inventory problems methods of MDP are used implicitly, and inventory problems have remained since then a major application of MDPs. Much influence came from the book by Howard (1960), where the important method of policy iteration for the computation of V and of a maximizer of LV was introduced; cf. Howard and Nemhauser (1968) for information about the origin of his work.

A breakthrough in the theory of countable state MDPs was the paper by Blackwell (1962), who also gave the rigorous foundation for MDPs with arbitrary transition laws in Blackwell (1965).

From the vast theoretical and applied literature we mention a few examples which have had much influence on the development on the field: Dynkin (1965) (non-stationary countable-state Bayesian models); Hinderer (1970) (theoretical foundations for non-stationary infinite-stage models, based in particular on Blackwell 1965 and Dynkin 1965); Hordijk (1974) (a theoretical treatment of the countable state infinite-stage MDP, stressing methods of Markov chains); Dynkin and Yushkevich (1979) (rigorous theory intermingled with applications); Bertsekas and Shreve (1978) (the most advanced theoretical treatise).

Among the textbooks which influenced the development are the following: Ross (1970), Bertsekas (1976, 1995), Heyman and Sobel (1984, Vol.II), Whittle (1982), Ross (1983), and Puterman (1994), a largely expanded version of Bertsekas (1976).

Special topics within MDPs are treated, for example, in Berry and Fristedt (1985) and Hernández-Lerma (1989). Information on literature of more specialized topics are given in the supplements of the corresponding sections.

Numerical methods for MDPs are contained in Powell (2007), Chang et al. (2007) and Almudevar (2014).

Chapter 13
Examples of Markovian Decision Processes with Finite Transition Law

In this chapter we investigate several examples and models with finite transition law: an allocation problem with random investment, an inventory problem, MDPs with an absorbing set of states, MDPs with random initial state, stopping problems and terminating MDPs. Finally, stationary MDPs are generalized to non-stationary MDPs.

13.1 Examples with Explicit Solutions

Example 13.1.1 (Deterministic allocation with random investment opportunities) Consider the allocation problem from Example 2.4.3 with continuous input and with the modification that in each period an investment opportunity arises only with probability $p \in (0, 1]$, independent of opportunities in other periods. (The extreme case $p = 1$ brings us back to the original example.) As an illustration one could think of a company which obtains in each period with probability p a proposal for some project. The allocation of capital a results in an immediate utility $u(a) \in \mathbb{R}_+$ with $u(0) = 0$. How much capital a_t should one allocate at time $0 \leq t \leq N-1$ to such a project in order to maximize the expected discounted N-stage utility? The randomness of the investment opportunities in period t is modeled by i.i.d. random variables η_{t+1}. Here η_{t+1} equals 1 or 0 according to whether or not at time t there arises an investment opportunity, hence $\eta_{t+1} \sim \mathrm{Bi}(1, p)$.

We use the following CM: s_t and a_t denote the available capital in period t and the capital consumed in period t, respectively; hence $S = A = [0, K]$ for some $K \in \mathbb{R}^+$ and $D(s) = [0, s]$, $Z = \{0, 1\}$;

$$T(s, a, z) = \gamma s - az, \quad r_Z(s, a, z) = u(a) \cdot z,$$

K. Hinderer et al., *Dynamic Optimization*, Universitext,
DOI 10.1007/978-3-319-48814-1_13

for some $\gamma \in [1, \infty)$; thus $r(s,a) = \mathbb{E}\, r_Z(s,a,\eta_1) = p \cdot u(a)$; $V_0 \in \mathbb{R}_+^S$ is arbitrary with $V_0(0) = 0$; $\beta < 1$.

From the Basic Theorem 12.1.5 for the adjoint MDPad, which by Proposition 12.1.11 is equivalent to the CM we obtain the following two results:

The VI holds and it has, using $q := 1 - p$, the following form:

$$V_n(s) = q\beta \cdot V_{n-1}(\gamma s) + p \cdot \sup_{0 \le a \le s} [u(a) + \beta \cdot V_{n-1}(\gamma s - a)],\ s \in [0, K], n \ge 1.$$

In particular, $V_n(0) = 0$ for $n \ge 0$.

The policy $(f_n)_N^1$ is optimal for CM_N if $f_n(s)$ is a maximum point of

$$a \mapsto u(a) + \beta \cdot V_{n-1}(\gamma s - a),\ s \in [0, K].$$

♦

Example 13.1.2 (An explicitly solvable discrete inventory problem)

(i) The integer stock s_t is restricted to $S = \{0, 1, 2\}$. Thus, if a denotes the inventory *after* ordering, we have $A = S$, $D(s) := \{s, s+1, \ldots, 2\}$. (Many realistic inventory models have much larger state space. The present model may be applied, for example, for the inventory of a costly and large engine.)

(ii) η_t is the demand between time t and $t+1$, and the η_t's are i.i.d. with discrete density $q(\cdot) = (0.1, 0.7, 0.2)$ on $Z := \{0, 1, 2\}$.

(iii) Excess demand is lost, i.e. $s_{t+1} = (a_t - \eta_{t+1})^+$, and thus $T(s,a,z) := (a-z)^+$.

(iv) The ordering costs per unit are one. The inventory costs per unit of inventory at the end of a period are one and the penalty costs per unit of unsatisfied demand are three. Thus the one-stage costs are

$$c_Z(s,a,z) := a - s + (a-z)^+ + 3 \cdot (a-z)^-.$$

Thus we have a CM where the one-stage costs depend on the momentary disturbance z. We know from Proposition 12.1.11 that this model is equivalent to an MDP with $c(s,a) := \mathbb{E}\, c_Z(s,a,\eta_1)$. Using $\mathbb{E}(a-\eta_1)^- = \mathbb{E}(a-\eta_1)^+ - \mathbb{E}(a-\eta_1)$, we obtain

$$\begin{aligned} c(s,a) &= a - s + \mathbb{E}(a-\eta_1)^+ + 3 \cdot \mathbb{E}(a-\eta_1)^- \\ &= -2a - s + 4 \cdot \mathbb{E}(a-\eta_1)^+ + 3 \cdot \mathbb{E}\,\eta_1. \end{aligned}$$

Observing that $\cdot\, \mathbb{E}\,\eta_1 = 1.1$ and $\mathbb{E}(a-\eta_1)^+ = (a + 7(a-1)^+)/10$, we obtain

$$10c(s,a) = -10s + 33 - 16a + 28(a-1)^+.$$

(v) The terminal costs are $C_0(s) := -d \cdot s$ for some $d \in [0,\ 1.5]$. Thus d is the scrap value per unit.

(vi) There is no discounting, i.e. $\beta = 1$.

We know from the minimization version of Theorem 11.6 that the VI has the form

$$10 \cdot C_{n+1}(s) = 10 \cdot \min_{s \le a \le 2} \{c(s,a) + \mathbb{E}\, C_n((a-\eta_1)^+)\} \quad (13.1)$$
$$= -10s + 33 + \min_{s \le a \le 2} H_{n+1}(a), \quad n \in \mathbb{N}_0,$$

where for $n \in \mathbb{N}_0$

$$H_{n+1}(a) := -16a + 28(a-1)^+ + C_n(a) + 7C_n((a-1)^+) + 2C_n(0).$$

(a) We show by induction on n that

$$s \mapsto f(s) := \begin{Bmatrix} 1, & \text{if } s \in \{0,1\}, \\ 2, & \text{if } s = 2, \end{Bmatrix} = \max\{1, s\}$$

is a minimizer at each stage n. Note that f prescribes to order 1 piece if the stock is empty, and nothing, otherwise. By Theorem 11.6 we know that $(f)_N^1$ is a stationary optimal N-stage policy.

Put $a_n := C_n(0)$, $b_n := C_n(1)$, $c_n := C_n(2)$, $n \ge 0$. Then

$$H_{n+1}(a) = \begin{cases} 10a_n, & \text{if } a = 0, \\ -16 + 9a_n + b_n, & \text{if } a = 1, \\ -4 + 2a_n + 7b_n + c_n, & \text{if } a = 2. \end{cases}$$

As $D(2) = \{2\}$, f is a minimizer at stage $n+1$ if and only if $a := 1$ is a minimum point of $W_{n+1}(s, \cdot)$ for $s \in \{0, 1\}$. This holds by (13.1) if and only if

$$H_{n+1}(1) \le \min\{H_{n+1}(0), H_{n+1}(2)\},$$

or equivalently if and only if

$$7b_n - 112 \le 7a_n \le 6b_n + c_n + 12. \quad (13.2)$$

As $a_0 = 0$, $b_0 = -d$, $c_0 = -2d$ and $d \le 1.5$, (13.2) holds for $n = 0$. Now assume that for some $n \in \mathbb{N}$ the decision rule f is a minimizer at stages $1 \le k \le n$. We show that (13.2) holds, i.e. that f is a minimizer at stage $n+1$. From the induction hypothesis and (13.1) we know that $10C_k(s) = -10s + 33 + H_k(f(s))$ for $s \in S$ and $1 \le k \le n$, which yields for $1 \le k \le n$

$$10a_k = 17 + 9a_{k-1} + b_{k-1}, \quad (13.3)$$

$$b_k = a_k - 1, \quad (13.4)$$

$$10c_k = 9 + 2a_{k-1} + 7b_{k-1} + c_{k-1}.$$

Firstly, (13.4) with $k = n$ implies the first inequality in (13.2). The second one holds by (13.4) with $k = n$ if and only if $10a_n \leq 10c_n + 60$. Substituting a_n and c_n by (13.3) and (13.4) with $k = n$, respectively, yields the condition $7a_{n-1} \leq 6b_{n-1} + c_{n-1} + 52$, which holds by the second inequality in (13.2) with n replaced by $n-1$. Therefore (13.2) is true, hence f is a minimizer at stage $n+1$.

The proof also shows that f is a minimizer at each stage whenever it is a minimizer at stage 1, and this holds by (13.2) with $n = 0$ if and only if

$$7C_0(1) - 112 \leq 7C_0(0) \leq 6C_0(1) + C_0(2) + 12.$$

(b) There is also an explicit representation of $C_n(s)$. In fact, from (13.3) with $k = n$, (13.4) with $k = n-1$ and $a_0 = 0$, $b_0 = -d$, one easily obtains

$$C_n(0) = C_n(1) + 1 = 1.6n + 0.1(1-d), \quad n \geq 1,$$

$C_1(2) = c_1 = 0.9(1-d)$ and

$$c_{n+1} = \alpha + \gamma n + \delta c_n, \quad n \geq 1, \tag{13.5}$$

where $\alpha := 0.29 - 0.09d$, $\gamma := 1.44$, $\delta := 0.1$. The solution of the difference equation (13.5) can be obtained using a computer. One obtains for $n \geq 2$

$$C_n(2) = (7.555 - 8d) \cdot 0.1^n + 1.6n - 1.455 - 0.1d,$$

hence $C_n(2) \sim 1.6n - 1.455 - 0.1d$ for large n. Thus for each initial state the increase of costs per period is asymptotically equal to 1.6.

(c) The structure of the solution of the present problem is typical for many CMs with finite state and action space, in the following sense.

(c1) Often there exists a turnpike horizon (f^*, N^*). Here we have $f^* := f$ and even $N^*(f) = 1$.

(c2) If $\beta = 1$, then the functions $n \mapsto C_n(s)$, $s \in S$, often behave asymptotically as in the present example, i.e. there exists some g (in our example $g = 1.6$), independent of s, such that $C_n(s) - gn$ has a finite limit for $n \to \infty$. ♦

13.2 MDPs with an Absorbing Set of States

The second version of the marketing problem from Example 12.1.6 with $V_0(0) = 0$ showed the special feature that under any policy the decision process $(\zeta_t)_1^N$ stops as soon as the state $s = 0$ is reached. More generally we define in analogy to Definition 3.3.1:

Definition 13.2.1 A nonempty proper subset J_0 of the state space S is called an ***absorbing set*** for an MDP if we have

- $\sum_{s' \in J_0} p(s,a,s') = 1$ for all $s \in J_0$, $a \in D(s)$.
- $r(s,a) = 0$ for all $s \in J_0$, $a \in D(s)$.
- $V_0(s) = 0$ for all $s \in J_0$.

In applications J_0 often consists of a single point, Condition (i) means that the decision process $(\zeta_t)_1^N$ under any $\pi \in \mathbb{F}^N$ entering J_0 stays there until time N (i.e. is absorbed in J_0). Conditions (ii) and (iii) ensure that staying in J_0 does not earn anything and that no terminal reward is paid if the decision process really enters J_0 at some time $t \le N$, respectively. Note that (i) holds if $p(s,a,s) = 1$ for $s \in J_0$, $a \in D(s)$. Moreover, (i) implies that $r(s,a) = \sum_{s' \in J_0} r_S(s,a,s')$ for $s \in J_0$, $a \in D(s)$, hence (ii) holds if $r_S(s,a,s') = 0$ for $s, s' \in J_0$, $a \in D(s)$. From (i)–(iii) and the VI we obtain by induction on n: In an MDP with absorbing set J_0 we have $V_n = 0$ on J_0 for all $n \in \mathbb{N}_0$. Thus we must only care about the states in the ***essential state space*** $J := S - J_0$. More precisely, the following holds:

(a) The VI assumes the form

$$\begin{aligned} V_n(s) &= \sup_{a \in D(s)} \left[r(s,a) + \beta \cdot \sum_{s' \in J} p(s,a,s') \cdot V_{n-1}(s') \right] \\ &= \sup_{a \in D(s)} W_n(s,a), \quad n \ge 1,\ s \in J. \end{aligned} \tag{13.6}$$

(b) Any decision rule f which maximizes $a \mapsto W_n(s,a)$ for $s \in J$ is a maximizer at stage n.

In some examples with an absorbing set J_0 one obtains in addition to the one-stage reward a termination reward $r_{\mathrm{term}}(s') \in \mathbb{R}$ as soon as the decision process enters a state $s' \in J_0$. We can subsume this model under our previous one by using the new one-stage reward function $\tilde{r}_S$ with $\tilde{r}_S(s,a,s') := 0$ for $s \in J_0$ and

$$\tilde{r}_S(s,a,s') := r_S(s,a,s') + \beta \cdot r_{\mathrm{term}}(s') \cdot 1_{J_0}(s'), \quad s \in J, a \in D(s),\ s' \in S. \tag{13.7}$$

If condition (ii) for absorbing sets holds for $r(s,a)$, then

$$\tilde{r}(s,a) := p\tilde{r}_S(s,a) = r(s,a) + \beta \cdot \sum_{s' \in J_0} p(s,a,s') \cdot r_{\mathrm{term}}(s'), \quad s \in J,\ a \in D(s).$$

MDPs can be treated by means of an $\widehat{\mathrm{MDP}}$ with an absorbing set as follows: enlarge the state space S to $\hat{S} := S + \{\hat{s}\}$, $\hat{s}$ arbitrary, and put $\hat{D}(\hat{s}) := A$, $\hat{r}(\hat{s},a) = \hat{V}_0(\hat{s}) = 0$ and

$$\hat{p}(s,a,s') := \begin{cases} p(s,a,s'), & \text{if } s, s' \in S, \\ 0, & \text{if } s \in S, s' = \hat{s}, \\ 1, & \text{if } s = s' = \hat{s}. \end{cases}$$

Then $\widehat{\text{MDP}}$ has the absorbing set $J_0 := \{\hat{s}\}$ and the essential state space $J := S$. Therefore classical results about MDPs without an absorbing set are contained in results about MDPs with an absorbing set.

Example 13.2.2 (A stochastic batch-size problem) A customer orders $x \in \mathbb{N}$ pieces of a certain product which is produced in batches. The sizes of the batches can be chosen from a finite subset A of $\mathbb{N}$. Not more than N batches are allowed. As the production equipment does not work perfectly a produced batch will usually contain some defective pieces. Let $g(a,j)$ denote the probability that a batch of size a contains k good pieces, $0 \le k \le a$, $\sum_{k=0}^{a} g(a,k) = 1$ for all a. It is convenient to define $g(a,k) := 0$ for $k \notin \mathbb{N}_{0,a}$. Production stops as soon as at least x good pieces have been produced or N batches have been produced. There are production costs $\alpha \in \mathbb{R}_+$ per unit, a set-up cost $K \in \mathbb{R}_+$ for each batch, $K + \alpha > 0$, and penalty costs of amount $d \cdot s$, $d \in \mathbb{R}_+$, if at time N there is a shortage of s good pieces. Each good piece produced in excess above x has a scrap value of amount $e \in \mathbb{R}_+$. We want to choose the batch sizes such that the expected N-stage production costs are minimized.

We model the problem as an MDP with infinite state space, but finite transition law as follows. Let $s_t \in J := \mathbb{N}_m$ denote the number of pieces which must still be produced at time t and the initial state is $s_0 := x$. Let $s_t < 0$ mean that up to time t there is an excess production of $|s_t|$ pieces. Therefore we assume $J_0 := \{0, -1, -2, \ldots\}$. There is no restriction on the batch-size, hence $D(s) = A$ for all s. As the process stops as soon as it reaches J_0, we have $p(s, a, s') := g(a, s - s')$ for all s, a, s'. The one-stage costs c_1 have the form (13.7) with $-r_S(s, a, s') := (\alpha a + K) \cdot 1_J(s)$ and $-r_{\text{term}}(s') := -es'$. The terminal costs are $C_0(s) = d \cdot s^+$. We assume that there is no discounting, i.e. $\beta = 1$. It follows from the minimization version of (13.6) that the minimal expected n-stage costs $C_n(s)$ for $s \in J$ and $n \ge 1$ satisfy

$$C_n(s) = \min_{a \in A} \left[\alpha a + K + c_1(s,a) + \sum_{j=0}^{s-1} g(a,j) \cdot C_{n-1}(s-j) \right],$$

where

$$c_1(s,a) = e \cdot \sum_{s'<0} g(a, s-s')s'.$$

Obviously J_0 is absorbing, hence $C_n = 0$ on J_0 for all $n \ge 0$. Table 13.1 contains numerical results for the special case of a binomial transition law, where each item produced has the probability $p = 0.9$ of being good. Then $g(a, \cdot)$ is the discrete density of $\text{Bi}(a, 0.9)$, i.e.

$$g(a,j) = \binom{a}{j} \cdot 0.9^j \cdot 0.1^{a-j}, \quad 0 \le j \le a.$$

Table 13.1 Results for the batch-size problem from Example 13.2.2: upper part for $A = \{2, 3, 5\}$, lower part for $A = \{0, 2, 3, 5\}$

	$f_n(s)$					
n	$s = 1$	$s = 2$	$s = 3$	$s = 4 - 10$	$C_n(1)$	$C_n(10)$
1	2	2	3	5	8.240	42.500
2	2	3	3	5	8.272	35.000
3	2	3	3	5	8.273	36.042
4	2	3	3	5	8.273	36.202
5, 6	2	3	3	5	8.273	36.204
1	0	2	3	5	5.000	42.500
2	0	2	3	5	5.000	35.000
$3 - 6$	0	2	3	5	5.000	34.787

The other parameters are $s_0 = x = 10$, $\alpha = 2$, $K = d = 5$, $e = 1$. The upper part of the table refers to the case where the possible batch-sizes are 2, 3, and 5. In accordance with intuition the optimal batch-size $f_n(s)$ is increasing in s. On the other hand it seems strange that $C_n(s)$ is not decreasing in n (e.g. $C_4(10) = 36.202 > C_3(10) = 36.042$) as more production periods should allow for better choices of the batch-sizes. The reason is that the choice of the action space $A = \{2, 3, 5\}$ forbids stopping the production process at low states $s > 0$ where the minimal future production costs exceed the penalty costs for an immediate stop (e.g. $C_1(1) = 8.240 > C_0(1) = 5$). Therefore one expects that the minimal production costs may be reduced by admitting in addition the batch-size $a = 0$. The results for this second version are given in the lower part of Table 13.1. Now C_n is in fact decreasing in n. This follows by induction on n as $C_1 \leq C_0$. The gain relative to the first model depends on n and s. For $n = 6$ and $s = 10$ it is about 4.

The inspection of Table 13.1 leads to the following conjectures: In the first version the function $n \mapsto C_n(10)$ has the minimum point $n = 2$. In the second version: (a) $n \mapsto C_n(10)$ is decreasing, (b) $C_n(10) = C_4(10)$ for $n \geq 4$, and (c) $(1, f^*)$ with $f^* = (0, 2, 3, 5)$ is a turnpike. ♦

13.3 MDPs with Random Initial State

In some problems (see e.g. Example 13.4.1 below) we need an MDP with random initial state, defined as follows.

Definition 13.3.1 Given an MDP with finite state space and horizon $N \in \mathbb{N}$ we assume that

- the initial state ζ_0 is random and stochastically independent of the states ζ_t, $1 \leq t \leq N$.
- p_0 is the discrete density of ζ_0 on S.

Then we call the following maximum problem an ***MDP$_N$* with random initial state** (***MDP$_N$*(p_0)** for short).

This situation is described for $\pi \in \mathbb{F}^N$ by the joint discrete density of the coordinate random variables $(\zeta_t)_0^N$ on the sample space S^{N+1}, given by (cf. (12.2))

$$(s_t)_0^N \mapsto p_0(s_0) \cdot p_\pi(s_0, s^N) := p_0(s_0) \cdot p_{\pi_0}(s_0, s_1) \cdots p_{\pi_{N-1}}(s_{N-1}, s_N).$$

(The case where p_0 gives all mass to a fixed point s_0 brings us back to problem $MDP_N(s_0)$.) Let $V_{N\pi}$ and V_N be defined on S as in Chap. 12. The N-stage expected reward $V_{N\pi}(p_0)$ for initial discrete density p_0 is defined by

$$V_{N\pi}(p_0) := p_0\, V_{N\pi} := \sum_{s_0 \in S} p_0(s_0)\, V_{N\pi}(s_0). \tag{13.8}$$

The sum exists and is finite for all N and π since S is finite.

Problem $MDP_N(p_0)$ then means to maximize $\pi \mapsto V_{N\pi}(p_0)$, i.e. to find

- $V_N(p_0) := \sup_{\pi \in \mathbb{F}^N} V_{N\pi}(p_0)$, the ***maximal expected N-stage reward for initial discrete density* p_0**,
- *an N-stage policy* $\pi^* \in \mathbb{F}^N$, *which is* ***p_0-optimal*** in the sense that it maximizes $\pi \mapsto V_{N\pi}(p_0)$ on $\mathbb{F}^N$.

Proposition 13.3.2 (Solution of MDPs with random initial state) *If $\pi^* \in \mathbb{F}^N$ is optimal for MDP_N it is also p_0-optimal for $MDP_N(p_0)$ for each discrete density p_0 on S, and $V_N(p_0) = p_0 V_N$.*

Proof For each $\pi \in \mathbb{F}^N$ we get from (13.8)

$$V_N(p_0) \ge V_{N\pi^*}(p_0) = \sum_{s \in S} p_0(s) \cdot V_{N\pi^*}(s) \ge \sum_{s \in S} p_0(s) \cdot V_{N\pi}(s) = V_{N\pi}(p_0).$$

Now the assertion follows by taking the supremum over π. □

13.4 Stopping Problems

Example 13.4.1 (Finite-stage stopping of an uncontrolled Markov chain) Consider an N-stage Markov chain with finite state space J and transition matrix $q = (q(s, s'),\ s, s' \in J)$. We firstly treat the case where the chain starts in a given state s_0, and afterwards the case where it starts according to a discrete density p_0 on J. We want either to stop the chain at one of the N time points $0 \le t \le N-1$ or to continue up to time N. If the chain is at time $0 \le t \le N-1$ not yet stopped and in state $s \in J$, one obtains $g(s) \in \mathbb{R}$ if it is stopped (action $a = 1$), and $h(s) \in \mathbb{R}$ if it is not stopped (action $a = 0$). If the chain is not stopped before time N and if it is in state $s \in J$ at time N, one gains $g(s)$. According to which policy do we obtain the maximal expected gain $V_N(s_0)$? We use the following MDP with absorbing set $\{\bar{s}\}$, hence $S = J + \{\bar{s}\}$; $s_t \in J$ denotes the state at time t, if the chain has not yet been stopped, while $s_t = \bar{s}$ means that the chain is stopped before time t; $A = D(s) = \{0, 1\}$; for $s \in J$ we have $p(s, 0, s') = q(s, s')$, $s' \in J$, $p(s, 1, \bar{s}) = 1$,

$$r(s,a) = \begin{cases} g(s), & \text{if } a = 1, \\ h(s), & \text{if } a = 0, \end{cases}$$

and $V_0(s) = g(s)$. Since $\{\bar{s}\}$ is absorbing, $V_n(\bar{s}) = 0$ for all $n \geq 0$. Put

$$v_n(s) := qV_{n-1}(s) := \sum_{j \in J} q(s,j) \cdot V_{n-1}(j), \; n \geq 1, \; s \in J.$$

Here and in the sequel we denote the multiplication of the matrix q or of the vector p_0 with a J-vector v by qv and by $p_0 v$, respectively. Then the VI (13.6) implies that on J

$$V_n = \max\{h + \beta v_n, g\}, \quad n \geq 1. \tag{13.9}$$

In addition, an optimal policy $\pi^* = (f_n)_N^1$ for MDP_N is given by the largest maximizers f_n at the stages n, determined by $f_n(\bar{s}) := 1$ and

$$f_n(s) := \text{ stop if and only if } \beta \cdot v_n(s) \leq g(s) - h(s), \; s \in J.$$

The optimal policy π^* prescribes to stop at stage n in state $s \in S$ (provided we have not stopped earlier) if and only if $g(s) \geq V_n(s)$ if and only if $g(s) = V_n(s)$. We call the set S_n of states $s \in S$ where $a = 1$ is optimal at stage $n \geq 1$ the ***stopping set*** at stage n. Obviously

$$B_n = [g \geq V_n] = [g = V_n], \quad n \geq 1.$$

(Note that $\bar{s}$ belongs to each stopping set.) The optimal policy π^* tells us to stop at the earliest time t where the momentary stopping reward $g(s_t)$ equals at least the momentary reward $h(s_t)$ for not stopping plus the expected discounted reward for the remaining periods. Thus one should stop in problem MDP_N at the time

$$\begin{aligned} \tau_N := \tau_N((s_t)_0^{N-1}) &:= \min\{0 \leq t \leq N-1 : s_t \in B_{N-t}\} \\ &= \min\{0 \leq t \leq N-1 : \beta \cdot v_{N-t}(s_t) \leq g(s_t) - h(s_t)\}, \end{aligned}$$

where $\min \emptyset := N$. We call τ_N the ***optimal stopping time*** of the stopped MDP_N. It gives for each possible path $(s_t)_0^{N-1}$ of the Markov chain the smallest time point $0 \leq t \leq N$ at which one should stop. The optimal stopping time τ_N in Example 13.4.1 has the remarkable property that for $0 \leq t \leq N-1$ the set $[\tau = t] := \{(s_0^{N-1}) \in S^N : \tau_N(s_0^{N-1}) = t\}$ does not depend on $s_{t+1}, s_{t+2}, \ldots, s_{N-1}$; in fact, $[\tau_N = t] =$

$\{(s_0^{N-1}) \in S^N : s_i \notin B_{N-i}, 0 \le i \le t-1, s_t \in B_{N-t}\}$. In other words: if at any time $0 \le t \le N-1$ one has observed $s_0, s_1, \ldots, s_t$ and if one has not yet stopped then it only depends on s_t whether or not one should stop at time t. ♦

Part (a) of the next result shows how V_n, $n \ge 1$, and hence B_n, can be computed recursively without maximization (except for the application of the operation $x \mapsto x^-$). Moreover, for fixed $N \ge 1$ isotonicity of $t \mapsto B_{N-t}$ in part (b) means that at a fixed state s one should stop the rather the fewer stages $N-t$ are still ahead.

Proposition 13.4.2 (The solution of the stopping problem from Example 13.4.1) *The following holds:*

(a) $V_n = g + e_n^-$, $n \ge 0$, *where* $e_0 :\equiv 0$, $e_1 := g - h - \beta \cdot qg$ *and* $e_{n+1} := e_1 - \beta \cdot q e_n^-$.
(b) V_n *is increasing in* n*; thus there exists* $V := \lim_n V_n$ *and* B_n *is decreasing in* n.
(c) Assume $\beta < 1$ *and put* $B := \cap_1^\infty B_n$, *i.e.* $B = [V = g]$. *Then* V *is the unique solution of the optimality equation,* $V \le \max h^+/(1-\beta) + \max g^+ < \infty$ *and* f_B *is asymptotically optimal.*
(d) Assume that B_1 *is quasi-absorbing in the sense that* $P(s, 0, B_1) = 1$ *for* $s \in B_1$. *Then* $V_n = g$ *on* B_1, $n \ge 1$, *and all stopping sets* B_n *equal* B_1.

Proof

(a) We prove the assertion (I_n) by induction on $n \ge 0$. (I_0) holds trivially. Assume that (I_n) holds for some $n \ge 0$. From (13.9) we infer that

$$V_{n+1} = g + \max\{0, -g + h + \beta \cdot qV_n\} = g + (g - h - \beta \cdot qV_n)^-. \quad (13.10)$$

From this equation the assertion follows by applying (I_n).
(b) As by (a) $V_1 \ge g = V_0$, it follows from (13.6) by induction on $n \ge 0$ that $V_{n+1} \ge V_n$.
(c) The upper bound for V holds since induction on $n \ge 0$ shows that $V_n \le \sigma_n(\beta) \cdot \max h^+ + \max g^+$ on J, $n \ge 1$. The other two assertions follow from Theorem 12.2.3(a) and (f), observing that 1_B is a maximizer of LV.
(d1) The assertion (I_n) that $V_n = g$ on B_1 holds for $n = 0$. Assume (I_n) for some $n \ge 0$. Then for s in the quasi-absorbing set B_1 we have

$$\begin{aligned} W_{n+1}(s,0) &= g(s) + \beta \sum_{s' \in B_1} q(s,s')\, V_n(s') \\ &= g(s) + \beta \sum_{s' \in B_1} q(s,s')\, g(s') = W_1(s,0). \end{aligned}$$

Thus $V_{n+1}(s) = V_1(s) = g$ by (13.9).
(d2) It follows from (d1) that $B_1 \subset B_n$. Now (d) follows from (b). □

Remark 13.4.3 (Example 13.4.1)

(i) If all sets B_n equal B_1, then the optimal stopping time for MDP_N equals

$$(s_t)_0^{N-1} \mapsto \tau_N((s_t)_0^{N-1}) = \min\{0 \le t \le N-1 : s_t \in B_1\},$$

which is called the *first entrance time of the set* B_1. It is also called a *one-step look-ahead (OSLA) stopping rule* for the following reason: in MDP_N one stops at time $0 \le t \le N-1$ in state s_t if and only if one stops in MDP_{t+1} at time t, i.e. one step before the end.

(ii) It follows from the proof of (c) that $V \le g^+ < \infty$ if $h \le 0$.

(iii) Assume $\beta \le 1$ and $h \le 0$. Then we see from (b) that each maximum point of g belongs to each set B_n. In particular, if $g \equiv \alpha \in \mathbb{R}$, then $B_n = S$ and $V_n = \alpha$, $n \ge 1$.

(iv) In the case where the chain starts according to a discrete density p_0 on J the maximal expected discounted N-stage reward is $d_{N+1} := V_N(p_0)$, $N \ge 1$. Set $d_1 := p_0 g$. From Propositions 13.3.2 and 13.4.2 one easily derives, using $p_0(\bar{s}) := 0$, the following results:

(a) $d_{n+1} = p_0 g + p_0 e_n^-$, $n \ge 0$.
(b) The policy π^* is p_0-optimal for each p_0.
(c) d_n is increasing in n. Thus there exists $d := \lim_n d_n$.
(d) The bounds in Proposition 13.4.2(c) hold with V replaced by d. ◇

Example 13.4.4 (Selling an asset) You possess an asset and obtain successively at most $N+1 \ge 2$ i.i.d. offers at times $t = 0, 1, \ldots, N$. Each offer takes values in a finite subset J of $\mathbb{R}$ and has a discrete density p_0. (Negative offers may occur as a price for selling waste.) When an offer arrives you must decide whether you accept it, upon which the process stops. Otherwise you continue. If none of the offers at times $0 \le t \le N-1$ is accepted, you must accept the offer s_N at time N. Each offer costs you $\gamma \in \mathbb{R}_+$ units.

According to which policy do you obtain d_N, the maximal expected gain when at most $N+1$ offers arise? This is that special case of Example 13.4.1 where $q(s, s') = p_0(s')$, independent of s and where $g(s) := s - \gamma$ and $h(s) := -\gamma$, $s \in J$. Since $v_n = p_0 V_{n-1} = d_n$, we see that v_n is independent of s and equal to the constant d_n.

Put $H(x) := \sum_{s \in J} p_0(s) \cdot \max\{\beta \cdot x, s\} - \gamma$, $x \in \mathbb{R}$. Now (13.10) and Proposition 13.4.2 show:

(a) $d_1 = \mathbb{E}\,\zeta_0 - \gamma$, $d_{n+1} = H(d_n)$, $n \ge 1$.
(b) The policy $\pi^* = (f_n)_N^1$ where $s \mapsto f_n(s) =$ stop if and only if $s \ge \beta \cdot d_n$, is p_0-optimal. Thus π^* is a control-limit policy.
(c) If $\beta \le 1$ then $d \le (\max J - \gamma)^+$, and $d = H(d)$. If $\beta < 1$ $(= 1)$ then d is the unique (the smallest) solution x of the equation $x = H(x)$. ♦

In accordance with intuition it follows that (i) $d_n = d_n(\gamma)$ is decreasing in γ and that (ii) the larger γ, the more likely an offer is accepted. This property is also exhibited in Fig. 13.1 where for the special case of throwing a die (see

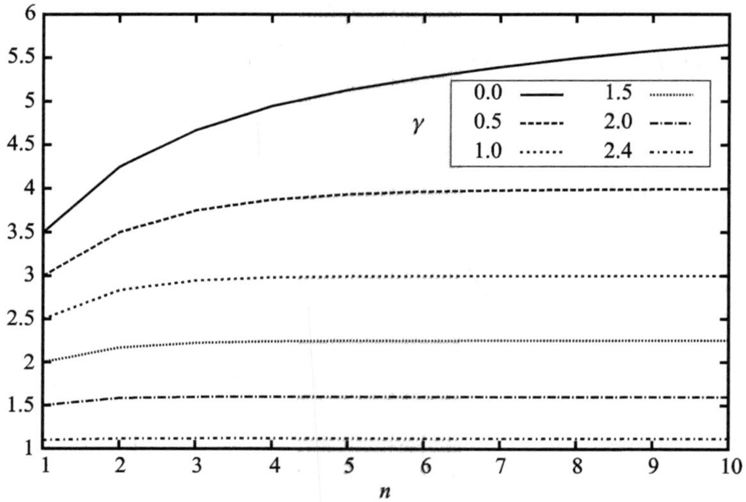

Fig. 13.1 Values d_n of Example 13.4.5

Example 13.4.5 below) the numbers $d_1, \ldots, d_{10}$ for various values of γ are plotted. (Points belonging to the same value of γ are joined by straight lines.)

Example 13.4.5 (Throwing a die) We treat the special case of Example 13.4.4 where the offers are the throws of a fair die, hence $\zeta_0 \sim U(\mathbb{N}_6)$ and where $\beta = 1$. This problem admits a detailed analysis, as follows.

(i) One should accept a throw of size s at stage n if and only if $s \geq \lceil d_n \rceil$. This holds as induction on $n \geq 1$ shows that $d_n \leq 6 - \gamma \leq 6$.

(ii) If each throw costs one unit (i.e. if $\gamma = 1$), induction on n yields the following explicit solution

$$d_{n+1} = 3 - \frac{1}{2 \cdot 3^n}, \; n \in \mathbb{N}_0. \tag{13.11}$$

Moreover, as d_n is strictly increasing from 2.5 to 3, the following stationary policy is optimal for each N: accept the first throw which is at least 3.

(iii) There exists

$$d^* := \lim_n d_n,$$

and $3.5 - \gamma \leq d^* \leq 6 - \gamma$. In fact, existence of d^* follows from isotonicity of (d_n), and the bounds for d^* are obvious. The general computation of d^* we leave as a problem; as an example, d^* equals 6, 4 and 3 if $\gamma = 0, 0.5$ and 1, respectively.

(iv) There exists a decision rule f^* (with $f^*(\infty)$ arbitrary) which has a finite turnpike horizon, i.e. f^* is a maximizer for all stages large enough. In fact, it follows from (iv) that

$$N^* := \min\{n \in \mathbb{N} : d_n > \lceil d^* \rceil - 1\} \tag{13.12}$$

is finite. As $\lceil d_n \rceil = \lceil d^* \rceil$ if and only if $n \geq N^*$, N^* is a turnpike horizon for any decision rule f^* such that for $s \in \mathbb{N}_6$ we have $f^*(s) := 1$ if and only if $s \geq \lceil d^* \rceil$. Figure 13.2 shows N^* as a function of γ. It seems that there are four numbers γ_1 to γ_4 such that the function $\gamma \mapsto N^*(\gamma)$ is increasing to infinity in a left neighborhood and constant in a right neighborhood of γ_i, $1 \leq i \leq 4$. This is indeed true and $\gamma_1 = 1/6$, $\gamma_2 = 0.5$, $\gamma_3 = 1$ and $\gamma_4 = 5/3$.

(v) The explicit representation (13.11) of d_n in case $\gamma = 1$ is typical for all $\gamma \in [0, 2.5)$. In fact, for those γ we have

$$d_n = d^* - (d^* - d_{N^*}) \cdot \alpha^{n-N^*}, \; n \geq N^*, \tag{13.13}$$

where $\alpha := (\lceil d^* \rceil - 1)/6 \in \mathbb{N}_5$. In particular, $(d_n)_1^\infty$ converges exponentially fast to d^*. A proof of (13.13) runs as follows. Obviously $\lfloor d_n \rfloor = \lceil d^* \rceil - 1 =: k$ for $n \geq N^*$. Now the recursion

$$d_{n+1} = 3.5 - \gamma + \lfloor d_n \rfloor (2d_n - \lfloor d_n \rfloor - 1)/12, \quad n \geq 0, \tag{13.14}$$

implies that $d_{n+1} = \alpha \cdot d_n + \delta$, $n \geq N^*$, for some δ. Applying Lemma 4.1.5 to $b_i := d_{N^*+i}$ shows that $b_i = \delta \cdot \sigma_i(\alpha) + b_0 \cdot \alpha^i$, $i \geq 1$. Letting i tend to ∞ yields

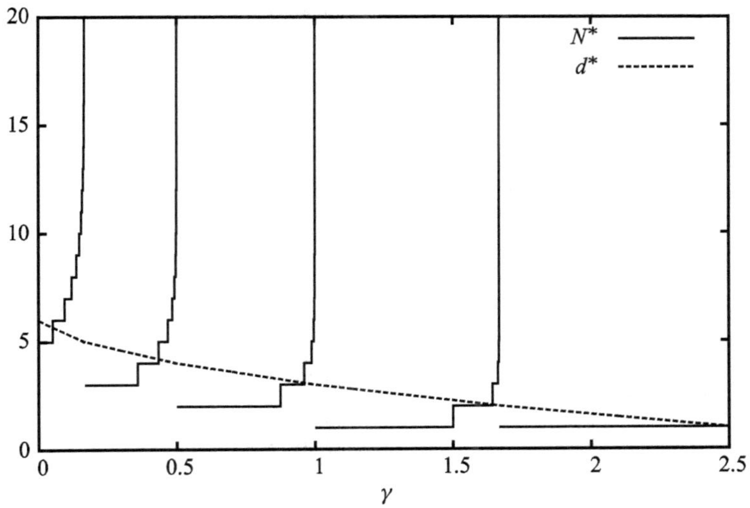

Fig. 13.2 Values of N^* and d^* for Example 13.4.5

$\delta = (1 - \alpha) \cdot d^*$, and then simple algebraic manipulations verify (13.13). — The explicit solution (13.13) requires the computation of d^* and of d_{N^*}, which allows us to compute N^* from (13.12). Finally we obtain d_{N^*} from (13.14).

(vi) The sequence of value functions V_n converges to

$$V(s) := \begin{cases} \max\{d^*, s\} - \gamma, & \text{if } s > 0, \\ 0, & \text{if } s = 0. \end{cases}$$

It can also be shown that V is the unique solution v of the optimality equation (cf. Theorem 12.2.3) with $v(0) = 0$. ♦

Example 13.4.6 (Selling m copies of an asset) The following is an extension of Example 13.4.4 to the case of selling more than one copy of an asset. We have to sell within N periods $s_0 = m \leq N$ copies of an asset. At each time $0 \leq t \leq N - 1$ we receive, unless all copies are sold, an offer of amount η_t, $0 \leq t \leq N - 1$, which costs $d \geq 0$. We assume that the η_t's are i.i.d. and take values in a compact interval Z. If $k \in \mathbb{N}_0$ copies remain unsold at time N, we obtain a scrap value of amount $V_0(k)$ with $V_0(0) = 0$. We want to maximize the expected discounted sum of offers accepted at times $0 \leq t \leq N - 1$. Thus $s \in S := \mathbb{N}_{0,m}$ is the number of units not yet sold; $t_1(s, z) = (s - 1)^+$, $t_0(s, z) = s$,

$$g(s, z, a) = \begin{cases} z - d, & \text{if } s > 0 \text{ and } a = 1, \\ -d, & \text{if } s > 0 \text{ and } a = 0, \\ 0, & \text{if } s = 0. \end{cases}$$

Obviously g and v_0 are bounded. The set $J_0 := \{0\} \times Z$ is absorbing, hence $V_n(0) = 0$ for all n. Next, we have $G := J := X - J_0$. Now we obtain $v_N(s) := \mathbb{E}\, V_N(s, \eta_1)$ for $s > 0$ by the recursion

$$v_n(s) = -d + \beta \cdot v_{n-1}(s) + \mathbb{E}\,[\eta_1 + \beta \cdot (v_{n-1}(s-1) - v_{n-1}(s)]^+ \,,\; 1 \leq s \leq m.$$

Moreover, the smallest maximizer f_n at stage n has for $s > 0$ the form

$$f_n(s, z) := 1 \text{ if and only if } z > \beta \cdot (v_{n-1}(s) - v_{n-1}(s-1)).$$ ♦

Example 13.4.7 (Cayley's lottery problem from 1875) In the 1875 edition of "The Educational Times", p. 189, the mathematician and lawyer Arthur Cayley (1821–1895; known above all for his work in algebra and projective geometry) posed the following entertaining problem, which seems to be the first publication of a stochastic DP problem.

> A lottery is arranged as follows: There are n tickets representing a, b, c, etc. pounds respectively. A person draws once; looks at his ticket, and if he pleases, draws again (out of the remaining $n - 1$ tickets), looks at his ticket, and if he pleases draws again (out of the remaining $n - 2$ tickets), and so on, drawing in all not more than k times. He receives

the value of the last drawn ticket. Supposing that he regulates his drawings in the manner most advantageous to him according to the theory of probabilities, what is the value of his expectation?

Cayley gave an informal solution on p. 237 of the same journal (with numerical values for $n = 4$ and $(a, b, c, d) = (1, 2, 3, 4)$), which reads in modern terminology as follows. (We replace n by m and k by $N \leq m$ in order to be consistent with our terminology.) Let $\mathbb{N}_m$ be the set of tickets j with values b_j, $1 \leq j \leq m$. For each non-empty subset $B \subseteq \mathbb{N}_m$ denote by $M_n(B)$ the maximal expected reward of the last drawn ticket in at most $n \leq |B|$ drawings without replacement from the tickets in B. Then the desired value of $M_n(\mathbb{N}_m)$ is obtained recursively by

$$M_n(B) = |B|^{-1} \cdot \sum_{i \in B} \max\{b_i, M_{n-1}(B - \{i\})\}, \quad 2 \leq n \leq k, \tag{13.15}$$

$$M_1(B) := |B|^{-1} \cdot \sum_{i \in B} b_i. \tag{13.16}$$

We are going to derive (13.15) and (13.16) by modeling the problem as an MDP as follows:

(i) $s_t = (i_t, B_t)$, where i_t is the number of the t-th ticket drawn and $B_t \neq \emptyset$ the set of tickets remaining after the t-th draw, so that $i_t \notin B_t$. Moreover, the state $(i_t, \emptyset)$ means that drawing has stopped after at most t draws and i_t was the ticket drawn last. Thus

$$S = \{(i, B) \in \mathbb{N}_m \times \mathfrak{P}(\mathbb{N}_m) : i \notin B\}.$$

(ii) $a_t = 1$ or 0 if the t-th draw is accepted (and the game ends) or not. Thus $A = D(s) = \{0, 1\}$ for all s.

(iii) If we are in state $s = (i, B)$ with $B \neq \emptyset$ and if we reject ticket i, the next drawing is from the set B with uniform distribution while in the case of acceptance of i the next state is $(i, \emptyset)$. Thus

$$p((i, B), a, (j, C)) = \begin{cases} 1/|B|, & \text{if } B \neq \emptyset, a = 0, j \in B, C = B - \{j\}, \\ 1, & \text{if } B = C = \emptyset, j = i, \\ 1, & \text{if } a = 1, j = i, C = \emptyset, \\ 0, & \text{else.} \end{cases}$$

(iv) The reward r is zero, while $V_0(i, B) = b_i$ and $\beta = 1$.

It follows from (i)–(iv) that $V_n(i, B)$ is the maximal expected reward if the ticket drawn first is i and if at most further $n \wedge |B|$ drawings from the set B are allowed. Obviously the set $J_0 := \mathbb{N}_m \times \{\emptyset\} \subseteq S$ is absorbing. Therefore $V_n(i, \emptyset) = V_0(i, \emptyset) =$

b_i for all n and i and the VI assumes for $B \neq \emptyset$ the form

$$\begin{aligned} V_n(i,B) &= \max_{a\in\{0,1\}} \left\{ \sum_{(j,C)\in S} p((i,B),a,(j,C))V_{n-1}(j,C) \right\} \\ &= \max\left\{ |B|^{-1} \cdot \sum_{j\in B} V_{n-1}(j,B-\{j\}), b_i \right\}. \end{aligned} \tag{13.17}$$

$M_n(B)$ is obtained from $V_n(i, B - \{i\})$ according to (13.8), by averaging over the possible tickets i in B, hence

$$M_n(B) = |B|^{-1} \cdot \sum_{j\in B} V_n(j, B - \{j\}).$$

Thus $M_1(B)$ is as in (13.16) and it follows now from (13.17) that (13.15) holds. Moreover, we realize from (13.15) that an optimal N-stage policy $(f_n)_N^1$ is given by

$$f_n(i,B) := 1 \text{ if and only if } b_i \geq M_{n-1}(B). \qquad \blacklozenge$$

Example 13.4.8 (A stochastic scheduling problem) This example uses the reduction of the state space by means of the reachable sets. Three jobs $i = 1, 2, 3$ arrive sequentially in this order at a service station and carry with them i.i.d. random gains η_1, η_2, η_3, respectively, which take values in a finite set $Z \subset \mathbb{R}$ with positive probabilities q_z, $z \in Z$. For processing the jobs one assigns to each of the jobs one of three machines having processing coefficients $b_1 > b_2 > b_3$. This means that assigning machine a to job i yields the reward $b_a \cdot \eta_i$, $1 \leq a, i \leq 3$. How should one assign the machines in order to maximize the expected sum of rewards? (The case $b_a := 10^{3-a}$, $Z := \mathbb{N}_{0,9}$ models a popular bowling game, called Hausnummern-Kegeln in German, where the number of fallen pins in three throws are used to form a 3-digit number.)

One expects that there exists an optimal policy of the following structure:

(i) For appropriate real numbers $y_1 < y_2$ one should assign to job 1 machine 3 if $\eta_1 < y_1$, machine 2 if $y_1 \leq \eta_1 < y_2$ and machine 1 if $\eta_1 \geq y_2$.
(ii) If a machine has been assigned to the first job, if i and j are the machines not yet assigned and if (without loss of generality) $b_i > b_j$, then one should assign machine j to job 2 if and only if $\eta_2 < y_3(i,j)$ for appropriate reals $y_3(i,j)$.
(iii) After machines have been assigned to the first two jobs, the remaining machine must be assigned to the last job.

We are going to verify this conjecture and compute explicitly the numbers y_1, y_2 and $y_3(i,j)$. Remarkably, it turns out that y_1 and y_2 are symmetric with respect to $\mathbb{E}\,\eta_1$ and $y_3(i,j)$ does not depend on (i,j).

We use the following CM: The states have the form $s = (x, \varepsilon) \in S := Z \times \{0,1\}^3$; x is the gain of that job which is momentarily arriving; $\varepsilon := (\varepsilon_1, \varepsilon_2, \varepsilon_3)$ where $\varepsilon_a = 0$ if and only if at the moment machine a is not yet assigned, $1 \leq a \leq 3$; for simplicity of notation the vector $(\varepsilon_1, \varepsilon_2, \varepsilon_3)$ is denoted by the codeword $\varepsilon_1\varepsilon_2\varepsilon_3$; $a \in A := \{1, 2, 3\}$ denotes the machine which is assigned to the momentarily arriving job; $D(x, \varepsilon) := D(\varepsilon) := \{a \in \mathbb{N}_3 : \varepsilon_a = 0\}$ for $\varepsilon \neq 111$, while $D(x, 111)$ is arbitrary; $T(x, \varepsilon, a, z) := (z, \varepsilon + e_a)$ (with e_a the a-th unit vector) if $\varepsilon \neq 111$, while $T(x, 111, a, z) := (z, 111)$; $r_Z(x, \varepsilon, a, z) = r(x, \varepsilon, a) = x \cdot b_a$ if $\varepsilon \neq 111$, while $r(x, 111, a) = 0$; $\beta := 1$; $V_0 :\equiv 0$. It follows that $J_0 := Z \times \{111\}$ is absorbing.

Consider the probability distribution on S with discrete density $p(x, \varepsilon) := q_x$, $x \in Z$, if $\varepsilon = 000$, and $p(x, \varepsilon) := 0$ otherwise. We want to find

(a) $V_3(p) := \max_{\pi \in \mathbb{F}^3} \mathbb{E}\, V_{3\pi}(\eta_1, 000)$,
(b) a p-optimal policy $\pi^* \in \mathbb{F}^3$.

We shall use that $V_3(p) = \mathbb{E}\, V_3(\eta_1, 000)$ according to (13.8) and that $\pi^* \in \mathbb{F}^3$ is p-optimal if it is s_0-optimal for all $s_0 \in Z \times \{000\}$.

Put $v_n(\varepsilon) := \mathbb{E}\, V_n(\eta_1, \varepsilon)$, hence $v_0(\varepsilon) = 0$. The VI holds by Theorem 12.1.5 and it has the form, using $J = S - J_0$,

$$\begin{aligned} V_n(x, \varepsilon) &= \max_{a \in D(\varepsilon)} \left[x \cdot b_a + v_{n-1}(\varepsilon + e_a)\right], \\ &=: \max_{a \in D(\varepsilon)} W_n(x, \varepsilon, a), \;\; 1 \leq n \leq 3, \; (x, \varepsilon) \in J, \end{aligned} \tag{13.18}$$

starting with $v_0 \equiv 0$. From the form of T we see that for $s_0 := (x, 000)$ the reachable state spaces do not depend on x and are

$$\begin{aligned} R_1(s_0) &= \cup_{z \in Z} \{(z, 001), (z, 010), (z, 100)\} =: S_1, \\ R_2(s_0) &= \cup_{z \in Z} \{(z, 011), (z, 101), (z, 110)\} =: S_2, \\ R_3(s_0) &= J_0. \end{aligned}$$

From our discussion of reachable state spaces we know, as $p(x, \varepsilon) = 0$ unless $\varepsilon = 000$, that we can find $V_3(x, 000)$ for all $x \in Z$ (and hence $V_3(p) = v_3(000)$) from the VI by computing V_1 only on $\cup_{x \in Z} R_2(x, 000) = S_2$ and V_2 only on $\cup_{x \in Z} R_1(x, 000) = S_1$. Moreover, a policy $(f_n)_3^1$ is p-optimal as soon as $f_n(z, \varepsilon)$, $1 \leq n \leq 3$, maximizes $W_n(z, \varepsilon, \cdot)$ for $(z, \varepsilon) \in \cup_{x \in Z} R_{3-n}(x, 000) = S_{3-n}$.

We are going to show, using $\mu := \mathbb{E}\, \eta_1$ and $\gamma := \mathbb{E}(\mu - \eta_1)^+$:

The conjecture above holds with

$$y_3 := \mu, \qquad y_1 := \mu - \gamma, \qquad y_2 := \mu + \gamma.$$

Moreover, the maximal expected reward is

$$v_3(000) = 3b_2 \cdot \mu + (b_1 - b_2) \cdot \mathbb{E} \max\{\eta_1, y_2\} - (b_2 - b_3) \cdot \mathbb{E} \min\{\eta_1, y_1\}.$$

Proof Let f_n be the smallest maximizer at stage n.

(a) Firstly we obtain from (13.18) as $V_0 \equiv 0$

$$v_1(\varepsilon) = \begin{cases} \mu \cdot b_1, & \text{if } \varepsilon = 011, \\ \mu \cdot b_2, & \text{if } \varepsilon = 101, \\ \mu \cdot b_3, & \text{if } \varepsilon = 110. \end{cases}$$

(b) Using (a), we see from (13.18) that

$$V_2(x, 001) = \max \{x \cdot b_1 + \mu \cdot b_2,\ x \cdot b_2 + \mu \cdot b_1\}.$$

It follows that $f_2(x, 001) = 2$ if and only if $x < \mu$. This implies, using $m := \mathbb{E} \min\{\eta_1, \mu\} = \mu - \gamma$ and $M := \mathbb{E} \max\{\eta_1, \mu\} = \mu + \gamma$, that

$$\begin{aligned} v_2(001) &= \mathbb{E}\, V_2(\eta_1, 001) \\ &= \sum_{x<\mu} q_x \cdot (x \cdot b_2 + \mu \cdot b_1) + \sum_{x \geq \mu} q_x \cdot (x \cdot b_1 + \mu \cdot b_2) = b_2 m + b_1 M. \end{aligned}$$

In the same way we see that $f_2(x, 010) = 3$ if and only if $f_2(x, 100) = 3$ if and only if $x < \mu$, and that

$$v_2(010) = b_3 m + b_1 M, \qquad v_2(100) = b_3 m + b_2 M.$$

In particular, the assertion about y_3 holds.

(c) We see from (13.18) that

$$\begin{aligned} V_3(x, 000) &= \max \{x \cdot b_1 + v_2(100),\ x \cdot b_2 + v_2(010),\ x \cdot b_3 + v_2(001)\} \\ &= \max_{1 \leq a \leq 3} W_3(x, 000, a). \end{aligned}$$

Now $f_3(x, 000) = 3$ if and only if both $W_3(x, 000, 1) < W_3(x, 000, 3)$ and $W_3(x, 000, 2) < W_3(x, 000, 3)$. Using (b) these two conditions assume the form

$$x(b_1 - b_3) < \Lambda := (b_2 - b3)m + (b_1 - b_2)M$$

and $x < m$, respectively. As $m \leq M$, the second of these conditions implies the first one. Thus $f_3(x, 000) = 3$ if and only if $x < m$. Next, $f_3(x, 000) = 1$ if and only if both $W_3(x, 000, 2) \leq W_3(x, 000, 1)$ and $W_3(x, 000, 3) \leq W_3(x, 000, 1)$. Using (b) these two conditions assume the form $x \geq M$ and $\Lambda \leq x(b_1 - b_3)$, respectively. As $m \leq M$, the first of these conditions implies the second one. Thus $f_3(x, 000) = 1$ if and only if $x \geq M$. Now the assertions about y_1 and y_2 are verified.

(d) From (c) and (13.8) we see that

$$v_3(000) = \mathbb{E}\, V_3(\eta_1, 000) = \sum_{x<m} q_x \cdot W_3(x, 000, 3)$$
$$+ \sum_{m\le x<M} q_x \cdot W_3(x, 000, 2) + \sum_{x\ge M} q_x \cdot W_3(x, 000, 1).$$

Now the value of $v_3(000)$ given above is obtained by writing the middle sum as

$$\sum_{m\le x<M} (\cdots) = \sum_{x\in Z}(\cdots) - \sum_{x<m}(\cdots) - \sum_{x\ge M}(\cdots)$$

and by inserting the values of $v(001)$, $v(010)$ and $v(100)$ from (b). □

As special case consider the Hausnummern-Kegeln and make the assumption that the discrete density q_z, $-1 \le z \le 10$, of the number η_1 of pins fallen is triangular and symmetric to 4.5 and with $q_{-1} = q_{10} = 0$. Then

$$f_3(x, 000) = \begin{cases} 3, & \text{if } 0 \le x \le 3, \\ 2, & \text{if } 4 \le x \le 5, \\ 1, & \text{if } 6 \le x \le 9, \end{cases}$$

and the maximal expected number generated by the number of pins fallen in three throws equals $v_3(000) = 642.5$. On the other hand, if one assigned the number of pins fallen randomly to the three digits, one would obtain on average only the reward $\mathbb{E}(100\eta_1 + 10\eta_2 + \eta_3) = 499.5$. ♦

13.5 Terminating MDPs

We now generalize some of the results in Sect. 5.1 to MDPs with finite S and an absorbing set $J_0 \notin \{\emptyset, S\}$. Recall from Chap. 12 the definition $J := S - J_0$ and the definition of the set $R(s) := R_1(s)$ of states reachable from $s \in S$ in one step.

Definition 13.5.1 Let h be a function h from S into $\mathbb{N}_0$ such that $[h = 0] := \{s \in S : h(s) = 0\} = J_0$ and such that

$$R(s) \subset [h \le h(s) - 1], \quad s \in J. \tag{13.19}$$

Then h is called a *horizon function* and the MDP is called ***h*-terminating** with *horizon* $H := \max_{s\in S} h(s)$.

Note that $H \geq 1$ since $J \neq S$. Intuitively, $h(s)$ measures the distance of the state s from J_0. Moreover, for each $s \in S$ the sets $R_t(s)$, $t \geq 1$, of states reachable in t steps from s drift towards J_0 in the sense that $R_t(s)$ is contained in $[h \leq (h(s) - t)^+]$, in particular $R_t(s) \subset J_0$ for $t \geq N$. Note that S is the disjoint union of the sets $[h = k]$, $1 \leq k \leq N$, that $s \in [h = h(s)]$ and that (13.19) holds if and only if for each $1 \leq k \leq N$ we have $R(s) \subset [h \leq k - 1]$ for all $s \in [h = k]$.

Example 13.5.2 (Optimal weighing with a balance beam) This problem is treated in Cairns (1963) and Ahlswede and Wegener (1979, p. 115). Among $k \geq 2$ coins there is exactly one which is heavier than the other ones. (The two weights involved are not known.) What is the minimal number of weighings required to detect the heavier coin using a balance beam? We model the problem by the following h-terminating MDP: $s_t \in S = \mathbb{N}_{0,k}$ denotes the number of coins still to be examined at time $t \geq 0$, hence $s_0 = k$; $a_t \in A := \mathbb{N}_k$, $t \geq 0$, denotes the number of coins you put on each side of the beam when weighing at time t; thus $D(s) := \{a \in A : a \leq s/2\}$ if $s \geq 2$ and $D(0) = D(1) = A$. At each time $t \geq 0$ the selection of the set B of $2a$ coins for the momentary weighing from the set of $s \geq 2$ coins not yet inspected is made randomly. This means that, independent of the previous states and actions, the heavier coin is contained in B with probability

$$p(s, a, a) := |\mathcal{F}|/|\mathcal{A}| = \binom{s-1}{2a-1} \Big/ \binom{s}{2a}.$$

Here $\mathcal{F}$ is the system of favorable sets B (determined by its $2a - 1$ coins of smaller weight) and $\mathcal{A}$ is the system of all sets B. It follows that $p(s, a, a) = 2a/s$ and that the heavier coin is not contained in B with probability $p(s, a, s - 2a) = 1 - 2a/s$. Moreover, $p(s, a, s) = 1$ for $s \in J_0 := \{0, 1\}$; $c(s, a) = 1$ for $s \geq 2$ and $c(0) = c(1) = 0$; $C_0 \equiv 0$; $\beta = 1$.

Obviously J_0 is absorbing. The MDP is h-terminating for $h(s) := s$, $s \geq 2$, $h(0) = h(1) = 0$, since $R(s) = \bigcup \{\{a\} \cup \{s - 2a\} : 1 \leq a \leq s/2\} \subset [h \leq s - 1] = \mathbb{N}_{0,s-1}$ for $s \geq 2$. Moreover, $H = k$. ♦

For an h-terminating MDP with absorbing set J_0 the problem ***MDP***$_{(s)}$:= $MDP_{H_0}(s)$, $s \in J$, means to maximize the expected sum of discounted rewards until any of the decision processes, which starts in $s \in J$, enters J_0. Thus $V(s) := V_{H_0}(s)$ is the maximal expected sum of discounted rewards earned by the decision processes starting in s until they enter the absorbing set J_0. Obviously $V|J_0 \equiv 0$ since $V_n|J_0 = 0$ for all $n \geq 1$. We call $s \mapsto V(s)$ the ***value function*** of the h-terminating MDP. Note that V does not depend on V_0 and on how $D(s)$ is defined for $s \in J_0$.

Problem $MDP(s)$ can be solved by applying the Basic Theorem 12.1.5 to MDP_{H_0}. As a rule, the following result is easier to apply.

Theorem 13.5.3 (Basic Theorem for terminating MDPs) *For each h-terminating MDP with absorbing set J_0 the following holds.*

(a) *The value function V satisfies the* **optimality equation** *(***OE** *for short)*

$$V(s) = \sup_{a\in D(s)} \left[r(s,a) + \beta \cdot \sum_{s'\in R(s)} p(s,a,s') \cdot V(s') \right] \tag{13.20}$$
$$=: \sup_{a\in D(s)} W(s,a), \quad s \in J.$$

The OE can be solved recursively: For fixed $s_0 \in J$ *compute successively V on* $[h = 1]$, $[h = 2]$, …, $[h = h(s_0)]$, *starting with* $V|J_0 \equiv 0$.

(b) *The* **optimality criterion:** *If f is a maximizer of W, then for each* $s_0 \in J$ *the stationary policy* $(f)^1_{H_0}$ *is* s_0*-optimal in MDP*(s_0).

Proof Firstly, the VI holds by Theorem 12.1.5 since S is finite.

(a1) Intuitively one expects that (I_k) holds for $1 \le k \le H$: If $s \in [h \le k]$ then $V_n(s) = V_k(s)$ for $n \ge k$. A proof by induction on $1 \le k \le N_0$ runs as follows. (I_1) is true since $R(s) \subset J_0$ for $s \in [h \le 1]$ and since $V_n|J_0 \equiv 0$ for $n \ge 1$. Now assume that (I_k) holds for some $1 \le k < H$. Select $s \in [h \le k+1]$, hence $R(s) \subset [h \le k]$. It follows that $V_n(s) = UV_{n-1}(s) = UV_k(s) = V_{k+1}(s)$ for $n \ge k+1$.

(a2) For $s \in S = [h \le H]$ and (a1) for $k := H$ and $n := H+1$ we get $V(s) = V_N(s) = V_{N+1}(s) = UV_N(s) = UV(s)$.

(b) We prove by induction on $1 \le k \le H$ that (I_k) holds: $V_k(s) = V_{kf}(s)$ for $s \in [h \le k]$. (Then (I_H) proves Theorem 13.5.3(b).) Firstly, consider (I_1). If $s \in [h \le 1]$ then (a1) for $n := H$ and (13.20) entail that $V_1(s) = V(s) = r_f(s) + \beta \cdot \sum_{s'\in R(s)} p_f(s,s')\, V(s')$. Now (I_1) follows since on $R(s) \subset [h = 0] = J_0$ we have $V = V_0$, hence $V_1(s) = U_f V_0(s) = V_{1f}(s)$ by the RI. Now assume (I_k) for some $1 \le k < H$. Select $s \in [h \le k+1]$, hence $R(s) \subset [h \le k]$. From (I_k), from $V(s') = V_k(s')$ for $s' \in R(s)$ by (a1) with $n := H$ and from the RI we get

$$V_{k+1}(s) = V(s) = r_f(s) + \beta \cdot \sum_{s\prime\in R(s)} p_f(s,s\prime)\, V(s\prime)$$
$$= r_f(s) + \beta \cdot \sum_{s\prime\in R(s)} p_f(s,s\prime)\, V_{kf}(s\prime) = U_f V_{kf}(s) = V_{k+1,f}(s). \qquad \square$$

Example 13.5.4 (Continuation of Example 13.5.2) It follows from (13.20) that $C(k)$ can be found recursively from

$$C(s) = 1 + \max_{1\le a\le s/2} [2a/s \cdot C(a) + (1 - 2a/s) \cdot C(s - 2a)]$$
$$=: \max_{1\le a\le s/2} W(s,a), \quad 2 \le s \le k,$$

starting with $C(0) = C(1) = 0$. Moreover, if f is a maximizer of W, then the stationary policy $(f)_H^1$ is s_0-optimal. Intuitively one expects that $C(s)$ is increasing in s. However, $C(7) = 13/2 < C(6) = 2$. ♦

Remark 13.5.5

(a) Consider an MDP with finite S and with a non-empty absorbing set J_0. Then there exists a horizon function h if and only if $N(s) := \inf\{t \geq 0 : R_t(s) \subset J_0\}$ (with $\inf \emptyset := \infty$) is finite for all $s \in S$. And then $N(\cdot)$ is the pointwise smallest horizon function. Note that $N(s)$ is the earliest time t such that all states which are reachable from s at time t are absorbing.

Proof If h is a horizon function and $s \in S$ then, as shown above, $R_t(s) \subset J_0$ for $t := h(s)$. Thus $N(s) \leq h(s) < \infty$. On the other hand, let $N(s) < \infty$, $s \in S$. Then $N(s) = 0$ if and only if $R_0(s) \subset J_0$ if and only if $s \in J_0$. Moreover, if $s \in [N = k]$ for some $k \geq 1$ then $R_k(s) = \cup_{s' \in R(s)} R_{k-1}(s')$ shows that $s' \in [N \leq k-1]$ for $s' \in R(s)$. □

(b) In Example 13.5.2 with $k := 4$ we have $h(s) - N(s) = 1, 2, 2$ for $s = 2, 3, 4$. The use of $N(\cdot)$ as horizon function requires the mostly burdensome computation of the sets $R_t(s)$, $t \geq 0$, $s \in S$. Therefore, for the computation of $V(s_0)$, often the use of another *simpler* horizon function h is advantageous, although this in general requires the computations of $C(s')$ for more states s'. ◇

Example 13.5.6 (Optimal sampling from an urn without replacement) An urn initially contains i_0 red and j_0 black balls. One draws balls successively without replacement one at a time. One can stop at any time before the urn is empty. Each time one draws a red ball (a black ball) one wins one unit (loses one unit). It is allowed to stop with zero reward immediately, i.e. not to sample at all. What is the maximal expected reward $V(i_0, j_0)$ for optimal stopping? (We have $V(i_0, j_0) \geq 0$ as immediate stopping is allowed.) By intuition one may be led to the following conjectures:

(a) Continue sampling as long as the maximal expected reward $V(i, j)$ for the future is positive and stop otherwise. (Here i and j are the number of momentarily remaining red and black balls, respectively.)
(b) $V(i, j) = 0$ if and only if $j \geq i$.
(c) $V(i, j)$ is increasing in i and decreasing in j. (The latter implies that $V(i, j) = 0$ for all $j \geq k$ whenever $V(i, k) = 0$.)

Table 13.2 below disproves (b), as, for example, $V(5, 7) > 0$, and it supports (c). Below we verify (a), (c) and a weaker form of (b): $V(i, j) = 0$ for $j \geq 2i$.

We use the following terminating MDP: $s = (i, j) \in S = \mathbb{N}_0^2$, where i and j denote the number of red and black balls respectively, before the momentary drawing; $s_t = (0, 0)$ means that before time $t \geq 0$ sampling has stopped (either voluntarily or since the urn has become empty); $a = 1$ or 0 whether one stops drawing or continues drawing; thus $A = D(s) = \{0, 1\}$, if $s \neq (0, 0)$, and $D((0, 0)) = \{1\}$; the action $a = 1$ leads from (i, j) to $(0, 0)$ while under action $a = 0$ a transition takes place

Table 13.2 Rounded values of $V(i,j)$ for Example 13.5.6

	i										
j	0	1	2	3	4	5	6	7	8	9	10
10	0.0	0.0	0.0	0.0	0.0	0.0	0.0	0.0	0.4	0.9	1.6
9	0.0	0.0	0.0	0.0	0.0	0.0	0.0	0.3	0.8	1.5	2.3
8	0.0	0.0	0.0	0.0	0.0	0.0	0.2	0.8	1.4	2.2	3.0
7	0.0	0.0	0.0	0.0	0.0	0.2	0.7	1.3	2.1	3.0	3.8
6	0.0	0.0	0.0	0.0	0.1	0.5	1.2	2.0	2.9	3.7	4.7
5	0.0	0.0	0.0	0.0	0.4	1.1	1.9	2.8	3.7	4.6	5.5
4	0.0	0.0	0.0	0.3	1.0	1.8	2.7	3.6	4.5	5.4	6.4
3	0.0	0.0	0.2	0.8	1.7	2.5	3.5	4.4	5.3	6.3	7.3
2	0.0	0.0	0.7	1.5	2.4	3.3	4.3	5.3	6.2	7.2	8.2
1	0.0	0.5	1.3	2.3	3.2	4.2	5.1	6.1	7.1	8.1	9.1
0	0.0	1.0	2.0	3.0	4.0	5.0	6.0	7.0	8.0	9.0	10.0

from (i,j) with $i+j>0$ to $(i-1,j)$ with probability $i/(i+j)$ and to $(i,j-1)$ with probability $j/(i+j)$; thus $P(i,j,1,\cdot,\cdot) = \delta_{(0,0)}$, and

$$P(i,j,0,\cdot,\cdot) = \frac{i}{i+j}\cdot\delta_{(i-1,j)} + \frac{j}{i+j}\cdot\delta_{(i,j-1)};$$

the reward is given by

$$r_S(i,j,a,i',j') = \begin{cases} 1, & \text{if } a=0, i'=i-1\geq 0, j'=j, \\ -1, & \text{if } a=0, i'=i, j'=j-1\geq 0, \\ 0, & \text{else,} \end{cases}$$

hence, (using $0/0 := 0$), $r(i,j,a) = (i-j)/(i+j)\cdot\delta_{a,0}$; $V_0 \equiv 0$, $\beta = 1$.

It is obvious that the MDP has the absorbing set $J_0 := \{(0,0)\}$. It follows from

$$R(i,j) = \begin{cases} \{(0,0),(i-1,j),(i,j-1)\}, & \text{if } i,j\geq 1, \\ \{(0,0),(i-1,0)\}, & \text{if } 0=j<i, \\ \{(0,0),(0,j-i)\}, & \text{if } 0=i<j, \end{cases}$$

that the MDP is h-terminating, hence $V(0,0)=0$. From Theorem 13.5.3(a) with $h(i,j) := i+j$ we obtain for $i,j\geq 1$

$$\begin{aligned} V(i,j) &= (i+j)^{-1}\max\{i-j+i\cdot V(i-1,j)+j\cdot V(i,j-1),0\} \\ &=: \max\{W(i,j,0),W(i,j,1)\}, \qquad (13.21) \\ V(i,0) &= (1+V(i-1,0))^+,\ V(0,j) = (-1+V(0,j-1))^+. \end{aligned}$$

In particular, $V \geq 0$. It is obvious (and also follows trivially by induction on i and j) that $V(i,0) = i$, $V(0,j) = 0$, $i, j \geq 0$. Moreover, the largest maximizer f of W, which determines according to Theorem 13.5.3(b) for each s_0 an s_0-optimal policy is given by

$$f(i,j) := 0 \text{ if and only if } V(i,j) > 0,$$

since $W(i,j,0) > W(i,j,1) = 0$ if and only if $V(i,j) > 0$ which verifies the conjecture (i) above. The recursive solution of (13.21) according to Theorem 13.5.3(a) is easily implemented on a PC, observing that $[h = k] = \{(i,j) \in \mathbb{N}^2 : i + j = k\}$; cf. Table 13.2.

Proposition 13.5.7 (Properties of the value function of Example 13.5.6) *In Example 13.5.6 we have:*

(a) $2 \cdot V(i,j) \leq (2i - j)^+$ *for all* $i, j \geq 0$. *In particular,* $V(i,j) = 0$ *if* $j \geq 2i$.
(b) $\sup_{i,j \geq 1}[V(i,j-1) - V(i-1,j)] = 2$.
(c) $V(i,j)$ *is increasing in* i *and decreasing in* j.
(d) $i - j \leq V(i,j) \leq i$ *for all* i, j, *hence* $V(i,j)/i \to 1$ *for* $i \to \infty$ *and all* j.

Proof

(a) We use induction on $k := i + j \geq 0$. The case $k = 0$ is trivial, as $V(0,0) = 0$. Assume that for some $k \geq 0$ the assertion holds for all $i, j \geq 0$ with $i + j = k$. Now select some i, j with $i + j = k + 1$. The cases $i = 0$ and $j = 0$ are trivial. Thus assume $i, j \geq 1$. Then we obtain from (13.21) and the induction hypothesis that

$$2 \cdot (i + j) \cdot V(i,j) \leq L(i,j) := \left[2i - 2j + i(2i - j - 2)^+ + j(2i - j + 1)^+\right]^+.$$

Thus it suffices to show that $L(i,j) \leq R(i,j) := (i + j)(2i - j)^+$.
Case α: $2i - j \leq 0$. Then $L(i,j) = 0 = R(i,j)$.
Case β: $2i - j = 1$. Then $L(i,j) = j + 1 \leq (3j + 1)/2 = R(i,j)$ as $j \geq 1$.
Case γ: $2i - j \geq 2$. Then

$$L(i,j) = [(i + j)(2i - j) - j]^+ \leq R(i,j).$$

(b) One easily derives from (13.21) by induction on i that $V(i,1) = i^2/(i + 1)$, $i \geq 0$. Therefore $V(i,0) - V(i-1,1) = 2 - 1/i$. Thus $M := \sup_{i,j \geq 1}[V(i,j-1) - V(i-1,j)] \geq 2$. That $M \leq 2$ holds will now be shown jointly with (c).

(c) Astonishingly, this intuitively obvious property seems to require an involved proof. As $V(0,j) = 0 = V(0,j-1)$, $j \geq 1$, and as $V(i,0) = i \geq V(i-1,0)$, $i \geq 1$, it suffices to show that

$$V(i-1,j) \leq V(i,j) \leq V(i,j-1) \leq 2 + V(i-1,j), \; i,j \geq 1, \tag{13.22}$$

which also proves $M \leq 2$ and thus completes the proof of (b). For the proof of (13.22) we use induction on $k := i+j \geq 2$. The assertion holds trivially for $k = 2$. Assume that for some $k \geq 2$ it holds for all $i, j \in N$ such that $i + j \leq k$. Select some $i, j \in \mathbb{N}$ with $i + j = k + 1$.

(c1) Firstly we prove the last inequality in (13.22). It holds for $i = 1$ (and $j = k$) as (a) implies $V(1, j-1) \leq (1-(j-1)/2)^+ \leq 2 = 2 + V(0,j)$. It also holds for $j = 1$ (and $i = k$) as $V(i-1, 1) = (i-1)^2/i \geq i-2 = V(i, 0) - 2$. Thus we can assume $i, j \geq 2$. Put $v_1 := V(i-1, j-1)$, $v_2 := V(i-2, j)$. By (13.21) we have to show that

$$B := [i-j+1+i\cdot v_1 + (j-1)\cdot V(i,j-2)]^+$$
$$\leq C := 2(i+j-1) + [i-j-1+(i-1)\cdot v_2 + j\cdot v_1]^+ .$$

The induction hypothesis implies $V(i,j-2) \leq 2 + v_1$. Hence, as $x \mapsto x^+$ is isotone on $\mathbb{R}$, we have

$$B \leq [i-j+1+i\cdot v_1 + (j-1)\cdot(2+v_1)]^+$$
$$= [2(i+j-1) + i-j-1+(i-1)\cdot v_2 + j\cdot v_1 + (i-1)\cdot(v_1 - v_2 - 2)]^+$$
$$\leq C.$$

The induction hypothesis also yields $v_1 \leq 2 + v_2$. Now we obtain, using that $(x+y+z)^+ \leq x^+ + y^+ + z^+$ for x, y, $z \in \mathbb{R}$

$$B \leq 2(i+j-1) +$$
$$[i-j-1+(i-1)\cdot v_2 + j\cdot v_1]^+ + (i-1)\cdot[v_1 - v_2 - 2]^+ \leq C.$$

(c2) Next we prove the first inequality in (13.22). It holds trivially for $i = 1$. Thus assume $i \geq 2$. Put $v_3 := V(i-1, j)$, $v_4 := V(i, j-1)$. By (13.21) we have to show that

$$D := (i+j)\cdot[i-j-1+(i-1)\cdot V(i-2,j) + j\cdot V(i-1,j-1)]^+ \leq K^+,$$

where $K := (i+j-1)(i-j+i\cdot v_3 + j\cdot v_4)$. From (c1) we know that $v_4 - v_3 - 2 \leq 0$. Now the induction hypothesis yields

$$D \leq (i+j)\cdot[i-j-1+(i-1)\cdot v_3 + j\cdot v_4]^+$$
$$= [K + j\cdot(v_4 - v_3 - 2)]^+ \leq K^+ + j\cdot[v_4 - v_3 - 2]^+ \leq K^+.$$

(c3) The proof of the second inequality in (13.22) is similar to (c2).

(d) For $s := (i,j)$ we have $N(s) = i + j$, and $V_{i+j,\pi}(s) \leq i$ for all $\pi \in \mathbb{F}^{i+j}$, hence $V(i,j) \leq i$. On the other hand, the policy $\sigma \in \mathbb{F}^{i+j}$, which prescribes always to continue sampling, yields $V_{i+j,\sigma}(s) = i - j$, hence $V(i,j) \geq i - j$. □

13.6 Non-stationary MDPs and CMs

Stationary MDPs can easily be generalized to non-stationary ones, except that the transition functions T_t must be replaced by stochastic transition laws p_t and the one-stage rewards $r_t(s,a)$ are replaced by $r_{S_t}(s,a,s')$. We consider the case of forward indexing of data, since it has the advantage of using the time-direction in which the decision process evolves. Possible time dependent discount factors are implicitly included in the reward functions; the explicit treatment of discount factors is indicated below. Altogether we use the

Definition 13.6.1 A **non-stationary N-stage MDP$_N$ with finite transition law** is a tuple

$$((S_t)_0^N,\ (A_t)_0^{N-1},\ (D_t)_0^{N-1},\ (p_t)_0^{N-1},\ (r_{S_t})_0^{N-1},\ v_N)$$

with the following meaning, where the index t refers to data in force at time t:

- S_t is the state space.
- A_t is the action space.
- D_t is the constraint set.
- p_t is the finite transition law.
- r_{S_t} is the one-stage reward function.
- v_N is the terminal reward function.

A decision rule at time t, $0 \le t \le N-1$, is a mapping f from S_t into A_t such that $f(s) \in D_t(s)$ for all $s \in S_t$. Denote by $\mathbb{F}_t$ the set of decision rules at time t. Then $\Pi_N := \times_{i=0}^{N-1}\mathbb{F}_i$ is the set of N-stage policies. We know from the beginning of Chap. 12 that there exists a non-homogeneous Markov chain $\zeta^N := (\zeta_t)_1^N$ on a discrete probability space $(\Omega, \mathbb{P}_{\pi s_0})$, starting in s_0 and controlled by a policy $\pi = (\pi_t)_0^{N-1}$, such that the stochastic $S_t \times S_{t+1}$-matrix $p_{t\pi} := (p_t(s, \pi_t(s), s'))$, $s \in S_t$, $s' \in S_{t+1}$, is the transition law at time t. The ***total random N-stage reward*** for initial state s_0 and policy π is defined as

$$R_{0\pi}(s_0, \zeta^N) := \sum_{t=0}^{N-1} r_{S_t}(\zeta_t, \pi_t(\zeta_t), \zeta_{t+1}) + v_N(\zeta_N).$$

Put

$$p_\pi(s_0, s^N) := p_{0\pi_0}(s_0, s_1) \cdot p_{1\pi_1}(s_1, s_2) \cdots p_{N-1,\pi_{N-1}}(s_{N-1}, s_N),\ s_0 \in S, s^N \in S^N.$$

Since the transition laws p_t are finite, the ***expected total N-stage reward***

$$v_{0\pi}(s_0) := \mathbb{E}_{\pi s_0} R_{0\pi}(s_0, \zeta^N) = \sum_{s^N \in \prod_1^N S_t} R_{0\pi}(s_0, s^N) \cdot p_\pi(s_0, s^N)$$

exists and is finite; here $\mathbb{E}_{\pi s_0}$ is the expectation with respect to $\mathbb{P}_{\pi s_0}$. The problem consists in computing for initial state $s_0 \in S_0$ the ***maximal expected N-stage reward***

$$v_0(s_0) := \sup\{v_{0\pi}(s_0) : \pi \in \Pi_N\}$$

and, if possible, an optimal policy, i.e. a maximum point of $\pi \mapsto v_{0\pi}(s_0)$ for all $s_0 \in S_0$. The model MDP$_N$ defines in an obvious way for $0 \le t \le N-1$ a non-stationary *submodel MDP$_{N-t}$* whose data consist of those which are valid between time t and time N. Denote by v_t the value function for *MDP$_{N-t}$*. (Since discounting is included in the reward functions, v_t constitutes the value function which is discounted back to time $t = 0$.) For $0 \le t \le N-1$ and a function v on S_{t+1} we use the abbreviation

$$p_t v(s,a) := \sum_{s' \in S_{t+1}} p_t(s,a,s') \cdot v(s'), \quad (s,a) \in D_t.$$

A decision rule $f \in \mathbb{F}_t$ is called a ***maximizer at time t*** if for all $s \in S_t$ the action $f(s)$ is a maximum point of

$$a \mapsto w_t(s,a) := r_t(s,a) + p_t \cdot v_{t+1}(s,a).$$

Here $r_t := p_t r_{S_t}$. Exactly as in Chap. 12 we get the RI and then the following

Theorem 13.6.2 (Basic Theorem for a non-stationary MDP$_N$ with finite transition law)

(a) The optimality criterion holds: If π_t is a maximizer at time t, $0 \le t \le N-1$, then the policy $(\pi_t)_0^{N-1}$ is optimal.
(b) v_0 may be computed by the (forward) value iteration

$$v_t(s) = \sup_{a \in D_t(s)} \Big[r_t(s,a) + p_t \cdot v_{t+1}(s,a) \Big], \; 0 \le t \le N-1, \; s \in S_t. \qquad (13.23)$$

Time-dependent discount factors $\beta_t > 0$ for period $0 \le t \le N$ (with $\beta_0 := 1$) and undiscounted reward functions $\hat{r}_{S_t}$ and $\hat{v}_N$ can easily be treated by choosing $r_{S_t} := \beta_0\beta_1 \cdots \beta_t \cdot \hat{r}_{S_t}$, $0 \le t \le N-1$, and $v_N := \beta_0\beta_1 \cdots \beta_N \cdot \hat{v}_N$. Then $\hat{v}_t := (\beta_0\beta_1 \cdots \beta_t)^{-1} \cdot v_t$ is the undiscounted value function for the submodel MDP$_{N-t}$. In particular $v_0 = \hat{v}_0$, hence the two approaches with and without explicit discounting are *equivalent*. Now (13.23) yields the following VI for the undiscounted value functions $\hat{v}_t$:

$$\hat{v}_t(s) = \sup_{a \in D_t(s)} \big[\hat{r}_t(s,a) + \beta_{t+1} \cdot p_t \cdot \hat{v}_{t+1}(s,a) \big], \; 0 \le t \le N-1, \; s \in S_t.$$

The concept of an absorbing set J_0 and of a termination reward function from Sect. 13.2 carries over as follows. For simplicity we assume that S_t, A_t and D_t do not depend on t. We say that a non-empty set of states $J_0 \neq S$ is **absorbing set for**

the non-stationary MDP$_N$ if for $s \in J_0$, $a \in D(s)$ and $0 \leq t \leq N-1$ we have

$$\sum_{s' \in J_0} p_t(s, a, s') = 1, \quad r_t(s, a) = 0, \quad v_N(s) = 0.$$

As in Sect. 13.2 one sees that $v_t = 0$ on J_0, $0 \leq t \leq N$, and that the value of the maximizers in states $s \in J_0$ does not matter. Again we call $J := S - J_0$ the essential state space. Now the VI assumes the form

$$v_t(s) = \sup_{a \in D(s)} \left[r_t(s, a) + p_t \cdot v_{t+1}(s, a) \right], \; 0 \leq t \leq N-1, \; s \in J.$$

We now define non-stationary CMs similarly as non-stationary MDPs.

Definition 13.6.3 A **non-stationary N-stage CM$_N$ with finite disturbance space** is a tuple

$$((S_t)_0^N, (A_t)_0^{N-1}, (D_t)_0^{N-1}, (Z_t)_1^N, (T_t)_0^{N-1}, (Q_t)_0^{N-1}, (r_{Z_t})_0^{N-1}, v_N)$$

having the following meaning, where the index t refers to data in force at time t:

- S_t, A_t, D_t and v_N have the same meaning as in a non-stationary MDP$_N$.
- Z_t is the finite disturbance space.
- T_t is the transition function.
- Q_t is the probability distribution of the disturbance η_t.
- r_{Z_t} is the one-stage reward function.

Again discounting is included in the reward functions. The notions decision rule at time t, $0 \leq t \leq N-1$, and set Π_N of N-stage policies $\pi = (\pi_t)_0^{N-1}$ is defined as for non-stationary MDPs. Similarly as in Chap. 11 we construct the canonical probability space $(\Omega, \mathbb{P})$ with coordinate variables η_t, $1 \leq t \leq N$, where $\Omega := \times_1^N Z_t$ and where $\mathbb{P}$ is the probability distribution on Ω with discrete density

$$z^N := (z_t)_1^N \mapsto \prod_{t=1}^N q_t(z_t), \quad z^N \in \Omega.$$

Here $z \mapsto q_t(z) := \mathbb{P}(\eta_t = z)$ is the discrete density of η_t. Put $\eta^t := (\eta_i)_1^t$, $1 \leq t \leq N$. The random states $\zeta_t := \zeta_t(\pi, s_0, z^N)$, $0 \leq t \leq N$, are defined by $\zeta_0 \equiv s_0$ and ζ_{t+1} for $0 \leq t \leq N-1$. The ***total random N-stage reward*** for initial state s_0 and policy π is defined as

$$G_{0\pi}(s_0, \eta^N) := \sum_{t=0}^{N-1} r_{Z_t}(\zeta_t, \pi_t(\zeta_t), \eta_{t+1}) + v_N(\zeta_N).$$

As Z_t is finite, the ***expected total N-stage reward***

$$V_{0\pi}(s_0) := \mathbb{E}\, G_{0\pi}(s_0, \eta^N) = \sum_{z^N \in \times_1^N M_t} G_{0\pi}(s_0, z^N) \cdot \prod_{t=1}^{N} q_t(z_t)$$

exists and is finite; here $\mathbb{E}$ is the expectation with respect to $\mathbb{P}$. The optimization problem is literally the same as for non-stationary MDPs. The model CM_N defines in an obvious way for $1 \leq t \leq N-1$ a *submodel* CM_{N-t} whose data are taken from the last $N-t$ stages of the CM_N. Denote by v_t the value function for CM_{N-t}. The definition of a ***maximizer at time t*** is the same as for a non-stationary MDP, using now

$$r_t(s,a) := \mathbb{E}\, r_{Z_t}(s,a,\eta_{t+1}),\ 0 \leq t \leq N-1,\ (s,a) \in D_t,$$

$$w_t(s,a) := r_t(s,a) + \mathbb{E}\, v_{t+1}(T_t(s,a,\eta_{t+1})),\ 0 \leq t \leq N-1,\ (s,a) \in D_t.$$

One derives exactly as in Chap. 11 the RI, which implies that the Basic Theorem 13.6.2 for non-stationary MDPs remains literally true. In particular, the VI has the form

$$v_t(s) = \sup_{a \in D_t(s)} \Big[r_t(s,a) + \mathbb{E}\, v_{t+1}(T_t(s,a,\eta_{t+1})) \Big], \quad 0 \leq t \leq N-1,\ s \in S_t.$$

Moreover, the RI shows that the CM_N is equivalent to the non-stationary MDPad_N which differs from the CM_N only by the one-stage reward r_t and by the transition law

$$p_t(s,a,s') := \mathbb{P}(T_t(s,a,\eta_{t+1}) = s'), \quad 0 \leq t \leq N-1,\ (s,a) \in D_t, s' \in S_{t+1}.$$

Example 13.6.4 (Gambling with discrete stakes and terminal utility)

The model. Assume you gamble in a casino at most N times by choosing an integer stake a between zero and your momentary capital $s \in \mathbb{N}$. You win with known probability $p \in (0,1)$ upon which you gain d times your stake, $d \in \mathbb{N}$, $d \geq 2$. (The usual case is $d = 2$.) Otherwise you gain nothing. At the beginning your capital is $s_0 \in \mathbb{N}$. Denoting by $\zeta_{t\pi}$ the capital at time t under policy $\pi \in \mathbb{F}^N$, your aim consists in maximizing your expected terminal utility $\mathbb{E}\, V_0(\zeta_{N\pi})$ for some function V_0 on $\mathbb{N}_0$. Of course, the game stops if you are ruined before the N-th play.

The solution. The problem may be formulated as a CM_N as follows. The state space is $S = \mathbb{N}_0$, and $s_t = 0$ means that ruin has occurred at some time $k \leq t$. The stake at time t is your action, hence $A = \mathbb{N}_0$ and $D(s) = \{0, 1, \ldots, s\}$. The disturbance η_{t+1} equals 1 if you win at time t, and this occurs with probability $p \in (0,1)$. Otherwise η_{t+1} is zero. Thus $Z = \{0,1\}$ and $Q = \mathrm{Bi}(1,p)$. Your capital at time $t+1$ equals your capital at time t minus the stake at time t plus d times your stake if you win. Therefore the state equation has the form $T(s,a,z) = s - a + daz$. As you want to maximize the expected terminal utility the one-stage reward function

r_Z vanishes. Moreover, $\beta = 1$. We also make the natural assumption that $V_0(0) = 0$. It follows from Theorem 11.6(b) that the VI holds, that $V_n(0) = 0$ for all $n \geq 0$ and that, using $q := 1 - p$, we have for $n \geq 1$ and $s \in \mathbb{N}$

$$V_n(s) = \max_{0 \leq a \leq s} \left[p \cdot V_{n-1}(s - a + da) + q \cdot V_{n-1}(s - a) \right]. \tag{13.24}$$

Moreover, as the sets $D(s) = \{0, 1, \ldots, s\}$ are finite, there exists at each stage n a smallest maximizer f_n, and by Theorem 11.6(a) the policy $(f_n)_1^N$ is optimal. Note also that the solution depends only on p, d, and V_0.

Structure of the solution. For large N and/or s_0 the computing time and also the memory requirements become quickly prohibitive. Therefore it is useful to search for reasonable conditions under which structural properties of the solution decrease the computational load. This problem constitutes a considerable part of the present treatise. It will be pursued in later sections, in particular in Chaps. 17 and 19, in a general context. In order to give a flavor of the problems and techniques considered we now attack a few questions by ad hoc methods. ♦

A decision rule f is called ***bang-bang*** if $f(s) \in \{0, s\}$ for all s, i.e. if it prescribes at each state either to stake nothing or to stake all available capital.

Proposition 13.6.5 (Structural properties of the solution of the gambling problem from Example 13.6.4) *There following hold:*

(a) *$V_n(s)$ is increasing in n.*
(b) *$V_n(s)$ is increasing in s if $V_0(s)$ is increasing in s.*
(c) *If V_0 is convex, V_n is convex for all n, there exists a bang-bang maximizer f_n at each stage n, and*

$$V_n(s) = \max\{V_{n-1}(s), p \cdot V_{n-1}(ds)\}, \quad n \geq 1, s \geq 1.$$

(d) *If $p \geq 1/d$ and if V_0 is increasing and convex, then bold play is optimal, which means that it is optimal to stake at each stage n the total capital available i.e. to use the decision rule $s \mapsto f_n(s) := s$.*
(e) *If bold play is optimal for all N, then*

$$V_N(s) = p^N \cdot V_0(d^N s), \quad N, s \geq 1.$$

In particular, if $V_0(s) = s$ for all s, if $d = 2$ and $p \geq 1/2$, then bold play is optimal for all N, and $V_N(s) = (2p)^N \cdot s$.
(f) *If $d = 2$ and if V_0 is concave, then for all n the value functions V_n are concave and the smallest maximizer f_n at stage n belongs to LIP(1), i.e. satisfies $|f_n(s + 1) - f_n(s)| \leq 1$ for all s.*

Proof We use that

$$Lv(s,a) = \mathbb{E}\, v(s-a+da\cdot\eta_1) = p\cdot v(s-a+da) + q\cdot v(s-a). \qquad (13.25)$$

(a) This follows from the VI since $V_n(s) \geq LV_{n-1}(s,0) = V_{n-1}(s)$ for all s.
(b) The assertion is true by assumption for $n = 0$. Assume that $v := V_{n-1}$ is increasing for some $n \geq 1$. For each s we have $D(s) \subset D(s+1)$. Then (13.25) shows that

$$Lv(s,a) \leq Lv(s+1,a) \leq Uv(s+1) = V_n(s+1), \quad a \in D(s).$$

Therefore $V_n(s) = \max_a Lv(s,a) \leq V_n(s+1)$.
(c) We use properties of discretely convex functions from Sect. 7.3. Firstly, V_0 is convex by assumption. Assume that $v := V_{n-1}$ is convex for some $n \in \mathbb{N}$. As $a \mapsto s+da$ and $a \mapsto s-a$ are affine, $Lv(s,\cdot)$ is convex on $D(s)$. It follows that $Lv(s,\cdot)$ assumes its maximum either at $a = 0$ or at $a = s$. This shows that one can select $f_n(s) \in \{0,s\}$ and that by the VI (13.24)

$$V_n(s) = \max\{Lv(s,0), Lv(s,s)\} = \max\{v(s), p\cdot v(ds)\}.$$

As $s \mapsto v(ds)$ is convex and as the maximum of two convex functions is convex, V_n is convex. This holds then for all n. Moreover, the existence of a bang-bang maximizer at stage n follows from the preceding reasoning.
(d) Fix n and put $v := V_{n-1}$, which is increasing and convex by (b) and (c), respectively. It suffices to show that $Lv(s,s) \geq Lv(s,0)$. Let $\tilde{v}$ denote the largest convex minorant of v, i.e. the function on $\mathbb{R}_+$ which coincides with v on $\mathbb{N}_0$ and which is affine on each interval $[k, k+1]$, $k \in \mathbb{N}_0$. Then $\tilde{v}$ is convex and increasing. Therefore, as $pd \geq 1$

$$\begin{aligned} Lv(s,s) &= p\cdot v(ds) + (1-p)\cdot v(0) = p\cdot \tilde{v}(d\,s) + (1-p)\cdot \tilde{v}(0) \\ &\geq \tilde{v}(pd\,s + (1-p)\cdot 0) \geq \tilde{v}(s) = v(s) = Lv(s,0). \end{aligned}$$

(e) This follows easily from $V_n(s) = p\cdot V_{n-1}(ds)$.
(f) The proof uses induction on the substitution $b := s-a$ and the following reasoning. Let v be a concave function on $\mathbb{N}_0$, and put

$$w(s,b) := p\cdot v(2s-b) + q\cdot v(b), \quad s \in \mathbb{N}_0,\ b \in \mathbb{N}_{0,s}.$$

Then the largest maximizer g of w belongs to *ILIP*(2) by Lemma 8.3.5 with $\lambda := 2$, $D_\lambda = \{(s,b) \in \mathbb{N}_0^2 : s \leq b \leq 2s\}$, $K := q\cdot v$ and $u := p\cdot v$. □

Remark 13.6.6 (Comments on Proposition 13.6.5)

(i) The properties (a) and (b) conform to intuition, which says that larger maximal expected n-stage reward is obtained under more chances to play and (under the natural assumption that larger terminal capital implies larger terminal reward) under larger initial capital. One also expects that $V_n(s, p)$ is increasing in p, for which we give a proof in Example 18.2.13.
(ii) If p is too small then there is a large chance of getting ruined before N plays have been made. Hence it may pay to stop staking anything after a few plays. ◇

Example 13.6.7 (Planning the sale of a facility) You own a facility, e.g. a machine, which at time $t = 0$ is in a working condition (*age* for short) $x_0 \in M_X := \mathbb{N}_{0,m}$ and at time $1 \le t \le N-1$ in a random age X_t taking values in M_X. Here $X_t = 0$ means a new facility and $X_t = m$ means an irreparable defective facility. The age changes in a Markovian manner during one period from $x \in M_X$ to $x' \ge x$ with probability $t(x, x')$, hence $t(m, m) = 1$. At each time $0 \le t \le N-1$ you may sell the facility for the random prize Y_t, taking vales in the finite set $M_Y \subset \mathbb{R}_+$. If the facility has at time t the age x, then the prize Y_t has the discrete density $y \mapsto q(x, y)$ on M_Y. More precisely we assume that the sequence of two-dimensional vectors (X_t, Y_t), $0 \le t \le N-1$, of ages and prizes is Markovian with state space $M_X \times M_Y$, a random initial state (x_0, Y_0) for some $x_0 \in M_X$ and transition probabilities $t(x, x') \cdot q(x', y')$ for the transition from (x, y) into (x', y').

As long as the facility is not sold it earns, when having the age $x < m$, a reward per period of amount $g(x) \in \mathbb{R}_+$. When the age reaches m, the facility becomes worthless. At the latest at time $N-1$ the facility must be sold.

When the initial age is $x_0 < m$ and Y_0 is random, according to which policy should one sell the facility in order to achieve the maximal expected discounted reward $d_N(x_0)$?

We use the following $MDP_{N-1}(p_0)$ with the understanding that a sold facility is considered to have age m. The states are $s = (x, y) \in S := M_X \times M_Y$; here $s = (m, y)$ means that either the age m has been reached or that the facility has been sold previously at some age smaller than m; $a = 1$ or $a = 0$ if the facility is sold or not, respectively, hence $A = D(s) = \{0, 1\}$ for all s; $V_0(x, y) = y$ for $x < m$, and $V_0(m, y) = 0$;

$$p(x, y, a, x', y') = \begin{cases} t(x, x') \cdot q(x', y'), & \text{if } a = 0, \\ \delta_{x',m} \cdot q(m, y'), & \text{if } a = 1; \end{cases}$$

$r(m, y, a) = 0$, and for $x < m$

$$r(x, y, a) = \begin{cases} g(x), & \text{if } a = 0, \\ y, & \text{if } a = 1. \end{cases}$$

The discrete density μ of the initial state (X_0, Y_0) is $(x, y) \mapsto \delta_{x,x_0} \cdot q(x_0, y)$.

Obviously $J_0 := \{m\} \times M_Y$ is absorbing, hence $V_n(m,0) = 0$ for all n, y. The VI holds and yields for $0 \le x < m$ and $y \in M_Y$

$$\begin{aligned} V_n(x,y) &= \max\left\{ g(x) + \beta \cdot \sum_{x \le x' \le m-1} t(x,x') \sum_{y' \in M_Y} q(x',y') \cdot V_{n-1}(x',y'),\ y \right\} \\ &=: \max\{W_n(x,y,0), y\}. \end{aligned}$$

Moreover, the smallest maximizer at stage $n \ge 1$ is

$$(x,y) \mapsto f_n(x,y) := \begin{cases} 0, & \text{if } W_n(x,y,0) \ge y, \\ 1, & \text{else.} \end{cases}$$

We want to maximize $\pi \mapsto V_{N-1,\pi}(p_0) := \mathbb{E}[V_{N-1,\pi}(X_0, Y_0)|X_0 = x_0]$ on $\mathbb{F}^N$. Now we obtain from the MDP-version of Proposition 13.3.2, with p_0 as defined above, the following results:

(a) $d_N(x_0)$ equals

$$V_{N-1}(p_0) = \sum_{y \in M_Y} q(x_0, y) \cdot V_{N-1}(x_0, y) =: V_{N-1}(x_0, q),\ x_0 < m,$$

and can be found by the recursion

$$V_n(x,q) = \sum_{y \in M_Y} q(x,y) \cdot \max\left\{ g(x) + \beta \cdot \sum_{x \le x' \le m-1} t(x,x') \cdot V_{n-1}(x',q),\ y \right\},$$

starting with $V_0(x,q) = \mathbb{E}[Y_1|X_0 = x]$.

(b) For each $N \ge 1$ the policy $(f_n)_{N-1}^1$ is p_0-optimal for the problem with at most N periods for sale. ♦

Example 13.6.8 (Solving the deterministic problem of an optimal multiplication of matrices by means of a terminating CM) Consider a sequence of matrices $M_1, M_2, \ldots, M_k$, $k \ge 3$, of compatible sizes. Let M_i be a $d_i \times d_{i+1}$-matrix, $1 \le i \le k$. The computation of the product of the matrices in the given order is done by setting parentheses such that each factor consists either of two consecutive matrices or of a single matrix. This can be done in several ways. It is an important problem to find the minimal number $C(1,k)$ of multiplications of matrix elements and of finding an optimal *multiplication policy*, as formally defined below. An efficient (yet theoretically involved) method is based on the idea of reducing the computation of $C(1,k)$ to the computation of the corresponding numbers $C(i,j)$ for all subproducts $M_{ij} := M_i \cdot M_{i+1} \cdot \cdots \cdot M_j$ of $d_i \times d_{j+1}$-matrices, $1 \le i < j \le k$. The number of multiplications of matrix elements required by the product $M_{ia} \cdot M_{a+1,j}$ equals $c(i,j,a) := d_i \cdot d_{a+1} \cdot d_{j+1}$ and is called the *multiplication cost* of the product

$M_{ia} \cdot M_{a+1j}$. In this product the *multiplication point a* determines four parentheses, namely $(M_{ia}) \cdot (M_{a+1,j})$. (Note that in contrast to the usual setting single matrices may also be enclosed in parentheses.) Let $W(i,j,a)$, $1 \le i \le a < j \le k$, be the cost resulting when firstly M_{ia} and M_{a+1j} are computed in an optimal way and then when M_{ij} is found as $M_{ia} \cdot M_{a+1j}$. Thus, using $C(i,i) := 0$, $1 \le i \le k$, we have

$$W(i,j,a) = c(i,j,a) + C(i,a) + C(a+1,j), \quad 1 \le i \le a < j \le k.$$

Intuitively one expects that

$$C(i,j) = \min_{i \le a < j} W(i,j,a), \quad 1 \le i < j \le k. \tag{13.26}$$

This would yield a numerical procedure for computing $C(1,k)$, starting with $C(i,i) := 0$. Unfortunately (13.26) cannot be justified within the framework of an absorbing DP since under action a a transition not to a single new state but to either of the two new states $(i',j') \in \{(i,a),(a+1,j)\}$ occurs, a situation which reminds us of MDPs. In fact we now show that the problem can be formalized by the h-terminating MDP of CM-type for the following CM $(S, A, D, Z, T, c, C_0, \beta)$: $S = \{(i,j) \in \mathbb{N}_k^2 : i \le j\}$; $A = \mathbb{N}_k$; $D(i,j) = \{i, i+1, \dots, j-1\}$ if $i < j$ and $D(i,i)$ arbitrary, e.g. $= A$; $Z = \{0,1\}$; $T(i,i,a,z) = (i,i)$, and if $i < j$ then

$$T(i,j,a,z) = \begin{cases} (i,a), & \text{if } z = 0, \\ (a+1,j), & \text{if } z = 1; \end{cases}$$

$q(0) = q(1) = 0.5$; c as above; $C_0 \equiv 0$ and $\beta = 2$. Obviously $J_0 := \{(i,i) : 1 \le i \le k\}$ is absorbing and the CM is h-terminating for $h(i,j) := j - i$, since $R(i,j) = \cup_{i \le a < j} \{(i,a),(a+1,j)\} \subset [h \le j - i - 1]$ for $i < j$. Thus the horizon is $N := k - 1$.

Returning to our multiplication problem, the formal definition of $C(i_0, j_0)$, $(i_0, j_0) \in S$, by means of an optimal choice of multiplication points is given, using arbitrary policies $\pi = (\pi_t)_0^{N-1} \in \mathbb{F}^N$ from the CM as follows: the first multiplication point selected is $a_0 := \pi_0(i_0, j_0)$; then, if after the t-th setting of quadruples of parentheses, $0 \le t \le N-1$, there remains the task of finding the minimal cost $C(i_t, j_t)$, one selects $a_t := \pi_t(i_t, j_t)$. Denote by $C_{N\pi}(i,j)$, $(i,j) \in S$, the cost in our multiplication problem when the initial state is (i,j) and when the multiplication policy $\pi \in \mathbb{F}^N$ is applied. Then the goal is to compute $C(i,j) := \min_{\pi \in \mathbb{F}^N} C_{N\pi}(i,j)$, which in particular yields $C(1,k)$. Obviously the functions $C_{n\pi}$, $\pi \in \mathbb{F}^N$, $1 \le n \le N$, which are defined analogously as $C_{N\pi}$, satisfy the recursion

$$C_{n,(f,\sigma)}(i,j) = c_f(i,j) + C_{n-1,\sigma}(T_f(i,j,0)) + C_{n-1,\sigma}(T_f(i,j,1)),$$
$$1 \le n \le N, (f,\sigma) \in \mathbb{F} \times \mathbb{F}^{n-1}, (i,j) \in S,$$

where $C_{0\pi} :\equiv 0$. This is exactly the RI in our CM. Therefore the OE of Theorem 13.5.3(a) for the CM yields (13.26), and from the optimality criterion in Theorem 13.5.3(b) we obtain an optimal multiplication policy. ♦

13.7 Problems

Problem 13.7.1 Assume in the gambling problem from Example 13.6.4 that V_0 is convex and that $V_0(0) = 0$.

(a) If $V_0(s) \leq p \cdot V_0(ds)$ for all s, then $V_n(s) = p^n \cdot V_0(d^n s)$, for all n and s, and bold play is optimal.
(b) If $V_0(s) \geq p \cdot V_0(ds)$ for all s, then $V_n = V_0$ for all n, and it is optimal never to play.
(c) Apply (a) and (b) to the case $V_0(s) = s^\gamma$ for $\gamma \in [1, \infty)$.

Problem 13.7.2 If it is optimal in the N-stage gambling problem from Example 13.6.4 to stake nothing at some time $k < N$ and at all states s, then the same holds for all times $\nu < k$, and $V_{N-k} = V_{N-k-1} = \cdots = V_N$.

The next three problems concern Example 13.4.5.

Problem 13.7.3

(a) It is optimal to accept the first throw (irrespective of its value) for some horizon $N \geq 1$ if and only if it is optimal to accept the first throw for each horizon $N \geq 1$ if and only if $\gamma \geq 2.5$.
(b) If $\gamma \in [0, 2.5)$, then it is never optimal to accept a throw of amount one.

Problem 13.7.4 (Computation of d^*)

(a) If $\gamma = 0$, then $d^* = 6$.
(b) If $\gamma \in (0, 2.5)$, then $(k-1)k < 12\gamma \leq k(k+1)$ for a unique $k \in \mathbb{N}_5$, and then

$$d^* = (13-k)/2 - 6\gamma/k.$$

$$H(d) := 3.5 + \mathbb{E}(d - \eta_1)^+ - d, \quad d \in [0, 6],$$

is affine on each interval $[k, k+1]$, $k \in \mathbb{N}_{0,5}$, H is continuous and strictly decreasing (figure!) and $H(k) = 3.5 - k(13-k)/12$ for $k \in \mathbb{N}_{0,6}$. Moreover, $H(d^*) = \gamma$. Verify that d^* equals 4 and 3 if $\gamma = 0.5$ and 1, respectively.

Problem 13.7.5 The condition $\gamma < 1/6$ is necessary and sufficient for the existence of a horizon $N_0 \geq 1$ with the following property: For all horizons $n \geq N_0$ it is optimal to accept the first throw if and only if it is a six.

Problem 13.7.6 Consider the stochastic scheduling problem from Example 13.4.8 and prove the following assertions:

(a) If the probability distribution of η_1 is symmetric with respect to $\mu := \mathbb{E}\,\eta_1$, then

$$V_3(p) = (2b_1 + b_2)\mu - (b_1 - b_3) \cdot \mathbb{E} \min\{\eta_1, y_1\}.$$

(b) Solve the Hausnummern-Kegeln problem when the probability distribution of η_1 has the symmetric triangular discrete density q_z, $-1 \le z \le 10$, with $q_{-1} = q_{10} = 0$.
(c) Solve the scheduling problem when $\eta_1 \sim U(\mathbb{N}_6)$.

Problem 13.7.7 Consider a terminating MDP with finite transition law and absorbing set J_0. Fix $s \in J$ and prove that:

(a) For any $s' \in S$ and $1 \le t \le N(s)$ we have $s' \in R_t(s)$ if and only if there exists some $\pi \in F^{N(s)}$ with $\mathbb{P}_{\pi s}(\zeta_t = s') > 0$.
(b) For all $\pi \in F^{N(s)}$ we have $\mathbb{P}_{\pi s}(\zeta_t \in J_0) = 1$ for $t = N(s)$, but not for $t < N(s)$.

Problem 13.7.8 Consider the discrete version of the allocation problem from Example 13.1.1 with $\gamma := 1$, $u(a) \in \mathbb{R}$ for all a and prove the following assertions:

(a) If V_0 is increasing, all value functions are increasing.
(b) If $\beta = 1$ and $u(0) \ge 0$ then $V_n(s)$ is increasing in n.
(c) If $\beta < 1$ then

$$V(s) \ge p \cdot \max\left\{\frac{u(0)}{1-\beta}, \frac{u(s)}{1-\beta q}\right\}, \; s > 0.$$

Problem 13.7.9 Consider the special continuous case of the allocation problem from Example 13.1.1 where the utility for investing the amount x is $u(x) := \sqrt{x}$ and where the terminal reward is $V_0(s) := e_0\sqrt{s}$ for some $e_0 \in \mathbb{R}_+$. One easily shows by induction on n that

$$V_n(s) = e_n\sqrt{s},\; n \in \mathbb{N}_0,\; s \in \mathbb{R}_+,$$

where

$$e_n = p\sqrt{1 + \beta^2\gamma e_{n-1}^2} + q\beta e_{n-1}\sqrt{\gamma},\; n \in \mathbb{N}.$$

Moreover,

$$s \mapsto f_n(s) := s/(1 + \beta^2\gamma e_{n-1}^2)$$

is a maximizer at stage $n \in \mathbb{N}$. Figure 13.3 shows the graph of $n \mapsto e_n$ for several values of $c := \beta\sqrt{\gamma}$, e_0 and p.

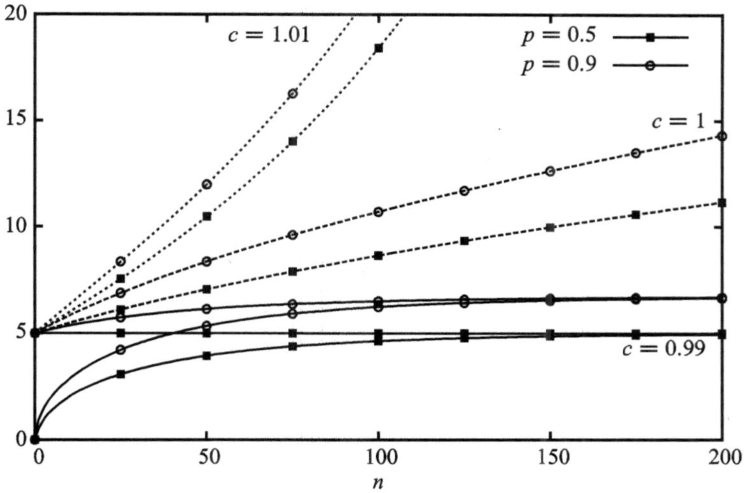

Fig. 13.3 Examples of e_n

13.8 Supplement

Gambling with discrete capital, discrete stakes and arbitrary terminal reward is rarely dealt with in the literature, which concentrates on what we call *target gambling*. Historically, gambling lies at the roots of probability theory, but it seems that gambling strategies were not discussed until the paper by Coolidge (1909); cf. Dubins and Savage (1976, p. 4).

Chapter 14
Markovian Decision Processes with Discrete Transition Law

We consider MDPs with countable state space and variable discount factor. The discount factor may depend on the state and the action. Under minimal assumptions we prove the reward iteration and formulate a structure theorem for MDPs. Also the useful notion of a bounding function is introduced.

14.1 The Finite Horizon Model

The approach of Chap. 12 is modified in order to deal with the problem of existence of expectations which arise in the case of a countably infinite state space and a so-called discrete transition law. In Chap. 12 the expected N-stage reward

$$V_{N\pi}(s_0) = \mathbb{E}_{\pi s_0} R_{N\pi}(s,\zeta^N) = \sum_y R_{N\pi}(s_0, y) \cdot \mathbb{P}_{\pi s_0}(\zeta^N = y)$$

was a finite sum as from each state s under action a only finitely many states s' could be reached with positive probability. If the number of such states s' is countably infinite, the expectation becomes an infinite series and we have to ensure the existence of $V_{N\pi}(s_0)$ for all N, π and s_0 by appropriate assumptions, as otherwise the optimization problem does not make sense at all. For applications such questions are certainly not of predominant importance. However, it is not satisfying to get rid of the problem by just assuming that all occurring random variables have expectations. It may well be worth having available easily checkable conditions which ensure that $V_{N\pi}(s_0)$ exists, i.e. that $V_{N\pi}(s_0)$ can really be used as a measure of performance of the policy π.

K. Hinderer et al., *Dynamic Optimization*, Universitext,
DOI 10.1007/978-3-319-48814-1_14

Firstly we comment on the appropriate definition of the expectation

$$\mathbb{E}\, F(\xi) := \sum_{i\in I} p(i) \cdot F(i) \tag{14.1}$$

of a random variable $F(\xi)$, where the discrete random variable ξ takes values in a countable set $I \subset \mathbb{R}$ with probabilities $p(\cdot)$ and where F is an arbitrary function on I.

The usual approach requires that the series in (14.1) converges just assuming that all occurring random variables have expectations. It may be well absolutely (with I in any order and hence to the same value), in which case $\mathbb{E}\, F(\xi)$ is finite. In the optimization context this definition sometimes entails cumbersome considerations about finiteness of series, a property not needed for many theoretical issues. A more elegant approach allows infinite values of series of the form $\sum_{i\in I} a_i$ of numbers $a_i \in \overline{\mathbb{R}}$ while ensuring that the most important properties of absolutely convergent series remain true. The minimal requirement for this to hold is the unconditional convergence of $\sum_{i\in I} a_i$ in $\overline{\mathbb{R}}$, (i.e. for each arrangement of the terms a_i), and hence to the same value. (In case $I := \mathbb{N}$ unconditional convergence to a finite value implies convergence in the usual sense, but the converse does not hold.) However, in order to obtain a theory as flexible as for absolutely convergent series (and in order to be completely consistent with integration theory as used from Chap. 16 on) we must strengthen unconditional convergence to the associative law for series; cf. (Q4) below.

Our approach is based on the following facts and definitions.

(Q1) A series $\sum a_i$ of numbers $a_i \in [0, \infty]$ is unconditional convergent in $[0, \infty]$. Thus, $\mathbb{E}\, F(\xi)$ is well-defined by (14.1) in case $F \geq 0$.

(Q2) A series $\sum a_i$ of numbers $a_i \in \overline{\mathbb{R}}$ is called **quasi-convergent** (in $\overline{\mathbb{R}}$) if at least one of the numbers $\sum a_i^+$, $\sum a_i^-$ (which are well-defined according to (Q1)) is finite. Then $\sum a_i$ is defined as the extended real number

$$\sum_{i\in I} a_i := \sum_{i\in I} a_i^+ - \sum_{i\in I} a_i^-.$$

(Q3) We see from

$$\sum_{i\in I} |a_i| = \sum_{i\in I} a_i^+ + \sum_{i\in I} a_i^-$$

that $\sum_{i\in I} a_i$ is quasi-convergent with finite value if and only if $\sum_{i\in I} a_i$ is absolutely convergent in $\mathbb{R}$.

(Q4) If $\sum a_i$ is quasi-convergent then $\sum a_i$ is unconditional convergent in $\overline{\mathbb{R}}$. Even the *associative law for series* holds in the following form: If $I = \sum_{j\in J} I_j$ then

$$\sum_{i\in I} a_i = \sum_{j\in J} \sum_{i\in I_j} a_i,$$

each of the series on the right-hand side being quasi-convergent. (This associative law is the discrete counterpart of Fubini's Theorem, a very important tool for the study of MDPs with general transition law in Chap. 16 below.) In particular, if $I = J \times K$ then the case $I_j = \{j\} \times K$ implies: If $\sum_{(j,k)} a_{jk}$ is quasi-convergent then $\sum_k a_{jk}$ is quasi-convergent for all j, $\sum_j \sum_k a_{jk}$ is quasi-convergent and

$$\sum_{(j,k)} a_{jk} = \sum_j \sum_k a_{jk} .$$

(Q5) According to (Q1), there are defined for an arbitrary extended real random variable $F(\xi)$ the extended real numbers

$$\mathbb{E}\,|F(\xi)| = \sum_i p(i)|F(i)|, \quad \mathbb{E}\,F(\xi)^{\pm} = \sum_i p(i)F^{\pm}(i).$$

We say that the expectation $\mathbb{E}\,F(\xi)$ *exists*, or that $F(\xi)$ is *quasi-integrable* with respect to p, if at least one of the numbers $\mathbb{E}\,F(\xi)^+$, $\mathbb{E}\,F(\xi)^-$ is finite, i.e. if $\sum_i p(i)F(i)$ is quasi-convergent. Then $\mathbb{E}\,F(\xi)$ is defined as

$$\mathbb{E}\,F(\xi) := \mathbb{E}\,F(\xi)^+ - \mathbb{E}\,F(\xi)^- = \sum_{i \in I} p(i)F(i).$$

The associative law for the latter series is crucial for the validity of the usual rules for expectations. The random variable $F(\xi)$ is called *integrable* if $\mathbb{E}\,F(\xi)$ exists and is finite. As $\mathbb{E}\,|F(\xi)| = \mathbb{E}\,F(\xi)^+ + \mathbb{E}\,F(\xi)^-$, $F(\xi)$ is integrable if and only if $\mathbb{E}\,|F(\xi)| < \infty$ and also if and only if both numbers $\mathbb{E}\,F(\xi)^+$, $\mathbb{E}\,F(\xi)^-$ are finite.

Series of the form

$$(pv)(b) := \sum_{s' \in S} p(b, s') \cdot v(s') \text{ for } v : S \to \overline{\mathbb{R}}$$

are only used if they are quasi-convergent. We shall use the abbreviations

$$pv(s, a) := \sum_{s' \in S} p(s, a, s') \cdot v(s'),$$

$$p_f(s, s') := p(s, f(s), s'),$$

$$p_f v(s) := \sum_{s' \in S} p_f(s, s') \cdot v(s') .$$

Definition 14.1.1 A **Markovian decision process with discrete transition law** is a tuple $(S, A, D, p, r_S, V_0, \beta)$ of the following kind:

- S is countable.
- A, D, V_0 and β have the same meaning as in the CM.
- $(s, a, s') \mapsto p(s, a, s')$ from $D \times S$ into $\mathbb{R}_+$ is called a **transition matrix** from D into S.
- $r(s, a) := \sum_{s' \in S} r_S(s, a, s') \cdot p(s, a, s')$ is assumed to exist and to be finite for all $(s, a) \in D$.

In the rest of this section *MDP* means *MDP with discrete transition law.*

Before proceeding to the further explanation of the problem we digress a bit for a slight generalization, namely MDPs with variable discount factors. As these models rarely occur in the present chapter, one can skip for first reading everything in the present chapter which refers to MDPvars. It suffices to return to this model later when studying MDPvars in the context of so-called Markov renewal programs. Treating MDPvars here saves a lot of repetitions of similar results without much impediment of the general line of presentation. This is mainly due to the fact that often assumptions and results can be formulated in terms of the operators L, U_f and U.

Definition 14.1.2 A **Markovian Decision Process with variable discount factor and with discrete transition law**, (**MDPvar** for short) is defined as an MDP with discrete transition law, except that $\beta: D \to \mathbb{R}_+$ is a bounded function, not a constant. β is called the ***discount function.***

The interpretation of the discount function is as follows: The reward $r_S(s_t, a_t, s_{t+1})$ is obtained at the beginning of period $t \in \mathbb{N}_0$. For $t \geq 1$ it is discounted back to the beginning of period $t-1$ by means of $\beta(s_{t-1}, a_{t-1})$. Thus the reward in period $t \geq 1$ is discounted back to time $t = 0$ by means of

$$\beta_t := \prod_{i=0}^{t-1} \beta(s_i, a_i), \ t \in \mathbb{N}.$$

In particular, if β is a constant β_0, then $\beta_t = \beta_0^t$ and the MDPvar is an MDP. In applications we often have $\beta \leq 1$. We allow $\beta(s, a) = 0$ for modeling *absorption.* All notions defined in the present section so far, such as decision rule and N-stage policy remain unchanged for MDPvars. The reward earned under policy π and initial state s_0 in period $t \geq 1$ is discounted back to time 0 by means of the bounded random factor

$$\beta_{t\pi}(s_0, \zeta^{t-1}) := \prod_{i=0}^{t-1} \beta(\zeta_i, \pi_i(\zeta_i)) \text{ with } \zeta_0 :\equiv s_0.$$

Many results about MDPs remain true for MDPvars when, roughly spoken, the constant β is replaced by the function $(s, a) \mapsto \beta(s, a)$ on D. However, our presentation is centered around MDPs. Whenever changes in definitions or results are necessary for MDPvars, this will be clearly indicated. Necessary changes in proofs are only indicated when they require more than just a replacement of the constant β by the function $\beta(\cdot,\cdot)$ on D. Moreover, we use the following device (cf. e.g. Lemma 14.1.3 below): *Whenever results about MDPs hold literally also for MDPvars this is indicated by "[Var]" at the beginning of the result for MDPs.*

Now we return to the development for MDPs. As in Chap. 12 we construct for given N-stage policy and initial state s_0 a discrete probability space $(\Omega, \mathbb{P}_{\pi s_0})$ and an N-stage Markov chain $\zeta^N := (\zeta_t)_1^N$ starting in $\zeta_0 := s_0$ and with transition matrices $p_{\pi_t}(s,s') := p(s,\pi_t(s),s')$, $0 \le t \le N-1$. In particular, $\mathbb{P}_{\pi s}$ is defined as in (12.2). The N-stage reward under initial state s_0 and policy π is defined as in (12.4) as the real random variable

$$R_{N\pi}(s_0,\zeta^N) := \sum_{t=0}^{N-1} \beta^t \cdot r_S(\zeta_t, \pi_t(\zeta_t), \zeta_{t+1}) + \beta^N \cdot V_0(\zeta_N). \tag{14.2}$$

For MDPvars one must replace in (14.2) the constants β^t, $0 \le t \le N$, by $\beta_{t\pi}(s_0,\zeta^{t-1})$, where $\beta_{0\pi}(s_0,\zeta^{-1}) := 1$.

As the ζ_t's may assume countably many values, the expectations $\mathbb{E}_{\pi s_0} R_{N\pi}(s_0,\zeta^N)$ need not exist. Therefore, the optimization problem is defined if and only if we make the following

first minimal assumption (MA1) for an MDP: For all $N \in \mathbb{N}$, $\pi \in \mathbb{F}^N$ and $s_0 \in S$ the ***expected N-stage reward*** $V_{N\pi}(s_0) := \mathbb{E}_{\pi s_0} R_{N\pi}(s_0,\zeta^N)$ exists.

This means that $R_{N\pi}(s_0,\zeta^N)$ is quasi-integrable with respect to $\mathbb{P}_{\pi s_0}$. Under (MA1) there is defined for each N the ***N-stage value function***

$$s \mapsto V_N(s) := \sup\left\{V_{N\pi}(s) : \pi \in \mathbb{F}^N\right\},\ s \in S.$$

Note that both $V_{N\pi}$ and V_N may assume the values $+\infty$ and $-\infty$. Obviously (MA1) is satisfied in the following two cases:

- ***(EP)***: r_S and V_0 are bounded below.
- ***(EN)***: r_S and V_0 are bounded above.

(EP) and (EN) means *essentially positive* and *essentially negative*, respectively. Obviously under (EP) [under (EN)] the functions $V_{N\pi}$ and V_N are bounded below [above]. Case (EN) is important for cost minimization problems; it holds in particular if $c_S := -r_S \ge 0$ and $C_0 := -V_0 \ge 0$.

One expects that under (MA1) or appropriate stronger assumptions the value iteration (VI for short) holds and has the same form as in Chap. 12:

$$V_n(s) = \sup_{a \in D(s)} \left[r(s,a) + \beta \cdot \sum_{s' \in S} p(s,a,s') V_{n-1}(s') \right] =: \sup_{a \in D(s)} W_n(s,a).$$

(For MDPvars one has to replace β by $\beta(s,a)$.) These equations do not make sense unless the MDP satisfies the following

second minimal assumption (MA2) for an MDP: (MA1) holds and all functions V_n, $n \in \mathbb{N}_0$, belong to the set $\mathfrak{V}_0$ of those functions v on S for which $pv(s,a)$, $(s,a) \in D$, exists.

Both (EN) and (EP) imply (MA2). (MA2) is not implied by (MA1) but (MA2) is only slightly stronger than (MA1): (MA1) implies (MA2) if for each $n \in \mathbb{N}$ there is some $\pi \in \mathbb{F}^n$ such that $V_{n\pi} > -\infty$.

Obviously $\mathfrak{V}_0$ is the largest subset of functions v on S on which we can define the operators L, U_f and U as in Chap. 12. For $v \in \mathfrak{V}_0$ the functions $U_f v$ and Uv need not belong to $\mathfrak{V}_0$, so $U_g U_f v$ for $g, f \in \mathbb{F}$, and $U^2 v$ need not be defined.

Lemma 14.1.3 *[Var] Under (MA1) the functions V_0 and $V_{n\pi}$, $n \in \mathbb{N}$, $\pi \in \mathbb{F}^n$, belong to $\mathfrak{V}_0$ and they may be computed by the* ***reward iteration (RI*** *for short)*

$$V_{1f} = U_f V_0,\, f \in \mathbb{F},\; V_{n(f,\sigma)} = U_f V_{n-1,\sigma},\, n \geq 1,\; (f,\sigma) \in \mathbb{F} \times \mathbb{F}^{n-1}. \tag{14.3}$$

Proof We assume $n \geq 2$; the case $n = 1$ is similar. Fix $s \in S$ and $\pi = (f,\sigma) \in \mathbb{F} \times \mathbb{F}^{n-1}$. For $(s_t)_1^n \in S^n$ put $x := (s_t)_2^n$. We have as in (12.6) the recursion

$$R_{n\pi}(s,(s_t)_1^n) = R_{n\pi}(s,s_1,x) = r_S(s,f(s),s_1) + \beta \cdot R_{n-1,\sigma}(s_1,x).$$

As $R_{n\pi}(s,\zeta^n)$ is $\mathbb{P}_{\pi s}$-quasi-integrable by (MA1), the same holds for $R_{n-1,\sigma}(\zeta_1,(\zeta_t)_2^n)$. As $p_\pi(s,s_1,x) = p_f(s,s_1)\,p_\sigma(s_1,x)$ by (12.3), we obtain

$$\begin{aligned} V_{n\pi}(s) &= \sum_{s_1,x} p_\pi(s,s_1,x)\,R_{n,\pi}(s,s_1,x) \\ &= r_f(s) + \beta \cdot \sum_{s_1,x} p_f(s,s_1)\,p_\sigma(s_1,x)\,R_{n-1,\sigma}(s_1,x). \end{aligned}$$

The series are quasi-convergent by (MA1). By the associative law (Q4) the last series equals

$$\sum_{s_1} p_f(s,s_1) \sum_x p_\sigma(s_1,x)\,R_{n-1,\sigma}(s_1,x) = \sum_{s_1} p_f(s,s_1)\,V_{n-1,\sigma}(s_1),$$

all series being quasi-convergent. This verifies (14.3). Quasi-convergence of the series on the right-hand side implies that $V_{n-1,\sigma}$ belongs to $\mathfrak{V}_0$. □

The following properties of operators are useful for lucid, succinct and unified proofs.

Lemma 14.1.4 (Properties of the operators L, U_f and U) *There following hold:*

(a) Each of the operators L, U_f and U is isotone.
(b) If $v \in \mathfrak{V}_0$ and $\alpha \in \mathbb{R}$ then $v + \alpha \in \mathfrak{V}_0$ and $B(v + \alpha) = Bv + \alpha\beta$ for $B \in \{L, U_f, U\}$.
(c) If (MA1) holds then $V_1 = UV_0$, and $V_n \leq UV_{n-1}$ for all $n \geq 1$ for which $V_{n-1} \in \mathfrak{V}_0$.
(d) For MDPvars (a) and (c) remain true, while (b) must be replaced by

$$L(v + \alpha) = Lv + \alpha\beta, \quad U_f(v + \alpha) = U_f v + \alpha \cdot \beta_f.$$

Proof Everything is shown as in the proof of Lemma 11.5, using for (c) the RI and observing that L, U_f and U are only defined on $\mathfrak{V}_0$. □

Assume (MA1) and that the recursion

$$v_n = Uv_{n-1},\ 1 \leq n \leq N,\ v_0 := V_0,$$

has a solution $v_n \in \mathfrak{V}_0$. Contrary to intuition this does not ensure that the functions v_n equal the value functions V_n, and thus the VI_N (i.e. the value iteration up to stage N) need not hold. However, everything goes well if in addition there exist maximizers f_n of $Lv_{n-1} = L(U^{n-1}V_0)$, $1 \leq n \leq N$. This condition can be sequentially checked and may be rephrased as: there exists a maximizer f_1 of LV_0 and maximizers f_{n+1} of LV_{n,π^n}, $1 \leq n \leq N-1$, where $\pi^n := (f_t)_n^1$. (Note that $V_{n,\pi^n} \in \mathfrak{V}_0$ by Lemma 14.1.3.) Such policies $\pi = \pi^N$ are called ***maximal.*** Now we obtain the desired result as follows.

Proposition 14.1.5 (Maximizers yield the VI and optimal policies) *[Var] Under (MA1) for each $N \geq 1$ the following* ***optimality criterion OC_N*** *holds: If there exists a maximal N-stage policy π (in particular, if $D(s)$ is finite for all s), then VI_N holds and π is optimal for MDP_N.*

Proof The proof is nearly a copy of the proof of the Basic Theorem 11.6(a). We prove the assertion, denoted by (I_N), by induction on N. For $n = 1$ we obtain from Lemma 14.1.4(c) $V_1 = UV_0 = U_{f_1}V_0 = V_{1f_1} \in \mathfrak{V}_0$ by Lemma 14.1.3. Thus (I_1) is true. Assume (I_N) to be true for some $N \geq 1$. Then $\sigma := (f_n)_N^1$ is optimal for MDP_N and $V_N \in \mathfrak{V}_0$ by Lemma 14.1.3. Now Lemma 14.1.4(c) for $f := f_{N+1}$ implies (I_{N+1}) as

$$V_{N+1} \leq UV_N = U_f V_N = U_f V_{N\sigma} = V_{N+1,(f,\sigma)} \leq V_{N+1}. \square$$

The subsequent Structure Theorem 14.1.6 is easier to apply than Proposition 14.1.5 as it ensures (MA2) and the existence of maximizers at each stage in a sequential manner. Often it not only verifies the VI, but also a certain structure of the value functions, e.g. monotonicity. Therefore we call it the *Structure Theorem.*

Theorem 14.1.6 (Structure Theorem for MDPs with discrete transition law) *[Var] Assume (MA1). Let $\mathfrak{V} \subset \mathfrak{V}_0$ be a set of functions on S which contains V_0 and has the following properties:*

(S1) For each $v \in \mathfrak{V}$ there exists a maximizer f_v of Lv,
(S2) $Uv \in \mathfrak{V}$ for all $v \in \mathfrak{V}$.

Then the following holds:

(a) $V_n \in \mathfrak{V}$ for all $n \geq 1$ and (MA2) is fulfilled.
(b) For each $n \geq 1$ the decision rule $f_n := f_{V_{n-1}}$ is a maximizer at stage n, and $(f_n)_N^1$ is an optimal policy for each $N \geq 1$.
(c) The VI holds: $V_n = UV_{n-1}$, $n \geq 1$.

Proof We verify by induction on $n \geq 1$ the assertion (I_n):

$V_{n-1} \in \mathfrak{V}$ (hence (MA2) is true), f_n is a maximizer at stage n, and $V_n = UV_{n-1}$.

(I_1) is true. In fact, $V_0 \in \mathfrak{V}$, by assumption, then f_1 is a maximizer at stage 1 by (S1), and $V_1 = UV_0$ holds by Lemma 14.1.4(c). Assume for some $n \geq 1$ that (I_ν) holds for $1 \leq \nu \leq n$. This implies (I_{n+1}) as follows. $V_n \in \mathfrak{V}$, holds by (S2) since $v := V_{n-1} \in \mathfrak{V}$, and since $V_n = UV_{n-1}$ by (I_n). Next, f_{n+1} is a maximizer at stage $n+1$ by (S1), since $v := V_n \in \mathfrak{V}$. Since f_ν is a maximizer at stage $1 \leq \nu \leq n$, $\sigma := (f_\nu)_n^1$ is optimal for MDP_n. Now Lemma 14.1.4(c) and the RI with $f := f_{n+1}$ yield

$$UV_n \geq V_{n+1} \geq V_{n+1,(f,\sigma)} = U_f V_{n\sigma} = U_f V_n = UV_n.$$

That $(f_n)_N^1$ is an optimal policy for each $N \geq 1$ follows from Proposition 14.1.5(a). □

In applications the choice of the set $\mathfrak{V}$ must balance (S1) against (S2): In order to fulfill (S1) a small $\mathfrak{V}$ is desirable while (S2) asks for a large $\mathfrak{V}$. If (S1) has been verified, one may use $Uv = U_f V$ for checking (S2).

In several contexts, finiteness of the value functions is required, e.g. in connection with convexity. Then, if r_S and V_0 not both are bounded, so-called bounding functions are useful. They are also helpful for studying the large and infinite-horizon case. Moreover, they ensure the VI (cf. Proposition 14.1.9 below) in some cases where the Structure Theorem is not applicable.

Our goal consists in finding conditions which ensure that for some function $b \geq 0$ on S and some constants $d_n = d_n(b) \geq 0$ we have

$$|V_n(s)| \leq d_n \cdot b(s), \quad n \geq 1, \ s \in S.$$

The approach is similar (but more complex since here the VI does not hold from the outset) to the one in Chap. 3, except that the condition for the transition function T is replaced by a corresponding condition for the transition law p. To motivate the conditions introduced below, assume that for some MDPvar we already know that

$|V_n| \le d_n \cdot b$ for some $n \ge 0$ and that $V_{n+1} = UV_n$. Then $p|V_n| \le d_n \cdot pb$ and

$$|V_{n+1}(s)| \le \sup_{a \in D(s)} [p|r_S|(s,a) + \beta(s,a) \cdot p|V_n|(s,a)]$$
$$\le \sup_a p|r_S|(s,a) + d_n \cdot \sup_a(\beta(s,a) \cdot pb(s,a)).$$

Now we see that $|V_{n+1}| \le d_{n+1} \cdot b$ for some $d_{n+1} \ge 0$ if we require that for some constant $\delta \ge 0$ and all (s,a)

$$p|r_S|(s,a) \le \delta \cdot b(s), \text{ and } \beta(s,a) \cdot pb(s,a) \le \delta \cdot b(s).$$

These two conditions are now written in a compact form using for functions v on S and w on D the b-norm (weighted sup-norm):

$$\|v\|_b := \sup_{s \in S} |v(s)|/b(s), \quad \|w\|_b := \sup_{(s,a) \in D} |w(s,a)|/b(s).$$

Note that we use the notation $\|v\|_b$ and $\|w\|_b$ whether or not these numbers are finite, and that then $|v| \le \|v\|_b \cdot b$ and $|w| \le \|w\|_b \cdot b$. Recall that B_b denotes the Banach space of functions on S with finite b-norm and that $0/0 := 0$.

Definition 14.1.7 A function $b\colon S \to \mathbb{R}_+$ is called a ***bounding function of an MDPvar*** if

- $\|p|r_S|\|_b < \infty$.
- $\|V_0\|_b < \infty$.
- $\beta_b := \|\beta \cdot pb\|_b = \sup_{(s,a)\in D}[\beta(s,a) \cdot pb(s,a)/b(s)] < \infty$.

In general there exist several bounding functions. In particular, if b is a bounding function then $s \mapsto b(s) + 1$ is also a bounding function, provided $\overline{\beta} := \sup_{(s,a)\in D} \beta(s,a) \in (0, \infty]$ is finite. Therefore in the literature bounding functions $b(\cdot)$ are often assumed to have the property $b \ge 1$. We do not make this assumption since the bound (14.5) below may be better for $b(\cdot)$ than for $b(\cdot) + 1$. For a first reading it is advised to assume that b is positive since otherwise much care must be given to divisions by $b(s)$ if $b(s) = 0$.

Remark 14.1.8 Consider some MDPvar and an arbitrary function $b \ge 0$ on S. Put $\gamma := \sup_{(s,a)\in D} pb(s,a)/b(s) \in [0, \infty]$. Then:

(a) In MDPs $\overline{\beta}$ equals the constant discount factor, say $\beta' > 0$, and $\beta_b = \beta' \cdot \gamma$.
(b) We have $\beta_b \le \overline{\beta} \cdot \gamma$. Thus β_b is finite if both $\overline{\beta}$ and γ are finite.
(c) If r_S and V_0 are bounded and if $\overline{\beta} < \infty$, then the MDPvar has the trivial bounding function $b \equiv 1$, $\|\cdot\|_b$ is the sup-norm, and $\beta_b = \overline{\beta}$.
(d) If $\|p|r_S|\|_b < \infty$, then $r = pr_S$ exists and is finite (as required in the definition of an MDPvar) since $p|r_S| \le \|p|r_S|\|_b \cdot b < \infty$. In addition, we have $|r| \le \|p|r_S|\|_b \cdot b$. In particular, $r(s,a)$ is bounded in a. (This yields examples of MDPs which do not have a bounding function.) The condition $\|p|r_S|\|_b < \infty$ holds if

$\sup_{(s,a,s_1)\in D\times S} |r_S(s,a,s_1)|/b(s) < \infty$. Moreover, if r_S is independent of s_1, then $\|p|r_S|\|_b = \|r\|_b$.

(e) If b is a bounding function then $B_b \subset \mathfrak{V}_0$, and the operators U_f and U map B_b into itself.

For the proof firstly note that $B_b \subset \mathfrak{V}_0$ since $\beta \cdot p|v| \le \beta \cdot \|v\|_b \cdot pb \le \|v\|_b \cdot \beta_b \cdot b < \infty$ for $v \in B_b$. Moreover, as $|r| \le \|p|r_S|\|_b \cdot b$ by (d), we have, using $\beta_f(s) := \beta(s,f(s))$ for $s \in S$,

$$\begin{aligned}|Uv(s)| &= |\sup_f U_f v(s)| \le \sup_f |U_f v(s)| \\ &\le \sup_f |r_f(s)| + \sup_f(\beta_f(s) \cdot p_f|v|(s)) \\ &\le \|p|r_S|\|_b \cdot b(s) + \|v\|_b \cdot \sup_f(\beta_f(s) \cdot p_f b(s)) \\ &\le [\|p|r_S|\|_b + \beta_b \cdot \|v\|_b] \cdot b(s).\end{aligned}$$ ◇

Proposition 14.1.9 (Bounds and the VI for MDPvars with discrete transition law and with a bounding function) *If the MDPvar has a bounding function b then the following holds:*

(a) $V_n \in B_b$, $n \ge 0$, *i.e. there exist constants* $d_n = d_n(b) \in \mathbb{R}_+$ *such that*

$$|V_n(s)| \le d_n \cdot b(s), \quad n \ge 0, s \in S. \tag{14.4}$$

(b) The VI holds.

Proof

(a) We verify the bounds (14.4) and (14.5). For this purpose we consider the MDPvar′ which differs from the MDPvar by $r'_S := |r_S|$ and $V'_0 := |V_0|$. In MDPvar′ we are in case (EP), hence (MA1) holds, which implies the RI by Lemma 14.1.3. Thus we get

$$0 \le V'_{n+1,(f,\sigma)} = U_f V'_{n\sigma} \le \left(\|p|r_S|\|_b + \beta_b \cdot \|V'_{n\sigma}\|_b\right) \cdot b,$$

hence

$$\|V'_{n+1,(f,\sigma)}\|_b \le \|p|r_S|\|_b + \beta_b \cdot \|V'_{n\sigma}\|_b.$$

Now induction on $n \ge 0$ shows that $\|V'_{n\pi}\|_b \le d_n$ with d_n from (14.5). This implies for all s

$$\mathbb{E}_{\pi s}|R_{n\pi}(s,\zeta^n)| \le V'_{n\pi}(s) \le d_n \cdot b(s) < \infty.$$

Therefore $V_{n\pi}(s) := \mathbb{E}_{\pi s} R_{n\pi}(s, \zeta^n)$ exists, i.e. (MA1) holds for the MDPvar. Now the assertion follows from

$$|V_n(s)| \leq \sup_{\pi \in \mathbb{F}^n} |V_{n\pi}(s)| = \sup_{\pi} |\,\mathbb{E}_{\pi s} R_{n\pi}(s, \zeta^n)|$$
$$\leq \sup_{\pi} \mathbb{E}_{\pi s} |R_{n\pi}(s, \zeta^n)| \leq d_n \cdot b(s).$$

(b) Firstly, (MA2) holds as $p|V_n| \leq d_n \cdot pb < \infty$ by (a). Now the VI is valid by Theorem 14.1.6(c) if $UV_n < \infty$ for $n \geq 0$. This holds since by the proof of Remark 14.1.8(e)

$$|UV_n| \leq [\|p|r_S|\|_b + \beta_b \cdot \|V_n\|_b] \cdot b \leq [\|p|r_S|\|_b + \beta_b \cdot d_n] \cdot b < \infty. \qquad \square$$

Remark 14.1.10 The above proof shows that (14.4) holds with

$$d_n := \sigma_n(\beta_b) \cdot \|p|r_S|\|_b + \beta_b^n \cdot \|V_0\|_b < \infty. \tag{14.5}$$

$\diamond$

Definition 14.1.11 A **control model with i.i.d. disturbances and countable disturbance space** (**CM** for short) is a tuple $(S, A, D, Z, Q, T, r_Z, V_0, \beta)$ of the following kind:

- S and A are arbitrary.
- D, T, r_Z, V_0 and β have the same meaning as in the CM of Chap. 11.
- The disturbance space Z may be countably infinite.
- Q is the discrete probability distribution of each of the i.i.d. disturbances.

The ***first minimal assumption (MA1) for a CM*** means that there exists for all $s \in S$, $N \geq 1$ and $\pi \in \mathbb{F}^N$ the expected N-stage reward $V_{N\pi}$, the expectation of $G_{N\pi}(s, \eta^N)$ from (11.1).

The N-stage value function V_n of the CM and the problem $CM_N(s)$ and CM_N are defined as in Chap. 11.

Definition 14.1.12 A function $b: S \to \mathbb{R}_+$ is called a ***bounding function for a CM*** if

- $\|\,\mathbb{E}\,|r_Z(\cdot, \cdot, \eta_1)|\,\|_b < \infty$.
- $\|V_0\|_b < \infty$.
- $\beta_b := \|\beta \cdot \mathbb{E}\, b(T(\cdot, \cdot, \eta_1))\|_b < \infty$.

The elementary facts in Remark 14.1.8 remain true for CMs with obvious changes. Validity of (MA2) for a CM is ensured by the existence of a bounding function. This follows similarly as in the proof of Proposition 14.1.9. Obviously

$$\|r\|_b = \|\,\mathbb{E}\, r_Z(\cdot, \cdot, \xi)\|_b \leq \|\,\mathbb{E}\,|r_Z(\cdot, \cdot, \xi)|\,\|_b.$$

It follows that a bounding function b of a CM is also a bounding function of the MDPad, and then the CM is equivalent to MDPad.

Example 14.1.13 (Discrete state replacement problem without failure state) We consider the variant of the replacement problem from Example 11.9 where the age of the system may be arbitrary large, i.e. where $S := \mathbb{N}_0$. We allow arbitrary large increases of age. Thus $A = D(s) = \{0, 1\}$ for $s \in \mathbb{N}_0$, $Z := \mathbb{N}_0$ and $T(s, a, z) = (s+z) \cdot \delta_{a0}$. Most statements in Example 11.9, in particular the existence of a control-limit policy, remain true when just putting $m := \infty$. However, one also needs to check whether the proofs carry over, which will be sketched below. We again assume that both g and V_0 are decreasing, hence r and V_0 are bounded above. Thus we are in case (EN), and (MA2) holds. ♦

If in the next proposition $s_n^* = 0$ one has to replace at all states s, whereas in case $s_n^* = \infty$ one does not replace at all.

Proposition 14.1.14 (Structure of the solution of replacement Example 14.1.13) *Assume that the functions g and V_0 are decreasing. Then:*

(a) The VI holds.
(b) V_n is finite and decreasing for all n.
(c) For $n \in \mathbb{N}$ put

$$s_n^* := \inf\{s \in \mathbb{N}_0 : \beta \cdot \mathbb{E}\, V_{n-1}(s + \xi) < \beta \cdot V_{n-1}(0) - \gamma\},$$

(with $\inf \emptyset := \infty$*), and*

$$f_n(s) = \begin{cases} 0, & \text{if } s < s_n^*, \\ 1, & \text{if } s \geq s_n^*. \end{cases} \tag{14.6}$$

Then f_n from (14.6) is the smallest maximizer at each stage n and $(f_n)_N^1$ is an optimal control-limit policy for CM_N, $N \in \mathbb{N}$.

Proof We only sketch the proof.

(a) Maximizers exist at all stages as $D(s) = \{0, 1\}$ for all s. Thus the VI holds by Proposition 14.1.5(b).
(b1) (EN) implies $V_n < \infty$ for all n. Now finiteness of V_n for all n follows by induction on n, as $V_{n-1} > -\infty$ implies the same for V_n, due to

$$V_n(s) \geq W_n(s, 1) = g(s) - \gamma + \beta \cdot V_{n-1}(0),\ s \in \mathbb{N}_0.$$

(b2) and (c). These parts of the proof of Example 11.9 remain unchanged. Note, however, that $\mathbb{E}\, V_{n-1}(s + \xi) = -\infty$ for some n and s is not excluded. □

14.2 Large and Infinite Horizon

The set-up for MDPs (and MDPvars) with infinite horizon (**MDP$_\infty$** and **MDPvar$_\infty$** for short) is the same as in Chap. 12. There also the expected infinite-stage reward $V_{\infty\pi}(s)$ earned under policy $\pi = (\pi_t)_0^\infty \in \Pi := \mathbb{F}^{\mathbb{N}_0}$ and initial state s is defined as well as $V_\infty(s)$ and optimality and stationarity of an infinite-stage policy. Theorems 12.2.3 and 12.3.2 about MDPs with large and with infinite horizon remain true without change of proof in the case where both r_S and V_0 are bounded. Using bounding functions we mention a more general result as follows.

Theorem 14.2.1 (Solution of the large horizon problem and of the infinite-stage MDP) *[Var] Assume that the MDP has a bounding function b with $\beta_b < 1$. Then:*

(a) $(V_n)_0^\infty$ converges for each choice of V_0 in B_b in b-norm to the same function $V \in B_b$, and

$$\|V\|_b \le \|r\|_b/(1-\beta_b).$$

(b) For all $\pi \in \Pi$ and $s \in S$ the total reward $R_{\infty\pi}(s,\zeta^\infty)$ exists $\mathbb{P}_\pi(s,\cdot)$-almost surely, $V_{\infty\pi}$ and V_∞ exist and belong to B_b. Moreover, $(V_{n\pi})_0^\infty$, $\pi \in \Pi$, converges for each choice of V_0 in B_b in b-norm towards $V_{\infty\pi}$.

(c) V equals V_∞ and is the unique solution of the optimality equation within B_b. Moreover,

$$\|V - V_N\|_b \le \|V - V_0\|_b \cdot \beta_b^N,\; N \in \mathbb{N}.$$

(d) If f is a maximizer of LV, then f is asymptotically optimal, f^∞ is a stationary optimal policy and $(V_{nf})_1^\infty$ converges in b-norm towards V. Moreover,

$$0 \le V_N - V_{Nf} \le 2\beta_b^N \cdot \|V - V_0\|_b \cdot b,\; n \in \mathbb{N}.$$

Proof The proof is nearly the same as for Theorem 20.1.3 below. Even some simplifications occur as here no measurability problems arise and as the existence of a bounding function implies (MA2) and the VI by Proposition 14.1.9.

Each MDPvar can be reduced to the following tuple $(S', A, D', p', r', V_0', \beta')$ (MDP′ in short) with constant discount factor β' according to the following idea: If the VI holds for the MDPvar it obviously has the form

$$V_n(s) = \sup_{a\in D(s)} \left[r(s,a) + \beta' \cdot \sum_{s'\in S} p'(s,a,s')\, V_{n-1}(s') \right], \quad s \in S,$$

where $\beta' := \sup_{(s,a)\in D} \beta(s,a) \in (0,\infty)$, and

$$p'(s,a,s') := \beta(s,a) \cdot p(s,a,s')/\beta',\; (s,a) \in D.$$

For $(s,a) \in D$, the sum $\sum_{s\in S} p'(s,a,s') = \beta(s,a)/\beta' \le 1$ need not equal one. Therefore we introduce an additional state $\hat{s}$ which absorbs the remaining probability $1 - \beta(s,a)/\beta'$. More precisely we define $(S', A, D', p', r', V_0', \beta')$ by: $S' := S + \{\hat{s}\}$; $D'(s)$, $r'(s,a)$ and $V_0'(s)$ for $s \in S$ are as in the MDP; $D'(\hat{s}) := A$; $p'(s,a,\hat{s}) := 1 - \beta(s,a)/\beta'$, $s \in S$ and $p'(\hat{s},a,\hat{s}) := 1$; $r'(\hat{s},a) := V_0'(\hat{s}) := 0$; β' as above. Then MDP′ has the absorbing set $J_0 := \{\hat{s}\}$, the essential state space $J := S$, and $V_n'(\hat{s}) = 0$ for $n \ge 0$.

As a rule structural properties (such as concavity) of the function V are proved by showing (i) that all value functions have the property and (ii) that the property is preserved under the pointwise limit of sequences of functions. In the proof of (i) difficulties may arise when applying the VI if $\beta(s,a)$ is not constant. Then the preceding reduction of an MDPvar to an MDP′ may be helpful.

Definition 14.2.2 Given an MDPvar $(S, A, p, r_S, V_0, \beta)$ with the following properties:

(i) We have $\beta' := \sup_{(s,a)\in D} \beta(s,a) < 1$, or equivalently, the function $D \ni (s,a) \mapsto \Delta(s,a) := \beta(s,a)/(1-\beta(s,a))$ is bounded,
(ii) the MDPvar has a bounding function b with $\beta_b < 1$.

Then the following ***MDP′*** $(S, A, D, p', r_S', V_0, \beta')$ with

- $\Delta(s,a) := \beta'(s,a)/(1-\beta'(s,a))$, $\Delta' := \beta'/(1-\beta')$,
- $p'(s,a,s') := \frac{\Delta(s,a)}{\Delta'} \cdot p(s,a,s') + \left(1 - \frac{\Delta(s,a)}{\Delta'}\right) \cdot \delta_{s,s'}$, $(s,a,s') \in D \times S$,
- $r_S'(s,a,s') := r'(s,a) := \frac{1+\Delta(s,a)}{1+\Delta'} \cdot r(s,a)$, $(s,a,s') \in D \times S$

is called a ***uniformization*** of the MDPvar.

Theorem 14.2.3 (Uniformization of an MDPvar) *For the MDPvar from Definition 14.2.2:*

(a) The function b is also a bounding function of MDP′ and $\beta_b' < 1$. Thus there exists V', independent of $V_0' \in B_b$.
(b) We have $V = V'$, hence V is the unique solution $v \in B_b$ of $U'v = v$.
(c) The decision rule $f \in \mathbb{F}$ is a maximizer of LV if and only if it is a maximizer of $L'V$, and then f^∞ is a stationary optimal policy for MDPvar.

Proof

(a) Firstly, $p'|r_S'| = |r'| \le |r| \le p|r_S|$ implies $\|p'|r_S'|\|_b \le \|p|r_S|\|_b < \infty$. Next we show that $\beta_b' := \beta' \cdot \|p'b\|_b < 1$. From $\beta = 1 - 1/(1+\Delta)$ we see that $\beta' = \Delta'/(1+\Delta')$. A simple computation yields

$$\Delta' \cdot p'v = \Delta \cdot pv + (\Delta' - \Delta) \cdot v, \quad v \in B_b. \tag{14.7}$$

From (14.7) with $v := b$ we obtain, using $\beta \cdot pb \le \beta_b \cdot b$, that

$$(1+\Delta') \cdot \beta' \cdot p'b = (1+\Delta) \cdot \beta \cdot pb + (\Delta' - \Delta) \cdot b$$
$$\le [(1+\Delta) \cdot \beta_b + \Delta' - \Delta] \cdot b \le (\beta_b + \Delta') \cdot b.$$

Thus $\beta'_b \le (\beta_b + \Delta')/(1+\Delta') < 1$.

From (14.7) we get the basic equality

$$(1+\Delta') \cdot (L'v - v) = (1+\Delta) \cdot (Lv - v), \ v \in B_b. \tag{14.8}$$

Now (14.8) implies for $v := V$, since $LV \le UV = V$, that

$$(1+\Delta') \cdot (L'V - V) = (1+\Delta) \cdot (LV - V) \le 0,$$

hence $L'V(s,a) \le V(s)$ for all (s,a), hence $U'V \le V$. Now induction on $n \ge 1$ shows that $(U')^n V \le V$. As $(U')^n V$ converges for $n \to \infty$ to V' by Theorem 14.2.1, we have $V' \le V$. In the same way one obtains from (14.8) for $v := V'$ that $V \le V'$, hence $V = V'$.

(b) The first part of the assertion follows from (14.8), and the second one from Theorem 14.2.1.

(c) The assertions follow directly from (b) and Theorem 14.2.1(d). □

14.3 Problems

The first three problems refer to the replacement problem from Example 14.1.13. If $v: \mathbb{N}_0 \to \mathbb{R}$ is decreasing, we put $v(\infty) := \lim_{s\to\infty} v(s)$.

Problem 14.3.1 (a) Let $f :\equiv 0$ and $h :\equiv 1$ be decision rules and put $T_\nu := \sum_1^\nu \eta_i$. Put $V_{Nh} := V_{N\pi}$ for $\pi := (h)_0^{N-1}$. Then

$$V_{Nf}(s) = \sum_0^{N-1} \beta^\nu \cdot \mathbb{E}\, g(s + T_\nu) + \beta^N \cdot \mathbb{E}\, V_0(s + T_N),$$

$$V_{Nh}(s) = g(s) - g(0) + (g(0) - \gamma) \cdot \sigma_n(\beta) + \beta^n V_0(0).$$

In case $g(s) = V_0(s) = e^{-s}$ for all s one can express $V_{Nf}(s)$ in terms of the generating function of η_1.

(b) The value functions have the following bounds:

$$\max\{V_{Nf}(s), V_{Nh}(s)\} \le V_N(s) \le g(s) + g(0) \cdot \sigma_{N-1}(\beta) + \beta^N \cdot V_0(0).$$

Thus V_N is bounded for all N if and only if $g(\infty) > -\infty$.

Problem 14.3.2 If the replacement costs γ are large enough, namely if

$$\beta \cdot [g(0) - g(\infty)] \leq (1-\beta)\gamma \text{ and } \beta \cdot [V_0(0) - V_0(\infty)] \leq \gamma,$$

then it is optimal not to replace at each stage and at each state.

Problem 14.3.3 If the replacement costs γ are small enough, namely if

$$\gamma \leq \beta \cdot [g(0) - \mathbb{E}\, g(\eta_1)] \text{ and } \gamma \leq \beta \cdot [V_0(0) - \mathbb{E}\, V_0(\eta_1)],$$

then it is optimal to replace at each stage and at each state.

Problem 14.3.4 In the following deterministic DP there do not exist maximizers at any stage, but for each $\delta > 0$ a smallest δ-maximizer exists at each stage: $S = A = D(s) = \mathbb{N}$; $T(s,a) = s + a$; $r(s,a) = -1/a$; $V_0(s) = 1/s$, $\beta = 1$.

Problem 14.3.5 ((MA1) for a CM does not imply (MA1) for the adjoint MDP) Consider the uncontrolled CM with $S = Z = \mathbb{N}$, $|A| = 1$, $q(z) = \delta/z^3$, where $\delta := \left(\sum \frac{1}{z^3}\right)^{-1} > \left(\sum \frac{1}{z^2}\right)^{-1}$, $r_Z(s,z) = s^2 z$, $T(s,z) = s + s^2 z$, $V_0(s) = -s$, $\beta = 1$. Fix N and put $\eta_\nu^s := \zeta_{\nu\pi}^s$. Then:

(a) $\eta_{\nu+1}^s = s + \sum_{i=0}^{\nu} (\eta_i^s)^2 \cdot \eta_{i+1}$, $0 \leq \nu \leq N-1$.
(b) $G_{N\pi}(s, (\eta_\nu)_1^N) = -s$, hence (MA1) holds for the CM.
(c) For the adjoint MDPad we have for $s = 1$ and $N = 2$: $R_{2\pi}(s, \eta_1^s, \eta_2^s) = m - 1 - \eta_1 + (m - \eta_2)(1 + \eta_1)^2$, where $m := \mathbb{E}\, \eta_1 > 1$. Therefore $\mathbb{E}\, R_{2\pi}^{\pm}(s, \eta_1^s, \eta_2^s) = \infty$, hence (MA1) does not hold for the adjoint MDP.

Problem 14.3.6 ((MA1) for a CM does not follow from (MA1) for the MDPad) Consider the CM with $S = \mathbb{N}_0$; A and D are arbitrary; $Z = \mathbb{Z}$, $q(z) = 3/(\pi z)^2$ for $z \neq 0$ and $q(0) = 0$; $T(s,a,z) = |z|$; $\tilde{r}(s,a,z) = sz$ for $z = \pm 1$ and zero otherwise; $V_0 \equiv 0$; $\beta = 1$. One easily shows:

(a) r exists and equals zero, hence (MA1) holds for MDPad.
(b) As $\mathbb{E}\, |\eta_1| = \infty$, $V_{2\pi}(0)$ does not exist, (MA1) fails for the CM.

14.4 Supplements

Supplement 14.4.1 (Validity of the VI) Assume that (MA1) holds and that for each $n \in \mathbb{N}$ there exists some $\pi \in \mathbb{F}^n$ such that $V_{n\pi} > -\infty$. Then (MA2) holds. If $V_{n\pi} > -\infty$ for each $n \in \mathbb{N}$ and each $\pi \in \mathbb{F}^n$ then the VI holds.

Supplement 14.4.2 (If $v \in \mathfrak{V}_0$, then $U_f v$ need not belong to $\mathfrak{V}_0$) Consider the MDP with $S = \mathbb{Z} - \{0\}$; $A = D(s) = \{1,2\}$; $p(s,1,\pm 1) = 1/2$, $p(s,2,s') = 6/(\pi s')^2$ if $s \cdot s' > 0$, and $p(s,a,s') = 0$, else; $\tilde{r} \equiv 0$; $V_0(s) = s$; $\beta = 1$. Then $V_0 \in \mathfrak{V}_0$, but $U_f V_0$ does not belong to $\mathfrak{V}_0$ for $f \equiv 2$, as $p(U_f V_0)^{\pm}(s,1) = \infty$.

Supplement 14.4.3 ((MA1) does not imply (MA2)) Consider the following MDP: $S = \mathbb{Z} - \{0\}$; $A = D(s) = \mathbb{N}$, $s \in S$; $r \equiv 0$; $V_0(s) = s$ for all s, $\beta = 1$ and

$$P(s, a, ds') = \begin{cases} 4^{-1} \cdot \delta_{-a} + 3 \cdot 4^{-1} \cdot \delta_a, & \text{if } s \geq 0, \\ 6\pi^{-2} \cdot \sum_1^\infty \delta_{-i}/i^2, & \text{if } s < 0. \end{cases}$$

For given N, $s > 0$ and $\pi = (\pi_t)_0^{N-1} \in \mathbb{F}^N$ define the positive numbers $t_0, t_1, \ldots, t_N$ by $t_0 := s$, $t_{\nu+1} := \pi(t_\nu)$, $0 \leq \nu \leq N - 1$. It follows from the definition of P that t_ν is the unique positive element in the support of the probability distribution of ζ_ν, $\nu \leq N$, with respect to $\mathbb{P}_{\pi s}$. Thus $\mathbb{E}_{\pi s} V_0^+(\zeta_N) < \infty$. Therefore (MA1) holds. On the other hand, V_1 does not belong to $\mathfrak{V}_0$, hence (MA2) is not fulfilled.

Supplement 14.4.4 (Historical remarks) Weak conditions for (MA1) and (MA2) were first introduced in Hinderer (1971). The notion of a bounding function is due to Wessels (1977), although similar ideas were used before, cf. e.g. Veinott (1969) and Lippman (1974/1975). The extension to one-sided bounding functions was made by Rieder (1975b). An investigation on the validity of the value iteration is due to Hinderer (1977).

Chapter 15
Examples with Discrete Disturbances and with Discrete Transition Law

In this chapter we apply the general theorems from Chap. 14 to special examples. In particular, we consider a production-inventory problem with backlogging and delivery lag and a queueing model with arrival control.

Example 15.1 (A production, sale and inventory problem with a non-trivial bounding function) A firm produces, sells and stores a certain product in integer quantities, for which there is a demand of amount $\eta_\nu \geq 0$ in period ν with finite mean. The random demands, occurring at the end of the periods, are assumed to be i.i.d. and integer-valued. Unsatisfied demand is backlogged. In each period one can produce at most $m \in \mathbb{N}$ units of the product. There is unlimited storage capacity. The gain and cost parameter are as follows:

$$\begin{aligned} R > 0 &= \text{reward for selling one unit} \\ d > 0 &= \text{production cost per unit} \\ h > 0 &= \text{cost for storing one unit for one period} \\ d_0 > 0 &= \text{gain for selling one unit of a terminal positive inventory} \\ p > 0 &= \text{penalty cost per unit of unsatisfied demand.} \end{aligned}$$

How much should be produced in each period in order to maximize the expected N-stage reward? Obviously the problem can be modeled as a CM with the following data: s and a denotes the inventory before and after ordering, respectively, hence $a-s$ is the order; $S = A = \mathbb{Z}$, $Z = \mathbb{N}_0$ and $D(s) = \{s, s+1, \dots, s+m\}$; $T(s, a, z) = a-z$; $V_0(s) = d_0 s^+$;

$$r_Z(s, a, z) := R \cdot \min\{z, a^+\} - d \cdot (a-s) - ha^+ - p \cdot (a-z)^-.$$

K. Hinderer et al., *Dynamic Optimization*, Universitext,
DOI 10.1007/978-3-319-48814-1_15

(a) As r_Z is unbounded below and as V_0 is unbounded above, $b \equiv 1$ cannot be a bounding function, and also neither (EN) nor (EP) holds. However, $s \mapsto 1 + |s|$ is a bounding function. This follows from the following inequalities:

$$\mathbb{E}\,|r_Z(s, a, \eta_1)| \leq R \cdot \mathbb{E}\,\eta_1 + dm + h \cdot (|s| + m) + p \cdot (|s| + m + \mathbb{E}\,\eta_1)$$

$$|V_0(s)| \leq d_0|s|$$

$$\mathbb{E}\,b(T(s, a, \eta_1)) = 1 + \mathbb{E}\,|a - \eta_1| \leq 1 + \mathbb{E}\,\eta_1 + m + |s|.$$

(b) Now it follows from Proposition 14.1.9 that the VI holds, and that for $n \in \mathbb{N}$, $s \in \mathbb{Z}$

$$V_n(s) = ds + \max_{s \leq a \leq s+m} \big[R \cdot \mathbb{E} \min\{\eta_1, a^+\} - da - ha^+ - p \cdot \mathbb{E}(a - \eta_1)^- + \beta \cdot \mathbb{E}\, V_{n-1}(a - \eta_1) \big].$$

Moreover, there exists at each stage a smallest maximizer, and these yield optimal policies according to Proposition 14.1.5. ♦

Example 15.2 (An inventory model with backlogging having an optimal policy of a special type) (Cf. Proposition 15.3(b) below.) We consider the variant of Example 13.1.2 where unsatisfied demand is backlogged. Thus states, inventory capacity $B \in \mathbb{N}$, actions, sets of admissible actions, random demand and ordering costs are as in Example 13.1.2, but now we define the discrete interval $S = A = (-\infty, B] := \{B, B-1, \ldots\}$ and $T(s, a) = s - a$. We now prove an analogue of Example 13.1.2 under partially weaker assumptions. However, in order to ensure (MA1) despite the unboundedness of S, we now require that C_0 is lower bounded. ♦

Proposition 15.3 (Structure of the solution of Example 15.2) *Put* $G(a) := \mathbb{E}\, g(a - \eta_1)$ *and assume that*

(i) $\mathbb{E}\,\eta_1 < \infty$, $g \geq 0$ *is convex and*

$$c_1 a + G(a) \to \infty \text{ for } a \to -\infty, \tag{15.1}$$

(ii) C_0 *is lower bounded and convex, and* $\mathbb{E}\, C_0(s - \eta_1)$ *is finite for all* $s \in S$.

Then we have

(a) All functions C_n *are finite and convex, and the VI holds and has the form*

$$C_n(s) = -c_1 s + \min_{s \leq a \leq B} [c_1 a + G(a) + \beta \cdot \mathbb{E}\, C_{n-1}(a - \eta_1)]\,.$$

(b) *There exists on* $(-\infty, B]$ *a smallest minimum point* S_n^* *of the function*

$$a \mapsto H_n(a) := c_1 \cdot a + G(a) + \beta \cdot \mathbb{E}\, C_{n-1}(a - \eta_1),\ n \geq 1,$$

and $s \mapsto f_n(s) := s \vee S_n^*$ *is the smallest minimizer at stage n.*

(c) *The function* $s \mapsto C_n(s) + c_1 s$ *is constant on* $(-\infty, S_n^*]$ *and increasing on* $[S_n^*, B]$.

Using the minimizer $s \mapsto f_n(s) := s \vee S_n^*$ means that it is optimal to order in case $s < S_n^*$ the amount $S_n^* - s$, i.e. to order *fill the inventory up to* S_n^*, and to order *nothing*, otherwise. Note that convexity of g implies convexity of G, as convexity is preserved under integration. If $\Delta H_n(B-1)$ is negative, then $S_n^* = B$. If $\Delta H_n(B-1) \geq 0$, then S_n^* is the smallest $x \in (-\infty, B-1]$ such that $\Delta H_n(x) \geq 0$. Condition (15.1) is equivalent to $\Delta G(-\infty) := \lim_{x\to-\infty} \Delta G(x) > -c_1$.

Proof As the VI is not available at the outset, we now use Theorem 14.1.6 with $\mathfrak{V} \subset \mathfrak{V}_0$ as the set of lower bounded convex functions v on S for which $\mathbb{E}\, v(a - \eta_1)$ is finite for all a.

(S1) For $v \in \mathfrak{V}$, we have $0 \leq Lv(s, a) < \infty$ for all (s, a), as

$$Lv(s, a) + c_1 \cdot s = H(a) := c_1 \cdot a + G(a) + \beta \cdot \mathbb{E}\, v(a - \eta_1).$$

Moreover, as v is lower bounded, (15.1) implies that $H(a) \to \infty$ for $a \to -\infty$. Moreover, convexity of v yields convexity of H. Thus H has a smallest minimum point S^*. Convexity of the function H implies that it is increasing on $[S^*, B]$. Therefore $Lv(s, \cdot)$ has on $D(s) = [s, B]$ the smallest minimum point $f_v(s) := s \vee S^*$. This shows that f_v is the smallest minimizer of Lv.

(S2) Select $v \in \mathfrak{V}$. Firstly Uv is lower bounded as $Uv \geq \beta \cdot \inf v$. Convexity of Uv follows from convexity of H since

$$Uv(s) = -c_1 s + \begin{cases} H(S_n^*), & \text{if } -\infty < s \leq S_n^*, \\ H(s), & \text{if } S_n^* < s \leq B. \end{cases}$$

The preceding equation also shows that $Uv(s) \leq -c_1 s + H(B)$. Therefore

$$\mathbb{E}\, Uv(a - \eta_1) \leq -c_1\, (a - \mathbb{E}\, \eta_1) + H(B) < \infty.$$

This completes the proof that $Uv \in \mathfrak{V}$. □

The *proportional cost case* is defined by the assumption that $g(y) = h \cdot y^+ + p \cdot y^-$, $y \in S$. The positive constants h and p are the holding cost factor and the penalty cost factor, respectively. Then the assumptions in Proposition 15.3 hold if

$$\mathbb{E}\, \eta_1 < \infty \text{ and } c_1 < p.$$

(The latter assumption is natural as it says that the penalty cost factor is larger than the ordering cost factor.) In fact, $\mathbb{E}\,\eta_1 < \infty$ implies $G(a) < \infty$ since $(a-z)^+ \leq |a-z| \leq |a| + z$. Moreover,

$$c_1 \cdot a + G(a) = (p+h) \cdot \mathbb{E}(\eta_1 - a)^- + p \cdot \mathbb{E}\,\eta_1 + (c_1 - p) \cdot a$$
$$\geq (c_1 - p) \cdot a \to \infty \text{ for } a \to -\infty.$$

Example 15.4 (A backlogging finite-stage model with a stationary optimal policy) Let s_t and a_t denote the inventory for a single product at the beginning of period $t \geq 0$ before and after ordering, respectively. Hence $a_t - s_t$ is the quantity ordered. There is a finite capacity bound $B \in \mathbb{N}$ for inventory. We allow backlogging of unsatisfied demand. Therefore s_t can assume negative values, in which case $|s_t|$ is the backlogged demand. There is no delivery lag, i.e. orders are available immediately. The random demand in period $t \geq 0$ is an $\mathbb{N}_0$-random variable η_{t+1}. The sequence $(\eta_t)_1^N$ is assumed to be i.i.d. and $\mathbb{E}\,\eta_1 < \infty$. There are two kinds of cost: (i) the cost $\gamma \in \mathbb{R}_+$ for ordering one unit and (ii) the holding and shortage costs per period, denoted by $g(a,z) \geq 0$. We assume $\mathbb{E}\,g(a,\eta_1) < \infty$ for all a. If after N periods there remains unsatisfied demand $s_N < 0$, it must be purchased at cost $C_0(s_N)$, whereas a positive stock $s_N \geq 0$ can be sold for a reward of $-C_0(s_N)$. We assume C_0 to be bounded below. The goal consists in minimizing the N-stage expected discounted costs.

Here we treat jointly the case where the stock is measured in the discrete scale; as shown in Chap. 14, the case of the continuous scale requires only minor changes. Intervals in $\mathbb{Z}$ will be denoted in the same manner as intervals in $\mathbb{R}$. The inventory problem is described by a CM as follows: $S = A = (-\infty, B]$; $D(s) = [s, B]$; $Z = [0, \infty)$. As backlogging is allowed, the transition function is $T(s,a,z) := a - z$. The one-stage costs are $c_Z(s,a,z) := \gamma \cdot (a-s) + g(a,z)$, hence

$$Lv(s,a) = \gamma \cdot (a-s) + \mathbb{E}\,g(a,\eta_1) + \beta \cdot \mathbb{E}\,v(a-\eta_1).$$

Note that c is finite as $\mathbb{E}\,g(a,\eta_1) < \infty$ for all a. As $c \geq 0$ and as C_0 is bounded below, (EN) holds.

We want to find conditions under which there exist minimizers f_n of the following simple *below s_n^* fill up to S_n^* type* (which can be expected under natural assumptions)

$$f_n(s) := \begin{cases} S_n^*, & \text{if } s < s_n^*, \\ s, & \text{if } s \geq s_n^*. \end{cases}$$

Policies consisting of such minimizers play an outstanding role in inventory theory. They are easy to implement since each of its decision rules is characterized by just two numbers. Our present model admits under mild assumptions the policy *below s_n^* fill up to S_n^*-minimizers* where even S_n^* equals s_n^* and is independent of n, so that

$$f_n(s) = s \vee s^* =: f^*(s), \; n \in \mathbb{N}, \; s \in S.$$

Here we encounter the rare situation where for finite horizon N a stationary policy (namely $(f^*)_N^1$) is optimal. (It is crucial for this simple form of f_n that the ordering costs are proportional to the quantity ordered.) The optimal quantity ordered is $f^*(s) - s = (s^* - s)^+$. Under f^* the inventory – once it has dropped below s^* – is filled up to s^* in each of the remaining periods. Moreover, if $s_0 \le s^*$, we have $s_t = s^* - \eta_t$ for all $1 \le t \le N$.

The application of the Structure Theorem requires a judicious choice of the set $\mathfrak{V}$. In order to give the reader a feeling of how results like Proposition 15.5 below can be detected, we shall now deviate from the usual theorem-proof scheme. We are looking for assumptions which guarantee the existence of some $s^* \in S$ such that the Theorem 14.1.6 is applicable and such that $s \mapsto s \vee s^*$ is a minimizer of Lv for all $v \in \mathfrak{V}$. (Notice that $v \in \mathfrak{V}_0$ if and only if $\mathbb{E}\, v(a - \eta_1)$ exists for all $a \in A$.) For this purpose we define for each $t \in S$ the decision rule $s \mapsto f_t(s) := s \vee t$. For fixed $v \in \mathfrak{V}_0$ such that $\mathbb{E}\, v(a - \eta_1)$ is finite for all a we can separate variables in $Lv(s, a)$ as $Lv(s, a) = -\gamma s + H_v(a)$, where

$$H_v(a) := \gamma a + \mathbb{E}\, g(a, \eta_1) + \beta \cdot \mathbb{E}\, v(a - \eta_1),\ a \in (-\infty, B].$$

(i) Obviously f_t is a minimizer of Lv if and only if for each $s \in S$ the restriction of H_v on $[s, B]$ has the minimum point $t \vee s$, i.e. if and only if H_v belongs to the set $\mathfrak{V}_1(t) := \{g \in \mathbb{R}^S : g$ has the minimum point t and isincreasing on $[t, B]\}$.
(ii) Now assume that $H_v \in \mathfrak{V}_1(t)$. Then $Uv(s) + \gamma s = H_v(s \vee t)$ for all s tells us that Uv belongs to

$$\mathfrak{V}(t) := \text{set of functions } v \text{ on } S \text{ such that } s \mapsto v(s) + \gamma s \text{ is constant on } (-\infty, t] \text{ and increasing on } [t, B].$$

We have $\mathfrak{V}(t) \subset \mathfrak{V}_0$ since $v \in \mathfrak{V}(t)$ implies $v \in \mathfrak{V}_0$ by boundedness of $s \mapsto v(s) + \gamma s$ and by $\mathbb{E}\, \eta_1 < \infty$.

Now we want to find some specific $t =: s^*$ such that the Structure Theorem 14.1.6 becomes applicable with $\mathfrak{V} := \mathfrak{V}(s^*)$.

(S1) Let t again be arbitrary. Select $v \in \mathfrak{V}(t)$, hence $v(s) = -\gamma s + u(s)$, $s \in S$, for some function u on S which is constant on $(-\infty, t]$ and increasing. We know from (i) that f_t is a minimizer of Lv if and only if $H_v \in \mathfrak{V}_1(t)$. Now we obtain

$$H_v(a) = G(a) + \beta \cdot \mathbb{E}\, u(a - \eta_1) + \gamma\beta \cdot \mathbb{E}\, \eta_1,$$

where

$$G(a) := \gamma \cdot a(1 - \beta) + \mathbb{E}\, g(a, \eta_1),\ a \in (-\infty, B]. \tag{15.2}$$

As u is increasing, $a \mapsto \mathbb{E}\, u(a - \eta_1)$ is also increasing. Therefore H_v is increasing on $[t, B]$ as soon as

(α) G is increasing on $[t, B]$.

Next, as u is constant on $(-\infty, t]$ and as $\eta_1 \geq 0$, $a \mapsto \beta \cdot \mathbb{E}\, u(a - \eta_1)$ is also constant on $(-\infty, t]$. Therefore H_v has the minimum point t as soon as

(β) G has the minimum point t.

Altogether, (S1) holds if (α) and (β) are true.

(S2) We have already seen under (ii) above that $Uv \in \mathfrak{V}(t)$ for $v \in \mathfrak{V}(t) = \mathfrak{V}(t) \cap \mathfrak{V}_0$ if f_t is a minimizer of Lv. ♦

Now the Structure Theorem 14.1.6 can be applied to

Proposition 15.5 (Structure of the solution of Example 15.4) *Assume that*

(i) the demand η_1 and the holding and shortage costs $g(a, Z)$, $a \in (-\infty, B]$, are integrable,
(ii) the function G from (15.2) has a minimum point $s^* \in (-\infty, B]$ and is increasing on $[s^*, B]$,
(iii) the function $s \mapsto C_0(s) + \gamma s$ is constant on $(-\infty, s^*]$ and increasing on $[s^*, B]$.

Then the VI holds and $s \mapsto f^*(s) := s \vee s^*$ is a minimizer at each stage. Moreover, the property stated in (iii) for C_0 holds for all minimal cost functions C_n.

Remark 15.6

(a) Condition (ii) holds for convex G if $G(a) \to \infty$ for $a \to -\infty$.
(b) Assumption (ii) is satisfied by $C_0(s) = -\gamma s$, $s \in S$, but not by $C_0 \equiv 0$, unless $s^* = B$.
(c) Although the minimizer f^* does not depend on the stage it is in general not myopic. ◇

Now we are going to discuss the important *proportional cost case,* where the holding and shortage costs are proportional to the inventory and the shortage, respectively. Denote by $h \geq 0$ and $p \geq$ the holding costs per unit and the shortage costs per unit, respectively. Assume that the costs are calculated on the basis of the inventory level at the end of the time period. Then

$$g(a, z) = h \cdot (a - z)^+ + p \cdot (a - z)^-, \tag{15.3}$$

and therefore

$$L(a) := \mathbb{E}\, g(a, \eta_1) = h \cdot \mathbb{E}(a - \eta_1)^+ + p \cdot \mathbb{E}(a - \eta_1)^-.$$

Because of $\mathbb{E}(a - \eta_1)^\pm \leq \mathbb{E}\, |a - \eta_1| \leq a + \mathbb{E}\, |\eta_1|$ assumption (i) of Proposition 15.3 is satisfied if η_1 is integrable. Moreover, as $\mathbb{E}\, g(a, \eta_1)$ and hence $G(a)$ from (15.2) is convex in a, assumption (ii) of Proposition 15.5 is fulfilled if G has a minimum point s^* in $(-\infty, B]$; figure! This reasoning leads to the next result.

Proposition 15.7 (Structure of the solution of the proportional cost case of Example 15.4) *Assume that η_1 is integrable, that the holding and shortage costs are of the form (15.3) with $p + h > 0$ and that $\alpha(1 - \beta) > p$. Let F denote the distribution function of η_1 and put*

$$\alpha := (p - \gamma(1 - \beta))/(p + h).$$

Then the conditions (i) and (ii) of Example 15.4 are fulfilled with the choice

$$s^* := \begin{cases} \text{an arbitrary } \alpha\text{ -quantile } q_\alpha \text{ of } F, & \text{if } 0 < \alpha < F(B-1), \\ B, & \text{if } \alpha \geq F(B-1). \end{cases}$$

Thus, if $s \mapsto C_0(s) + \gamma s$ is constant on $(-\infty, s^*]$ and increasing on $[s^*, B]$, then $s \mapsto s \vee q_\alpha$ is a minimizer at each stage in case $0 < \alpha < F(B - 1)$; and in case $\alpha \geq F(B - 1)$ it is optimal to fill up the inventory completely at each stage.

Proof Firstly, we already showed above that condition (i) of Proposition 15.5 holds because of the integrability of η_1. Next we notice that the function $a \mapsto (a - \eta_1)^\pm$ from $\mathbb{Z}$ into R_+ is convex so that $a \mapsto \mathbb{E}(a - \eta_1)^\pm$ is convex. This shows convexity of L and hence of

$$a \mapsto G(a) = (1 - \beta) \cdot \gamma\, a + L(a),$$

considered for the moment as a function on $\mathbb{Z}$. In particular, G is convex on $(-\infty, B]$. Next we are going to compute the right- and left-hand difference of G, denoted by $G'_+(a) := G(a + 1) - G(a)$, $G'_-(a) := G(a) - G(a - 1) = G'_+(a - 1)$. Since $(a - \eta_1)^\pm = (\eta_1 - a)^\mp$ and $(\eta_1 - a)^+ = (\eta_1 - a) + (\eta_1 - a)^-$, we get

$$G(a) = (p + h)[\mathbb{E}(\eta_1 - a)^- - \alpha a] + p\, \mathbb{E}\, \eta_1.$$

It is well-known that $\mathbb{E}(\eta_1 - a)^- = \int_{-\infty}^a F\, d\lambda$. Thus we get

$$G' + \pm(a) = (p + h)\, [\int_a^{a+1} F\, d\lambda - \alpha] = (p + h)\, [F(a) - \alpha], \;\; a \in \mathbb{Z}.$$

If $F(B-1) \leq \alpha$ then $G'_- \leq 0$ on $(-\infty, B)$, hence G is decreasing on $(-\infty, B]$. Thus $G|(-\infty, B]$ has the minimum point $s^* = B$.

Now assume $0 \leq \alpha < F(B - 1)$. We know that $b \in \mathbb{Z}$ is a minimum point of $G: \mathbb{Z} \to \mathbb{R}$ if and only if $G'_-(b) \leq 0 \leq G'_+(b)$ i.e. if and only if $F(b-1) \leq \alpha \leq F(b)$, i.e. if and only if b is an α-quantile of F. Each such b must belong to $(-\infty, B - 1]$ as $F(b - 1) \leq \alpha < F(B-)$. Thus $b < B$ and the proof is complete. □

Remark 15.8 The critical inventory level s^* is always positive as $\eta_1 \geq 0$ and $\alpha > 0$ implies $q_\alpha > 0$. ◇

Example 15.9 (An inventory model with backlogging and delivery lag) Consider Example 15.4, but without the assumption that g and the terminal cost function are convex. Thus we have the state space $S = \{B, B-1, \ldots\}$, proportional ordering cost $c_1 \geq 0$ per unit, a holding and shortage cost $g(t) \geq 0$ when t is the stock on hand at the end of the period and a lower bounded terminal cost, denoted by $C_0'(t)$, for a terminal inventory of amount t. In practice often orders are not delivered instantaneously but with a certain time lag, possibly of random length. We restrict ourselves to the simpler case of a deterministic lag of $\lambda \in \mathbb{N}$ periods. (This assumption is appropriate, for example, if the good has a fixed production and transportation time.) Thus orders made at times $t = 0, 1, 2, \ldots, N-1$ are received at times $\lambda, \lambda+1, \ldots, \lambda+N-1$, respectively. Orders at times $t \geq \lambda$ are made after the arrival of the order made at time $t - \lambda$. It is meaningful to assume that the inventory is in operation not only for N but for $N + \lambda$ periods and that also random demands $\eta_{N+1}, \eta_{N+2}, \ldots, \eta_{N+\lambda}$ arise in periods $N+1, N+2, \ldots, N+\lambda$. We assume that $\eta_1, \eta_2, \ldots, \eta_{N+\lambda}$ are i.i.d. In the last λ periods no ordering decisions are made but the arriving orders are used for meeting backlogged demand and/or demand arising in these periods. We assume that orders are paid upon delivery. Thus ordering costs arise at times $t = \lambda, \lambda+1, \ldots, \lambda+N-1$ while holding and shortage costs arise at all times $t = 0, 1, \ldots, N+\lambda-1$ and the terminal cost arises at time $N+\lambda$. We want to minimize the expected sum of discounted costs in all $\mathbb{N} + \lambda$ periods.

One must carefully choose the state s_t. Of course, s_0 is the initial stock at hand. For $1 \leq t \leq N-1$ the state s_t must contain all information on which the order decision at time t is based. It is intuitively clear that one should take as s_t not the stock on hand, but the stock on hand plus the orders not yet delivered at time t. With this choice we call s_t the *inventory position before ordering*. As in other inventory models, we choose as action a_t not the order but – equivalently – the *inventory position after ordering*. More precisely, select some policy $\pi = (\pi_t)_0^{N-1}$ and put $a_t := \pi_t(s_t)$, where s_t is the random state at time t (depending on s_0 and π), defined recursively by the state equation

$$s_{t+1} = a_t - \eta_{t+1}, \quad 0 \leq t \leq N-1. \tag{15.4}$$

Moreover, $b_t := a_t - s_t$ is the order at time t, $0 \leq t \leq N-1$. We also use $b_t := 0$ for $t < 0$ and for $t = N$. Put $\eta_{ij} := \sum_{t=i}^{j} \eta_t$ for $1 \leq i \leq j \leq N+\lambda$ and $\eta_{ij} := 0$ for $i > j$. Then

$$y_t := s_0 + \sum_{i=0}^{t-\lambda} b_i - \eta_{1t} \quad 0 \leq t \leq N+\lambda, \tag{15.5}$$

is the stock on hand (including in case $t \geq \lambda$ the order $b_{t-\lambda}$) at time t. Obviously we have

$$s_t = y_t + \sum_{i=t-\lambda+1}^{t-1} b_i = s_0 + \sum_{i=0}^{t-1} b_i - \eta_{1t}, \; 1 \leq t \leq N, \tag{15.6}$$

which implies

$$a_k = s_k + b_k = s_0 + \sum_{i=0}^{k} b_i - \eta_{1k},\ 1 \le k \le N-1.$$

Now we see from (15.5) that

$$y_t = s_0 - \eta_{1t} \text{ for } 0 \le t \le \lambda - 1,$$

$$y_{k+\lambda} = s_0 + \sum_{i=0}^{k} b_i - \eta_{1,k+\lambda} \text{ for } 0 \le k \le N.$$

Therefore and as $a_0 = s_0 + b_0$, we get

$$y_{k+\lambda} = a_k - \eta_{k+1,k+\lambda},\ 0 \le k \le N-1.$$

Moreover, the case $t := N$ in both (15.6) and (15.5) shows, using $b_N = 0$, that $y_{N+\lambda}$, the terminal stock on hand, equals $s_N - \eta_{N+1,N+\lambda}$.

Now we can compute the discounted sum of costs as

$$\begin{aligned}
&H_{N\pi}(s_0, \eta^{N+\lambda}) \\
&:= \sum_{t=0}^{N-1} \beta^{t+\lambda} \cdot c_1 \cdot (a_t - s_t) + \sum_{t=0}^{N+\lambda-1} \beta^t \cdot g(y_t - \eta_{t+1}) + \beta^{N+\lambda} \cdot C_0'(y_{N+\lambda}) \\
&= \sum_{t=0}^{N-1} \beta^{t+\lambda} \cdot c_1 \cdot (a_t - s_t) + \sum_{t=0}^{\lambda-1} \beta^t \cdot g(s_0 - \eta_{1,t+1}) \\
&+ \sum_{k=0}^{N-1} \beta^{k+\lambda} \cdot g(a_k - \eta_{k+1,\lambda+k+1}) + \beta^{N+\lambda} \cdot C_0'(s_N - \eta_{N+1,N+\lambda}).
\end{aligned}$$

It follows that

$$H'_{N\pi}(s_0, \eta^{N+\lambda}) := \left(K_{N\pi}(s_0, \eta^{N+\lambda}) - \sum_{t=0}^{\lambda-1} \beta^t \cdot g(s_0 - \eta_{1,t+1}) \right) \Big/ \beta^\lambda \tag{15.7}$$

$$= \sum_{k=0}^{N-1} \beta^k \left[c_1 \cdot (a_k - s_k) + g(a_k - \eta_{k+1,\lambda+k+1}) \right] + \beta^N \cdot C_0'(s_N - \eta_{N+1,N+\lambda}). \blacklozenge$$

This looks similar to the N-stage random cost in a CM, but $g(a_k - \eta_{k+1})$ and $C_0(s_N)$ are replaced by $g(a_k - \eta_{k+1,\lambda+k+1})$ and $C_0'(s_N - \eta_{N+1,N+\lambda})$, respectively. This difficulty can be overcome and yields the next result. Note also that $\mathbb{E}\, H_{N\pi}(s)$

exists for all N, π and s (possibly with value ∞) as $c_1 \geq 0$, $g \geq 0$ and as C_0' is lower bounded.

Proposition 15.10 (Reduction of the Example 15.9 with delivery lag to a CM) *Assume that* $\mathbb{E}\, g(t - \sum_1^t \eta_i) < \infty$ *for* $1 \leq t \leq \lambda + 1$ *and* $\mathbb{E}\, C_0'(t - \sum_1^\lambda \eta_i) < \infty$ *for all* $t \leq B$. *Then Example 15.9 with delivery lag* λ *has the same optimal policies as the CM with one-stage cost*

$$c_Z(s, a, z) = c(s, a) := c_1 \cdot (a - s) + \mathbb{E}\, g(a - \sum_1^{\lambda+1} \eta_i)$$

and terminal cost $C_0(s) := \mathbb{E}\, C_0'(t - \sum_1^\lambda \eta_i)$.

Moreover, the minimal cost $K_N(s)$ *of the original problem and the minimal cost* $C_N(s)$ *of the CM are related by*

$$K_N(s) = \beta^\lambda \cdot C_N(s) + \sum_{t=0}^{\lambda-1} \beta^t \cdot \mathbb{E}\, g(s - \sum_1^{t+1} \eta_i),\ s \in S. \tag{15.8}$$

Proof Firstly we obtain by induction on t from (15.4) with $a_t = \pi_t(s_t)$ that s_k and hence a_k, $1 \leq k \leq N-1$, in (15.7) depend on $\eta^{N+\lambda} := (\eta_i)_1^{N+\lambda}$ only via η^k. Therefore $a_k = a_k(\eta^k)$ and $\eta_{k+1,\lambda+k+1}$ are independent. Next, denote for fixed $1 \leq k \leq N-1$ by q_k the discrete density of η^k and by p the discrete density of $\eta_{k+1,\lambda+k+1}$, which is also the discrete density of $\eta_{1,\lambda+1}$. Now we obtain for $G(a) := \mathbb{E}\, g(a - \eta_{1,\lambda+1})$,

$$\begin{aligned}\mathbb{E}\, g(a_k(\eta^k) - \eta_{k+1,k+1+\lambda}) &= \sum_{z^k \in \mathbb{N}_0^k} \sum_{u \in \mathbb{N}_0} q_k(z^k) \cdot p(u) \cdot g(a_k(z^k) - u)\\ &= \mathbb{E}\, G(a_k(\eta^k)).\end{aligned}$$

In the same way one shows that

$$\mathbb{E}\, C_0'(s_N(\eta^N) - \eta_{N+1,N+\lambda}) = \mathbb{E}\, C_0(s_N(\eta^N)).$$

Now the assertion follows, observing that $\mathbb{E}\, g(a_0 - \eta_{1,\lambda+1}) = \mathbb{E}\, G(a_0)$ and putting $a_0(\eta^0) := a_0$, $s_0(Z^0) := s_0$, from

$$\mathbb{E}\, H_{N\pi}'(s_0) = \mathbb{E}\left(\sum_{k=0}^{N-1} \beta^k \Big[c_1 \cdot (a_k(\eta^k) - s_k(\eta^k)) + G(a_k(\eta^k))\Big] + \beta^N \cdot C_0(s_N(\eta^N))\right).$$

The right-hand side coincides, observing that $G(a_k) < \infty$ by assumption, with the expected costs $C_{N\pi}(s_0)$ of the CM. Finally, (15.8) follows from (15.7), using finiteness of $\mathbb{E}\, g(t - \eta_{1,t})$, $t \leq B$, $1 \leq t \leq \lambda$. □

In the case of proportional holding and shortage costs, i.e. if $g(t) = h \cdot t^+ + p \cdot t^-$, the assumption $\mathbb{E}\, g(t - \sum_1^t \eta_i) < \infty$ for $1 \le t \le \lambda + 1$ holds if $\mathbb{E}\,\eta_1 < \infty$.

As a consequence of Proposition 15.10, the previous Proposition 15.5 for the inventory problem without delivery lag remains true if $\mathbb{E}\, g(t - \eta_1)$ and $C_0'(t)$ are replaced by $\mathbb{E}\, g(t - \sum_1^{\lambda+1} \eta_i)$ and $\mathbb{E}\, C_0'(t - \sum_1^{\lambda} \eta_i)$, provided the latter two expectations are finite.

Example 15.11 (A time slot queueing model with arrival control) (See Sennott 1999, p. 16 and 51.) We consider a queueing system consisting of a single service station and an infinite buffer (also called waiting room); together they form the queueing system. At time $0 \le t \le N-1$ there arrives at the system a batch of random size $Y_{t+1} \ge 0$ of packets (or customers), which are available for service at time $t+1$. The period $[t, t+1)$, $t \in \mathbb{N}_0$, is called *slot t*. A controller has to decide at time $0 \le t \le N-1$ whether to admit the arriving batch ($a = 1$) or to reject it ($a = 0$). In slot t, if at least one packet is waiting for service, one of the packets is served, and its service is completed just before time $t+1$ with probability $\mu \in (0, 1)$. (This corresponds to a total service time per packet which has a geometric distribution.) Put X_{t+1} equal to zero if no service is completed at time $t+1$, and equal to one, otherwise. We assume that the sequence $(\eta_t)_1^N := ((X_t, Y_t))_1^N$ is i.i.d. and that also X_t and Y_t are independent of each other for all t. Thus

$$\mathbb{P}(X_t = 1, Y_t = y) = \mu \cdot q(y),\ \mathbb{P}(X_t = 0, Y_t = y) = (1 - \mu) \cdot q(y),\ y \in \mathbb{N}_0,$$

where q is the discrete density of Y_1. There is a holding cost $h(s) \in \mathbb{R}_+$ per slot when $s \in \mathbb{N}_0$ packets are currently in the system. Each rejection of a batch is penalized by $R \in \mathbb{R}^+$. If s_N packets are still in the system at time N, they obtain a special service at cost $C_0(s_N) \in \mathbb{R}_+$. It is natural to assume that both h and C_0 are increasing and that without loss of generality $h(0) = C_0(0) = 0$. We want to minimize the N-stage expected discounted cost.

We use a CM with the following data: $S = \mathbb{N}_0$; $A = D(s) = \{0, 1\}$ for all s; $Z = \{0, 1\} \times \mathbb{N}_0$; the state s_{t+1} equals s_t minus X_{t+1} (if $s_t > 0$) plus Y_{t+1}, provided the batch arriving at time t is admitted. Thus we have

$$s_{t+1} = T(s_t, a_t, (X_{t+1}, Y_{t+1})) = (s_t - X_{t+1})^+ + Y_{t+1} \cdot \delta_{a_t,1}, \quad t \ge 0.$$

Since S is countable and since C_0 and $c(s, a) = h(s) + R \cdot \delta_{a0}$ are non-negative, (MA2) holds. Moreover, finiteness of A ensures by Proposition 14.1.5 the validity of the VI and the existence of minimizers at each stage. Thus for $n \ge 1$ and $s \in \mathbb{N}_0$ we have

$$C_n(s) - h(s) = \min\left\{R + \beta \cdot \mathbb{E}\, C_{n-1}((s - X_1)^+), \beta \cdot \mathbb{E}\, C_{n-1}((s - X_1)^+ + Y_1)\right\}. \tag{15.9}$$

If h and C_0 are increasing, induction on $n \ge 0$ shows easily that C_n is increasing and finite for all $n \ge 0$. For finiteness one uses that from the isotonicity of

C_{n-1} we obtain

$$0 \leq C_n(s) \leq W_n(s, 0) = h(s) + R + \beta \cdot C_{n-1}(s), \; n \geq 1, s \in \mathbb{N}_0,$$

which yields the finite upper bound $C_n(s) \leq (h(s) + R) \cdot \sigma_n(\beta) + \beta^n \cdot C_0(s)$.

For numerical computations one will write the VI in the form

$$C_n(s) - h(s) = \min \Big\{R + \beta \cdot \big[(1 - \mu) \cdot C_{n-1}(s) + \mu \cdot C_{n-1}((s-1)^+)\big],$$
$$\beta \cdot \sum_{y=0}^{\infty} q(y) \cdot \big[(1 - \mu) \cdot C_{n-1}(s + y) + \mu \cdot C_{n-1}((s-1)^+ + y)\big]\Big\}.$$

♦

Example 15.12 (A time slot queueing model with service control) (A similar model was considered by Sennott 1999, p. 207.) We consider a queueing system consisting of a single service station and an infinite buffer. At time $0 \leq t \leq N-1$ there arrives at the system a batch of random size $Y_{t+1} \geq 0$ of customers which are available for service at time $t+1$. The random variables $Y_1, Y_2, \ldots, Y_N$ are i.i.d. with discrete density $y \mapsto q(y)$.

At each time t where no customer is under service and at least one customer is waiting in the buffer, one of these is admitted to service. The probability distribution of its service time is selected by a controller from a finite set, determined by discrete densities u_a on $\mathbb{N}$, $a \in A$. Thus $u_a(j)$ is the probability that the service will last j time slots. When action a is taken it is not yet known which realization j of the random service time arises. The selection of u_a costs $d(a) \in \mathbb{R}_+$. The state of the system equals $(i, k) \in S = \mathbb{N} \times \mathbb{N}_0 + \{(0, 0)\}$, where i is the number of customers in the system. Moreover, $k \geq 1$ means that there is a customer under service which still requires k slots until completion, while $k = 0$ means that no customer is under service. Thus genuine actions are made if and only if $(i \geq 1) \wedge (k = 0)$. For modeling the problem it is useful to admit formally also all actions in the other states. There is a holding cost $h(i) \in \mathbb{R}_+$ per slot when i customers are in the system, and a service cost $e \in \mathbb{R}_+$ per slot when a customer is momentarily served, i.e. if $i \geq 1$. If (i_N, k_N) is the final state, the customers still in the system obtain a special service at cost $C_0((i_N, k_N) \in \mathbb{R}_+$. It is natural to assume that h and C_0 are increasing and that $h(0) = C_0(0, 0) = 0$. We want to minimize the N-stage expected discounted cost.

We use an MDP with the following data: S is as above; the state $(0, 0)$ means that no customer is in the system; $D(s) = A$ for all s; for $i, k \in \mathbb{N}_0$ we have $c(i, k, a) = (d(a) \cdot \delta_{k0} + h(i) + e) \cdot 1_{\mathbb{N}}(i)$; for $i, k, j \geq 1$ and $y \geq 0$

$$p((0, 0), a, (y, 0)) = q(y);$$
$$p((i, 0), a, (i + y - \delta_{j1}, j - 1)) = q(y) \cdot u_a(j);$$
$$p((i, k), a, (i + y - \delta_{k1}, k - 1)) = q(y).$$

Since S is countable and since C_0 and c are non-negative, (MA2) holds. Moreover, finiteness of A implies by Proposition 14.1.5 the validity of the VI and the existence of minimizers at each stage. Moreover, for $n \geq 1$ the VI has the form

$$C_n(0,0) = \beta \cdot \sum_{y=0}^{\infty} q(y) \cdot C_{n-1}(y,0),$$

$$C_n(i,0) = h(i) + e + \min_{a \in A} \left[d(a) + \beta \cdot \sum_{y=0}^{\infty} q(y) \sum_{j=1}^{\infty} u_a(j) \cdot C_{n-1}(i + y - \delta_{j1}, j - 1) \right],$$

$$C_n(i,k) = h(i) + e + \beta \cdot \sum_{y=0}^{\infty} q(y) \cdot C_{n-1}(i + y - \delta_{k1}, k - 1), \; i, k \geq 1.$$

By a somewhat tedious induction on $n \geq 0$ one shows that C_n is increasing if h and C_0 are increasing. ♦

Chapter 16
Models with Arbitrary Transition Law

In this chapter we consider MDPs with general measurable state and action spaces. We extend the results of Chaps. 12 and 14 to the present model. Under minimal assumptions we state the reward iteration and the structure theorem. Binary MDPs and continuous versions of examples illustrate the results. A useful generalization of MDPs are MDPs with random environment. The random environment (such as economic factors) evolves within a set of states as an uncontrolled Markov chain.

16.1 The Models MDP and CM

In MDPs with arbitrary transition law (specified by a transition probability P, cf. below) we encounter the following new problems:

(i) In order that the N-stage total reward $\omega \mapsto R_{N\pi}(s_0, \zeta^N(\omega))$ for policy π and initial state s_0 is a random variable, we must ensure that it depends measurably on ω. This is easily achieved by assuming that all data and all decision rules are measurable.

(ii) If $R_{N\pi}(s_0, \zeta^N)$ is measurable, we must ensure that its expectation $V_{N\pi}(s_0)$ exists. This problem will be approached exactly as in Chap. 14 via assumptions like (EN) and (EP), or via bounding functions.

(iii) If we have ensured the existence of $V_{N\pi}(s_0)$ for all N, π and s_0 and hence of the value functions V_N, the value iteration (cf. (16.4) below) makes sense only if the value functions are measurable (which may not hold if $\mathbb{F}$ is uncountable) and if they are measurable, that they are also integrable with respect to the transition law P. While the latter problem can be treated as in Chap. 14, the former requires new techniques called *selection problems*. We deal with these subtle measurability problems only to such an extent that we can treat most applications in a rigorous way.

K. Hinderer et al., *Dynamic Optimization*, Universitext,
DOI 10.1007/978-3-319-48814-1_16

Firstly we turn to the measurability problems. We freely use elementary facts from measure-theoretical probability; these are collected in Appendix B. In treating applied models in a unified and transparent way the notions of measure, transition probability, abstract integral etc. can be of genuine help. On the other hand, the notion of measurability with respect to a σ-algebra $\mathfrak{M}$ on a set M, though being unavoidable for a rigorous treatment, will usually play only a minor role due to the following fact. Choosing a σ-algebra for the description of a random experiment means that one selects the sets which are of interest. In applications there are nearly always natural σ-algebras, which on the one hand are large enough to contain all sets of interest and which on the other hand are small enough to admit the construction of suitable probability measures on them. Nearly always we shall use such natural σ-algebras. If M is a Borel subset of a Euclidean space, we always take for $\mathfrak{M}$ the *natural* σ-algebra $\mathfrak{B}_M$ of Borel subsets of M. If M is countable we always take for $\mathfrak{M}$ (usually without mention) the *natural* σ-algebra $\mathfrak{P}(M)$. Notice that $\mathfrak{B}_M = \mathfrak{P}(M)$ if M is a countable subset of an Euclidean space. If $(M_1, \mathfrak{M}_1)$ and $(M_2, \mathfrak{M}_2)$ are measurable spaces then measurability of sets in $M_1 \times M_2$ or functions on $M_1 \times M_2$ are understood with respect to the product σ-algebra $\mathfrak{M}_1 \otimes \mathfrak{M}_2$, unless stated otherwise.

In the sequel, if a mapping g from a measurable space $(M_1, \mathfrak{M}_1)$ into a measurable space $(M_2, \mathfrak{M}_2)$ is $(\mathfrak{M}_1, \mathfrak{M}_2)$-measurable, we say that g is $\mathfrak{M}_1$-measurable or measurable when it is clear which σ-algebra is used in the domain or in both the domain and range of g, respectively. Thus, for example, a $\mathfrak{M}_1$-measurable function is a $(\mathfrak{M}_1, \mathfrak{B})$-measurable function from M_1 into $\mathbb{R}$. If $(M_1, \mathfrak{M}_1)$ and $(M_2, \mathfrak{M}_2)$ are measurable spaces, then measurability of subsets of $M_1 \times M_2$ and of functions on $M_1 \times M_2$ are understood with respect to the product σ-algebra $\mathfrak{M}_1 \otimes \mathfrak{M}_2$.

Sometimes we shall meet measures μ on $\mathfrak{M}$, which are defined (e.g. as images of other measures) in a complicated way. Then it is often simpler to specify the form of the integrals $\int v \, d\mu$ for arbitrary measurable $v \geq 0$ rather than the form of the probabilities $\mu(C)$ for $C \in \mathfrak{M}$. Notice that $\mu(C)$ is obtained from $\int v \, d\mu$ by $v := 1_C$.

Now we turn to the extension of the MDP of Chap. 14 to the case of an arbitrary transition law. Firstly, we choose suitable σ-algebras $\mathfrak{S}$ and $\mathfrak{A}$ on S and A, respectively. We assume that D belongs to $\mathfrak{S} \otimes \mathfrak{A}$. D is endowed with the trace $\mathfrak{D}$ of the product σ-algebra $\mathfrak{S} \otimes \mathfrak{A}$ on D. The state at time t, $1 \leq t \leq N$, of an N-stage decision process with initial state s_0 under policy $\pi = (\pi_t)_0^{N-1}$ is an S-valued random variable $\omega \mapsto \zeta_t(\omega)$ on a probability space $(\Omega, \mathfrak{F}, \mathbb{P}_\pi(s_0, d\omega))$, to be defined below in a canonical way. Then the N-stage total reward is defined as before by

$$R_{N\pi}(s_0, \zeta^N) := \sum_{t=0}^{N-1} \beta^t r_S(\zeta_t, \pi_t(\zeta_t), \zeta_{t+1}) + \beta^N \cdot V_0(\zeta_N),$$

$$\zeta_0 :\equiv s_0, \ \zeta^N := (\zeta_\nu)_1^N. \tag{16.1}$$

$R_{N\pi}(s_0, \zeta^N)$ is $\mathfrak{F}$-measurable if r_S and V_0 are measurable and if only measurable decision rules π_t are allowed.

This follows by the following observations. A vector of measurable mappings is measurable with respect to the product σ-algebra in the domain of the mappings and the composition of measurable mappings is again measurable. Measurability of $\beta^t \cdot r_S(\zeta_t, \pi_t(\zeta_t), \zeta_{t+1})$ and of $\beta^N \cdot V_0(\zeta_N)$ follows as multiplication with the constant β^t preserves measurability. Now the assertion follows as the sum of measurable functions is measurable.

The transition law p from Chap. 14, which was a stochastic $D \times S$-matrix, must now be replaced by a family $P(s, a, ds')$ of probability measures on $\mathfrak{S}$, indexed by $(s, a) \in D$. Of course, $P(s, a, B)$ must be chosen as the probability that the system moves during one period from the state s into the set B under the influence of action a, independent of the preceding states and actions. (In applications S is often a Borel set in a Euclidean space and then $P(s, a, ds')$ is usually defined by a Lebesgue density $x \mapsto p(s, a, x)$ on S.) In order that the joint distribution of $\zeta_1, \ldots, \zeta_N$ can be described by P (cf. (16.2) below) we assume that P is a transition probability according to

Definition 16.1.1 A function $P: D \times \mathfrak{S} \to \mathbb{R}_+$ is called a **transition probability** from D into S if

- $D \ni (s, a) \mapsto P(s, a, B)$ is $\mathfrak{D}$–measurable for all $B \in \mathfrak{S}$.
- $B \mapsto P(s, a, B)$ is a probability measure on $\mathfrak{S}$ for all $(s, a) \in D$.

Altogether, we are led to the following definition.

Definition 16.1.2 A **Markovian Decision Process with arbitrary transition law** (**MDP** for short) is a tuple $(S, A, D, P, r_S, V_0, \beta)$ of the following kind:

- $(S, \mathfrak{S})$ is a measurable space.
- $(A, \mathfrak{A})$ is a measurable space.
- D belongs to the product σ-algebra $\mathfrak{S} \otimes \mathfrak{A}$ and contains the graph of an $(\mathfrak{S}, \mathfrak{A})$-measurable mapping from S into A. D is endowed with the trace $\mathfrak{D}$ of the σ-algebra $\mathfrak{S} \otimes \mathfrak{A}$ on D.
- P is a transition probability from D into S.
- The one-stage reward function $r_S: D \to S$ is such that

$$r(s, a) := \int P(s, a, ds')\, r_S(s, a, s'),\ (s, a) \in D,$$

  exists and is finite.
- The terminal reward function $V_0: (S, \mathfrak{S}) \to \mathbb{R}$ is measurable.
- $\beta \in \mathbb{R}^+$ is the discount factor.

Remark 16.1.3

(a) The set $D(s)$ of admissible actions at state $s \in S$ belongs to $\mathfrak{A}$ as sections of sets in a product σ-algebra are measurable.

(b) In the early literature it was sometimes assumed that $\mathfrak{S}$ contains all singletons in order to guarantee that for $v \in \mathfrak{V}_0$

$$Uv(s) = \sup_{f \in \mathbb{F}} U_f v(s), \; s \in S.$$

However, this holds without the assumption stated above. For the proof fix $s \in S$. Firstly, $\sup_{f \in \mathbb{F}} U_f v(s) \leq Uv(s)$ holds as for all $f \in \mathbb{F}$

$$U_f v(s) = Lv(s,f(s)) \leq \sup_{a \in D(s)} Lv(s,a) = Uv(s).$$

Fix $a \in D(s)$ and select some $h \in \mathbb{F}$. Then the mapping

$$s' \mapsto g(s') := \begin{cases} a, & \text{if } s' \in D_a, \\ h(s'), & \text{if } s' \notin D_a, \end{cases}$$

is measurable, hence belongs to $\mathbb{F}$, and $g(s) = a$. Therefore

$$Lv(s,a) = U_g v(s) \leq \sup_{f \in \mathbb{F}} U_f v(s).$$

As this holds for all a, we have $Uv(s) \leq \sup_{f \in \mathbb{F}} U_f v(s)$. ◇

Example 16.1.4 (Measurability properties of D) In the following cases D has the measurability properties required in Definition 16.1.2 which cover many applications.

(a) $D(s) = A$ for all s and hence $D = S \times A$. (For each $a \in A$, D contains the graph of the measurable mapping $f :\equiv a$.)
(b) $A = \mathbb{R}$, and there are two measurable functions $\underline{d}, \overline{d}$ on S such that

$$D(s) = [\underline{d}(s), \overline{d}(s)] \text{ for all } s, \text{ and } \underline{d} \leq \overline{d},$$
$$D(s) = [\underline{d}(s), \infty) \text{ for all } s,$$
$$D(s) = (-\infty, \overline{d}(s)] \text{ for all } s.$$

(c) $A = \mathbb{Z}$, and there are two measurable functions $\underline{d}, \overline{d}$ from S into $\mathbb{Z}$ such that

$$D(s) = \{\underline{d}(s), \underline{d}(s)+1, \ldots, \overline{d}(s)\} \text{ for all } s, \text{ and } \underline{d} \leq \overline{d},$$
$$D(s) = \{\underline{d}(s), \underline{d}(s)+1, \ldots\} \text{ for all } s,$$
$$D(s) = \{\overline{d}(s), \overline{d}(s)-1, \ldots\} \text{ for all } s.$$

◆

Definition 16.1.5 Two important terms for the sequel are:

- A measurable mapping f from S into A such that graph $f \subset D$ (i.e. $f(s) \in D(s)$ for all s) is called a ***decision rule.***
- A sequence $\pi = (\pi_t)_0^{N-1}$ of N decision rules is called an ***N-stage policy.***

The assumption that D contains the graph of a measurable mapping from S into A guarantees that there exists at least one decision rule and therefore at least one N-stage policy for each N. As before, $\mathbb{F}$ and $\mathbb{F}^N$ denote the set of decision rules and of N-stage policies, respectively. As before we use for $f \in \mathbb{F}$ and mappings $h: D \times M \to M'$ for arbitrary sets M, M' the abbreviation

$$h_f(s,x) := h(s,f(s),x), \; s \in S, \; x \in M.$$

This is used, for example, for P_f and r_{Sf}. For $f \in \mathbb{F}$ the mapping $s \mapsto (s,f(s))$ from S into D is measurable. As the composition of measurable mappings is measurable, for $B \in \mathfrak{S}$ and $f \in \mathbb{F}$ the functions

$$(s,s') \mapsto r_S(s,f(s),s') \text{ and } s \mapsto P_f(s,B) := P(s,f(s),B)$$

are measurable. In particular, P_f is a transition probability from S into S. It describes the transition during a period under the influence of the decision rule f from the momentary to the next state.

As we want to carry over the results from Chaps. 12 and 14 to the present situation we must construct for given $N \in \mathbb{N}$, $s_0 \in S$ and $\pi \in \mathbb{F}^N$ the probability space $(\Omega, \mathfrak{F}, \mathbb{P}_{N\pi}(s_0, d\omega))$, on which the decision process $(\zeta)_1^N$ is defined, in analogy to the theory in Chap. 12. This means that the decision process must be a Markov chain $(\zeta_t)_1^N$ with general measurable state space $(S, \mathfrak{S})$. It is advantageous to define the latter concept by prescribing the form of expectations of arbitrary non-negative measurable functions v of $(\zeta_t)_1^N$ rather than prescribing directly the probability distribution of the sequence $(\zeta_t)_1^N$. The latter is obtained from the former by choosing for v an indicator function.

Definition 16.1.6 Let $(S, \mathfrak{S})$ be an arbitrary measurable space and let P_t, $0 \le t \le N-1$, be transition probabilities from S into S. A sequence of $(S, \mathfrak{S})$-valued random variables (ζ_t^N) on an arbitrary probability space is said to be a (N-stage) ***Markov chain*** with state space $(S, \mathfrak{S})$, initial state s_0 and transition probabilities P_t, $0 \le t \le N-1$, if the probability distributions $\mathbb{P}_N(s_0, d\omega)$ of the random vector $(\zeta_t)_1^N$ satisfy

$$\begin{aligned}\int \mathbb{P}_N(s_0, d\omega) \cdot v(\omega) &= \int P_0(s_0, ds_1) \int P_1(s_1, ds_2) \cdots \\ &\cdots \int P_{N-1}(s_{N-1}, ds_N) \cdot v((s_t)_1^N) \qquad (16.2)\end{aligned}$$

for all measurable $v \ge 0$ on S^N.

This definition describes again the evolution of the sequence $(\zeta_t)_1^N$ in an intuitive way: The process starts in s_0 and moves in the first step to some state s_1, selected according to the probability distribution $P_1(s_0, \cdot)$. In the second step it moves to some state s_2, selected according to the probability distribution $P_2(s_1, \cdot)$, etc. Finally it stops in some state s_N, selected according to $P_{N-1}(s_{N-1}, \cdot)$. Note that the selection of s_{t+1} is made independent of $s_0, s_1, \ldots, s_{t-1}$, and this is the core of the Markovian nature of the sequence $(\zeta_t)_1^N$.

The requirement that the decision process $(\zeta_t)_1^N$ be a Markov chain with state space $(S, \mathfrak{S})$, initial state s_0 and transition probabilities $P_t := P_{\pi_t}, 0 \le t \le N-1$ can be achieved by the following canonical construction: The sample space is $\Omega := S^N$ with the generic point $\omega = (s_t)_1^N$, and $\zeta_t((s_1, s_2, \ldots, s_N)) := s_t$, $1 \le t \le N$. Of course, as $\mathfrak{F}$ we take the product σ-algebra $\bigotimes_1^N \mathfrak{S}$. Finally, $\mathbb{P}_{N\pi}(s_0, d\omega)$ is defined by (16.2) with $P_t := P_{\pi_t}$, $0 \le t \le N-1$. The dependence of Ω, $\mathfrak{F}$, and ζ_t on N is made explicit in the notation only when needed.

It is crucial for the derivation of the reward iteration in Lemma 16.1.9 below that besides $R_{N\pi}(s_0, \zeta^N)$ also $\mathbb{P}_{N\pi}(s_0, d\omega)$ has a sequential structure. Indeed, from (16.2) one immediately obtains: If $v \ge 0$ is measurable, then

$$\int \mathbb{P}_{N(f,\sigma)}(s_0, d\omega) v(\omega) = \int P_f(s_0, ds') \int \mathbb{P}_{N-1,\sigma}(s', dy)\, v(s', y),$$

$$N \in \mathbb{N},\ (f, \sigma) \in \mathbb{F} \times \mathbb{F}^{N-1},\ s_0 \in S. \tag{16.3}$$

Definition 16.1.7 A ***Markovian Decision Process with variable discount factor*** (**MDPvar** for short), is defined as in Chap. 14, except that the discount function $(s, a) \mapsto \beta(s, a)$ must also be measurable.

Measurability of β and π_i for all i ensures measurability of

$$\beta_{t\pi}(s_0, \zeta^{t-1}) := \prod_{i=0}^{t-1} \beta(\zeta_i, \pi_i(\zeta_i)),$$

since products of measurable functions are measurable.

Recall from Chap. 14 the following convention: Whenever results about MDPs hold literally also for MDPvars this is indicated by [Var] at the beginning of the result for MDPs.

Now we return to the main stream of development. We use expectations of extended real random variables Y (i.e. functions) on a probability space $(M, \mathfrak{M}, \mu)$, defined as the *abstract* integral $\int Y\, d\mu$, for which we also write $\int Y(x)\, \mu(dx)$ or $\int \mu(dx)\, Y(x)$. We remind the reader of a few definitions. Firstly, $\mathbb{E}\, Y$ is defined for all non-negative random variables Y. For arbitrary Y the expectations of the non-negative random variables Y^+, Y^- and $|Y|$ are defined. We say that $\mathbb{E}\, Y$ *exists*, or that Y is *quasi-integrable* with respect to μ, if at least one of the numbers $\mathbb{E}\, Y^+$, $\mathbb{E}\, Y^-$ is finite, and then $\mathbb{E}\, Y$ is defined as $\mathbb{E}\, Y := \mathbb{E}\, Y^+ - \mathbb{E}\, Y^-$. We call Y *integrable*, if $\mathbb{E}\,Y$ exists and if this number is finite. $\mathbb{E}\,Y$ is integrable if and only if $\mathbb{E}\,|Y| < \infty$ if and only if both numbers $\mathbb{E}\,Y^+$, $\mathbb{E}\,Y^-$ are finite.

From now on the initial state will usually be denoted by s. The **expected N-stage reward** for initial state s and policy $\pi \in \mathbb{F}^N$ is defined as

$$V_{N\pi}(s_0) := \mathbb{E}_{\pi s}\, R_{N\pi}(s_0, \zeta^N) := \int R_{N\pi}(s_0, \omega)\mathbb{P}_{N\pi}(s_0, d\omega),$$

provided the integral exists. The minimal requirement that our optimization problem makes sense is the following

First minimal assumption (MA1) The expectations $V_{n\pi}(s)$ exist (i.e. $R_{n\pi}(s, \zeta^n)$ is $\mathbb{P}_{n\pi}(s, d\omega)$-quasi-integrable) for all $n \in \mathbb{N}$, $\pi \in \mathbb{F}^n$ and $s \in S$.

(MA1) is satisfied in the following two cases:

- **(EN)**: if both r_S and V_0 are bounded above, as then $R_{N\pi}(s_0, \zeta^N)$ is bounded above,
- **(EP)**: if both r_S and V_0 are bounded below, as then $R_{N\pi}(s_0, \zeta^N)$ is bounded below.

Another condition is given in Lemma 16.1.18 below. Case (EN) is important for cost minimization problems; it holds in particular if $c_S \geq 0$ and $C_0 \geq 0$.

Under (MA1) the ***N*-stage value function** is defined as

$$s \mapsto V_N(s) := \sup\{V_{N\pi}(s) : \pi \in \mathbb{F}^N\}.$$

It may assume both of the values $+\infty$ and $-\infty$. If (MA1) holds, then $V_{N\pi}(\cdot)$ is measurable. However, $V_N(\cdot)$ need not be measurable since the supremum of an uncountable family of measurable functions need not be measurable.

One expects that under appropriate assumptions the **value iteration** (**VI** for short) holds in the form

$$\begin{aligned} V_n(s) &= \sup_{a \in D(s)} \left[r(s,a) + \beta \cdot \int P(s,a,ds')\, V_{n-1}(s') \right], \\ &=: \sup_{a \in D(s)} W_n(s,a), \; n \geq 1, \; s \in S. \end{aligned} \qquad (16.4)$$

(For MDPvars the constant β must be replaced by $\beta(s,a)$.) The question whether the VI holds does not make sense unless the MDP satisfies the

Second minimal assumption (MA2): (MA1) holds and all value function V_n, $n \in \mathbb{N}_0$, belong to the set $\mathfrak{V}_0$ of those measurable functions v on S for which Pv exists.

Moreover, we say that the VI holds if (MA2) and (16.4) hold.

Note that a measurable function v belongs to $\mathfrak{V}_0$ if and only if $P_f v$ exists for all $f \in \mathbb{F}$. Obviously $\mathfrak{V}_0$ is the largest set of functions on which we can define the operator $Lv := r + \beta \cdot Pv$, or explicitly

$$Lv(s,a) := r(s,a) + \beta \cdot \int P(s,a,ds')v(s').$$

(For MDPvars one has to replace β by $\beta(s,a)$.) The operators U_f and U on $\mathfrak{V}_0$ are obtained from L as before as

$$U_f v(s) := Lv(s, f(s)), \;\; Uv(s) := \sup\{Lv(s,a) : a \in D(s)\}.$$

Notice that $U_f v$ is the expected reward in a one-stage MDP with policy $\pi := f$ and terminal reward function v. Lv and $U_f v$ are measurable, but $U_f v$ need not belong to $\mathfrak{V}_0$. Uv need not even be measurable as the following remark shows.

Remark 16.1.8 There exists a Borel set $B \subset \mathbb{R}^2$ whose projection B' onto $\mathbb{R}$ is not Borel. Put $S = A = D(s) = \mathbb{R}$, $\mathfrak{S} = \mathfrak{A} = \mathfrak{B}$, $r_S \equiv 0$, $V_0(s) = s$, and

$$P(s,a,dt) = \begin{cases} N(1,1), & \text{if } (s,a) \in B, \\ N(0,1), & \text{if } (s,a) \notin B. \end{cases}$$

Then $UV_0 = \beta \cdot 1_{B'}$ is not $\mathfrak{B}$-measurable. ◇

Recall from Remark 16.1.3(b) that $Uv(s) := \sup\{U_f v(s) : f \in \mathbb{F}\}$.

Lemma 16.1.9 (Reward iteration) *[Var] Under (MA1) the functions V_0 and $V_{n\pi}$, $n \in \mathbb{N}$, $\pi \in \mathbb{F}^n$, belong to $\mathfrak{V}_0$ and may be computed by the* ***reward iteration (RI*** *for short)*

$$V_{1f} = U_f V_0, f \in \mathbb{F},$$
$$V_{n(f,\sigma)} = U_f V_{n-1,\sigma}, \;\; n \geq 2, \;\; (f,\sigma) \in \mathbb{F} \times \mathbb{F}^{n-1}.$$

Proof The reasoning is very similar to the proof for Lemma 14.1.3; only quasi-convergence of series must be replaced by quasi-integrability of functions. Assume $n \geq 2$; the proof for $n = 1$ is similar and simpler. Fix s and $\pi = (f,\sigma) \in \mathbb{F} \times \mathbb{F}^{n-1}$. Put $s' := s_1$ and $x := (s_i)_2^n$. Firstly (16.1) implies

$$R_{n\pi}(s,(s_i)_1^n) = R_{n\pi}(s,s',x) = r_S(s,f(s),s') + \beta \cdot R_{n-1,\sigma}(s',x). \tag{16.5}$$

The recursive property (16.3) of $\mathbb{P}_{N\pi}(s,d\omega)$ allows the application of Fubini's Theorem (cf. Appendix B.2.3). Thus, as $V_{n\pi}$ and $V_{n-1,\sigma}$ exist by (MA1), we obtain from (16.5) and (16.3) with $\mu(ds') := P_f(s,ds')$, as $\mathbb{P}_{N\pi}(s,\cdot) = \mu \otimes \mathbb{P}_{N-1,\sigma}$,

$$\begin{aligned} V_{n\pi}(s) &= \int \mu(ds') \int \mathbb{P}_{n-1,\sigma}(s',dx)\, R_{n\pi}(s,s',x) \\ &= \int \mu(ds') \left[r_S(s,f(s),s') + \beta \cdot \int \mathbb{P}_{n-1,\sigma}(s',dx)\, R_{n-1,\sigma}(s',x) \right] \\ &= \int \mu(ds') \left[r_S(s,f(s),s') + \beta \cdot V_{n-1,\sigma}(s') \right] =: \int (g+h)\, d\mu. \end{aligned}$$

By assumption $r := \int g\,d\mu$ exists and is finite. Now the existence of $V_{n\pi}(s) = \int (g + h)\,d\mu$ and the additivity of the integral imply that $\int [(g + h) - g]\,d\mu = \beta \cdot P_f(V_{n-1,\sigma})(s)$ exists and equals

$$\int (g + h)\,d\mu \;-\; \int g\,d\mu = V_{n\pi}(s) - r_f(s).$$

Therefore $V_{n\pi} = U_f V_{n-1,\sigma}$. Moreover, as $P_f V_{n-1,\sigma}(s)$ exists for arbitrary f and s, $V_{n-1,\sigma}$ belongs to $\mathfrak{V}_0$. □

Lemma 14.1.4 and its proof remain literally true and yield

Lemma 16.1.10 (Properties of the operators *L*, U_f and *U*)

(a) Each of the operators L, U_f and U is isotone.
(b) If $v \in \mathfrak{V}_0$ and $\alpha \in \mathbb{R}$ then $v + \alpha \in \mathfrak{V}_0$ and

$$B(v + \alpha) = Bv + \alpha \cdot \beta \text{ for } B \in \{L, U_f, U\}.$$

(c) Assume (MA1). Then $V_1 = UV_0$, and $V_n \le UV_{n-1}$ for all n for which $V_{n-1} \in \mathfrak{V}_0$.
(d) For MDPvars (a) and (c) remain true, while (b) must be replaced by

$$L(v + \alpha) = Lv + \alpha \cdot \beta, \quad U_f(v + \alpha) = U_f v + \alpha \cdot \beta_f.$$

Under (MA1) optimality of a policy is defined, and also the notion *maximality of a policy* is defined as in Chap. 14. Due to Lemma 16.1.10 the proof of the OC_N Proposition 14.1.5 carries over literally.

Proposition 16.1.11 (Maximizers yield the VI and optimal policies) *[Var] Under (MA1) for each $N \ge 1$ the* ***optimality criterion OC_N*** *holds: If there exists a maximal N-stage policy π (in particular, if D(s) is finite for all s), then VI_N holds and π is optimal for MDP_N.*

In applying Proposition 16.1.11 we must ensure measurability of the decision rules. Below we give in Lemma 16.1.22 sufficient conditions under which this difficulty can be overcome. However, the method we usually apply is the Structure Theorem which holds literally as in Theorem 14.1.6, and which ensures (MA2) in a sequential manner. Because of its importance we repeat the formulation of Theorem 14.1.6 as follows.

Theorem 16.1.12 (Structure Theorem for MDPs with arbitrary transition law) *[Var] Assume (MA1). Let $\mathfrak{V} \subset \mathfrak{V}_0$ be a set of functions on S which contains V_0 and has the following properties:*

(S1) For each $v \in \mathfrak{V}$ there exists a maximizer f_v of Lv,
(S2) $Uv \in \mathfrak{V}$ for all $v \in \mathfrak{V}$.

Then the following holds.

(a) $V_n \in \mathfrak{V}$ *for all* $n \in \mathbb{N}$ *and (MA2) is fulfilled.*
(b) *For each* n *the decision rule* $f_n := f_{V_{n-1}}$ *is a maximizer at stage* n *and* $(f_n)_N^1$ *is an optimal policy for each* N *by Proposition* 16.1.11.
(c) *The VI holds.*

The proof is literally the same as the proof of Theorem 14.1.6.

Note that (S1) implies (S2) if $\mathfrak{V} = \mathfrak{V}_0$. This follows from Lemma 16.1.9 with $\mathfrak{V}_0 = \mathfrak{V}$ since $Uv = U_{f_v}$.

Our first application of the Structure Theorem 16.1.12 will be to the stochastic linear-quadratic system, which is a control model in the following sense.

Definition 16.1.13 A **control model CM with arbitrary i.i.d. disturbances** is a tuple $(S, A, D, Z, Q, T, r_Z, V_0, \beta)$ of the following kind:

- S, A, D and V_0 satisfy the measurability conditions of an MDP, stated in Definition 16.1.2.
- The disturbance space $Z \neq \emptyset$ is arbitrary and endowed with a σ-algebra $\mathfrak{Z}$.
- the probability distribution Q of the i.i.d. disturbances is arbitrary.
- T is a measurable mapping from $D \times Z$ into S.
- r_Z is a measurable function on $D \times Z$ such that $r(s,a) := \mathbb{E}\, r_Z(s, a, \eta_1)$ exists and is finite for all $(s,a) \in D$.

The CMs from Chap. 14, where no σ-algebras on S, A, D and Z are needed, are contained in the present model by taking as σ-algebras the corresponding powersets. Of course, $G_{N\pi}$, $V_{N\pi}$, V_N and problem CM_N are defined as in Chap. 11. By the

first minimal assumption (MA1) for a CM we understand that $V_{N\pi}$ exists for all N and $\pi \in \mathbb{F}^N$.

As in Chap. 14 we utilize results about MDPs for CMs by adjoining to the latter an MDPad, which has the one-stage reward function r and otherwise differs from the CM only in the transition law P, defined by

$$P(s, a, B) := Q(T(s, a, \eta_1) \in B), \quad (s,a) \in D, B \in \mathfrak{S}.$$

Thus the operator L of the MDPad has again the form (12.8). A proof similar to the one for the RI Lemma 16.1.9 for MDPs yields the RI for CMs with arbitrary disturbance space. Now it follows that each CM is equivalent to its MDPad, provided (MA1) holds in both models.

Example 16.1.14 (Stochastic linear-quadratic system) Consider the variant of the linear-quadratic system of Example 4.1.7 where the transition at time t is disturbed by a real random variable η_t. We assume that the disturbances are i.i.d., have mean zero and finite variance $\mathbb{E}\,\eta_1^2$. Thus we have a CM with the following data: $S = A = D(s) = Z = \mathbb{R}$; $T(s,a,z) = gs - ha + z$ for some real h and $g \neq 0$; $c(s,a) = \lambda \cdot s^2 + \gamma \cdot a^2$ and $C_0(s) = d_0 \cdot s^2$ for some $\lambda, d_0 \in \mathbb{R}_+$, $\gamma \in \mathbb{R}^+$. As $c \geq 0$ and $C_0 \geq 0$ we are in case (EN), so that (MA1) for the CM and for the adjoint MDP hold. For the solution we apply the Structure Theorem 16.1.12 with $\mathfrak{V}$ as the set of all functions $s \mapsto \delta \cdot s^2 + \varepsilon \cdot \mathbb{E}\,\eta_1^2$ for arbitrary $\delta, \varepsilon \in \mathbb{R}_+$, hence $\mathfrak{V} \subset \mathfrak{V}_0$. As $C_0 \in \mathfrak{V}$, we only have to verify conditions (S1) and (S2). For this purpose select from $\mathfrak{V}$ an arbitrary function $s \mapsto v(s) := d \cdot s^2 + \varepsilon \cdot \mathbb{E}\,\eta_1^2$. A simple computation yields

$$Lv(s,a) = \lambda s^2 + \gamma \cdot a^2 + \beta\delta \cdot (gs - ha)^2 + \beta \cdot \mathbb{E}\,\eta_1^2 \cdot (\delta + \varepsilon).$$

It follows easily that $Lv(s,\cdot)$ has the unique minimum point

$$f_v(s) = \frac{\beta gh \cdot \delta}{\gamma + \beta h^2 \cdot \delta} \cdot s, \quad s \in \mathbb{R},$$

and that

$$Uv(s) = \left(\lambda + \frac{\beta\gamma g^2 \cdot \delta}{\gamma + \beta h^2 \cdot \delta}\right) \cdot s^2 + \beta \cdot \mathbb{E}\,\eta_1^2 \cdot (\delta + \varepsilon),$$

so that Uv belongs to $\mathfrak{V}$. As the linear function f is measurable, f is the unique minimizer of Lv. Therefore the assumptions of the Structure Theorem 16.1.12 are fulfilled. Thus the VI holds and $C_n \in \mathfrak{V}$ for all n. It follows by comparing the two coefficients of the quadratic functions in v and Uv above that our problem has the following solution:

$$C_n(s) = d_n s^2 + \varepsilon_n \cdot \mathbb{E}\,\eta_1^2,$$
$$f_n(s) = \frac{h \cdot (d_n - \lambda)}{\gamma g} \cdot s, \quad n \geq 1,\ s \in \mathbb{R}.$$

Here (d_n) and (ε_n) can be computed recursively by $\varepsilon_0 := 0$ and

$$d_{n+1} = \lambda + \frac{\beta\gamma g^2 \cdot d_n}{\gamma + \beta h^2 \cdot d_n},$$
$$\varepsilon_{n+1} = \beta \cdot (d_n + \varepsilon_n),\ n \geq 0,$$

hence

$$\varepsilon_n = \sum_{t=1}^{n} \beta^t d_{n-t}.$$ ♦

For MDPs with discrete transition laws we saw in Proposition 14.1.5 and in Proposition 14.1.5(c) that there exists at each stage a maximizer if all sets $D(s)$ are finite. This fact also holds in the present situation in case A is countable, but the proof in Chap. 14 needs additional considerations. This is due to the fact that now measurability of decision rules has to be proved: for f to be of a maximizer at stage n it is not sufficient that $f(s)$ is a maximum point $f_n(s)$ of $W_n(s,\cdot)$ for each $s \in S$;

$f_n(s)$ must be selected from the set of maximum points of $W_n(s, \cdot)$ in such a way that f_n becomes measurable. This is the so-called **measurable selection problem.** In the case where $\mathfrak{S} = \mathbb{P}(S)$, in particular when S is countable, the measurable selection problem does not arise as then each mapping from S into A is measurable. In other cases problems may arise, even if A consists only of two points. Many general theorems concerning the measurable selection problem are known, most of them requiring sophisticated topological and measure-theoretical tools. We consider in Lemmas 16.1.15, 16.1.22 and Theorem 17.5 below three special cases, having elementary proofs and covering a lot of examples.

Lemma 16.1.15 (First elementary Selection Theorem) *Assume that $A \subset \mathbb{Z}$ and that $D(s)$ is finite for all s. Let w be a measurable function on D. Then w has a smallest and a largest maximizer, and $s \mapsto w^*(s) := \max_{a \in D(s)} w(s, a)$ is measurable.*

Proof There exists for each s a smallest maximum point of $f(s)$ of $w(s, \cdot)$. As measurability of f implies measurability of $s \mapsto w^*(s) = w(s, f(s))$, we only have to verify measurability of f. This holds, when $\{s \in S : f(s) = b\}$ is measurable for all b in the essential action space $A_0 = \text{pr}(D \to A)$. Fix $b \in A_0$ and put $B := \{a \in A_0 : D_a \cap D_b \neq \emptyset\}$. The sets D_a are measurable for all a. For $a \in B$ measurability of w implies measurability of the restriction of $w(\cdot, a)$ to the measurable and non-empty set $D_a \cap D_b$. Therefore for $a \in B$ the sets

$$S(a, b) := \{s \in D_a \cap D_b : w(s, a) < w(s, b)\}$$

and

$$S'(a, b) := \{s \in D_a \cap D_b : w(s, a) \leq w(s, b)\}$$

are measurable. Finally the assertion follows from

$$\{s \in S : f(s) = b\} = \left[\cap_{a<b} S(a, b)\right] \bigcap \left[\cap_{a>b} S'(a, b)\right].$$

Here the intersections are taken only over $a \in B$. □

We now obtain the following counterpart to Proposition 14.1.5(c).

Theorem 16.1.16 (The VI and existence of optimal policies if $A \subset \mathbb{Z}$) *[Var] Let $(S, \mathfrak{S})$ be arbitrary, let $A \subset \mathbb{Z}$ and let $D(s)$ be finite for all s. Assume (MA1).*

Then the VI holds, $s \mapsto f_n(s) := \min D(s)$ is the smallest maximizer at stage $n \geq 1$, and for each $N \geq 1$ the policy $(f_n)_N^1$ is optimal.

Proof We use the Structure Theorem 16.1.12 with $\mathfrak{V}$ as the set of all functions $V_{n\pi}$, $n \geq 0$, $\pi \in \mathbb{F}^n$ with $V_{0\pi} := V_0$. Then $V_0 \in \mathfrak{V}$ and $\mathfrak{V} \subset \mathfrak{V}_0$ by the RI.

(S1) This condition holds by Theorem 16.1.12(a) with $w := Lv$.

(S2) Select $v = V_{n\pi} \in \mathfrak{V}$, and let f be the smallest maximizer of Lv. Then $Uv = U_f V_{n\pi} = V_{n+1,(f,\pi)}$ belongs to $\mathfrak{V}$ by the RI. □

In Example 16.1.14 the application of the Structure Theorem 16.1.12 was possible as condition (EN) holds. For the application of Theorem 16.1.12 under weaker conditions the notion of an *bounding function b of the MDP* is important. We refer to Chap. 14 for the definition of $\|v\|_b$ and $\|w\|_b$ for functions v on S and w on D, respectively. Denote by $\boldsymbol{B_b}$ [MB_b] the Banach space of [measurable] functions on S with finite b-norm. Moreover, we use for functions v on $D \times S$ the abbreviation

$$Pv(s,a) := \int P(s,a,s') \cdot v(s,a,s'), \;\; (s,a) \in D, s' \in S.$$

Definition 16.1.17 A measurable function $b: S \to \mathbb{R}_+$ is called a **bounding function for an MDP [MDPvar]** if

$$\|P|r_S|\|_b, \; \|V_0\|_b \text{ and } \beta_b := \|\beta \cdot Pb\|_b \text{ are finite.}$$

Lemma 16.1.18 (Bounds for value functions) *[Var] If the MDP has a bounding function b, then (MA1) holds and $V_n \in B_b$, i.e.*

$$|V_n(s)| \leq d_n \cdot b(s), \quad n \geq 0, s \in S, \tag{16.6}$$

for appropriate constants $d_n \geq 0$. In particular, one may use d_n from (14.5).

Definition 16.1.19 A measurable function b from S into $\mathbb{R}_+$ is called a ***bounding function for a CM*** if

$$\| \mathbb{E}\, |r_Z(\cdot,\cdot,\eta_1)| \|_b, \; \|V_0\|_b \text{ and } \beta_b \text{ are finite.}$$

If a CM has a bounding function b then b is also a bounding function of the adjoint MDP, and then the CM is reducible to its adjoint MDPad.

Example 16.1.20 (Binary MDPs) In many applications only the two actions $a = 0$ and $a = 1$ are possible, where $a = 1$ means an action like overhaul, replacement, selling or harvesting while $a = 0$ means do nothing. Another application is the two-armed bandit where a means to pull arm a. We call an MDP **binary** if $A = \{0, 1\}$ and if there exist states s and s' in S such that $D(s) = \{0\}$ and $D(s') = \{1\}$. (This condition excludes trivial cases.) We shall use for the general binary MDP the maintenance interpretation. Here the state s in the arbitrary state space S is interpreted as the state of a technical system and $a = 1$ means an overhaul. Then the states in $S_0 := \{s \in S : D(s) = \{0\}\}$ are so good that overhaul is forbidden and the states in $S_1 := \{s \in S : D(s) = \{1\}\}$ are so bad that overhaul is enforced. Thus only in the states in $S_2 := S - S_0 - S_1 = \{s \in S : D(s) = \{0, 1\}\}$ do decisions have to be made. (It is easy to see that S_0, S_1 and S_2 are measurable.) As an example consider a technical system consisting of k components i of momentary age $s^i \in \mathbb{N}_0$, $1 \leq i \leq k$. Then $S = \mathbb{N}_0^k$ and S_0 and S_1 may have the form $S_0 = \{(s^i)_1^k \in \mathbb{N}_0^k : s^i \leq m \text{ for all } i\}$ and $S_1 = \{(s^i)_1^k \in \mathbb{N}_0^k : s^i \geq M \text{ for some } i\}$, respectively, for some $m < M$; figure for $k = 2$! Define for $s \in S_a + S_2$ and $a \in \{0, 1\}$

$$P_a(s, ds') := P(s, a, ds'), \; r_a(s) := r(s, a). \qquad \blacklozenge$$

Proposition 16.1.21 (The VI and existence of optimal policies in binary MDPs) *If the binary MDP has a bounding function then we have:*

(a) The VI holds and has the form

$$V_n = r_0 + \beta \cdot P_0 V_{n-1} \text{ on } S_0,$$
$$V_n = \max\{r_0 + \beta \cdot P_0 V_{n-1}, r_1 + \beta \cdot P_1 V_{n-1}\} \text{ on } S_2,$$
$$V_n = r_1 + \beta \cdot P_1 V_{n-1} \text{ on } S_1.$$

(b) Define on S_2 the function

$$\Delta_n = r_0 - r_1 + \beta \cdot [P_0 V_{n-1} - P_1 V_{n-1}], \; n \geq 1.$$

Then the decision rule

$$s \mapsto f_n(s) := \begin{cases} 0, & \text{if } s \in S_0 \text{ or } (s \in S_2 \text{ and } \Delta_n(s) \geq 0), \\ 1, & \text{else,} \end{cases}$$

is the smallest maximizer f_n at stage n. Thus for each N the policy $(f_n)_N^1$ is optimal.

A case where the measurable selection problem can be solved relatively easily is based on the following result, which is contained in Lemma 9.2.4(b2).

Lemma 16.1.22 (Second elementary Selection Theorem) *Assume that*

(i) $S \subset \mathbb{R}^d$ and $A = \mathbb{R}$,
(ii) D has the interval form $[\underline{d}(\cdot), \overline{d}(\cdot)]$ with continuous $\mathbb{R}$-functions $\underline{d}(\cdot)$ and $\overline{d}(\cdot)$,
(iii) w is a continuous function on D.

Then the smallest [largest] maximizer f of w (which exists by Lemma 9.2.3(a)) is lsc [usc], hence measurable.

When applying the Structure Theorem 16.1.12 to an MDP with $\mathfrak{V}$ as a set of continuous functions, one encounters the problem that continuity of $v \in \mathfrak{V}$ must be preserved under integration with respect to $P(s, a, ds')$. This problem is deferred to Chap. 17. The corresponding problem for CMs is simpler, as follows.

Theorem 16.1.23 (The VI and existence of optimal policies for a CM when the sets $D(s)$ are compact intervals) *[Var] Assume*

(i) $S \subset \mathbb{R}^d$ *and* $A = \mathbb{R}$,
(ii) *D has the interval form* $[\underline{d}(\cdot), \overline{d}(\cdot)]$ *with continuous* $\mathbb{R}$*-functions* $\underline{d}(\cdot)$ *and* $\overline{d}(\cdot)$,
(iii) $(s, a) \mapsto T(s, a, z)$ *is continuous for* Q*-almost all* $z \in Z$,
(iv) $(s, a) \mapsto r(s, a) := \mathbb{E}\, r_Z(s, a, \eta_1)$ *and* V_0 *are continuous,*
(v) *the CM has a continuous bounding function b, and*

$$b(T(s, a, z)) \le \delta \cdot b(s) \text{ for all } (s, a) \text{ and } z \text{ and some } \delta \in \mathbb{R}_+, \tag{16.7}$$

(vi) *in case of an MDPvar the discount function* $(s, a) \mapsto \beta(s, a)$ *is continuous.*

Then

(a) *The VI holds.*
(b) *The value functions exist, are finite and continuous.*
(c) *For each n there exists a smallest [largest] maximizer* f_n *at stage n. Thus for each N the policy* $(f_n)_N^1$ *is optimal.*
(d) *Moreover,* f_n *is lsc [usc], and hence continuous if unique.*

Proof We use the Structure Theorem 16.1.12 with $\mathfrak{V}$ as the set of continuous functions with finite b-norm. Obviously $\mathfrak{V}$ contains V_0 as V_0 is continuous and as b is a bounding function. Notice that $\mathfrak{V} \subset MB_b \subset \mathfrak{V}_0$.

(S1) Select $v \in \mathfrak{V}$.
(α) Firstly $(s, a) \mapsto H(s, a, z) := v(T(s, a, z))$ is continuous for Q-almost all $z \in Z$. Moreover, $\mathbb{E}\, H(s, a, \eta_1)$ is finite as

$$|H(s, a, z)| \le \|v\|_b \cdot b(T(s, a, \eta_1)) \le \delta \cdot \|v\|_b \cdot b(s).$$

Select $(s, a) \in D$ and a sequence of points $(s_n, a_n) \in D$ converging to (s, a). Then $|H(s_n, a_n, z)| \le \delta \cdot \|v\|_b \cdot \sup_k b(s_k)$ for all n and z. As the latter term is finite by continuity of b, the Bounded Convergence Theorem can be applied and shows that $(s, a) \mapsto \mathbb{E}\, H(s, a, \eta_1)$ is continuous. In the same way one shows that $(s, a) \mapsto r(s, a) = \mathbb{E}\, r_Z(s, a, \eta_1)$ is finite and continuous. Thus $w := Lv$ is continuous.
(β) Now the existence of a smallest and lsc [largest and usc] maximizer of Lv follows from Lemma 16.1.22.
(S2) We have $Uv \in B_b$ as $v \in MB_b$. Moreover, Uv is continuous. □

Remark 16.1.24

(i) Continuity of r follows from continuity of $(s, a) \mapsto r_Z(s, a, z)$ for Q-almost all $z \in Z$. This is shown in the same way as the continuity of $(s, a) \mapsto \mathbb{E}\, H(s, a, \eta_1)$ in the proof of Theorem 16.1.23.

(ii) Continuity of b in Theorem 16.1.23 can be weakened to the condition that b is *locally bounded* in the sense that each $s \in S$ has a neighborhood on which b is bounded.
(iii) Assumption (16.7) strengthens the condition $\| \mathbb{E}\, b(T(\cdot, \cdot, \eta_1))\|_b < \infty$ of the bounding function b.
(iv) Theorem 16.1.23 and its proof remain true if (S, ρ_S) is a separable metric space and if on $S \times A$ a metric is used which induces the product topology. ◇

Example 16.1.25 (Splitting problem with compact state space and continuous reward functions) Consider the problem of splitting a resource into two parts, as defined in Example 6.4.1, with the following data: $S = A = [0, B]$ for some $B \in \mathbb{R}^+$; $D(s) = [0, s]$; Z is arbitrary; $(s, a) \mapsto T(s, a, z)$ is continuous for all z; $r_Z(s, a, z) = u_1(a, z) + u_2(s - a, z)$ for bounded and measurable u_1 and u_2 such that $u_1(\cdot, z)$ and $u_2(\cdot, z)$ are continuous for Q-almost all z; V_0 is continuous. It follows, as r_Z and V_0 are bounded, that $b \equiv 1$ is a bounding function. Now we obtain from Theorem 16.1.23: The VI holds, the value functions are continuous and there exists a smallest [largest] maximizer f_n at each stage n, and f_n is lsc [usc]. ♦

Definition 16.1.26 An N-stage **non-stationary MDP$_N$** is a tuple

$$((S_t)_0^N, A, (D_t)_0^{N-1}, (P_t)_0^{N-1}, (r_{S_t})_0^{N-1}, \tilde{V}_N, \beta)$$

as in Chap. 13, except for the following changes:

- $(S_t, \mathfrak{S}_t)$ are measurable spaces.
- $(A, \mathfrak{A})$ is a measurable space.
- $D_t \in \mathfrak{S}_t \otimes \mathfrak{A}$, and D_t is endowed with the σ-algebra $D_t \cap (\mathfrak{S}_t \otimes \mathfrak{A})$.
- The stochastic matrices p_t from Chap. 13 are replaced by the transition probabilities P_t from D_t into S_{t+1}. Denote by $\mathbb{F}_t$ the set of decision rules at time t. For given policy $\pi \in \Pi_N := \times_{i=0}^{N-1} \mathbb{F}_i$ and initial state s_0 the probability distribution $\mathbb{P}_{\pi s_0}$ is constructed on $\otimes_1^N \mathfrak{S}_t$ so that the sequence $\zeta^N := (\zeta_t)_1^N$ of coordinate random variables form a non-homogeneous Markov chain with state spaces S_t and transition probabilities P_t.
- The functions r_{S_t} and the decision rules π_t are measurable, and $r_t(s, a) := \int P_t(s, a, ds') \cdot r_{S_t}(s, a, s')$ is assumed to exist and to be finite for all t and (s, a).
- The function $\tilde{V}_N$ is measurable.
- $\beta \in \mathbb{R}^+$.

As in Chap. 13 we define $R_{0\pi}(s_0, \eta^N)$ and $\tilde{V}_{0\pi}$, $\pi \in \Pi_N$, and also $\tilde{V}_0$. The model MDP$_N$ defines in an obvious way for $0 \le t < N$ a *subproblem MDP$_{N-t}$* whose data are taken from the last $N - t$ stages of MDP$_N$. Denote by $\tilde{V}_{t\pi}$, $\pi \in \times_{i=t}^{N-1} \mathbb{F}_i$, and by $\tilde{V}_t$ the total expected reward and the maximal total expected reward in *MDP$_{N-t}$*, respectively. We call the functions $\tilde{V}_t$ the value functions for MDP$_N$. Condition (MA1) means that all functions $\tilde{V}_{t\pi}$, $0 \le t \le N-1$, exist, which implies the RI and hence that $\tilde{V}_N \in \mathfrak{V}_0$. Condition (MA2) means that $P_t \tilde{V}_{t+1}$ exists for $0 \le t \le N-1$.

Denote by $\mathfrak{V}_{0,t+1}$, $0 \le t \le N-1$, the set of those measurable functions v on S_{t+1} for which $P_t v$ exists. Obviously $\mathfrak{V}_{0,t+1}$ is the largest set of functions on S_{t+1} on which the operator $L_t v := r_t + \beta \cdot P_t v$ is defined. (For MDPvars one has to replace β by $\beta(s,a)$.) The operator U_t on $\mathfrak{V}_{0,t+1}$ is obtained from L_t as

$$U_t v(s) := \sup\{L_t v(s,a) : a \in D_t(s)\},\ 0 \le t \le N-1,\ s \in S_t.$$

We say that the VI holds if (MA2) is true and if $\tilde{V}_t = U_t \tilde{V}_{t+1}$, $0 \le t \le N-1$. Under (MA1) the OC remains true, but the VI need not hold. For the proof of the Structure Theorem 16.1.28 below we need the following non-stationary version of Lemma 16.1.10(c) and Proposition 16.1.11(a), whose proofs carry over easily to the present situation.

Lemma 16.1.27 *[Var] Assume (MA1) for the non-stationary* MDP_N.

(a) We have $\tilde{V}_{N-1} = U_{N-1}\tilde{V}_N$, *and* $\tilde{V}_t \le U_t \tilde{V}_{t+1}$ *for all* $0 \le t \le N-1$ *for which* $\tilde{V}_{t+1} \in \mathfrak{V}_{0,t+1}$.
(b) The OC holds: If $\tilde{V}_{t+1} \in \mathfrak{V}_{0,t+1}$ *and if there exists a maximizer* π_t *at all times* $0 \le t \le N-1$ *(in particular, if* $D_t(s)$ *is finite for all s and t) then* $(\pi_t)_0^{N-1}$ *is optimal for the non-stationary* MDP_N.

Theorem 16.1.28 (Structure Theorem for a non-stationary MDP_N) *[Var] Let* $\mathfrak{V}_t$ *be an arbitrary set of functions in* $\mathfrak{V}_{0t}$, $1 \le t \le N$. *Assume (MA1),* $\tilde{V}_N \in \mathfrak{V}_N$, *and that the following holds:*

(S1) For each $v \in \mathfrak{V}_{t+1}$, $0 \le t \le N-1$, *there exists a maximizer* $f_t(v,\cdot)$ *of* $L_t v$,
(S2) $U_t v \in \mathfrak{V}_t$ *for all* $v \in \mathfrak{V}_{t+1}$, $0 \le t \le N-1$.

Then:

(a) $\tilde{V}_t \in \mathfrak{V}_t$ *for* $1 \le t \le N$, *and (MA2) is fulfilled.*
(b) For each $0 \le t \le N-1$ *the decision rule* $\pi_t := f_t(\tilde{V}_{t+1},\cdot)$ *is a maximizer at time t, and* $(\pi_t)_0^{N-1}$ *is an optimal policy by Lemma* 16.1.27*(b).*
(c) The VI holds.

Proof The proof runs along the lines the proof of Theorem 16.1.12 and uses Lemma 16.1.27. □

16.2 Models with Random Environment

We consider a useful extension of MDPs, so-called *MDPs with random environment* or *MDPs with uncontrollable component*. Intuitively, it is defined as an MDP—whose states are called *core states*—and whose data in each period depend on some random environment (such as economic factors). The latter evolves within some set I of *environmental states* as an uncontrolled Markov chain. Moreover, the transition law depends on the environment both at the beginning and at the end of the period.

Definition 16.2.1 An **MDP with random environment** (**MDP-RE** for short) is an MDP $(X, A, D, P_X, r_X, V_0, \beta)$ of the following kind:

- The state space X with states $x = (i, s)$ is the Cartesian product $I \times S$ of the set I of environmental states i and the set S of core states s. The sets I, S and X are endowed with σ-algebras $\mathfrak{I}$, $\mathfrak{S}$ and $\mathfrak{X} := \mathfrak{I} \otimes \mathfrak{S}$, respectively.
- $(A, \mathfrak{A})$ is a measurable space.
- D belongs to the product σ-algebra $\mathfrak{S} \otimes \mathfrak{A}$ and contains the graph of an $(\mathfrak{S}, \mathfrak{A})$-measurable mapping from S into A. D is endowed with the trace $\mathfrak{D}$ of the σ-algebra $\mathfrak{S} \otimes \mathfrak{A}$ on D.
- The transition law P_X has a decomposition of the form

$$P_X(i, s, a, d(i', s')) = Q_I(i, di') \otimes P_I(i, s, a, i', ds'), \; (i, s, a) \in D, \tag{16.8}$$

  where Q_I and P_I are given transition probability distributions from I into I and from $D \times I$ into S, respectively.
- The one-stage reward function $r_X: D \to X$ is such that

$$r(x, a, x') := \int P_X(x, a, dx')\, r_X(x, a, x'), \; (x, a) \in D,$$

  exists and is finite.
- The terminal reward function $V_0: (X, \mathfrak{X}) \to \mathbb{R}$ is measurable.
- $\beta \in \mathbb{R}^+$ is the discount factor.

  For each measurable $v \geq 0$ on X we have

$$\begin{aligned} P_X v(i, s, a) &:= \int P_X(i, s, a, d(i', s')) \cdot v(i', s') \\ &= \int Q_I(i, di') \int P_I(i, s, a, i', ds') \cdot v(i', s'), \; (i, s, a) \in D. \end{aligned}$$

This formula can be used for an explicit representation of $Lv = r + \beta \cdot P_X v$, where $r := P_X r_X$. The random states in the MDP-RE are $X_t := (I_t, \zeta_t)$, where I_t and ζ_t denote the random environmental states and the random core states, respectively. It follows from (16.8) that for each $N \geq 1$, initial state $x_0 = (i_0, s_0)$ and $\pi \in \mathbb{F}^N$ the sequence $(I_t)_1^N$ is a stationary Markov chain with initial state i_0 and transition probability distribution P_I, which is *uncontrolled*, i.e. independent of π. On the other hand, the sequence $(\zeta_t)_1^N$ with initial core state s_0 need not be Markovian, and $(I_t)_1^N$ and $(\zeta_t)_1^N$ need not be stochastically independent.

Two special cases are of interest:

(i) If $|I| = 1$, we obtain an MDP; thus MDPs and MDP-REs are equivalent.
(ii) If $Q_I(i, \cdot) = \delta_i$ for all i, we obtain an MDP whose data depends on the parameter $i \in I$.

We now introduce the corresponding generalization of a CM, namely:

Definition 16.2.2 A ***CM with random environment*** $(X, A, D, Z, Q, T, r_Z, V_0, \beta)$ (**CM-RE** for short) (or *CM with uncontrolled component*) is defined by

- X, A, D, Z, V_0 and β are as in the CM.
- Q is a transition probability distribution from I into $I \times Z$.
- The random variables I_t are assumed to be defined on the same probability space $(\Omega', \mathfrak{F}', \mathbb{P})$ as the disturbance variables η_t. The disturbances need not be i.i.d.; instead we assume that for given initial state (i_0, z_0) the sequence $((I_t, \eta_t))_1^N$ is an (uncontrolled) stationary Markov chain with the transition probability distribution Q(i,d(i',z')), the latter being independent of the present disturbance z.
- The transition function $T: D \times Z \times I \to S$ is measurable.
- The one-stage reward $r_Z: D \times Z \times I \to \mathbb{R}$ is measurable.

The above conditions imply: The probability distribution of $((\eta_t, I_t))_1^N$ is independent of z_0 and the sequence of environmental states I_t is also Markovian with initial state i_0 and transition probability distribution $P_I(i, B) := Q(i, B \times Z)$. However, the sequence of disturbances need not be Markovian. The core states depend on the disturbance and the environment, both at the beginning and at the end of the period, via the transition function T:

$$\zeta_{t+1} := T(I_t, \zeta_t, \pi_t(I_t, \zeta_t), \eta_{t+1}, I_{t+1}),\ 0 \leq t \leq N-1.$$

As T does not determine the whole new state (i', s') but only the core state s', a CM-RE cannot be defined as a special CM; it is a new model. Now we can define $V_{n\pi}$, V_n and (MA1) as for CMs. We adjoin to the CM-RE an MDPad which differs from the CM-RE by the transition law P' and the one-stage reward r'_X, as follows. The transition law P' is determined by two steps: in the first one we select (i', z') according to the probability distribution $Q(i, d(i', z'))$, and then the system moves deterministically to $s' := T(i, s, a, i', z')$; moreover

$$r'_X(i, s, a, i', s') = r'(i, s, a) := \int Q(i, d(i', z')) \cdot r_Z(i, s, a, i', z').$$

It follows that the operator L in the adjoint MDP has the form

$$Lv(i, s, a) = r'(i, s, a) + \beta \cdot \int Q(i, d(i', z')) \cdot v(i', T(i, s, a, z', i')).$$

By deriving the RI for the CM-RE, one easily sees: If $r_Z(i, s, a, z')$ does not depend on z, then the CM-RE is equivalent to MDPad; otherwise it can be reduced to it. Therefore the CM-RE can be solved in either case by means of the MDPad.

Three special cases are of interest:

(i) If $|I| = 1$, we obtain a CM.
(ii) A CM-RE with i.i.d. disturbances is defined by the property that Q has a decomposition of the form $Q(i, d(i', z')) = Q_Z(dz') \otimes Q_I(i, di')$.
(iii) If in (ii) $Q_I(i, \cdot) = \delta_i$ for all i, we obtain a CM whose data depend on the parameter $i \in I$.

16.3 Continuous Versions of Some Examples

Most examples from Chaps. 2, 3, 4, 5 and 6 have continuous versions. When the former can be solved with the aid of the Structure Theorem 14.1.6, usually the latter can be solved with minor changes with the aid of the Structure Theorem 16.1.12. In this context the verification of (S1) the second elementary selection theorem Lemma 16.1.22 or Theorem 16.1.23 are often useful. We now sketch the necessary changes for a few examples.

Example 16.3.1 (Continuous version of the production, sale and inventory problem from Example 15.1*)* Assume that the product can be sold, ordered and stored in arbitrary real non-negative quantities, that the upper bound m for the orders is positive real and that the demands η_t are $\mathbb{R}_+$-valued. The CM describing the problem differs from the CM in Example 15.1 only by the following data: $S = A = \mathbb{R}$, $M = \mathbb{R}_+$ and $D(s) = [s, s + m]$. It follows as in Example 15.1 that $s \mapsto 1 + |s|$ is a bounding function. Therefore (MA1) is true by Lemma 16.1.18. We assert:

The value functions are continuous, (MA2) is true, the VI holds and has the same form as in Example 15.1 (except that the minimization is over the interval $[s, s+m]$), and for each n there exists a smallest [largest] maximizer f_n at stage n, and f_n is lsc [usc].

For the proof it suffices to observe that the CM can be reduced to an MDP and to check the assumptions in Theorem 16.1.23. The assumptions (i)–(iii) and (v) and continuity of V_0 obviously hold. Continuity of r is equivalent to continuity of $a \mapsto \mathbb{E}\, g(\eta_1, a)$ on $(0, \infty)$, where $g(z, a) := \min\{z, a^+\}$. Now continuity of g at some point $a_0 > 0$ follows from the Bounded Convergence Theorem as $g(z, a) \leq a_0 + 1$ for all $z \in \mathbb{R}_+$ and all $a \in (0, a_0 + 1]$. ♦

Example 16.3.2 (Continuous version of the inventory Example 15.4*)* We consider the continuous case of Example 15.4 with the modification that the inventory capacity B equals ∞ and that $c_1 > 0$. By checking the proof of Proposition 15.3 one easily sees that this result remains true when $[S_n^*, B]$ is replaced by the real interval $[S_n^*, \infty)$. ♦

Example 16.3.3 (Continuous version of the inventory Example 15.4 *with proportional ordering costs)* Proposition 15.7 remains true, except that $F(B - 1)$ must be replaced by $F(B-)$. The necessary changes in the proof are as follows. Firstly, we get as in the proof of Proposition 15.7 that

$$G(a) = (p+h) \cdot \left[\int_{-\infty}^{a} F \, d\lambda - \alpha a \right] + p \cdot \mathbb{E}\, \eta_1, \; a \in \mathbb{R}.$$

Now we get for the right- and left-hand derivative of $G'_{\pm}$ of G

$$G'_{\pm}(a) = (p+h)\,[F(a\pm) - \alpha], \; a \in \mathbb{R}.$$

If $F(B-) \leq \alpha$ then $G'_{-} \leq 0$ on $(-\infty, B)$. Then G is decreasing on $(-\infty, B]$ and by continuity of G. Thus $G|(-\infty, B]$ has the minimum point $s^* = B$.

Now assume $0 \leq \alpha < F(B-)$. Then $b \in \mathbb{R}$ is a minimum point of $G\colon \mathbb{R} \to \mathbb{R}$ if and only if $G'_{-}(b) \leq 0 \leq G'_{+}(b)$, i.e. if and only if $F(b-) \leq \alpha \leq F(b)$, i.e. if and only if b is an α-quantile of F. Each such b must belong to $(-\infty, B)$ as $F(b-) \leq \alpha < F(B-)$. Thus $b < B$ and the proof is complete. ♦

Example 16.3.4 (A CM with a bounding function)

(a) Consider the following CM (which models a splitting problem): $S = A = Z = \mathbb{R}_+$; $D(s) = [0, s]$; $T(s, a, z) = G(s, a) \cdot z$ for a measurable function G on D.

We assume: (i) $\mathbb{E}\, \eta_1 < \infty$; (ii) $r_Z(s, az)$ is independent of z; (iii) $0 \leq G(s, a) \leq c_1 + c_2 s$ for some $c_1, c_2 \in \mathbb{R}_+$; $0 \leq r(s, a) \leq c_1 + c_2 s$; $0 \leq V_0(s) \leq c_1 + c_2 s$.

Then obviously $s \mapsto b(s) := 1 + s$ is a bounding function, hence all functions V_n and W_n, $n \geq 1$, are finite.

(b) The assumptions in (a) are fulfilled for the special case where V_0 is non-negative and concave and where

$$0 \leq G(s, a) = d_1 \cdot G_1(a) + d_2 \cdot G_2(s - a),$$
$$0 \leq r(s, a) = d_1 \cdot u_1(a) + d_2 \cdot u_2(s - a), \; 0 \leq a \leq s,$$

for concave functions G_1, G_2, u_1 and u_2 on $\mathbb{R}_+$. For the proof one uses that by Appendix D.2.4 concave functions on $\mathbb{R}_+$ have affine upper bounds. ♦

Example 16.3.5 (Maximizing the terminal utility of a two asset portfolio with borrowing admitted) At time $t = 0$ you possess some wealth $s_0 \in \mathbb{R}_+$. If $s_t \in \mathbb{R}_+$ is your your wealth at time t, $0 \leq t \leq N-1$, you may invest the fraction $a_t \in \mathbb{R}_+$ in a risky asset which has a random return of $\eta_{t+1} \in \mathbb{R}_+$ per unit. (Here we have an example where the continuous version is much easier to solve than the discrete one. In the latter the action a must be the amount rather than the fraction of wealth invested. Therefore one cannot hope to find maximizers of the form $f_n(s) = d_n \cdot s$ for appropriate constants d_n.) The sequence $\eta_1, \eta_2, \ldots, \eta_N$ is assumed to be i.i.d. with finite expectation. If $a_t < 1$, you loan the amount $s_t - a_t \cdot s_t$ to a bank as riskless asset with constant return $\gamma_+ > 1$, i.e. with interest rate $\gamma_+ - 1$; if $a_t > 1$, you borrow the amount $a_t \cdot s_t - s_t$ with interest rate $\gamma - 1 \geq \gamma_+ - 1$; if $a_t = 1$, you invest all your wealth without loaning and without borrowing. Your investment at times $0 \le t \le N-1$ is restricted by the requirement that your wealth at time $t+1$ must be non-negative with certainty. The final wealth s_N has a utility $u(s_N)$, where $u\colon \mathbb{R}_+ \to \mathbb{R}$ is concave and increasing. How should you invest in the risky asset in order to maximize the expected utility of the final wealth?

We are going to solve explicitly this problem for the utility function $u(s) := s^\lambda$, where $0 < \lambda < 1$. We exclude what has been named by Hakansson (1970/1971) the *easy-money-case* where $\eta_1 \ge \gamma$ almost sure, as then one can reach arbitrary large expected utility of the final wealth by investing always a sufficiently large fraction.

We use the CM with the following data: $S = A = \mathbb{R}_+$; note that $a > 1$ means borrowing and $a \le 1$ means loaning; $Z \subset \mathbb{R}_+$ is chosen as the support of the probability distribution Q of η_1; if action a is taken in state s the new state is $s' = T(s,a,z) = T(a,z) \cdot s$, where

$$T(a,z) := az + (1-a)^+ \cdot \gamma_+ - (1-a)^- \cdot \gamma;$$

$r_Z \equiv 0$, $V_0(s) = s^\lambda$; as we have only a terminal reward, β is of minor importance, and for simplicity we choose $\beta = 1$. Finally the set $D(s)$ is determined by the requirement that at each time t your wealth s_{t+1} must be non-negative with certainty, i.e. that $P(T(s,a,\eta_1) \ge 0) = 1$, or equivalently that $P(T(a,\eta_1) \ge 0) = 1$. This holds trivially if $0 \le a \le 1$. Now it is easy to see that we have to admit besides $a \in [0,1]$ exactly those $a > 1$ for which $az + (1-a) \cdot \gamma \ge 0$ for all $z \in Z$. This holds, as $a > 0$, if and only if $a \cdot \min Z + (1-a) \cdot \gamma \ge 0$. Finally we obtain, as the exclusion of the easy-money-case implies $\min Z < \gamma$, that $D(s) = [0, \overline{a}]$, where

$$\overline{a} := \gamma/(\gamma - \min Z) \ge 1.$$

Note that $D(s)$ is independent of s, and that borrowing is allowed if and only if $\min Z > 0$, i.e. if and only if the random return η_1 is almost sure positive. As $r_Z = 0$ and $V_0 \ge 0$, (MA1) holds. Now we are ready for the solution of our problem and insert the following proposition.

Proposition 16.3.6 (The solution of the portfolio problem from Example 16.3.5) *Assume that* $\mathbb{E}\,\eta_1 < \infty$ *and that* $\eta_1 < \gamma$ *with positive probability (the no-easy-money condition). Then:*

(a) The function

$$K(a) := \mathbb{E}\, T(a,\eta_1)^\lambda, \quad a \in [0,\overline{a}],$$

is finite and concave, and it has a smallest maximum point a^*.

(b) $V_N(s) = (K(a^*))^N \cdot s^\lambda$ *for all N and s, and the stationary policy* $(f)_N^1$ *with* $f :\equiv a^*$ *is optimal. Thus it is optimal to invest, when the momentary wealth is s, the amount* $a^* \cdot s$ *in the risky asset.*

(c) (c1) It is optimal to invest nothing in the risky asset (i.e. $a^* = 0$*) if and only if* $\mathbb{E}\,\eta_1 \le \gamma_+$*, and then* $V_N(s) = (\gamma_+^N \cdot s)^\lambda$.

(c2) If $\overline{a} > 1$ then $a^* = 1$ if and only if

$$\gamma_+ \cdot \mathbb{E}\,\eta_1^{\lambda-1} < \mathbb{E}\,\eta_1^{\lambda} \le \gamma \cdot \mathbb{E}\,\eta_1^{\lambda-1},$$

and then $V_N(s) = (\mathbb{E}\,\eta_1^{\lambda})^N \cdot s^{\lambda}$.

(c3) We have $a^* = \overline{a}$ if and only if $g(a) > 0$ for all $a \in (0, \overline{a})$, where

$$g(a) := \mathbb{E}\left[T(a, \eta_1)^{\lambda-1} \cdot \left(\eta_1 - \gamma_+ \cdot 1_{[0,1)}(a) - \gamma \cdot 1_{[1,\overline{a})}(a)\right)\right],\ 0 < a < \overline{a},$$

which exists and is finite.

(c4) If neither of the cases (c1)–(c3) holds, then a^* is the smallest solution of $g(a) = 0$ in $(0, \overline{a}) - \{1\}$.

Proof

(a) Firstly, $K(a)$ is finite as

$$\begin{aligned} 0 \le T(a, \eta_1)^{\lambda} &\le (\eta_1 \cdot \overline{a} + \gamma_+)^{\lambda} \le (2 \cdot \max\{\eta_1 \cdot \overline{a}, \gamma_+\})^{\lambda} \\ &\le 2^{\lambda} \cdot \left[\eta_1^{\lambda} \cdot \overline{a}^{\lambda} + \gamma_+^{\lambda}\right] =: h(\eta_1), \end{aligned} \tag{16.9}$$

and as $\mathbb{E}\,\eta_1 < \infty$ and $0 < \lambda < 1$ imply $\mathbb{E}\,\eta_1^{\lambda} < \infty$. The function $t(\cdot, z)$ is concave on $[0, \overline{a}]$ as $\gamma_+ \le \gamma$; figure! As $s \mapsto s^{\lambda}$ is increasing and concave, K is concave on $[0, \overline{a}]$. Finally, as $T(a, z)^{\lambda}$ is continuous in a on $[0, \overline{a}]$ and has by (16.9) the Q-integrable majorant h, K is continuous. The continuous function K on $[0, \overline{a}]$ has a smallest maximum point a^*.

(b) We apply the Structure Theorem 16.1.12 with the set $\mathfrak{V} \subset \mathfrak{V}_0$ of functions of the form $v(s) = c \cdot s^{\lambda}$ for some $c \in \mathbb{R}^+$. Obviously $V_0 \in \mathfrak{V}$. As $Lv(s, a) = c \cdot s^{\lambda} \cdot K(a)$, Lv has by (a) the maximizer $f :\equiv a^*$ and $Uv(s) = c \cdot s^{\lambda} \cdot K(a^*)$, hence $Uv \in \mathfrak{V}$. Thus (S1) and (S2) are true. Now the Structure Theorem tells us that the VI holds, from which we obtain $V_N(s) = (K(a^*))^N \cdot s^{\lambda}$ by induction.

(c) (c0) The concave function K is left and right differentiable on $(0, \overline{a})$. Using D'_+ as right derivative symbol and the obvious fact that the chain rule also holds for one-sided derivatives, we obtain

$$K'_+(a) = \mathbb{E}\,D'_+ T(a, \eta_1)^{\lambda} = \lambda \cdot g(a), \quad 0 < a < \overline{a}. \tag{16.10}$$

(c1) Firstly we prove that $K'_+(0) = \lambda \cdot \gamma_+^{\lambda-1} \cdot (\mathbb{E}\,\eta_1 - \gamma_+)$. For $0 < a < 0.5$ we have $T(a, z) \ge \gamma_+/2$, hence

$$G(a, z) := T(a, z)^{\lambda-1} \cdot (z - \gamma_+) \le \frac{|z| + \gamma_+}{(\gamma_+/2)^{1-\lambda}} =: h_1(z).$$

As $G(a, z) \to (z - \gamma_+) \cdot \gamma_+^{\lambda-1}$ for $a \downarrow 0$ and as G has the Q-integrable majorant h_1, $\mathbb{E}\,G(a, \eta_1) \to \mathbb{E}(\eta_1 - \gamma_+) \cdot \gamma_+^{\lambda-1}$, which proves the formula for $K'_+(0)$. Now the characterization of the case $a^* = 0$ follows, and the formula for $V_N(s)$ is obvious.

(c2) As in (c0) one shows that $K'_-(1) = \lambda \cdot \mathbb{E}\left[\eta_1^{\lambda-1} \cdot (\eta_1 - \gamma_+)\right]$. Now the assertion follows, using $K'_+(1) = \lambda \cdot \mathbb{E}\left[\eta_1^{\lambda-1} \cdot (\eta_1 - \gamma)\right]$, since $a^* = 1$ if and only if $K'_+(1) \leq 0 < K'_-(1)$.

(c3) This is obvious from (16.10).

(c4) A proof as in (c0), using that $t(\cdot, z)^\lambda$ is differentiable on $(0, 1) + (1, \overline{a})$, shows that K' exists there and equals g. Now the assertion follows. □

In general a^* can be found only numerically. A simple example which is explicitly solvable is as follows: $\lambda = 0.5$, $\gamma_+ = 1.1$, $\gamma = 1.2$ and $Q = (p \cdot \delta_{0.7} + q \cdot \delta_{1.3})$ for some $p \in (0, 1)$ with $p + q = 1$, hence $\overline{a} = 2.4$. Then we obtain e.g.

$$\begin{cases} a^* \approx 1.56, & \text{if } p = 0.1, \\ a^* = 1, & \text{if } p = 0.2, \\ a^* \approx 0.533, & \text{if } p = 0.3, \\ a^* = 0, & \text{if } p = 0.4. \end{cases}$$ ♦

Remark 16.3.7

(a) The proof shows that K is differentiable on $[0, \overline{a}]$, except for $a = 1$ in case $\overline{a} > 1$.

(b) We have $a^* < 1$ if $K'_+(1) < 0$ and (in case $\overline{a} > 1$) $a^* > 1$ if and only if $K'_+(1) > 0$. Therefore the optimal policy prescribes to loan at each stage t (the amount $s_t \cdot (1 - a^*)$) if $\mathbb{E}[\eta_1^{\lambda-1}(\eta_1 - \gamma)] < 0$, and to borrow at each stage t (the amount $s_t \cdot (a^* - 1)$) if and only if $\mathbb{E}[\eta_1^{\lambda-1}(\eta_1 - \gamma)] > 0$. ◇

Example 16.3.8 (Maximizing the terminal utility of a three asset portfolio without borrowing) (Two assets are treated by Hakansson 1970/1971). We consider a problem which differs from Example 16.3.5 only in the following respect: (i) One can choose between three assets, (ii) borrowing is not allowed.

The model has the following data. The action $a = (b_1, b_2)$ consists of the portions b_1 and b_2 of the momentary capital $s \in S := \mathbb{R}_+$ which are invested into asset $i = 1$ and $i = 2$, respectively; thus $1 - b_1 - b_2$ is that portion of s which is invested into asset $i = 3$; $D(s) = A = \{(b_1, b_2) \in [0, 1]^2 : b_1 + b_2 \leq 1\}$; the random return per unit from asset i is $X_i \geq 0$, and $\eta_1 := (X_1, X_2, X_3)$ has an arbitrary probability distribution; if asset $i = 3$ is a riskless loan with interest factor $\gamma > 1$ then $X_3 \sim \delta_\gamma$; we exclude the trivial case where $X_i \sim \delta_0$ for some i, as then one would never invest into asset i;

$$T(s, b_1, b_2, X_1, X_2, X_3) = s \cdot [b_1 X_1 + b_2 X_2 + (1 - b_1 - b_2) \cdot X_3];$$

$r_Z \equiv 0$, $V_0(s) = s^\lambda$ for some constant $\lambda \in (0, 1)$, and $\beta = 1$. ♦

It is easy to prove the following analogue of Proposition 16.3.6.

Proposition 16.3.9 (The solution of the portfolio problem from Example 16.3.8) *If* $\mathbb{E}X_i^\lambda < \infty$ *for* $i = 1, 2, 3$ *then:*

(a) *The function*

$$K(b_1, b_2) := \mathbb{E}\left[b_1X_1 + b_2X_2 + (1 - b_1 - b_2) \cdot X_3\right]^\lambda, \quad (b_1, b_2) \in A,$$

is finite and has a maximum point $a^* = (b_1^*, b_2^*)$. *Moreover,* $K(a^*) \geq \max\{\mathbb{E}X_1^\lambda, \mathbb{E}X_2^\lambda, \mathbb{E}X_3^\lambda\} > 0$.

(b) $V_N(s) = (K(a^*))^N \cdot s^\lambda$ *for all* N, s, *and the stationary policy* $(f)_N^1$ *with* $f :\equiv a^*$ *is optimal. Thus it is optimal to invest at any stage, when the momentary wealth is* s, *the amount* $b_1^* \cdot s$, $b_2^* \cdot s$ *and* $(1 - b_1^* - b_2^*) \cdot s$ *in asset* $i = 1, 2$ *and* $i = 3$, *respectively.*

Example 16.3.10 (Finite-stage stopping of a controlled Markov chain) We now generalize Example 13.4.1 of stopping an N-stage uncontrolled Markov chain in several respects: the chain may be controlled and the state space and V_0 may be arbitrary. Stopping now means that the decision process $(\zeta_t)_0^N$ is transferred by some action $\bar{a}$ (but not by other actions) almost surely to a single absorbing state $\bar{s}$. We call such a problem a ***stopped*** $\boldsymbol{MDP_N}$. For simplicity we assume that stopping is nowhere forbidden and nowhere enforced, i.e. that $\bar{a} \in D(s) \neq \{\bar{a}\}$, $s \in S$. Formally the stopped MDP_N is determined by the following properties, using $J := S - \{\bar{s}\}$:

$$P(\bar{s}, a, J) = r(\bar{s}, a) = V_0(\bar{s}) = 0, \; a \in D(\bar{s}),$$
$$P(s, \bar{a}, \{\bar{s}\}) = P(s, a, J) = 1, \; s \in J, a \in D(s) - \{\bar{a}\}.$$

Assume that the VI holds. Then $V_n(\bar{s}) = 0$ for all n and

$$V_n(s) = \max\{g(s), \bar{V}_n(s)\}, \; s \neq \bar{s}, \tag{16.11}$$

where $g := r(\cdot, \bar{a})$ and $\bar{V}_n(s) := \sup\{W_n(s, a) : a \in D(s), a \neq \bar{a}\}$. If a maximizer f_n at stage n exists, it tells us

(i) whether to stop, i.e. whether $f_n(s) = \bar{a}$,
(ii) which action $f_n(s) \neq \bar{a}$ to take if one should not stop.

While a maximizer at stage n need not exist, we can always decide whether to stop at stage n in state $s \in S$ (provided we have not stopped earlier), namely if and only if $g(s) \geq \bar{V}_n(s)$, i.e. if and only if $g(s) = V_n(s)$. (Here we use the convention to stop in case of indifference.) We call the set S_n of states $s \in S$ where $\bar{a}$ is optimal at stage $n \geq 1$ the ***stopping set*** at stage n. Obviously

$$B_n = \{s \in S : g(s) = V_n(s)\} = \{s \in S : g(s) \geq V_n(s)\}, \; n \geq 1. \tag{16.12}$$

(Note that $\bar{s}$ belongs to each stopping set.) Thus one should stop in problem MDP_N at the time

$$\tau_N := \min\{0 \le t \le N-1 : \zeta_t \in B_{N-t}\},$$

where $\min \emptyset := N$. We call τ_N, which is a random variable on $(S^N, \otimes_1^N \mathfrak{S})$, the *optimal stopping time* of the stopped MDP_N. It has the remarkable property that for $0 \le t \le N-1$ the event $[\tau_N = t]$ does not depend on $\zeta_{t+1}, \zeta_{t+2}, \ldots, \zeta_N$; in fact, $[\tau_N = t] = (\cap_{i=0}^{t-1}[\zeta_i \notin B_{N-i}]) \cap [\zeta_t \in B_{N-t}]$. In other words: if at any time $0 \le t \le N-1$ one has observed $\zeta_0, \zeta_1, \ldots, \zeta_t$ and if one has not yet stopped then it only depends on ζ_t whether or not one should stop at time t.

Obviously the computation of τ_N requires the computation of $(V_n)_1^{N-1}$. ♦

If, however, τ_N is determined for all N by the single function V_1 we call the model a ***stopped MDP***.

Proposition 16.3.11 (Stopped MDP) *Assume for a stopped MDP that*

(i) $g \ge V_0$,
(ii) the VI holds.

Then:

(a) $n \mapsto V_n$ *is increasing.*
(b) The sequence $(B_n)_1^\infty$ *of stopping sets is decreasing.*
(c) Assume in addition that B_1 *is quasi-absorbing in the sense that*

$$P(s, a, B_1) = 1 \textit{ for } s \in B_1 \textit{ and } a \in D(s), a \neq \bar{a}.$$

Then
(c1) $V_n = V_0$ *on* B_1.
(c2) All stopping sets B_n, $n \ge 1$, *equal* B_1.

Proof

(a) As $V_1(\bar{s}) = 0 = V_0(\bar{s})$ and as $V_1(s) \ge g(s) \ge V_0(s)$ for $s \neq \bar{s}$ by (16.11), it follows by induction on $n \ge 0$ that $V_{n+1} \ge V_n$.
(b) follows from (a) and from (16.12).
(c1) The assertion (I_n) that $V_n = V_0$ on B_1 holds for $n = 0$. Assume (I_n) for some $n \ge 0$. Then we obtain for $s \in B_1$ and $a \in D(s)$, $a \neq \bar{a}$, as B_1 is absorbing, that

$$\begin{aligned} W_{n+1}(s,a) &= r(s,a) + \beta \int_{B_1} P(s,a,ds')\, V_n(s') \\ &= r(s,a) + \beta \int_{B_1} P(s,a,ds')\, V_0(s) = W_1(s,a). \end{aligned}$$

Thus $\bar{V}_{n+1}(s) = \bar{V}_1(s)$, hence $V_{n+1}(s) = V_1(s) = V_0(s)$ by (16.11).
(c2) It follows from (c1) that $B_1 \subset B_n$. Now (c2) follows from (b). □

If all sets B_n equal B_1, then the optimal stopping time for MDP_N equals

$$\tau_N = \min\{0 \le t \le N-1 : \zeta_t \in B_1\}.$$

It is called the *first entrance time of the set* B_1 and also a *one-step look-ahead (OSLA) stopping rule*. The reason for the latter name is the following: in MDP_N one stops at time $0 \le t \le N-1$ if and only if one stops in MDP_{t+1} at time t, i.e. one step before the end.

Example 16.3.12 (Multiple asset selling) We introduce a model which comprises many special asset selling and assignment problems such as Examples 13.4.1 and 13.4.6. We move in a set S of internal states. If we are at time $0 \le t \le N-1$ in state s_t we receive a random offer η_t, where $(\eta_i)_0^N$ are i.i.d. random variables (the offer η_N is required only for formal reasons); if $x := (s,z) \in X := S \times Z$ belongs to a given non-empty measurable set $G \subset X$ we have the choice of either accepting the offer upon receiving the reward $g(x,1)$ and move to the new internal state $s' := t_1(x)$ or we may reject the offer, receive the reward $g(x,0) \ge 0$ and move to $s' := t_0(x)$. In case $x \notin G$ only rejection is allowed. We assume that g is measurable. The terminal reward $V_0(s,z) =: V_0(s)$ is assumed to be independent of z; this is natural, as no decision about the offer η_N must be made. For simplicity we assume that g and V_0 are bounded, so that (MA1) holds. The problem consists in maximizing the expected sum of rewards (for given initial state s) obtained in periods $0 \le t \le N-1$.

The traditional approach uses a CM with data as follows. The offer is included in the state, i.e. the states are the pairs $x = (s,z) \in X$; X is endowed with the product σ-algebra; the actions are $a = 1$ (acceptance of the offer) and $a = 0$ (rejection), hence $A = \{0,1\}$; $D(x) := \{0,1\}$ if $x \in G$, and $D(x) := \{0\}$, else; $T(x,a,z') := (t_a(x), z')$; $r(x,a) := g(x,a)$. As A is finite, Proposition 16.1.11 holds. Thus, for $x \in G$ we have

$$V_n(x) = \max_{a=1,2} \left[g(x,a) + \beta \cdot \mathbb{E}\, V_{n-1}(t_a(x), \eta_1)\right], \tag{16.13}$$

while in case $x \notin G$

$$V_n(x) = g(x,0) + \beta \cdot \mathbb{E}\, V_{n-1}(t_0(x), \eta_1).$$

We know from Remark 11.7 that $\pi \mapsto \mathbb{E}\, V_{N\pi}(s, \eta_1)$ is maximized by optimal policies for CM_N and that the maximal value is $v_N(s) := \mathbb{E}\, V_N(s, \eta_1)$. Now we obtain, using $g_a := g(\cdot, a)$, from (16.13) the recursion

$$v_n(s) = Eg_0(s,\eta_1) + \beta \cdot Ev_{n-1}(t_0(s,\eta_1)) + \mathbb{E}(\Delta_n^+(s,\eta_1) \cdot 1_G(s,\eta_1)),$$

where

$$\Delta_n := g_1 - g_0 + \beta \cdot (v_{n-1} \circ t_1 - v_{n-1} \circ t_0).$$

(Note that Δ_n is defined as boundedness of g and of V_0 imply the same for V_n and hence finiteness of v_n.) Moreover,

$$s \mapsto f_n(s,z) := \begin{cases} 1, & \text{if } (s,z) \in G \text{ and } \Delta_n(s,z) > 0, \\ 0, & \text{else} \end{cases}$$

is the smallest maximizer at stage n. ♦

Chapter 17
Existence of Optimal Policies

In this chapter the state and action spaces are assumed to be separable metric spaces. We apply the measurable selection theorem of Kuratowski/Ryll-Nardzewski in order to prove the existence of maximizers. We present different sets of assumptions under which optimal policies for MDPs exist.

Recall from Chap. 10 that for a function $b \geq 0$ on S we denote by B_b the set of functions on S with finite b-norm $\|v\|_b := \sup_s(|v(s)|/b(s))$. For functions w on D we put $\|w\|_b := \sup_{(s,a)\in D} |(w(s,a)|/b(s))$. Denote by B_b^+ the set of functions v on S whose positive part v^+ belongs to B_b, i.e. such that $v(s) \leq \gamma \cdot b(s)$ for all s and some $\gamma = \gamma_v \in \mathbb{R}_+$.

In this section we assume for the MDP considered, unless stated otherwise:

- State and action space are *separable* metric spaces (S, ρ_S) and (A, ρ_A), endowed with the σ-algebras $\mathfrak{S}$ and $\mathfrak{A}$ of Borel sets, respectively. (Note that neither S nor A need be a Borel space.)
- D is endowed with an arbitrary product metric ρ (cf. Appendix C.1), e.g. the taxicab metric, and with the σ-algebras of Borel sets. Note that each ρ induces the product topology.
- The MDP has an upper bounding function b. This means that $b \geq 0$ on S is measurable and that the numbers

$$\|Pb\|_b, \quad \|Pr_S^+\|_b, \quad \|V_0^+\|_b$$

are finite. This implies that $Uv \in B_b^+$ for each measurable $v \in B_b^+$.

Usc functions and lsc functions v on S or on A are measurable (cf. Appendix B.1.8(d)). Also usc and lsc functions on D are measurable, due to

K. Hinderer et al., *Dynamic Optimization*, Universitext,
DOI 10.1007/978-3-319-48814-1_17

separability of S and of A (cf. Dudley 1989). Moreover, the latter property ensures that the σ-algebra of Borel sets on $S \times A$ equals the product σ-algebra $\mathfrak{S} \otimes \mathfrak{A}$.

In Chaps. 15 and 16 we were able to establish the existence of maximizers in a number of applications without recourse to general theorems. Such theorems will now be treated, similarly as for deterministic DPs in Chap. 9. We prove the existence of so-called strongly optimal policies for MDPs with an upper bounding function under two different sets of assumptions: In the first one we require upper [lower] semicontinuity of the data in (s, a) and weak continuity of $(s, a) \mapsto P(s, a, ds')$, and we infer upper [lower] semicontinuity of the value functions; in the second one we require upper [lower] semicontinuity of the data in a and strong continuity of $a \mapsto P(s, a, ds')$, and we infer measurability of the value functions. We use the Structure Theorem 16.1.12. When applying the latter, e.g. under the first set of assumptions, we need conditions

(i) under which (semi-)continuity of v is preserved under integration with respect to $P(s, a, ds')$,
(ii) under which one can select maximum points $f(s)$ of $a \mapsto Lv(s, a)$, $s \in S$, such that f is measurable.

Definition 17.1 Let P be a transition law. Then:

- P is called (weakly) ***b*-continuous** if Pv is usc for each usc $v \in B_b^+$.
- P is called strongly ***b*-continuous** in a if $Pv(s, a)$ is usc in a for each measurable $v \in B_b^+$.

Lemma 17.2 (*b*-continuous transition laws) *The following holds for continuous b:*

(a) If P is a (weakly) b-continuous transition law then Pb and Pv are continuous for all continuous $v \in B_b$.
(b) If P is a strongly b-continuous transition law then $Pb(s, a)$ and $Pv(s, a)$ are continuous in a for all measurable $v \in B_b$.

Proof For the proof of the first assertion note that $\pm v$ is usc and belongs to B_b^+. Hence $\pm Pv$ is usc, hence Pv is continuous. The second assertion is proven in the same way. □

Remark 17.3

(a) Strong b-continuity in a does not imply (weak) b-continuity.
(b) If $\inf b > 0$, then the converse of the first assertion of Lemma 17.2 holds. In particular, if $b \equiv 1$ then P is b-continuous if and only if for each $(s, a) \in D$ and each sequence $((s_n, a_n))_1^\infty$ in D converging to (s, a) the sequence of probability measures $P(s_n, a_n, ds')$ is weakly convergent to $P(s, a, ds')$.
(c) P is strongly 1-continuous in a if and only if $P(s, a, B)$ is continuous in a for each measurable $B \subset S$. ◇

Lemma 17.4 (Criteria for b-continuity of P)

(a) *If Pb is usc, then P is b-continuous and strongly b-continuous in a and Pb is continuous under each of the following conditions:*

(a1) *P has a transition density $(s,a,s') \mapsto p(s,a,s')$ with respect to some σ-finite measure $\mu(ds')$ on $\mathfrak{S}$ such that $(s,a) \mapsto p(s,a,s')$ is lsc for μ-almost all $s' \in S$.*

(a2) *P belongs to the MDP to which a CM can be reduced, i.e. $P(s,a,ds')$ is the image of some probability distribution Q on $\mathfrak{Z}$ under $z \mapsto T(s,a,z)$, and $(s,a) \mapsto T(s,a,z)$ is continuous for Q-almost all $z \in Z$.*

(b) *If P is strongly 1-continuous in a and if $Pb(s,a)$ is continuous in a, then P is strongly b-continuous in a.*

Proof

(a) Since $Pv(s,a)$ is usc in a if it is usc in (s,a), it suffices to show that Pv is usc for each measurable $v \in B_b^+$. For the proof fix some measurable $v \in B_b^+$, and denote the elements (s,a) of D by x.

(a1) Firstly assume $v \leq 0$. Then we define $w(x,y) := p(x,y)v(y) \leq 0$ for $(x,y) \in D \times S$. Now $x \mapsto w(x,y)$ is usc for μ-almost all $y \in S$. For (a2) we put $\mu := Q$ and define $w(x,y) := v \circ T(x,y) \leq 0$ for $(x,y) \in D \times Z$. Then $x \mapsto w(x,y)$ is usc for μ-almost all y by Lemma 9.1.4(b). Select x in D and a sequence (x_n) in D converging to x. Then we obtain in both cases from Fatou's Lemma

$$\limsup_n Pv(x_n) = \limsup_n \int w(x_n,y)\,\mu(dy)$$
$$\leq \int \limsup_n w(x_n,y)\,\mu(dy) \leq \int w(y)\,\mu(dy) = Pv(x).$$

Thus Pv is usc. In particular, $P(-b)$ is usc, hence Pb is continuous as Pb was assumed to be usc.

(a2) Now let $v \in B_b^+$ be arbitrary measurable. Then $v \leq \gamma b < \infty$ for some $\gamma \in \mathbb{R}_+$. Now it follows from (a1), applied to $v - \gamma b \leq 0$ instead of v, that for $x_n \to x$

$$\begin{aligned} Pv(x) &= P(v-\gamma b)(x) + \gamma \cdot Pb(x) \\ &\geq \limsup_n P(v-\gamma b)(x_n) + \gamma \cdot Pb(x) \\ &= \limsup_n Pv(x_n) - \gamma \cdot \limsup_n Pb(x_n) + \gamma \cdot Pb(x), \end{aligned}$$

which equals $\limsup_n Pv(x_n)$ by continuity of Pb. Thus Pv is usc.

(b) We refer to Hernández-Lerma and Lasserre (1999, p. 48). □

It is easily seen that the second elementary Selection Lemma 16.1.22 and its proof remain true if (S, ρ_S) is a separable metric space. Using this fact in the Structure Theorem 16.1.12 with $\mathfrak{V}$ as the set of continuous functions in B_b and using Lemma 17.2(a), we obtain the next result; cf. Theorem 16.1.23.

Theorem 17.5 (Existence of optimal policies and continuity of the value functions when the sets $D(s)$ are intervals) *Assume*

(i) *The MDP has a bounding function b,*
(ii) $A = \mathbb{R}$, *and* $D(\cdot) = [c(\cdot), d(\cdot)]$ *for continuous* $c(\cdot)$ *and* $d(\cdot)$,
(iii) *P is b-continuous,*
(iv) *r and* V_0 *are continuous.*

Then we have:

(a) *The VI holds.*
(b) *All value functions belong to* B_b *and are continuous.*
(c) *There exists at each stage n a smallest [largest] maximizer* f_n*. Thus for each N the policy* $(f_n)_N^1$ *is optimal. Moreover,* f_n *is lsc [usc] and hence continuous, if unique.*

In addition to auxiliary results from Chap. 9 we need preparations concerning so-called measurable selectors. Recall the definition of a (quasi-continuous) correspondence from Chap. 9, and that

$$\rho_A(a, B) := \inf_{b \in B} \rho_A(a, b) \text{ for } B \subset A.$$

Definition 17.6 Let $\Gamma(\cdot)$ be a correspondence from S into A. Then:

- A mapping f from S into A such that $f(s) \in \Gamma(s)$ for all s is called a ***selector of the correspondence*** $\boldsymbol{\Gamma(\cdot)}$.
- The correspondence $\Gamma(\cdot)$ is called ***measurable*** if $s \mapsto \rho_A(a, \Gamma(s))$ is measurable for all $a \in A$.

Lemma 17.7 (Measurability of correspondences) *A correspondence* $\Gamma(\cdot)$ *is measurable if there exists a decreasing sequence of supersets* Γ_n *of* Γ, $n \geq 1$, *such that*

(i) *all s-sections of* Γ_n *are open,*
(ii) *all a-sections of* Γ_n *are measurable,*
(iii) *for each* $s \in S$ *each sequence of points* $a_n \in \Gamma_n(s)$, $n \geq 1$, *has a cluster value in* $\Gamma(s)$.

For the proof of Lemma 17.7 and the following Theorem 17.8 we refer to Kuratowski and Ryll-Nardzewski (1965).

Theorem 17.8 (Measurable Selection Theorem of Kuratowski/Ryll-Nardzewski) *A compact-valued and measurable correspondence has a measurable selector.*

We now apply Theorem 17.8 to the correspondence $s \mapsto D^*(s) :=$ set of maximum points of a function $w(s, \cdot)$.

Proposition 17.9 (Existence of maximizers of w) *Let $D(\cdot)$ be compact-valued and let w be a measurable function on D.*

Then:

(a) If $w(s,a)$ is usc in a, then w has a maximizer, and $s \mapsto w^*(s) := \sup_{a \in D(s)} w(s,a)$ is measurable.
(b) If $D(\cdot)$ is quasi-continuous and if w is usc, then $s \mapsto D^*(s) :=$ set of maximum points of $w(s,\cdot)$ is a compact-valued measurable correspondence.

Proof

(a) We refer to Schäl (1975) and Rieder (1978).

(b1) We know from Lemma 9.2.4 that

$$s \mapsto w^*(s) := \sup\{w(s,a) : a \in D(s)\}$$

is usc and that $D^*(s) \neq \emptyset$ for all s. In particular, $D^*(\cdot)$ is a correspondence. Moreover,

$$D^*(s) := \{a \in D(s) : w(s,a) \geq w^*(s)\}$$

is a closed subset of $D(s)$, as $w(s,\cdot)$ is usc. Since $D(s)$ is compact, $D^*(s)$, $s \in S$, is also compact.

(b2) Now we verify measurability of $D^*(\cdot)$ by verifying the assumptions (i)–(iii) of Lemma 17.7 for $\Gamma := D^*$. By Lemma 9.1.4(g) the usc function w is the limit of a decreasing sequence of continuous functions w_n on D. Consider the sets

$$D_n := \{(s,a) \in D : w_n(s,a) > w^*(s) - 1/n\},\ n \geq 1.$$

It is obvious that $D_n \supset D^*$ and that the sequence (D_n) is decreasing.

(i) By continuity of $w_n(s,\cdot)$ the sets $D_n(s)$ are open for all n and s.
(ii) It suffices to consider $a \in A_0$. Fix $n \geq 1$. The a-section D_a of the measurable set D is measurable and the restrictions of $w_n(\cdot,a)$ and w^* on D_a are measurable. Therefore the a-section of D_n is measurable.
(iii) Fix $s \in S$ and select $a_n \in D_n(s) \subset D(s)$ for $n \geq 1$. As $D(s)$ is compact, there exists a subsequence (a_{n_k}) of (a_n) converging to some $a \in D(s)$. Now we obtain for $k \geq 1$ and $m < n_k$

$$w_m(s,a_{n_k}) \geq w_{n_k}(s,a_{n_k}) > w^*(s) - 1/n_k.$$

Letting k tend to ∞ yields $w_m(s,a) \geq w^*(s)$ by continuity of w_m. Letting m tend to ∞ implies $w(s,a) \geq w^*(s)$. Therefore $a \in D^*(s)$. □

We are now ready for the proof of the next three results.

Theorem 17.10 (Existence of optimal policies and upper [lower] semicontinuity of the value functions) *Assume*

(i) *the MDP has an upper bounding function b [a bounding function b],*
(ii) *$D(\cdot)$ is quasi-continuous [continuous],*
(iii) *P is (weakly) b-continuous,*
(iv) *r and V_0 are upper semicontinuous [continuous].*

Then:

(a) *All value functions are upper semicontinuous [continuous] and belong to B_b^+, hence to $\mathfrak{V}_0$.*
(b) *The VI holds.*
(c) *There exists for each $N \geq 1$ a strongly optimal policy for MDP_N.*

Proof We give a proof for usc value functions. (The proof for continuity is similar.) We use the Structure Theorem 16.1.12 with $\mathfrak{V}$ as the set of usc functions in B_b^+. Firstly, V_0 belongs to $\mathfrak{V}$ by (i) and (iv). Next, if $v \in \mathfrak{V}$, then Pv is usc by (iii). Then Lv is usc by Lemma 9.1.4(a) as r is usc. Because of (ii) it follows from Proposition 17.9(a) that $w := Lv$ has a maximizer. Furthermore, it follows from Lemma 9.2.4(a) that Uv is usc. Moreover, (16.6) with $n := 1$ and $V_0 := v$ shows that $Uv \in B_b^+$. Now the assertion follows from Theorem 16.1.12. □

Corollary 17.11 (Upper [lower] semicontinuity of maximizers) *Assume that the MDP has a bounding function b, that $D(\cdot)$, r and V_0 are continuous and that P is (weakly) b-continuous; thus there exists by Theorem* 17.10 *a maximizer f_n at each stage n. Then the following holds:*

(a) *If f_n is unique, then f_n is continuous.*
(b) *If $A \subset \mathbb{R}$, then there exists a smallest [largest] maximizer at stage n, and it is lower [upper] semicontinuous.*

Proof From Theorem 17.10 we know that $w := V_{n-1}$ is continuous. Now the assertion follows from Proposition 17.9(b). □

Theorem 17.12 (Existence of optimal policies and upper semicontinuity of the value functions) *Assume that*

(i) *the MDP has an upper bounding function b,*
(ii) *$D(\cdot)$ is compact-valued,*
(iii) *P is strongly b-continuous in a,*
(iv) *$r(s, a)$ is upper semicontinuous in a.*

Then:

(a) *All value functions are measurable and belong to B_b^+, hence to $\mathfrak{V}_0$.*
(b) *The VI with an upper semicontinuous value function holds.*
(c) *There exists for each $N \geq 1$ a strongly optimal policy for MDP_N.*

Proof We use the Structure Theorem 16.1.12 with $\mathfrak{V}$ as the set of measurable functions in B_b^+. Firstly, V_0 belongs to $\mathfrak{V}$ by (i). Next, if $v \in \mathfrak{V}$, then $Pv(s, a)$ is usc in a by (iii). Then $Lv(s, a)$ is usc in a by Lemma 9.1.4(a) as $r(s, a)$ has this property, and Lv is measurable. Because of (ii) it follows from Proposition 17.9(a) that $w := Lv$ has a maximizer, hence Uv is measurable. Moreover, (16.6) with $n := 1$ and $V_0 := v$ shows that $Uv \in B_b^+$. Now the assertion follows from Theorem 16.1.12. □

Chapter 18
Stochastic Monotonicity and Monotonicity of the Value Functions

We consider MDPs and CMs with structured state space and arbitrary transition law. We assume that the minimal assumption (MA1) holds. As in Chap. 6 we are looking for conditions under which the value functions are monotone in the initial state s. This is easy for CMs, but requires a thorough treatment of the notion of stochastic monotonicity for MDPs. Unless stated otherwise, a result holds for both MDPs and CMs.

It is convenient to separate the two problems of monotonicity and of the validity of the VI. Therefore we usually include the latter in our assumptions.

18.1 Monotonicity

Recall from Chap. 6 the following notions. A **structured set** $(M, \leq^M)$ is a non-empty set M with a relation $\leq^M$; sometimes $\leq^M$ will be dropped from $(M, \leq^M)$ when no confusion is possible. For a first reading one should think of M as being a subset of $\mathbb{R}^k$ or even of $\mathbb{R}$, endowed with the usual ordering. Admitting only orderings instead of arbitrary relations on M would result in no simplification. Arbitrary orderings on $\mathbb{R}^k$ instead of $\leq^k$ allow a joint treatment of all 2^k kinds of componentwise orderings.

If $\mathfrak{M}$ is a σ-algebra on M we call $(M, \leq^M, \mathfrak{M})$ a **structured measurable space**. The set of all probability distributions on $\mathfrak{M}$ is denoted by $\mathbb{P}(\mathfrak{M})$. A probability distribution μ on $\mathfrak{M}$ is also called a probability distribution on $(M, \leq^M, \mathfrak{M})$. Subsets M of $\mathbb{R}^k$ (of $\overline{\mathbb{R}}$) are always endowed with the usual ordering $\leq^k$ ($\leq := \leq^1$) unless stated otherwise. A function v on M is said to be increasing [decreasing] if $x \leq^M y$ implies $v(x) \leq v(y)$ [$v(x) \geq v(y)$]. A subset B of the structured set $(M, \leq^M)$ is called **increasing** [**decreasing**] if $x \in B$ implies $y \in B$ for all $y \geq^M x$ [$y \leq^M x$]. As an example, $B \subset \mathbb{R}$ is increasing if and only if B is a right unbounded interval. $B \subset M$ is increasing if and only if B^c is decreasing. Recall the following convention:

K. Hinderer et al., *Dynamic Optimization*, Universitext,
DOI 10.1007/978-3-319-48814-1_18

If v is a mapping from a $M \subset X \times Y$ into M', where X and M' are structured, we say that $v(x, y)$ is *increasing in x* if $x \mapsto v(x, y)$ is increasing on the y-section of M for all $y \in \mathrm{pr}(M \to Y)$. For additional notions concerning relations and monotone functions, see Chap. 6.

In this subsection the space S is assumed to be structured; cf. Chap. 6. For the moment denote by $\mathfrak{V}\uparrow$ the set of functions on S which are increasing with respect to $\leq^S$. Obviously V_n in an MDP or a CM is increasing for each $n \geq 1$ if the VI holds, if V_0 is increasing and if $\mathfrak{V}\uparrow$ is *invariant* under U in the sense that $Uv \in \mathfrak{V}\uparrow$ for all $v \in (\mathfrak{V}\uparrow) \cap \mathfrak{V}_0$. Keeping in mind that for MDPs

$$Uv(s) = \sup_{a \in D(s)} [r(s, a) + \beta \cdot Pv(s, a)], \quad s \in S,$$

and for CMs

$$Uv(s) = \sup_{a \in D(s)} [r(s, a) + \beta \cdot \mathbb{E}\, v(T(s, a, \eta_1))], \quad s \in S,$$

invariance under U of an arbitrary set $\mathfrak{V}$ of functions on S for MDPs [for CMs] is usually shown according to the following *scheme*: Find a set $\mathfrak{W}$ of functions on D with the following properties:

(i) $Pv \in \mathfrak{W}$ [$\mathbb{E}\, v(T(\cdot, \cdot, \eta_1) \in \mathfrak{W}$] for all $v \in \mathfrak{V} \cap \mathfrak{V}_0$,
(ii) $r \in \mathfrak{W}$,
(iii) $\mathfrak{W}$ is closed under the multiplication with a non-negative constant and under the addition of two function, at least one of which (in our case r) is finite,
(iv) for all $w \in \mathfrak{W}$ the function

$$s \mapsto w^*(s) := \sup_{a \in D(s)} w(s, a) \quad \text{belongs to } \mathfrak{V}.$$

As conditions (ii) through (iv) were already treated in Chap. 2, we must now concentrate on (i). It is customary to take such desirable integral properties as definitions for concepts such as stochastic isotonicity. As a rule, these integral properties are difficult to check directly. Therefore there arises the problem of finding simpler sufficient conditions; cf. Theorem 18.2.9(b) and (c) below.

In the present section the scheme is used when $\mathfrak{V}$ is the set $\mathfrak{V}\uparrow$. Then we take for $\mathfrak{W}$ the set of functions $(s, a) \mapsto w(s, a)$ on D which are increasing in s. Now (iii) is obviously true, (ii) will be assumed and (iv) holds by Proposition 6.3.7 whenever $D(\cdot)$ is increasing. It remains to ensure condition (i). By the isotonicity of the integral, (i) holds for CMs if $T(s, a, z)$ is increasing in s for each $a \in A_0$ and $z \in Z$. This proves already the following stochastic generalization of Theorem 6.3.5.

Theorem 18.1.1 (Increasing value functions in CMs) *If* $(S, \leq^S)$ *is structured then* $V_n(s)$ *is increasing [decreasing] in s for all* $n \geq 0$ *under the following conditions:*

(i) The correspondence $s \mapsto D(s)$ *is increasing [decreasing],*
(ii) $T(s, a, z)$ *is increasing in s for all* $a \in A_0$, $z \in Z$,
(iii) the functions $r(s, a)$ *and* $V_0(s)$ *are increasing [decreasing] in s for all* $a \in A_0$,
(iv) the VI holds.

Example 18.1.2 (Splitting of a resource into two parts) In the CM for Example 16.1.25 we have $S = [0, B]$, $s \mapsto D(s) = [0, s)$ is increasing, and

$$T(s, a, z) = t_1(s - a, z) + t_2(a, z)$$

for two measurable functions t_1 and t_2 from $S \times \mathbb{R}_+$ into $\mathbb{R}_+$ such that $T(s, a, z) \in S$ for all s, a, z. (In the water reservoir interpretation we have $t_1 \equiv 0$ and $t_2(a, z) = \min\{B, a + z\}$, where z is the inflow during one period.) The reward function r_Z is assumed to be of the form

$$r_Z(s, a, z) = u_1(s - a, z) + u_2(a, z)$$

for two measurable functions u_1 and u_2 on $S \times \mathbb{R}_+$ such that $\mathbb{E}\, u_1(s - a, \eta_1)$ and $\mathbb{E}\, u_2(a, \eta_1)$ exist and are finite. We assume that the VI holds. Now it follows from Theorem 18.1.1 that V_n is increasing if $t_1(s, a)$, $u_1(s, a)$ and $V_0(s)$ are increasing in s. ♦

18.2 Stochastic Monotonicity

In this section we look for sufficient conditions for monotonicity of the value functions of MDPs, given in terms of stochastic monotonicity.

Definition 18.2.1 Let $(M, \leq^M, \mathfrak{M})$ be a structured measurable space.

- The relation $\leq_{\text{st}}^M$ ($\leq_{\text{st}}$ for short) on $\mathbb{P}(\mathfrak{M})$ is defined by

  $P_1 \leq_{\text{st}}^M P_2 :\Longleftrightarrow \int v\, dP_1 \leq \int v\, dP_2$ for all increasing measurable functions v on M such that $\int v\, dP_1$ and $\int v\, dP_2$ exist.

  We then say that P_1 is ***stochastically smaller*** (with respect to $\leq^M$) than P_2. In the special case $(M, \leq^M, \mathfrak{M}) = (\mathbb{R}^n, \leq^n, \mathfrak{B}_n)$ we use $\leq_{\text{st}}^n$ instead of $\leq_{\text{st}}^M$, and $\leq_{\text{st}}$ instead of $\leq_{\text{st}}^1$.
- Let $(I, \leq^I)$ be a structured set. A family $(P_i, i \in I)$ of probability distributions on $\mathfrak{M}$ is said to be ***stochastically increasing*** (with respect to $(\leq^I, \leq^M)$) if $i \leq^I j$ implies $P_i \leq_{\text{st}}^M P_j$.
- A family of n-dimensional random vectors is called stochastically increasing if the corresponding probability distributions have this property.

Remark 18.2.2

(a) It is easily seen that a family $(P_i, i \in I)$ (or (P_i) for short) is stochastically increasing if and only if $i \mapsto \int v\, dP_i$ is increasing for all measurable increasing v on the subset I_v of I of those $i \in I$ for which $\int v\, dP_i$ exists. Moreover, it follows from Proposition 18.2.11(a) below, that a family (P_i) is stochastically increasing if and only if $i \mapsto \int v\, dP_i$ is increasing for all measurable increasing v for which $\int v\, dP_i$ exists for all i.

(b) It follows from Proposition 18.2.11 below that requiring non-negativity of the functions v leads to an equivalent and for theoretical derivations more elegant property. However, the set of non-negative increasing measurable functions v is still too large for checking stochastic isotonicity of specific families of probability distributions. Below we give several easier checkable criteria for this case; cf. Theorem 18.2.9(b) if $M = \mathbb{R}$. ◇

Remark 18.2.3 By reversing the relation in I one defines stochastically decreasing for families of probability distributions and of random vectors. Obviously (P_i) is stochastically decreasing if and only if for each $i \le^I j$ and each increasing measurable function v on M such that $\int v\, dP_i$ and $\int v\, dP_j$ exist we have $\int v\, dP_i \ge \int v\, dP_j$. ◇

Remark 18.2.4 The relation $\le_{\text{st}}^M$ is obviously a preordering on $\mathbb{P}(\mathfrak{M})$. It follows from the uniqueness theorem for measures that $\le_{\text{st}}^n$ is even an ordering on $\mathbb{P}(\mathfrak{B}_n)$. ◇

Remark 18.2.5 A two-parameter family $(P_{ij}, (i,j) \in I \times J)$ of probability distributions on $\mathfrak{M}$ is stochastically increasing with respect to $(\le^I \times \le^J, \le^M)$ if the family is stochastically increasing in i and in j. The converse holds if $\le^I$ and $\le^J$ are reflexive. ◇

Directly from the definition of stochastic monotonicity we obtain: $X \le_{\text{st}} Y$ implies $\mathbb{E}\, X^k \le \mathbb{E}\, Y^k$ for $k = 1, 3, 5, \ldots$ whenever the expectations exist, and even $\mathbb{E}\, X^\alpha \le \mathbb{E}\, Y^\alpha$ for $\alpha \in \mathbb{R}^+$ if $X \ge 0$ and $Y \ge 0$.

The property $X \le_{\text{st}} Y \Longrightarrow \mathbb{E}\, X \le \mathbb{E}\, Y$ is useful for checking the correctness of proofs of stochastic monotonicity. As an example, if $X \sim \Gamma_{\alpha,b}$, then $\mathbb{E}\, X = b/\alpha$. Thus, if $\Gamma_{\alpha,b}$ is stochastically monotone at all in α (in b) then it must be stochastically decreasing in α (stochastically increasing in b).

The next result, which is useful in several contexts (e.g. for Bayes models), shows how monotonicity can be passed to integrals.

Lemma 18.2.6 (Isotonicity of Integrals) *Let (P_i) be a stochastically increasing family of probability distributions on $\mathfrak{M}$ and let v_i be measurable functions on M such that $\int v_i\, dP_i$ exists for all i. If $v_i(x)$ is increasing both in i and in x, then $\int v_i\, dP_i$ is increasing in i.*

Proof Let $i \le j$. Then v_i, v_j are increasing and $v_i \le v_j$. We have to show that $\int v_i\, dP_i \le \int v_j\, dP_j$. This holds trivially if $\int v_i\, dP_i = -\infty$. Otherwise $\int v_j\, dP_i$ exists since $\int v_j^-\, dP_i \le \int v_i^-\, dP_i < \infty$. Then $\int v_i\, dP_i \le \int v_j\, dP_i$ as $v_i \le v_j$. Finally $\int v_j\, dP_i \le \int v_j\, dP_j$ since v_j is increasing and $P_i \le_{\text{st}}^M P_j$. □

Note that in Lemma 18.2.6 isotonicity of $v_i(x)$ both in i and in x can be replaced by isotonicity in (i, x) if both relations $\leq^I$ and $\leq^M$ are reflexive.

Now we obtain the following basic result.

Theorem 18.2.7 (Increasing value functions) *If $(S, \leq^S)$ is structured then $s \mapsto V_n(s)$ is increasing [decreasing] for all n under the following conditions:*

(i) The correspondence $D(\cdot)$ is increasing [decreasing],
(ii) $P(s, a, ds')$ is stochastically increasing in s for all $a \in A_0$,
(iii) the functions $s \mapsto r(s, a)$, $a \in A_0$, and V_0 are increasing [decreasing],
(iv) the VI holds.

Remark 18.2.8 The crucial condition (ii) in Theorem 18.2.7 means that $a \in A_0$, s, $s' \in D_a$, $s \leq^S s'$ implies $P(s, a, B) \leq P(s', a, B)$ for all increasing $B \in \mathfrak{S}$. ◇

Of course, the preceding result becomes useful only when we can find easily checkable conditions for stochastic monotonicity of $s \mapsto P(s, a, ds')$. This is our next topic.

A set $B \subset M$ is increasing (with respect to $\leq^M$) if and only if 1_B is increasing. Therefore $\mu \leq_{\text{st}}^M \nu$ implies that $\mu(B) \leq \nu(B)$ for all measurable increasing subsets B of M. It is useful that the converse also holds and that for the important case of probability distributions on the real line stochastic monotonicity can be checked by means of distribution functions, as follows.

Theorem 18.2.9 (Stochastically increasing families of distribution functions) *Let (P_i) be a family of probability distributions on a structured measurable space $(M, \leq^M, \mathfrak{M})$. Then:*

(a) The family (P_i) is stochastically increasing if (and only if) $P_i(B)$ is increasing in i for each measurable increasing subset B of M.
(b) A family (P_i) of probability distributions on $(\mathbb{R}, \leq, \mathfrak{B})$ with distribution functions F_i is stochastically increasing if and only if $\overline{F}_i(x) := 1 - F_i(x)$ is increasing in i for all $x \in \mathbb{R}$.
(c) If P_i from (b) has a [discrete] density f_i on a right unbounded interval M in $\mathbb{R}$ [in $\mathbb{Z}$], then (P_i) is stochastically increasing if and only if $\int_x^\infty f_i \, d\lambda$ [if and only if $\sum_{j=x}^\infty f_i(j)$] is increasing in i for all $x \in M$.

Proof

(a) Let $i \leq^I j$ and let v be an increasing and measurable function on M such that $\int v \, dP_i$ and $\int v \, dP_j$ exist.

$$\int v \, dP_i = \int H_i(x) \, \lambda(dx) \tag{18.1}$$

$$:= \int_{\mathbb{R}_+} \Big[P_i(v^{-1}((x, \infty])) + P_i(v^{-1}((-x, \infty])) - 1 \Big] \lambda(dx),$$

and a corresponding representation holds for $\int v\,dP_j$ with H_j. It is easy to see that $v^{-1}(B) \subset M$ is an increasing measurable subset of M for each increasing (hence measurable) subset B of $\overline{\mathbb{R}}$, in particular for $B := (x, \infty]$ and for $B := (-x, \infty]$, $x \in \mathbb{R}$. Now it follows from (18.1) that $H_i \le H_j$, hence $\int v\,dP_i = \int H_i\,d\lambda \le \int H_j\,d\lambda = \int v\,dP_j$.

(b) The condition is necessary by (a) since $\overline{F}_i(x) = P_i((x, \infty))$ for all x and since the intervals (x, ∞) are increasing sets. In order to show that the condition is also sufficient we note that the only other increasing sets in $\mathbb{R}$ are of the form $B = \mathbb{R}$ and $B = [x, \infty)$, $x \in \mathbb{R}$. Moreover,

$$P_i((-n, \infty)) \to P_i(\mathbb{R}), \quad P_i((x - 1/n, \infty)) \to P_i([x, \infty))$$

for $n \to \infty$ by continuity of P_i from below and from above, respectively. Now the assertion follows as isotonicity of $i \mapsto P_i((-n, \infty))$ and of $i \mapsto P_i((x - 1/n, \infty))$ is preserved under the limit for $n \to \infty$.

(c) This is an easy consequence of (b). □

Example 18.2.10 (Stochastically monotone families of probability distributions)

(a) From Table 12.1 for the second version of the marketing problem from Example 12.1.6 we see that

$$\sum_{j=0}^{s'} p(s, 1, j) \le \sum_{j=0}^{s'} p(s, 0, j) \quad \text{for } 0 \le s,\, s' \le 3.$$

It follows from Theorem 18.2.9(c) that $P(s, 1, \cdot) \ge_{\text{st}} P(s, 0, \cdot)$ for all s. The other conditions in Theorem 18.2.7 are also fulfilled. Thus all value functions are increasing, and the same also holds for the first version.

(b) Each one-dimensional location parameter family $(P_m, m \in \mathbb{R})$ of probability distributions on $\mathfrak{B}$ is stochastically increasing in m. This holds as $m \mapsto \overline{F}_m(x) = \overline{F}_0(x - m)$ is increasing in m. In particular, the family of normal distributions $N(m, \sigma^2)$ is stochastically increasing in $m \in \mathbb{R}$. On the other hand, $\sigma \mapsto N(m, \sigma^2)$, $\sigma \in \mathbb{R}^+$, is neither stochastically increasing nor stochastically decreasing.

(c) Each one-dimensional scale parameter family $(P_b, b \in \mathbb{R}^+)$ of probability distributions on $\mathfrak{B}(\mathbb{R}^+)$ is stochastically increasing in b. This holds as $b \mapsto \overline{F}_b(x) = \overline{F}_1(x/b)$ (for $x \in \mathbb{R}_+$) is increasing in b. It follows that the family of gamma distributions $\Gamma_{\alpha,b}$ (in particular the family of exponential distributions $\text{Exp}(\alpha)$) is stochastically decreasing in α.

(d) The family of geometric distributions $\text{Geo}(p)$, $p \in (0, 1)$, is stochastically decreasing in p since $\sum_{j=x}^{\infty} f_p(j) = q^{x-1}$ is decreasing in p for all $x \in Z := \mathbb{N}$.

For later reference we collect several examples, all of which are treated in the present chapter, in Table 18.1. ♦

Table 18.1 Stochastically monotone families of probability distributions

Probability distribution	Increasing in	Decreasing in
$N(m, \sigma^2)$	m	–
$\Gamma_{\alpha,b}$	b	α
χ^2_n	n	–
$\text{Exp}(\alpha)$	–	α
$\text{Be}(\alpha_1, \alpha_2)$	α_1	α_2
$\text{Geo}(p)$	–	p
$\text{Bi}(n, p)$	n and p	–
$\text{Poi}(\alpha)$	α	–

The next result is useful, for example, when $\mathfrak{F}$ is the set of measurable increasing functions on M which are non-negative [which are bounded].

Proposition 18.2.11 (Stochastically monotone probability distributions) *Let $(M, \leq^M, \mathfrak{M})$ be a structured measurable space. Then:*

(a) Let $\mathfrak{F}$ denote a set of increasing measurable functions on M which contains the indicators of all increasing measurable sets in M, and let P_1 and P_2 be probability distributions on $\mathfrak{M}$. Then $P_1 \leq^M_{\text{st}} P_2$ if and only if $\int v\, dP_1 \leq \int v\, dP_2$ for all $v \in \mathfrak{F}$ for which both integrals exist.

(b) Stochastic monotonicity is preserved under increasing mappings: Let g be an increasing and measurable mapping from M into a structured measurable space $(M', \leq^{M'}, \mathfrak{M}')$, and let $(P_i, i \in I)$ be a stochastically increasing family of probability distributions on $\mathfrak{M}$. Then the family $((P_i)_g, i \in I)$ of images of P_i under g is stochastically increasing with respect to $(\leq^I, \leq^{M'})$.

Proof

(a) The first condition trivially implies the second one. In order to prove the converse, it suffices by Theorem 18.2.9(a) to show that $P_1(B) \leq P_2(B)$ for each measurable and increasing set B. This is obvious since $P_i(B) = \int 1_B\, dP_i$, $i = 1$, 2 and since $1_B \in \mathfrak{F}$.

(b) By (a) it suffices to show that $\int v\, d(P_i)_g$ is increasing for all $v \in \mathfrak{F}(M')$, the set of non-negative measurable increasing functions on M'. This follows from (a) with $\mathfrak{F} := \mathfrak{F}(M)$ since $v \circ g$ is measurable and increasing and since $\int v\, d(P_i)_g = \int v \circ g\, dP_i$. □

In CMs the transition law $Q(dz) = Q(y, dz)$ often depends on a parameter y in a structured set Y. The next result, which is easily verified by induction on $n \geq 0$, gives conditions under which $V_n(s, y)$ is increasing in y.

Proposition 18.2.12 (Monotone dependence of value functions of CMs on a parameter) *Assume that $Q(y, dz)$ depends on a parameter y in a structured set Y. Then $V_n(s, y)$, $n \geq 1$, is increasing in y if*

(i) $V_n(s, y)$, $n \geq 1$, and $V_0(s)$ are increasing in s,
(ii) the family $Q(y, dz)$ is stochastically increasing,

(iii) $T(s,a,z)$ *and* $r_Z(s,a,z)$ *are increasing in* z,
(iv) the VI holds.

Example 18.2.13 (The gambling Example 13.6.4*)* We confirm intuition by showing that the maximal expected n-stage reward $V_n(s,p)$ is increasing in the probability p of winning, provided V_0 is increasing.

For the proof recall the data from the CM in Example 13.6.4: $S = \mathbb{N}_0$; $A = \mathbb{N}_0$; $D(s) = \{0, 1, \ldots, s\}$; $Z = \{0, 1\}$; $\eta_t \sim Q(p, dz) := \mathrm{Bi}(1, p)$; $T(s, a, z) = s - a + \gamma a z$; $r_Z \equiv 0$, $V_0(s)$ is increasing and $V_0(0) = 0$; $\beta = 1$. Now the assertion follows since the assumptions of Proposition 18.2.12 are fulfilled; for (i) use Theorem 18.1.1. ◆

Example 18.2.14 (Cf. Example 13.4.4 *of selling an asset)* We confirm intuition by showing that $d_n = d_n(Q)$, the maximal expected sum of discounted costs and of the accepted offer when at most n offers arise, is increasing in Q in the sense that $Q \le_{\mathrm{st}} Q'$ implies $d_n(Q) \le d_n(Q')$ for all $n \ge 1$.

In fact, this follows by induction on $n \ge 1$

$$d_{n+1}(Q) = \int Q(dz)\, \max\{\beta \cdot d_n(Q), \gamma - \gamma' + z\} - \gamma, \quad n \ge 1,$$

starting with $d_1 = \int Q(dz)\, z - \gamma'$. ◆

Theorem 18.2.9(b) does not extend to arbitrary probability distributions on multidimensional Euclidean spaces. This is due to the fact that there are many more increasing Borel sets in $\mathbb{R}^m$, $m \ge 2$, than only $\mathbb{R}^m$ and the intervals of the form (x, ∞) or $[x, \infty)$, $x \in \mathbb{R}^m$. On the other hand, for products of probability distributions (i.e. for random vectors with independent components) stochastic isotonicity is equivalent to the same property of the marginals, cf. Proposition 18.2.18(a) below. In the case of random vectors with dependent components the following result can be helpful. It can be extended by induction on m to the case of probability distributions on the product of m measurable spaces.

Proposition 18.2.15 (Stochastically monotone probability distributions and probability densities) *Let* $(I, \le^I)$ *and* $(J, \le^J)$ *be a structured set and let* $(M_1, \le_1, \mathfrak{M}_1)$ *and* $(M_2, \le_2, \mathfrak{M}_2)$ *be structured measurable spaces with reflexive relations* $\le_1$ *and* $\le_2$*, respectively. Let* P_i, $i \in I$*, be probability distributions on* $\mathfrak{M}_1$ *and* Q_j, $j \in J$*, transition probabilities from* M_1 *into* M_2*. Then the family* $(P_i \otimes Q_j, (i,j) \in I \times J)$ *on* $M_1 \otimes M_2$ *is stochastically increasing with respect to* $(\le^I \times \le^J, \le_1 \times \le_2)$ *if the following holds:*

(i) The family (P_i) *is stochastically increasing,*
(ii) $Q_j(x, dy)$ *is stochastically increasing in* j *as well as in* x.

Proof Let $v \ge 0$ be an increasing measurable function on $M_1 \times M_2$. Reflexivity of $\le_1$ and $\le_2$ ensures that $v(x, \cdot)$ and $v(\cdot, y)$ are also increasing. Then $w_j(x) := \int Q_j(x, dy) \cdot v(x, y)$ is increasing in j by (ii). It follows from the isotonicity of the integral that $\int w_j \cdot dP_i$, which by Fubini's Theorem equals $\int v \cdot d(P_i \otimes Q_j)$, is increasing in j. Next, keeping j fixed, it follows from (ii) and Lemma 18.2.6 with $v_i(y)$ replaced by $v(x, y)$ and $P_i(dy)$ replaced by $Q_j(x, dy)$, that $w_j(x)$ is not only increasing in j, but also in x. Finally (i) and Lemma 18.2.6 with $v_i(x)$ replaced by $w_j(x)$ shows that $\int v \cdot d(P_i \otimes Q_j)$ is also increasing in i. Now isotonicity of $\int v \cdot d(P_i \otimes Q_j)$ in (i, j) follows from reflexivity of $\leq^I$ and $\leq^J$. □

Remark 18.2.16 The special case where $Q_j(x, dy)$ does not depend on x yields a sufficient condition for stochastic isotonicity of the family of product probability distributions $P_i \times Q_j$, $(i, j) \in I \times J$. ◇

Remark 18.2.17 In the special case of Proposition 18.2.15 where $I = J$ the condition (ii) with $j := i$, together with (i), is sufficient for stochastic isotonicity of the family $P_i \otimes Q_i$, $i \in I$, in particular for stochastic isotonicity of the family of product probability distributions $P_i \times Q_i$, $i \in I$. ◇

Proposition 18.2.18 (Stochastically increasing convolutions) *Let (P_i) and (Q_i) be families of probability distributions on $\mathfrak{B}_n$ and $\mathfrak{B}_m$, respectively. Denote convolution of probability distributions by $*$. Then:*

(a) The family $i \mapsto P_i \times Q_i$ is stochastically increasing if and only if both $i \mapsto P_i$ and $i \mapsto Q_i$ are stochastically increasing.
(b) If $m = n$ and if both (P_i) and (Q_i) are stochastically increasing then $(P_i * Q_i)$ is stochastically increasing.
(c) If $I = \mathbb{N}$ or $I = \mathbb{R}^+$, if P_i is concentrated on $\mathbb{R}^n_+$ and if (P_i) is a convolution semigroup, i.e. if

$$P_{i+j} = P_i * P_j \textit{ for } i, j \in I, \tag{18.2}$$

then (P_i) is stochastically increasing.

Proof

(a) The if-part is that special case of Remark 18.2.17 where $Q_i(x, dy)$ does not depend on x. The only-if-part follows from Proposition 18.2.11(b).
(b) This part follows from (a) and from Proposition 18.2.11(b) as $P_i * Q_i$ is the image of $P_i \times Q_i$ under the measurable increasing mapping $(x, y) \mapsto x + y$, x, $y \in \mathbb{R}^n$.
(c) Assume $i < j$ and let δ_0 be the probability distribution concentrated at $0 \in \mathbb{R}^n$. Then obviously $\delta_0 \leq^n_{\text{st}} P_{j-i}$. Now it follows from (b) that

$$P_i = P_i * \delta_0 \leq^n_{\text{st}} P_i * P_{j-i} = P_j.$$ □

Remark 18.2.19 Part (a) and its proof remain true for an arbitrary relation on $\mathbb{R}^n$. ◇

Remark 18.2.20 Many common families of probability distributions, such as $(\text{Bi}(n, p), n \in \mathbb{N})$, $(\text{Poi}(\alpha), \alpha \in \mathbb{R}^+)$, $(\Gamma_{\alpha,b}, \eta \in \mathbb{R}^+)$ and $(\chi^2_n, n \in \mathbb{N})$ have the semigroup property (18.2) and thus are stochastically increasing. ◇

Example 18.2.21 We assert that the family of two-dimensional probability distributions R_i, $i := (\alpha, \beta, b, c) \in (\mathbb{R}^+)^4$ with density

$$(x, y) \mapsto \begin{cases} \frac{\alpha^b \beta^c}{\Gamma(b)\Gamma(c)}, \cdot x^{b-1}(y-x)^{c-1} e^{-[\alpha-\beta)x+\beta y]}, & \text{if } 0 < x < y, \\ 0, & \text{otherwise,} \end{cases}$$

is stochastically decreasing in α and β and stochastically increasing in b and c. For the proof one could use Proposition 18.2.15 with $i := (\alpha, b)$ and $j := (\beta, c)$ since $R_i = P_i \otimes Q_i$ with $P_i = \Gamma_{\alpha,b}$ and $Q_i(x, \cdot)$ is $\Gamma_{\beta,c}$ shifted x units to the right. ♦

18.3 Further Concepts of Stochastic Monotonicity

In the preceding subsection we saw that it is mostly difficult to check stochastic monotonicity in the multivariate case, i.e. for families of probability distributions on $\mathbb{R}^d$ for $d > 1$. We now consider a relation $\leq_{\text{tp}}$ between probability distributions which is stronger than $\leq_{\text{st}}$, but sometimes easier to check for the following reason: properties of the set function $B \mapsto P(B)$ needed for stochastic monotonicity are taken over by properties of functions on Cartesian products in Euclidean spaces. The relation $\leq_{\text{tp}}$ is particularly useful for Bayesian models. Throughout this section we consider ordered measurable spaces $(M, \leq^d, \mathfrak{M})$ where $M \subset \mathbb{R}^d$ is the Cartesian product of Borel sets in $\mathbb{R}$, and where $\mathfrak{M}$ is the σ-algebra of Borel subsets of M. Often the usual ordering $\leq^d$ is denoted by $\leq$ if no confusion is possible. We know that M is a ***lattice***, i.e. that $x, y \in M$ implies $x \wedge y, x \vee y \in M$. Recall that $x \wedge y = (x_i \wedge y_i)_1^d$, $x \vee y = (x_i \vee y_i)_1^d$ for $x = (x_i)_1^d \in \mathbb{R}^d$ and $y = (y_i)_1^d \in \mathbb{R}^d$.

Definition 18.3.1 Let M be a Cartesian product in $\mathbb{R}^d$.

- The **tp-relation** $\leq_{\text{tp}}$ on the set of non-negative functions on M is defined as follows: $g_1 \leq_{\text{tp}} g_2$ if

$$g_1(x) \cdot g_2(y) \leq g_1(x \wedge y) \cdot g_2(x \vee y) \quad \text{for all } x, y \in M. \tag{18.3}$$

- A function $g \geq 0$ on M is called a ***TP*$_2$-function** or ***TP*$_2$** on M or totally positive function of order two if $g \leq_{\text{tp}} g$ on M, i.e. if

$$g(x) \cdot g(y) \leq g(x \wedge y) \cdot g(x \vee y) \quad \text{for all } x, y \in M. \tag{18.4}$$

Although in general the relation $\leq_{\text{tp}}$ is neither reflexive, nor transitive nor antisymmetric, it is in the literature often called the tp-ordering.

Examples of a family of TP$_2$-functions: $x \mapsto \exp(g(\sum_1^d \alpha_i \cdot x_i))$, $x = (x_i)_1^d \in \mathbb{R}_+^d$, is TP$_2$ if g is convex on $\mathbb{R}_+$ and if $\alpha_i \in \mathbb{R}_+$ for all i. Note that each function $g \geq 0$ on $M \subset \mathbb{R}$ is TP$_2$.

Definition 18.3.2 Let $M \subset \mathbb{R}^k$ be the Cartesian product of Borel sets in $\mathbb{R}$.

- Let P_1 and P_2 be probability distributions on $\mathfrak{M}$. We say that P_1 is **smaller** than P_2 in the sense of the tp-relation (**$P_1 \leq_{tp} P_2$** for short), if P_1 and P_2 have densities g_1 and g_2, respectively, with respect to the same σ-finite product measure (possibly dependent on P_1, P_2) on $\mathfrak{M}$ such that $g_1 \leq_{tp} g_2$.
- Let $(I, \leq^I)$ be structured. A family $(P_i, i \in I)$ of probability distributions on $\mathfrak{M}$ is said to be ***$\leq_{tp}$-increasing*** if $P_i \leq_{tp} P_j$ whenever $i \leq^I j$.

Remark 18.3.3 In case $d = 1$ the relation $\leq_{tp}$ is also called the ***likelihood ratio relation*** and denoted by $\leq_{lr}$. In this case $\leq_{lr}$-increasing families $(P_i, i \in I)$ are also called families with *increasing likelihood ratio*. The name increasing likelihood ratio stems from the fact that in case $d = 1$ the condition (18.3) holds if and only if g_2/g_1 (with $x/0 := \infty$ for $x \in \mathbb{R}^+$) is increasing on the set $[g_1 > 0] \cup [g_2 > 0]$. ◇

Remark 18.3.4 Obviously (18.4) holds trivially for all pairs x, y for which $(g_1(x) = 0) \vee (g_2(y) = 0)) \vee (x \leq y)$. ◇

Remark 18.3.5 In applications often $\leq_{tp}$-increasing families (P_i) have the even stronger property that all P_i have densities g_i with respect to a fixed σ-finite measure ν (mostly Lebesgue measure or counting measure) such that $g_i \leq_{tp} g_j$ for $i \leq^I j$. Then the functions g_i are TP_2, if $\leq^I$ is reflexive. ◇

Proposition 18.3.6 ($\leq_{tp}$-increasing families of probability distributions) *If $(P_i, i \in I)$ is $\leq_{tp}$-increasing then it is also stochastically increasing; the converse does not hold.*

Proof We give here a proof for the case $n = 1$. A proof for the general case is given in Karlin and Rinott (1980, Theorem 2.2). It suffices to assume $I := \{1, 2\}$. Let each P_i, $i \in I$, have a μ-density g_i with respect to some σ-finite measure. We apply the characterization of stochastic monotonicity given in Proposition 18.2.11(a) with $\mathfrak{F}$ as the set of bounded, increasing and measurable functions on $\mathbb{R}$. Put $G := [g_1 > 0] \cup [g_2 > 0]$, $B := [g_2 \leq g_1] \cap G$ and $C := [g_2 > g_1] \cap G$. We assert that $b := \sup v|B \leq c := \inf v|C$ for each $v \in \mathfrak{F}$. In fact, if $x \in B$ and $y \in C$ then $g_2(x)/g_1(x) \leq 1 < g_2(y)/g_1(y)$. As g_2/g_1 is increasing on $B + C = G$ the relation $y \leq x$ cannot hold. Thus $x \leq y$, hence $v(x) \leq v(y)$. This shows that $b \leq c$. Now

$$P_i((B + C)^c) = P_i(G^c) \leq P_i(g_i = 0) = 0$$

for $i = 1, 2$ implies

$$\begin{aligned}
&\int v\, dP_1 - \int v\, dP_2 \\
&= \int_{B+C} v\, dP_1 - \int_{B+C} v\, dP_2 = \int_B v \cdot (g_1 - g_2)\, d\mu - \int_C v \cdot (g_2 - g_1)\, d\mu \\
&\leq b \cdot \int_B (g_1 - g_2)\, d\mu - c \cdot \int_C (g_2 - g_1)\, d\mu \leq b \cdot \int_{B+C} (g_1 - g_2)\, d\mu \\
&= b[P_1(B + C) - P_2(B + C)] = 0.
\end{aligned}$$

□

Example 18.3.7 (Monotone likelihood ratio)

(a) The family of binomial distributions $\mathrm{Bi}(n,p)$, $0 \le p \le 1$ and n fixed, has increasing likelihood ratio. In fact, for $0 \le p_1 < p_2 \le 1$ denote by $q(x)$, $x \in TR := [p_1 > 0] \cup [p_2 > 0]$ the ratio of the discrete densities of $\mathrm{Bi}(n,p_2)$ and $\mathrm{Bi}(n,p_1)$. Put $q_1 = 1 - p_1$ and $q_2 = 1 - p_2$ and observe $0^0 := 1$.

Case 1: If $(0 < p_1) \wedge (p_2 < 1)$, then $x \mapsto q(x) = (q_2/q_1)^n\,(p_2q_1/p_1q_2)^x$ is increasing on $TR = \mathbb{N}_{0,n}$ since $p_2q_1/p_1q_2 > 1$.
Case 2: If $(p_1 = 0) \vee (p_2 = 1)$, then q is increasing on $TR = \{0, n\}$ since $q(0) = 0$ and $q(n) = \infty$.

(b) Many stochastically monotone families of probability distributions on the real line (cf. Table 18.1) even have monotone likelihood ratio. In some cases the latter property is easier to verify. As an example, let $I := (\mathbb{R}^+)^2$, endowed with the SE-ordering. We assert that for each probability distribution ν on $(0, 1)$ the family $P_{\alpha,\gamma}(dx)$ with ν-density proportional to $x^{\alpha-1}\cdot(1-x)^{\gamma-1}\,\nu(dx)$ has increasing likelihood in $i = (\alpha,\gamma)$, hence is stochastically increasing in α and stochastically decreasing in γ. (These two properties hold in particular for the family of beta distributions $\mathrm{Be}(\alpha,\gamma)$, as seen by choosing for ν the Lebesgue measure on $(0, 1)$.)

For the proof select $x \ge y$, $i = (\alpha,\gamma)$ and $j = (\alpha',\gamma')$ with $i \le^{\mathrm{SE}} j$, i.e. with $\alpha \le \alpha'$, $\gamma \ge \gamma'$. Put $g_{\alpha,\gamma}(x) := x^{\alpha-1}\,(1-x)^{\gamma-1}$, $x \in (0, 1)$. A simple calculation shows that $g_i(x)\cdot g_j(y) \le g_i(y)\cdot g_j(x)$ is equivalent to

$$(y/x)^{\alpha'-\alpha} \le ((1-y)/(1-x))^{\gamma-\gamma'}.$$

This holds since $y/x \le 1 \le (1-y)/(1-x)$ and since the exponents are non-negative. ♦

Lemma 18.3.8 (Properties of TP$_2$-functions) *Let $X \subset \mathbb{R}^k$ and $Y \subset \mathbb{R}^m$ be Cartesian products.*

(a) If f is TP$_2$ on X then $(x,y) \mapsto g(x,y) := f(x)$ is TP$_2$ on $X \times Y$.
(b) If $f \le_{\mathrm{tp}} g$ on X then $c\cdot f \le_{\mathrm{tp}} d\cdot g$ for all c, $d \in \mathbb{R}_+$. In particular, if g is TP$_2$ then $c\cdot g$ is TP$_2$ for each $c \in \mathbb{R}_+$.
(c) If $f_1 \le_{\mathrm{tp}} g_1$ on X and $f_2 \le_{\mathrm{tp}} g_2$ on X, then $f_1\cdot f_2 \le_{\mathrm{tp}} g_1\cdot g_2$ on X. In particular, each finite product of TP$_2$-functions on X is TP$_2$ on X.
(d) If $f_1 \le_{\mathrm{tp}} g_1$ on X and if $f_2 \le_{\mathrm{tp}} g_2$ on Y, then $f_1 \times f_2 \le_{\mathrm{tp}} g_1 \times g_2$ on $X \times Y$, where $(f_1 \times f_2)(x,y) := f_1(x)\cdot f_2(y)$. Moreover, each product $\times_1^k f_i$ of k TP$_2$-functions on X is TP$_2$ on X^k.
(e) If g is TP$_2$ on $X \times Y$ then $g(\cdot,y) \le_{\mathrm{tp}} g(\cdot,y')$ on X for y, $y' \in Y$ with $y \le y'$. (In particular, $g(\cdot,y)$ is TP$_2$ on X for all $y \in Y$.) The converse holds in case $d = 1$.
(f) Let $X \subset \mathbb{R}^k$ and $Y \subset \mathbb{R}^m$ be Cartesian products of Borel sets in $\mathbb{R}$, and endow X with the σ-algebra $\mathfrak{X}$ of its Borel sets. Let $g_1 \ge 0$ and $g_2 \ge 0$ be functions on $X \times Y$, measurable in x, with $g_1 \le_{\mathrm{tp}} g_2$.

(f1) If κ is a σ-finite measure on $\mathfrak{X}$ then

$$\int \kappa(dx)\, g_1(x,\cdot) \leq_{\text{tp}} \int \kappa(dx)\, g_2(x,\cdot) \text{ on } Y,$$

provided all these integrals are finite.

(f2) If P_1 and P_2 are probability distributions on $\mathfrak{X}$ such that $P_1 \leq_{\text{tp}} P_2$, then

$$\int P_1(dx)\, g_1(x,\cdot) \leq_{\text{tp}} \int P_2(dx)\, g_2(x,\cdot) \text{ on } Y,$$

provided all these integrals are finite.

The proof of (a)–(e) is straightforward and therefore omitted. For (f1) we refer to Karlin and Rinott (1980, 2.4).

For (f2) we note that $P_1(dx) = f_1(x) \cdot \kappa(dx)$ and $P_2(dx) = f_2(x) \cdot \kappa(dx)$ for some functions $f_1 \geq 0, f_2 \geq 0$ on X with $f_1 \leq_{\text{tp}} f_2$ and some σ-finite measure κ on $\mathfrak{X}$. Then $h_1 := f_1 \cdot g_1$ and $h_2 := f_2 \cdot g_2$ satisfy $h_1 \leq_{\text{tp}} h_2$ on $X \times Y$ by (a) and (c). Now the assertion follows from (f1) with g_1 replaced by h_1 and g_2 replaced by h_2. □

Example 18.3.9 As an application of Lemma 18.3.8(d) let $(I, \leq^I)$ be structured, let P_{ij}, $i \in I$, $1 \leq j \leq n$, be probability distributions on Cartesian products $M_j \in \mathfrak{B}_{n_j}$. If $i \mapsto P_{ij}$ has increasing likelihood for each j then $i \mapsto \times_{j=1}^n P_{ij}$ has increasing likelihood.

For the proof we select $i \leq^I i'$. By assumption the probability distributions P_{ij} and $P_{i'j}$ have densities g_{ij} and $g_{i'j}$, respectively, with respect to some σ-finite measure ν_j such that $g_{ij} \leq_{\text{tp}} g_{i'j}$. Then the probability distributions $\times_{j=1}^n P_{ij}$ and $\times_{j=1}^n P_{i'j}$ have the densities $\prod_{j=1}^n g_{ij}$ and $\prod_{j=1}^n g_{i'j}$, respectively, with respect to the σ-finite measure $\times_{j=1}^n \nu_j$. Now the assertion follows from Lemma 18.3.8(d). ♦

Chapter 19
Concavity and Convexity of the Value Functions and Monotone Maximizers

We now extend some of the results in Chaps. 7 and 8 to CMs and MDPs. The latter are considerably more difficult to deal with than CMs.

19.1 Concave and Convex Value Functions

Consider a CM (or an MDP) as defined in Chap. 16 and where the restriction set D (and hence S) is a convex set in a Euclidean space. Before looking for conditions under which the value functions are concave or convex, we must ensure that they are finite, since we only consider finite convex or concave functions. We separate this problem and also the validity of the VI from the present study by using the following requirement (where F stands for *finite*):

Assumption (VIF): The VI holds, and the functions V_n and W_n are finite for all $n \geq 1$.

The validity of the VI is treated under several types of assumptions in Chap. 16. The finiteness requirement is fulfilled if the CM (or the MDP) has a bounding function, in particular if r_Z (or r_S) and V_0 are bounded. Note also that by Appendix D.2.4 concave functions have affine upper bounds.

As a rule one verifies (strict) concavity of the V_n's by induction on $n \geq 0$. In the induction step one ensures that (strict) concavity of $v := V_{n-1}$ implies the same property for $(s, a) \mapsto \mathbb{E}\, v(T(s, a, \eta_1))$. This problem is typically solved by combining assumptions which ensure (strict) concavity of $(s, a) \mapsto v(T(s, a, z))$ for Q-almost all z with Appendix D.2.14 about the preservation of convexity by integration.

After these preparations it is rather easy to extend most results in Chap. 7 to the present situation. Often we omit proofs and refer to the corresponding deterministic version only. The reader should have no difficulty in extending more results from Chap. 7.

K. Hinderer et al., *Dynamic Optimization*, Universitext,
DOI 10.1007/978-3-319-48814-1_19

Theorem 19.1.1 (Concavity of value functions; cf. Proposition 7.2.1) *Consider a CM which fulfills (VIF) and for which:*

(i) D is convex,
(ii) there exists a relation $\leq_S$ on S such that $T(s,a,z)$ is $\leq_S$-convex in (s,a) for Q-almost all $z \in Z$,
(iii) r and V_0 are concave,
(iv) all functions V_n, $n \geq 0$, are $\leq_S$-decreasing.

Then the following holds:

(a) V_n and $W_n(s,\cdot)$, $s \in S$, are concave for all $n \geq 1$.
(b) If in addition to (i)–(iv) the function $r(s,\cdot)$, $s \in S$, is strictly concave then $W_n(s,\cdot)$, $n \geq 1$, $s \in S$, is strictly concave, and at each stage there exists at most one maximizer.
(c) Assume that in addition to (i)–(iv) the mapping $T(\cdot,\cdot,z)$ is injective for Q-almost all $z \in Z$, V_0 is strictly concave and that there exists at each stage $n \geq 1$ a maximizer f_n. Then V_n and $a \mapsto W_n(s,a)$, $n \geq 1$, $s \in S$, are strictly concave, and the f_n's are unique.

The *minimization version* of Theorem 19.1.1 has the following form:

Theorem 19.1.2 (Convexity of value functions; cf. Proposition 7.2.1) *Consider a CM which fulfills (VIF) and with the following properties:*

(i) D is convex,
(ii) there exists a relation $\leq_S$ on S such that $T(s,a,z)$ is $\leq_S$-convex in (s,a) for Q-almost all $z \in Z$,
(iii) c and C_0 are convex,
(iv) all functions C_n, $n \geq 0$, are $\leq_S$-increasing.

Then C_n and $W_n(s,\cdot) := c(s,\cdot) + \beta \cdot \mathbb{E}\, C_{n-1} \circ T(s,\cdot,\eta_1)$ are convex for all $n \geq 1$, $s \in S$.

Example 19.1.3 (Minimization of the stochastic version of the continuous allocation problem) Here $S = A = \mathbb{R}_+$, $D(s) = [0,s]$, $T(s,a,z) = z \cdot (s-a)$ and

$$c(s,a) := u_0(s) + u_1(a) + u_2(s-a)$$

with increasing, convex and non-negative u_0, u_1, u_2 and C_0. Finiteness of C_n for all n follows from non-negativity of c and C_0. Using the transformation $a' := a/s$, one proves that C_n is increasing for all n. We cannot use this transformation for verifying convexity of C_n as $T'(s,a') := s \cdot (1-a')$ is not concave. On the other hand, the minimization version of Theorem 19.1.1 above leads immediately to the desired result. ♦

Theorem 19.1.4 (Convexity of value functions when $D(s)$ is independent of s; cf. Proposition 7.2.1) *Assume that a CM fulfills (VIF) and that*

(i) S is convex, A is an arbitrary set and $D(s) = A$ for all s,
(ii) $T(s, a, z)$ is convex [concave] in s for all $a \in A$ and Q-almost all z,
(iii) $r(s, a)$ and $V_0(s)$ are convex in s,
(iv) V_n is increasing [decreasing] for all $n \geq 0$.

Then V_n is convex for all $n \geq 1$.
The same holds if (ii) $\wedge$ (iv) are replaced by

(v) $T(s, a, z)$ is affine in s for all a and Q-almost all z.

Theorem 19.1.5 (Convexity of value functions in invariant MDPs when $D(s)$ is independent of s) *Consider an MDP which fulfills (VIF) and which satisfies*

(i) S is convex, A is an arbitrary set and $D(s) = A$ for all s,
(ii) the MDP is invariant, i.e. $P(s, a, ds')$ does not depend on s,
(iii) $r(s, a)$ and $V_0(s)$ are convex in s.

Then V_n is convex for all $n \geq 1$.

The results in Chap. 7 about CMs with discretely concave or discretely convex value functions can be extended in a similar way.

For the next result recall that the set of vertices of a polytope A is denoted by $\mathrm{vert}(A)$.

Proposition 19.1.6 (Existence of bang-bang maximizers; cf. Theorem 7.2.3) *Consider a CM with the following properties:*

(i) S is convex and $D(s) = A$, $s \in S$, for a polytope A,
(ii) $T(s, a, z)$ is affine in s for Q-almost all z and affine in a for Q-almost all z,
(iii) $r(s, a)$ is convex in s and in a, and V_0 is convex,
(iv) the CM has a bounding function b such that $\mathbb{E}\, b(T(s, a, \eta_1)) < \infty$ for all $(s, a) \in D$.

Then the (VI) holds and for $n \geq 1$ we have:

(a) V_n is convex and

$$V_n(s) = \max_{a \in \mathrm{vert}(A)} W_n(s, a), \; s \in S.$$

(b) If A is endowed with any total ordering (e.g. the lexicographic ordering), then $s \mapsto f_n(s) :=$ smallest maximum point of $W_n(s, \cdot)$ on $\mathrm{vert}(A)$ is a bang-bang maximizer at stage n.

Proof

(a) (MA1) holds since the CM has a bounding function. Now we use the Structure Theorem 16.1.12 with $\mathfrak{V}$ as the set of Borel-measurable and convex functions v

on S such that $|v| \leq \delta b$ for some $\delta \in \mathbb{R}_+$, possibly depending on v. By (iv) we have $\mathfrak{V} \subset \mathfrak{V}_0$, and $V_0 \in \mathfrak{V}$ by (iii), hence $\mathbb{E}\,|v(T(s,a,\eta_1)| \leq \delta \cdot b(s)$.

The Structure Theorem 16.1.12 yields the assertion, provided (S1) and (S2) are fulfilled. For checking (S2) select $v \in \mathfrak{V}$. Then $Lv(s,a) = r(s,a) + \beta \cdot \mathbb{E}\,v(T(s,a,\eta_1))$ is measurable in (s,a), convex in s and convex in a by (ii). Now finiteness of vert(A) implies that $s \mapsto Uv(s) = \max_{a \in \mathrm{vert}(A)} Lv(s,a)$ is measurable and convex. Moreover, $Uv \in \mathfrak{V}$ since

$$|Uv(s)| \leq \max_a \left[|r(s,a)| + \beta \cdot |\mathbb{E}\,v(T(s,a,\eta_1))|\right]$$
$$\leq \max_a \left(\delta \cdot b(s) + \beta\delta \cdot b(s)\right) \leq c \cdot (1+\beta) \cdot b(s).$$

(S1) also holds since Lv has the maximizer $s \mapsto f_v(s) :=$ smallest maximum point of $Lv(s,\cdot)$ on vert(A).

(b) This part follows obviously from part (a). □

19.2 Monotone Maximizers

For notions such as monotonely complete correspondences or functions with increasing differences the reader should consult Chap. 8.

Assume that $(S, \leq_S)$ is structured, that $(A, \leq_A)$ is totally ordered, that D is monotonely complete and that (VIF) holds. Assume further that the σ-algebras $\mathfrak{S}$ and $\mathfrak{A}$ are such that increasing mappings f from S into A with $f(s) \in D(s)$ for all s, are measurable. (This property holds, for example, when S and A are Borel sets in $\mathbb{R}$. It does not suffice that S and A are Borel sets in $\mathbb{R}^d$ for $d > 1$.)

Proposition 19.2.1 (Increasing maximizers in invariant MDPs) *Assume that an MDP fulfills (VIF) and that*

(i) S and A are Borel sets in $\mathbb{R}$,
(ii) D is monotonely complete,
(iii) the MDP is invariant,
(iv) r has increasing differences,
(v) $W_n(s,\cdot)$ has a smallest [largest] maximum point $f_n(s)$ for all s and n.

Then f_n is an increasing maximizer at stage $n \geq 1$.

Proof For the proof see Theorem 8.2.9. □

Example 19.2.2 (Concavity of the minimal cost functions in the time slot queueing model Example 15.11*)* By (15.9) the VI in Example 15.11 has the form

$$C_n(s) - h(s) = \min\{R + \beta \cdot \mathbb{E}\,C_{n-1}((s-X_1)^+),\ \beta \cdot \mathbb{E}\,C_{n-1}((s-X_1)^+ + Y_1)\}. \quad (19.1)$$

Is it true that $C_n(s)$ is concave in s if h and C_0 are concave? It follows easily from (19.1) by induction on $n \geq 0$, observing that the minimum of two concave functions is concave, that $C_n(s)$ is concave in s on $\{n, n+1, \ldots\}$, but in general the convex term $s \mapsto (s - X_1)^+$ in (19.1) excludes concavity of C_n on the whole state space $\mathbb{N}_0$.

Intuitively one expects that one should always accept the arriving batch if the penalty R for rejection is very large. We now prove a result of this type which also includes concavity of the minimal cost functions on the whole state space. ♦

Proposition 19.2.3 (Solution of the queueing problem from Example 15.11 for large penalty cost) *Assume that*

(i) $0 < \mathbb{E}\, Y_1 < 1$,
(ii) h *and* C_0 *are increasing and concave,*
(iii) $\beta < 1$, $d := 2h(1) - h(2) > 0$ *and*

$$R \geq \beta \cdot \mathbb{E}\, Y_1 \cdot \max\left(C_0(1), \frac{h(1)}{1-\beta}\right),$$

$$\beta \leq d \cdot \min\left(\frac{1}{h(1)+d}, \frac{1}{C_0(1)}\right).$$

Then each minimal cost function C_n, $n \geq 0$, *is concave on* $\mathbb{N}_0$, *and it is optimal for each* N*-stage problem to accept all arriving batches.*

Proof Let f_n be the largest minimizer at stage n. Put $X := X_1$ and $Y := Y_1$.

(a) We have seen in Example 15.11 that C_n is increasing and finite for all $n \geq 0$ since h and C_0 are increasing.
(b) Consider for $n \geq 0$ the condition
(A_n) $\beta \cdot \mathbb{E}\, Y \cdot (C_n(1) - C_n(0)) \leq R$.
Fix $n \geq 1$ and assume (A_{n-1}) and that $v := C_{n-1}$ is concave. We assert that then $f_n \equiv 1$. For the proof we note that concavity of v yields

$$v(z+y) - v(z) = \sum_{i=z}^{z+y-1} (v(i+1) - v(i)) \leq y \cdot (v(1) - v(0)), \; y, z \in \mathbb{N}_0. \tag{19.2}$$

Now we obtain, using (19.1) and (19.2) with $y := Y$, $z := (s - X)^+$ and (A_{n-1}) that for $n \geq 1$

$$\begin{aligned} &W_n(s, 1) - h(s) \\ &= \beta \cdot \mathbb{E}\, v((s-X)^+ + Y) \leq \beta \cdot [\mathbb{E}\, Y \cdot (v(1) - v(0)) + \mathbb{E}\, v((s-X))^+] \\ &\leq R + \beta \cdot \mathbb{E}\, v((s-X))^+ = W_n(s, 0) - h(s), \quad s \geq 0. \end{aligned}$$

(c) Consider for $n \geq 0$ the condition
(B_n) $\beta \cdot (C_n(1) - C_n(0)) \leq d$.
We prove by induction on $n \geq 0$ the assertion
(I_n): C_n is concave and $(A_n) \wedge (B_n)$ holds, hence $f_{n+1} \equiv 1$ by (b).
Firstly, (I_0) is ensured by (ii), (iii) and $C_0(0) = 0$.
Assume (I_{n-1}) for some $n \geq 1$, hence $f_n \equiv 1$. Put $v := C_{n-1}$.

(c1) We show that (A_n) holds. In fact, isotonicity and concavity of v and (19.1) yield

$$\begin{aligned} C_n(1) - C_n(0) &= h(1) + \beta \cdot \mathbb{E}[v(1 - X + Y) - v(Y)] \\ &\leq h(1) + \beta \cdot \mathbb{E}[v(Y + 1) - v(Y)] \leq h(1) + \beta \cdot [v(1) - v(0)]. \end{aligned} \tag{19.3}$$

This implies by (iii) and (A_{n-1}) that

$$\beta \cdot \mathbb{E}\, Y \cdot (C_n(1) - C_n(0)) \leq (1 - \beta) \cdot R + \beta \cdot R = R.$$

(c2) Next, (B_n) holds since (19.3), (iii) and (B_{n-1}) yield $\beta \cdot (C_n(1) - C_n(0)) \leq \beta \cdot (h(1) + d) \leq d$.

(c3) Fix $n \geq 1$. We show that C_n is concave. Firstly the VI shows that concavity of v on $\mathbb{N}_0$ implies concavity of C_n on $\mathbb{N}$. Thus it suffices to show that $2 \cdot C_n(1) \geq C_n(0) + C_n(2)$. This holds since $f_n \equiv 1$, if and only if

$$2h(1) + 2\beta \cdot \mathbb{E}\, v(1 - X + Y) \geq \beta \cdot \mathbb{E}\, v(Y) + h(2) + \beta \cdot \mathbb{E}\, v(2 - X + Y),$$

if and only if

$$\beta \cdot \mathbb{E}[v(2 - X + Y) - 2 \cdot v(1 - X + Y) + v(Y)] \leq d.$$

The latter inequality holds, since X assumes only the values zero and one, if for all $y \in \mathbb{N}_0$ we have $\beta \cdot [v(2 + y) - 2 \cdot v(1 + y) + v(y)] \leq d$ and $\beta \cdot [v(1 + y) - v(y)] \leq d$. These inequalities follow from $d \geq 0$, concavity of v and from (B_{n-1}). □

Chapter 20
Markovian Decision Processes with Large and with Infinite Horizon

We now extend the investigation in Sect. 12.2 to MDPs with arbitrary transition law. The general issues arising in the case of a large or infinite horizon are described in detail at the beginning of Chap. 10 and in Sect. 10.5.

20.1 Large Horizon

Recall that the problem of the validity of the VI has been treated in Chaps. 14 and 16. We now prove a stochastic counterpart to Proposition 10.1.5(b).

Proposition 20.1.1 (Validity of the optimality equation) *Assume that (MA2), the VI and one of the following two conditions hold.*

(a) $V_0 \geq v$ *for some measurable* v *on* S *such that* $P|v| < \infty$, *and* $V_1 \geq V_0$.
(b) The value functions are finite and converge uniformly.

Then V *exists, belongs to* $\mathfrak{V}_0$ *and satisfies the optimality equation.*

Proof Existence of V is obvious as $(V_n)_0^\infty$ is increasing. It follows under condition (a) and (b), respectively, from the Monotone Convergence Theorem that $V \in \mathfrak{V}_0$ and that $PV_n \uparrow PV$ and $PV_n \to PV$, respectively, for $n \to \infty$. Therefore $W_n \to W = LV$. For fixed s we obtain under condition (a) from Lemma 10.1.4(a1) with $g_n(a) := W_n(s, a)$, as $W_n \uparrow W$, that

$$V_n(s) = \sup_a W_n(s, a) \to \sup_a W(s, a) = UV(s),$$

hence $V = UV$. Under condition (b) we use Lemma 10.1.4(a2), observing that (g_n) converges uniformly as

$$|W_n - W| \leq \beta\, |P(V_n - V)| \leq \beta\, P|V_n - V| \leq \beta \sup |V_n - V|. \qquad \square$$

K. Hinderer et al., *Dynamic Optimization*, Universitext,
DOI 10.1007/978-3-319-48814-1_20

We need the following preparation. Concepts related to bounding functions were introduced in Chap. 15.

Lemma 20.1.2 (*U* **is Lipschitz on** MB_b **with constant** β_b) *If b is a bounding function of the MDP, then U and* $U_f, f \in \mathbb{F}$, *map* MB_b *into* B_b *and*

$$\|Uv - Uw\|_b \leq \sup_{f \in \mathbb{F}} \|U_f v - U_f w\|_b \leq \beta_b \, \|v - w\|_b \text{ for } v, w \in MB_b\,.$$

Proof The proof is nearly the same as for Lemma 10.1.9. One uses

$$\beta \cdot |Pv| \leq \beta P|v| \leq \beta_b \cdot \|v\|_b \cdot b(\cdot) \text{ for } v \in MB_b\,.$$

□

Theorem 20.1.3 (The optimality equation and asymptotically optimal decision rules) *Assume that the MDP has a bounding function b with* $\beta_b < 1$. *Then we have:*

(a) *The sequence of value functions* V_N, $N \in \mathbb{N}_0$, *converges in b-norm to* $V \in B_b$, *and V is independent of* $V_0 \in MB_b$. *Moreover,*

$$\|V\|_b \leq \|r\|_b / (1 - \beta_b)$$

and

$$\|V - V_N\|_b \leq \Big[\|r\|_b/(1 - \beta_b) + \|V_0\|_b\Big] \beta_b^N, \; N \in \mathbb{N}. \tag{20.1}$$

(b) *Assume that (MA2) and the VI hold.*

(b1) *V is the unique solution of the optimality equation within* MB_b *and*

$$\|V - V_N\|_b \leq \|V - V_0\|_b \cdot \beta_b^N, \; N \in \mathbb{N}. \tag{20.2}$$

(b2) *The decision rule f is asymptotically optimal if and only if it is a maximizer of LV. And then* $(V_{Nf})_1^\infty$ *converges in b-norm towards V, and*

$$\|V_N - V_{Nf}\|_b \leq 2\, \beta_b^N \cdot \|V - V_0\|_b, \; n \in \mathbb{N}.$$

Proof

(a1) By Lemma 16.1.18 the value functions exist and belong to the Banach space B_b. We prove the existence of *V* by showing that (V_n) is a Cauchy sequence.

(a11) Consider the MDP′ which differs from MDP by $r_S' :\equiv 0$ and $V_0' := b$. Then *b* is also a bounding function of MDP′ and $V_{\nu\pi}' = \beta^\nu \cdot \mathbb{E}_{\pi s}\, b(\zeta_\nu) \leq V_\nu'$. Now we obtain from Lemma 16.1.18, using $\|b\| = 1$, that

$$\beta^\nu \cdot \mathbb{E}_{\pi s}\, b(\zeta_\nu) \leq \beta_b^\nu \cdot b(s) \text{ for } s \in S \text{ and } \nu \in \mathbb{N}_0. \tag{20.3}$$

(a12) For $n \leq m$ and $\pi = (\pi_\nu)_0^{m-1}$ define $\sigma = (\pi_\nu)_0^{n-1}$ and $V_{n\pi} := V_{n\sigma}$. Then

$$V_{m\pi}(s) - V_{n\pi}(s) = \mathbb{E}\left[\sum_{\nu=n}^{m-1} \beta^\nu r_{\pi_\nu}(\zeta_\nu) + \beta^m V_0(\zeta_m) - \beta^n V_0(\zeta_n)\right].$$

Therefore, using (20.3), we obtain

$$\begin{aligned}|V_{m\pi}(s) - V_{n\pi}(s)| &\leq \sum_n^{m-1} \beta^\nu \, \mathbb{E}\,|r_{\pi_\nu}(\zeta_\nu)| + \beta^m \, \mathbb{E}\,|V_0(\zeta_m)| + \beta^n \, \mathbb{E}_{\pi s}\,|V_0(\zeta_n)| \\ &\leq \Big[\|r\| \sum_n^{m-1} \beta_b^\nu + \|V_0\|(\beta_b^m + \beta_b^n)\Big] b(s).\end{aligned}$$

It follows that

$$\|V_{m\pi} - V_{n\pi}\| \leq \Big[\|r\|/(1-\beta_b) + \|V_0\|\Big]\beta_b^m + \|V_0\|\beta_b^n) =: d(m,n).$$

As

$$|V_m(s) - V_n(s)| \leq \sup_\pi |V_{m\pi}(s) - V_{n\pi}(s)| \leq d(m,n) \cdot b(s), \tag{20.4}$$

and as $d(m,n) \leq d(n,n)$ for $m \geq n$, we obtain

$$\|V_m - V_n\| \leq d(n,n) \to 0 \text{ for } n \to \infty.$$

Thus (V_n) is a Cauchy sequence. Moreover, letting m tend to ∞ in (20.4), we obtain (20.1) and also independence of V from V_0.

(b1) We have for all n, using Lemma 20.1.2,

$$\|UV - V_{n+1}\| = \|UV - UV_n\| \leq \beta_b \|V - V_n\|.$$

Therefore

$$\|UV - V\| \leq \|UV - V_{n+1}\| + \|V_{n+1} - V\| \leq \beta\|V - V_n\| + \|V_{n+1} - V\|.$$

By (20.1) the latter sum tends to 0 for $n \to \infty$. Thus V is a fixed point of U. It is easily seen to be the unique fixed point of U within MB_b as $\beta_b < 1$. The inequality (20.2) is easily derived using Lemma 20.1.2.

(b2) The proof is exactly the same as for (12.13). □

Remark 20.1.4 The assumption in Theorem 20.1.3(b) is fulfilled if the conditions (S1)–(S2) of the Structure Theorem 16.1.12 hold for an arbitrary set $\mathfrak{V}$ of functions on S. ◇

Example 20.1.5 (The stochastic linear-quadratic system from Example 16.1.14*)* We exclude the deterministic version by requiring positive variance $\mathbb{E}\,\eta_1^2 > 0$. Recall that we assumed $\mathbb{E}\,\eta_1 = 0$, $\lambda, \gamma \in \mathbb{R}^+$, $g, h \neq 0$ and $C_0(s) = d_0 s^2$, $s \in \mathbb{R}$, for some $d_0 \in \mathbb{R}_+$. We use an ad hoc method, as follows. In Example 16.1.14 we saw that

$$C_n(s) = d_n s^2 + e_n \cdot \mathbb{E}\,\eta_1^2, \quad f_n(s) = \frac{h \cdot (d_n - \lambda)}{\gamma g} \cdot s, \quad n \geq 1,\ s \in \mathbb{R}.$$

Here (d_n) and (e_n) can be computed recursively by $e_0 := 0$ and

$$d_{n+1} = \lambda + \frac{\beta\gamma g^2 \cdot d_n}{\gamma + \beta h^2 \cdot d_n}$$

$$e_{n+1} = \beta \cdot (d_n + e_n), \tag{20.5}$$

$$= \sum_{t=1}^{n+1} \beta^t d_{n+1-t}, \quad n \geq 0. \tag{20.6}$$

In Example 10.3.8 we have seen that d_n is monotonely converging to a positive finite limit d for each $\beta \in \mathbb{R}^+$, and that d can be computed as the unique positive solution of the quadratic equation (10.20). We now show for the stochastic version of Example 16.1.14:

(a) There exists $C(s) := \lim_{n\to\infty} C_n(s)$, $s \in \mathbb{R}$, and

$$C(s) = \begin{cases} ds^2 + \frac{\beta d}{1-\beta} \cdot \mathbb{E}\,\eta_1^2, & \text{if } \beta < 1, \\ \infty, & \text{if } \beta \geq 1. \end{cases}$$

(b) If $\beta < 1$, the limit of the decision rules f_n, i.e. the decision rule

$$s \mapsto f(s) := h(d - \lambda)s/(\gamma g)$$

is the unique minimizer of LC and also asymptotically optimal.

Note the difference with the deterministic case, where by Example 10.3.8 for all $\beta \in \mathbb{R}^+$ the function C is finite and f is asymptotically optimal.

Proof

(a) Since d_n converges to d, it suffices to consider the convergence of e_n for $n \to \infty$.

Case 1: $\beta \geq 1$. From (20.5) and (20.6) we see that $e_n \geq \beta \cdot (\lambda + e_{n-1}) \geq \lambda + e_{n-1}$ for $n \geq 2$. Now induction on $n \geq 2$ yields $e_n \geq (n-1)\lambda \to \infty$ for $n \to \infty$.

Case 2: $\beta < 1$. For $\varepsilon > 0$ there exists an $n_0 = n_0(\varepsilon) \geq 1$ such that $|d - d_n| \leq \varepsilon$ for $n \geq n_0$. From (20.5) we get by induction on $n \geq 0$ that

$$e_{n_0+n} \leq \beta \cdot (d + \varepsilon) \cdot \sigma_n(\beta) + \beta^n \cdot e_{n_0}.$$

Thus $\limsup_n e_n \leq \beta \cdot (d + \varepsilon)/(1 - \beta)$ for all $\varepsilon > 0$, hence $\limsup_n e_n \leq \beta \cdot d/(1 - \beta)$. In a similar way one shows that $\liminf_n e_n \geq \beta \cdot d/(1 - \beta)$. This proves (a).

(b) It is easy to see that $U_f C = C$. Put $\delta := g - h^2 \cdot (d - \lambda)/(\gamma g)$, hence $gs - hf(s) = \delta s$. From the proof of Example 10.3.8 we know that $\mu := \beta\delta^2 < 1$. Now induction on $N \geq 0$, using that $C - C_{N+1f} = U_f C - U_f C_{Nf}$, yields

$$|C(s) - C_{Nf}(s)| \leq |d - d_0| \cdot \left[\mu^N \cdot s^2 + N \cdot (\beta \vee \mu)^N \cdot \mathbb{E}\,\eta_1^2\right] + \beta^N \alpha \cdot \mathbb{E}\,\eta_1^2,$$

where $\alpha := \beta d/(1 - \beta)$. Since β, μ and hence also $\beta \vee \mu$ are smaller than 1, $C(s)(s) - C_{Nf}(s)$ converges to zero for $N \to \infty$. Now the assertion follows from (a) since

$$C_N(s) - C_{Nf}(s) = [C_N(s) - C(s)] + [C(s) - C_{Nf}(s)]. \qquad \square\blacklozenge$$

20.2 Infinite Horizon

MDPs with infinite horizon (MDP_∞) are defined in the same way as MDPs except that no terminal reward function V_0 is required. For the definition of the expected ∞-stage reward earned under policy $\pi = (\pi_\nu)_0^\infty \in \pi := \mathbb{F}^{\mathbb{N}_0}$ and initial state s we need an infinite sequence of state random variables ζ_ν, $\nu \in \mathbb{N}$, on some probability space $(\Omega, \mathfrak{F}, \mathbb{P}_\pi(s, \cdot))$ which describes the infinite-stage decision process. This is the same situation as in Sect. 12.3, except that now the state space is arbitrary. Again we can use the Theorem of Ionescu Tulcea, cf. Appendix B.3.5 and find in a canonical way $(\Omega, \mathfrak{F}, \mathbb{P}_\pi(s, \cdot))$ and $\zeta := (\zeta_\nu)_1^\infty$ such that for each $N \in \mathbb{N}$ the probability distribution of $\zeta^N := (\zeta_\nu)_1^N$ has the form given in (12.2). This means that ζ is a non-homogeneous Markov chain with state space S, initial probability distribution $P_{\pi_0}(s, ds_1)$ and transition probability distributions P_{π_ν}, $\nu \in \mathbb{N}$. The quantities $R_{\infty\pi}(s, \zeta)$, $V_{\infty\pi}(s)$ and $V_\infty(s)$ as well as optimality and stationarity of an infinite-stage policy are defined as in Chap. 12. As in Chap. 12 one shows that $V_{n\pi}(s) = \mathbb{E}_{\pi s} R_{n\pi}(s, \zeta^n)$.

Theorem 20.2.1 (Infinite-stage MDPs) *Assume that the MDP has a bounding function with $\beta_b < 1$; hence V_N, $N \in \mathbb{N}$, and V exist and belong to B_b. Then:*

(a) For all $\pi \in \Pi$ and $s \in S$ the total reward $R_{\infty\pi}(s, \zeta)$ exists $\mathbb{P}_\pi(s, \cdot)$-almost surely, $V_{\infty\pi}$ and V_∞ exist and belong to B_b and $V_\infty = V$. Moreover, we have for $N \in \mathbb{N}$

$$\|V_\infty - V_N\|_b \leq \sup_{\pi \in \Pi} \|V_{\infty\pi} - V_{N\pi}\|_b \leq \left[\|r\|_b/(1 - \beta_b) + \|V_0\|_b\right]\beta_b^N. \qquad (20.7)$$

In particular, $V_{N\pi}$ converges for each choice of $V_0 \in MB_b$ in b-norm and uniformly with respect to $\pi \in \Pi$ towards $V_{\infty\pi}$.

(b) *Assume that (MA2) and the VI hold. If f is a maximizer of LV, then $(f)_0^\infty$ is a stationary optimal policy.*

Proof Drop b in $\|\cdot\|_b$.

(a1) For $R'_{\infty\pi}(s,\zeta) := \sum_{\nu=0}^\infty \beta^\nu |r_{\pi_\nu}(\zeta_\nu)|$ we obtain, using (20.3)

$$\mathbb{E}_{\pi s} R'_{\infty\pi}(s,\zeta) = \sum_{\nu=0}^\infty \beta^\nu \, \mathbb{E}_{\pi s} |r_{\pi_\nu}(\zeta_\nu)| \le \|r\| \sum_0^\infty \beta^\nu \, \mathbb{E}_{\pi s} b(\zeta_\nu)$$

$$\le \|r\| \sum_0^\infty \beta_b^\nu \, b(s) = \|r\| \, b(s)/(1-\beta_b).$$

$R'_{\infty\pi}(s,\zeta(\omega)) < \infty$ for $\mathbb{P}_\pi(s,\cdot)$-almost all ω yields that $R_{\infty\pi}(s,\zeta)$ is defined $\mathbb{P}_\pi(s,\cdot)$-almost surely. Moreover,

$$\mathbb{E}_{\pi s} |R_{\infty\pi}(s,\zeta)| \le \mathbb{E}_{\pi s} \mathbb{R}'_{\infty\pi}(s,\zeta) \le \|r\| b(s)/(1-\beta_b) < \infty.$$

Therefore $V_{\infty\pi}$ exists and

$$|V_{\infty\pi}(s)| = |\, \mathbb{E}_{\pi s} R_{\infty\pi}(s,\zeta)| \le \mathbb{E}_{\pi s} |R_{\infty\pi}(s,\zeta)| \le \|r\| b(s)/(1-\beta_b).$$

Thus $V_{\infty\pi} \in B_b$. Finally, also $V_\infty \in B_b$ as

$$|V_\infty(s)| = |\sup_\pi V_{\infty\pi}(s)| \le \sup_\pi |V_{\infty\pi}(s)| \le \|r\|/(1-\beta_b)\, b(s).$$

It follows that $V_\infty = V$ and that $V_{n\pi} \to V_{\infty\pi}$ in b-norm.

(a2) The relation (20.7) follows as in part (b) of the proof of Theorem 12.3.2, using (20.3).

(b) This part is proved exactly as part (c) of Theorem 12.3.2. □

Remark 20.2.2 For the proof of Theorem 20.2.1(a) we do not need the assumption that (MA2) and the VI hold, but we cannot show that $V = V_\infty$ is measurable (and then belongs to MB_b) or even solves the optimality equation. Further results on the models $\lim MDP_n$ and MDP_∞ can be found in Schäl (1975). ◇

Part III
Generalizations of Markovian Decision Processes

Chapter 21
Markovian Decision Processes with Disturbances

In Part II we treated two basic models of Stochastic Dynamic Programming: Control Models with independent disturbances and Markovian Decision Processes. In the first of these models the state $\zeta_t = \zeta_{t\pi}$ in period t and the disturbance η_{t+1} in the same period are stochastically independent of each other. However, there are other important models in which η_{t+1} depends on ζ_t. This holds in particular in some of the Markov renewal programs treated in Chap. 22 below, where a random time η_{t+1} elapses between the t-th and the $(t + 1)$-st decision; another relevant model is the Markovian control model MCM introduced below. We now present a model, called an MDP *with disturbances*, which due to a very flexible transition law comprises all these models. As a further important generalization, necessary for the Markov renewal programs, we allow the discount factor in each period to be a function of the disturbance.

21.1 The Model MDPD

Definition 21.1.1 A ***Markovian Decision Process with Disturbances*** (**MDPD** for short) is a tuple $(S, A, D, Z, K, \tilde{r}, V_0, \tilde{\beta})$ of the following kind:

- State space S, action space A and constraint set D with corresponding σ-algebras $\mathfrak{S}$, $\mathfrak{A}$ and $\mathfrak{D} := D \cap (\mathfrak{S} \otimes \mathfrak{A})$ and the measurable terminal reward function V_0 have the same meaning as in the MDPs of Chap. 16.
- $Z \neq \emptyset$ is the set of disturbances, endowed with a σ-algebra $\mathfrak{Z}$.
- The **transition law** $K(s, a, d(z, s'))$ is a transition probability from $D \to Z \times S$. Its first marginal Q, called the **disturbance transition law**, and second marginal P, called the **state transition law**, are given by

$$Q(s, a, dz) := K(s, a, dz \times S), \quad (s, a) \in D,$$
$$P(s, a, ds') := K(s, a, Z \times ds'), \quad (s, a) \in D.$$

K. Hinderer et al., *Dynamic Optimization*, Universitext,
DOI 10.1007/978-3-319-48814-1_21

- The one-stage reward function $\tilde{r}$ is a measurable function on $D \times Z \times S \to \mathbb{R}$ such that

$$r(s, a) := \int K(s, a, d(z, s')) \cdot \tilde{r}(s, a, z, s')$$

exists and is finite. We call r the **expected one-stage reward.** If $\tilde{r}(s, a, z, s')$ does not depend on (z, s'), we just denote it by $r(s, a)$.
- The **discount function** $z \mapsto \tilde{\beta}(z)$ is a non-negative measurable function on Z such that $0 < \sup_{(s,a)\in D} \beta(s, a) < \infty$, where

$$\beta(s, a) := \int Q(s, a, dz) \cdot \tilde{\beta}(z), \quad (s, a) \in D. \tag{21.1}$$

We call $(s, a) \mapsto \beta(s, a)$ the **expected discount function.**

Decision rules and N-stage policies are defined as for MDPs; in particular, decision rules at time t depend only on the momentary state s_t. For $f \in \mathbb{F}$ we use the abbreviation

$$K_f(s, d(z, s')) := K(s, f(s), d(z, s')), \ s \in S,$$

and Q_f, P_f, r_f and $\tilde{\beta}_f$ are defined similarly. Obviously K_f is a transition probability from S into $Z \times S$ which describes the transition from the momentary state under the decision rule f to the next pair of disturbance and state, and similar interpretations hold for Q_f and P_f.

Remark 21.1.2

(a) Although our new model is more complicated than MDPs and CMs, its analysis is very much the same. (Moreover, it can be reduced to an MDP′; cf. Theorem 21.1.10(a) below.) This is due to the fact that under each probability distribution $\mathbb{P}_{N\pi}(s_0, \cdot)$, constructed below, the sequence $((\eta_t, \zeta_t))_1^N$ is Markovian and of a special type: The transition law at time $t \geq 1$ is a version of the conditional probability distribution

$$K(s_t, \pi_t(s_t), G) = \mathbb{P}_{N\pi}(s_0, (\eta_{t+1}, \zeta_{t+1}) \in G \mid (\eta_t, \zeta_t) = (z_t, s_t)).$$

It describes the transition from (z_t, s_t) under policy π to (z_{t+1}, s_{t+1}), and it is special since it does not depend on the disturbance z_t.

(b) In the special case where for all (s, a) the probability distribution $K(s, a, d(z, s'))$ is the product of the probability distributions $Q(s, a, dz)$ and $P(s, a, ds')$ we say that *disturbances and states are conditionally independent*. An example, often occurring in queueing models (cf. Chap. 22) is $K(s, a, d(z, s')) = \mathrm{Exp}(\Lambda(s, a)) \times P(s, a, ds')$ for a positive function Λ on D and $\tilde{\beta}(z) = e^{-\alpha z}$ for some $\alpha \in \mathbb{R}_+$. Then $\beta(s, a) = \Lambda(s, a)/(\alpha + \Lambda(s, a))$ and $\sup_{s,a} \beta(s, a) \leq 1$.

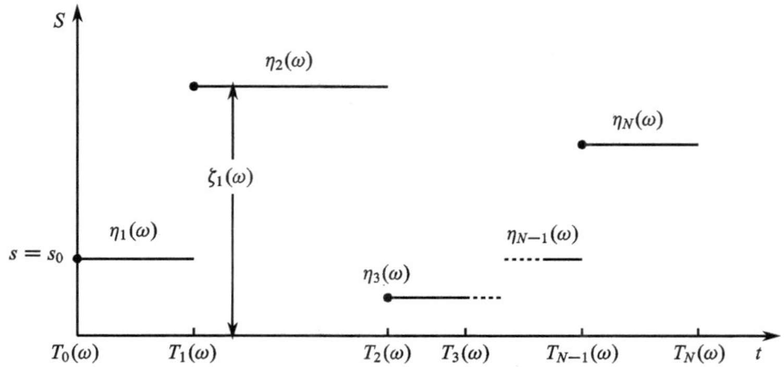

Fig. 21.1 Path of a decision process for an N-stage Markov renewal program

(c) The evolution of the state process $(\zeta_t)_1^N$ can be imagined as follows; cf. Fig. 21.1 which shows a path for the case where the η_t's are random times between successive decisions: The system starting in the state s_0 moves in the first period to a new (possibly the same!) state s_1, where the pair (z_1, s_1) is selected according to the probability distribution $K_{\pi_0}(s, d(z_1, s_1))$; then the system moves to a state s_2, where (z_2, s_2) is selected according to $K_{\pi_1}(s_1, d(z_2, s_2))$, irrespective of z_1, etc.
(d) The reward $\tilde{r}(s_t, a_t, z_{t+1}, s_{t+1})$ is obtained at the beginning of period $t \geq 0$. For $t \geq 1$ it is discounted to the beginning of period zero by means of $\prod_{\nu=1}^{t} \tilde{\beta}(z_t)$. This number takes over the role of β^t in MDPs.
(e) In this section we interpret $\tilde{\beta}(z)$ as a discount factor for periods of unit length, which varies probabilistically, e.g. due to a random environment z. Another important interpretation is given in Chap. 22. ◇

Remark 21.1.3 The second marginal of $K(s, a, d(z, s'))$ is

$$P(s, a, ds') = K(s, a, Z \times ds').$$

It is easy to see that P and Q are transition probabilities from D into S and Z, respectively. They describe the transition to the next state and next disturbance, respectively. The functions $r(\cdot, \cdot)$ and $\beta(\cdot, \cdot)$ are measurable. ◇

Remark 21.1.4 In applications the transition from ζ_t to $(\eta_{t+1}, \zeta_{t+1})$ is often made in two stages by firstly selecting η_{t+1} and then ζ_{t+1}, or vice versa. This is modeled by the following two *factorizations* of K:

$$\begin{aligned} K(s, a, d(z, s')) &= Q(s, a, dz) \otimes K_{12}(s, a, z, ds') \\ &= P(s, a, ds') \otimes K_{21}(s, a, s', dz). \end{aligned} \tag{21.2}$$

Here $K_{12}(s, a, z, ds')$ is a conditional distribution of ζ_{t+1}, given $[\zeta_t = s, \eta_{t+1} = z]$ and $K_{21}(s, a, s', dz)$ is a conditional distribution of η_{t+1}, given $[\zeta_t = s, \zeta_{t+1} = s']$, both with respect to the probability distribution $\mathbb{P}_{N\pi}(s_0, dy^N)$ constructed below. The latter factorization is the usual one in Markov renewal processes. ◇

Remark 21.1.5 In applications often the state space is countable. Then one need not care about measurability with respect to states or actions, but if Z is uncountable (e.g. $Z = \mathbb{R}_+$) measurability with respect to $\mathfrak{Z}$ must be kept in mind. ◇

For measurable functions v on $D \times Z \times S$ we define the measurable function Kv on D by

$$Kv(s,a) := \int K(s,a,d(z,s')) \cdot v(s,a,z,s'),$$

provided these integrals exists. Similar notations are used for other transition probabilities such as Q or P. As an example, (21.1) may be written as $\beta = Q\tilde{\beta}$.

Besides the Markov renewal programs in Chap. 22, the following special MDPD with constant discount function is of interest. It extends CMs as now the disturbance transition law Q may depend on (s,a). Thus $Q(s,a,dz)$ is a transition probability from D into Z, while all other data have the meaning as in a CM.

Definition 21.1.6 A **Markovian control model** $(S, A, D, Z, Q, T, r_Z, V_0, \beta)$ (**MCM** for short) is defined as the special MDPD $(S, A, D, Z, K, \tilde{r}, V_0, \tilde{\beta})$ with the following three properties:

- K is determined by a random transition from D into the disturbance space Z according to $Q(s,a,dz)$, followed by a deterministic transition from (s,a,z) into $s' = T(s,a,z)$.
- $\tilde{r}(s,a,z,s') = r_Z(s,a,z)$, independent of s'.
- $\tilde{\beta} \equiv \beta$.

Remark 21.1.7

(a) If a model M (e.g. an MCM) is defined by means of another model $\hat{M}$ (e.g. an MDPD), we understand by problem M_N the problem $\hat{M}_N$.
(b) In an MCM we have (cf. (21.2)) $K(s,a,d(z,s')) = Q(s,a,dz) \otimes \delta_{T(s,a,z)}(ds')$, this ensures for all measurable $v \geq 0$ on $Z \times S$ that

$$Kv(s,a) = \int Q(s,a,dz) \cdot v(z, T(s,a,z)), \ (s,a) \in D. \tag{21.3}$$

(c) After having defined below the functions $V_{n\pi}$ for an MDPD, it is intuitively clear that an MDP $(S, A, D, P, r_S, V_0, \beta)$ as defined in Chap. 16 is equivalent (in the sense of Chap. 12) to an MDPD if (MA1) holds in both models. In fact, a proof similar to the proof in Chap. 12 of the equivalence of a CM and its adjoint MDP, shows that the MDP is equivalent to the special MCM $(S, A, D, Z, Q, T, r_Z, V_0, \beta)$ where $Z := S$, $z := s'$, $T(s,a,s') := s'$ for all (s,a,s'), $Q := P$ and $r_Z(s,a,s',s'') := r_S(s,a,s')$, independent of s''.

(d) An MDP is also equivalent to the special MDPD where Z is a singleton (hence one can drop the disturbance z from all data), where $K := P$, $\tilde{r} := r_S$ and $\tilde{\beta} \equiv \beta$, provided, (MA1) holds in both models. ◇

Example 21.1.8 (Choice of production modes) N identical machines must be produced, one at a time. We have the option among m production modes a, $1 \leq a \leq m$, with costs $g_a \geq 0$ per unit time. The random production time under mode a has the probability distribution $Q(a, dz)$. A reward $R \in \mathbb{R}^+$ is obtained upon the completion of the production of a piece. If it takes $s_N \in \mathbb{R}_+$ time units to produce all N pieces then there arises a penalty cost of amount $d_0 \cdot s_N$, $d_0 > 0$. A constant discount factor per unit time of the form $e^{-\alpha}$ for some $\alpha > 0$ is assumed. In general stochastically small production times will be coupled with high costs. Thus one must balance the speed and the costs of production in order to maximize the discounted expected reward.

We model the problem as an MCM as follows: s_t is the time point when the production of the $(t+1)$-st piece begins; thus $S = Z = \mathbb{R}_+$; the initial state is $s_0 = 0$; $A = D(s) = \mathbb{N}_m$; $Q(s, a, dz)$ is independent of s; $T(s, a, z) = s + z$; $r_Z(s, a, z) = (R - g_a z) \cdot e^{-\alpha z}$; $V_0(s) := -d_0 \cdot s$; $\tilde{\beta}(z) = e^{-\alpha z}$. Existence and finiteness of r is shown as follows. The Laplace transform $\hat{Q}_a$ of $Q(a, dz)$ has on $\mathbb{R}^+$ the finite derivative

$$\hat{Q}_a'(x) = -\int_0^\infty z \cdot e^{-xz}\, Q_a(dz).$$

Now we obtain, since $\alpha > 0$,

$$Q|r_Z|(s, a) = \int e^{-\alpha z} \cdot |R - g_a z|\, Q_a(dz) \leq R \cdot \hat{Q}_a(\alpha) - g_a \cdot \hat{Q}_a'(\alpha) < \infty. \qquad \blacklozenge$$

Put $y_t := (z_t, s_t) \in Y := Z \times S$ and $y^t := (z_1, s_1, z_2, s_2, \ldots, z_t, s_t)$, $t \geq 1$. Fix an initial state s_0 and an N-stage policy $\pi = (\pi_t)_0^{N-1} \in \mathbb{F}^N$. We use the canonical probability space $(Y^N, \mathfrak{Y}_N, \mathbb{P}_{N\pi}(s_0, dy^N))$ where the sample space Y^N has the generic element y^N, and $\mathfrak{Y}_N := \otimes_1^N (\mathfrak{Z} \otimes \mathfrak{S})$. We also use the coordinate random variables $y^N \mapsto (\eta_t(y^N), \zeta_t(y^N)) := (z_t, s_t) = y_t$, $1 \leq t \leq N$. They describe the random disturbance time η_t and the random state ζ_t at the beginning of period t. Define $\mathbb{P}_{N\pi}(s_0, dy^N)$ in such a way (cf. Chap. 16) that the sequence $((\eta_t, \zeta_t))_1^N$ is Markovian with state space Y, initial probability distribution $K_{\pi_0}(s_0, d(z_1, s_1))$ and transition probabilities K_{π_t}, $1 \leq t \leq N-1$. According to Chap. 16 this means that $\mathbb{P}_{1f}(s_0, dy_1) = K_f(s_0, dy_1)$ and that for $n \geq 2$ and $(f, \sigma) \in \mathbb{F} \times \mathbb{F}^{n-1}$

$$\mathbb{P}_{n(f,\sigma)}(s_0, dy^n) = \int K_f(s_0, dy_1) \int \mathbb{P}_{n-1,\sigma}(s_1, d(y_t)_2^n). \tag{21.4}$$

The N-stage reward is the random variable

$$y^N \mapsto R_{N\pi}(s_0, y^N) := \tilde{r}(s_0, \pi_0(s_0), y_1)$$
$$+ \sum_{t=1}^{N-1} \prod_1^t \tilde{\beta}(z_\nu) \cdot \tilde{r}(s_t, \pi_t(s_t), y_{t+1}) + \prod_1^N \tilde{\beta}(z_\nu) \cdot V_0(s_N). \tag{21.5}$$

The **expected N-stage reward** for policy π and initial state s_0 and policy π is

$$V_{N\pi}(s_0) := \int \mathbb{P}_{N\pi}(s_0, dy^N) \cdot R_{N\pi}(s_0, y^N), \tag{21.6}$$

provided the integral exists. From now on we denote the initial state by s rather than s_0.

The problem of maximizing $\pi \mapsto V_{N\pi}(s)$ for all s and $N \geq 1$ makes sense if and only if the MDPD satisfies the

first minimal assumption (MA1) for an MDPD: for all $N \in \mathbb{N}$, $\pi \in \mathbb{F}^N$ and $s_0 \in S$ the expected N-stage reward $V_{N\pi}(s_0) := \mathbb{E}_{\pi s_0} R_{N\pi}(s_0, \zeta^N)$ exists.

Then the ***N-stage value function*** $V_N: S \to \overline{\mathbb{R}}$ is defined by

$$V_N(s) := \sup\{V_{N\pi}(s) : \pi \in \mathbb{F}^N\}.$$

It follows as in Chap. 16 that $V_{N\pi}(\cdot)$ is $\mathfrak{S}$-measurable, but $V_N(\cdot)$ need not be $\mathfrak{S}$-measurable. Obviously (MA1) holds under the following

assumption (LUBF): The model has either a lower or an upper bounding function.

Another condition by means of bounding functions is given in Theorem 21.1.13 below.

We show in Lemma 21.1.9 below that (MA1) implies the RI in the form

$$V_{n(f,\sigma)}(s) = r_f(s) + \int K_f(s, d(z, s')) \cdot \tilde{\beta}(z) \cdot V_{n-1,\sigma}(s'). \tag{21.7}$$

In view of (21.7) we define $\mathfrak{V}_0$ as the set of those measurable functions v on S for which there exists

$$K(\tilde{\beta}v)(s, a) = \int K(s, a, d(z, s')) \cdot \tilde{\beta}(z) \cdot v(s') \text{ for all } (s, a). \tag{21.8}$$

As $K(\tilde{\beta}v^{\pm})(s, a) \leq \sup v^{\pm} \cdot \sup \beta$, the set $\mathfrak{V}_0$ contains all measurable v which are bounded from above or from below.

The integral in (21.7) can be written more succinctly as

$$\int \psi_f(s, ds') \cdot V_{n-1,\sigma}(s'),$$

for the bounded measure

$$\psi(s,a,B) := \int K(s,a,d(z,s')) \cdot \tilde{\beta}(z) \cdot 1_B(s'), \quad B \in \mathfrak{S}.$$

In fact, since $\psi(s,a,\cdot)$ is the second marginal of the measure on $\mathfrak{S}$ which has the $K(s,a,d(z,s'))$-density $(z,s') \mapsto \tilde{\beta}(z)$, we have

$$\psi v(s,a) := \int \psi(s,a,ds') \cdot v(s') = K(\tilde{\beta} v)(s,a), \; (s,a) \in D$$

for all measurable $v \geq 0$. Since $(s,a) \mapsto \psi(s,a,C)$ is measurable for all C, ψ is a bounded transition probability from D into S. We call ψ the ***discounted state transition law.*** Note that $\beta(s,a) = \psi(s,a,S)$ for all (s,a), that ψ is substochastic (i.e. $\psi(s,a,S) \leq 1$) if $\tilde{\beta} \leq 1$, and that $v \in \mathfrak{V}_0$ if and only if $\psi_f v$ exists for all $f \in \mathbb{F}$.

Lemma 21.1.9 *If (MA1) holds for an MDPD then V_0 and $V_{n\pi}$ belong to $\mathfrak{V}_0$ for $n \geq 1$ and $\pi \in \mathbb{F}^n$, and the **reward iteration (RI** for short) holds in the form* (21.7), *which may be written as*

$$V_{1f} = r_f + \psi_f V_0,$$
$$V_{n(f,\sigma)} = r_f + \psi_f V_{n-1,\sigma}, \quad n \geq 2, f \in \mathbb{F}, \sigma \in \mathbb{F}^{n-1}.$$

The proof is omitted since it is very similar to the one given for Lemma 16.1.9.

Again it is crucial that $\mathbb{P}_{N\pi}$ and $R_{N\pi}$ from (21.5) have a recursive structure; this follows for $\mathbb{P}_{N\pi}$ from (21.4), and for $R_{N\pi}$ since $R_{1f}(s_0,y_1) = \tilde{r}_f(s_0,y_1) + \tilde{\beta}(z_1) \cdot V_0(s_1)$ and since for $n \geq 2$

$$R_{n(f,\sigma)}(s_0,(y_t)_1^n) = \tilde{r}_f(s_0,y_1) + \tilde{\beta}(z_1) \cdot R_{n-1,\sigma}(s_1,(y_t)_2^n).$$

$\mathfrak{V}_0$ is the largest set of functions on which we can define the operator

$$Lv := r + K(\tilde{\beta} v) = r + \psi v, \tag{21.9}$$

i.e. for which ψv exists. The operators U_f and U on $\mathfrak{V}_0$ are obtained from L as before by

$$U_f v(s) := Lv(s,f(s)), \quad Uv(s) := \sup\{Lv(s,a) : a \in D(s)\}. \tag{21.10}$$

In an MCM (where β is constant) we have $v \in \mathfrak{V}_0$ if and only if $Q(v \circ T)$ exists and then $Lv = r + \beta \cdot Q(v \circ T)$.

If disturbances and states are conditionally independent we have

$$Lv(s,a) = r(s,a) + \beta(s,a) \cdot Pv(s,a), \quad (s,a) \in D. \tag{21.11}$$

This representation of L holds for general MDPDs, when P is replaced by the probability distribution

$$P_0(s,a,ds') := \begin{cases} \psi(s,a,ds')/\beta(s,a), & \text{if } \beta(s,a) > 0, \\ P(s,a,ds'), & \text{else.} \end{cases}$$

Clearly $P_0(s,a,ds')$ is a transition probability from D into S. We call P_0 the ***normalized state transition law.*** From $\psi(s,a,ds') = \beta(s,a) \cdot P_0(s,a,ds')$ for all (s,a) we see that $v \in \mathfrak{V}_0$ if and only if $P_0 v$ exists, and then for all (s,a)

$$\beta(s,a) \cdot P_0 v(s,a) = \psi v(s,a) = K(\tilde{\beta} v)(s,a).$$

Moreover, P_0 coincides with P if $\tilde{\beta}$ is constant or if disturbances and states are conditionally independent.

In analogy to Part II we define the

second minimal assumption (MA2) for an MDPD: (MA1) holds and all value functions V_n belong to $\mathfrak{V}_0$.

Then UV_{n-1} exists and maximizers at stage $n \geq 1$ are defined. We say that the VI holds for the MDPD, if (MA2) holds and if for all s and $n \geq 1$

$$V_n(s) = UV_{n-1}(s) = \sup_{a \in D(s)} [r(s,a) + \psi V_{n-1}(s,a)] \tag{21.12}$$

$$= \sup_{a \in D(s)} [r(s,a) + K(\tilde{\beta} V_{n-1})(s,a)] = \sup_{a \in D(s)} [r(s,a) + \beta(s,a) \cdot P_0 V_{n-1}(s,a)].$$

An MDPD can be reduced to the following ***adjoint MDP'*** $(S', A, D', P', r', V_0', \beta')$ with an absorbing point $x \notin S$: $S' = S + \{x\}$; $D'(s) = D(s)$, $D'(x) = A$; $P'(s,a,B) = \psi(s,a,B)/\beta'$ for $s \in S$, $B \subset S$; $P'(s,a,\{x\}) := 1 - P'(s,a,S)$ and $P'(x,a,\{x\}) = 1$; $r'(s,a) = r(s,a)$ and $V_0'(s) = V_0(s)$ for $(s,a) \in D$; $r'(x,a) = V_0'(x) = 0$; $\beta' = \sup_{(s,a) \in D} \beta(s,a)$. Then the set $\mathfrak{V}_0$ and by (21.9) also the operator L in MDPD and in MDP' coincide, hence the RI Lemma 21.1.9 for the MDPD has the same form as for the MDP'. This proves part (a) of the next result, and then (b) follows from Theorem 16.1.12 and Proposition 16.1.11.

Theorem 21.1.10 (Basic Theorem for MDPDs)

(a) If (MA1) holds both in the MDPD and in the adjoint MDP', then the MDPD is reducible in the sense of Definition 12.1.10 *to the MDP'.*

(b) (b1) Under (MA1) the Structure Theorem 16.1.12 *remains true.*

(b2) Under (MA2) the OC from Proposition 16.1.11 *remains true.*

(b3) The VI holds in the MDPD if and only if it holds in the MDP'; this is in particular the case when there exist maximizers at each stage $n \geq 1$.

Example 21.1.11 (Choice of production modes of Example 21.1.8, *continuation)* As both r_Z and V_0 are bounded above, (MA1) holds both in the MDPD and the adjoint MDP'. As $D(s)$ is finite for all s, it follows from Theorem 21.1.10 that the

VI holds and that there exists a smallest maximizer at each stage and hence an optimal policy for $MDPD_N$ for all N. Moreover, from (21.4) we see that

$$Lv(s,a) = R \cdot \hat{Q}_a(\alpha) + K_a \cdot \hat{Q}'_a(\alpha) + \int_0^\infty e^{-\alpha z} v(s+z)\, Q_a(dz). \qquad \square$$

Definition 21.1.12 A measurable function $b\colon S \to \mathbb{R}_+$ is called a ***bounding function of the MDPD*** if

$$\|K|\tilde{r}|\,\|_b,\ \|V_0\|_b \text{ and } \beta_b := \|\psi b\|_b = \|\beta \cdot P_0 b\|_b \text{ are finite.}$$

We collect a few *properties* of bounding functions:

- b is a bounding function if and only if for some $\delta \geq 0$ and all (s,a) we have

$$K|\tilde{r}(s,a)| \leq \delta \cdot b(s),\ |V_0(s)| \leq \delta \cdot b(s),\ \psi b(s,a) \leq \delta \cdot b(s).$$

  The first of these conditions holds if $|\tilde{r}(s,a,z,s')| \leq \delta \cdot b(s)$ for all s, a, z, s'.
- Finiteness of $\|K|\tilde{r}|\|_b$ implies existence and finiteness of r.
- If $\tilde{r}$ and V_0 are bounded then $b :\equiv 1$ is a bounding function and $\beta_b = \sup_{(s,a)} \beta(s,a)$.
- A bounding function b of the MDPD is also a bounding function of the adjoint MDP′ and the two numbers β_b coincide in both models.

Now one can copy the proof of Theorem 14.1.6 with $\beta \cdot p_f v$ replaced by $\psi_f v$ in order to obtain

Theorem 21.1.13 (MDPDs with bounding functions) *Assume that the MDPD has a bounding function b. Then:*

(a) (MA1) and the OC hold.
(b) If either S or A is countable the VI holds.
(c) If A is countable and endowed with a total ordering and if $D(s)$ is finite for all s, then the smallest maximizer f_n at each stage n is measurable, and for each $N \in \mathbb{N}$ the policy $(f_n)_N^1$ is optimal.

Sometimes one encounters a situation which looks like an MDPD except that the data and the decisions at time $t \geq 1$ also depend on the preceding disturbance z_t. This situation should not be modeled by an MDPD but by an MDPvar with augmented state space $X := S \times Z$ and states $x_t = (s_t, z_t)$; z_0 must be interpreted as an additional initial disturbance. Denoting the momentary disturbance by z' (rather than by z as before), the data have the form $P(x,a,dx')$, $r_X(x,a,x')$, $\tilde{\beta}(x,a,x')$, and $V_0(x')$.

For MDPDs the problem of large horizon can be analyzed with the aid of the adjoint MDP′ as only $V_{n\pi}$ and V_n for finite n are involved; some of the results are incorporated into Theorem 21.1.15 below. On the other hand, some comments are necessary for the case of infinite horizon, as follows.

MDPDs with ***infinite horizon*** (***MDPD***$_\infty$ for short) are defined as MDPDs except that no terminal reward function V_0 is required. We must use infinite-stage policies $\pi = (\pi_t)_0^\infty \in \Pi := \mathbb{F}^\infty$ and their sections $\pi_N := (\pi_t)_0^{N-1}$, $N \geq 1$. Given an initial state s we use the infinite sequence of two-dimensional coordinate random vectors (η_t, ζ_t), $t \geq 1$ on the canonical probability space $(Y^\infty, \mathfrak{Y}_\infty, \mathbb{P}_\pi(s, dy))$ where $Y^\infty := (Z \times S)_1^\infty$ with generic element $y := ((z_t, s_t))_1^\infty$, $\mathfrak{Y}_\infty := (\mathfrak{Z} \otimes \mathfrak{S})^\infty$ and $(\eta_t(y), \zeta_t(y)) := (z_t, s_t)$. This is the same situation as in Sect. 20.2, except that the state space is now replaced by the Cartesian product of the disturbance space and the state space. By the Theorem of Ionescu Tulcea we find in a canonical way the probability distribution $\mathbb{P}_\pi(s, dy)$ such that $((\eta_t, \zeta_t))_1^\infty$ is a non-homogeneous Markov chain with state space $Z \times S$, initial distribution $K_{\pi_0}(s, d(z_1, s_1))$ and transition probabilities K_{π_t}, $t \in \mathbb{N}$. The connection to the functions $V_{N\sigma}$, $\sigma \in \mathbb{F}^N$, is as follows: the definition of $\mathbb{P}_\pi(s, dy)$ implies that for all N, π and s

$$V_{N\pi_N}(s) = \int \mathbb{P}_\pi(s, dy) \cdot R_{N\pi}(s, y^N) =: V_{N\pi}(s), \tag{21.13}$$

and hence $V_N = \sup_{\pi \in \mathbb{F}^N} V_{N\pi}$. Equation (21.13) becomes important below when comparing MDPD$_\infty$ with an MDPD(V_0), defined by the data of the MDPD$_\infty$, augmented by some terminal reward function V_0. (The corresponding functions $V_{n\pi}$ and the value functions V_n then depend on V_0.)

For MDPD$_\infty$ we use the following

Definition 21.1.14

- The **first minimal assumption** (***MA*1**)$_\infty$ means that for all $\pi \in \mathbb{F}^N$ and $s_0 \in S$ there exist $\mathbb{P}_\pi(s, dy)$-almost surely

$$R_{\infty\pi}(s_0, y) := \tilde{r}(s_0, \pi_t(s_0), z_1, s_1) + \sum_{t=1}^{\infty} \prod_1^t \tilde{\beta}(z_t) \cdot \tilde{r}(s_t, \pi_t(s_t), z_{t+1}, s_{t+1}),$$

$$V_{\infty\pi}(s) := \int \mathbb{P}_\pi(s, dy)\, R_{\infty\pi}(s, y).$$

- Under (MA1)$_\infty$ there is defined $V_\infty(s) := \sup_{\pi \in \Pi} V_{\infty\pi}(s)$, $s \in S$.
- $\pi^* \in \Pi$ is called **optimal** if $V_{\infty\pi^*} = V_\infty$, and π is called **stationary** if $\pi = (f)_0^\infty =: f^\infty$ for some decision rule f.

An MDPD$_\infty$ can be analyzed by means of its adjoint MDP$'_\infty$ since under (MA1)$_\infty$ for both the MDPD$_\infty$ and the MDP$'_\infty$ the MDPD$_\infty$ is reducible in the sense of Chap. 20 to the MDP$'_\infty$.

Recall that MB_b is the set of measurable functions on S with finite b-norm.

Theorem 21.1.15 (Reduction Theorem for MDPDs with large and with infinite horizon) *Assume that the MDPD has a bounding function with* $\beta_b < 1$. *Then* $(MA1)_\infty$ *holds for the MDPD*$_\infty$ *and its adjoint MDP*$'_\infty$, *and the former is reducible* *to the latter. As a consequence, the results from Chap.* 20 *about* $MDP'_\infty s$ *carry over to* $MDPD_\infty s$. *In particular:*

(a) The sequence $(V_n)_0^\infty$ *of value functions converges in b-norm to* V, V *belongs to* MB_b, *and* V *is independent of* V_0 *in* MB_b.
(b) There exists V_∞, *and* $V_\infty = V$.
(c) Assume that VI holds.

(c1) V *is within* MB_b *the unique solution of the optimality equation*

$$\begin{aligned} V(s) &= \sup_{a \in D(s)} [r(s,a) + \beta(s,a) \cdot P_0 V(s,a)] \\ &= \sup_{a \in D(s)} [r(s,a) + \psi V(s,a)], \quad s \in S. \end{aligned}$$

(c2) The decision rule f *is asymptotically optimal if and only if it is a maximizer of* LV. *And then* f^∞ *is a stationary optimal policy.*

The proof is similar to the one given for Proposition 14.1.5 and Theorem 20.2.1, and is omitted.

Example 21.1.16 (A time slot queueing problem with service control) We consider a queueing system similar to the one treated as a CM in Example 15.11. The system consists of a single server and an infinite buffer. At time $0 \le t \le N-1$ there arrives a batch of random size $Y_{t+1} \ge 0$ of customers. A controller has to decide at the beginning of each time slot where at least one customer is present, which one of a finite set $A \subset [0,1)$ of service rates a to use. This means that service is completed in the present slot with probability a. Thus the service rate may change from slot to slot during the total service time of a specific customer. When $a = 0$, no service is provided, and this is the only choice if no customer is present.

Put X_t equal to zero if no service is completed in slot $t-1$, and equal to one, otherwise. We assume that the sequence $(\eta_t)_1^N := ((X_t, Y_t))_1^N$ is i.i.d. There is a holding cost $h(s) \in \mathbb{R}_+$ per slot when $s \in \mathbb{N}_0$ packets (including the one being served) are waiting, and there is a service cost $d(k) \in \mathbb{R}_+$ when the customer being served still requires k slots for service. If s_N packets are still waiting at time N, they obtain a special service at cost $C_0(s_N) \in \mathbb{R}_+$. It is natural to assume that both h and C_0 are increasing and that $h(0) = C_0(0) = 0$. We want to minimize the N-stage expected discounted cost.

Since $X_t \sim \mathrm{Bi}(1,a)$ we cannot use a CM, but we must use an MCM with the following data: $S = \mathbb{N}_0$; $D(s) = A$ from above for $s \ge 1$ and $D(0) := \{0\}$; $Z = \{0,1\} \times \mathbb{N}_0$; the state s_{t+1} equals s_t minus X_{t+1} (if $s_t > 0$) plus Y_{t+1}. Thus

$$s_{t+1} = T(s_t, a_t, (X_{t+1}, Y_{t+1})) = (s_t - X_{t+1})^+ + Y_{t+1}, \quad t \ge 0.$$

The transition law $Q(s, a, dz)$ has the discrete density

$$(x, y) \mapsto q(s, a, (x, y)) := \begin{cases} a \cdot p_y, & \text{if } s \geq 1,\ x = 1, \\ (1-a) \cdot p_y, & \text{if } s \geq 1,\ x = 0, \\ p_y, & \text{if } s = a = 0. \end{cases}$$

Since S is countable and since C_0 and $c(s, a) = (h(s) + d(a)) \cdot 1_{\mathbb{N}}(s)$ are non-negative, (MA2) holds. Moreover, finiteness of A implies by Proposition 14.1.5 the validity of the VI and the existence of minimizers at each stage. From (21.3) with $v(s, z) := C_{n-1}(s)$ we see that

$$KC_{n-1}(s, a) = \sum_{x=0}^{1} \sum_{y=0}^{\infty} q(s, a, (x, y)) \cdot C_{n-1}((s-x)^+ + y).$$

Now the VI has by (21.12) the form

$$\begin{aligned} C_n(0) &= \beta \cdot \mathbb{E}\, C_{n-1}(Y_1), \\ C_n(s) &= h(s) + \max_{a \in A} \big[d(a) + \\ &\qquad \beta \cdot \{a \cdot \mathbb{E}\, C_{n-1}(s-1+Y_1) + (1-a) \cdot \mathbb{E}\, C_{n-1}(s+Y_1)\} \big],\ s \geq 1. \end{aligned}$$

Assume that h and C_0 are increasing. Then one easily shows by induction on $n \geq 0$:

(i) All minimal cost functions C_n, $n \geq 0$, are increasing.
(ii) All functions C_n are finite if both h and C_0 belong to the set $\mathfrak{V}$ of functions $v \geq 0$ on $\mathbb{N}_0$ for which there exists a $\gamma \in \mathbb{R}^+$ such that $\mathbb{E}\, v(s + Y_1) \leq \gamma \cdot v(s)$, $s \in \mathbb{N}_0$.

Note that, for example, $s \mapsto s^\alpha$ and $s \mapsto \exp(\alpha \cdot s)$, $\alpha \in \mathbb{R}^+$, belong to $\mathfrak{V}$, provided $\mathbb{E}\, Y_1^\alpha < \infty$ or $\mathbb{E} \exp(Y_1) < \infty$, respectively, and that the assumption in (ii) implies that both $\mathbb{E}\, h(Y_1)$ and $\mathbb{E}\, C_0(Y_1)$ are finite. ♦

21.2 Problems

Problem 21.2.1 Consider an MDPD starting in s_0. The sequence $(\zeta_\nu)_1^N$ of states is Markovian with initial distribution $P_{\pi_0}(s_0, ds')$ and transition probability distributions P_{π_ν}, $1 \leq \nu \leq N-1$. The sequence of disturbances $(\eta_\nu)_1^N$ is in general not Markovian.

Problem 21.2.2 Assume that the MDPD$_\infty$ has a bounding function b with $\beta_b < 1$ and that the VI holds for some $V_0 \in M_b$. If $\overline{\beta} := \sup_{(s,a)} \beta(s, a) < 1$ and if f is an

ε-maximizer of LV for some $\varepsilon \geq 0$ (i.e. $LV_f \geq LV - \varepsilon$), then the stationary policy $(f)_0^\infty$ is $\varepsilon/(1-\overline{\beta})$-optimal.

21.3 Supplements

Supplement 21.3.1 (MDPD modeled as an MDPvar′) We can also model an MDPD as a special case of the following MDPvar′ with augmented state space $X := S \times Z$. This models the case where the data and the decisions at time $t \geq 1$ also depend on the preceding disturbance z_t, as $x_t = (s_t, z_t)$; z_0 can be interpreted as an additional initial disturbance. Denote the momentary disturbance not as before by z but by z'. The special case which models the MDPD is obtained if all data at time t do not depend on z_t, i.e. $P'(s, z, a, d(z', s')) := K(s, a, d(z', s'))$, $r'_X(s, z, a, s', z') := \tilde{r}(s, a, z', s')$, $\tilde{\beta}'(s, z, a, s', z') := \tilde{\beta}(s, a, z')$, and $V'_0(s, z) := V_0(s)$. It turns out that in this special case $V_n(s, z)$ and $W_n(s, z, a)$ (but not $V_{n\pi}(s, z)$ for $\pi \in F^n$) do not depend on z. Hence a maximizer at stage n, if it exists, is independent of z, and is in fact a maximizer for the MDPD. A policy consisting of maximizers is then an optimal policy for the MDPD.

Supplement 21.3.2 (MDPDs with random environment) We now consider an MDPD with a useful structure, so-called ***MDPDs with random environment*** (***MDPD-RE*** for short); cf. Sect. 16.2. Intuitively, it is defined as an MDPD (whose states are called *core states*) and whose data in each period depends on some random environment (such as economic factors). The latter evolves within some set I of *environmental* states such that they form together with the disturbances an uncontrolled Markov chain, independent of the core states. Moreover, the transition to the next state depends on the environment both at the beginning and at the end of the period. More precisely, an MDPD-RE is an MDPD $(X, A, D, K, \tilde{r}, V_0, \tilde{\beta})$ of the following kind:

- The state space is X with states $x = (i, s)$ is the Cartesian product $I \times S$ of the set I of environmental states i and the set S of core states s. I, S and X are endowed with σ-algebras $\mathfrak{I}$, $\mathfrak{S}$ and $\mathfrak{X} := \mathfrak{I} \otimes \mathfrak{S}$, respectively.
- The transition law K has a decomposition of the form

$$K(i, s, a, d(z, i', s')) = P_I(i, d(z, i')) \otimes K_S(i, s, a, z, i', ds'), \; (i, s, a) \in D, \tag{21.14}$$

where P_I and K_S are given transition probabilities from I into $Z \times I$ and from $D \times Z \times I$ into S, respectively.

The random states in the MDPD-RE are denoted by $X_t := (I_t, \zeta_t)$. Here I_t and ζ_t denote the random environmental states and the random core states, respectively. It follows from (21.14) that for each $N \in \mathbb{N}$, initial state $x_0 = (i_0, s_0)$ and $\pi \in$

$\mathbb{F}^N$ the sequence $(\eta_t, I_t)_1^N$ is an uncontrolled stationary Markov chain with initial probability distribution $P(i_0, d(z_1, i_1))$ and transition probability P_I. This implies that the sequence of environmental states is also Markovian with initial state i_0 and transition probability $Q_I(i, B) := P(i, Z \times B)$.

Supplement 21.3.3 (History-dependent policies) We treat a problem which beyond its own interest is useful for the proof of the Basic Theorems in Theorems 25.1.10 and 26.2.3: Markovian policies suffice.

(a) So far we used policies $(\pi_t)_0^{N-1} \in \mathbb{F}^N$. They are called *Markovian policies* since at any time t their decision rule π_t depends only on the present state s_t and not on the other components of the so-called history h_t up to time t, consisting of the preceding states $s_0, s_1, \ldots, s_t$ and the preceding disturbances $z_1, \ldots, z_t$. (No dependence of π_t on the preceding actions $a_0, a_1, \ldots, a_{t-1}$ is needed since these are functions of $s_0, s_1, \ldots, s_{t-1}$ via $\pi_0, \pi_1, \ldots, \pi_{t-1}$.) Does the use of history-dependent policies (which includes the Markovian policies) increase the maximal expected N-stage reward? Since at any time t we can only control the sum of future rewards, which does not depend on states and disturbances before time t, one expects a negative answer. (This is supported by the following simple fact: In order to maximize a function $x \mapsto g(x)$ nothing is gained by adding another variable y and maximizing $(x, y) \mapsto g(x)$.) We are going to confirm this in Sect. 21.3 below under conditions sufficient for most applications. (For weaker conditions cf. Bertsekas and Shreve 1978.)

(b) We need some definitions, as follows.

- The elements $h_0 := s_0 \in H_0 := S$ and $h_t := (s_0, y^t) := (s_0, z_1, s_1, \ldots, z_t, s_t) \in H_t := S \times Y^t$, $t \geq 1$, are called the ***history*** at time zero and at time t, respectively.
- A measurable mapping π_t from H_t into A is called a ***history-dependent decision rule*** at time $t \geq 0$ if $\pi_t(h_t) \in D(s_t)$ for all $h_t \in H_t$. The set of all history-dependent decision rules at time t is denoted by $\mathbb{F}_t$. Note that $\mathbb{F}_0 = \mathbb{F}$.
- A sequence $\pi = (\pi_t)_0^{N-1}$ of decision rules $\pi_t \in \mathbb{F}_t$ is called a ***history-dependent N-stage policy***. Thus $\Delta_N := \times_{t=0}^{N-1} \mathbb{F}_t$ is the set of such policies.

The definition of $R_{N\pi}(s_0, y^N)$ given in (21.5) extends from Markovian policies to history-dependent policies $\pi \in \Delta_N$ by replacing in (21.5) the action $\pi_t(s_t)$ by $\pi_t(h_t)$. For $\pi \in \Delta_N$ the probability distribution $\mathbb{P}(s_0, dy^N)$ on $\mathfrak{Y}_N$ is now defined *evolutionary* by

$$B \mapsto \mathbb{P}(s_0, B) := \int K(s_0, \pi_0(s_0), dy_1) \int K(s_1, \pi_1(s_0, y^1), dy_2)$$
$$\cdots \int K(s_{N-1}, \pi_{N-1}(s_0, y^{N-1}), dy_N)\, 1_B(y^N). \qquad (21.15)$$

Moreover, $V_{N\pi}(s_0)$ for $\pi \in \Delta_N$ is defined as in (21.6) as the expectation of $RN\pi(s_0, \cdot)$ with respect to $\mathbb{P}(s_0, dy^N)$. Obviously $V_{N\pi}(s)$ exists under (LUBF).

(c) As preparation for the proof of Sect. 21.3 below we need the following lemma, namely the RI for history-dependent policies $\pi =: (f, \sigma) \in \Delta_N, f \in \mathbb{F}, N \geq 2$. This requires more effort than for Markovian ones since σ no longer belongs to Δ_{N-1}, as, for instance, π_1 depends not only on s_1 but also on s_0 and z_1, hence $\pi_1 \notin \mathbb{F}_0$. On the other hand, for fixed (s_0, z_1) the (s_0, z_1)-section $\pi_t(s_0, z_1, \cdot)$ of π_t, $t \geq 1$, i.e. the mapping

$$(s_1, z_2, s_2, \ldots, z_t, s_t) \mapsto \pi_t(s_0, z_1, s_1, z_2, \ldots, z_t, s_t)$$

belongs to $\mathbb{F}_{t-1}$. Therefore the (s_0, z_1)-section of σ defined by

$$\sigma_{(s_0,z_1)} := (\pi_{t+1}(s_0, z_1, \cdot))_{t=0}^{N-2} \tag{21.16}$$

belongs to Δ_{N-1}. Since $\sigma_{(s_0,z_1)}$ is that part of π which is used for the last $N-1$ periods of the decision process determined by (f, σ) and (s_0, z_1), the following result is intuitively clear.

Lemma (The reward iteration for history-dependent policies) *Assume (LUBF) for an MDPD. Then for* $f \in \mathbb{F}$, $(f, \sigma) \in \Delta_N$ *and* $s \in S$

$$V_{1f}(s) = r_f(s) + \int K_f(s, d(z, s'))\, \tilde{\beta}(z) \cdot V_0(s')$$

$$V_{N(f,\sigma)}(s) = r_f(s) + \int K_f(s, d(z, s'))\, \tilde{\beta}(z) \cdot V_{N-1,\sigma_{(s,z)}}(s').$$

Idea of *proof.* The proof of the first assertion is obvious. The basic idea for the proof of the second assertion is the recursive structure of both $R(s, y^N)$ and of $\mathbb{P}(s, dy^N)$.

In fact, (21.5) with $\pi_t(s_t)$ replaced by $\pi_t(s, y^t)$ shows that for $y^N \in Y^N$ and using $z := z_1$, $s' := s_1$, we have

$$R_{N(f,\sigma)}(s, y^N) = \tilde{r}_f(s, z, s') + \tilde{\beta}(z) \cdot R_{N-1,\sigma_{(s,z)}}(s', (y_t)_2^N). \tag{21.17}$$

Moreover, in analogy to (21.4) it follows from (21.15) that

$$\mathbb{P}_{N(f,\sigma)}(s, dy^N) = K_f(s, d(z, s')) \otimes \mathbb{P}_{N-1,\sigma_{(s,z)}}(s', d(y_t)_2^N). \tag{21.18}$$

Now we obtain Sect. 21.3 from (21.17) and (21.18), since

$$\int \mathbb{P}_{N-1,\sigma_{(s,z)}}(s', d(y_t)_2^N)\, R_{N-1,\sigma_{(s,z)}}(s', (y_t)_2^N) = V_{N-1,\sigma_{(s,z)}}(s').$$

Proposition (Markovian policies suffice) *Assume (LUBF) for the MDPD and that the VI holds. Then*

$$\sup_{\pi \in \Delta_N} V_{N\pi}(s) = V_N(s) := \sup_{\sigma \in \mathbb{F}^N} V_{N\sigma}(s), \quad N \geq 1, s \in S.$$

Proof Since $\mathbb{F}^N \subset \Delta_N$, we only have to verify the assertion (I_N) that $V_{N\pi} \leq V_N$ for $\pi \in \Delta_N$. We use induction on $N \geq 1$. Assertion (I_1) holds trivially since $\Delta_1 = \mathbb{F}$. Assume (I_N) for some $N \geq 1$. Then we obtain from the RI Sect. 21.3 for $f \in \mathbb{F}$, $(f, \sigma) \in \Delta_{N+1}$

$$\begin{aligned} V_{N+1,(f,\sigma)}(s) &= r_f(s) + \int K_f(s, d(z, s')) \cdot \tilde{\beta}(z) \cdot V_{N,\sigma_{(s,z)}}(s') \\ &\leq r_f(s) + \int K_f(s, d(z, s')) \cdot \tilde{\beta}(z) \cdot V_N(s') \\ &= U_f V_N(s) \leq U V_N(s) = V_{N+1}(s). \end{aligned}$$

□

Chapter 22
Markov Renewal Programs

In the models of Part II we usually assumed a constant discount factor. This is only appropriate if the decisions are taken at equally spaced time points. However, there are many situations where the times between successive decisions are random variables whose probability distribution depends in a Markovian manner on the momentary state and action. Prominent examples are the optimal control of queueing systems where decisions (e.g. about the speed of service or the admission of customers) are often taken at the arrival or departure times of customers. Such situations can be described by special MDPDs where the disturbance variable η_{t+1} is the random time between the t-th and the $(t+1)$-st decision, $t \geq 0$.

Besides the horizon $N \in \overline{\mathbb{N}}$, which now does not denote time but the total number of stages, we have in reality a time limit $t_0 \in (0, \infty]$ beyond which the decision process must stop. In our opinion, often the most realistic (but also the most complicated) model is the case where both $N = \infty$ and $t_0 < \infty$. This case and the case $N < \infty$ and $t_0 < \infty$ is treated briefly in Sect. 22.3, while the other two cases with infinite time-horizon t_0 are the topic of the present section. The two cases with $N < \infty$ are often used as a theoretical tool for studying the structure of $V := \lim_{n \to \infty} V_n$, but they are also of independent interest; cf. Example 21.1.8. The discount factor per unit time is a constant which by tradition is not denoted by β but by $e^{-\alpha}$.

22.1 The Finite Horizon Model

Definition 22.1.1 A **Markov Renewal Program** $(S, A, D, Z, K, \tilde{r}, V_0, \alpha)$ (**MRP** for short) is an MDPD $(S, A, D, Z, K, \tilde{r}, V_0, \tilde{\beta})$ such that

- $Z = \mathbb{R}_+$ (continuous time case) or $Z = \mathbb{N}_0$ (discrete time case), and $\mathfrak{Z} := \mathfrak{B} \cap Z$,
- $\tilde{\beta}(s, a, z) = e^{-\alpha z}$ for some constant $\alpha \in \mathbb{R}_+$, called the ***discount rate.***

K. Hinderer et al., *Dynamic Optimization*, Universitext,
DOI 10.1007/978-3-319-48814-1_22

Markov renewal programs are often called ***Semi-Markov Decision processes.***

The non-negative random variable η_{t+1}, $0 \le t \le N-1$, is called the ***sojourn time*** in state ζ_t. Thus

$$T_t := \sum_{i=1}^{t} \eta_i,\ 0 \le t \le N,$$

(with $T_0 :\equiv 0$) is in case $t < N$ the time of the t-th decision, called the t-th ***decision epoch.*** Moreover, T_N is the time when the N-stage process ends. The time interval $[T_t, T_{t+1})$, $0 \le t \le N-1$, is called the t-th ***decision period.*** (From the context no confusion of T_t with the transition function in a CM will be possible.) The realizations of the random variable T_t are denoted by τ_t.

Choose an initial state s_0 and some N-stage policy $\pi = (\pi_t)_0^{N-1}$. Recalling from Chap. 21 the meaning of K_{π_t}, $t \ge 1$, as

$$K_{\pi_t}(s_t, G) = \mathbb{P}_{N\pi}(s_0, (\eta_{t+1}, \zeta_{t+1}) \in G \,\big|\, (\eta_i, \zeta_i) = (z_i, s_i),\ 1 \le i \le t),$$

the evolution of the state process $(\zeta_t)_1^N$ can be imagined as follows; cf. Fig. 21.1. The system stays in the initial state s_0 for a period of length z_1, then it jumps to another (or the same!) state s_1, where the pair (z_1, s_1) is selected according to the probability distribution $K_{\pi_0}(s, d(z_1, s_1))$; next, the system stays in s_1 for a period of length z_2 and then jumps to a state s_2, where (z_2, s_2) is selected according to $K_{\pi_1}(s_1, d(z_2, s_2))$, irrespective of z_1, etc. The process stops at time T_N after having made N jumps. The one-stage reward earned at time τ_t (or more generally, the reward earned during the t-th period and discounted to time τ_t) must be discounted to time zero with the factor $e^{-\alpha\tau_t}$, and similarly for the terminal reward.

According to (21.5) the N-stage reward for initial state s_0 and policy π is given by

$$R(s_0, y^N) = \sum_{t=0}^{N-1} e^{-\alpha\tau_t} \cdot \tilde{r}(s_t, \pi_t(s_t), z_{t+1}, s_{t+1}) + e^{-\alpha\tau_N} \cdot V_0(s_N).$$

An MRP has the following properties:

(a) The expected discount function

$$\beta(s,a) := \int Q(s,a,dz)\, e^{-\alpha z}$$

equals the Laplace transform $x \mapsto \hat{Q}(s,a,x)$ of $Q(s,a,dz)$ at point $x = \alpha$.

(b) $\mathfrak{V}_0$ is the set of measurable functions v on S such that $\int K(s,a,d(z,s'))\, e^{-\alpha z} \cdot v(s')$ exists for all (s,a). And then

$$\beta(s,a) \cdot P_0 v(s,a) = \psi v(s,a) = \int K(s,a,d(z,s'))\, e^{-\alpha z} \cdot v(s').$$

(c) Using $r := K\tilde{r}$ the operator L of the adjoint MDPvar has the form

$$Lv(s,a) = r(s,a) + \int K(s,a,d(z,s'))\, e^{-\alpha z} \cdot v(s')$$
$$= r(s,a) + \beta(s,a) \cdot \int P_0(s,a,ds')\, v(s'). \qquad (22.1)$$

Remark 22.1.2

(i) In the discrete time case $Q(s,a,dz)$ has a discrete density.
(ii) In many applications the state space is countable; in particular, in the control of queueing systems the state s often denotes the number of customers in the system. Then the state transition law $P(s,a,ds')$ has a discrete density. ◇

Remark 22.1.3 If the sojourn times are equal to a constant $c \in \mathbb{R}^+$ then the MRP reduces to an MDP where the transition law equals the state transition law P of the MRP and where the constant discount factor is the constant $\beta := e^{-\alpha c}$. ◇

Special cases:

(a) In many applications $Q(s,a,dz)$ is an exponential distribution with parameter $\Lambda(s,a) \in \mathbb{R}^+$. We call such an MRP an ***exponential MRP.*** (In the literature it is sometimes also called a *controlled continuous-time Markov chain.*) It follows that

$$\beta(s,a) = \frac{\Lambda(s,a)}{\alpha + \Lambda(s,a)},$$

and hence by (22.1)

$$Lv(s,a) = r(s,a) + \frac{\Lambda(s,a)}{\alpha + \Lambda(s,a)} \cdot P_0 v(s,a).$$

If Λ is a constant $\lambda > 0$, i.e. if $Q(s,a,dz) = \mathrm{Exp}(\lambda)$ for all (s,a), then the adjoint MDPvar is even an MDP with transition law P_0 and constant discount factor $\beta := \lambda/(\alpha + \lambda)$.

(b) If the sojourn times η_{t+1} and the states ζ_t are conditional independent given ζ_t, i.e. if $K(s,a,d(z,s')) = Q(s,a,dz) \times P(s,a,ds')$, and if (ii) $\tilde{r}(s,a,z,s') =: r_S(s,a,s')$ does not depend on z, we call the MRP a ***Renewal Markovian Decision Process*** (***RMDP*** for short). It models the same problem as described by the MDP $(S, A, D, P, r_S, V_0, \beta)$ with $\beta := e^{-\alpha}$, except that the times between decisions are random variables whose interaction with the MDP is described by the conditional probability distribution $Q(s,a,dz)$ of the sojourn time given the momentary state s and action a. It follows that $r(s,a)$ from the MDPD and from the MDP coincide. The operator L in the MDPD assumes by (21.11) the form

$$Lv(s,a) = r(s,a) + \beta(s,a) \cdot Pv(s,a).$$

This shows that L of the RMDP is obtained from L of the embedded MDP simply by replacing in the latter the constant $e^{-\alpha}$ by $\beta(s,a) = \int Q(s, adz) \cdot e^{-\alpha z}$.

For modeling purposes it is often useful to represent $\tilde{r}$ in the form

$$\tilde{r}(s,a,z,s') = r_0(s,a,z) + r_3(s,a,z,s') \cdot e^{-\alpha z} \tag{22.2}$$

for measurable functions r_0 on $D \times Z$, and r_3 on $D \times Z \times S$. (Here $\tilde{r}$ depends on the discount rate α.)

Note that (22.2) is no restriction of generality, as the choice $r_0 :\equiv 0$, $r_3(s,a,z,s') := e^{\alpha z} \cdot \tilde{r}(s,a,z,s')$ yields the original $\tilde{r}$. However, for applications it is useful to keep the terms separated, due to the following interpretations: $r_0(s,a,z)$ is typically a reward which accumulates up to time z and discounted to the beginning of the decision period, while $r_3(s,a,z,s')$ is obtained at the end of the decision period. Moreover, in the continuous time case, $r_0(s,a,z)$ often has the form:

$$r_0(s,a,z) = r_1(s,a) + \int_0^z e^{-\alpha x} \cdot r_2(s,a,x)\, dx,\ z \in \mathbb{R}_+, \tag{22.3}$$

provided $x \mapsto e^{-\alpha x} r_2(s,a,x)$ is integrable on $[0,z]$ for all $z \in \mathbb{R}_+$. Here $r_2(s,a,x)$ is a *reward rate,* possibly depending on time x, which is continuously discounted to the beginning of the decision period. In the discrete time case it is appropriate to assume that a reward (not reward rate!) $r_2(s,a,x)$ is obtained at each of the z time units $x = 0$, $x = 1, \ldots, x = z-1$ of the decision period, and that $r_2(s,a,x)$ is discretely discounted to the beginning of the decision period. This yields

$$r_0(s,a,z) := r_1(s,a) + \sum_{x=0}^{z-1} e^{-\alpha x} \cdot r_2(s,a,x),\ z \in \mathbb{N}. \tag{22.4}$$

It follows from (22.2) that $r = Qr_0 + \psi r_3$, provided the two integrals exist and are finite.

Consider the special case when $r_2(s,a,\cdot)$ is constant during the decision period. If r_0 has the form (22.3) or (22.4), then Qr_0 exists, is finite, and for $(s,a) \in D$

$$Qr_0(s,a) = r_1(s,a) + r_2(s,a) \cdot (1 - \beta(s,a))/c,$$

where $c := \alpha$ in the first case and $c := 1 - e^{-\alpha}$ in the second case.

We have seen above how an MDP may be extended to a Renewal MDP by means of a transition probability $Q(s,a,dz)$ from D into $Z \in \{\mathbb{R}_+, \mathbb{N}_0\}$ which models the time z between decisions, in such a way that the sojourn times η_{t+1} and states ζ_{t+1} are conditional independent given ζ_t. An analogous extension of a CM to a ***Renewal CM*** (***RCM*** for short) can be obtained as follows. Since z denotes the sojourn time, the disturbances in the CM is here denoted by x. Thus the CM is determined by a tuple $(S, A, D, X, Q_X, T, r_X, V_0, \beta)$ with $\beta := e^{-\alpha}$; in particular, $Q_X(dx)$ is the probability distribution of the disturbances. The extension RCM is simply RMDPad, where MDPad is the MDP adjoint to the CM. The transition law P in MDPad is determined by $Pv(s,a) := \int Q_X(dx) \cdot v(T(s,a,x))$ for measurable $v \geq 0$. Thus K in the RCM is determined by $K(s,a,d(z,s'))\, v(z,s') = \int Q(s,a,dz) \int Q_X(dx)\, v(z, T(s,a,x))$ for measurable $v \geq 0$. Moreover, $\tilde{r}(s,a,z,s') = r(s,a) = \int Q_X(dx)\, r_X(s,a,x)$. The operator L in the RCM assumes the form

$$Lv(s,a) = r(s,a) + \beta(s,a) \cdot \int Q_X(dx) \cdot v(T(s,a,x)).$$

Thus the operator L of the RCM is obtained from L of the CM simply by replacing there the constant $e^{-\alpha}$ by $\beta(s,a) = \int Q(s,a,dz)\, e^{-\alpha z}$.

Example 22.1.4 (Selling an asset with i.i.d. random times between successive offers) Consider the CM in Example 13.4.4 which models the selling of an asset. Let $Q(dz)$ and $M(dx)$ be the probability distribution of the offers η_t and of the i.i.d. random times between successive offers, respectively. Put $\beta' := \int M(dx)\, e^{-\alpha x}$. Then it follows from the preceding reasoning and from Example 13.4.4 that the maximal expected N-stage reward d_N for at most $N \geq 1$ offers can be found recursively, using $d_1 = \mathbb{E}\, \eta_1 - \gamma$, by

$$d_{n+1} = \mathbb{E} \max\{\beta' \cdot d_n, \eta_1\} - \gamma, \quad n \geq 1. \tag{22.5}$$

Moreover, it is optimal to accept an offer of amount z at stage n if $z \geq \beta' \cdot d_n$.

As an example, if $M = \text{Exp}(\lambda)$ and if $\mathbb{E}\, \eta_1 =: 1/\lambda \geq \gamma$, then it follows from (22.5) by induction on n that $d_n \geq d_1 \geq 0$, and the intuitively obvious fact that d_n is decreasing with $\mathbb{E}\, \eta_1$. — One may criticize that the preceding problem is not appropriately modeled. In fact, in general one shall not wait for offers until a given number $N + 1$ of offers arise, but one will wait for a given time $t_0 > 0$. This problem is treated in Chap. 22. ♦

22.2 The Infinite Horizon Model

In this section we consider an MDP with infinite horizon.

Definition 22.2.1 A **Markov Renewal Program with infinite horizon** $(S, A, D, Z, K, \tilde{r}, \alpha)$ (**MRP**$_\infty$ for short) is defined as an MRP except that no terminal reward function V_0 is required. The process runs up to time $T_\infty := \lim_n T_n$.

In many applications we have $T_\infty \equiv \infty$. This happens in particular under the following *drive* assumption (DR) on the distribution of the sojourn time.

(DR) There exist constants $\varepsilon > 0$ and $\eta > 0$ such that

$$K(t,s,a,s',[\eta,\infty)) \geq \varepsilon \text{ for all } (t,s,a,s').$$

Assumption (DR) means that under each initial state (t, s) and each admissible action a and next core state s' the momentary sojourn time z is larger than η with positive probability, uniformly in (t, s, a, s'). If $K(t, s, a, s', dz)$ does not depend on (t, s, a, s'), then we have i.i.d. sojourn times η_ν with probability distribution $K(dz)$ and the jump time process $(T_\nu)_1^\infty$ is a renewal process. Then right continuity of the distribution function $z \mapsto K((-\infty, z])$ implies that (DR) holds if and only if $K(\{0\}) < 1$.

Of course, all notions and general results about MDPD$_\infty$s also apply to MRP$_\infty$s. For treating the control of queueing systems, we need

Lemma 22.2.2 *Let A_1, A_2, ..., A_k be independent random variables on some probability space $(\Omega', \mathfrak{F}, \mathbb{P})$ with $A_i \sim \text{Exp}(\mu_i)$. Then:*

(a) $\min_{1\le i\le k} A_i$ *has the probability distribution* $\text{Exp}(\sum_{i=1}^k \mu_i)$.
(b) $\mathbb{P}(A_j = \min_{1\le i\le k} A_i)) = \mu_j/(\sum_{i=1}^k \mu_i)$, $1 \le j \le k$.

Proof

(a) follows from

$$\mathbb{P}(\min_i A_i > z) = \prod_i \mathbb{P}(A_i > z) = \exp(-\sum_i \mu_i \cdot z),\ z \in \mathbb{R}_+.$$

(b) Fix k and j. The random variable $B := \min_{i\ne j} A_i$ is independent of A_j and $B \sim \text{Exp}(\mu)$ with $\mu := \sum_{i\ne j} \mu_i$ by (a). Now the assertion follows as

$$\mathbb{P}(A_j = \min_i A_i) = \mathbb{P}(A_j \le B) = \int_0^\infty \mathbb{P}_{A_j}(dz)\, \mathbb{P}(B > z)$$
$$= \int_0^\infty \text{Exp}(\mu_j)(dz)\, e^{-\mu z} = \mu_j/(\mu + \mu_j). \qquad \square$$

Example 22.2.3 (Controlling the service rate in an M/M/1-queue at arrival and departure times) (The notation M/M/1 is an abbreviation for *Markov/Markov/one server.*) A system consists of a single server and unlimited waiting room. Each arriving customer joins the queue unless he is served immediately. We use the following assumptions:

(i) Customers arrive according to a Poisson process with (uncontrolled) intensity $\lambda \in \mathbb{R}^+$, called the *arrival rate*. Thus, if A_1 denotes the time of arrival of the first customer and A_i, $i \ge 2$, the time between the arrival of the $(i-1)$-st and the i-th customer, then the A_i's are i.i.d. random variables on some probability space $(\Omega', \mathfrak{F}', \mathbb{P})$ with $A_1 \sim \text{Exp}(\lambda)$. As the mean time between arrivals is $1/\lambda$, a high arrival rate implies short mean interarrival times. Customers are served in the order of their arrival. The number of customers in the system at a certain time point are those which either are waiting or are served.

(ii) The server selects at the t-th decision epoch, i.e. at time T_t, $t \geq 0$, a *service rate* a which is used during the t-th decision period (which begins at time T_t). Decision epochs are: time zero, each time epoch when either a customer departs (because his service is completed) or when a customer arrives. If at time T_t, $t \geq 1$, the service of the customer being served in the $(t-1)$-th decision period is not yet completed, it must be continued. At decision epochs the server has the following choices: (a) If the system is empty, only the action $a = 0$, which means that service is switched off, is admissible. (b) If the system is non-empty he may either choose $a = 0$ or a service rate $a > 0$ from a finite non-empty set $A^+ \subset \mathbb{R}^+$. (It may happen that the same customer is served in several successive periods with different or the same service rates.)

(iii) Let the random variable $B_{a,t+1}$ on $(\Omega', \mathfrak{F}', \mathbb{P})$ denote the service time in period $t \geq 0$ in case the server is busy and uses the service rate $a \in A^+$. It is assumed that $B_{at} \sim \text{Exp}(a)$ for $a \in A^+$ and $t \in \mathbb{N}_0$. The family of all random variables A_i, B_{at}, $i, t \in \mathbb{N}$, $a \in A^+$, is assumed to be independent. When using service rate $a \in A^+$ during a decision period the mean service time during that period is $1/a$. Thus high service rates means *fast* service.

(iv) If at the beginning of a decision period $s \geq 1$ customers are in the system and if the server works with service rate $a \in A^+$, there arise during the decision period waiting costs and service costs with cost *rates* (i.e. costs per unit time) $g(s) \in \mathbb{R}_+$ and $h(a) \in \mathbb{R}_+$, respectively. At each time of departure we obtain a reward of amount $R \in \mathbb{R}_+$ for completing the service of a customer. Costs and rewards are discounted to time 0 with a discount rate $\alpha \in \mathbb{R}^+$. The goal consists in minimizing the infinite-stage expected discounted costs.

We describe the problem by means of an exponential MRP_∞ with conditional independent sojourn times and states as follows. The state $s_t \in S = \mathbb{N}_0$ is the number of customers in the system at time T_t, after the arrival or departure of a customer; $D(s) = A := A^+ + \{0\}$ for $s \geq 1$, and $D(0) = \{0\}$.

We comment on difficulties in modeling the transition law. We call *model I* the model described in terms of the random variables A_i and B_{at}. This model seems to be a first rate candidate if one is interested in estimating the data of the model. Assume that an initial state s_0 and some policy $\pi = (\pi_t)_0^\infty$ are given. Then model I requires

(i) the proof that $((\eta_t, \zeta_t))_0^\infty$ is a Markov renewal process, using the probability distributions of A_1 and B_{a1} and the assumption that the family $(A_i, B_{at}, i, t \in \mathbb{N}, a \in A^+)$ is independent,

(ii) the computation of $\beta(s, a)$ and $P_0(s, a, ds')$ from the assumptions of model I.

This seems to be a very difficult task. In our opinion the usual argument of invoking the lack of memory of the exponential distribution is far from a rigorous mathematical proof. Fortunately the situation is not as bad as it might look at first sight since in each applied problem one cannot do more than decide on an individual basis whether a proposed model is convincing or not. Here also the following model II becomes convincing.

Assume for model II that after the t-th decision, which takes place at time T_t, we are in state s and take action $a := \pi_t(s)$. It is plausible that due to the lack of memory of the exponential distribution the Poisson arrival process starts anew at time T_t. Then $Q(s, 0, dz)$, $s \geq 0$, equals $\text{Exp}(\lambda)$, the probability distribution of A_1, and $Q(s, a, dz)$, $s \geq 1$, $a \in A^+$ equals the probability distribution of $A_1 \wedge B_{a1}$. Thus we get from Lemma 22.2.2(a) for $s \geq 1$, $a \in A^+$

$$Q(s, a, dz) = \text{Exp}(\lambda + a) =: Q_a,$$

$$\Lambda(s, a) = \lambda + a, \quad \beta(s, a) = \frac{\lambda + a}{\alpha + \lambda + a} =: \beta_a. \tag{22.6}$$

Next, from

$$K_{12}(s, a, z, \{s'\}) := \mathbb{P}_\pi(s_0, [\zeta_{t+1} = s'] \big| \zeta_t = s, \eta_{t+1} = z)$$

we get $K_{12}(s, 0, z, \{s + 1\}) = 1$, $s \geq 1$, and for $s \geq 1$, $a \in A^+$

$$K_{12}(s, a, z, \{s'\}) = \begin{cases} \mathbb{P}(A_1 = \min\{A_1, B_{a1}\}), & \text{if } s' = s + 1, \\ \mathbb{P}(B_{a1} = \min\{A_1, B_{a1}\}), & \text{if } s' = s - 1, \\ 0, & \text{else.} \end{cases}$$

As $K_{12}(s, a, z, \{s'\})$ is independent of z, sojourn times and states are conditionally independent. Hence $P_0(s, a, ds') = P(s, a, ds') = K_{12}(s, a, z, ds')$; cf. Chap. 21. From Lemma 22.2.2(b) we obtain for all (s, a)

$$P(s, a, \cdot) = \frac{a}{\lambda + a} \cdot \delta_{s-1} + \frac{\lambda}{\lambda + a} \cdot \delta_{s+1}.$$

Obviously the one-stage costs, discounted to the beginning of the period, have the form (22.2) with $r_1 :\equiv 0$, $r_2(s, a, x) := -(g(s) + h(a))$ and $r_3(s, a, z, s') := R \cdot \delta_{s,s-1}$. Thus we have for all (s, a)

$$\tilde{c}(s, a, z, s') = [g(s) + h(a)] \cdot (1 - e^{-\alpha z})/\alpha - Re^{-\alpha z} \cdot \delta_{s',s-1}.$$

An elementary computation, using $v(-1) := 0$, yields for $(s, a) \in D$

$$Lv(s, a) = \frac{g(s) + h(a) - a \cdot R}{\alpha + \lambda + a} + \frac{a \cdot v(s - 1) + \lambda \cdot v(s + 1)}{\alpha + \lambda + a}.$$

Proposition 22.2.4 (Properties of Example 22.2.3) *Assume*

(i) A^+ *is finite,*
(ii) $g(s) \leq K \cdot d^s$, $s \in \mathbb{N}_0$, *for some* $K \in \mathbb{R}^+$ *and some* $d \in [1, (\alpha + \lambda)/\lambda)$,
(iii) the service cost rate h *is increasing,*
(iv) $\alpha > 0$.

Then we have:

(a) $s \mapsto b(s) := d^s$ *is a bounding function of the* MRP_∞ *with* $\beta_b < 1$.
(b) C_∞ *exists and is within* B_b *the unique solution* v *of the OE* $v = Uv$.
(c) *If* f *is the smallest minimizer of* LC_∞*, then* f^∞ *is a stationary optimal policy.*

Proof

(a1) We show that $\beta_b < 1$. For $s \geq 1$ and $a \in A$ we obtain, using the discrete density $p(s,a,\cdot)$ of $P(s,a,ds')$

$$(\alpha + \lambda + a) \cdot \beta_a \cdot pb(s,a) = \lambda \cdot b(s+1) + a \cdot b(s-1) = d^{s-1} \cdot (\lambda \cdot d^2 + a).$$

Therefore $\beta(s,a) \cdot pb(s,a)/b(s) = H(a)/d$, where

$$H(a) := (\lambda \cdot d^2 + a)/(\alpha + \lambda + a) = 1 + (\lambda \cdot d^2 - \alpha - \lambda)/(\alpha + \lambda + a).$$

As H is monotone and as

$$\beta_0 \cdot pb(0,0)/b(0) = \lambda \cdot d/(\alpha + \lambda) = H(0)/d < 1,$$

we have $\beta_b < 1$ if $H(\bar{a}) < d$, where $\bar{a} := \max A$. Now

$$H(\overline{a}) < [d \cdot (\alpha + \lambda) + \overline{a}]/(\alpha + \lambda + \overline{a}) < d.$$

(a2) Isotonicity of h implies, as $d^s \geq 1$, that for all s

$$|\tilde{c}(s,a,z,s')| \leq (g(s) + h(\bar{a}))/\alpha + R \leq [(K + h(\bar{a}))/\alpha + R] \cdot d^s.$$

This proves finiteness of $\|K \cdot |\tilde{c}|\|_b$.
(b) This follows from (a) and the minimization version of Theorem 14.2.1.
(c) As $D(s)$ is finite for all s, Lv has a smallest minimizer for all $v \in B_b$. Now the assertion follows again from the minimization version of Theorem 14.2.1. □

Remark 22.2.5 If the waiting cost rate is a constant $\gamma \in \mathbb{R}^+$ per person, i.e. if $g(s) := \gamma \cdot s$, then assumption (ii) is fulfilled for any $d \in (1, (\alpha + \lambda)/\lambda)$ as for K large enough

$$K \cdot d^s = K \cdot e^{\log d \cdot s} \geq K \cdot \log d \cdot s \geq \gamma s. \qquad \diamond$$

Assume that we want to prove for some MRP_∞ a structural property of V_∞. If the property is preserved under the pointwise limit of sequences of functions, then one will succeed if one can show by means of the VI the property for all value functions V_n in the adjoint MDPvar (starting with an appropriate V_0, often $V_0 :\equiv 0$), provided $(V_n)_0^\infty$ converges pointwise to V_∞. Here difficulties may arise when applying the VI as $\beta(s,a)$ in the adjoint MDPvar is in general not constant. This problem can often be overcome by the next result which gives weak conditions under which V_∞

is also the limit of the value functions V'_n in a certain MDP′ with a constant discount factor β'. Moreover, Part (b) of the next result is useful for establishing structural properties of a maximizer f of LV_∞ (which yields the stationary optimal policy f^∞).

Theorem 22.2.6 (Uniformization of an MDPvar) *Consider an MDPvar with countable state space. Let $p(s,a,\cdot)$ be the discrete density of $P(s,a,ds')$, $(s,a) \in D$. Assume:*

(i) $\beta' := \sup_{(s,a)\in D} \beta(s,a) < 1$, or equivalently, the function $\Delta := \beta/(1-\beta)$ is positive and bounded,

$$\beta(s,a) = \frac{\Delta(s,a)}{1+\Delta(s,a)}, \quad (s,a) \in D,$$

(ii) the MDPvar$_\infty$ has a bounding function b with $\beta_b < 1$, so that V exists by Theorem 14.2.1 and belongs to $\mathbb{B}_b$.

Put $\Delta' := \sup \Delta$, select some $V_0 \in \mathbb{B}_b$ and consider the MDP′(V_0), called a ***uniformization*** *of the MDPvar, which has the same S, A, D as MDPvar, the constant discount factor β' and for all (s,a,s')*

$$p'(s,a,s') := \frac{\Delta(s,a)}{\Delta'} \cdot p(s,a,s') + \left(1 - \frac{\Delta(s,a)}{\Delta'}\right) \cdot \delta_{s,s'},$$

$$r'(s,a) := \frac{1+\Delta(s,a)}{1+\Delta'} \cdot r(s,a).$$

Then we have:

(a) The function b is also a bounding function of MDP′(V_0), and $\beta'_b < 1$. The sequence $(V'_n)_0^\infty$ converges exponentially fast in b-norm to $V \in \mathbb{B}_b$. V is the unique solution $v \in \mathbb{B}_b$ of the OE for MDP′, i.e. of $U'v = v$.

(b) The decision rule $f \in \mathbb{F}$ is a maximizer of LV if and only if it is a maximizer of $L'V$, and then f^∞ is a stationary optimal policy for MDPvar$_\infty$.

Proof

(a1) Firstly, $|r'| \le |r|$ implies $\|r'\|_b < \infty$, and $\|V_0\|_b < \infty$ holds as $V_0 \in \mathbb{B}_b$. It remains to show that $\beta'_b := \beta' \cdot \|p'b\|_b < 1$. A simple computation yields

$$\Delta' \cdot p'v = \Delta \cdot pv + (\Delta' - \Delta) \cdot v, \; v \in \mathbb{B}_b. \tag{22.7}$$

From (22.7) with $v := b$, we obtain, using $\beta \cdot pb \le \beta_b \cdot b$, that

$$(1+\Delta') \cdot \beta' \cdot p'b = (1+\Delta) \cdot \beta \cdot pb + (\Delta' - \Delta) \cdot b \le (\beta_b + \Delta') \cdot b.$$

Thus $\beta'_b \le (\beta_b + \Delta')/(1+\Delta') < 1$.

(a2) Of basic importance is the equality

$$(1 + \Delta') \cdot (L'v - v) = (1 + \Delta) \cdot (Lv - v), \; v \in B_b, \tag{22.8}$$

which follows from (22.7). From (22.8) we obtain for $v := V$, since $LV \leq UV = V$, that

$$(1 + \Delta') \cdot (L'V - V) = (1 + \Delta) \cdot (LV - V) \leq 0,$$

hence $L'V(s, a) \leq V(s)$ for all (s, a), hence $U'V \leq V$. Now induction on $n \in \mathbb{N}$ shows for MDP′(V) that $V_n' = (U')^n V \leq V$. As V_n' converges for $n \to \infty$ to V' by Theorem 14.2.1, we have $V' \leq V$. In the same way one obtains from (22.8) for $v := V'$ that $V \leq V'$, hence $V = V'$.

(b) The first part of the assertion follows from (22.8), and the second one from Theorem 14.2.1. □

Remark 22.2.7

(i) From $\beta = 1 - 1/(1 - \Delta)$ we see that $\beta' = \Delta'/(1 - \Delta')$.

(ii) The proof of Theorem 22.2.6 shows that $\beta_b' \leq \hat{\beta} := (\beta_b + \Delta')/(1 + \Delta')$. On the other hand, in general $\beta_b < \hat{\beta}_b'$, and then the bound

$$\|V - V_N'\|_b \leq \min\left\{\|V_1' - V_0\|_b, \|r'\|_b + (1 - \hat{\beta})) \cdot \|V_0\|_b\right\} \cdot \frac{\hat{\beta}^N}{1 - \hat{\beta}}, \tag{22.9}$$

which is obtained from Theorem 20.1.3 is worse than the corresponding bound for $\|V_\infty - V_N\|_b$, which is obtained from (22.9) by replacing V_1' by V_1, r' by r and $\hat{\beta}$ by β_b.

(iii) $P'(s, a, ds')$ is a mixture of $P(s, a, ds')$ and of δ_s. ◇

Remark 22.2.8 If the MDPvar is adjoint to an exponential MRP with positive discount rate α and $Q(s, a, dz) = \text{Exp}(\Lambda(s, a))$ for a bounded function Λ, then assumption (i) holds with $\Delta(s, a) := \Lambda(s, a)/\alpha$. ◇

Remark 22.2.9 The proof of Theorem 22.2.6 shows that $\beta_b' \leq \hat{\beta} := (\beta_b + \Delta')/(1 + \Delta')$. One can find V_∞ approximately either by successive approximation in MDPvar or in MDP′. In general the latter is easier to implement as L' often has a simpler form than L, and an additional acceleration occurs when one succeeds in showing that the maximizers at each stage in MRP′ are increasing (e.g. in Example 22.2.3). On the other hand, in general $\beta_b < \hat{\beta}_b'$, and then the bound

$$\|V - V_N'\|_b \leq \min\left\{\|V_1' - V_0\|_b, \|r'\|_b + (1 - \hat{\beta})) \cdot \|V_0\|_b\right\} \cdot \frac{\hat{\beta}^N}{1 - \hat{\beta}}, \tag{22.10}$$

which is obtained from Theorem 20.1.3 is worse than the corresponding bound for $\|V_\infty - V_N\|_b$, which is obtained from (22.10) by replacing V_1' by V_1, r' by r and $\hat{\beta}$ by β_b. ◇

Remark 22.2.10 It follows from the proof of Theorem 22.2.6 that the theorem remains true if Δ' is any positive constant such that

$$(1 - p(s, a, s)) \cdot \Delta(s, a) \leq \Delta' \text{ for } (s, a) \in D.$$

(This holds in particular for all $\Delta' \geq \sup \Delta$.) This variation only affects the speed of convergence in (22.10). It is highest if Δ' is as small as possible, as $\hat{\beta} = (\beta_b + \Delta')(1 + \Delta')$ is increasing in Δ'. ◇

Proposition 22.2.11 (Structural properties of the solution of Example 22.2.3) *Under the assumptions (i)–(iv) in Proposition* 22.2.4 *we have:*

(a) Put $\bar{a} := \max A$. C_∞ *is within* B_b *the unique solution* v *of the equation*

$$\begin{aligned} \lambda \cdot v(s) = g(s) + \lambda \cdot v(s+1) \\ + \min_{a \in A} \left[h(a) + a \cdot (v(s-1) - v(s) - R)\right], \; s \in \mathbb{N}_0. \end{aligned} \tag{22.11}$$

(b) If $f^*(s)$ *is the smallest minimum point of*

$$a \mapsto h(a) + a \cdot \left[C_\infty(s-1) - C_\infty(s) - R\right], \; s \in \mathbb{N}_0, \tag{22.12}$$

then f^* *is the smallest minimizer of* LC_∞, *and* $f^{*\infty}$ *is a stationary optimal policy for Example* 22.2.3.

(c) If $h(a) \geq a \cdot R + g(0) - g(1) + h(0)$ *for all a, and if g is increasing then* C_∞ *is increasing.*

(d) If g is convex then C_∞ *is convex.*

(e) If the assumptions of (b) and (c) hold then f^* *is increasing.*

Proof

(a) and (b): We consider the MDPvar which is adjoint to the MRP$_\infty$. By (22.6) we have $\beta(s, a) = \Delta(s, a)/(1 + \Delta(s, a))$ for $\Delta(s, a) := (\lambda + a)/\alpha$. According to Theorem 22.2.6 the uniformization MDP$'$ of MDPvar is determined by $\Delta' = (\lambda + \bar{a})/\alpha$, $\beta' := (\lambda + \bar{a})/(\alpha + \lambda + \bar{a})$ and

$$\begin{aligned} P'(s, a, \cdot) &:= \frac{a}{\alpha + \lambda + \bar{a}} \cdot \left[a \cdot \delta_{s-1} + (\alpha + \bar{a} - a) \cdot \delta_s + \lambda \cdot \delta_{s+1}\right] / (\Delta + \bar{a}), \\ c'(s, a) &:= (g(s) + h(a) - a \cdot R)/(\alpha + \lambda + \bar{a}). \end{aligned}$$

We also assume $V_0' :\equiv 0$. We know from Proposition 22.2.4 that b is a bounding function of MRP$_\infty$. Now Theorem 22.2.6 tells us that C_∞ is within B_b the unique solution v of $v = U'v$, and, observing the proof of Theorem 22.2.6(b), that f^* is the smallest minimizer of $L'C_\infty$. Now a simple computation yields both assertions.

(c) and (d): As the value functions C'_n of MDP′ exist, belong to B_b and converge in b-norm towards C_∞, and as the two properties under discussion are preserved under pointwise convergence, it suffices to prove them for all functions C'_n. Moreover, as $C'_0 :\equiv 0$ has both properties and as the VI holds for MDP′, it suffices to prove that if $v \in B_b$ has one of the two properties then $U'v$ has the same property.

For (c) we show that $U'v$ is increasing whenever $v \in B_b$ is increasing. From Theorem 22.2.6 we obtain

$$\begin{aligned}&(\alpha + \lambda + \overline{a}) \cdot L'v(s,a) \qquad (22.13)\\ &= g(s) + a \cdot v(s-1) + (\alpha + \bar{a} - a) \cdot v(s) + \lambda \cdot v(s+1) + h(a) - a \cdot R.\end{aligned}$$

Thus, as g and v are increasing and as $\alpha + \bar{a} - a \geq 0$, $s \mapsto L'v(s,a)$ is increasing on $\mathbb{N}$. As $D(s) = A$ for all $s > 0$, $U'v$ is increasing on $\mathbb{N}$. It remains to show that $U'v(0) \leq U'v(1)$, i.e. that $L'v(0,0) \leq L'v(1,a)$ for all $a \in A$. From (22.13) we have

$$\begin{aligned}&(\alpha + \lambda + \overline{a}) \cdot L'v(1,a)\\ &= g(1) + h(a) - a \cdot R + \lambda \cdot v(2) + a \cdot v(0) + (\alpha + \bar{a} - a) \cdot v(1)\\ &\geq g(0) + h(0) + \lambda \cdot v(1) + a \cdot v(0) + (\alpha + \bar{a} - a) \cdot v(0)\\ &= (\alpha + \lambda + \overline{a}) \cdot L'v(0,0).\end{aligned}$$

For (d) we put, using $v(-1) := 0$,

$$J(s,a) := h(a) - a \cdot R + a \cdot v(s-1) + (\bar{a} - a) \cdot v(s), \;\; (s,a) \in D,$$

and $I(s) := \min_{a \in D(s)} J(s,a)$.

Then by (22.13)

$$(\alpha + \lambda + \overline{a}) \cdot U'v(s) = I(s) + g(s) + \lambda \cdot (v(s+1) - v(s)).$$

Let $f(s)$ be a minimum point of $J(s,\cdot)$ for $s \in \mathbb{N}_0$. As g and $s \mapsto v(s+1)$ are convex and as sums of convex functions are convex, it suffices to show that $s \mapsto I(s) = J(s,f(s))$ is convex on $\mathbb{N}$. Now we have for $s \in \mathbb{N}_0$, observing that $f(0) = 0$,

$$\begin{aligned}&I(s+2) - I(s+1) - [I(s+1) - I(s)]\\ &\geq J(s+2,f(s+2)) - J(s+1,f(s+2)) - J(s+1,f(s)) + J(s,f(s))\\ &= (\bar{a} - f(s+2)) \cdot [v(s+2) + v(s) - 2v(s+1)]\\ &\quad + f(s) \cdot [v(s+1) + v(s-1) - 2v(s)] \geq 0.\end{aligned}$$

(e) Let $v \in B_b$ be convex and let f be the smallest minimizer of $L'v$. We show that f is increasing on $\mathbb{N}$ (hence also on $\mathbb{N}_0$ as $f(0) = 0$). Then the assertion follows, as C_∞ is convex by (d). As $D(s) = A$ for all $s > 0$, the correspondence $s \mapsto D(s)$, $s \in \mathbb{N}$, is NE-complete by Example 8.2.4(a). Therefore it suffices to show that $L'v$ has decreasing differences on $\{(s,a) \in D : s > 0\}$. Now we see from (22.13) that it suffices by Theorem 8.2.9 to show that

$$(s,a) \mapsto w(s,a) := a \cdot [v(s-1) - v(s)]$$

has decreasing differences i.e. that $a \mapsto w(s+1,a) - w(s,a)$ is decreasing. This follows from convexity of v. □

We now comment on the use of (22.11) and (22.12) for computing C_∞ and f^*, respectively. Assume that we want to compute $C_\infty(s)$ for $0 \le s \le K$, where $K \in \mathbb{N}$ is given. Obviously this could be done exactly, according to (22.11) by recursion in state space, if $C_\infty(0)$ were known. However, we do not know of a method to compute $C_\infty(0)$ exactly. (In particular we do not know how to exploit the fact that C_∞ is within B_b the unique solution of (22.11).) Fortunately an approximate numerical solution of $C_\infty(s)$, $0 \le s \le K$, with an error not larger than a given $\varepsilon > 0$ is possible as follows: Firstly select N so large that the bound (22.10) yields $\|V_\infty - V'_N\|_b \le \varepsilon/d^K$. As $V_0 \equiv 0$ and as

$$\|C'_1\|_b = \sup_s \frac{|\min_a c'(s,a)|}{b(s)} \le \sup_{(s,a)} \frac{|c'(s,a)|}{b(s)} = \|c'\|_b$$

this condition assumes the form

$$\|C'_1\|_b \cdot \hat{\beta}^N \le \varepsilon \cdot (1 - \hat{\beta}).$$

Then we can compute $C_\infty(s)$, $0 \le s \le K$, with an error not larger than ε by the VI in MDP$'$. This requires, as seen from (22.13), only the computation of the finitely many numbers $C'_n(s)$, $1 \le n \le N-1$, $1 \le s \le K+n$. Moreover, one can show that under the assumptions in Proposition 22.2.11(c) and (d) the minimizers at each stage n in MDP$'$ are increasing, which accelerates the VI.

22.3 Infinite Stage Markov Renewal Programs with Finite Time Horizon

In reality the decision process of an infinite-stage MRP determined by an initial state s_0 and a policy, cannot be run indefinitely, but must stop after some finite positive time t_0, called *initial time horizon*. This concerns, for example, the many models of control of queues in cases where the servers only work during an interval of finite length; another example is the selling of an asset where one cannot wait infinitely long. The existence of an initial time horizon t_0 means that the decision process runs up to time $\min\{T_\infty, t_0\}$. Thus it runs up to time t_0 if $T_\infty = \infty$ almost surely under any policy, for which condition (DR) in Definition 22.2 is sufficient. Also finite-stage MRPs with finite time horizon are of interest, but they are not treated here.

Definition 22.3.1 The **finite time horizon Markov Renewal Program** of an $(S, A, D, K, \tilde{r}, \alpha)$ (***MRP***^{t_0} for short), $t_0 \in \mathbb{R}^+$, is defined by the following data:

- MRP $(S, A, D, Z, K, \tilde{r}, V_0, \alpha)$.
- $t_0 \in \mathbb{R}^+$, called the **initial time horizon** (also: initial residual time).
- $\tilde{h}: \mathbb{R}_+ \times D \times \mathbb{R}_+ \times S \to \mathbb{R}_+$ measurable, called the **termination reward function.**

To explain the meaning of $\tilde{h}$, let us have a momentary initial time-horizon $t \geq 0$ and a momentary sojourn time z. Denote the state at the end of the period by u rather than by s' as before. If $z \leq t$, one earns immediately $\tilde{r}(s, a, z, u)$ and the process continues for at least one more period. Otherwise, i.e. if $z > t$, one obtains immediately the reward $\tilde{h}(t, s, a, z, u)$ and the process stops at time t_0. (Obviously it suffices to know $\tilde{h}(t, s, a, z, u)$ only for $t < z$.)

All problems MRP^{t_0}, $t_0 > 0$, can be modeled by a single infinite time-horizon model MRP$'$, say, with extended states $s' = (t, s)$ in the new state space $S' := \mathbb{R} \times S$ in the following sense.

Definition 22.3.2 The **infinite time horizon model** $(S', A', D', Z', K', \tilde{r}', \alpha')$ (***MRP***$'$ for short) corresponding to the MRP $(S, A, D, Z, K, \tilde{r}, \alpha)$ is defined by the following data:

- $X' := \mathbb{R} \times S$ with states $s' = (t, s)$, s called the **momentary core state.**
- $A' := A$.
- $D'(t, s) := D(s)$.
- $K'(t, s, a, d(z, t', u')) := K(s, a, d(z, u)) \otimes \delta_{t-z}(dt')$.
- $\tilde{r}'(t, s, a, z, u) := 1_{\mathbb{R}_+}(t) \left[\tilde{r}(s, a, z, u) 1_{[0,t]}(z) + \tilde{h}(t, s, a, z, u) 1_{(t,\infty)}(z)\right]$.
- $\alpha' := \alpha$.

Negative values of t, which are only needed for a convenient modeling, mean that the elapsed time has surpassed t_0 by the amount $|t|$. As initial time horizon only positive t_0 are of interest. The decision rules $f' \in F'$ depend on both the core state s and the initial time-horizon t; a transition occurs from $x = (t, s)$ to the next state $(t - z, u)$ according to $K'(t, s, a, d(z, t', u'))$. Obviously $J_0 := (-\infty, 0) \times S$ is absorbing for MRP$'$.

In applications it is useful to assume that $\tilde{r}$ has the form (22.2), i.e.

$$\tilde{r}(s, a, z, u) = r_0(s, a, z) + e^{-\alpha z} \cdot r_3(s, a, z, u). \tag{22.14}$$

Here r_0 may be as in Chap. 22 the sum of a reward r_1 earned at the beginning of the period and a reward earned continuously during the period, while r_3 is a reward earned at the end of the period.

If $\tilde{r}$ has the form (22.14), we use as a natural form of $\tilde{h}$

$$\tilde{h}(t,s,a,z,u) = r_0(s,a,t) + e^{-\alpha t} \cdot R(t,s,a,z,u). \tag{22.15}$$

Here $R(t,s,a,z,u)$ is an appropriately chosen reward, earned when the process stops.

In this section we assume that there exists a measurable function $b \geq 0$ on S and a measurable function $h \geq 0$ on $D \times \mathbb{R}_+ \times S$ with $|\tilde{h}(t,s,a,z,u)| \leq h(s,a,z,u)$ for all (t,s,a,z,u), such that $\beta_b < 1$ and $\|K|\tilde{r}|\|_b$ and $\|Kh\|_b$ are finite. Recall that

$$\beta_b = \sup_{s,a} \frac{\int K(s,a,d(z,u)) \cdot e^{-\alpha z} b(u)}{b(s)}.$$

It is easy to see that b and $(t,s) \mapsto b'(t,s) := b(s)$ are bounding functions for MRP and MRP′, respectively, and $\beta_{b'} = \beta_b < 1$. Thus (MA1) holds and by Theorem 21.1.15 the value functions and their limits V and V' exist and belong to the set B_b of function on S with finite b-norm and to the set $B_{b'}$ of functions v' on S' with finite b'-norm, respectively. Moreover, the functions r, r', V_∞ and V'_∞ exist, and $V = V_\infty$, $V' = V'_\infty$.

As $\tilde{r}'$ is independent of t' we have

$$r^t(s,a) := r'(t,s,a) = \int K(s,a,d(z,u)) \cdot \tilde{r}'(t,s,a,z,u).$$

Thus $r^t(s,a) := 0$ for $t < 0$ and for $t \geq 0$

$$r^t(s,a) = r(s,a) + \int K(s,a,d(z,u)) \cdot 1_{(t,\infty)}(z) \left[\tilde{h}(t,s,a,z,u) - \tilde{r}(s,a,z,u)\right].$$

For $\pi' = (\pi'_\nu)_0^\infty \in (F^\infty)'$ and initial state $(\tau_0, \zeta_0) := (t,s)$ we have

$$V'_{\infty\pi'}(t,s) = \mathbb{E}_{\pi',(t,s)}\left[\sum_{\nu=0}^{\infty} e^{-\alpha T_\nu} \cdot \tilde{r}'(\tau_\nu, \zeta_\nu, \pi'_\nu(\tau_\nu,\zeta_\nu), \eta_{\nu+1}, \zeta_{\nu+1})\right]. \tag{22.16}$$

Here $\tau_\nu := t - T_\nu$ denotes the random time horizon at time T_ν.

All results from Chaps. 21 and 22 are applicable to MRP′. In particular, if the OE holds, it has the form

$$V^t(s) := V'_\infty(t,s) \tag{22.17}$$
$$= \sup_{a \in D(s)} \left[r^t(s,a) + \int K(s,a,d(z,u)) \cdot e^{-\alpha z} V^{t-z}(u) 1_{[0,t]}(z)\right], \; t \in \mathbb{R}_+, s \in S.$$

Moreover, (22.16) implies that $V^t(s) = 0$ for $t < 0$.

In general the OE (22.17) can be solved numerically only after a discretization of time (and possibly of the state space), which entails a discretization error. This difficulty does not arise if $Q(s, a, dz)$ is concentrated on $\mathbb{N}$ (or more generally on $\{h, 2h, \ldots\}$ for some $h > 0$), e.g. if $Q(s, a, dz) := \text{Geo}(p)$. We then speak of a *discrete-time model.* It follows that $K(s, a, d(z, u))$ is concentrated on $\mathbb{N} \times S$. Therefore the OE allows us to compute $V^k(\cdot)$ for some $k \in \mathbb{N}$ as soon as $V^j(\cdot)$ is known for $0 \leq j \leq k-1$. Thus the OE of the infinite-stage time-discrete model can be solved numerically by recursion in state space, starting with $k = 0$ and observing that $V^k \equiv 0$ for $k < 0$.

Example 22.3.3 (An optimal stopping problem) Consider a homogeneous Markov chain with finite state space $I \subset \mathbb{R}_+$ and with transition matrix $(p(i,j))$. Let $Q(dz)$ denote the probability distribution of the times between jumps of the Markov chain, and assume that Q is concentrated on $\mathbb{N}$. Interpret the states $s \in I$ as offers which occur at the jump times. Upon accepting an offer ($a = 1$) the process stops and goes to state $s = \infty$. If an offer is rejected ($a = 0$), one waits for the next offer. As long as the process is not stopped there is a cost $c_0 > 0$ per unit time. If the process is not stopped during the time interval $[0, t_0]$ for some given $t_0 \in \mathbb{N}$, an additional reward $R_0 \in \mathbb{R}$ is paid at time t_0. Payoffs are discounted with rate $\alpha > 0$. When should one stop in the infinite-stage model in order to maximize the expected reward?

We model the problem by an MRP$'$, as follows: $S = I + \{\infty\}; A = D(s) = \{0, 1\}$ for all s; using the notation $K(s, a, z, s') := K(s, a, \{(z, s')\})$ we have $K(s, 0, z, s') = q(z) \cdot p(s, s')$ if $s, s' < \infty$, where q is the discrete density q of Q; $K(s, a, z, \infty) = q(z)$ if $(s = \infty) \vee (a = 1)$; we use $\tilde{r}$ as in (22.14), with $r_0(\infty, a, z) = 0$ and, using $\beta := e^{-\alpha}$, for $s < \infty$

$$r_0(s, a, z) = a \cdot s - c_0 \cdot (1 - a) \cdot \sum_{x=0}^{z-1} \beta^x = a \cdot s - (1 - a) \cdot c_0 \cdot (1 - \beta^z)/(1 - \beta);$$

we use $\tilde{h}$ from (22.15) with $R(\infty, a, z, s') = 0$ and $R(s, a, z, s') = (1 - a) \cdot R_0$ for $s < \infty$.

Obviously $b \equiv 1$ is a bounding function, and $\beta_b = \sum_1^\infty q(z) \cdot \beta^z < 1$ as $\beta < 1$. Moreover, it follows from Theorem 21.1.15 that the OE holds. It follows from (22.17) that $V^k(\infty) = 0$ for all k and that for $s < \infty$

$$\begin{aligned} V^k(s) &= \max\left\{ r^k(s, 0) + \sum_{z=1}^{k} q(z) \cdot \beta^z \sum_{j \in I} p(s, j) \cdot V^{k-z}(j), s \right\} \\ &=: \max\left\{W^k(s, 0), s\right\}, \end{aligned} \tag{22.18}$$

which can be solved numerically by recursion in state space, starting with $k = 0$ and observing that $V^k \equiv 0$ for $k < 0$. Moreover, it is optimal to accept the first offer s for which $s \geq W^k(s, 0)$. The explicit form of $r^k(s, 0)$ is a bit involved. Values of $V^k(s)$ may be computed from (22.18) by recursion in state space. ♦

Chapter 23
Bayesian Control Models

We present the Bayesian approach to solving control models with i.i.d. disturbances where the reward functions and the distributions depend measurably on some unknown parameter. We introduce the associated MDP′ and state the basic theorem for Bayesian control models. Proofs of the results are involved and postponed to Chap. 25. A gambling problem illustrates the results.

23.1 The Model BCM

Example 23.1.1 (The Bayesian version of the gambling problem from Example 13.6.4*)* Recall Example 13.6.4: You enter a casino with a capital $s_0 \in \mathbb{N}$ and gamble at most N times by choosing each time an integer stake a between zero and your momentary capital $s \in \mathbb{N}$. The stake is kept by the casino. You win with known probability $p \in (0, 1)$ upon which you gain d times your stake, where $d \in \mathbb{N}$, $d \geq 2$, is constant. Otherwise you gain nothing. Your aim consists in maximizing the expectation of your terminal utility $V_0(s_N) \geq 0$. The game stops if you are ruined before the N-th play.

In practice often the probability of winning will not be known, but an unknown number $\vartheta \in \Theta := (0, 1)$. This means that the probability distribution $\mathrm{Bi}(1, p)$ of the i.i.d. disturbances in Example 13.6.4 must be replaced by $Q(\vartheta, dz) := \mathrm{Bi}(1, \vartheta)$. As a consequence, the value functions $s \mapsto V_N(\vartheta, s)$ and the minimizers $s \mapsto f_n(\vartheta, s)$ as given in Example 13.6.4 depend on the unknown ϑ and hence are useless. An obvious way out, known from statistical decision theory, is the *Bayes approach:* one assumes that from experience one knows a so-called ***prior,*** i.e. a probability distribution $\mu_0(d\vartheta)$ (e.g. a beta distribution $\mathrm{Be}(\alpha_1, \alpha_2)$, $\alpha_1, \alpha_2 \in \mathbb{R}^+$) according to

K. Hinderer et al., *Dynamic Optimization*, Universitext,
DOI 10.1007/978-3-319-48814-1_23

which the parameter ϑ changes in repetitions of the experiment. Now it is natural to maximize

$$\pi \mapsto \int \mu_0(d\vartheta)\, V_{N\pi}(\vartheta, s), \quad s \in S, \tag{23.1}$$

the expected N-stage reward for initial state s and policy $\pi \in \mathbb{F}^N$ (whose decision rules do not depend on ϑ) averaged over $\vartheta \in \Theta$ according to μ_0. ♦

Prima facie it seems that the preceding problem cannot be attacked by DP methods, since $\int \mu_0(d\vartheta)\, V_{N\pi}(\vartheta, s)$ is defined in a different way than an expected N-stage reward in a CM. However, it turns out that our problem can be solved by DP methods, provided we admit at each time $0 \le t \le N-1$ decision rules π_t which not only depend on the present state s_t but also on the *disturbance history* $z^t := (z_1, z_2, \ldots, z_t) \in Z^t$. In fact, we can solve the N-stage problem for fixed prior μ_0 as soon as we know the solution of all $(N-1)$-stage problems for arbitrary priors μ. This recursive method can be framed into an MDP′ having the pairs (μ, s) as states.

Proofs for the results in this section are involved. In order to allow a quick access to concrete examples, proofs are postponed to Chap. 25, where they are given for a more general model. Here our framework is a family $CM(\vartheta)$, $\vartheta \in \Theta$, of control models with i.i.d. disturbances where not only Q but also r_Z and V_0 (but not T!) may depend measurably on ϑ.

Definition 23.1.2 A **Bayesian control model** (or **BCM** for short) is a tuple $(\Theta, S, A, D, Z, Q, T, r_Z, V_0, \beta)$ of the following kind:

- The set $\Theta \neq \emptyset$ is called the ***parameter space*** and is endowed with a σ-algebra $\mathfrak{T}$. We denote by $\mathbb{P}(\mathfrak{T})$ the set of all prior distributions, i.e. of all probability distributions on $\mathfrak{T}$.
- The sets S, A, D, Z (with their σ-algebras $\mathfrak{S}$, $\mathfrak{A}$, $\mathfrak{D}$ and $\mathfrak{Z}$, respectively) and also T and β have the same meaning as in the CM of Chap. 16.
- The transition law Q is a transition probability from Θ into Z.
- The measurable function $(\vartheta, s, a, z) \mapsto r_Z(\vartheta, s, a, z)$ on $\Theta \times D \times Z$ is the one-stage reward function. The integrals

$$r(\vartheta, s, a) := Qr_Z(\vartheta, s, a) := \int Q(\vartheta, dz)\, r_Z(\vartheta, s, a, z)$$

  are assumed to exist and to be finite for all (ϑ, s, a).
- The measurable function $(\vartheta, s) \mapsto V_0(\vartheta, s)$ on $\Theta \times S$ is the terminal reward function.

From each CM we obtain a *Bayesian version* by replacing $Q(dz)$ by $Q(\vartheta, dz)$ and by keeping r_Z and V_0 unchanged. Note, however, that then r may depend on ϑ. BCMs generalize CMs: the latter is obtained from the former by taking for Θ a singleton.

In CMs we used policies in $\mathbb{F}^N$. These are called Markovian, since their decision rules at any time t do not depend on states or disturbances before time t. We must extend the definition (23.1) to the case of ***disturbance history-dependent policies,*** i.e. to policies $\pi = (\pi_t)_0^{N-1}$ where each decision rule π_t depends not only on the present state s_t but also on the disturbance history z^t, while $\pi_0 = \pi_0(s_0)$. The set of such policies is denoted by Δ_{NZ}. For fixed $s_0 \in S$ we use the sample space $\Theta \times Z^N$ with σ-algebra $\mathfrak{F} := \mathfrak{T} \bigotimes \mathfrak{Z}^N$ with coordinate random variables $\tau(\vartheta, z^N) := \vartheta$ and $\eta_t(\vartheta, z^N) := z_t$, $1 \le t \le N$. Then the N-stage reward under policy $\pi \in \Delta_{NZ}$ is defined as the random variable

$$(\vartheta, z^N) \mapsto G_{N\pi}(\vartheta, s_0, z^N) := r_Z(\vartheta, s_0, \pi_0(s_0), z_1) \quad (23.2)$$
$$+ \sum_{t=1}^{N-1} \beta^t \cdot r_Z(\vartheta, \zeta_{t\pi}, \pi_t(z^t, \zeta_{t\pi}), z_{t+1}) + \beta^N \cdot V_0(\vartheta, \zeta_{N\pi}),$$

where $\zeta_{0,\pi}(z^0, s_0) := s_0$ and $\zeta_{t+1,\pi}(s_0, z^{t+1}) := T(\zeta_{t\pi}, \pi_t(z^t, \zeta_{t\pi}, z_{t+1})), 0 \le t \le N-1$. The disturbances are assumed to be conditional i.i.d. with probability distribution $Q(\vartheta, dz)$, conditional on $[\tau = \vartheta]$. Thus we use on $\mathfrak{F}$ the probability distribution

$$B \mapsto \mathbb{Q}_N^{\mu}(B) := \int \mu(d\vartheta) \int Q(\vartheta, dz_1) \cdots \int Q(\vartheta, dz_N)\, 1_B(\vartheta, z^N). \quad (23.3)$$

In order to ensure the existence of $v_{N\pi}(\mu, s)$, defined in (23.4) below, we make the following assumption (LUBF). It implies that there exists a function b on S which either is a lower bounding function or an upper bounding function for each $CM(\vartheta)$.
Assumption (LUBF). There exists a measurable function $b\colon S \to \mathbb{R}_+$ and a constant $\delta \in \mathbb{R}_+$ such that for all $(\vartheta, s, a, z) \in \Theta \times D \times Z$ we have $\int Q(\vartheta, dz)\, b(T(s, a, z)) \le \delta \cdot b(s)$ and either

$$r_Z(\vartheta, s, a, z) \le \delta \cdot b(s), \quad V_0(\vartheta, s) \le \delta \cdot b(s), \quad \text{or}$$
$$r_Z(\vartheta, s, a, z) \ge -\delta \cdot b(s), \quad V_0(\vartheta, s) \ge -\delta \cdot b(s).$$

Remark 23.1.3 (LUBF) Note that (LUBF) holds with $b \equiv 1$ and $\delta = 1$ if both r_Z and V_0 are non-negative, and in minimization problems if both c_Z and C_0 are non-negative. ◇

Under (LUBF) there exists for $N \ge 1$, $\pi \in \Delta_{NZ}$ and $\mu \in \mathbb{P}(\mathfrak{T})$

$$v_{N\pi}(\mu, s) := \int \mathbb{Q}_N^{\mu}(d(\vartheta, z^N))\, G_{N\pi}(\vartheta, s, z^N) = \int \mu(d\vartheta)\, V_{N\pi}(\vartheta, s). \quad (23.4)$$

We call $v_{N\pi}(\mu, s)$ the N-stage ***Bayes reward of policy*** π with respect to prior μ and initial state s.

A BCM and some prior μ_0 define the N-stage ***Bayesian control model*** $\boldsymbol{BCM_N(\mu_0)}$, $N \geq 1$, with prior μ_0 as follows:

(i) Find for each initial state $s \in S$ the ***maximal N-stage Bayes reward*** within the set of history-dependent policies, i.e.

$$v_N(\mu_0, s) := \sup\{v_{N\pi}(\mu_0, s) : \pi \in \Delta_{NZ}\}.$$

(ii) Find a policy $\pi^* \in \Delta_{NZ}$ which is ***Bayes-optimal*** for $BCM_N(\mu_0)$ in the sense that it maximizes $\pi \mapsto v_{N\pi}(\mu_0, s)$ on Δ_{NZ} for all $s \in S$.

Both the process of gaining information about ϑ and of making decisions influence the decision process and hence each other. Therefore in general one has to balance precise and costly information about ϑ with the goal to obtain maximal expected reward.

We now motivate the introduction of the MDP′ with state space $S' := \mathbb{P}(\mathfrak{T}) \times S$ which solves problem $BCM_N(\mu_0)$ for arbitrary but fixed prior μ_0. We must ensure that $v_n(\mu, s) = V_n'(\mu, s) := \sup\{V_{n\pi'}(\mu, s) : \pi' \in (\mathbb{F}')^n\}$ for $1 \leq n \leq N$ and all $(\mu, s) \in S'$. Before giving the formal definition of MDP′ we now motivate it by considering the simplest case $N = 1$. Put $\mathbb{Q}^\mu(d(\vartheta, z)) = \mu(d\vartheta) \otimes Q(\vartheta, dz)$. Thus $\mathbb{Q}^\mu$ is given in factorized form, and the notation indicates that Q remains fixed while μ varies in $\mathbb{P}(\mathfrak{T})$. Then we have for each decision rule f' by (23.2) and (23.4) for $(\mu, s) \in S'$

$$v_{1f'}(\mu, s) = \int \mathbb{Q}^\mu(d(\vartheta, z))\, r_Z(\vartheta, s, f'(s), z) + \beta \cdot \int \mathbb{Q}^\mu(d(\vartheta, z))\, V_0(\vartheta, T_{f'}(s, z)).$$

From this it is clear that we must define $r'_{S'}(\mu, s, a, \mu', s') = r'(\mu, s, a) := \int \mathbb{Q}^\mu(d(\vartheta, z))\, r_Z(\vartheta, s, a, z)$, independent of (μ', s'). Also it is natural to define $V_0'(\mu, s) := \int \mu(d\vartheta)\, V_0(\vartheta, s)$. As a crucial point there remains the definition of the transition law P'. For this purpose consider for fixed μ the probability space $(\Theta \times Z, \mathfrak{T} \bigotimes \mathfrak{Z}, \mathbb{Q}^\mu)$ with coordinate random variables τ and η_1. It follows that μ is the first marginal $\mathbb{Q}^\mu_\tau$ of $\mathbb{Q}^\mu$ and that Q is a version of the conditional distribution $\mathbb{Q}^\mu_{\eta_1|\tau}$ of η_1 conditioned on τ, i.e. $\mathbb{Q}^\mu = \mathbb{Q}^\mu_\tau \otimes \mathbb{Q}^\mu_{\eta_1|\tau}$. The crucial idea for our aim consists in factorizing the probability distribution $\mathbb{Q}^\mu$ in the opposite order as $\mathbb{Q}^\mu(d(z, \vartheta)) = \mathbb{Q}^\mu_{\eta_1}(dz) \otimes \Phi^\mu(z, d\vartheta)$. Here $\mathbb{Q}^\mu_{\eta_1}$ is the first marginal of $\mathbb{Q}^\mu$ and $\Phi^\mu(z, d\vartheta)$ is any version of the conditional distribution $\mathbb{Q}^\mu_{\tau|\eta_1}$ of τ conditioned on η_1. Then

$$\int \mathbb{Q}^\mu(d(\vartheta, z))\, v(\vartheta, z) = \int \mathbb{Q}^\mu_{\eta_1}(dz) \int \Phi^\mu(z, d\vartheta)\, v(\vartheta, z) \tag{23.5}$$

for all measurable $v \geq 0$ on $\Theta \times Z$. (Note that (23.5) is the integral formulation of the well-known formula $\mathbb{E}\, v(\tau, \eta_1) = \mathbb{E}(\mathbb{E}[v(\tau, \eta_1)|\eta_1])$.) In this context $\Phi^\mu(z, d\vartheta)$ is called a version of the ***posterior*** (probability distribution) of the prior (probability distribution) μ. It is useful to imagine $\Phi^\mu(z, d\vartheta)$ as an update of $\mu(d\vartheta)$ after the observation of a realization z of η_1. If Θ and Z are countable, then the existence of $\Phi^{\mu}(z, d\vartheta)$ is trivial. In all practical cases at least one version of $\mathbb{Q}^{\mu}_{\tau|\eta_1}$ exists, and in general several versions exist. However, as we need only property (23.5), all versions serve the same purpose.

For notational convenience we put $Q'(\mu, dz) := \mathbb{Q}^{\mu}_{\eta_1}(dz)$ and write $\Phi(\mu, z, d\vartheta)$ instead of $\Phi^{\mu}(z, d\vartheta)$. Now we obtain from (23.5) with $v(\vartheta, z) := V_0(\vartheta, T_f(s, z))$, s fixed,

$$\begin{aligned} v_{1f}(\mu, s) &= r'_f(\mu, s) + \beta \cdot \int Q'(\mu, dz) \int \Phi(\mu, z, d\vartheta)\, V_0(\vartheta, T_f(s, z)) \\ &= r'_f(\mu, s) + \beta \cdot \int Q'(\mu, dz)\, V'_0(\Phi(\mu, z), T_f(s, z)). \end{aligned}$$

Now we see that we should choose $P'(\mu, s, a, d(\mu', s'))$ as the image of $Q'(\mu, dz)$ under the mapping $z \mapsto (\Phi(\mu, z), T(s, a, z))$, and it is plausible that the VI for the MDP′ (if it holds) must have the form

$$\begin{aligned} V'_n(\mu, s) = \sup_{a \in D(s)} \left[r'(\mu, s, a) + \beta \cdot \int Q'(\mu, dz)\, V'_{n-1}(\Phi(\mu, z), T(s, a, z)) \right], \\ n \geq 1, (\mu, s) \in \mathbb{P}(\mathfrak{T}) \times S. \end{aligned} \tag{23.6}$$

This is confirmed in Theorem 23.1.17 below. Our motivation is complete, and we can continue our formal exposition.

A mapping Φ from $\mathbb{P}(\mathfrak{T}) \times Z$ into $\mathbb{P}(\mathfrak{T})$, which transforms each prior μ under the disturbance z into a version of its posterior

$$\begin{aligned} \Phi(\mu, z, d\vartheta) &:= \mathbb{Q}^{\mu}_{1,\tau|\eta_1}(z, d\vartheta) \\ &= \text{the conditional probability distribution of } \tau \text{given } \eta_1 = z, \end{aligned}$$

is called a version of the ***Bayes operator (generated by the transition law Q)***. Note that a mapping Φ from $\mathbb{P}(\mathfrak{T}) \times Z$ into $\mathbb{P}(\mathfrak{T})$ such that $\Phi(\mu, z, B)$ is measurable in z is a version of the Bayes operator if and only if for $\mu \in \mathbb{P}(\mathfrak{T})$ and all measurable $v \geq 0$ on $\Theta \times Z$ we have

$$\int \mathbb{Q}^{\mu}_{\tau,\eta_1}(d(\vartheta, z))\, v(\vartheta, z) = \int \mathbb{Q}^{\mu}_{\eta_1}(dz) \int \Phi(\mu, z, d\vartheta')\, v(\vartheta', z),$$

i.e. if and only if the following important equation holds

$$\int \mu(d\vartheta) \int Q(\vartheta, dz)\, v(\vartheta, z) = \int \mu(d\vartheta) \int Q(\vartheta, dz) \int \Phi(\mu, z, d\vartheta')\, v(\vartheta', z). \tag{23.7}$$

Each version $B \mapsto \mu_t(\mu_0, z^t, B)$ of the conditional probability distribution $\mathbb{Q}^{\mu_0}_N(\tau|\eta_t = z^t)$, $z^t \in Z^t$, is called ***a posterior*** at time $t \geq 1$ of the prior μ_0 under the observation history z^t. We shall need posteriors in Theorem 23.1.17 for the computation of the value functions and of Bayes-optimal policies in the MDP′ to which the BCM is reduced.

It is easily seen that each version Φ of the Bayes operator determines a sequence of posteriors recursively by $\mu_1(\mu_0, z_1) := \Phi(\mu_0, z_1)$ and

$$\mu_{t+1}(\mu_0, z^{t+1}) := \Phi(\mu_t(\mu_0, z^t), z_{t+1}), \ t \geq 1. \tag{23.8}$$

In MDP′ the states could be chosen as the pairs $(\mu, s) \in \mathbb{P}(\mathfrak{T}) \times S$. However, for most applications $\mathbb{P}(\mathfrak{T})$ is unnecessarily large for the following two reasons:

(a) The VI (23.6) shows that the computation of $v_N(\mu_0, s_0)$ requires $v_n(\mu, s)$, $0 \leq n \leq N-1$, $s \in S$, only for $\mu = \mu_{N-n}(\mu_0, z^{N-n})$ for all $z^{N-n} \in Z^{N-n}$.
(b) In applications with given prior μ_0 often there exists some family $\hat{\mu} = (\hat{\mu}(i, \vartheta), i \in I)$ of probability distributions on $\mathfrak{T}$ with the following properties:

  (b1) The family $\hat{\mu}$ contains μ_0, i.e. $\mu_0 = \hat{\mu}(i_0)$ for some $i_0 \in I$.
  (b2) The index set I is simple, e.g. a Cartesian product in $\mathbb{R}^2$; see Table 23.1 below.
  (b3) There exist posteriors μ_t, $t \geq 1$, of μ_0, which belong again to the family $\hat{\mu}$, i.e. $\mu_t(\mu_0, z^t) = \hat{\mu}(i_t(i_0, z^t))$ for mappings i_t from $I \times Z^t$ into I.

It follows easily that (b3) holds as soon as some version of $\mu_1(\mu_0, z_1)$ belongs to $\hat{\mu}$, $z_1 \in Z$. This motivates the next definition.

Definition 23.1.4 Let BCM be a Bayesian control model, and let $(i, B) \mapsto \hat{\mu}(i, B)$ be a transition probability from the measurable space $(I, \mathfrak{I})$ into Θ. A measurable mapping φ from $I \times Z$ into I is called a ***sufficient statistic*** with respect to Q for the family $\hat{\mu}$ of probability distributions $\hat{\mu}(i) := \hat{\mu}(i, d\vartheta)$, $i \in I$, if

$$\Phi(\hat{\mu}(i), z, d\vartheta) = \hat{\mu}(\varphi(i, z), d\vartheta), \ \text{for all } (i, z) \in I \times Z,$$

for some version Φ of the Bayes operator. Then we also call the triple $(BCM, \hat{\mu}, \varphi)$ a ***Bayesian control model.***

Note that Φ and hence sufficiency of φ with respect to Q for $\hat{\mu}$ does not depend on D, T, r_Z, V_0 and β, but only on $(\Theta, \mathfrak{T})$, $(Z, \mathfrak{Z})$ and Q.

For a particular $BCM_N(\mu_0)$ one sometimes gets a hint for an appropriate choice of I, $\hat{\mu}$, φ with $\mu_0 \in \hat{\mu}$ by simply computing the first few posteriors $\mu_1(\mu_0, z_1) = \Phi(\mu_0, z_1)$, $\mu_2(\mu_0, z^2)$, etc. by Lemma 23.1.6 below. In most applications Q has a transition density with respect to some σ-finite measure on $\mathfrak{Z}$. Then sufficiency for an appropriately chosen family $\hat{\mu}$ can mostly be established via Lemma 23.1.6 below. In particular, this provides us with the trivial sufficient statistic $\varphi := \Phi$ in Lemma 23.1.6(b) and also leads to the very useful criterion Proposition 23.1.20 (cf. DeGroot 1970, p. 159).

One easily derives from Theorem 23.1.17 below that the solution of a particular problem $BCM_N(\mu_0)$ does not depend on the choice of I, $\hat{\mu}$, φ.

In part (b) of the next result, where $I \subset \mathbb{P}(\mathfrak{T})$, we need a σ-algebra $\mathfrak{J}$ on I such that $i \mapsto \hat{\mu}(i) := i$ becomes a probability distribution from I into Θ, i.e. such that all functions $i \mapsto i(B)$, $B \in \mathfrak{T}$, are $\mathfrak{J}$-measurable. This leads to a natural choice for $\mathfrak{J}$ according to the next definition, and this choice also ensures the measurability of the sufficient statistic $\varphi: I \times Z \to I$ in Lemma 23.1.6(b) below.

Definition 23.1.5 The ***-σ-algebra** on $I \subset \mathbb{P}(\mathfrak{T})$ is the smallest σ-algebra $\mathfrak{F}$ on I such that for each $B \in \mathfrak{T}$ the function $i \mapsto i(B)$ on I is $\mathfrak{F}$-measurable.

In applications one mostly needs the Bayes operator when Z is a Euclidean set with positive Lebesgue measure and when $Q(\vartheta, dz)$ has a Lebesgue density. This case is covered by the next result, which can be motivated by the special case where Z is countable. Then simple computations with elementary conditional probabilities suffice. In fact, if $z \mapsto q(\vartheta, z)$, $\vartheta \in \Theta$, is the discrete density of $Q(\vartheta, dz)$ then $\Phi(\mu, z, d\vartheta)$ has the μ-density $\vartheta \mapsto q(\vartheta, z)/\int \mu(d\vartheta')\, q(\vartheta', z)$ (to be understood as 1 if the denominator vanishes). Note that in the next result the measure ρ may not depend on ϑ.

Lemma 23.1.6 (Computation of the Bayes operator) *Assume that Q has a transition density $q: \Theta \times Z \to \mathbb{R}_+$ with respect to some σ-finite measure ρ on $\mathfrak{Z}$. For $\mu \in \mathbb{P}(\mathfrak{T})$ and $z \in Z$ put $q_\mu(z) := \int \mu(d\vartheta')\, q(\vartheta', z)$. Put*

$$\Phi(\mu, z, d\vartheta) := \text{the probability distribution on } \mathfrak{T} \text{ with } \mu\text{-density}$$
$$\vartheta' \mapsto q(\vartheta, z)/q_\mu(z), \ \text{if } 0 < q_\mu(z) < \infty,$$

and with μ-density $\equiv 1$ (i.e. $\Phi(\mu, z, d\vartheta) := \mu(d\vartheta)$), otherwise. Then:

(a) Φ is a version of the Bayes operator.
(b) Assume $\hat{\mu}(i) := i$, $i \in I \subset \mathbb{P}(\mathfrak{T})$, and use as σ-algebra $\mathfrak{J}$ on I the $*$-σ-algebra. Then $\varphi := \Phi$ is sufficient for $\hat{\mu}$ if $\Phi(i, z) \in I$ for all $(i, z) \in I \times Z$, in particular, if $I = \mathbb{P}(\mathfrak{T})$.

Proof

(a1) Firstly one sees, using measurability of q_μ, that $(z, B) \mapsto \Phi(\mu, z, B)$ is a transition probability from Z into Θ. Put $\Lambda := \mu \times \rho$. According to (23.7) it suffices to verify for measurable $v \geq 0$ and $w(z) := \int \mu(d\vartheta')\, g_\mu(\vartheta', z) \cdot v(\vartheta', z)$ that for $z \in Z$

$$\int \Lambda(d(\vartheta, z))\, q(\vartheta, z) \cdot v(\vartheta, z) = \int \Lambda(d(\vartheta, z))\, q(\vartheta, z) \cdot w(z). \tag{23.9}$$

(a2) Obviously (23.9) is true if $0 < q_\mu(z) < \infty$. In order to cover the general case, we note that for $u \geq 0$ on $\Theta \times Z$

$$\int \Lambda(d(\vartheta, z))\, q(\vartheta, z) \cdot v(\vartheta, z) = \int \Lambda(d(\vartheta, z))\, q(\vartheta, z) \cdot 1_{[0<q_\mu<\infty]}(z) \cdot u(\vartheta, z). \tag{23.10}$$

This is true since $q_\mu(z) = 0$ implies $q(\cdot, z) = 0$ μ-almost surely and since $q_\mu < \infty$ ρ-almost surely due to $\int \rho(dz)\, q_\mu(z) = \int \mu(d\vartheta) \int \rho(dz)\, q(\vartheta, z) = 1 < \infty$.

Now (23.10) with $u := w$ shows that the right-hand side of (23.9) equals

$$\int \rho(dz)\, q_\mu(z) \cdot 1_{[0<q_\mu<\infty]}(z) \int \mu(d\vartheta')\, g_\mu(\vartheta', z) \cdot v(\vartheta', z)$$

$$= \int \rho(dz)\, 1_{[0<q_\mu<\infty]}(z) \int \mu(d\vartheta')\, q(\vartheta', z) \cdot v(\vartheta', z)$$

$$= \int \Lambda(d(\vartheta', z))\, 1_{[0<q_\mu<\infty]}(z) \cdot q(\vartheta', z) \cdot v(\vartheta', z),$$

which by (23.10) with $u := v$ equals the left-hand side of (23.9).

(b) is obvious from (a) and from the definition of sufficiency. □

Remark 23.1.7 Φ from Lemma 23.1.6(a) is not unique since $Q(\vartheta, dz)$ may have several transition density functions with respect to ρ. In the sequel, when q has been selected, then Φ from Lemma 23.1.6(a) and the posteriors determined by Φ will be called *the* Bayes operator and *the* posteriors, respectively. ◇

Example 23.1.8 (First continuation of the gambling problem from Example 23.1.1*)* Let the prior be the beta distribution $\mathrm{Be}(\alpha_1, \alpha_2)$. Then $\mu_0 := \mathrm{Be}(\alpha_1, \alpha_2)$ is embedded by the choice $i_0 := (0, 0)$ into the family $\hat{\mu}(k, m) := \mathrm{Be}(k+\alpha_1, m+\alpha_2)$, $i := (k, m) \in I := \mathbb{N}_0^2$. We have $q(\vartheta, z) := \vartheta^z \cdot (1 - \vartheta)^{1-z}$, where $z = 1$ means success and $z = 0$ means failure. An easy computation, using Lemma 23.1.6 with $\mu := \hat{\mu}(k, m)$ shows that $\Phi(\hat{\mu}(k, m), z, d\vartheta) = \mathrm{Be}(k + z + \alpha_1, m + 1 - z + \alpha_2) = \hat{\mu}(k + z, m + 1 - z)$. Therefore $(i, z) \mapsto i + (z, 1 - z)$ is sufficient for $\hat{\mu}$. The same result holds for a much more general family $\hat{\mu}$; cf. the second continuation below. ◆

A model $(BCM, \hat{\mu}, \varphi)$ determines a family of Bayesian Problems $BCM_N(\hat{\mu}(i), \varphi)$, $i \in I$. In such a problem we identify $\hat{\mu}(i)$ with i, and we only need care about those priors which belong to $\hat{\mu}$. This leads to the following modification of our previously defined MDP′, which will solve all Bayesian Problems $BCM_N(\hat{\mu}(i))$, $i \in I$, and hence $BCM_N(\mu_0)$ for a specific μ_0, provided it is contained in $\hat{\mu}$.

Definition 23.1.9 (The associated MDP′) Consider some Bayesian control model $(BCM, \hat{\mu}, \varphi)$ which satisfies assumption (LUBF). The Markovian decision process MDP′ $(S', A, D', P', r', V_0', \beta)$ is called *associated* to $(BCM, \hat{\mu}, \varphi)$ if the following holds:

- The states are $(i, s) \in S' := I \times S$ with σ-algebra $\mathfrak{J} \otimes \mathfrak{S}$.
- $D'(i, s) := D(s)$ is independent of i.
- $P'(i, s, a, d(i', s'))$ is the image of the probability distribution $Q'(i, dz) := \int \hat{\mu}(i, d\vartheta)\, Q(\vartheta, dz)$ under the mapping $z \mapsto (\varphi(i, z), T(s, a, z))$. Thus P' is

determined by the property that for all (i, s, a) and all measurable $v \geq 0$ on S'

$$P'v(i,s,a) := \int Q'(i,dz)\, v(\varphi(i,z), T(s,a,z)).$$

- $r'(i,s,a) := \int \hat{\mu}(i,d\vartheta)\, r(\vartheta,s,a) = \int \hat{\mu}(i,d\vartheta) \int Q(\vartheta,dz)\, r_Z(\vartheta,s,a,z)$ is finite.
- $V_0'(i,s) := \int \hat{\mu}(i,d\vartheta)\, V_0(\vartheta,s)$ is finite.

Remark 23.1.10

(a) The functions r' and V_0' exist by (LUBF) and are measurable. If r_Z doesn't depend on ϑ, we have $r'(i,s,a) = \int Q'(i,dz)\, r_Z(s,a,z)$.
(b) The form of P' is intuitively clear: If we take action a in the augmented state (i,s), and if the disturbance z occurs, then the next index i' is $\varphi(i,z)$ and the next core state s' is $T(s,a,z)$. This also shows that the MDP′ is equivalent to the MCM $(S', A, D', Z, Q', T', r', V_0', \beta)$ with $Q'(i,s,a,dz)$ independent of (s,a) and with $T'(i,s,a,z) := (\varphi(i,z), T(s,a,z))$.
(c) From (23.7) with $\mu := \hat{\mu}(i)$ and $v(\vartheta,z) := \int Q(\vartheta,dz')\, u(z')$ for some arbitrary measurable $u \geq 0$ on Z we obtain the remarkable formula

$$\int Q'(i,dz)\, u(z) = \int Q'(i,dz) \int Q'(\varphi(i,z),dz')\, u(z'), \tag{23.11}$$

which will be useful in Chap. 24 for studying the structure of solutions of BCMs.

Example 23.1.11 (Second continuation of the gambling problem from Example 23.1.1*)* Let us identify the data of the associated MDP′, assuming that μ_0 is an *arbitrary* prior. Put $M(k,m) := \int \vartheta^k(1-\vartheta)^m \cdot \mu_0(d\vartheta)$, $(k,m) \in I := \mathbb{N}_0^2$. Then μ_0 belongs by $(k,m) := (0,0)$ to the family $\hat{\mu}(k,m,d\vartheta) := \vartheta^k(1-\vartheta)^m \cdot \mu_0(d\vartheta)/M(k,m)$, $(k,m) \in I$. Then $\varphi(k,m,z) := (k+z, m+1-z)$ is a sufficient statistic by no. 3 of Table 23.1. In view of the data from the CM in Example 13.6.4 we have: $S = \mathbb{N}_0$; $A = \mathbb{N}_0$; $D(s) = \{0,1,\ldots,s\}$; $Z = \{0,1\}$; $\eta_t \sim Q(\vartheta,dz) := \mathrm{Bi}(1,\vartheta)$; $T(s,a,z) = s - a + daz$; $Q'(i,dz) = \mathrm{Bi}(1,p'(i))$, where $p'(k,m) := \int \hat{\mu}(k,m,d\vartheta)\, \vartheta = M(k+1,m)/M(k,m)$ and $q'(k,m) := 1 - p'(k,m) = M(k,m+1)/M(k,m)$; $r' \equiv 0$; $V_0'(k,m,s) = V_0(s) \geq 0$ is arbitrary with $V_0(0) = 0$; $\beta = 1$. Note that $V_0 \geq 0$ implies that $v_n(i,s) \geq 0$ for all n, i, s. ♦

In MDP′ the sets $\mathbb{F}'$ and $(\mathbb{F}')^N$ denote the set of decision rules and the set of N-stage Markovian policies, respectively. Moreover, by (LUBF) the function $(i,s) \mapsto b(s)$ is either an upper bounding function or a lower bounding function for MDP′, so that $V_{N\pi'}'$, defined as in Chap. 16, exists for $\pi' \in (\mathbb{F}')^N$. By $V_N'(i,s)$ we denote the maximal expected N-stage reward for initial state (i,s) within the set $(\mathbb{F}')^N$ of Markovian policies in *MDP*$'_N$.

Part (b) of our basic result Theorem 23.1.17 below shows that in maximizing $\pi \mapsto v_{N\pi}(i_0,s)$ on Δ_{NZ} for all s one can in general restrict oneself to those history-dependent policies $\pi = (\pi_t)_0^{N-1}$ where $\pi_t(z^t, s_t)$ depends on the disturbance history

z^t only via the momentary posterior index, defined below. Policies π with this property are generated by Markovian policies π' in MDP$'$ as follows.

Definition 23.1.12 (Posterior index mappings and generation of disturbance-dependent policies in $BCM_N(\hat{\mu}(i))$ from Markovian policies in MDP'_N)

- The **posterior index mappings** $i_t: I \times Z^t \to I$, $0 \le t \le N$, generated by $i \in I$, are defined recursively by

$$i_{t+1}(i, z^{t+1}) := \varphi(i_t(i, z^t), z_{t+1}), \quad i_0 :\equiv i. \tag{23.12}$$

- The ***disturbance-dependent policy*** $\pi^i = (\pi_t^i)_0^{N-1} \in \Delta_{NZ}$, ***generated by*** $i \in I$ ***and a Markovian policy*** $\pi' = (\pi_t')_0^{N-1} \in (\mathbb{F}')^N$ is defined by $\pi_0^i(s_0) := \pi_0'(i, s_0)$, $s_0 \in S$, and for $1 \le t \le N-1$ by

$$\pi_t^i(z^t, s_t) := \pi_t'(i_t(i, z^t), s_t), \quad (z^t, s_t) \in Z^t \times S. \tag{23.13}$$

If φ is sufficient for $\hat{\mu}$, then obviously $\hat{\mu}(i_t(i, z^t))$ is the posterior in (23.8) at time t for prior $\hat{\mu}(i)$, $t \ge 1$, $i \in I$, determined by Φ.

Example 23.1.13 (Third continuation of the gambling problem from Example 23.1.1*)* We assume an arbitrary prior μ_0. As seen in the second continuation, we have $\mu_0 = \hat{\mu}(0,0)$ for a certain family $\hat{\mu}(i)$, $i \in \mathbb{N}_0^2$, which has the sufficient statistic $\varphi(i, z) := i + (z, 1 - z)$. It follows that for $i_0 := (0, 0)$ the posterior index i_t at time $t \ge 1$ equals $i_t(i_0, z^t) = (\sum_{j=1}^t z_j, t - \sum_{j=1}^t z_j)$, the number of successes and of failures in the disturbance vector z^t. Moreover, the policy $\pi = (\pi_t)_0^{N-1} \in \Delta_{NZ}$ generated by i_0 and by some policy $\pi' \in (\mathbb{F}')^N$ is given by $\pi_0 = \pi_0'(i_0, \cdot)$ and for $1 \le t \le N-1$ by

$$\pi_t(i_0, z^t, s_t) = \pi_t'\left(\sum_{j=1}^t z_j, t - \sum_{j=1}^t z_j, s_t\right), \quad (z^t, s_t) \in \{0,1\}^t \times \mathbb{N}_0. \qquad \blacklozenge$$

For simplicity of notation we often write $v_{N\pi}(i, s)$ and $v_N(i, s)$ instead of $v_{N\pi}(\hat{\mu}(i), s)$ and $v_N(\hat{\mu}(i), s)$, respectively. Thus (23.4) shows that for $\pi \in \Delta_{NZ}$ and all N, i, s

$$v_{N\pi}(i, s) = \int \mathbb{Q}_N^{\hat{\mu}(i)}(d(\vartheta, z^N))\, G_{N\pi}(\vartheta, s, z^N),$$

$$v_N(i, s) = \sup_{\pi \in \Delta_{NZ}} v_{N\pi}(i, s).$$

We call $v_N: I \times S \to \overline{\mathbb{R}}$ the *N-stage* ***Bayes value function.***

For $f \in \Delta_1$, $\pi := (f, \sigma) \in \Delta_N$ and $z_1 \in Z$ the z_1-section of π_t, $t \ge 1$, is the mapping $(z_2, z_3, \ldots, z_t, s_t) \mapsto \pi_t(z^t, s_t)$, which belongs to $\mathbb{F}^{t-1}$. Thus the z_1-section of σ, defined by $\sigma_{z_1} := (\pi_t(z_1, \cdot))_{t=1}^{N-1}$ belongs to $\Delta_{N-1,Z}$.

The result of Proposition 23.1.16 is the crucial point in the proof of our Basic Theorem 23.1.17. Its proof needs the following two auxiliary Lemmas.

Lemma 23.1.14 *In each $(BCM, \hat{\mu}, \varphi)$ we have $v_{1f} = V'_{1f}$ for $f \in \mathbb{F}$. Moreover, for $N \geq 2$, $(f, \sigma) \in \Delta_{NZ}$ and $(i, s) \in I \times S$*

$$v_{N(f,\sigma)}(i, s) = r'_f(i, s) + \beta \cdot \int Q'(i, dz)\, v_{N-1,\sigma_z}(\varphi(i, z), T_f(s, z)).$$

Proof We omit the proof of $v_{1f} = V'_{1f}$ since it is similar and simpler than the proof of the other assertion.

(a) Put $\pi := (f, \sigma)$. Using $v(\vartheta, s, z) := V_{N-1,\sigma_z}(\vartheta, T_f(s, z))$ we have for fixed ϑ

$$V_{N\pi}(\vartheta, s) = r(\vartheta, s) + \beta \cdot \int Q(\vartheta, dz)\, v(\vartheta, s, z). \tag{23.14}$$

In fact, one can either copy the proof of Sect. 21.3 or one can derive (23.14) from Sect. 21.3 since $\Delta_{NZ} \subset \Delta_N$ and since $CM(\vartheta)$ is a special MDPD(ϑ).

(b) Put $\mu := \hat{\mu}(i)$. Now the second assertion in Lemma 23.1.14 follows, since our convention $v_n(i, s) := v_n(\hat{\mu}(i), s)$ yield

$$(v_{N\pi}(i, s) - r'(i, s))/\beta = \int \mu(d\vartheta) \int Q(\vartheta, dz)\, v(\vartheta, s, z)$$

$$= \int \mu(d\vartheta) \int Q(\vartheta, dz) \int \Phi(\mu, z, d\vartheta')\, V_{N-1,\sigma_z}(\vartheta', T_f(s, z))$$

$$= \int Q'(i, dz)\, v_{N-1,\sigma_z}(\Phi(\mu, z), T_f(s, z))$$

$$= \int Q'(i, dz)\, v_{N-1,\sigma_z}(\varphi(i, z), T_f(s, z)).$$

□

Lemma 23.1.15 *Fix $i \in I$ and $N \geq 2$. If $\pi = (f, \sigma) \in \Delta_{NZ}$ is generated by i and $(f', \sigma') \in \mathbb{F}' \times (\mathbb{F}')^{N-1}$ then the z-section σ_z of σ is generated by $i' := \varphi(i, z)$ and σ', $z \in Z$.*

Proof Put $H_{tZ} := Z^t \times S$ and let $z_1 := z$ vary in Z. Now (23.13) implies for $1 \leq t \leq N - 1$, $(z^t, s_t) \in H_{tZ}$ that

$$\pi_{t+1}(z^{t+1}, s_{t+1}) = \pi'_{t+1}(i_{t+1}(i', z^{t+1}), s_{t+1}). \tag{23.15}$$

Since $\sigma_t = \pi_{t+1}$ and $\sigma'_t = \pi'_{t+1}$, $0 \leq t \leq N - 2$, we have to show, using $z_2^{t+1} := (z_\nu)_2^{t+1} \in Z^t$, that

$$\pi_{t+1}(z^{t+1}, s_{t+1}) = \pi'_{t+1}(i_t(i', z_2^{t+1}), s_{t+1}), \quad (z^{t+1}, s_{t+1}) \in H_{t+1,Z}.$$

By (23.15) this is true if for $0 \leq t \leq N-2$ the assertion (I_t) holds:

$$i_t(i, z^{t+1}) = i_t(i', z_2^{t+1}), \quad z^{t+1} \in Z^{t+1}.$$

Firstly, (I_0) holds since $i_i(1, z_1) = \varphi(i, z_1) = i' = i_0(i')$. Moreover, (I_t) for some $t \geq 0$ implies (I_{t+1}) since

$$i_{t+2}(i, z^{t+2}) = \varphi(i_{t+1}(i, z_t^{t+1}), z_{t+2}) = \varphi(i_t(i, z_2^{t+1}), z_{t+2}) = i_{t+1}(i', z_2^{t+2}). \qquad \square$$

For the following Proposition 23.1.16(a) note that in BCM each observation-dependent policy $(\pi_t)_0^{N-1} \in \Delta_{NZ}$ can be regarded as a history-dependent policy $\pi' = (\pi'_t)_0^{N-1}$ in MDP′, where π'_t does not depend on $i_0, i_1, \ldots, i_t$. Thus $\Delta_{NZ} \subset \Delta'_N$; in particular, $V'_{N\pi}$ is defined for $\pi \in \Delta_{NZ}$.

Proposition 23.1.16 *In each (BCM, $\hat{\mu}, \varphi$) we have for all $N \geq 1$:*

(a) $v_{N\pi} = V'_{N\pi}$ *for all* $\pi \in \Delta_{NZ}$.
(b) If $\pi^i \in \Delta_{NZ}$ *is generated by i and* $\pi' \in (\mathbb{F}')^N$ *then* $v_{N\pi^i}(i, s) = V'_{N\pi'}(i, s)$ *for all s.*

Proof

(a) We verify the assertion (I_N) by induction on $N \geq 1$. Firstly, (I_1) is contained in Lemma 23.1.14. Assume (I_N) for some $N \geq 1$. Then for $f \in \mathbb{F}$, $(f, \sigma) \in \Delta_{N+1,Z}$ and $z \in Z$ we have $\sigma_z \in \Delta_{NZ}$. Thus $v_{N,\sigma_z} = V'_{N,\sigma_z}$ by (I_N). Now fix (i, s). From Lemma 23.1.14 we obtain

$$\begin{aligned} v_{N+1,(f,\sigma)}(i, s) &= r'_f(i, s) + \beta \cdot \int Q'(i, dz)\, v_{N,\sigma_z}(\varphi(i, z), T_f(s, z)) \\ &= r'_f(i, s) + \beta \cdot \int Q'(i, dz)\, V'_{N,\sigma_z}(\varphi(i, z), T_f(s, z)). \end{aligned}$$

By Sect. 21.3 applied to MDP′, the latter expression equals $V'_{N+1,(f,\sigma)}(i, s)$.

(b) We prove the assertion by induction on $N \geq 1$. Firstly, (I_1) holds by (a). Assume (I_N) for some $N \geq 1$. Then for $f' \in \mathbb{F}'$, $\pi' := (f', \sigma') \in \Delta_{N+1,Z}$ and $z \in Z$ we have $\sigma_z \in \Delta_{NZ}$. Since π is generated by (i, π'), we know from Lemma 23.1.15 that σ_z is generated by $i' := \varphi(i, z)$ and σ'. Therefore $v_{N,\sigma_z}(i', s) = V'_{N,\sigma'}(i', s)$, $s \in S$. Finally Lemma 16.1.9 and the Markovian policy π' and Lemma 23.1.14 for $\pi \in \Delta_{NZ}$ yield

$$\begin{aligned} V'_{N+1,\pi'}(i, s) &= r'_{f'}(i, s) + \beta \cdot \int Q'(i, dz)\, V'_{N\sigma'}(\varphi(i, z), T_f(s, z)) \\ &= r'_{f'}(i, s) + \beta \cdot \int Q'(i, dz)\, v_{N\sigma_z}(\varphi(i, z), T_f(s, z)) = v_{N+1,\pi}(i, s). \end{aligned} \qquad \square$$

In the next basic result the validity of the VI for MDP′ can often be established by means of the Structure Theorem 16.1.12.

Theorem 23.1.17 (Basic Theorem for a Bayesian control model) *If the VI holds for the associated MDP′, then we have for all $N \geq 1$:*

(a) The Bayes value function v_N equals the value function V'_N in MDP′, hence can be found recursively by the value iteration

$$v_n(i,s) = \sup_{a \in D(s)} \left[r'(i,s,a) + \beta \cdot \int Q'(i,dz)\, v_{n-1}(\varphi(i,z), T(s,a,z)) \right],$$

$$n \geq 1, (i,s) \in I \times S \tag{23.16}$$

starting with $v_0(i,s) := \int \hat{\mu}(i,d\vartheta)\, V_0(\vartheta,s) = V'_0(i,s)$.

(b) Assume that at any stage $n \geq 1$ of the associated MDP′ a maximizer f'_n exists. Then for each $i_0 \in I$ the disturbance-dependent policy $\pi^* \in \Delta_{NZ}$, generated by i_0 and the Markovian policy $\pi' := (f'_n)_N^1$, is Bayes-optimal for problem $BCM_N(\hat{\mu}(i_0))$.

Proof

(a) From Proposition 23.1.16(a) and Sect. 21.3, applied to the MCM′, which is equivalent to MDP′, we get (21.3) and $\Delta_{NZ} \subset \Delta'_{NZ}$,

$$\begin{aligned} v_N &= \sup\{v_{N\sigma} : \sigma \in \Delta_{NZ}\} = \sup\{V'_{N\sigma} : \sigma \in \Delta_{NZ}\} \\ &\leq \sup\{V'_{N\pi} : \pi \in \Delta'_{NZ}\} = V'_N. \end{aligned}$$

On the other hand, for given $\pi' \in (\mathbb{F}')^N$ and $(i,s) \in S'$ let π^i be generated by (i,π'). Then Proposition 23.1.16(b) yields $v_N(i,s) \geq v_{N\pi^i}(i,s) = V'_{N\pi'}(i,s)$. Since π' is arbitrary, we get $v_N(i,s) \geq V'_N(i,s)$. Thus $v_n = V'_N$.

(b) From Theorem 16.1.12(b) we know that π' is optimal for MDP'_N. Now we obtain from Proposition 23.1.16(b) and from (a) above for each $(i,s) \in S'$

$$V'_{N\pi'}(i,s) = v_{N\pi^i}(i,s) \leq v_N(i,s) = V'_N(i,s) = V'_{N\pi'}(i,s).$$ □

For solving a problem $BCM_N(\mu_0)$ for a given μ_0 by means of Theorem 23.1.17, one must find a family $\hat{\mu}$ containing μ_0 and having a sufficient statistic φ. Many results below for specific problems are proven under the assumption that $\hat{\mu}$ is arbitrary with a sufficient statistic φ. Then the results hold for *all* problems $BCM_N(\mu_0)$ for which the prior μ_0 belongs to $\hat{\mu}$. Note that while the VI in MDP′ gives the Bayes value functions for all problems $BCM_N(\hat{\mu}(i))$, $i \in I$, the policy π^* in Theorem 23.1.17(b) depends on i_0. In general no policy $\pi \in \Delta_{NZ}$ is Bayes-optimal for *all* problems $BCM_N(\hat{\mu}(i))$, $i \in I$.

Remark 23.1.18

(a) For the numerical computation of $v_N(i_0, s_0)$ by the VI (23.16) for fixed (i_0, s_0) one needs $v_{N-t}(i, s)$, $1 \le t \le N-1$, $s \in S$, not for all $i \in I$ but only for the set of posterior indexes $i_t(i_0, z^t)$, $z^t \in Z^t$. However, the set $R_1(i, s)$ of states in MDP′ which are reachable in one step from $(i, s) \in I \times S$ equals $\{(\varphi(i, z), T(s, az)) : a \in D(s), z \in Z\}$, $(i, s) \in I \times S$. Thus the exact numerical computation of $v_{n-1}(i, s)$ is in general impossible unless S, $D(s)$ and Z are finite.
(b) The implementation (e.g. on a computer) of π^* in Theorem 23.1.17(b) does not require the explicit form of the decision rules of π^*. It suffices to observe the states (i_t, s_t) which arise during the control, and to take the actions $a_t^* := f'_{N-t}(i_t, s_t)$, $0 \le t \le N-1$.
(c) Note that $\pi^* = (\pi_t^*)_0^{N-1} \in \Delta_{NZ}$ is given by

$$\pi_0^*(s_0) := f'_N(i_0, s_0), \; \pi_t^*(z^t, s_t) := f'_{N-t}(i_t(i_0, z^t), s_t), \; 1 \le t \le N-1.$$

Example 23.1.19 (Solution of the gambling problem from Example 23.1.1*)* Assume an arbitrary prior μ_0. As seen in the second continuation, we have $\mu_0 = \hat{\mu}(0, 0)$ for a certain family $\hat{\mu}(i)$, $i \in \mathbb{N}_0^2$, which has the sufficient statistic $\varphi(i, z) := i + (z, 1-z)$. The assumptions of the Basic Theorem 23.1.17 are fulfilled. The VI yields $v_n(i, 0) = 0$ for all n and i, and for $n \ge 1$, $s \in \mathbb{N}$ and all i

$$v_n(i, s) = \max_{0 \le a \le s} \left[p'(i) \cdot v_{n-1}(i + (1, 0), s - a + da) + q'(i) \cdot v_{n-1}(i + (0, 1), s - a) \right], \tag{23.17}$$

starting with $v_0(i, s) := V_0(s)$. It follows by induction on $n \ge 0$ that $v_n \ge 0$ is finite for all n. Note that although we are basically only interested in the N-stage Bayes reward $v_N(\mu_0, s_0) = v_N(0, 0, s_0)$ for some fixed $s_0 \in \mathbb{N}$, its computation via the VI (23.17) also requires $v_n(i, s)$ for $1 \le n \le N-1$, but not for all i from the infinite set $I = \mathbb{N}_0^2$. In fact, one needs $v_{N-1}(i, s)$ for $i = (0, 1), (1, 0)$, $v_{N-2}(i, s)$ for $i = (0, 2), (1, 1), (2, 0)$, etc. In Proposition 24.1.9 we study structural properties of the solution. ♦

For the application of the Basic Theorem 23.1.17 it is advantageous if one can embed μ_0 into a family $(\hat{\mu}(i), i \in I)$, having a sufficient statistic φ and where I is a subset of a low-dimensional Euclidean space. In Tables 23.1 and 23.2 we collect such results. The main tool is the next result.

Proposition 23.1.20 (Neyman criterion for sufficiency in BCMs) *Let $(I, \mathfrak{I})$ be a measurable space. Assume that Q has a transition density function q with respect to a σ-finite measure on $\mathfrak{Z}$. Assume further that there exist measurable functions $l \ge 0$ on $\Theta \times I$ and $H > 0$ on $I \times Z$ and a measurable mapping φ from $I \times Z$ into I such that for all ϑ, i, z*

$$l(\vartheta, i) \cdot q(\vartheta, z) = l(\vartheta, \varphi(i, z)) \cdot H(i, z). \tag{23.18}$$

Table 23.1 Sufficient statistics φ for families $\hat{\mu}_\nu(i, d\vartheta) \propto l(\vartheta, i) \cdot \nu(d\vartheta)$, $\nu \in \mathbb{P}_e$ and transition probabilities $Q(\vartheta, dz)$ (for notations cf. comments below)

$Q(\vartheta, dz)$	Θ	i	I	$l(\vartheta, i)$	$\mathbb{P}_e$	$\varphi(i, z)$
1. $\Gamma_{\vartheta,b}$, $b \in \mathbb{R}^+$	$\mathbb{R}^+$	(x, y)	$\mathbb{R}_+ \times \mathbb{R}_+$	$\vartheta^x e^{-y\vartheta}$	$\mathbb{P}_{>0}(\mathfrak{T})$	$i + (b, z)$
2. $\mathrm{Exp}(\vartheta)$	$\mathbb{R}^+$	(x, y)	$\mathbb{R}_+ \times \mathbb{R}_+$	$\vartheta^x e^{-y\vartheta}$	$\mathbb{P}_{>0}(\mathfrak{T})$	$i + (1, z)$
3. $\mathrm{Bi}(n, \vartheta)$	$(0, 1)$	(k, m)	$\mathbb{N}_0^2$	$\vartheta^k \cdot (1 - \vartheta)^m$	$\mathbb{P}(\mathfrak{T})$	$i + (z, n - z)$
4. $N(0, 1/\vartheta)$	$\mathbb{R}^+$	(x, y)	$(0.5 \cdot \mathbb{N}_0) \times \mathbb{R}_+$	$\vartheta^x e^{-y\vartheta}$	$\mathbb{P}_{>0}(\mathfrak{T})$	$i + (0.5, 0.5z^2)$
5. $U(0, \vartheta)$	$\mathbb{R}^+$	(k, x)	$\mathbb{N}_0 \times \mathbb{R}_+$	$1_{(x,\infty)}(\vartheta)/\vartheta^k$	$\mathbb{P}_{>-\infty}(\mathfrak{T})$	$(x \vee z, k + 1)$
6. $\mathrm{Geo}(\vartheta)$	$(0, 1)$	(k, m)	$\mathbb{N}_0^2$	$\vartheta^k \cdot (1 - \vartheta)^m$	$\mathbb{P}(\mathfrak{T})$	$i + (1, z - 1)$
7. $\mathrm{NBi}(n, \vartheta)$	$(0, 1)$	(k, m)	$\mathbb{N}_0^2$	$\vartheta^k \cdot (1 - \vartheta)^m$	$\mathbb{P}(\mathfrak{T})$	$i + (n, z)$
8. $\mathrm{Poi}(\vartheta)$	$\mathbb{R}^+$	(k, m)	$\mathbb{N}_0^2$	$\vartheta^k \cdot e^{-m\vartheta}$	$\mathbb{P}_{>0}(\mathfrak{T})$	$i + (z, 1)$

Table 23.2 The family $\hat{\mu}_\nu(i, d\vartheta) \propto l(\vartheta, i) \cdot \nu(d\vartheta)$, with l and Q from Table 23.1 and for special choices of $\nu := \hat{\mu}(0, 0)$

$Q(\vartheta, dz)$	i	I	$\nu := \hat{\mu}_\nu(0, 0)$	$\hat{\mu}_\nu(i)$
1. $\Gamma_{\vartheta,b}$	(x, y)	$\mathbb{R}_+ \times \mathbb{R}_+$	$\Gamma_{\alpha,\delta}$	$\Gamma_{y+\alpha,x+\delta}$
2. $\mathrm{Exp}(\vartheta)$	(x, y)	$\mathbb{R}_+ \times \mathbb{R}_+$	$\Gamma_{\alpha,\delta}$	$\Gamma_{y+\alpha,x+\delta}$
3. $\mathrm{Bi}(n, \vartheta)$	(k, m)	$\mathbb{N}_0^2$	$\mathrm{Be}(\alpha_1, \alpha_2)$	$\mathrm{Be}(k + \alpha_1, m + \alpha_2)$
4. $N(0, 1/\vartheta)$	(x, y)	$(0.5 \cdot \mathbb{N}_0) \times \mathbb{R}_+$	$\Gamma_{\alpha,\varepsilon}$	$\Gamma_{y+\alpha,x+\varepsilon}$
5. $\mathrm{Geo}(\vartheta)$	(k, m)	$\mathbb{N}_0^2$	$\mathrm{Be}(\alpha_1, \alpha_2)$	$\mathrm{Be}(k + \alpha_1, m + \alpha_2)$
6. $\mathrm{NBi}(n, \vartheta)$	(k, m)	$\mathbb{N}_0^2$	$\mathrm{Be}(\alpha_1, \alpha_2)$	$\mathrm{Be}(k + \alpha_1, m + \alpha_2)$
7. $\mathrm{Poi}(\vartheta)$	(k, m)	$\mathbb{N}_0^2$	$\Gamma_{\alpha,\varepsilon}$	$\Gamma_{m+\alpha,k+\varepsilon}$

Then for each σ-finite measure $\nu \not\equiv 0$ on $\mathfrak{T}$ such that

$$l_\nu(i) := \int \nu(d\vartheta)\, l(\vartheta, i)$$

is finite and positive for all i, φ is sufficient for the family

$$\hat{\mu}_\nu(i, d\vartheta) := l(\vartheta, i) \cdot \nu(d\vartheta)/l_\nu(i), \quad i \in I. \tag{23.19}$$

Proof Fix i, z and put $\mu := \hat{\mu}_\nu(i)$. Recall that $q_\mu(z) := \int \mu(d\vartheta')\, q(\vartheta', z)$. We have to show that $\Phi(\hat{\mu}(i), z) = \hat{\mu}_\nu(\varphi(i, z))$.

For $B \in \mathfrak{T}$ we obtain from Lemma 23.1.6, whether or not $0 < q_\mu(z) < \infty$,

$$\Phi(\mu, z, B) \cdot q_\mu(z) = \int_B \mu(d\vartheta)\, q(\vartheta, z) \tag{23.20}$$

$$= \int_B \nu(d\vartheta)\, l(\vartheta, i) \cdot q(\vartheta, z)/l_\nu(i) = H(i, z) \cdot \int_B \nu(d\vartheta)\, l(\vartheta, \varphi(i, z))/l_\nu(i)$$

$$= \hat{\mu}_\nu(\varphi(i, z), B) \cdot H(i, z) \cdot l_\nu(\varphi(i, z))/l_\nu(i) =: \hat{\mu}_\nu(\varphi(i, z), B) \cdot \Lambda(i, z).$$

Equation (23.20) for $B := \Theta$ shows that $q_\mu(z) = \Lambda(i, z)$, with $\Lambda(i, z)$ being finite and positive, as both H and l_ν have this property. Inserting $q_\mu(z) = \Lambda(i, z)$ into (23.20) yields the assertion. □

One often encounters transition probabilities μ from a set X into a set Y of the form

$$\mu(x, dy) := g(x, y) \cdot \nu(dy) \Big/ \int \nu(dy') \cdot g(x, y'), \quad x \in X, \tag{23.21}$$

for some measurable function $g \geq 0$ on $X \times Y$. As the denominator of the right-hand side of (23.21) is determined by the numerator, it is customary to write (23.21) in the simpler form

$$\mu(x, dy) \propto g(x, y) \cdot \nu(dy), \quad x \in X.$$

The proportionality symbol $\propto$ is pronounced as *proportional to*. As an example, (23.19) may be written as $\hat{\mu}_\nu(i, d\vartheta) \propto l(\vartheta, i) \cdot \nu(d\vartheta)$.

Example 23.1.21 (Sufficient statistics for BCMs)

(a) Assume that $Q(\vartheta, dz)$ is the exponential distribution $\text{Exp}(\vartheta)$ for $\vartheta \in \Theta := \mathbb{R}^+$, and $Z := \mathbb{R}^+$. Then condition (23.18) of the Neyman criterion is fulfilled for $l(\vartheta, k, y) := \vartheta^k \cdot e^{-y\vartheta}$, $i := (k, y) \in I := \mathbb{N}_0 \times \mathbb{R}_+$, $H \equiv 1$ and $\varphi(i, z) := i + (1, z)$. It is easy to see that $0 < l_\nu(k, y) < \infty$ for some $\nu \in \mathbb{P}(\mathfrak{T})$ if and only if ν belongs to the set $\mathbb{P}_{>0}(\mathfrak{T})$ of those priors which have finite moments of all orders in $(0, \infty)$. It follows from the Neyman criterion that for each $\nu \in \mathbb{P}_{>0}(\mathfrak{T})$ the mapping φ is sufficient for the family

$$\hat{\mu}_\nu(k, y, d\vartheta) \propto \vartheta^k \cdot e^{-y\vartheta} \cdot \nu(d\vartheta), \quad (k, y) \in \mathbb{N}_0 \times \mathbb{R}_+.$$

In particular, the choice $\nu := \text{Exp}(\delta)$, $\delta > 0$ fixed, shows that φ is sufficient for the family $\hat{\mu}_\nu(k, y) := \Gamma_{y+\delta, k+1}$.

From (23.12) we obtain by induction on $t \geq 0$ that for each $i_0 \in I$ the posterior index at time $t \geq 1$ equals $J_t^{i_0}(z^t) = i_0 + (t, \sum_{j=1}^t z_j)$.

(b) Example (a) and many more examples from Table 23.1 are contained in the case where $Q(\vartheta, dz)$ is a one-parameter exponential family, i.e. where

$$Q(\vartheta, dz) = g(\vartheta) \cdot \exp[p(\vartheta) \cdot h(z)] \cdot b(z) \cdot \rho(dz).$$

Here g and $b > 0$ are measurable functions on Θ and on Z, respectively and ρ is some σ-finite measure on Z; since $Q(\vartheta, \cdot)$ is a probability distribution also $g > 0$. (As an example, $Q(\vartheta, dz) := \text{Exp}(\vartheta)$, $\vartheta \in \mathbb{R}^+$, is an exponential family with $g(\vartheta) := p(\vartheta) := \vartheta$, $h(z) := -z$, $b :\equiv 1$ and $\rho :=$ Lebesgue measure on $\mathbb{R}^+$.) Put $I := \mathbb{N}_0 \times \mathbb{R}$, $\varphi(i, z) := i + (1, z)$ and $l(\vartheta, k, y) = g(\vartheta)^k \cdot \exp[p(\vartheta) \cdot y]$. Then the Neyman criterion with $H(i, z) := b(z)$ shows: If ν is a probability distribution on Θ such that $0 < \int \nu(d\vartheta)\, g(\vartheta)^k \cdot \exp[p(\vartheta)\cdot y] < \infty$ for all $(k, y) \in I$, then φ is sufficient for the family $\hat{\mu}(k, y, d\vartheta) \propto g(\vartheta)^k \cdot \exp[p(\vartheta)\cdot y] \cdot \nu(d\vartheta)$, $(k, y) \in I$.

(c) Assume that $Q(\vartheta, dz) = N(\vartheta, \sigma^2)$ for $\vartheta \in \Theta := \mathbb{R}$, and $Z := \mathbb{R}$. As prior μ_0 we use the normal distribution $\mu_0 := N(m, b^2)$ with given $m \in \mathbb{R}$ and $b \in \mathbb{R}^+$. Then μ_0 is contained by the choice $i := (0, 0)$ in the family $\hat{\mu}(x, k) := N(M(x, k), \Sigma^2(k))$, $i = (x, k) \in I := \mathbb{R} \times \mathbb{N}_0$, where

$$M(x, k) := \frac{\sigma^2 m + b^2 x}{\sigma^2 + b^2 k}, \quad \Sigma^2(k) := \frac{\sigma^2 b^2}{\sigma^2 + b^2 k}.$$

We assert that $\hat{\mu}$ has the sufficient statistic $\varphi(i, z) := i + (z, 1)$. Firstly, we have $N(M(x, k), \Sigma^2(k), d\vartheta) \propto l(\vartheta, i) \cdot \lambda(d\vartheta)$, where

$$l(\vartheta, x, k) := \exp\left(-\frac{(\vartheta - M(x, k))^2}{2 \cdot \Sigma^2(k)}\right),$$

and λ is Lebesgue measure on $\mathfrak{B}$. A simple yet tedious calculation shows that condition (23.18) of the Neyman criterion is fulfilled. Now the assertion follows from Proposition 23.1.20 with $\nu := \lambda$. Note that $0 < l_\nu(i) < \infty$ since $\vartheta \mapsto l(\vartheta, i)$ equals, except for a multiplicative constant, the λ-density of a normal distribution. From (23.12) we obtain by induction on $t \geq 0$ that for each $i_0 \in I$ the posterior index at time $t \geq 1$ equals $i_t(i_0, z^t) = i_0 + (\sum_{j=1}^t z_j, t)$. ♦

All results of the next table can be verified by the Neyman criterion.

Comments on Table 23.1:

1. We use $0.5 \cdot \mathbb{N}_0 := \{0, 0.5, 1, 1.5, \ldots\}$. $\mathbb{P}_e$ is the set of probability distributions ν on $\mathfrak{T}$ with $0 < l_\nu(i) < \infty$ for $i \in I$. By $\mathbb{P}_{>\alpha}(\mathfrak{T})$, $\alpha \in [-\infty, \infty)$, we denote the set of probability distributions on $\mathfrak{T}$ which have finite moments of all orders in (α, ∞). The condition $\nu \in \mathbb{P}_{>-\infty}$ is fulfilled, for example, if ν is concentrated on an interval $[b, \infty)$ with $b \in \mathbb{R}^+$; $\nu \in \mathbb{P}_{>0}$ is fulfilled, for example, if ν is a Gamma distribution.
2. Geo(ϑ) is that version of the geometric distribution which gives probability zero to the point $z = 0$; NBi(n, ϑ) denotes the negative binomial distribution; in particular, NBi($1, \vartheta$) is that version of the geometric distribution which gives probability ϑ to the point $z = 0$.

Remark 23.1.22

(a) In many cases of Table 23.1 we have $l(\vartheta, i_0) = 1$, $\vartheta \in \Theta$, for some $i_0 \in I$. Then each $\nu \in \mathbb{P}_e$ may be used to model our preferences about ϑ since $\nu = \hat{\mu}(i_0)$, and then Theorem 23.1.17 may be applied for $\mu_0 := \nu$. Thus in each case of Table 23.1 one can apply the Basic Theorem 23.1.17 for each prior $\mu_0 := \nu = \hat{\mu}(0, 0)$.

(b) If the assumptions in Proposition 23.1.20 hold and if $I' \subset I$ is invariant under φ in the sense that $\varphi(i, z) \in I'$ for all $(i, z) \in I' \times Z$, then the assumptions in Proposition 23.1.20 hold when I is replaced by I'.

(c) From Table 23.1 one easily obtains more examples of sufficient statistics by making a change of the parameter ϑ as explained in Problem 23.3.4. ◇

23.2 The Model BCM with Large Horizon

Consider a model $(BCM, \hat{\mu}, \varphi)$ with associated MDP′. When the horizon N is large, one will try to use $v^* := \lim_{n\to\infty} v_n$, if it exists as finite limit, as approximate value for v_N. In addition one will ask for a decision rule $f' \in \mathbb{F}'$ which is asymptotically optimal for the model $(BCM, \hat{\mu}, \varphi)$ in the sense that for each $i_0 \in I$ the sequence of n-stage policies $\pi^n_{i_0} \in \Delta_n$, $n \geq 1$, generated according to (23.1.15) by i_0 and by the stationary policy $\pi' := (f')_0^{n-1} \in (\mathbb{F}')^n$ satisfies for all $s \in S$

$$v_n(i_0, s) - v_{n,\pi^n_{i_0}}(i_0, s) \to 0 \text{ for } n \to \infty.$$

Obviously the policies $\pi^n_{i_0}$ are relatively simple to compute from f'. Fortunately both problems can be settled by means of corresponding properties of the associated MDP′ since $v_n = V'_n$ for all $n \geq 1$ by the Basic Theorem 23.1.17. In fact, the following proposition holds:

Proposition 23.2.1 (Existence of asymptotically decision rules)

(a) *v^* exists if and only if $V' := \lim_{n\to\infty} V'_n$ exists, and then $v^* = V'$; this holds in particular if the sequence $(v_n)_0^\infty$ is monotone. And then v^* is a fixed point of U' if and only if V' is a fixed point of U'; this holds in particular if $(v_n)_0^\infty$ is increasing and if $v_0 \geq 0$.*

(b) *If v^* exists and if $f' \in \mathbb{F}'$ is asymptotically optimal for MDP′, then f' is asymptotically optimal for $(BCM, \hat{\mu}, \varphi)$.*

Proof In (a) only the last assertion needs a proof which follows from Proposition 20.1.1.

Assertion (b) holds since $v_{N\pi^n_{i_0}}(i_0, s) = V'_{nf'}(i_0, s)$ for all s by Proposition 25.1.9(b) below. □

Now one can combine (a) and (b) with large horizon results for MDPs. As an example, we now combine (a) and (b) with Theorem 20.1.3. For a measurable function $b\colon S \to \mathbb{R}_+$ denote by $MB(X)_b$ the set of measurable functions on $X = I \times S$ which have finite b-norm.

Theorem 23.2.2 (The optimality equation and asymptotically optimal decision rules) *Consider a model $(BCM, \hat{\mu}, \varphi)$ with associated MDP′, in which the VI holds. Assume that there exists a measurable function $b\colon S \to \mathbb{R}_+$ with*

$$\|r'\|_b < \infty, \quad \|V'_0\|_b < \infty,$$

$$\beta_b := \beta \cdot \sup_{(i,s,a)\in I\times D} \int Q'(i, dz)\, b(T(s, a, z))/b(s) < 1.$$

Then the following holds:

(a) *The sequence of Bayes value functions v_N, $N \geq 1$, converges in b-norm to some $v^* \in MB(X)_b$, and v^* is independent of V_0 in $MB(X)_b$.*
(b) *v^* is the unique solution of the optimality equation $U'v = v$ within $MB(X)_b$, and*

$$\|v^* - v_N\|_b \leq \beta_b^N \cdot \|v^* - v_0\|_b, \quad N \geq 1.$$

(c) *If f' is a maximizer of $L'v^*$ then it is asymptotically optimal for $(BCM, \hat{\mu}, \varphi)$. And then we have for all i_0 and s*

$$|v_N(i_0, s) - v_{\pi_{i_0}^n}(i_0, s)| \leq \mathrm{sp}((v^* - v_0)/b) \cdot \beta_b^N, \quad n \geq 1.$$

Proof One easily shows that $(i, s) \mapsto b'(i, s) := b(s)$ is a bounding function for MDP′ with $\beta_{b'} = \beta_b < 1$, $\|r'_X\|_{b'} = \|r'\|_b < \infty$ and $\|V'_0\|_{b'} = \|V'_0\|_b < \infty$. Now the assertions follow from Theorem 20.1.3 and Proposition 23.2.1. □

Remark 23.2.3

(a) The assumptions $\|r'\|_b < \infty$, $\|V'_0\|_b < \infty$ and $\beta_b < 1$ hold if $\|Qr_Z\|_b < \infty$, $\|V_0\|_b < \infty$ and if

$$\tilde{\beta}_b := \beta \cdot \sup_{(\vartheta,s,a)\in\Theta\times D} \int Q(\vartheta, dz)\, b(T(s, a, z))/b(s) < 1,$$

respectively; then $\beta_b \leq \tilde{\beta}_b$. In particular, for $b \equiv 1$ these three assumptions reduce to the requirement that both r_Z and V_0 are bounded and that $\beta < 1$.
(b) If v^* exists, then structural properties of v^*, preserved under the pointwise limit, are usually proven by verifying them for each v_n. ◇

Example 23.2.4 (Large horizons in the gambling problem from Example 23.1.1*)* We have:

(a) v^* exists.
(b) If $V_0 \geq 0$ then $U'v^* = v^*$.
(c) If $V_0(s) \leq d_0 \cdot s^\alpha$ for some $d_0, \alpha \in \mathbb{R}^+$ and if $\beta < (d-1)^{-\alpha}$, then the assumptions of Theorem 23.2.2 are fulfilled for $b(s) := s^\alpha$, $s \in \mathbb{N}_0$, and $\beta_b \leq (d-1)^{-\alpha}$.

Only (c) needs a proof. Since $Q'(i, dz) = \mathrm{Bi}(1, p'(i))$ and since $s \mapsto s^\alpha$ is increasing, we have

$$\begin{aligned}\max_{0\leq a\leq s} \int Q'(i, dz)\, b(T(s, a, z)) &= \max_{0\leq a\leq s} \left[p'(i) \cdot (s - a + da)^\alpha + q'(i) \cdot (s - a)^\alpha\right] \\ &\leq (d-1)^\alpha \cdot s^\alpha.\end{aligned}$$

Therefore $\beta_b \leq \beta \cdot (d-1)^\alpha < 1$. This proves the assertion. ◇

23.3 Problems

Problem 23.3.1 (Generalization of Example 23.1.21(a)) Assume that $Q(\vartheta, dz)$ is the Gamma distribution $\Gamma_{\vartheta,b}$, $\vartheta \in \mathbb{R}^+$, for a constant $b \in \mathbb{R}^+$.

(a) For each $\nu \in \mathbb{P}_\infty(\mathfrak{T})$ the mapping $(i, z) \mapsto \varphi(i, z) := i + (\eta, z)$ is sufficient for the family

$$\hat{\mu}_\nu(x, y, d\vartheta) \propto \vartheta^x \cdot e^{-y\vartheta} \cdot \nu(d\vartheta), \quad i := (x, y) \in I := \mathbb{R}_+^2.$$

(b) Compute $i_t(i_0, z^t)$ for given $i_0 \in I$.
(c) Compute $\hat{\mu}_\nu(i)$ for $\nu := \Gamma_{\alpha,\varepsilon}$, $\alpha, \varepsilon \in \mathbb{R}^+$.
(d) Compute a density of $Q'(i, d\vartheta)$ for ν as in (c).

Problem 23.3.2 (Special cases of the Neyman-criterion) Several examples of sufficient statistics are special cases of the following situation (cf. DeGroot 1970, p. 163).

Assume that Q has the Λ_Z-density

$$z \mapsto q(\vartheta, z) = g_1(\vartheta)^{h_1(z)} \cdot g_2(\vartheta)^{h_2(z)} \cdot H(z)$$

for some non-negative measurable functions g_1, g_2, h_1, h_2 and a positive function H. For non-negative measurable functions H_1 and H_2 put $l(\vartheta, i) := g_1(\vartheta)^{H_1(i)} \cdot g_2(\vartheta)^{H_2(i)}$, where $i \in I \subset \mathbb{R}^2$. Assume that $H_1(i) + h_1(z) = H_1(\varphi(i, z))$ and $H_2(i) + h_2(z) = H_2(\varphi(i, z))$ for some measurable mapping φ from $I \times Z$ into I. For any $\tilde{\mu} \in \mathbb{P}(\mathfrak{T})$ such that l_ν is positive and finite, φ is sufficient for $\hat{\mu}(i, d\vartheta) \propto l(\vartheta, i)\, \nu(d\vartheta)$.

Problem 23.3.3

(a) Assume that φ is sufficient for $(\hat{\mu}(i), i \in I)$. If $\emptyset \neq I_1 \subset I$ is invariant under φ (i.e. if $\varphi(i, z) \in I_1$ for $i \in I_1$ and all z), then $\varphi|I_1 \times S$ is sufficient for $\hat{\mu}|I_1$.
(b) Apply (a) to that special case of Problem 23.3.1 where $Q(\vartheta, dz)$ is the Erlang distribution $\Gamma_{\vartheta,k}$, $k \in \mathbb{N}$.
(c) In Problem 23.3.1 the set $I_2 := \mathbb{R}_+ \times \mathbb{R}^+$ is also invariant under φ. In this case one does not need a moment condition for ν since $0 < l_\nu(\vartheta, i) < \infty$, $\vartheta \in \Theta$, $i \in I_2$, for any $\nu \in \mathbb{P}(\mathfrak{T})$. However, ν is no longer contained in $\hat{\mu}|I_2$.

Problem 23.3.4 (Reparametrization preserves sufficiency) Consider a model $(BCM, \hat{\mu}, \varphi)$. Assume that Q has a transition density q with respect to a σ-finite measure such that $0 < \int q(\vartheta, z)\, \hat{\mu}(i, d\vartheta) < \infty$ for all i and z. Reparametrize Q by means of a bijection Ψ from a set Θ_1 onto Θ as follows: Endow Θ_1 with the σ-algebra $\mathfrak{T}_1 := \Psi^{-1}(\mathfrak{T})$, hence Ψ and Ψ^{-1} are measurable. For $\vartheta_1 \in \Theta_1$ put $Q_1(\vartheta_1, dz) := Q(\Psi(\vartheta_1), dz)$. Then φ is sufficient with respect to Q_1 for the family $\hat{\mu}_1(i) := (\hat{\mu}(i))_{\Psi^{-1}}$, $i \in I$.

Problem 23.3.5 (The inverse Gamma distribution) The inverse Gamma distribution $\Gamma^-_{\alpha,b}$, $\alpha, b \in \mathbb{R}^+$ on $\mathbb{R}^+$ is defined by the density $\vartheta \mapsto \alpha^b \cdot \vartheta^{-b-1} \cdot e^{-\alpha/\vartheta} / \Gamma(b)$.

(The name stems from the fact that $Y \sim \Gamma_{\alpha,b}$ implies $1/Y \sim \Gamma^{-}_{\alpha,b}$). Derive from Problem 23.3.4 with $\Theta_1 := \Theta := \mathbb{R}^+$ and $\Psi(\vartheta_1) := 1/\vartheta_1$ and from Tables 23.1 and 23.2, No. 2, that for $Q_1(\vartheta_1, dz) := \mathrm{Exp}(1/\vartheta_1)$, $\vartheta_1 \in \mathbb{R}^+$, the mapping $(x,y,z) \mapsto \varphi(x,y,z) := (x+1, y+z)$ on $I \times Z := \mathbb{R}_+ \times \mathbb{R}_+ \times \mathbb{R}^+$ is sufficient with respect to Q_1 for the family $\hat{\mu}_1(x,y) := \Gamma^{-}_{\alpha+y,\delta+x}$.

23.4 Supplements

Supplement 23.4.1 (Posterior distribution) To deal with the posterior distributions of ϑ we need the probability space $(\Theta \times \Omega_N, \mathfrak{T} \otimes \mathfrak{G}_N, \mathbb{P}_{N\pi}(\mu, s, d(\vartheta, g_N)))$ with

$$\mathbb{P}_{N\pi}(\mu, s, d(\vartheta, g_N)) := \mu(d\vartheta) \otimes \mathbb{P}_{N\pi}(\vartheta, s, dg_N).$$

Let τ, η_1, ζ_1, ..., η_N, ζ_N be the coordinate variables on this probability space. Put $P(s, d(\vartheta, g_N)) := \mathbb{P}_{N\pi}(\hat{\mu}(i), s, d(\vartheta, g_N))$. It is intuitively clear and can be proved that in case $\mu_0 = \hat{\mu}(i_0)$ the posterior distribution $\mu_t(\mu_0, h_t) := P_{\tau|((\eta_i,\zeta_i))_1^t}(h_t, d\vartheta)$ equals $\hat{\mu}(i_t(i_0, h_t))$. Note that the approach via $\hat{\mu}$ is no restriction, as also $I := \mathbb{P}(\mathfrak{T})$, $\hat{\mu}(i) := i$ for $i \in I$ and $\varphi(\mu, s, a, z, s') := \Phi(\mu, s, a, z, s')$ is allowed.

Supplement 23.4.2 (Solution of a problem $BCM_N(\mu_0)$) The solution of a problem $BCM_N(\mu_0)$ by means of Theorem 23.1.17 does not depend on the particular choice of I, $\hat{\mu}$ and φ in the following sense: If $\tilde{I}$, $\tilde{\mu}$ and $\tilde{\varphi}$ is another choice such that $\mu_0 = \hat{\mu}(i_0) = \tilde{\mu}(\tilde{i}_0)$ for some $i_0 \in I$ and $\tilde{i}_0 \in \tilde{I}$, then for $i \in I^* := I \cap \tilde{I}$:

(a) $v_n(i,s) = \tilde{v}_n(i,s)$ for $s \in S$.
(b) If there exist maximizers f'_n and $\tilde{f}'_n$ of $L'v_{n-1}$ and of $\tilde{L}'v_{n-1}$, $1 \le n \le N$, respectively, then the policies π^i and $\tilde{\pi}^i \in \Delta_{NZ}$, generated by $(f'_n)_N^1$ and $(\tilde{f}'_n)_N^1$, respectively, coincide and are Bayes-optimal for $BCM_N(\mu_0)$.

Proof For the proof one uses that for $(s,a) \in D$ and $z \in Z$ we have

$$r'(i,s,a) = \tilde{r}'(i,s,a), \quad v'_0(i,s) = \tilde{v}_0(i,s),$$
$$\varphi(i,z) = \tilde{\varphi}(i,z), \quad Q'(i,dz) = \tilde{Q}'(i,dz).$$

Now (a) follows from the VI (23.16) by induction on $n \ge 0$, which also implies (b). □

Supplement 23.4.3 (Historical remarks) The presentation here and in Chap. 24 owes much to the work of Rieder (1975a, 1988). In the latter Lecture Notes Bayesian MCMs (special cases of BMDPDs from Chap. 25) are treated in a similar way. Structural results for partially observed CMs are derived in Rieder (1991). Early treatments of the theory are contained in DeGroot (1970) and van Hee (1978). Other monographs include Davis and Vinter (1985), Girlich et al. (1990) and Presman and Sonin (1990).

Besides the Bayesian approach there exist other approaches for solving the family CM(ϑ), $\theta \in \Theta$, of control models. Well-known from statistical decision theory is the minimax (or worst case) approach. Sometimes these models are also called robust MDPs, see e.g. Nilim and El Ghaoui (2005) and Iyengar (2005). Adaptive control processes have been investigated by Hernández-Lerma (1989).

Chapter 24
Examples of Bayesian Control Models

In this chapter we study several examples: linear-quadratic and gambling problems, stopping problems and asset selling. Important applications of the Bayesian theory are those statistical decision problems (in particular problems of estimation and testing) which are not based on a fixed number of observations, but where sampling may be discontinued at a random time, determined by the observations made so far. An example is the sequential probability ratio test for two simple hypotheses.

24.1 Linear-Quadratic and Gambling Problems

We start with the following

Example 24.1.1 (The linear-quadratic system of Example 16.1.14) We want to solve the BCM for the linear-quadratic system of Example 16.1.14 with the following data: $S = A = D(s) = \mathbb{R}$; $C_0(\vartheta, s) := d_0 s^2$ for some constant $d_0 \in \mathbb{R}^+$. Since the new features of the solution become visible already in the special case $g = h = \lambda = \gamma = 1$ in Example 16.1.14, we assume this case, i.e.

$$T(s, a, z) := s - a + z, \quad c_Z(\vartheta, s, a, z) := s^2 + a^2.$$

We admit for $Q(\vartheta, dz)$, $\vartheta \in \Theta = \mathbb{R}^+$, any probability distribution on $Z = \mathbb{R}$ which has mean zero and variance $1/\vartheta$. We further assume a prior μ_0 on $\mathfrak{B}(\mathbb{R}^+)$ which is embedded in some family $\hat{\mu}(i)$, $i \in I$, having a sufficient statistic φ. Since $c_Z \geq 0$ and $C_0 \geq 0$, assumption (LUBF) is fulfilled. For the associated MDP′ we have

$$T'(i, s, a, z) = (\varphi(i, z), s - a + z)), \quad c'(i, s, a) = s^2 + a^2, \quad C_0'(i, s) = d_0 s^2.$$

As $c' \geq 0$ and $C_0' \geq 0$, (MA1) holds for MDP′. Obviously $Q'(i, dz)$ has mean zero and variance $\sigma^2(i) := \int \vartheta^{-1} \, \hat{\mu}(i, d\vartheta)$, which is measurable in i.

K. Hinderer et al., *Dynamic Optimization*, Universitext,
DOI 10.1007/978-3-319-48814-1_24

In Proposition 24.1.2 below we see that the solution of our problem is the same as the solution of the problem from Example 16.1.14 with known transition law $Q(dz)$, except that the variance $\mathbb{E}\eta_1^2$ of $Q(dz)$ there is replaced by the variance $\sigma^2(i)$ of $Q'(i, dz)$. Moreover, the solution has four remarkable features: (i) The Bayes value functions v_n are separable in i and s, quadratic in s and linear in $\sigma^2(i)$. (ii) $v_n(i, s)$ equals the minimal Bayes cost within the set of Markovian policies. (iii) The limit v^* of the Bayes value functions exists for all $\beta \in \mathbb{R}^+$, but is finite only for $\beta < 1$. (iv) In case $\beta < 1$ the unique minimizer of $L'v^*$ is not asymptotically optimal. ◆

Proposition 24.1.2 (Explicit solution of the linear-quadratic problem from Example 24.1.1) *Assume that the variance $\sigma^2(i)$ of $Q'(i, dz)$ is finite and positive for all $i \in I$. Then the following holds:*

(a) The minimal n-stage Bayes cost for prior $\hat{\mu}(i)$ and initial state s equals

$$v_n(i, s) = d_n \cdot s^2 + e_n \cdot \sigma^2(i), \quad n \geq 1.$$

Here the numbers d_{n+1} and e_n, $n \geq 0$, are given by

$$d_{n+1} = 1 + \frac{\beta \cdot d_n}{1 + \beta \cdot d_n}, \qquad e_n := \sum_{t=1}^{n} \beta^t d_{n-t}.$$

(b) For each choice of $i_0 \in I$ the Markovian policy $(f_n)_N^1 \in \Delta_N$ with $f_n(s) := (d_n - 1) \cdot s$, $s \in \mathbb{R}$, is Bayes-optimal for $BCM_N(\hat{\mu}(i_0))$.

(c) If I is structured and if $\hat{\mu}(i, d\vartheta)$ is stochastically increasing in i, then $v_n(i, s)$ is decreasing in i.

(d) There exists $v^*(i, s) := \lim_{n\to\infty} v_n(i, s)$, $(i, s) \in I \times \mathbb{R}$, and

$$v^*(i, s) = \begin{cases} ds^2 + \frac{\beta d}{1-\beta} \cdot \sigma^2(i), & \text{if } \beta < 1, \\ \infty, & \text{if } \beta \geq 1. \end{cases}$$

Here $d := \lim_n d_n = (2\beta - 1 + \sqrt{1 + 4\beta^2})/(2\beta)$.

(e) If $\beta < 1$, the decision rule $(i, s) \mapsto f'(i, s) := (d - 1) \cdot s$ is the unique minimizer of $L'v^*$ and asymptotically optimal.

Proof The proof runs along the lines of the proofs of Examples 16.1.14 and 20.1.5.

(a) We use the Structure Theorem 16.1.12 with $\mathfrak{V}$ as the set of functions $(i, s) \mapsto v(i, s) = \delta \cdot s^2 + \varepsilon \cdot \sigma^2(i)$ for constants $\delta, \varepsilon \geq 0$, possibly depending on v. Since $v \geq 0$ we have $\mathfrak{V} \subset \mathfrak{V}_0$. Moreover, from (23.11) with $u(z) := z^2$ we obtain

$$\int Q'(i, dz)\, \sigma^2(\varphi(i, z)) = \sigma^2(i).$$

Now we get for $v \in \mathfrak{V}$, since $L'v = r' + \beta \cdot P'v$,

$$(L'v(i,s,a) - s^2 - a^2)/\beta = \int Q'(i,dz)\, v(\varphi(i,z), s-a+z)$$

$$= \int Q'(i,dz)\, \left[\delta \cdot (s-a+z)^2 + \varepsilon \cdot \sigma^2(\varphi(i,z))\right]$$

$$= \delta \cdot (s-a)^2 + \delta \cdot \int Q'(i,dz)\, z^2 + \varepsilon \cdot \int Q'(i,dz)\, \sigma^2(\varphi(i,z))$$

$$= \delta \cdot (s-a)^2 + (\varepsilon + \delta) \cdot \sigma^2(i), \quad (i,s,a) \in I \times \mathbb{R}^2.$$

Easy computations show that $L'v$ has the unique minimizer $f_v(i,s) := \beta\delta{\cdot}s/(1+\beta\delta)$, and that for $(i,s) \in I \times \mathbb{R}$

$$U'v(i,s) = (1 + \beta \cdot \delta/(1 + \beta \cdot \delta)) \cdot s^2 + \beta \cdot (\varepsilon + \delta) \cdot \sigma^2(i). \tag{24.1}$$

Thus (S1) and (S2) from the Structure Theorem hold. Moreover, $C_0' \in \mathfrak{V}$. Now the Basic Theorem 23.1.17 tells us that the VI holds for the MDP′, that $v_n(i,s) = C_n'(i,s) = \delta_n \cdot s^2 + \varepsilon_n \cdot \sigma^2(i)$, $n \geq 1$, for some constants δ_n, ε_n, and that $(i,s) \mapsto f_n'(i,s) := \beta \cdot \delta_{n-1} \cdot s/(1 + \beta \cdot \delta_{n-1})$ is a minimizer for MDP′ at stage n. Moreover, (24.1) yields

$$\delta_{n+1} = 1 + \beta \cdot \delta_n/(1 + \beta \cdot \delta_n), \quad \varepsilon_{n+1} = \beta \cdot (\varepsilon_n + \delta_n), \quad n \geq 0,$$

with $\delta_0 := d_0$, $\varepsilon_0 := 0$. Now the assertion follows since obviously $\delta_n = d_n$, since the latter equation has the solution $\varepsilon_n = e_n$ and since $\beta \cdot \delta_{n-1}/(1 + \beta \cdot \delta_{n-1}) = \delta_n - 1 = d_n - 1$.

(b) By Theorem 23.1.17 the policy $\pi = (\pi_t)_0^{N-1} = (f_n)_N^1$, which is generated from i_0 and from $\pi' = (\pi_t')_0^{N-1} := (f_n')_N^1$ according to (23.1.15), is Bayes-optimal for $BCM_N(\hat{\mu}(i_0))$. Since $f_n'(i,s)$ is independent of i, we see from (23.1.15) that $\pi_t(z^t, s_t) = \pi_t'(s_t)$, hence f_n has the form as asserted in (b), and it is obviously Markovian.

(c) This follows from (a) since $\sigma^2(i) = \int \vartheta^{-1}\, \hat{\mu}(i, d\vartheta)$.

(d) The proof is exactly the same as for the corresponding result Example 20.1.5 for the problem with known transition law, except that $\mathbb{E}\eta_1^2$ there is replaced by $\sigma^2(i)$. Note that because of $\lambda = \gamma = g = h = 1$ the quadratic equation $I(y) = 0$ from (10.20) has the unique positive solution d as given above.

(e) From (a) with $\delta := d$, $\varepsilon := \beta d/(1-\beta)$ and $f_{v^*} := f'$ we see that $U_{f'}'v^* = v^*$, and that f' is the unique minimizer of $L'v^*$. Now we can copy the proof of Example 20.1.5(b) replacing $\mathbb{E}\eta_1^2$ there by $\sigma^2(i)$. Observe that in Example 23.1.21 we saw that f' is asymptotically optimal for $(BCM, \hat{\mu}, \varphi)$ if it is asymptotically optimal for MDP′. □

Remark 24.1.3 Consider the special case of Proposition 24.1.2 where $Q(\vartheta, dz) = N(0, 1/\vartheta)$, $\vartheta \in \mathbb{R}^+$. We see from no. 4 in Tables 23.1 and 23.2 that $\varphi(x,y) :=$

$(0.5, 0.5z^2)$, $i := (x, y) \in I := (0.5 \cdot \mathbb{N}_0) \times \mathbb{R}_+$ is sufficient for the family $\hat{\mu}(x, y) := \Gamma_{y+\alpha,x+\varepsilon}$, α, $\varepsilon \in \mathbb{R}^+$. Then $\sigma^2(i) < \infty$ for all i if and only if $\varepsilon > 1$, and $\sigma^2(x, y) = (y + \alpha)/(x + \varepsilon - 1)$. Thus in this special case $v_n(x, y, s)$ is decreasing in x and increasing in y. This is confirmed by Proposition 24.1.2(c) when choosing on I the SE-ordering, since $\Gamma_{y+\alpha,x+\varepsilon}$ is stochastically increasing in x and stochastically decreasing in y by Table 18.1. ◆

In examples it often happens that several properties of $v_n(i, s)$ as a function of n or s easily carry over from CMs to BCMs. The following proposition contains a general result in this direction concerning monotonicity.

Proposition 24.1.4 (Isotonicity of Bayes value functions in n and in s) *Assume that the VI holds in the MDP′ associated to a Bayesian control model (BCM, $\hat{\mu}$, φ).*

(a) If $v_1 \geq v_0$, then $v_n(i, s)$ is increasing in n.
(b) If S is structured and if $D(s)$, $T(s, a, z)$, $r_Z(\vartheta, s, a, z)$ and $V_0(\vartheta, s)$ are increasing in s, then $v_n(i, s)$ is increasing in s for all $n \geq 0$.

Proof

(a) This follows by induction on n from the VI (23.16).
(b) This is proved exactly as for the corresponding result Theorem 18.1.1 for CMs, observing that the assumptions about r_Z and V_0 imply that $r'(i, s, a)$ and $V_0'(i, s)$ are increasing in s. □

The problem of isotonicity of $v_n(i, s)$ in the information state i is more difficult. From Theorem 18.2.7, the criterion for isotonicity of $s \mapsto V_n(s)$ in MDPs, applied to MDP′, one can find conditions under which $v_n(i, s)$ is increasing in (i, s). We show in Proposition 24.1.6 below that under slightly stronger conditions we even obtain isotonicity of $v_n(i, s)$ in i and in s. Recall from Chap. 6 that for mappings v from a Cartesian product of structured sets into a structured set $(M, \leq^M)$ isotonicity is weaker than componentwise isotonicity, provided $\leq^M$ is transitive (in particular, if v is a function).

Lemma 24.1.5 *Assume that I, Θ and Z are structured. Then $Q'(i, dz)$ is stochastically increasing in i if both $\hat{\mu}(i, d\vartheta)$ and $Q(\vartheta, dz)$ are stochastically increasing (or both are stochastically decreasing) in i and in ϑ, respectively.*

Proof We consider the case where both $\hat{\mu}(i)$ and $Q(\vartheta, dz)$ are stochastically increasing. Let $v \geq 0$ be measurable and increasing on Z. Then $\int Q'(i, dz)\, v(z) = \int \hat{\mu}(i, d\vartheta) \int Q(\vartheta, dz)\, v(z)$. The inner integral is increasing in ϑ since $Q(\cdot, dz)$ is stochastically increasing, and then $\int Q'(i, dz)\, v(z)$ is increasing in i since $\hat{\mu}(\cdot)$ is stochastically increasing. □

Proposition 24.1.6 (Isotonicity in i and in s of Bayes value functions) *Consider a model (BCM, $\hat{\mu}$, φ) with structured sets Θ, I, S and Z. Assume that the VI holds for the associated MDP′. Then all Bayes value functions $v_n(i, s)$, $n \geq 1$, are increasing* *in i and in s under the following conditions:*

(i) $D(s)$ *is increasing in s,*
(ii) $T(s, a, z)$ *is increasing in s and in z,*
(iii) $\varphi(i, z)$ *is increasing in i and in z,*
(iv) $i \mapsto \hat{\mu}(i, d\vartheta)$ *and* $\vartheta \mapsto Q(\vartheta, dz)$ *are stochastically increasing,*
(v) $r_Z(\vartheta, s, a, z)$ *is increasing in* ϑ*, s and z, and* $V_0(\vartheta, s)$ *is increasing in* ϑ *and s.*

Proof

(a) We show that $r'(i, s, a)$ is increasing in i. (In a similar way one shows that $V_0'(i, s)$ is increasing in i.) Firstly we mention that $r(\vartheta, s, a) = \int Q(\vartheta, dz)\, r_Z(\vartheta, s, a, z)$ is increasing in ϑ by Lemma 18.2.6 since r_Z is increasing in ϑ and in z and since $Q(\vartheta, dz)$ is stochastically increasing in ϑ. Then the assertion follows from $r'(i, s, a) = \int \hat{\mu}(i, d\vartheta)\, r(\vartheta, s, a)$ as $\hat{\mu}(i, d\vartheta)$ is stochastically increasing in i.

(b) We know from Theorem 23.1.17 that $v_n(i, s) = V_n'(i, s)$, which is increasing in s by Proposition 24.1.4(b) and (i), (ii) and (v). Now we show by induction on $n \geq 0$ the assertion (I_n) that $v_n(i, s)$ is also increasing in i. (I_0) holds by (a) since $v_0 = V_0'$. Assume that (I_{n-1}) holds for some $n \geq 1$. By comparison with the proof for Theorem 18.2.7 about isotonicity of the value functions in arbitrary MDPs we see, using (i) and (a), that we only have to show that

$$w_n(i, s, a) := \int Q'(i, dz)\, v_{n-1}(T'(i, s, a, z))$$

is increasing in i. According to Lemmas 24.1.5 and 18.2.6 it suffices to show that $v_{n-1}(T'(i, s, a, z)) = v_{n-1}(\varphi(i, z), T(s, a, z))$ is increasing in i and in z. Isotonicity in i holds since $v_{n-1}(j, s)$ is increasing in j by (I_{n-1}) and since $\varphi(i, z)$ is increasing in i by (iii). Moreover, isotonicity of $v_{n-1}(\varphi(i, z), T(s, a, z))$ in z holds since $v_{n-1}(j, s')$ is increasing in j and in s' (hence also in (j, s')) by (I_{n-1}) and since both $\varphi(i, z)$ and $T(s, a, z)$ (hence also $T'(i, s, a, z)$) are increasing in z by (iii) and (ii), respectively. □

Remark 24.1.7 Assume that $\Theta \subset \mathbb{R}^d$ is the Cartesian product of Borel sets in $\mathbb{R}$, that $\leq^{\Theta} = \leq^d$ and that $\hat{\mu}(i, d\vartheta) = g_i(\vartheta) \cdot \nu(d\vartheta)$ for some σ-finite measure ν and some measurable function $(i, \mu) \mapsto g_i(\mu) \geq 0$ on $I \times \Theta$. Then $\hat{\mu}(i)$ is stochastically increasing in i if $g_i \leq_{\text{TP}} g_j$ whenever $i \leq^I j$. This result follows from the definition of increasing likelihood ratio (cf. Chap. 18), since the latter implies stochastic monotonicity by Proposition 18.3.6. ◇

Example 24.1.8 (Structure of the solution of the gambling problem from Example 23.1.19) Assume an arbitrary prior μ_0 and put $M(k, m) := \int \vartheta^k (1 - \vartheta)^m \cdot \mu_0(d\vartheta)$, $(k, m) \in I := \mathbb{N}_0^2$. Then $\mu_0 = \hat{\mu}(0, 0)$ for $\hat{\mu}(k, m, d\vartheta) := \vartheta^k (1 - \vartheta)^m \cdot \mu_0(d\vartheta)/M(k, m)$, $(k, m) \in I$, and $\varphi(k, m, z) := (k + z, m + 1 - z)$ is a sufficient statistic as seen in Example 23.1.19. Recall from (23.17) that the finite Bayes value functions satisfy the VI

$$v_n(i,s) = \max_{0\le a\le s}\big[p'(i)\cdot v_{n-1}(\varphi(i,1), s-a+da) \qquad (24.2)$$
$$+ q'(i)\cdot v_{n-1}(\varphi(i,0), s-a)\big],\quad n\ge 1,\ i\in\mathbb{N}_0^2,\ s\in\mathbb{N}_0,$$

starting with $v_0(i,s) := V_0(s)$. Since we admit an arbitrary prior μ_0, for many questions (such as monotonicity of $v_N(i,s)$ in s) it suffices to consider $v_N(\mu_0,\cdot) = v_N(0,0,\cdot)$. ♦

A decision rule f is called ***bang-bang*** in s if $f(i,s)\in\{0,s\}$ for all i, s, i.e. if it prescribes at each state either to stake nothing or to stake all available capital. The proof of part (b) of the next result exemplifies the fact that structural properties of the solution may considerably simplify computations.

Proposition 24.1.9 (Structural properties of the solution of the gambling problem from Example 24.1.8) *The following hold:*

(a) $v_n(\mu_0,s)$ is increasing in n.
(b) $v_n(\mu_0,s)$ is increasing in s if $V_0(s)$ is increasing in s.
(c) If V_0 is convex then $v_n(\mu_0,s)$ is convex in s, there exists at each stage n for the MDP′ a maximizer $(i,s)\mapsto f_n(i,s)$ which is bang-bang in s.
(d) If $d=2$ and if V_0 is concave, then for all n the Bayes value functions $v_n(\mu_0,s)$ are concave in s, and the smallest maximizer f_n' at stage n of the MDP′ belongs to LIP(1) with respect to s, i.e. $|f_n(i,s+1)-f_n(i,s)|\le 1$ for all i, s.
(e) If V_0 is increasing then $v_n(k,m,s)$ is increasing in k and s and decreasing in m.
(f) If $V_0(s)=s$ for $s\in\mathbb{N}_0$ then $v_n(i,s)=d_n(i)\cdot s$ for some $d_n(i)\in R_+$ and $n\ge 0$ and all i, s. If in addition the mean of μ_0 is $\ge 1/d$ then $f_n'(i,s)=s$ for all n, (i,s), hence $\pi_0'(i_0,s)=s$ and $\pi_t'(i_0,z^t,s)=s$ for $t\ge 1$. Moreover,

$$d_n(\mu_0) = d^n\cdot\int \vartheta^n\,\mu_0(d\vartheta).$$

Proof

(a) This follows from Proposition 24.1.4(a) since by (24.2)

$$v_1(i,s)\ge w_1(i,s,0) = p'(i)\cdot v_0(\varphi(i,1),s)$$
$$+ q'(i)\cdot v_0(\varphi(i,0),s) = V_0(s) = v_0(i,s).$$

(b) This follows directly from Proposition 24.1.4(b).

(c) and (d) have proofs that are exactly the same as the proof of Proposition 13.6.5(c) and (f), respectively.

(e) We apply Proposition 24.1.6 with $(I,\le^I) := (\mathbb{N}_0^2,\le^{\text{SE}})$. Assumption (iii) in Proposition 24.1.6 is satisfied, as mentioned above. The assumptions (i), (ii) and (v) hold trivially. Finally assumption (iv) holds since

$i \mapsto \hat{\mu}(i)$ and $\vartheta \mapsto \mathrm{Bi}(1, \vartheta)$ are stochastically increasing by Table 18.1 and Example 18.3.7(b), respectively.

(f1) We prove the first assertion (I_n) by induction on $n \geq 0$. Firstly, (I_0) holds with $d_0(i) = 1$ for all i. Now assume (I_n) for some $n \geq 0$. Put $\lambda_n(i) := p'(i) \cdot d_n(\varphi(i, 1))$, $\rho_n(i) := q'(i) \cdot d_n(\varphi(i, 0))$. Then the VI (24.2) yields

$$\begin{aligned} v_{n+1}(i, s) &= \max_{0 \leq a \leq s} [\lambda_n(i)(s - a + da) + \rho_n(i)(s - a)] \\ &= \left[\lambda_n(i) + \rho_n(i) + ((d - 1) \cdot \lambda_n(i) - \rho_n(i))^+\right] \cdot s \\ &=: d_{n+1}(i) \cdot s. \end{aligned} \tag{24.3}$$

Hence

$$d_{n+1}(i) = d\lambda_n(i) + ((d - 1)\lambda_n(i) - \rho_n(i))^-.$$

Note that $d_{n+1}(i) = v_{n+1}(i, 1) \geq 0$.

(f2) We prove the second assertion. Put $m(k) := \int \vartheta^k \, \mu_0(d\vartheta)$, $k \geq 0$. Fix $n \geq 1$ and $k \geq 0$. We have already used in part (e) of the proof that $\hat{\mu}(k, m)$ is stochastically increasing in k. Therefore $p'(k, 0) = \int \vartheta \, \hat{\mu}(k, 0, d\vartheta) \geq p'(0, 0) \geq 1/d$. From (e) we know that $v_n(k + 1, 0, s) \geq v_n(k, 1, s)$ for all s, hence $d_n(k + 1, 0) \geq d_n(k, 1)$. Since $p'(k, 0) \geq 1/d$ implies $q'(k, 0) = 1 - p'(k, 0) \leq (d - 1)/d$, we obtain

$$\begin{aligned} (d - 1) \cdot \lambda_n(k, 0) &= (d-1) \cdot p'(k, 0) \cdot d_n(k + 1, 0) \geq (d-1)/d \cdot d_n(k, 1) \\ &\geq q'(k, 0) \cdot d_n(k, 1) = \rho_n(k, 0). \end{aligned}$$

Since $p'(k, 0) = m(k + 1)/m(k)$, as seen in Chap. 23 after (23.11), it follows from (24.3) that

$$d_{n+1}(k, 0) = d \cdot \lambda_n(k, 0) = d \cdot m(k + 1) \cdot d_n(k + 1, 0)/m(k).$$

Now backward induction on n yields the second assertion. □

Example 24.1.10 (for Proposition 24.1.9(f)) If $\mu_0 = \mathrm{Be}(\alpha_1, \alpha_2)$ for some $\alpha_1, \alpha_2 \in \mathbb{R}^+$ with $d \geq (\alpha_1 + \alpha_2)/\alpha_1$, then

$$d_n(\mu_0) = d^n \cdot \prod_{j=0}^{n-1} \frac{\alpha_1 + j}{\alpha_1 + \alpha_2 + j}.$$

In particular, if μ_0 is the uniform probability distribution on (0, 1) with $d \geq 2$ then $d_n(\mu_0) = d^n/(n + 1)$. ◆

24.2 Optimal Stopping and Asset Selling

We now treat N-stage *stopping problems* (such as the asset selling problem from Example 13.4.4) when the probability distribution $Q(\vartheta, dz)$ of the i.i.d. observations $z_1, z_2, \ldots, z_N$ in Z (which in the asset selling problem are the offers) depends on an unknown parameter ϑ. At the latest the last offer must be accepted. We search for a policy under which we obtain $d_N(\mu_0)$, the maximal expected discounted gain when μ_0 is the prior. In contrast to our goal such problems in the literature are mostly treated in the simpler case where there is an additional *known* initial offer $s_0 \in Z$; then one looks for a policy under which one obtains $v_N(\mu_0, s_0)$, the maximal expected discounted gain when there is a known initial offer s_0. We now treat the often more realistic case where no known initial offer is available.

Example 24.2.1 (Optimal stopping of Bayesian models with i.i.d. offers and without known initial offer) We describe the problem by a binary BCM with known initial state, using a somewhat tricky choice of the state space and building upon the following idea: The state space S contains not only all observations z and a state $\hat{s}$ (in which the process is absorbed as soon as an offer is accepted) but also a special state $\overline{s}$. The data of the BCM are then chosen such that (i) the process which starts in $\overline{s}$ does not gain anything in the first period and the rewards gained at times $1 \leq t \leq N$ are discounted back to time $t = 0$, (ii) the next state s_1 equals the first offer z_1, (iii) the process evolves from z_1 at time $t = 1$ onward as it would do from time $t = 0$ onward with an offer $s_0 := z_1$ and (iv) $s_{t+1} = z_{t+1}$ if s_t is not accepted.

More precisely, we use the following BCM: the observation space Z is arbitrary; often, e.g. in the asset selling problem, Z equals a Borelian subset of $\mathbb{R}$; Θ is arbitrary and $Q(\vartheta, dz)$ is the probability distribution of each of the offers; $S = Z + \{\overline{s}, \hat{s}\}$; $s_0 \in Z$ denotes an initial offer; $s_t \in Z$ for some $1 \leq t \leq N$ means that no offer has been accepted before time t and that the offer at time t equals s_t; thus the next state s_{t+1} equals $\hat{s}$ or z_{t+1} if the offer s_t is accepted ($a_t = 1$) or rejected ($a_t = 0$), respectively; hence $s_t = \hat{s}$ for some $1 \leq t \leq N$ means that an offer has been accepted before time t; ($s_0 := \hat{s}$ occurs only for formal reasons and may be interpreted as meaning that no offers at all are received and no costs arise); obviously the system moves from a state $s_0 \in Z$ to states $s_1, s_2, \ldots, s_{t-1}$ within Z until either it is absorbed in $s_t := \hat{s}$ for some $1 \leq t \leq N$ or until some $s_N \in Z$ is reached (no absorption); $\overline{s}$ can occur only as initial state upon which a transition to some offer $z \in Z$ is enforced. We assume that an accepted offer $z \in Z$ yields a reward $\psi(\vartheta, z)$, where $\psi \geq 0$ is a measurable function on $\Theta \times Z$ (for asset selling we have $\psi(\vartheta, s) = s$); in addition, each offer (not only the rejected ones, as sometimes assumed in the literature) costs $c \geq 0$; the state $\overline{s}$ yields neither a gain nor a cost. Altogether we have for $(s, a) \in D$ and $z \in Z$

$$T(s, a, z) = \begin{cases} z \cdot \delta_{a,0} + \hat{s} \cdot \delta_{a,1}, & \text{if } s \in Z, \\ z, & \text{if } s = \overline{s}, \\ \hat{s}, & \text{if } s = \hat{s}; \end{cases}$$

$$r(\vartheta, s, a) = \begin{cases} \psi(\vartheta, s) \cdot \delta_{a,1} - c, & \text{if } s \in Z, \\ 0, & \text{if } s \in \{\overline{s}, \hat{s}\}; \end{cases}$$

$$V_0(\vartheta, s) = r(\vartheta, s, 1) = \begin{cases} \psi(\vartheta, s) - c, & \text{if } s \in Z, \\ 0, & \text{if } s \in \{\overline{s}, \hat{s}\}. \end{cases} \qquad \square$$

It is intuitively clear and follows formally from (23.4) that $v_N(\mu_0, \hat{s}) = 0$ and that $v_N(\mu_0, s_0)$ for $s_0 \in Z$ is the maximal expected N-stage discounted reward when the decision process starts with a known offer s_0. Moreover, inserting T and r into (23.2) and using (23.4), we see that

$$d_N(\mu_0) = v_N(\mu_0, \overline{s})/\beta, \; N \geq 2. \tag{24.4}$$

(Here the denominator β takes care of the fact that the rewards, when starting in $s_0 = \overline{s}$, are discounted to time $t = 1$.) The importance of (24.4) stems from the fact that we can use it for computing $d_N(\mu_0)$ within our BCM with known initial state. For this purpose we have to find (i) $v_N(\mu_0, \overline{s})$ (for which we also need $v_n(\mu_0, z)$, $1 \leq n \leq N-1, z \in Z$, cf. (24.7) below) and (ii) a policy π^* which is Bayes-optimal for $BCM_N(\mu_0)$, hence in particular maximizes $\pi \mapsto v_{N\pi}(\mu_0, \overline{s})$ on Δ_N. ♦

From now on we assume that μ_0 belongs to some family $(\hat{\mu}(i), i \in I)$ having a sufficient statistic φ. The data of the associated MDP′ is determined from $(BCM, \hat{\mu}, \varphi)$ as in Chap. 23.

The core of each policy $\pi \in \Delta_N$, when applied for initial state $\overline{s}$, may be described by its ***stopping time,*** i.e. the time at which π prescribes to accept the momentary offer. It is defined by

$$z^N \mapsto \tau_{N\pi}(z^N) := \min\{1 \leq t \leq N-1 : \pi_t(z^t, z_t) = 1\},$$

where $\min \emptyset := N$. (Note that $\tau_{N\pi}$ does not depend on π_0.) It is clear that for all N, $\pi \in \Delta_N$, $\vartheta \in \Theta$ and $z^N \in Z^N$, using $\tau := \tau_{N\pi}(z^N)$, we have

$$G_{N\pi}(\vartheta, \overline{s}, z^N) = \beta^\tau \cdot \psi(\vartheta, z_\tau) - c \cdot \beta \cdot \sigma_\tau(\beta). \tag{24.5}$$

Thus it suffices to find an ***optimal stopping time*** $\boldsymbol{\tau_N^*}$, defined as the stopping time $\tau_{N\pi^*}$ of any Bayes optimal $\pi^* \in \Delta_N$. We sketch a proof of (24.5): Fix z^N. From the definition of τ we get $\pi_\nu(z^\nu, z_\nu) = 0$ for $1 \leq \nu \leq m-1$, and $\pi_m(z^m, z_m) = 1$ if $m < N$. Next, the form of T implies that $\zeta_{\nu\pi}(\overline{s}, z^\nu) = z_\nu$, $1 \leq \nu \leq m$, and $\zeta_{\nu\pi}(\overline{s}, z^\nu) = \hat{s}$ for $m+1 \leq \nu \leq N$. Now the assertion follows from (23.2) and the form of r.

Note that $d_1(i)$ in the next result equals due to (23.7) the expected reward one gets under prior $\hat{\mu}(i)$ when one obtains, paying c, a single offer which is accepted.

Proposition 24.2.2 (The solution of the stopping problem from Example 24.2.1) *Consider the stopping model* $(BCM, \hat{\mu}, \varphi)$. *Assume that* $G(i,z) := \int \hat{\mu}(i, d\vartheta)\, \psi(\vartheta, z)$, $(i,z) \in I \times Z$, *is finite. Fix some* $i_0 \in I$.

(a) The maximal expected reward $d_N(i_0)$ *for prior* $\hat{\mu}(i_0)$ *can be obtained recursively, starting with* $d_1(i) := \int Q'(i,dz)\, G(\varphi(i,z), z) - c$, $i \in I$, *as follows: For* $n \geq 1$, $i \in I$

$$d_{n+1}(i) = \int Q'(i,dz) \cdot \max\{\beta \cdot d_n(\varphi(i,z)),\ G(\varphi(i,z), z)\} - c. \tag{24.6}$$

(b) For prior $\hat{\mu}(i_0)$ *the mapping*

$$\begin{aligned} z^N &\mapsto \tau_N^*(i_0, z^N) \\ &:= \min\{1 \leq t \leq N-1 : \beta \cdot d_{N-t}(i_t(i_0, z^t)) < G(i_t(i_0, z^t), z_t)\}, \end{aligned}$$

where $\min \emptyset := N$, *is an optimal stopping time.*

(c) $d_n(i,c)$ *is increasing in* $n \geq 1$ *and decreasing in c.*

(d) If Θ *and I are structured,* $d_n(i)$ *is increasing in i for* $n \geq 1$ *if* $G(i,z)$ *is increasing in i,* $\varphi(i,z)$ *is increasing in i and in z and if* $\hat{\mu}(i, d\vartheta)$ *and* $Q(\vartheta, dz)$ *are stochastically monotone in the same sense in i and in* ϑ, *respectively.*

Proof We apply Theorem 23.1.17 to the $(BCM, \hat{\mu}, \varphi)$.

(a) Firstly for $i \in I$ and $s \in Z$ we have $r'(i,s,1) = v_0(i,s) = G(i,s) - c$, $r'(i,s,0) = -c$, and $r'(i,\bar{s},a) = r'(i,\hat{s},a) = v_0(i,\bar{s}) = v_0(i,\hat{s}) = 0$, $a \in \{0,1\}$. From Theorem 23.1.17(a) we obtain $v_n(i,\hat{s}) = \beta \cdot \int Q'(i,dz)\, v_{n-1}(\varphi(i,z), \hat{s})$, for $n \geq 1$ and $i \in I$, which implies by induction on $n \geq 0$ that $v_n(i,\hat{s}) = 0$, $n \geq 0$. Now one obtains from (23.16) for $i \in I$ and $n \geq 1$, observing that $d_1(i) = \int Q'(i,dz)\, v_0(\varphi(i,z), z)$,

$$d_{n+1}(i) = v_{n+1}(i,\bar{s})/\beta = \int Q'(i,dz)\, v_n(\varphi(i,z), z)), \tag{24.7}$$

$$v_n(i,z) = \max\{\beta \cdot d_n(i), G(i,z)\} - c, \quad z \in Z.$$

The combination of these two equations yields (24.6).

(b) The proof of (24.7) via (23.16) also shows that for fixed $N \geq 1$ the mapping f_n', defined by $f_n'(i,\bar{s}) := f_n'(i,\hat{s}) := 0$ (= reject) and

$$f_n'(i,z) := 1 \text{ if and only if } \beta \cdot d_n(i) < G(i,z), \quad z \in Z,\ i \in I, \tag{24.8}$$

is the smallest maximizer of Lv_{n-1}, $1 \leq n \leq N$. Then the policy $\pi^* = (\pi_t^*)_0^{N-1} \in \Delta_N$, generated by i_0 and by $(f_n')_N^1$, is Bayes-optimal by Theorem 23.1.17(b). For $1 \leq t \leq N-1$ and $z^t \in Z^t$ we have

$\pi_t^*(z^t, z_t) = f'_{N-t}(i_t(i_0, z^t), z_t)$. Now the assertion follows from (24.8) and the definition $\tau_N^* := \tau_{N\pi^*}$.

(c) and (d) follow from (24.6) by induction on n. For (d) one uses Theorem 18.2.7 and that $Q'(i, dz)$ is stochastically increasing by Lemma 24.1.5. □

Example 24.2.3 (for part (d) of Proposition 24.2.2) If $\Theta = Z = \mathbb{R}^+$, $Q(\vartheta, dz) = \text{Exp}(\vartheta)$ and $\hat{\mu}(k, y) := \Gamma_{y+1,k+1}$, $i = (k, y) \in I := \mathbb{N}_0 \times \mathbb{R}_+$, then $d_n(k, y)$ is decreasing in k and increasing in y. (This conforms to intuition: Consider, for example, antitonicity in k. We know from Table 18.1 that $\hat{\mu}(k, y)$ is stochastically increasing in k, i.e. large values of k mean a tendency to large values of ϑ and hence via stochastic antitonicity of $\text{Exp}(\vartheta, dz)$ to small values of the offers z). For the proof we use the usual ordering on Θ. We know from no. 2 of Tables 23.1 and 23.2 with $\nu := \text{Exp}(1)$ that $\varphi(i, z) := i + (1, z)$ is sufficient for $\hat{\mu}$. Endow I with the NW-ordering, i.e. isotonicity in i means antitonicity in k and isotonicity in y; cf. Chap. 6. Obviously $\varphi(i, z)$ is increasing in i and in z. Moreover, $\hat{\mu}(i, d\vartheta)$ and $\text{Exp}(\vartheta, dz)$ are stochastically decreasing in i and ϑ, respectively. Now the assertion follows from (d). ♦

Example 24.2.4 (A Bayesian asset selling problem with i.i.d. offers and without known initial offer) We consider the following special case of the stopping problem from Example 24.2.1: $\Theta = Z = \mathbb{R}^+$, $Q(\vartheta, dz) := \text{Exp}(1/\vartheta)$; thus the unknown parameter ϑ equals the mean of the probability distribution of the offers. Moreover, $\psi(\vartheta, z) = z$ and $c = 0$, hence $G(i, z)$ from Proposition 24.2.2 equals z. As prior μ_0 we take the inverse Gamma distribution $\Gamma^-_{\alpha,\delta}$, $\alpha \in \mathbb{R}^+$, $\delta \in (1, \infty)$, introduced in Problem 23.3.5. The solution, consisting of $d_N(i_0)$ and of $\tau_N^*(i_0, z^N)$ for $i_0 := (0, 0)$ and the family $\hat{\mu}(x, y, d\vartheta) = \Gamma^-_{\alpha+y,\delta+x}(d\vartheta)$, $(x, y) \in \mathbb{R}^2_+$, is given in Proposition 24.2.7 below. Note that $\mu_0 = \Gamma^-_{\alpha,\delta}$ equals $\hat{\mu}(0, 0)$. First we provide auxiliary results in Lemmas 24.2.5 and 24.2.6. ♦

Lemma 24.2.5 *The following hold:*

(a) $\varphi(x, y, z) := (x + 1, y + z)$, $(x, y) \in \mathbb{R}^2_+$, *is sufficient for* $\hat{\mu}(x, y, d\vartheta) = \Gamma^-_{\alpha+y,\delta+x}(d\vartheta)$.

(b) For $\kappa \in \mathbb{R}^+$ *denote by* $\text{Be2}(\kappa, dz)$ *the special Beta distribution of the second kind on* $\mathbb{R}^+$ *with the density* $z \mapsto \kappa/(1 + z)^{\kappa+1}$. *Then*

(b1) $\text{Be2}(\kappa, dz)$ *has the distribution function*

$$z \mapsto (1 - 1/(1 + z)^{\kappa}) \cdot 1_{[0,\infty)}(z).$$

(b2) For $\kappa > 1$ *and* $a \in \mathbb{R}_+$ *we obtain*

$$\int_a^{\infty} \text{Be2}(\kappa, dz)\, z = \frac{1 + \kappa a}{(\kappa - 1)(1 + a)^{\kappa}}.$$

In particular, $\int_0^{\infty} \text{Be2}(\kappa, dz)\, z = 1/(\kappa - 1)$.

(c) For each measurable $g \geq 0$ on $\mathbb{R}^+$ we have

$$\int Q'(x, y, dz)\, g(z) = \int \text{Be2}(\delta + x, dz)\, g((\alpha + y) \cdot z), \quad (x, y) \in \mathbb{R}_+^2.$$

(d) We have $d_1(x, y) = \frac{\alpha+y}{\delta+x-1}$, and for $n \geq 1$ and $(x, y) \in I$ we have

$$d_{n+1}(x, y) = \int Q'(x, y, dz) \max\{\beta \cdot d_n(x + 1, y + z), z\}. \tag{24.9}$$

Proof

(a) This is stated in Problem 23.3.5.
(b1) follows by a simple computation, while (b2) is contained in tables of integrals.
(c) The proof starts with

$$\begin{aligned}\int Q'(x, y, dz)\, g(z) &= \int \hat{\mu}(x, y, d\vartheta) \int \vartheta^{-1} \cdot e^{-z/\vartheta} \cdot g(z)\, dz \\ &= \int \left[\int \Gamma^-_{\alpha+y,\delta+x}(d\vartheta)\, \vartheta^{-1} \cdot e^{-z/\vartheta}) \right] g(z)\, dz.\end{aligned}$$

Then the substitution $z = (\alpha + y) \cdot u$ in the integral implies the assertion since the integral over the density of $\Gamma^-_{(\alpha+y)(1+u),\delta+x}$ equals one.
(d1) For the proof of the assertion about d_1 we use (c) and Proposition 24.2.2(a) with $g(z) := z$. This implies $d_1(x, y) = (\alpha + y) \cdot \int \text{Be2}(\delta + x, dz)\, z$. Now the assertion follows from (b2) with $\kappa := \delta + x$.
(d2) The assertion about d_{n+1}, $n \geq 1$, follows from (a) and from Proposition 24.2.2(a). □

Lemma 24.2.6 *The following hold:*

(a)

$$d_n(x, y) = (\alpha + y) \cdot c_n(x), \quad n \geq 1, (x, y) \in \mathbb{R}_+^2, \tag{24.10}$$

where $c_1(x) := 1/(\delta + x - 1)$, and for $n \geq 1$

$$c_{n+1}(x) := \int \text{Be2}(\delta + x, dz) \max\{\beta \cdot (1 + z) \cdot c_n(x + 1), z\}. \tag{24.11}$$

(b) $d_N(0, 0) = \alpha \cdot c_N(0)$.
(c) $c_n(x)$ is increasing in n and decreasing in x.

Proof

(a) For $n = 1$ equation (24.10) holds by Lemma 24.2.5(d), while (24.10) for $n > 1$ follows by induction on $n \geq 1$, using Lemma 24.2.5(c) and (24.9).

(b) This is obvious from the definition of $c_N(0)$.

(c1) We show that $c_n(x)$ is increasing in n. Firstly from (24.11) and Lemma 24.2.5 (b2) we get $c_2(x) \geq \int \mathrm{Be2}(\delta + x, dz)\, z = c_1(x)$. Then it follows from (24.11) by induction on $1 \leq n \leq N-1$ that $c_n \leq c_{n+1}$.

(c2) We show that $c_n(x)$ is decreasing in x. This holds trivially for $n = 1$ since $c_1(x) = 1/(\delta + x - 1)$. Assume that it holds for some $n \geq 1$. Put $g(x,z) := \max\{\beta \cdot (1+z) \cdot c_n(x+1), z\}$, which is decreasing in x and increasing in z. Now we get from (24.11), since $\mathrm{Be2}(\delta + x, dz)$ is stochastically decreasing in x by Lemma 24.2.5(b1), that for $x \leq x'$

$$c_{n+1}(x') = \int \mathrm{Be2}(\delta + x', dz)\, g(x', z) \leq \int \mathrm{Be2}(\delta + x', dz)\, g(x, z)$$

$$\leq \int \mathrm{Be2}(\delta + x, dz)\, g(x, z) = c_{n+1}. \qquad \square$$

Proposition 24.2.7 (The algorithmic solution of the special asset selling problem from Example 24.2.4) *Fix $N \geq 2$. Put $e_{n+1} := e_{n+1}(N) := c_{N-n}(n)$, $0 \leq n \leq N-1$ with $c_{N-n}(n)$ from Lemma 24.2.6(a). Thus $d_N(0,0) = \alpha \cdot e_1$ by Lemma 24.2.6(b).*

(a) The sequence $n \mapsto e_n(N)$ is decreasing on $\mathbb{N}_N$.

(b) e_1 and hence $d_N(0,0)$ can be found from the backward recursion

$$e_n := \frac{1}{\delta + n - 2} \cdot \left[(\delta + n - 1) \cdot \beta e_{n+1} + [(1 - \beta e_{n+1})^+]^{\delta + n - 1}\right] \tag{24.12}$$

for $1 \leq n \leq N-1$ and starting with $e_N = 1/(\delta + N - 2)$.

(c) Using $\min \emptyset := N$, the following stopping time is optimal:

$$\tau_N^*((0,0), z^N) := \min\{1 \leq n \leq N-1 : \beta \cdot (\alpha + \sum_{\nu=1}^{n} z_\nu) \cdot e_{n+1} < z_n\}.$$

Proof

(a) It follows from Lemma 24.2.6(c) that for $1 \leq n \leq N-1$

$$e_{n+1}(N) = c_{N-n}(n) \leq c_{N-n}(n-1) \leq c_{N-n+1}(n-1) = e_n(N).$$

(b) Firstly, $e_N = c_1(N-1) = 1/(\delta + N - 2)$ follows from Lemma 24.2.6(a). Next, fix $1 \leq n \leq N-1$ and put $\kappa := \delta + n - 1$ and $\eta := \beta \cdot e_{n+1}(N)$. Then (24.11) implies

$$e_n = c_{N-n+1}(n-1) = \int \mathrm{Be2}(\kappa, dz)\, \max\{(1+z) \cdot \eta, z\}. \tag{24.13}$$

Put $n^*(N) := \max\{1 \leq n \leq N : \beta \cdot e_n \geq 1\}$ with $\max \emptyset := 0$.

Case 1: $n \leq n^* - 1$ with $n^* = n^*(N)$ from (c). Then $\eta \geq 1$ by (a), hence $(1+z) \cdot \eta \geq z$. Now (24.13) and Lemma 24.2.5(b) imply

$$e_n = \eta \cdot \int \text{Be2}(\kappa, dz)\,(1+z) = \eta \cdot \frac{\kappa}{\kappa - 1} = \frac{\delta + n - 1}{\delta + n - 2} \cdot \beta \cdot e_{n+1}.$$

Thus (24.12) holds in case 1.

Case 2: $n \geq n^*$, hence $\eta < 1$ by (a). This also holds if $n^*(N) = 0$. It follows that $\eta \cdot (1+z) < z$ if and only if $z > \lambda := \eta/(1-\eta)$. Thus, using (24.13), we get

$$e_n = \eta \cdot \frac{\kappa}{\kappa - 1} \cdot \text{Be2}(\kappa - 1, (-\infty, \lambda]) + \int_\lambda^\infty \text{Be2}(\kappa, dz)\, z.$$

Now (24.12) in case 2 follows by simple calculations, using Lemma 24.2.5(b).

(c) For $i_0 := (0,0)$ we get from (23.12) and Lemma 24.2.5(a) by induction on $t \geq 1$ that $i_t(i_0, z^t) = (t, \sum_{\nu=1}^t z_\nu)$. Now the form of $\tau_N^*(i_0, z^N)$ follows from Proposition 24.2.2(b), using Lemma 24.2.6(a) and that $G(i,z) = z$ by Example 24.2.4. □

24.3 Bayesian Sequential Statistical Decision Theory

Important applications of the Bayesian Theory are those statistical decision problems (in particular problems of estimation and testing) which are not based on a fixed number of observations, but where sampling may be discontinued at a random time, determined by the observations made so far and bounded above by some number N. This may result in a considerable saving of sampling costs relative to the procedure with a fixed number N of observations.

We consider the case of *N-stage sampling*. This means that we are allowed to take sequentially at most N i.i.d. observations $\eta_1, \eta_2, \ldots, \eta_N$ with values in some measurable space $(Z, \mathfrak{M})$ and having the (partially) unknown probability distribution $Q(\vartheta, dz)$, ϑ in some set Θ. At time $t = 0$ one has to decide, without having available any observation, whether to make at least one observation (*continuation action* $a_0 = \infty$) or to make a *statistical decision* a_0 (i.e. an inference about the unknown parameter ϑ, an estimate in estimation problems, a decision between hypotheses in testing problems) from some nonempty set A_{st}, not containing ∞. (In case $A_{st} = \Theta$ the statistician which takes action $a_0 \in \Theta$ contends a_0 to be the *true* parameter.) If one decides to sample at least once, one must decide at each time $1 \leq t \leq N$ at which one has not yet stopped sampling, on the basis of the observed values of $\eta_1, \eta_2, \ldots, \eta_t$, whether to continue (action $a_t := \infty$) or to stop sampling and take a statistical decision $a_t \in A_{\text{st}}$. At the latest after the observation of η_N a statistical decision must be made. The statistical decision $a \in A_{\text{st}}$ results in a *statistical* cost (e.g. in case $A = \Theta$ a penalty for having not found the correct ϑ) of amount $L(\vartheta, a) \geq 0$.

Each observation costs $\kappa(\vartheta) \geq 0$. Here L, the so-called *loss function,* and κ are measurable functions on $\Theta \times A_{\text{st}}$ and Θ, respectively. We also use a discount factor $\beta > 0$ where $\beta < 1$ may be appropriate when observations are time-consuming. The terminal cost is introduced below. Assuming that the prior μ_0 belongs to a given family $(\hat{\mu}(i), i \in I)$, with sufficient statistics φ_n, independent of n, we arrive at the following N-stage ***(Bayesian) sequential statistical decision problem:***

Minimize the expected sum of N-stage discounted sampling costs and statistical costs, averaged over Θ by means of a prior μ_0.

We include the case $N = 0$ where a statistical decision must be made without having available any observation. The problem is modeled (using a dummy observation η_{N+1}) as the $(N+1)$-stage Bayesian problem $BCM_{N+1}(\mu_0)$, $N \geq 0$, of the following model $(BCM, \hat{\mu}, \varphi)$:

- $S = \{0, 1\}$; $s = 0$ and $s = 1$ means that one has or has not yet stopped sampling, respectively; only the initial state $s_0 = 1$ is relevant.
- $A = A_{\text{st}} + \{\infty\} = D(s)$, $s \in S$.
- T is independent of z, $T(1, \infty) = 1$ and $T(s, a) = 0$, otherwise.
- $Q(\vartheta, dz)$ is arbitrary.
- $c_Z = c$ with $c(\vartheta, 0, a) = 0$ and

$$c(\vartheta, 1, a) = \begin{cases} L(\vartheta, a), & \text{if } a \in A_{\text{st}}, \\ \kappa(\vartheta), & \text{if } a = \infty. \end{cases}$$

- $C_0(\vartheta, 0) = 0$, $C_0(\vartheta, 1) = \infty$; the latter definition ensures that in the N-stage problem only a statistical decision is allowed after the observation of η_N, i.e. at stage $n = 1$.

According to Remark 23.1.3 condition (LUBF) is fulfilled since $c_Z = c \geq 0$, $C_0 \geq 0$. Thus problem $BCM_{N+1}(\hat{\mu}(i))$ and the MDP′, associated to $(BCM, \hat{\mu}, \varphi)$, are well-defined. For $i \in I$ we have $c'(i, 0, a) = 0$ and

$$c'(i, 1, a) = \begin{cases} \lambda(i, a) := \int \hat{\mu}(i, d\vartheta)\, L(\vartheta, a), & \text{if } a \in A_{\text{st}}, \\ \kappa'(i) := \int \hat{\mu}(i, d\vartheta)\, \kappa(\vartheta), & \text{if } a = \infty. \end{cases} \tag{24.14}$$

We assume finiteness of $\lambda(i, a)$ and $\kappa'(i)$ for all (i, a); then c' is finite. Since $J_0 := \{(i, 0) : i \in I\}$ is absorbing, we have $v_n(i, 0) = 0$ for all $n \geq 0$ and $i \in I$, and only actions in the states $(i, 1)$, $i \in I$, matter. Therefore we often write $\pi'_t(i)$ and $f'_n(i)$ instead of $\pi'_t(i, 1)$ and $f'_n(i, 1)$, respectively.

Of course, a major role is played by $\rho_N(i) := v_{N+1}(i, 1) \geq 0, N \geq 0$, the minimal expected sum of N-stage discounted sampling costs and statistical costs, averaged over ϑ by means of $\hat{\mu}(i)$. We call $\rho_N(i)$ the N-stage ***Bayes risk of the prior*** $\hat{\mu}(i)$.

We now show that the sequence $(\rho_n(\cdot))_0^\infty$ determines the solution of all N-stage problems.

Theorem 24.3.1 (Basic Theorem for Bayesian sequential statistical decision problems) *Assume that $\lambda(i,a)$ and $\kappa'(i)$ from (24.14) are finite and that there exists a measurable mapping f^* from I into $A_{\rm st}$ such that $f^*(i)$ is a minimum point of $a \mapsto \lambda(i,a)$ on $A_{\rm st}$ for all i. For each $i_0 \in I$ and $N \geq 1$ the following holds:*

(a) The N-stage Bayes risk $\rho_N(i_0)$ is finite and can be found by the recursion

$$\rho_n(i) = \min\{\rho_0(i), d_n(i)\}, \quad n \geq 1, i \in I, \tag{24.15}$$

starting with $\rho_0(i) = \lambda(i, f^*(i))$. Here

$$d_n(i) := \kappa'(i) + \beta \cdot \int Q'(i,dz)\, \rho_{n-1}(\varphi(i,z)) \geq 0, \quad n \geq 1,\ i \in I. \tag{24.16}$$

(b) The following policy $\pi' := (f_n')_{N+1}^1$ (with arbitrary $f_n'(i,0)$, measurable in i) is optimal for MDP'_{N+1}: $f_1'(i) = f^*(i)$, and for $2 \leq n \leq N+1$

$$f_n'(i) = \begin{cases} f^*(i), & \text{if } \rho_0(i) \leq d_n(i), \\ \infty, & \text{if } \rho_0(i) > d_n(i). \end{cases} \tag{24.17}$$

Thus the policy $\pi^* \in \Delta_{N+1}$, generated from i_0 and π', is optimal for $\mathrm{BCM}_{N+1}(\hat{\mu}(i_0))$.

(c) The optimal policy π^* prescribes to sample, when starting in $s_0 := 1$, up to time

$$N^* = N^*(N, i_0, z^N) := \min\{0 \leq t \leq N-1 : \rho_0(i_t(i_0, z^t)) = \rho_{N+1-t}(i_t(i_0, z^t))\},$$

where $i_0(i_0, z^0) := i_0$ and $\min \emptyset := N$, and then to take the statistical decision $f_{\rm st}(N, i_0, z^N) := f^*(i_{N^*}(i_0, z^{N^*}))$.

(d) The n-stage Bayes risk $\rho_n(i)$ and the numbers $d_n(i)$ are decreasing in n. Moreover, if Θ, Z and I are structured, $\rho_n(i)$ and $d_n(i)$ are increasing in i under the following conditions:

(d1) $i \mapsto \hat{\mu}(i, d\vartheta)$ and $\vartheta \mapsto Q(\vartheta, dz)$ are stochastically increasing,
(d2) $\kappa(\vartheta)$ and $L(\vartheta, a)$ are increasing in ϑ,
(d3) $\varphi(i,z)$ is increasing in i and in z.

Proof

(a1) We prove that in MDP′ the VI holds and that there exists a minimizer at each stage. For this purpose we verify the assumptions (S1) and (S2) of the Structure Theorem 16.1.12 with $\mathfrak{V}$ as the set of measurable $[0,\infty]$-functions v on $I \times \{0,1\}$ with $v(i,0) = 0$ for all i. The operator L' in MDP′ is given by

$L'v(i, 0, a) = 0$ and

$$L'v(i, 1, a) = \begin{cases} \lambda(i, a), & \text{if } a \in A_{\text{st}}, \\ \kappa'(i) + \beta \cdot \int Q'(i, dz)\, v(\varphi(i, z), 1), & \text{if } a = \infty. \end{cases}$$

(S1) is true since for each $v \in \mathfrak{V}$ a minimizer f'_v of L'_v is given as follows: $f'_v(i, 0)$ is arbitrary (measurable in i) and

$$f'_v(i, 1) := \begin{cases} f^*(i), & \text{if } \lambda(i, f^*(i)) \leq \kappa'(i), \\ \infty, & \text{else.} \end{cases}$$

(S2) is true, i.e. $U'v \in \mathfrak{V}$ for all $v \in \mathfrak{V}$. In fact, $U'v(i, 0) = 0$, $U'v(i, 1) = \min\{L'v(i, 1, \infty), L'v(i, 1, f^*(i))\} \in [0, \infty]$, and $i \mapsto U'v(i, 1)$ is measurable by measurability of $i \mapsto L'v(i, 1, \infty)$ and of $i \mapsto L'v(i, 1, f^*(i))$.

(a2) Because of (a1) we can apply the Basic Theorem 23.1.17(a). It tells us that $v_n = V'_n$ for $n \geq 1$ and that the VI holds in MDP$'$. Observing that ρ_{n-1} in (24.16) equals v_n, that $L'v_0(i, 1, \infty) = L'C_0(i, 1, \infty) = \infty$ and that $L'v_{n-1}(i, 1, \infty) = d_n(i)$, the VI has the form

$$\rho_0(i) = v_1(i) = \inf_a \lambda(i, a) = \lambda(i, f^*(i)) < \infty,$$

$$\rho_n(i) = v_{n+1}(i) = \min\{\inf_a \lambda(i, a), d_n(i)\} < \infty, \ n \geq 1.$$

(b) This follows from (a2) and Theorem 23.1.17(b).

(c) This follows from (b), observing that $\rho_0(i) \leq d_n(i)$ if and only if $\rho_0(i) = \rho_n(i)$, as seen from (24.15).

(d) The first assertion follows from (24.16) and (24.15) by induction on $n \geq 1$ for the assertion that $(\rho_n \leq \rho_{n-1}) \wedge (d_{n+1} \leq d_n)$. For the proof of the second assertion increasing here means increasing in i. We use generalized induction Sect. A.6 on $n \geq 1$ for (I_n): *$\rho_{n-1}(i)$ is increasing*, and for (J_n): *$d_n(i)$ is increasing*. Firstly, (I_1) holds. In fact, since $\hat{\mu}(i)$ is stochastically increasing and $L(\vartheta, a)$ is increasing in ϑ, $\lambda(i, a)$ is increasing in i, hence $\rho_0(i) = \inf_{a \in A_{\text{st}}} \lambda(i, a)$ is increasing. Assume (I_n) for some $n \geq 1$. We show (J_n) (which implies (I_{n+1}) by (24.15)). It follows from (d1) and (d2) that $\kappa'(i)$ is increasing. Moreover, $Q'(i, dz)$ is stochastically increasing in i by (d1) and Lemma 24.1.5. From (I_n) and (d3) we know that $\rho_{n-1}(\varphi(i, z))$ is increasing in i and in z. Now we obtain from (24.16) and Theorem 18.2.7 that $d_n(i)$ is increasing. □

Remark 24.3.2 The form of π' in (b) is intuitively clear: If at time $0 \leq t \leq N - 1$ a statistical decision has not yet been made and if i_t is the momentary posterior index, one takes another observation if and only if the minimal expected cost $\rho_0(i_t)$ for taking a statistical decision is now larger than $\rho_{N+1-t}(i_t)$, the Bayes risk for the remaining $N + 1 - t$ periods. It corresponds to intuition that $\rho_n(i)$ is decreasing in n. ◇

We now apply the Basic Theorem 24.3.1 to several special problems.

Example 24.3.3 (Sequential estimation with quadratic loss) We want to estimate the unknown mean $\vartheta \in \Theta := \mathbb{R}$ of a normal distribution $Q(\vartheta, dz) := N(\vartheta, \sigma^2)$, where σ^2 is known. Thus $A_{st} := \Theta = \mathbb{R}$ and $Z = \mathbb{R}$. We use the popular loss function $L(\vartheta, a) := (\vartheta - a)^2$ and a constant positive sampling cost κ_0. As prior μ_0 we use the normal distribution $\mu_0 := N(m, b^2)$ with given $m \in \mathbb{R}$ and $b \in \mathbb{R}^+$. Then μ_0 is contained by the choice $i := (0, 0)$ in the family $\hat{\mu}(x, k) := N(M(x, k), \Sigma^2(k))$, $i = (x, k) \in I := \mathbb{R} \times \mathbb{N}_0$, where

$$M(x, k) := \frac{\sigma^2 m + b^2 x}{\sigma^2 + b^2 k}, \quad \Sigma^2(k) := \frac{\sigma^2 b^2}{\sigma^2 + b^2 k}.$$

By Example 23.1.21(b) the family $\hat{\mu}$ has the sufficient statistic $\varphi(i, z) := i + (z, 1)$.

Fix $(x, k) \in \mathbb{R} \times \mathbb{N}_0$ and let $\tau \sim \hat{\mu}((x, k))$ be a random variable. Then the definition of λ in (24.14) yields

$$\lambda(x, k, a) = E(\tau - a)^2 = (M(x, k) - a)^2 + \Sigma^2(k). \tag{24.18}$$

Theorem 24.3.4 (Solution of the estimation problem with quadratic loss) *Consider the N-stage problem with prior $\mu_0 := N(m, b^2)$ and $0 < \beta \leq 1$. Let π^* be the policy from Theorem 24.3.1(b), generated by π' and $i_0 := (0, 0)$, which is optimal for problem $BCM_{N+1}(\mu_0)$.*

(a) The statistical decision $f_{st}(N, (0, 0), z^N)$ of π^* equals $M(\sum_{\nu=1}^{N^*} z_\nu, N^*)$.
(b) The Bayes risk equals

$$\rho_N(0, 0) = \min_{0 \leq \nu \leq N} \left(\kappa_0 \cdot \sigma_\nu(\beta) + \beta^\nu \cdot \frac{\sigma^2 b^2}{\sigma^2 + b^2 \nu} \right).$$

(c) If $\kappa_0 < \kappa^* := \frac{b^2 \cdot (b^2 + (1-\beta)\sigma^2)}{b^2 + \sigma^2}$ then $N^*(N) = \min\{N, \lceil x_0 \rceil\}$, where

$$x_0 := \sqrt{\sigma^2/\kappa_0 + \lambda^2} - \lambda - \sigma^2/b^2, \quad 2\lambda := 1 - (1 - \beta) \cdot \sigma^2/\kappa_0.$$

(d) If $\kappa_0 \geq \kappa^*$ then it is optimal to take no observation and to use m as estimate for the unknown mean ϑ in $N(\vartheta, \sigma^2)$. Moreover, the Bayes risk is b^2.

Note that (c) and (d) imply that $N^* = N^*(N, (0, 0), z^N)$ is independent of z^N.

Proof

(a) We see from (24.18) that for all (x, k) the function $a \mapsto \lambda(x, k, a)$ has the unique minimum point $f^*(x, k) := M(x, k)$ and that $\rho_0(x, k) = \Sigma^2(k)$. Since $i_t(i_0, z^t) = (\sum_{j=1}^t z_j, t)$ by Example 23.1.21(b), f_{st} has by Theorem 24.3.1(c) the form stated above.

(b) Equation (24.15) and induction on $n \geq 0$ yield

$$\rho_n(x,k) = \min_{0 \leq \nu \leq n} H_\nu(k), \quad n \geq 0, (x,k) \in \mathbb{R} \times \mathbb{N}_0, \tag{24.19}$$

where

$$H_\nu(k) := \kappa_0 \cdot \sigma_\nu(\beta) + \beta^\nu \cdot \Sigma^2(k+\nu), \; \nu \geq 0, k \geq 0. \tag{24.20}$$

(c1) Define $H_\nu(x)$ for all $x \in \mathbb{R}_+$ by (24.20). We show that for $x \in \mathbb{R}_+$ we have $H_0(x) \leq H_1(x)$ if and only if $x \geq x_0$. (Note that x_0 may be negative.) For the proof put $c := b^2/\kappa_0$ and $d := \sigma^2/\kappa_0$, and make the substitution $y := d/c + x$, hence $y > 0$. An elementary calculation shows that

$$H_0(x) \leq H_1(x) \Longleftrightarrow y^2 + 2\lambda \cdot y - d \geq 0 \Longleftrightarrow x \geq x_0.$$

(c2) We show that $\nu \mapsto H_\nu(k)$ increases on $\mathbb{N}_0$ for fixed $k \geq 0$ if and only if $H_0(k) \leq H_1(k)$. Obviously it suffices to show that $\nu \mapsto G_\nu(k) := \beta^{-\nu} \cdot (H_{\nu+1}(k) - H_\nu(k))$ is increasing on $\mathbb{N}_0$. For the proof we note that the function $M(x) := \beta/(d+cx+c) - 1/(d+cx), x \in \mathbb{R}_+$, is increasing since its derivative is positive. Now the assertion follows since $G_\nu(t) = \kappa_0 \cdot [1 + cd \cdot M(t+\nu)]$.

(c3) From Theorem 24.3.1(c) we know that N^* equals the smallest $0 \leq t \leq N-1$ such that $\rho_0(i_t(i_0, z^t)) = \rho_{N+1-t}(i_t(i_0, z^t))$, if such a t exists, and it equals N, otherwise. Now, if $0 \leq t \leq N-1$, then (24.19), (c1) and (c2) imply that $\rho_0(i_t(i_0, z^t)) = \rho_{N+1-t}(i_t(i_0, z^t))$ if and only if $t \geq x_0$. Moreover, a simple calculation shows that $x_0 > 0$ if and only if $\kappa_0 < \kappa^*$. Now we have the following three cases:

Case 1: If $x_0 \leq 0$, i.e. if $\kappa_0 \geq \kappa^*$, then $N^* = 0$.
Case 2: If $0 < x_0 \leq N-1$, then $N^* = \lceil x_0 \rceil$.
Case 3: If $x_0 > N-1$ then $N^* = N$.

(d) This is contained in (b) and in (c3), case 1. □♦

Example 24.3.5 (Sequential testing of two simple hypotheses) We want to test for arbitrary Z and arbitrary $Q(\vartheta, dz)$, $\vartheta \in \Theta := \{\vartheta_0, \vartheta_1\}$, the two simple hypotheses $H_0 : \vartheta = \vartheta_0$ $(a = 0)$ against $H_1 : \vartheta = \vartheta_1$ $(a = 1)$. Thus $A_{st} = \{0, 1\}$. Without loss of generality we relabel the elements of Θ by $\vartheta_0 := 0$, $\vartheta_1 := 1$, hence $\Theta = \{0, 1\}$. We assume that $Q(\vartheta, dz)$ has a density $z \mapsto q_\vartheta(z)$ with respect to some σ-finite measure $\Lambda(dz)$. Without loss of generality we assume that $q_0(z) + q_1(z) > 0$ for all z, since otherwise we may replace everywhere Z by $[q_0 + q_1 > 0]$. We use constant observation costs $\kappa_0 > 0$ and for constants $c_0, c_1 \in \mathbb{R}^+$ the loss function

$$L(\vartheta, a) := \begin{cases} c_0, & \text{if } \vartheta = 1, a = 0, \\ c_1, & \text{if } \vartheta = 0, a = 1, \\ 0, & \text{else.} \end{cases}$$

We admit as μ_0 any probability distribution on Θ, i.e. any binomial probability distribution $\hat{\mu}(i) := \mathrm{Bi}(1, i)$, $i \in I := [0, 1]$. Obviously

$$q_i(z) := i \cdot q_1(z) + (1 - i) \cdot q_0(z), \quad i \in [0, 1], z \in Z,$$

is positive for $0 < i < 1$. An easy computation using Lemma 23.1.6 shows that $\Phi(\hat{\mu}(i), z) = \mathrm{Bi}(1, \varphi(i, z)) = \hat{\mu}(\varphi(i, z))$, where for $z \in Z$

$$\varphi(i, z) := \begin{cases} \frac{i \cdot q_1(z)}{q_i(z)}, & \text{if } 0 < i < 1, \\ i, & \text{if } i = 0, 1. \end{cases}$$

Thus φ is sufficient and $Q'(i, dz)$ has the Λ-density q_i.

Only priors $\hat{\mu}(i_0)$ with $0 < i_0 < 1$ are of interest since, for example, $i_0 = 0$ means that the unknown ϑ equals zero with probability one, so that one can accept the hypothesis H_0 without any testing. On the other hand, if $0 < i_0 < 1$ it may well happen that for some $t \geq 1$ the posterior index i_t equals zero (or one), namely if $q_1(z_t) = 0 < q_0(z_t)$, i.e. if one observes some z_t which can occur only under H_0. (An example of this situation: if $Q(0, dz)$ and $Q(1, dz)$ are the uniform probability distributions on $(0, 2)$ and on $(0, 1)$, respectively, then H_1 can be accepted as soon as one observes some $z_t \in (1, 2)$.)

We have

$$\lambda(i, a) = \begin{cases} i \cdot c_0, & \text{if } a = 0, \\ (1 - i) \cdot c_1, & \text{if } a = 1. \end{cases}$$

From $\rho_0(i) = \lambda(i, 0) \wedge \lambda(i, 1)$ and from (24.15) and (24.16) we obtain for $i \in [0, 1]$

$$\begin{aligned} \rho_0(i) &= \min\{i \cdot c_0, (1 - i) \cdot c_1\}, \quad \rho_n(i) = \min\{\rho_0(i), d_n(i)\} \\ d_n(i) &= \kappa_0 + \beta \cdot \int \Lambda(dz)\, q_i(z) \cdot \rho_{n-1}(\varphi(i, z)), \quad n \geq 1. \end{aligned} \tag{24.21}$$

Proposition 24.3.6 (Solution of the sequential testing problem from Example 24.3.5) *The following hold:*

(a) The Bayes risk $\rho_n(i)$ and also $d_{n+1}(i)$ are decreasing in n on $\mathbb{N}_0$.
(b) $\rho_n(i)$ and $d_{n+1}(i)$ are concave and continuous in $i \in [0, 1]$, and $\rho_n(0) = \rho_n(1) = 0$, $d_{n+1}(0) = d_{n+1}(1) = \kappa_0$, $n \geq 0$.
(c) Fix $N \geq 0$ and let $\pi' := (f'_n)_{N+1}^1$ be the policy from Theorem 24.3.1(b) which is optimal for the MDP'_{N+1}. Then

$$f'_1(i) = \begin{cases} 0 \text{ (i.e. accept} H_0), & \text{if } 0 \leq i \leq c_1/(c_0 + c_1) =: i^*, \\ 1 \text{ (i.e. accept} H_1), & \text{if } i^* < i \leq 1. \end{cases} \tag{24.22}$$

Moreover, for $2 \leq n \leq k + 1$:

(c1) *If* $d_n(i^*) \geq i^* \cdot c_0$, *then* $f_n' = f_1'$.
(c2) *If* $d_n(i^*) < i^* \cdot c_0$, *then the equation* $d_n(i) = i \cdot c_0$ *has on* $(0, i^*)$ *a unique solution* $\underline{i}_n$ *and the equation* $d_n(i) = (1 - i) \cdot c_1$ *has on* $(i^*, 1)$ *a unique solution* $\overline{i}_n$, *and*

$$f_n'(i) = \begin{cases} 0 \text{ (i.e. accept} H_0), & \text{if } 0 \leq i \leq \underline{i}_n, \\ \infty \text{ (i.e. continue sampling)}, & \text{if } \underline{i}_n < i < \overline{i}_n, \\ 1 \text{ (i.e. accept} H_1), & \text{if } \overline{i}_n \leq i \leq 1. \end{cases}$$

Proof

(a) This follows from Theorem 24.3.1(d).
(b1) Fix $i \in \{0, 1\}$ and $n \geq 0$. Then $0 \leq \rho_n(i) \leq \rho_0(i) = 0$, and $d_{n+1}(0) = d_{n+1}(1) = \kappa_0$ follows from (24.21) since $\rho_n(i) = 0$.
(b2) For $n \geq 0$, $i \in [0, 1]$ put

$$g_n(i, z) := q_i(z) \cdot 1_{[q_i > 0]}(z) \cdot \rho_n(i \cdot q_1(z)/q_i(z)).$$

Then we obtain from (24.21) for all i and $n \geq 0$

$$d_{n+1}(i) = \kappa_0 + \beta \cdot \int \Lambda(dz)\, g_n(i, z).$$

(b21) Here and in (b22) concave means concave on $(0, 1)$. We show that concavity of ρ_n for some $n \geq 0$ implies concavity of d_{n+1}. The assertion follows from (24.16) and Theorem 19.1.1, provided $g_n(\cdot, z)$ is concave for each $z \in Z$. Fix $z \in Z$ and drop for the moment z in $q_i(z)$. Fix $i, j, \lambda \in (0, 1)$ and put $\lambda' := 1 - \lambda$. Note that $q_i > 0$, $q_{i'} > 0$ and that $q_{\lambda i + \lambda' j} = \lambda \cdot q_i + \lambda' \cdot q_j$. We have to show that $\lambda \cdot g_n(i, z) + \lambda' \cdot g_n(j, z) \leq g_n(\lambda i + \lambda' j, z)$. Put $\mu := \lambda q_i / (\lambda q_i + \lambda' q_j) \in (0, 1)$. Then our assertion reads

$$\mu \cdot \rho_n(iq_1/q_i) + (1 - \mu) \cdot \rho_n(jq_1/q_j) \leq \rho_n\left(\mu \cdot iq_1/q_i + (1 - \mu) \cdot jq_1/q_j\right),$$

which holds by concavity of ρ_n.
(b22) We show by induction on $n \geq 0$ the assertion (I_n) that ρ_n and d_{n+1} are concave (on $(0, 1)$). Firstly, (I_0) holds since obviously ρ_0 is concave as the minimum of two affine functions (figure!) and since concavity of d_1 then follows from (b21). Now assume that (I_n) holds for some $n \geq 0$. Then concavity of ρ_{n+1} follows since ρ_{n+1} is the minimum of the two concave functions ρ_0 and d_{n+1}. Moreover, concavity of d_{n+2} follows from (b21).
(b31) Continuity of ρ_n in $i = 0, 1$ follows from $0 \leq \rho_n(i) \leq \rho_0(i)$, $i \in [0, 1]$, from (b1) and from continuity of ρ_0 in $i = 0$ and $i = 1$, respectively.
(b32) We show that d_{n+1} is continuous in $i = 0$. (In the same way one shows continuity in $i = 1$.) Firstly we prove continuity of d_1 in $i = 0$. A simple

computation shows that $0 \le g_0(i,z) = q_1 \cdot \rho_0(iq_1/q_i) = \min\{c_0 i q_1, c_1 \cdot (1-i)q_0\}$, $0 < i < 1$, which converges to $g_0(0,z) = 0$ for $i \to 0$. Moreover, $0 \le g_0(i,z) \le (q_0(z) + q_1(z)) \cdot \sup \rho_0 \le (q_0(z) + q_1(z)) \cdot (c_0 + c_1) =: h(z)$ and $\int \Lambda(dz)\, h(z) = 2 \cdot (c_0 + c_1) < \infty$. Thus h is an integrable majorant of g_1. Now (24.16) and the theorem about continuity of integrals with parameters (cf. Appendix B.3.4) proves that d_1 is continuous in $i = 0$. For $n \ge 0$ we obtain from (24.21), since $g_n \ge 0$, and from (a), that $\kappa_0 \le d_{n+1}(i) \le d_1(i) \to \kappa_0$.

(b33) From (b22) we know that ρ_n and d_{n+1} are concave and continuous on $(0,1)$, and they are continuous in $i = 0$ and $i = 1$ by (b31) and (b32). It follows from Appendix D.2.2 that they are also concave on $[0,1]$.

(c0) Firstly, the form of f_1' follows from Theorem 24.3.1(b) since $a \mapsto \lambda(i,a)$ has the (smallest) minimum point $f^*(i) = f_1'(i)$, $i \in I$. Now assume $n \ge 2$. The function $h_n := d_n - \rho_0$ is continuous on $[0,1]$ by (b). We see from (24.17) that $f_n'(i) = f_1'(i)$ if $h_n(i) \ge 0$ and $f_n'(i) = \infty$ if $h_n(i) < 0$. Since d_n is concave by (b) and since $\rho_0(i) = c_0 \cdot i$ for $0 \le i \le i^*$ and $\rho_0(i) = c_1 \cdot (1-i)$ for $i^* \le i \le 1$, h_n is concave on $[0, i^*]$ and on $[i^*, 1]$. Moreover, from (b) we obtain $h_n(0) = h_n(1) = \kappa_0 > 0$.

(c1) By assumption we have $h_n(i^*) \ge 0$. Now concavity of h_n on $[0, i^*]$ yields for $0 \le \alpha \le 1$

$$h_n(\alpha \cdot i^*) = h_n(\alpha \cdot i^* + (1-\alpha) \cdot 0)) \ge \alpha \cdot h_n(i^*) + (1-\alpha) \cdot h_n(0) \ge (1-\alpha) \cdot \kappa_0 \ge 0.$$

Thus $h_n \ge 0$ on $[0, i^*]$, and in the same way one shows that $h_n \ge 0$ on $[i^*, 1]$. This proves the assertion by (c0).

(c2) By assumption we have $h_n(i^*) < 0$. Then $h_n(0) = \kappa_0 > 0$, and continuity of h_n implies that h_n has at least one zero in $(0, i^*)$. Assume that $j_1 < j_2$ are two zeros of h_n in $(0, i^*)$. Then concavity of h_n ensures that each point in $[j_1, j_2]$ is a local, hence (cf. Appendix A.4.12(b)) is also a global maximum point of h_n. However, this contradicts that $h_n(0) > 0$. Thus there exists a unique solution $\underline{i}_n \in (0, i^*)$ of the equation $d_n(i) = i \cdot c_0$ in $[0, i^*]$. Moreover, continuity of h_n ensures that $h_n \ge 0$ on $[0, \underline{i}_n]$ and $h_n < 0$ on $(\underline{i}_n, i^*]$. Now (c0) tells us that $f_n'(i) = 0$ if $0 \le i \le \underline{i}_n$ and $f_n'(i) = \infty$ if $\underline{i}_n < i \le i^*$.
In the same way one shows the existence of $\bar{i}_n$ and that $f_n'(i) = \infty$ if $i^* \le i < \bar{i}_n$ and $f_n'(i) = 0$ if $\bar{i}_n \le i \le 1$. □♦

We now determine in the preceding problem the $(N+1)$-stage policy $\pi^* = (\pi_t)_0^N$, generated by some $i_0 \in (0,1)$ and the policy π' from Proposition 24.3.6(c), and optimal for $BCM_{N+1}(\hat{\mu}(i_0))$ by Theorem 24.3.1(b). Intuitively one expects that π_t has the following form: If at some time t one has not yet made a statistical decision and if the probability that the observed value z^t occurs is much smaller [much larger] under H_1 than under H_0, one will accept H_0 [H_1]; otherwise one will continue sampling. We now confirm this form of π^*. For simplicity we assume

$q_0 > 0, q_1 > 0$, hence $q_i > 0$ for $0 \le i \le 1$. For $t \ge 1$ put

$$L_t(\vartheta, z^t) := \prod_{\nu=1}^{t} q_\vartheta(z_\nu), \quad R_t(z^t) := \frac{L_t(1, z^t)}{L_t(0, z^t)}, \quad (\vartheta, z^t) \in \{0, 1\} \times Z^t.$$

L_t is called the likelihood function at time t, and R_t the likelihood ratio function at time t. Obviously $L_t(\vartheta, z^t)$ is the probability that under ϑ the observation history is z^t. We write $\pi_t(z^t)$ instead of $\pi_t(z^t, 1)$, $1 \le t \le N$, and $\pi_0(z^0)$ instead of $\pi_0(1)$.

Corollary 24.3.7 (Sequential probability ratio test for two simple hypotheses) *Assume $q_0 > 0$ and $q_1 > 0$. With i^* from (24.22) and $\underline{i}_n$, $\overline{i}_n$ from Proposition 24.3.6(c2) define for $i_0 \in (0, 1)$*

$$B(i_0) := \frac{(1 - i_0) \cdot i^*}{i_0 \cdot (1 - i^*)}, \quad \underline{B}_n(i_0) := \frac{(1 - i_0) \cdot \underline{i}_n}{i_0 \cdot (1 - \underline{i}_n)}, \quad \overline{B}_n(i_0) := \frac{(1 - i_0) \cdot \overline{i}_n}{i_0 \cdot (1 - \overline{i}_n)}.$$

Fix $N \ge 0$. The $(N + 1)$-stage policy $\pi^* = (\pi_t)_0^N$ which is optimal for $BCM_{N+1}(\hat{\mu}(i_0))$ is given as follows:

$$\pi_N(z^N) = \begin{cases} 0 \text{ (i.e. accept} H_0), & \text{if } R_N(z^N) \le B(i_0), \\ 1 \text{ (i.e. accept} H_1), & \text{else,} \end{cases}$$

and for $0 \le t \le N - 1$ with d_n from (24.21):

(a) If $d_{N+1-t}(i^*) \ge i^* \cdot c_0$, then

$$\pi_t(z^t) = \begin{cases} 0 \text{ (i.e. accept} H_0), & \text{if } R_t(z^t) \le B(i_0), \\ 1 \text{ (i.e. accept} H_1), & \text{else.} \end{cases}$$

(b) If $d_{N+1-t}(i^*) < i^* \cdot c_0$, then

$$\pi_t(z^t) = \begin{cases} 0 \text{ (i.e. accept} H_0), & \text{if } R_t(z^t) \le \underline{B}_{N+1-t}(i_0), \\ \infty \text{ (continue sampling)}, & \text{if } \underline{B}_{N+1-t}(i_0) < R_t(z^t) < \overline{B}_{N+1-t}(i_0), \\ 1 \text{ (i.e. accept} H_1), & \text{if } \overline{B}_{N+1-t}(i_0) \le R_t(z^t). \end{cases}$$

Proof Firstly, induction on $t \ge 1$ shows that the posterior i_t is generated by i_0 via φ from

$$i_t(i_0, z^t) = \frac{i_0 \cdot R_t(z^t)}{i_0 \cdot R_t(z^t) + 1 - i}, \quad t \ge 1, z^t \in Z^t.$$

Now the assertion follows from Proposition 24.3.6(c), using that $i_t(i_0, z^t) \le x$ if and only if $R_t(z^t) \le [(1 - i_0) \cdot x]/[i_0 \cdot (1 - x)]$ for $x \in (0, 1)$. □

24.4 Problems

Problem 24.4.1 Consider the asset selling problem from Example 24.2.4 and assume that J is bounded and that $\beta \leq 1$. Then:

(a) The functions d_n are bounded, and $\inf J - c \leq d_n \leq (\sup J - c)^+$.
(b) We have $d := \lim_n d_n = H(d)$, where H is the operator from the set of measurable bounded functions on J into itself, defined by

$$Hv(i) := \int Q'(i, dz) \max\{\beta \cdot v(\varphi(i, z)), z\} - c.$$

(c) If $\beta < 1$ $(= 1)$, then d is the unique (the smallest) fixed point of H.

Problem 24.4.2 In the estimation problem from Theorem 24.3.4 with prior $N(m, b^2)$ the MCM$'$ solves the problem for a prior $N(m', (b')^2)$ if and only if $b \leq b'$ and $m'/(b')^2 \geq m/b'^2$.

Problem 24.4.3 Consider the sequential test from Example 24.3.5. Prove that $\underline{i}_n$ is decreasing in n, $\bar{i}_n$ is increasing in n and

$$\frac{\kappa_0}{c_0} \leq \underline{i}_n \leq i^* \leq \bar{i}_n \leq 1 - \frac{\kappa_0}{c_0}.$$

Chapter 25
Bayesian Models with Disturbances

The model of the present chapter generalizes the Bayesian control models (BCMs) from Chap. 23. The BCMs in Chap. 23 were essentially families of CMs, indexed by the unknown parameter $\vartheta \in \Theta$. We now consider a Bayesian MDPD (BMDPD for short), which essentially is a family of MDPDs (cf. Chap. 21), indexed by ϑ, where the transition law has the factorization property below (see Definition 25.1.2).

25.1 The Model BMDPD

Example 25.1.1 (The k-armed Bernoulli bandit) Consider a system consisting of k possible projects $a \in \mathbb{N}_k$. At times $0 \le t \le N-1$ you decide which one of the projects a to promote. The random result of such a decision may be success ($\eta_{t+1} = 1$) or failure ($\eta_{t+1} = 0$) with $\eta_1, \eta_2, \ldots, \eta_N$ being stochastically independent. The probabilities ϑ_a for success when project $a \in \mathbb{N}_k$ is promoted are unknown with a prior $\mu_0(d\vartheta)$ for the vector $\vartheta = (\vartheta_a)_{a=1}^k$. We want to maximize the expected number of successes within N trials with respect to μ_0.

Applications include medical trials and scheduling problems and—most popular—gambling at a bandit with k arms. Therefore we sometimes say *playing at arm a* instead of *promoting project a*. Since the probability distribution for the outcome η_t depends on a, we cannot model our problem by a BCM, but a Bayesian MCM (cf. Example 25.2.5 below). ♦

The motivation in the present section is the same as for BCMs in Chap. 23. Therefore we restrict ourselves mostly to the formal treatment. The main differences between a BCM and our new model BMDPD are as follows:

(i) At time $t \ge 1$ information about ϑ is not only contained in $z^t = (z_1, z_2, \ldots, z_t)$ and s_t as in Chap. 23, but also in $s_0, s_1, \ldots, s_{t-1}$. Therefore we admit as in Supplement 21.3.3 decision rules at time $t \ge 1$ which depend on the whole

K. Hinderer et al., *Dynamic Optimization*, Universitext,
DOI 10.1007/978-3-319-48814-1_25

history $h_t := (s_0, z_1, s_1, z_2, \ldots, z_t, s_t) \in H_t := S \times (Z \times S)^t$, while π_0 depends on $h_0 := s_0 \in H_0 := S$. The set of policies $\pi = (\pi_t)_0^{N-1}$ with decision rules of this type, called ***history-dependent policies***, is denoted by Δ_N.

(ii) Let $\mathbb{P}(\mathfrak{T})$ be the set of all probability distributions on $\mathfrak{T}$. Again we assume that one can embed the prior $\mu_0 \in \mathbb{P}(\mathfrak{T})$ as $\mu = \hat{\mu}(i_0)$ into a family $\hat{\mu} = (\hat{\mu}(i, d\vartheta), i \in I)$ of probability distributions on $\mathfrak{T}$ having a sufficient statistic φ defined formally below. But now φ may depend not only on (i, z) but also on (s, a).

(iii) We use a model $(BMDPD, \hat{\mu}, \varphi)$ defining for each $i_0 \in I$ problems $BMDPD_N(\hat{\mu}(i_0))$, $N \geq 1$, which are solved jointly by means of an associated MDPD′ with states $(i, s) \in I \times S$ (cf. Definition 25.1.4).

(iv) We admit a variable discount factor $\tilde{\beta}(z)$, $z \in Z$.

Definition 25.1.2 A ***Bayesian Markovian decision process with disturbances*** (***BMDPD*** for short) (or ***Bayesian model***) is a tuple $(\Theta, S, A, D, Z, K, \tilde{r}, V_0, \tilde{\beta})$ of the following kind:

- The parameter space $\Theta \neq \emptyset$ is endowed with a σ-algebra $\mathfrak{T}$.
- The sets S, A, D and Z with σ-algebras $\mathfrak{S}$, $\mathfrak{A}$, $\mathfrak{D}$ and $\mathfrak{Z}$, respectively, and $\tilde{\beta}: Z \to \mathbb{R}_+$ have the same meaning as in the MDPDs of Sect. 21.1.
- The transition law K is a transition probability from $\Theta \times D$ into $Z \times S$ and has a factorization of the form

$$K(\vartheta, s, a, d(z, s')) = Q(\vartheta, s, a, dz) \otimes K_S(s, a, z, ds'). \tag{25.1}$$

  Observe that K_S is independent of ϑ.
- The measurable function $\tilde{r}$ on $\Theta \times D \times Z \times S$ is the one-stage reward function. The integrals

$$r(\vartheta, s, a) := K\tilde{r}(\vartheta, s, a) := \int K(\vartheta, s, a, d(z, s'))\, \tilde{r}(\vartheta, s, a, z, s')$$

  are assumed to exist and to be finite for all (ϑ, s, a).
- The measurable function $(\vartheta, s) \mapsto V_0(\vartheta, s)$ on $\Theta \times S$ is the terminal reward function.

Keeping the parameter ϑ in the data of a BMDPD fixed, we obtain a ***model MDPD(ϑ)*** with transition law as in (25.1) and disturbance transition law $Q(\vartheta, s, a, dz)$. Note that although $\tilde{\beta}$ does not depend on ϑ, the expected discount factor $\beta(\vartheta, s, a) := \int Q(\vartheta, s, a, dz)\, \tilde{\beta}(z)$ may depend on ϑ.

Put $y_t := (z_t, s_t) \in Y := Z \times S$, $t \geq 1$, and $y_0 := s_0 \in S$. Then $h_t = (s_0, y^t) = (h_{t-1}, y_t)$, $t \geq 1$. For the sake of a unified notation, unless stated otherwise, we drop z_0, z^0, y^0 where these symbols occur. For an initial state s_0 the N-stage reward under policy $\pi \in \Delta_N$ is defined on the sample space $\Theta \times Y^N$ with points (ϑ, y^N) and with σ-algebra $\mathfrak{F} := \mathfrak{T} \otimes \left(\bigotimes (\mathfrak{Z} \otimes \mathfrak{S})^N\right)$ as the random variable

$$(\vartheta, y^N) \mapsto R_{N\pi}(\vartheta, s_0, y^N) := \tilde{r}(\vartheta, s_0, \pi_0(s_0), y_1)$$
$$+ \sum_{t=1}^{N-1} \prod_1^t \tilde{\beta}(z_\nu) \cdot \tilde{r}(\vartheta, s_t, \pi_t(h_t), y_{t+1}) + \prod_1^N \tilde{\beta}(z_\nu) \cdot V_0(\vartheta, s_N).$$

For $t \geq 0$ the transition from (z_t, s_t) under $a_t := \pi_t(h_t)$ to (z_{t+1}, s_{t+1}) is governed by

$$K_{\pi_t}(\vartheta, h_t, d(z_{t+1}, s_{t+1})) := K(\vartheta, s_t, \pi_t(h_t), d(z_{t+1}, s_{t+1})).$$

Therefore we construct for $\mu \in \mathbb{P}(\mathfrak{T})$ and $\pi \in \Delta_N$ the probability distribution $\mathbb{P}^\mu_{N\pi}(s_0, d\,(\vartheta, y^N))$ on $\mathfrak{F}$ as

$$\mathbb{P}^\mu_{N\pi}(s_0, B) := \int \mu(d\vartheta) \int K_{\pi_0}(\vartheta, h_0, d(z_1, s_1)) \int K_{\pi_1}(\vartheta, h_1, d(z_2, s_2))$$
$$\cdots \int K_{\pi_{N-1}}(\vartheta, h_{N-1}, d(z_N, s_N))\, 1_B(\vartheta, y^N)$$
$$=: \int \mu(d\vartheta)\, \mathbb{P}_{N\pi}(\vartheta, s_0, B), \quad B \in \mathfrak{F}.$$

In order to ensure the existence of $v_{N\pi}(\mu, s)$, defined in (25.2) below, we make throughout this chapter the following assumption (LUBF). It implies that there exists a function b on S which either is a lower bounding function or an upper bounding function for each MDPD(ϑ).

Assumption (LUBF): There exists a measurable function $b\colon S \to \mathbb{R}_+$ and a constant $\delta \in \mathbb{R}_+$ such that for all $(\vartheta, s, a) \in \Theta \times D$ we have $K(\tilde{\beta} b)(\vartheta, s, a) \leq \delta \cdot b(s)$ and either

$$r(\vartheta, s, a) \leq \delta \cdot b(s), \quad V_0(\vartheta, s) \leq \delta \cdot b(s) \quad \text{or}$$
$$r(\vartheta, s, a) \geq -\delta \cdot b(s), \quad V_0(\vartheta, s) \geq -\delta \cdot b(s).$$

Under (LUBF) there exists for $N \geq 1$, $\pi \in \Delta_N$

$$v_{N\pi}(\mu, s) := \int \mathbb{P}^\mu_N(s, d(\vartheta, y^N))\, R_{N\pi}(\vartheta, s, y^N), \quad \mu \in \mathbb{P}(\mathfrak{T}), s \in S, \tag{25.2}$$

the so-called N-stage Bayes reward of policy π with respect to prior μ and initial state s. The proof is essentially the same as for the BMDPs in Chap. 23.

A Bayesian model BMDPD and some prior μ_0 leads to the following

Definition 25.1.3 A ***Bayesian problem BMDPD***$_N$ (μ_0), $N \geq 1$, is defined as follows:

- Find for each initial state $s \in S$ the maximal N-stage **Bayes reward** with respect to μ_0 within the set of all history-dependent policies, i.e.

$$v_N(\mu_0, s) := \sup\{v_{N\pi}(\mu_0, s) : \pi \in \Delta_N\}.$$

- Find a policy $\pi^* \in \Delta_N$ which is ***Bayes-optimal*** for $BMDPD_N(\mu_0)$ in the sense that it maximizes $\pi \mapsto v_{N\pi}(\mu_0, s)$ on Δ_N for all $s \in S$.

Similarly as in Chap. 23 we solve this problem by means of a so-called associated MDPD′. In it the state s_t is augmented by a component μ_t which consists of the information about ϑ contained in (z_t, s_t). As in Chap. 23 the information is formulated as a posterior probability distribution of the random variable τ with realizations ϑ. We need some preparations. For fixed (μ, s, a) consider on the sample space $(\Theta \times Z, \mathfrak{T} \otimes \mathfrak{Z})$ the probability distribution

$$P^{\mu}(s, a, d(\vartheta, z)) := \int \mu(d\vartheta) \int Q(\vartheta, s, a, dz) = \mu(d\vartheta) \otimes Q(\vartheta, s, a, dz).$$

Let τ and η_1 be the coordinate random variables on $\Theta \times Z$. The definition of a version $\Phi: \mathbb{P}(\mathfrak{T}) \times D \times Z \to \mathbb{P}(\mathfrak{T})$ of the ***Bayes operator*** generated by the disturbance transition law Q as

$$(\mu, s, a, z) \mapsto \Phi(\mu, s, a, z, d\vartheta) := P^{\mu}_{\tau|\eta_1}(s, a, z, d\vartheta)$$

is essentially the same as for BCMs in Chap. 23. However, now Φ may also depend on $(s, a) \in D$. Note that a mapping Φ from $\mathbb{P}(\mathfrak{T}) \times D \times Z$ into $\mathbb{P}(\mathfrak{T})$ such that $\Phi(\mu, s, a, z, B)$, $B \in \mathfrak{T}$, is measurable in (s, a, z) is a version of the Bayes operator if and only if for $\mu \in \mathbb{P}(\mathfrak{T})$, $(s, a) \in D$ and all measurable $v \geq 0$ on $\Theta \times Z$ we have

$$\int P^{\mu}_{\tau,\eta_1}(s, a, d(\vartheta, z))\, v(\vartheta, z) = \int P^{\mu}_{\eta_1}(s, a, dz) \int \Phi(\mu, s, a, z, d\vartheta')\, v(\vartheta', z),$$

i.e. if and only if

$$\begin{aligned} &\int \mu(d\vartheta) \int Q(\vartheta, s, a, dz)\, v(\vartheta, z) \qquad (25.3)\\ &= \int \mu(d\vartheta) \int Q(\vartheta, s, a, dz) \int \Phi(\mu, s, a, z, d\vartheta')\, v(\vartheta', z). \end{aligned}$$

As in Chap. 23 we only treat the case where there exists a sufficient statistic in the following sense.

Definition 25.1.4 Let Q be the disturbance transition law of a BMDPD and let $(i, B) \mapsto \hat{\mu}(i, B)$ be a transition probability from the measurable space $(I, \mathfrak{I})$ into Θ. A measurable mapping φ from $I \times D \times Z$ into I is called a ***sufficient statistic*** (with respect to Q) for the family $\hat{\mu}$ of probability distributions $\hat{\mu}(i) := \hat{\mu}(i, d\vartheta)$, $i \in I$, if

$$\Phi(\hat{\mu}(i), s, a, z, d\vartheta) = \hat{\mu}(\varphi(i, s, a, z), d\vartheta) \quad \text{for } (i, s, a, z) \in I \times D \times Z. \tag{25.4}$$

(If φ is independent of some of its arguments, these are dropped in the notation tacitly.) Then we also call ***(BMDPD, $\hat{\mu}$, φ)*** a ***Bayesian model.*** It determines a family of *Bayesian problems* $BMDPD_N(\hat{\mu}(i))$, $N \geq 1$, $i \in I$.

The next definition is very similar to the corresponding one in Chap. 16.

Definition 25.1.5 (The *associated MDPD′*) Consider some Bayesian model $(BMDPD, \hat{\mu}, \varphi) = (\Theta, S, A, D, Z, Q, K_S, \tilde{r}, V_0, \tilde{\beta}, \hat{\mu}, \varphi)$ which satisfies assumption (LUBF). We associate to it the following Markovian decision process with disturbances MDPD′ $(I \times S, A, D', Z, K', r', V_0', \tilde{\beta})$:

- The states are $(i, s) \in I \times S$, with σ-algebra $\mathfrak{I} \otimes \mathfrak{S}$.
- $D'(i, s) := D(s)$ is independent of i.
- K' is determined as follows: for all measurable $v \geq 0$ on $Z \times I \times S$

$$\begin{aligned} K'v(i, s, a) &:= \int K'(i, s, a, d(z, i', s'))\, v(z, i', s')) \\ &= \int Q'(i, s, a, dz) \int K_S(s, a, z, ds')\, v(z, \varphi(i, s, a, z), s'). \end{aligned} \tag{25.5}$$

  Here Q' is the disturbance transition law Q of the MDPD′, given by

$$Q'(i, s, a, dz) := \int \hat{\mu}(i, d\vartheta)\, Q(\vartheta, s, a, dz). \tag{25.6}$$

- The one-stage reward does not depend on (z, i', s') and equals

$$r'(i, s, a) := \int \hat{\mu}(i, d\vartheta) \int Q(\vartheta, s, a, dz) \int K_S(s, a, z, ds')\, \tilde{r}(\vartheta, s, a, z, s'), \tag{25.7}$$

  which is assumed to be finite.
- The terminal reward equals $V_0'(i, s) := \int \hat{\mu}(i, d\vartheta)\, V_0(\vartheta, s)$.

The form of K' in the above definition is intuitively clear: If we take action a in the *augmented* state (i, s), and if the disturbance z occurs, then the next index i' is $\varphi(i, s, a, z)$. The functions r' and V_0' exist by (LUBF) and they are measurable.

In MDPD′ the sets $\mathbb{F}'$, $(\mathbb{F}')^N$ and Δ_N' denote the set of decision rules, the set of N-stage Markovian policies and the set of N-stage history-dependent policies, respectively. Note that $\Delta_1' = \mathbb{F}'$ and that $\pi_t' = \pi_t'(h_t')$, where $h_t' =$

$(i_0, s_0, (z_1, i_1, s_1), \ldots, (z_t, i_t, s_t))$. Moreover, $(i, s) \mapsto b(s)$ is an upper or lower bounding function for MDPD′, so that $V'_{N\pi'}$, defined as in (21.6), exists for $\pi' \in \Delta'_N \supset (\mathbb{F}')^N$. By $V'_N(i, s)$ we denote the maximal expected N-stage reward within the set $(\mathbb{F}')^N$ of Markovian policies in $MDPD'_N$. For $f' \in \mathbb{F}'$ put $\varphi_{f'}(i, s, z) := \varphi(i, s, f'(i, s), z)$ for all (i, s, z).

Part (b) of our basic result Theorem 25.1.10 below shows that in maximizing $\pi \mapsto v_{N\pi}(i_0, s)$ on Δ_N for all s one can in general restrict oneself to those history-dependent policies π where $\pi_t(h_t)$ depends on the history h_t only via the momentary posterior index, defined below. Policies π with this property are generated by Markovian policies π' in MDP′ as follows.

Definition 25.1.6 (Posterior index functions and generation of history-dependent policies in *BMDPD*$_N(\hat{\mu}(i))$ from Markovian policies for $MDPD'_N$**)**

- The ***posterior index functions*** $i_{t\pi'}(i, \cdot): H_{t-1} \times Z \to I$, $1 \le t \le N$, generated by $i \in I$ and $\pi' \in (\mathbb{F}')^N$, are defined recursively by

$$i_{t+1,\pi'}(i, h_t, z_{t+1}) := \varphi_{\pi'_t}(i_{t\pi'}(i, h_{t-1}, z_t), s_t, z_{t+1}), \quad i_{0\pi'}(i, \cdot) :\equiv i.$$

- The ***history-dependent policy in a BMDPD*** $\pi^i = (\pi^i_t)_0^{N-1} \in \Delta_N$, generated by $i \in I$ and a Markovian policy $\pi' = (\pi'_t)_0^{N-1} \in (\mathbb{F}')^N$ is defined by $\pi^i_0(s_0) := \pi'_0(i, s_0)$, $s_0 \in S$, and for $1 \le t \le N-1$ by

$$\pi^i_t(h_t) := \pi'_t(i_{t\pi'}(i, h_{t-1}, z_t), s_t), \quad h_t \in H_t.$$

From now on we write $v_{N\pi}(i, s)$ and $v_N(i, s)$ instead of $v_{N\pi}(\hat{\mu}(i), s)$ and $v_N(\hat{\mu}(i), s)$, respectively. We call $v_N: I \times S \to \overline{\mathbb{R}}$ the N-stage ***Bayes value function.***

The next proposition is the crucial point in the proof of our Basic Theorem 25.1.10. For Proposition 25.1.9(a) note that in BMDPD each history-dependent policy $(\pi_t)_0^{N-1} \in \Delta_N$ can be regarded as a history-dependent policy $(\pi'_t)_0^{N-1}$ in MDPD′, where π'_t does not depend on $i_0, i_1, \ldots, i_t$. Thus $\Delta_N \subset \Delta'_N$, in particular $\mathbb{F} \subset \mathbb{F}'$.

For the proof of the proposition we need the following two auxiliary lemmas which we mention without proof. The definition of the (s, z)-section $\sigma_{(s,z)} \in \Delta_{N-1}$ is given in (21.16). In the first lemma we need the factorization (25.1) of K.

Lemma 25.1.7 (Reward iteration for the functions $v_{n\pi}$) *In each $(BMDPD, \hat{\mu}, \varphi)$ we have $v_{1f} = V'_{1f}$, $f \in \mathbb{F}$. Moreover, for $N \ge 2$, $(f, \sigma) \in \Delta_N$ and $x = (i, s) \in I \times S$ we have*

$$v_{N,(f,\sigma)}(x) = r'(x, f(s)) + \int K'(x, f(s), d(z, x'))\, \tilde{\beta}(z) \cdot v_{N-1,\sigma_{(s,z)}}(x').$$

Lemma 25.1.8 *Assume $N \geq 2$, $i \in I, f \in \mathbb{F}$ and $(s,z) \in S \times Z$. If $\pi = (f,\sigma) \in \Delta_N$ is generated by i and $(f',\sigma') \in \mathbb{F}' \times (\mathbb{F}')^{N-1}$, then the (s,z)-section $\sigma_{(s,z)}$ of σ is generated by $i' := \varphi(i,s,f'(i,s),z)$ and by σ'.*

Proposition 25.1.9 *In each $(BMDPD, \hat{\mu}, \varphi)$ the following hold for all $N \geq 1$:*

(a) $v_{N\pi} = V'_{N\pi}$ for all $\pi \in \Delta_N$.
(b) For $i \in I$ the following holds: If $\pi^i \in \Delta_N$ is generated by i and by $\pi' \in (\mathbb{F}')^N$ then $v_{N\pi^i}(i,s) = V'_{N\pi'}(i,s)$ for all s.

Proof

(a) We verify the assertion (I_N) by induction on $N \geq 1$. Firstly, (I_1) is the first assertion in Lemma 25.1.7. Assume (I_N) for some $N \geq 1$. Then for $f \in \mathbb{F}$, $(f,\sigma) \in \Delta_{N+1}$ and $(s,z) \in S \times Z$ we have $\sigma_{(s,z)} \in \Delta_N$. Thus $v_{N,\sigma_{(s,z)}} = V'_{N,\sigma_{(s,z)}}$ by (I_N), hence Lemma 25.1.7 implies for $x = (i,s) \in I \times S$

$$v_{N+1,(f,\sigma)}(i,s) = r'_f(i,s) + \int K'_f(i,s,d(z',x'))\,\tilde{\beta}(z) \cdot V'_{N,\sigma_{(s,z)}}(x').$$

By Sect. 21.3 applied to MDPD$'$, the latter expression equals $V'_{N+1,(f,\sigma)}(i,s)$.

(b) We prove the assertion (I_N) by induction on $N \geq 1$. Fix $i \in I$. For the proof of (I_1) fix $s \in S$ and observe that $f^i = f'(i,\cdot) \in \Delta_1$. Now (a) for $N = 1$ and the RI in MDPD$'$ imply for $j \in I$

$$v_{1f^i}(j,s) = V'_{1f^i}(j,s) = r'(j,s,f'(i,s)) + \int K'(j,s,f'(i,s),d(z,x'))\,\tilde{\beta}(z) \cdot V'_0(x').$$

In particular, for $j := i$ we obtain, using again the RI in MDPD$'$,

$$v_{1f^i}(i,s) = r'_{f'}(i,s) + \int K'_{f'}(i,s,d(z,x'))\,\tilde{\beta}(z) \cdot V'_0(x') = V_{1f'}(i,s),$$

which verifies (I_1). Now assume (I_N) for some $N \geq 1$. Fix s, $\pi' = (f',\sigma') \in \mathbb{F}' \times (\mathbb{F}')^N$ and $\pi^i = (f^i,\sigma^i) \in \Delta_N$ with $f^i \in \mathbb{F}$. We know from Lemma 25.1.8 that $\sigma^i_{(s,z)}$, which belongs to Δ_N, is generated by $i'(z) := \varphi_{f'}(i,s,z)$, $z \in Z$, and σ'. Thus $v_{N,\sigma^i_{(s,z)}}(i'(z),s') = V'_{N,\sigma'}(i'(z),s')$, $s' \in S$, by (I_N). Finally the RI Lemma 21.1.9 for MDPD$'$ and for the Markovian policy π', (25.5) and Lemma 25.1.7 yield, observing that $f'(i,s) = f^i(s)$ and $r'_{f'}(i,s) = r'_{f^i}(i,s)$,

$$\begin{aligned}
&V'_{N+1,\pi'}(i,s)\\
&= r'_{f'}(i,s) + \int Q'_{f'}(i,s,dz) \int K_S(s,f'(i,s),z,ds')\,\tilde{\beta}(z) \cdot V'_{N\sigma'}(i'(z),s')\\
&= r'_{f^i}(i,s) + \int Q'_{f^i}(i,s,dz) \int K_S(s,f^i(s),z,ds')\,\tilde{\beta}(z) \cdot v_{N\sigma^i(s,z)}(i'(z),s')\\
&= v_{N+1,\pi^i}(i,s).
\end{aligned}$$

□

Theorem 25.1.10 (Basic Theorem for the Bayesian Model) *If the VI holds in the associated MDPD′ then for $N \geq 1$:*

(a) The Bayes value function v_N equals the value function V'_N in MDPD′, hence can be found recursively by value iteration in MDPD′:

$$v_n(i,s) = \sup_{a \in D(s)} [r'(i,s,a) + K'(\tilde{\beta} \cdot v_{n-1})(i,s,a)], \tag{25.8}$$

$$1 \leq n \leq N,\ (i,s) \in I \times S,\ v_0(i,s) := \int \hat{\mu}(i, d\vartheta)\, V_0(\vartheta, s).$$

(b) Assume that at each stage $1 \leq n \leq N$ a maximizer $f'_n \in \mathbb{F}'$ exists in MDPD′. Then for each $i \in I$ the history-dependent policy $\pi^i \in \Delta_N$, generated by i and by $\pi' := (f'_n)_N^1$, is Bayes-optimal for $BMDPD_N(\hat{\mu}(i))$.

Note that by (25.5) the term $K'(\tilde{\beta} \cdot v_{n-1})(i,s,a)$ in (25.8) equals

$$\int Q'(i,s,a,dz)\, \tilde{\beta}(z) \cdot \int K_S(s,a,z,ds')\, v_{n-1}(\varphi(i,s,a,z), s').$$

Proof

(a) From Proposition 25.1.9(a) and Sect. 21.3, applied to MDPD′ we get, since $\Delta_n \subset \Delta'_n$,

$$\begin{aligned} v_n &= \sup\{v_{n\sigma} : \sigma \in \Delta_n\} = \sup\{V'_{n\sigma} : \sigma \in \Delta_n\} \\ &\leq \sup\{V'_{n\pi} : \pi \in \Delta'_n\} = V'_n. \end{aligned}$$

On the other hand, for given $\pi' \in (\mathbb{F}')^n$ and $(i,s) \in I \times S$ let π^i be generated by (i, π'). Then Proposition 25.1.9(b) yields $v_n(i,s) \geq v_{n\pi^i}(i,s) = V'_{n\pi'}(i,s)$. Since π' is arbitrary, we get $v_n(i,s) \geq V'_n(i,s)$.

(b) From Theorem 21.1.10(b), applied to MDPD′, we know that π' is optimal for MDPD′. Now we obtain from Proposition 25.1.9(b) and from (a) for $(i,s) \in I \times S$

$$V'_{N\pi'}(i,s) = v_{N\pi^i}(i,s) \leq v_N(i,s) = V'_N(i,s) = V'_{N\pi'}(i,s). \qquad \square$$

To solve a problem $BMDPD_N(\mu_0)$ for given μ_0 by means of Theorem 25.1.10, one must find a family $\hat{\mu}$ containing μ_0 and having a sufficient statistic φ. Note that while the VI in MDP′ gives the Bayes value functions for all problems $BCM(\hat{\mu}(i))$, $i \in I$, in general no policy $\pi \in \Delta_N$ is Bayes-optimal for *all* problems $BMDPD(\hat{\mu}(i))$, $i \in I$.

The implementation (e.g. on a computer) of π^i in Theorem 25.1.10(b) does not require the explicit form of the policies π^i. It suffices for given initial state $s_0 \in S$ to observe the states (i_t, s_t) which arise at time t during the control, and to take the actions $a_t^i := f'_{N-t}(i_t, s_t)$, $0 \leq t \leq N-1$.

Of course, there exists a minimization version, an infinite-stage version and a non-stationary version of Theorem 25.1.10.

25.2 The Models BMCM and BMDP

We now consider the associated MDPD′ for the two important special cases of a Bayesian MCM and a Bayesian MDP. These are models where, roughly speaking, each model MDPD(ϑ), $\vartheta \in \Theta$, is an $MCM(\vartheta)$ (with $T(s,a,z)$ independent of ϑ) and a special $MCM(\vartheta)$, respectively. More precisely, we use the following definitions.

Definition 25.2.1

(a) A ***Bayesian MCM*** (***BMCM*** for short) $(\Theta, S, A, D, Z, Q, T, r_Z, V_0, \beta)$ is a BMDPD $(\Theta, S, A, D, Z, Q, K_S, \tilde{r}, V_0, \tilde{\beta})$ such that:

- $K_S(s,a,z,ds') = \delta_{T(s,a,z)}(ds')$ for some measurable mapping $T: D \to S$.
- $\tilde{r}(\vartheta, s, a, z, s') = r_Z(\vartheta, s, a, z)$, independent of s'.
- $\tilde{\beta} \equiv \beta$.

Note that each BMCM satisfies the factorization assumption (25.1) since $T(s,a,z)$ does not depend on ϑ. Obviously a BCM, treated in Chap. 23, equals that special BMCM where Q does not depend on (s,a).

(b) A ***Bayesian MDP*** (***BMDP*** for short) $(\Theta, S, A, D, P, r_S, V_0, \beta)$ is a BMCM $(\Theta, S, A, D, Z, Q, T, r_Z, V_0, \beta)$ such that:

- $Z = S$, $z = s'$.
- $Q(\vartheta, s, a, ds') = P(\vartheta, s, a, ds')$.
- $T(s,a,s') = s'$.
- $r_Z(\vartheta, s, a, s') = r_S(\vartheta, s, a, s')$.

Lemma 25.2.2 (The associated MDPD′ for special BMDPDs) *The following holds:*

(a) If the model (BMDPD, $\hat{\mu}, \varphi$) has a constant discount factor β the associated MDPD′ is equivalent to the MDP′ $(I \times S, A, D', P', r', V_0', \beta)$, called the associated MDP′, where P' is determined by the property that for all measurable $v \geq 0$ on $I \times S$

$$P'v(i,s,a) = \int Q'(i,s,a,dz) \int K_S(s,a,z,ds')\, v(\varphi(i,s,a,z), s').$$

Here Q' is defined by (25.6).

(b) *In the MDP′ associated to a Bayesian MCM we have*

$$r'(i,s,a) = \int \hat{\mu}(i,d\vartheta) \int Q(\vartheta,s,a,dz)\, r_Z(\vartheta,s,a,z),$$

and for all measurable $v \geq 0$ *on* $I \times S$ *we have*

$$P'v(i,s,a) = \int Q'(i,s,a,dz)\, v(\varphi(i,s,a,z), T(s,a,z)).$$

(c) *In the MDP′ associated to a Bayesian MDP we have*

$$r'(i,s,a) = \int \hat{\mu}(i,d\vartheta) \int P(\vartheta,s,a,ds')\, r_S(\vartheta,s,a,s'),$$

$$Q'(i,s,a,ds') = \int \hat{\mu}(i,d\vartheta)\, P(\vartheta,s,a,ds'),$$

and for all measurable $v \geq 0$ *on* $I \times S$ *we have*

$$P'v(i,s,a) = \int Q'(i,s,a,ds')\, v(\varphi(i,s,a,s'),s').$$

Proof

(a) Firstly, (MA1) holds in MDPD′ and in MDP′ since $(i,s) \mapsto b(s)$ is a lower or an upper bounding function in both models. Now the assertion holds since the operator L' is the same in both models.
(b) Follows from (a) since $K_S(s,a,z,ds') = \delta_{T(s,a,z)}(ds')$.
(c) Follows from (b) and (25.7) with $Z = S$, $T(s,a,s') = s'$, $Q = P$. □

Lemma 25.2.3 (Computation of the Bayes operator) *Assume that* Q *has a transition density* $q\colon \Theta \times D \times Z \to \mathbb{R}_+$ *with respect to some* σ*-finite measure on* $\mathfrak{Z}$. *For* $\mu \in \mathbb{P}(\mathfrak{T})$ *and* $(s,a,z) \in D \times Z$ *put* $q_\mu(s,a,z) := \int \mu(d\vartheta')\, q(\vartheta',s,a,z)$ *and let* $\Phi(\mu,s,a,z,d\vartheta)$ *be the probability distribution on* $\mathfrak{T}$ *with* μ*-density* $\vartheta \mapsto g_\mu(\vartheta,s,a,z)$, *where* $g_\mu(\vartheta,s,a,z) := q(\vartheta,s,a,z)/q_\mu(s,a,z)$ *if* $0 < q_\mu(s,a,z) < \infty$, *and* $g_\mu(\vartheta,s,a,z) := 1$, *otherwise. Then* Φ *is a version of the Bayes operator.*

Proof Keep $(s,a) \in D$ fixed and apply Lemma 23.1.6. □

Proposition 25.2.4 (Neyman criterion for sufficiency in BMDPs)

(a) *Assume that* Q *has a transition density* q *with respect to a* σ*-finite measure. Assume further that there exist measurable functions* $l \geq 0$ *on* $\Theta \times I$ *and* $H > 0$ *on* $I \times D \times Z$ *and a measurable mapping* φ *from* $I \times D \times Z$ *into* I *such that for all* ϑ, i, s, a, z

$$l(\vartheta,i) \cdot q(\vartheta,s,a,z) = l(\vartheta,\varphi(i,s,a,z)) \cdot H(i,s,a,z).$$

Then for each σ-finite measure κ on $\mathfrak{T}$ such that $l_\kappa(i) := \int \kappa(d\vartheta)\, l(\vartheta, i)$ is finite and positive for all i, φ is sufficient for the family

$$\hat{\mu}_\nu(i, d\vartheta) := l(\vartheta, i) \cdot \kappa(d\vartheta)/\, l_\kappa(i), \quad i \in I.$$

(b) The conclusion of (a) remains true for BMDPs, provided the assumptions of (a) hold with Q, z and Z replaced by the transition law P of the BMDP, by s' and by S, respectively.

Proof

(a) Keep $(s, a) \in D$ fixed and apply Proposition 23.1.20.
(b) This is obvious since a BMDP with transition law P is a BMDPD with $Z := S$ and $Q := P$. □

Example 25.2.5 (Continuation of the k-armed Bernoulli bandit problem from Example 25.1.1) We model the problem by a BMCM as follows: $\Theta = [0, 1]^k$ with elements $\vartheta = (\vartheta_a)_{a=1}^k$; S consists of a dummy variable s_0, which will usually be dropped; $A = D(s_0) = \mathbb{N}_k$; $Z = \{0, 1\}$; $Q(\vartheta, a, dz) = \mathrm{Bi}(1, \vartheta_a)$; $T(s_0, a, z) = s_0$; $r_Z(\vartheta, a, z) = z$, hence $r(\vartheta, a) = \int Q(\vartheta, a, dz)\, z = \vartheta_a$; $V_0 \equiv 0$; $\beta > 0$. Our goal is to maximize the expectation of $R_{N\pi}(\vartheta, s_0, ((z_t, s_t))_{t=1}^N) := \sum_1^N \eta_t$.

We now verify the intuitively obvious fact that the optimal decision at time t is based on the momentary number of successes and failures at the k arms. Put $I := \mathbb{N}_0^{2k}$ with elements $i = (m_j)_1^{2k}$, and $l(\vartheta, i) := \prod_{j=1}^k \vartheta_j^{m_j}(1 - \vartheta_j)^{m_{k+j}}$, $i \in I$. (It turns out from φ introduced below that the components m_j and m_{k+j} of the index $i_{t\pi'}(i, \eta^t)$ at time $t \geq 1$ are the number of successes and of failures, respectively, up to time t at arm j, observed via the disturbance vector z^t.) We exclude the very special case where the prior μ_0 is concentrated on the boundary $\partial[0, 1]^k$ of Θ. Then we can embed $\mu_0 =: \nu$ by the choice $i = \mathbf{0} := (0, 0, \ldots, 0)$ into the family

$$\hat{\mu}(i, d\vartheta) := \frac{l(\vartheta, i)\nu(d\vartheta)}{l_\nu(i)}, \quad i \in \mathbb{N}_0^{2k}, \tag{25.9}$$

since $l_\nu(i)$ is finite and positive for all i. In fact, we obviously have $l_\nu(i) \leq 1$. Moreover $l_\nu(i) > 0$ holds since ν is not concentrated on $\partial[0, 1]^k$ and since for fixed $\vartheta \in \Theta$ the number $l(\vartheta, i)$ is positive for all $i \in I$ if and only if $0 < \vartheta_j < 1$ for all j if and only if $\vartheta \notin \partial[0, 1]^k$.

Let e_j, $1 \leq j \leq 2k$, denote the j-th unit vector in $\mathbb{R}^{2k}$. It follows easily from Proposition 25.2.4(a) that $\varphi(i, a, z) := i + e_a z + e_{k+a}(1 - z)$ is sufficient for $\hat{\mu}$. From Lemma 25.2.2(a) we get for the MDP′ associated to the $(BMCM, \hat{\mu}, \varphi)$: $P'(i, a, di') = \mathrm{Bi}(1, p_a(i))$ for all (i, a), where

$$p_a(i) = \int \hat{\mu}(i, d\vartheta)\, \vartheta_a = \int \nu(d\vartheta)\, \vartheta_a \cdot l(\vartheta, i)/l_\nu(i),$$

and $r'(i,a) = p_a(i)$. Since A is finite, the VI holds in MDP′, and the maximal expected discounted number of successes in N trials equals $v_N(\mathbf{0})$. By the Basic Theorem 25.1.10 and by Lemma 25.2.2(a) we obtain $v_N(\mathbf{0})$ by the recursion

$$v_n(i) = \max_{1\le a\le k} \big[p_a(i) + \beta \cdot p_a(i) \cdot v_{n-1}(i+e_a) \\ + \beta \cdot (1 - p_a(i)) \cdot v_{n-1}(i+e_{k+a})\big],\ n \ge 1, i \in \mathbb{N}_0^{2k}, \qquad (25.10)$$

where $v_0 \equiv 0$. Note that the computation of $v_N(\mathbf{0})$ via (25.10) requires v_n, $1 \le n \le N-1$, only on the simplex $\{(m_j)_1^{2k} \in \mathbb{N}_0^{2k} : m_j \le N - n \text{ for } 1 \le j \le 2k\}$. ♦

The VI above solves the problem not only for the prior $\mu_0 := \nu$ but also for each prior $\mu_0 := \hat{\mu}(i)$, $i \in \mathbb{N}_0^{2k}$, where ν is fixed. One can interpret $\hat{\mu}(i)$ with $i = (m_j)_1^{2k}$ as information that some games before time $t = 0$ showed at arm j m_j successes and m_{k+j} failures. One will expect that in general $v_n(i)$ will increase when all numbers m_j of previous successes increase and all numbers m_{k+j} of previous failures decrease. We then say shortly that $v_n(i,a)$ is increasing in the number of successes and decreasing in the number of failures. This means that $i \mapsto v_n(i)$ increases with respect to the ordering $\preceq$, defined as the product of $\le^k$ and the inverse of $\le^k$. An answer to this problem is given in the next result.

Proposition 25.2.6 (Isotonicity of $v_n(i)$ in i in the bandit problem from Example 25.2.5) *Assume that $\hat{\mu}$ has the form (25.9) with arbitrary ν, not concentrated on $\partial[0,1]^k$. Then:*

(a) If $r'(i,a)$ is increasing in the number of successes and decreasing in the number of failures, then the same holds for $v_n(i)$ for all $n \ge 1$.
(b) The assumption in (a) holds if the arms are stochastically independent in the sense that ν is the product of k probability distributions on $(0,1)$.

Proof

(a) We prove the assertion (I_n) by induction on $n \ge 0$. Firstly, (I_0) holds trivially. Assume (I_n) for some $n \ge 0$. For $(i,a) \in \mathbb{N}_0^{2k} \times \mathbb{N}_k$ put

$$\begin{aligned} w(i,a) &:= r'(i,a) \cdot v_n(i+e_a) + (1 - r'(i,a)) \cdot v_n(i+e_{k+a}) \\ &= \int Q'(i,a,dz)\, v_n(i + e_a z + e_{k+a}(1-z)). \end{aligned}$$

Firstly, $i \mapsto Q'(i,a,dz)$ is stochastically increasing. In fact, since $r'(i,a)$ is increasing in i,

$$Q'(i,a,\{0\}) = \int \hat{\mu}(i,d\vartheta)\, Q(\vartheta,a,\{0\}) = \int \hat{\mu}(i,d\vartheta)\,(1-\vartheta_a) = 1 - r'(i,a)$$

is decreasing in i. Next, $v_n(i + e_a z + e_{k+a}(1-z))$ is increasing in i and in z, since $v_n(\cdot)$ is increasing by (I_n). Now Lemma 18.2.6 shows that $w(i,a)$ is increasing in

i. Observing that $r'(i,a)$ is increasing in i, (I_{n+1}) follows from Proposition 6.3.7 and (25.10).

(b) Let ν be the product of probability distributions $\nu_j(d\vartheta_j)$, $1 \le j \le k$, on $(0,1)$. Fix $a \in \mathbb{N}_k$. A simple computation shows that $r'((m_j)_1^{2k}, a) = \int \vartheta_a P_{m_a,m_{k+a}}(d\vartheta_a)$ where $P_{m_a,m_{k+a}}(d\vartheta_a) \propto \vartheta_a^{m_a}(1-\vartheta_a)^{m_{k+a}} \cdot \nu_a(d\vartheta_a)$. From Lemma 18.3.8(b) with $x := \vartheta_a$, $\alpha := m_a$ and $\gamma := m_{a+k}$ we see that the family $P_{m_a,m_{k+a}}(d\vartheta_a)$ of probability distributions on $(0,1)$ is stochastically increasing in m_a and stochastically decreasing in m_{k+a}. Now the assertion follows since $\vartheta_a \mapsto \vartheta_a$ is increasing. □

Definition 25.2.7 A BMCM is called **binary** if

- $\Theta = \Theta_1 \times \Theta_2$.
- S consists of a dummy variable s_0, which mostly is dropped.
- $D(s_0) = A := \{1,2\}$.
- $Q(\vartheta, a, dz)$ depends on ϑ only via ϑ_a, i.e. $Q(\vartheta,a,dz) = Q_a(\vartheta_a, dz)$ for a transition probability Q_a from Θ_a into Z.
- $r_Z(\vartheta, a, z) = r_a(\vartheta_a, z)$ for measurable functions r_a on $\Theta_a \times Z$.

As $T(s_0,a,z) = s_0$ and $K_{12}(s_0,a,z,\cdot) = \delta_{s_0}$ the terms T and K_{12} can be dropped.

Lemma 25.2.8 *Consider a binary BMCM and a family $\hat{\mu}$ satisfying the following conditions:*

(i) there exists a transition probability $(i,B) \mapsto \hat{\mu}(i,d\vartheta)$ from $I = I_1 \times I_2$ into Θ,

(ii) for all $i \in I$ we have $\hat{\mu}(i_1,i_2,d(\vartheta_1,\vartheta_2)) = \hat{\mu}_1(i_1,d\vartheta_1) \times \hat{\mu}_2(i_2,d\vartheta_2)$, for two families $\hat{\mu}_a$ of probability distributions on Θ_a. $a = 1,2$. In particular, for all $i \in I$ the random vector $\tau =: (\tau_1,\tau_2)$ has stochastically independent components with respect to $\hat{\mu}(i)$,

(iii) each family $\hat{\mu}_a$, $a = 1,2$, has a sufficient statistic φ_a with respect to Q_a, $a = 1$, 2. Note that φ_a maps $I_a \times Z$ into I_a.

Then $\hat{\mu}$ has with respect to Q the sufficient statistic

$$\varphi(i_1,i_2,1,z) = (\varphi_1(i_1,z), i_2), \quad \varphi(i_1,i_2,2,z) = (i_1, \varphi_2(i_2,z)).$$

We then say that (BMCM, $\hat{\mu}$, φ) has independent experiments.

Proof We give a proof for $a = 1$; the proof for $a = 2$ is similar. By (25.4) it suffices to show that (25.3) holds with μ, $Q(\vartheta,dz)$ and $\Phi(\mu,z,d\vartheta)$ replaced by $\hat{\mu}(i)$, $Q(\vartheta,s,a,dz)$ and $\hat{\mu}(\varphi(i,s,a,z),d\vartheta)$, respectively, i.e. that the following holds for all measurable $v \ge 0$ on $\Theta \times Z$

$$\int \hat{\mu}_2(i_2,d\vartheta_2) \int \hat{\mu}_1(i_1,d\vartheta_1) \int Q_1(\vartheta_1,dz)\, v(\vartheta_1,\vartheta_2,z) \tag{25.11}$$

$$= \int \hat{\mu}_1(i_1,d\vartheta_1) \int Q_1(\vartheta_1,dz) \int \hat{\mu}_1(\varphi_1(i_1,z),d\vartheta_1') \int \hat{\mu}_2(i_1,d\vartheta_2')\, v(\vartheta_1',\vartheta_2',z).$$

Since $\hat{\mu}_1$ has the sufficient statistic φ_1 with respect to Q_1, we know from (25.3) with Θ, Q and $\Phi(\mu, z, d\vartheta')$ replaced by Θ_1, Q_1 and $\hat{\mu}_1(\varphi_1(i_1, z), d\vartheta_1')$, respectively, that for all $w \geq 0$ on $\Theta_1 \times Z$ we have

$$\int \hat{\mu}_1(i_1, d\vartheta_1) \int Q_1(\vartheta_1, dz)\, w(\vartheta_1, z)$$
$$= \int \hat{\mu}_1(i_1, d\vartheta_1) \int Q_1(\vartheta_1, dz) \int \hat{\mu}_1(\varphi_1(i_1, z), d\vartheta_1')\, w(\vartheta_1', z).$$

Inserting this with $w(\vartheta_1, z) := v(\vartheta_1, \vartheta_2, z)$ for fixed ϑ_2 into the left-hand side of (25.11) yields the right-hand side of (25.11). □

The model MDP′ from Lemma 25.2.8 has the following properties:

(a) For $i = (i_1, i_2) \in I$ and $a = 1, 2$:

(a1) $r'(i, a) = r_a'(i_a) := \int \hat{\mu}_a(i_a, d\vartheta_a) \int Q_a(\vartheta_a, dz)\, r_a(\vartheta_a, z)$.
(a2) $Q'(i, a, dz) = Q_a'(i_a, dz) := \int \hat{\mu}_a(i_a, d\vartheta_a)\, Q_a(\vartheta_a, dz)$.

(b) By finiteness of S the VI in MDP′ holds and yields for $n \geq 1$, $i = (i_1, i_2) \in I$, using $v_0(i) = \int \hat{\mu}(i, d\vartheta)\, V_0(\vartheta)$,

$$v_n(i) = \max_{a \in \{1,2\}} \left[r_a'(i_a) + \beta \int Q_a'(i_a, dz)\, v_{n-1}(\varphi(i, a, z)) \right].$$

(c) There exists $v^* := \lim_n v_n$, and each maximizer of LV^* is asymptotically optimal for MDP′.

Our goal is to exhibit the structure of the smallest asymptotically optimal decision rule for MDP′. We need several preparations.

Example 25.2.9 (A binary stopped MDP with stopping reward L) Consider a binary stopped MDP^L $(S, A, D, P, r, V_0, \beta)$ with the absorbing set $\{\overline{s}\}$, essential state space $J := S - \{\overline{s}\}$ and action space $A = \{0, 1\} = D(s)$, $s \in S$, where $a = 0$ denotes stopping; cf. Example 16.3.10. Therefore we have

$$P(\overline{s}, a, J) = r(\overline{s}, a) = V_0(\overline{s}) = 0,\ a = 0, 1,$$
$$P(j, 0, \{\overline{s}\}) = P(j, 1, J) = 1,\ j \in J.$$

Thus under $a = 1$ the system stays in J. We assume:

(i) $r_1 := r(\cdot, 1)$ is bounded,
(ii) there is a stopping reward $L \in \mathbb{R}$, i.e. $r(s, 0) = L$, $s \in J$,
(iii) $V_0 \equiv 0$ and $\beta < 1$.

The value function for the n-stage problem is denoted by V_n^L. Finiteness of A and boundedness of r_1 imply by Lemma 16.1.18 the validity of the VI. We get for $n \geq 1$ that $V_n(\overline{s}) = 0$ and, using $V_0^L := V_0 \equiv 0$,

$$V_n^L(s) = \max\left\{L, r_1(s) + \beta \cdot PV_{n-1}^L(s,1)\right\}, \; s \in J. \tag{25.12}$$

Since $\beta < 1$ there exists the limit value function V^L by Theorem 20.1.3. Note that $V^L(\overline{s}) = 0$. ♦

Proposition 25.2.10 (Properties of the large horizon solution of the stopped MDP^L from Example 25.2.9) *For all $L \in \mathbb{R}$ the following holds:*

(a) $V^L(s)$ is increasing and convex in L for $s \in J$.
(b) $V^L(s) - L$ is decreasing in L for $s \in J$.
(c) For $s \in J$ there exists $\kappa(s) := \min\{x \in \mathbb{R} : V^x(s) = x\}$, and

$$V^L(s) = \begin{cases} L, & \text{if } \kappa(s) \leq L, \\ r_1(s) + \beta \cdot PV^L(s,1) \in (L, \kappa(s)), & \text{otherwise.} \end{cases}$$

(d) The decision rule $s \mapsto f^L(s) := 1_{(L,\infty)}(\kappa(s))$, $s \in J$, (which says to stop in state s if and only if $\kappa(s) \leq L$) is asymptotically optimal for MDP^L.

Proof

(a) One easily shows by induction on $n \geq 0$, using the VI (25.12), that $V_n^L(s)$ (and hence $V^L(s)$) is increasing and convex in L for $s \in J$. For convexity one uses that by Theorem 19.1.1 and boundedness of $s \mapsto V_{n-1}^L(s)$ the convexity of V_{n-1}^L in L implies convexity of $PV_n^L(\cdot, 1)$ in L and that the maximum of two convex functions is convex.
(b) This follows similarly as (a) since $\beta < 1$ and since for $s \in J$

$$V_n^L(s) - L = \max\left\{0, r_1(s) + \beta \cdot P(V_{n-1}^L - L)(s,1) - (1-\beta) \cdot L\right\}.$$

(c) It follows by induction on $n \geq 1$ that $V_n^L \equiv L$ on J (hence $V^L \equiv L$ on J) if $L \geq \max_{s \in J} r_1^+(s)/(1-\beta)$. Next, if $L < \alpha := -\max_{s \in J} r_1(s)^-/(1-\beta)$, it follows by induction on $n \geq 1$ that $V_n^L \geq \alpha > L$, hence $V^L \geq \alpha > L$. Finally, the assertion follows since $L \mapsto V_n^L(s) - L$ is decreasing by (b) and convex by (a), hence also continuous on $\mathbb{R}$.
(d) Put $LV^L(s,0) := L$, $LV^L(s,1) := r_1(s) + \beta \cdot PV^L(s,1)$. Obviously $f^L(s)$ maximizes $a \mapsto LV^L(s,a)$ for $s \in J$. Now the assertion follows from Theorem 20.1.3. □

Example 25.2.11 (A two-action stopped MDP) Consider the following stopped MDP^L := $(S, A, D, P, r, V_0, \beta)$ with the absorbing set $\{\overline{s}\}$, essential state space $J := S - \{\overline{s}\}$ and action space $A = \{0, 1, 2\} = D(s)$, $s \in S$, where $a = 0$ denotes stopping; cf. Example 16.3.10. Thus we have

$$P(\overline{s}, a, J) = r(\overline{s}, a) = V_0(\overline{s}) = 0,\ a = 0, 1,$$
$$P(j, 0, \{\overline{s}\}) = P(j, 1, J) = 1,\ j \in J.$$

We assume that the system has two components c_1 and c_2 which change only under the actions $a = 1$ and $a = 2$, respectively. More precisely, we assume:

(i) $J = J_1 \times J_2$. The states in J are denoted by $j = (j_1, j_2)$,
(ii) for transition probabilities P_a from J_a into J_a, $a = 1, 2$ we have

$$P(j, 1, dj') = P_1(j_1, dj_1') \times \delta_{j_2}(dj_2'), P(j, 2, dj') = \delta_{j_1}(dj_1') \times P_2(j_2, dj_2'),$$

(iii) $r(j, a) = r_a(j_a), j \in J, a = 1, 2$, for some bounded function r_a on J_a,
(iv) there is a stopping reward $L \in \mathbb{R}$, i.e. $r(j, 0) = L, j \in J$,
(v) $V_0 \equiv 0, \beta < 1$.

The value function for the n-stage problem is denoted by V_n^L. Finiteness of A and boundedness of r imply by Proposition 16.1.21 the validity of the VI. We get for $n \geq 1$ that $V_n(\overline{s}) = 0$ and, using $V_0^L :\equiv 0$,

$$V_n^L(j) = \max\left\{L, L_1 V_{n-1}^L(j), L_2 V_{n-1}^L(j)\right\},\ j \in J.$$

Here $L_1 v$, $L_2 v$ are defined for measurable $v \geq 0$ on J by

$$L_1 v(j) := r_1(j_1) + \beta \cdot \int P_1(j_1, dj_1')\, v(j_1', j_2),$$
$$L_2 v(j) := r_2(j_2) + \beta \cdot \int P_2(j_2, dj_2')\, v(j_1, j_2'),\ j = (j_1, j_2) \in J.$$

Since $\beta < 1$ we know from Theorem 20.1.3 that there exists $V^L := \lim_{n\to\infty} V_n^L$. Note that $V^L(\overline{s}) = 0$. ◆

The two-action stopped MDP^L defines two binary stopped models MDP$^L(a) := (S_a, A_a, D_a, \tilde{P}_a, \tilde{r}_a, V_0, \beta)$, $a = 1, 2$ as follows. (Here a is an index, not to be confused with the notation for actions.) The absorbing set is $\{\overline{s}\}$, $J := S - \{\overline{s}\}$ is the essential state space and $A = \{0, 1\} = D(s)$, $s \in S$, where $a = 0$ denotes stopping. Thus we have for $a = 1, 2$

$$\tilde{P}_a(\overline{s}, b, J) = \tilde{r}_a(\overline{s}, b) = V_0(\overline{s}) = 0,\ b = 0, 1,$$
$$\tilde{P}_a(j_a, 0, \{\overline{s}\}) = \tilde{P}_a(j_a, 1, J) = 1,\ j \in J.$$

In addition we define for $a = 1, 2$:

(i) $\tilde{P}_a(j_a, 0, \{\overline{s}\} := 1$ and $\tilde{P}_a(j_a, 1, dj'_a) := P_a(j_a, 1, dj'_a)$ with P_a from the MDP^L,
(ii) $\tilde{r}_a(j_a, 1) := r_a(j_a)$ with r_a from the MDP^L,
(iii) $L \in \mathbb{R}$ is a stopping reward, i.e. $r_a(j_a, 0) := L, j_a \in J_a$,
(iv) $V_0 :\equiv 0, \beta < 1$.

Since A_a is finite, r_a bounded and $\beta < 1$ we know from Proposition 16.1.21 and Example 16.1.20 that there exists the limit value function V_a^L in $\text{MDP}^L(a)$.

We now show how the large horizon solution of the two-action stopped MDP^L from Example 25.2.11 can be found using the limit value functions of $\text{MDP}^L(1)$ and $\text{MDP}^L(2)$.

Proposition 25.2.12 (Large horizon solution of the two-action stopped MDP^L from Example 25.2.11) *For all $L \in \mathbb{R}$ the following holds:*

(a) $V^L(j)$ is increasing and convex in L for $j \in J$.
(b) $V^L(j) - L$ is decreasing in L for $j \in J$.
(c) For $a = 1, 2$ and $j_a \in J_a$ there exists

$$\kappa_a(j_a) := \min\{x \in \mathbb{R} : V_a^x(j_a) = x\}.$$

(d) Let v^* be the limit value function in the (unstopped) MDP', and put $\kappa(j) := \max\{\kappa_1(j_1), \kappa_2(j_2)\}$, $j = (j_1, j_2) \in J$. Then for $j \in J$ we have

$$V^L(j) = \begin{cases} L, & \text{if } L \geq \kappa(j), \\ L_a V^L(j), & \text{if } L < \kappa(j) = \kappa_a(j_a),\ a = 1, 2, \\ v^*(j), & \text{if } L \leq -\max\{\|r_1\|, \|r_2\|\}/(1-\beta). \end{cases}$$

(e) The decision rule

$$j \mapsto f^L(j) := \begin{cases} 0, & \text{if } L \geq \kappa(j), \\ 1, & \text{if } L < \kappa(j) = \kappa_1(j_1), \\ 2, & \text{if } L < \kappa(j) = \kappa_2(i_2) < \kappa_1(j_1), \end{cases}$$

is asymptotically optimal for MDP^L.

Proof

(a) and (b) follow exactly as in the proof of Example 25.2.11 for the binary problem.
(c) Is obvious.
(d) Put $v(j, L) := V^L(j)$ and $v_a(j_a, L) := V_a^L(j_a)$. The derivative of $v(j, y)$ with respect to y exists and it holds (for a proof, see Tsitsiklis 1986)

$$\frac{\partial}{\partial y} v(j, y) = \frac{\partial}{\partial y} v_1(j_1, y) \cdot \frac{\partial}{\partial y} v_2(j_2, y).$$

By integration we obtain for $L \leq L_0$

$$V^L(j) = V^{L_0}(j) - \int_L^{L_0} \frac{\partial}{\partial y} v_1(j_1, y) \cdot \frac{\partial}{\partial y} v_2(j_2, y)\, dy.$$

For $L_0 \geq \max\{\|r_1\|, \|r_2\|\}/(1-\beta)$ we have $V^{L_0}(j) = L_0$.

Thus we define the function

$$w^L(j) := L_0 - \int_L^{L_0} \frac{\partial}{\partial y} v_1(j_1, y) \cdot \frac{\partial}{\partial y} v_2(j_2, y)\, dy, \quad j \in J,\ L \leq L_0.$$

By partial integration we obtain with the notation $Q_a(j, y) := \frac{\partial}{\partial y} v_b(j_b, y)$ for $b \neq a$

$$w^L(j) = L_0 - Q_a(j, y) v_a(j_a, y)\Big|_L^{L_0} + \int_L^{L_0} v_a(j_a, y)\, dQ_a(j, y).$$

According to Proposition 25.2.10 $Q_a(j, y)$ has the following properties:

$$0 \leq Q_a(j, y) \leq 1, \;\; j \mapsto Q_a(j, y) \text{ is increasing in } j, \;\; Q_a(j, y) = 1 \text{ for } y \geq \kappa(j).$$

For $L_0 \to \infty$ we obtain

$$w^L(j) = Q_a(j, L) v_a(j_a, L) + \int_L^{\infty} v_a(j_a, y)\, dQ_a(j, y).$$

Let us introduce the following functions

$$\Delta_a^L := w^L(j) - L_a w^L(j),$$
$$\delta^L(j_a) := v_a(j_a, L) - r_a(j_a) - \beta \int P_a(j_a, dj_a')\, v_a(j_a', L).$$

It can be shown that they are related as follows

$$\Delta_a^L(j_a) = \delta^L(j_a) \cdot Q_a(j, L) + \int_L^{\infty} \delta^y(j_a)\, dQ_a(j, y).$$

From Proposition 25.2.10 we conclude that $\delta^L(j_a) \geq 0$ and that for $L < \kappa_a(j_a)$ we have $\delta^L(j_a) = 0$. From these relations we obtain

$$L \geq \kappa(j) \quad \Longrightarrow \quad w^L(j) = L,$$
$$L < \kappa_a(j_a) = \kappa(j) \quad \Longrightarrow \quad \Delta_a^L(j) = 0.$$

Further $\Delta_a^L(j) \geq 0$ and $w^L(j) \geq L$ imply

$$L \geq \kappa_a(j_a) = \kappa(j) \quad \Longrightarrow \quad L \geq \max_a L_a w^L(j),$$

$$L < \kappa_a(j_a) = \kappa(j) \quad \Longrightarrow \quad L \leq L_a w^L(j).$$

From these conclusions we derive

$$w^L(j) = \max\{L, L_1 w^L(j), L_2 w^L(j)\}, \quad j \in J,$$

i.e. w^L is a fixed point of the optimality equation. Since the fixed point is unique we get $w^L(j) = V^L(j)$ which implies (d).

(e) From (d) it follows directly that f^L is the smallest maximizer of LV^L. Hence f^L is asymptotically optimal for MDP^L. □

Now we are ready for the main result of this subsection.

Theorem 25.2.13 (Asymptotic optimality of the so-called Gittins index) *In the MDP′ associated to the binary model (BMCM, $\hat{\mu}, \varphi$) with independent experiments from Definition 25.2.7, the decision rule*

$$(i_1, i_2) \mapsto f^*(i_1, i_2) := \begin{cases} 2, & \text{if } \kappa_2(i_2) \geq \kappa_1(i_1), \\ 1, & \text{otherwise}, \end{cases}$$

is asymptotically optimal.

Proof Letting $L \to 0$ the statement follows from Proposition 25.2.12(d). □

The main results can be extended easily from the binary to an m-action BMCM.

25.3 Problems

Problem 25.3.1 Assume that $P_{X,Y}^{\mu} := \mu \otimes P_{Y|X}$, and that $P_{Y|X}$ has a transition density function g with respect to some σ-finite measure $\Lambda(dy)$. Then:

(a) g is a $\mu \times \Lambda$-density of $P_{X,Y}^{\mu}$ and $y \mapsto g_\mu(y) := \int \mu(dx) \cdot g(x, y)$ is a Λ-density of Y.

(b) The Bayes operator Φ generated by $P_{Y|X}$ exists, and $\Phi(\mu, y, dx) \propto g(x, y)\, \mu(dx)$ for all y with $0 < g_\mu(y) < \infty$.

Problem 25.3.2 Assume that φ is sufficient for $(\hat{\mu}(i), i \in I)$. If $\emptyset \neq I_1 \subset I$ is invariant under φ (i.e. if $\varphi(i, s, a, z, s') \in I_1$ for $i \in I_1$ and all s, a, z, s'), then $\varphi|I_1$ is sufficient for $\hat{\mu}|I_1$.

Problem 25.3.3 Assume that K has a transition density function $g\colon \Theta \times D \times Z \times S \to \mathbb{R}_+$ with respect to some σ-finite measure on $\mathfrak{Z} \otimes \mathfrak{S}$. Put $g_\mu(s, a, z, s') := \int \mu(d\vartheta') \cdot g(\vartheta', s, a, z, s')$. Then Φ exists, and for all μ, s, a, z, s' such that $0 < g_\mu(s, a, z, s') < \infty$ we have

$$\Phi(\mu, s, a, z, s', d\vartheta) = \frac{g(\vartheta, s, a, z, s') \cdot \mu(d\vartheta)}{g_\mu(s, a, z, s')}.$$

Chapter 26
Partially Observable Models

POMs or models with incomplete information are more general than the Bayesian models from the previous chapter. We introduce the associated MDPD$'$ and state the basic theorem for POMs. A classical maintenance problem illustrates the solution technique.

26.1 The Models POM and POMDP

We start with a rough description of a classical example.

Example 26.1.1 (Machine maintenance problem)

(a) *The classical model.* In many applications modeled by an N-stage MDP with finite state and action space and with transition matrix p, at each time the state is unobservable in the sense that it is not completely observable, but one can observe some value, called the observable state, which contains information about the unobservable state. We denote the momentary unobservable and observable state by ϑ and by s, respectively, and the succeeding states by ϑ' and s', respectively. The sets Θ and S of these states are assumed to be finite. Of course, the set of admissible actions must not depend on ϑ, hence is assumed to be equal to A. We also assume that the information about ϑ' contained in s' when action a is taken, is given by known probabilities $p_S(a, \vartheta', s')$. In addition, we assume information about the initial unobservable state in form of a discrete density μ. Finally, we make the plausible assumption that the joint evolution of the pairs $(\vartheta_t, s_t)_0^N$ is Markovian with transition probabilities $p(\vartheta, s, a, \vartheta', s') := p_\Theta(\vartheta, a, \vartheta') \cdot p_S(a, \vartheta', s')$, independent of s. There is a one-stage reward $\hat{r}(\vartheta, a, \vartheta', s')$ (mostly assumed to be independent of (ϑ', s')), and some terminal reward $V_0(\vartheta, s)$ (mostly assumed to be independent of s). We want to maximize for each initial observable state the N-stage expected reward,

K. Hinderer et al., *Dynamic Optimization*, Universitext,
DOI 10.1007/978-3-319-48814-1_26

averaged over μ. This problem turns out to be solvable by methods very similar to those used for the Bayesian MDPDs in Chap. 25.

(b) One of the applications of the model in (a) is the following simple *machine maintenance problem.* A machine produces during each period one item, which is delivered at the end of the period and whose quality s' can be observed, namely $s' = 0$ (faulty) or $s' = 1$ (good). The unobservable state ϑ' at the end of the period is the working condition of the machine which may assume any of the following three values: $\vartheta' = 0$ (good), $\vartheta' = 1$ (medium) and $\vartheta' = 2$ (failed). There are two actions, namely $a = 0$ (do nothing) and $a = 1$ (replace the machine). Each good item produced yields a reward of $d \in \mathbb{R}^+$ units, and a replacement costs $e \in \mathbb{R}^+$ units. We assume that from experience one knows the discrete densities $\vartheta' \mapsto p_\Theta(\vartheta, a, \vartheta')$ and $s' \mapsto p_S(a, \vartheta', s')$. In general the latter will be independent of a, and $p_\Theta(2, 0, 2) = p_\Theta(\vartheta, 1, 0) = 1$, $\vartheta \in \Theta$. Moreover, $\hat{r}(\vartheta, a, \vartheta', s') = d \cdot \delta_{s',1} - e \cdot \delta_{a,1}$ for some non-negative constants d and e.

(c) Besides several variants of the problem in (a) the following other applications can be treated: (i) *Medical diagnostic.* Here the unobservable state is the patient's physiological condition, the observable state is the result of a diagnostic test, and the actions are possible therapies. (ii) *Learning models.* Here the unobservable state is the student's state of knowledge, the observable state is the student's response to questions, and the actions are possible questions.

We now generalize the model in Example 26.1.1(a).

Definition 26.1.2 A **partially observable model** $(\Theta \times S, A, D_S, Z, P, K_S, \tilde{r}, V_0, \tilde{\beta})$ (**POM** for short) (or **model with incomplete information**) is an MDPD $(\Theta \times S, A, D, Z, K, \tilde{r}, V_0, \tilde{\beta})$ of the following kind:

- The state space is $\Theta \times S$ with elements (ϑ, s) and σ-algebra $\mathfrak{T} \otimes \mathfrak{S}$. Here Θ and S are called the set of **unobservable states** and of **observable states**, respectively.
- $D(\vartheta, s) =: D_S(s)$ does not depend on ϑ. Thus $D = \Theta \times D_S$ where $D_S := \{(s, a) \in S \times A : a \in D_S(s)\}$.
- The transition law K is a transition probability from D into $Z \times \Theta \times S$, having a factorization of the form

$$K(\vartheta, s, a, d(z, \vartheta', s')) = P(\vartheta, s, a, d(\vartheta', z)) \otimes K_S(s, a, z, ds'), \qquad (26.1)$$

for transition probabilities P from D into $\Theta \times Z$ and of K_S from $D_S \times Z$ into S. Observe that K_S is independent of ϑ and ϑ'.
- The expected one-stage reward

$$\begin{aligned} r(\vartheta, s, a) &:= K\tilde{r}(\vartheta, s, a) \\ &:= \int P(\vartheta, s, a, d(z, \vartheta')) \int K_S(s, a, z, ds')\, \tilde{r}(\vartheta, s, a, z, \vartheta', s') \end{aligned}$$

is assumed to exist and to be finite for all (ϑ, s, a). If $\tilde{r}$ does not depend on (z, ϑ', s') it is denoted by r.

Remark 26.1.3

(a) Sometimes one uses a factorization of P of the form

$$P(\vartheta, s, a, d(\vartheta', z)) = P_\Theta(\vartheta, s, a, d\vartheta') \otimes P_{Z|\Theta}(\vartheta, s, a, \vartheta', dz)$$

in order to model via $P_{Z|\vartheta}$ the dependence of the disturbance z on the unobservable states (ϑ, ϑ') and hence via K_S also the dependence of the new observable state s' on (ϑ, ϑ').

(b) The disturbance transition law Q of the POM is defined by $Q(\vartheta, s, a, dz) := K(\vartheta, s, a, dz \times (\Theta \times S))$. It equals

$$Q(\vartheta, s, a, dz) := P(\vartheta, s, a, \Theta \times dz). \tag{26.2}$$

Definition 26.1.4 **A partially observable Markovian control model** (**POMCM** for short) $(\Theta \times S, A, D_S, Z, P, T, r_{\Theta\times Z}, V_0, \tilde{\beta})$ is a POM $(\Theta \times S, A, D, Z, P, K_S, \tilde{r}, V_0, \tilde{\beta})$ such that:

- $K_S(s, a, z, ds') := \delta_{T(s,a,z)}(ds')$ for some measurable mapping $T: D_S \times Z \to S$.
- $\tilde{r}(\vartheta, s, a, z, \vartheta', s') = r_{\Theta\times Z}(\vartheta, s, a, \vartheta', z)$, independent of s', for some measurable function $r_{\Theta\times Z}$ on $D \times \Theta \times Z$. (If $\tilde{r}$ depends on s', one can replace s' by $T(s, a, z)$.)

Thus

$$r(\vartheta, s, a) = \int P(\vartheta, s, a, d(\vartheta', z))\, r_{\Theta\times Z}(\vartheta, s, a, \vartheta', z).$$

Note that each POMCM satisfies the factorization assumption (26.1) since $T(s, a, z)$ does not depend on ϑ.

We provide a list of three submodels of POMCMs.

(a) A ***partially observable control model*** (***POCM*** for short) is a POMCM where P does not depend on (s, a).

(b) A ***partially observable Markovian decision process*** (***POMDP*** for short) $(\Theta \times S, A, D_S, P, r_{\Theta\times S}, V_0, \tilde{\beta})$ is a POMCM $(\Theta \times S, A, D_S, Z, P, T, r_{\Theta\times Z}, V_0, \tilde{\beta})$ with the following properties:

  - $Z = S$, $z = s'$; hence $P(\vartheta, s, a, d(\vartheta', z))$ is written as $P(\vartheta, s, a, d(\vartheta', s'))$ and $r_{\Theta\times Z}(\vartheta, s, a, \vartheta', z)$ is written as $r_{\Theta\times S}(\vartheta, s, a, \vartheta', s')$.
  - $T(s, a, s') = s'$.

(c) A ***simple POMCM*** $(\Theta, A, Z, P, r_{\Theta\times Z}, V_0, \tilde{\beta})$ is a POMCM $(\Theta\times S, A, D_S, Z, P, T, r_{\Theta\times Z}, V_0, \tilde{\beta})$ such that $|S| = 1$ and $D_S(s) = A$, $s \in S$. Then one can drop s everywhere. Thus

$$r(\vartheta, a) = \int P(\vartheta, a, d(\vartheta', z))\, r_{\Theta\times Z}(\vartheta, a, \vartheta', z).$$

In the literature on POMs the classical POMDPs (often with finite Θ and S) prevail. As an example, Example 26.1.1 (with the special forms of $\hat{r}$ and of $\hat{V}_0$) can be modeled by a classical POMDP.

Remark 26.1.5

(a) POMs often arise as a so-called partially observable version of an $\widehat{\text{MDPD}}$ $(\hat{S}, Z, \hat{K}, \hat{\tilde{r}}, \hat{V}_0, \tilde{\beta})$, called the *core process,* with non-observable states $\hat{s}$. This situation can be modeled by a POM as soon as a set S of observable states s and the dependence of the new observable state s' on the observable data (s, a, z) of the momentary period is given by a transition probability $\rho(s, a, z, ds')$ from $S\times A\times Z$ into S. The *partially observable version* (*PO-version* for short) *of the* $(\widehat{\text{MDPD}}, S, \rho)$ is then defined as the POM $(\Theta\times S, A, Z, P, K_S, \tilde{r}, V_0, \tilde{\beta})$ with the following specification:

(a1) $\Theta = \hat{S}$, $\vartheta = \hat{s}$,

(a2) $P(\vartheta, s, a, d(\vartheta', z)) := \hat{K}(\vartheta, a, d(\vartheta', z))$, independent of s, and $K_S := \rho$. Thus the following holds for measurable $v \geq 0$ on $Z\times\Theta\times S$

$$Kv(\vartheta, s, a) = \int \hat{K}(\vartheta, a, d(\vartheta', z)) \int \rho(s, a, z, ds')\, v(\vartheta', z, s').$$

(a3) $\tilde{r}(\vartheta, s, a, z, \vartheta', s') := \hat{\tilde{r}}(\vartheta, a, z, \vartheta')$, independent of s and s', hence

$$r(\vartheta, s, a) = \int P(\vartheta, s, a, d(\vartheta', z))\, \hat{\tilde{r}}(\vartheta, a, \vartheta', z), \quad \text{independent of } s.$$

(a4) $V_0(\vartheta, s) := \hat{V}_0(\vartheta)$, independent of s.

(b) By specializing the $\widehat{\text{MDPD}}$ in (a) to an $\widehat{\text{MCM}}$ $(\hat{S}, A, Z, \hat{Q}, \hat{T}, \hat{r}_Z, \hat{V}_0, \tilde{\beta})$ we obtain what we call the *PO-version of the* $\widehat{\text{MCM}}$. It is the POM $(\Theta\times S, A, Z, P, K_S, \tilde{r}, V_0, \tilde{\beta})$ with the following properties: (b1) $\Theta := \hat{S}$, $\vartheta := \hat{s}$, (b2) $P(\vartheta, s, a, d(\vartheta', z)) := \hat{Q}(\vartheta, s, a, dz) \otimes \delta_{\hat{T}(\vartheta,a,z)}(d\vartheta')$, $K_S := \rho$, (b3) $\tilde{r}(\vartheta, s, a, \vartheta', z, s') := \hat{r}_Z(\vartheta, a, z)$. This POM is an POMCM if $\rho(s, a, z, ds') = \delta_{T(s,a,z)}(ds')$ for some mapping $T: S\times A\times Z \to S$, i.e. if the new observable state s' depends deterministically on the observable data (s, a, z) of the momentary period.

(c) If $\hat{Q}$ in (b) does not depend on (s, a) we obtain the definition of the *PO-version* of a $\widehat{\text{CM}}$.

(d) We define the *PO-version* of an $\widehat{\text{MDP}} = (\hat{S}, A, \hat{P}, \hat{r}_S, \hat{V}_0, \beta)$ as the PO-version of the $\widehat{\text{MDPD}}$ which is equivalent to the $\widehat{\text{MDP}}$ by choosing Z as a singleton and $\tilde{\beta} \equiv \beta$, and by dropping z in the data of the $\widehat{\text{MDPD}}$. Thus the PO-version of the $\widehat{\text{MDP}}$ is the POM $(\Theta \times S, A, Z, P, K_S, \tilde{r}, V_0, \tilde{\beta})$ where $Q_{Z\vartheta}(\vartheta, s, a, d(\vartheta', z)) := \hat{P}(\vartheta, a, d\vartheta')$, $K_S := \rho$, $\tilde{r}(\vartheta, s, a, \vartheta', z, s') := \hat{r}_S(\vartheta, a, \vartheta')$. ◇

26.2 The Formal Treatment

As in Chap. 25 we use history-dependent policies π, i.e. the decision rules π_t depend on the history $h_t := (s_0, y^t) := (s_0, z_1, s_1, \ldots, z_t, s_t) \in H_t := S \times (Z \times S)^t$. All elements of h_t are observable at time t. Again the set of ***history-dependent policies*** is denoted by $\boldsymbol{\Delta}_N$. We also need at time $t \geq 1$ the histories in the MDPD defining the POM, called ***complete histories.*** They are given by $(\vartheta_0, s_0, \omega^t) = (\vartheta_0, s_0, \omega_1, \ldots, \omega_t) \in (\Theta \times S) \times \Omega^t$ where $\omega_\nu := (z_\nu, \vartheta_\nu, s_\nu) \in \Omega := Z \times \Theta \times S$.

For fixed s_0 the N-stage reward is defined on the sample space $\Theta \times \Omega^N$ with σ-algebra $\mathfrak{F} := \mathfrak{T} \otimes \left(\bigotimes_1^N (\mathfrak{Z} \otimes \mathfrak{T} \otimes \mathfrak{S})\right)$ as the random variable

$$\begin{aligned}(\vartheta_0, \omega^N) \mapsto R_{N\pi}(\vartheta_0, s_0, \omega^N) &:= \tilde{r}(\vartheta_0, s_0, \pi_0(h_0), \omega_1)\\ &+ \sum_{t=1}^{N-1} \prod_{\nu=1}^{t} \tilde{\beta}(z_\nu) \cdot \tilde{r}(\vartheta_t, s_t, \pi_t(h_t), \omega_{t+1}) + \prod_{\nu=1}^{N} \tilde{\beta}(z_\nu) \cdot V_0(\vartheta_N, s_N).\end{aligned}$$

For $t \geq 0$ the transition from $\omega_t = (z_t, \vartheta_t, s_t)$ under $a_t := \pi_t(h_t)$ to $\omega_{t+1} = (z_{t+1}, \vartheta_{t+1}, s_{t+1})$ is governed by

$$K_{\pi_t}(\vartheta_t, h_t, d\omega_{t+1}) := K(\vartheta_t, s_t, \pi_t(h_t), d\omega_{t+1}).$$

Therefore we construct for $\mu \in \mathbb{P}(\mathfrak{T})$ and $\pi \in \Delta_N$ on $\mathfrak{F}$ the probability distribution

$$\begin{aligned}\mathbb{P}^{\mu}_{N\pi}(s_0, B) &:= \int \mu(d\vartheta_0) \int K_{\pi_0}(\vartheta_0, h_0, d\omega_1) \int K_{\pi_1}(\vartheta_1, h_1, d\omega_2)\\ &\quad \cdots \int K_{\pi_{N-1}}(\vartheta_{N-1}, h_{N-1}, d\omega_N)\, 1_B(\vartheta_0, \omega^N)\\ &=: \int \mu(d\vartheta_0)\, \mathbb{P}_{N\pi}(\vartheta_0, s_0, B).\end{aligned}$$

Assumption (LUBF) has the same form as in Chap. 25. Then one shows with essentially the same proof as for BCMs in Chap. 23 that there exists for $N \geq 1$, $\pi \in \Delta_N$

$$v_{N\pi}(\mu, s) := \int \mathbb{P}^{\mu}_{N\pi}(s, d(\vartheta_0, \omega^N))\, R_{N\pi}(\vartheta_0, s, \omega^N),\ \mu \in \mathbb{P}(\mathfrak{T}), s \in S,$$

the so-called *N-stage expected reward of policy* π with respect to prior μ and initial state s.

A partially observable model POM and some prior μ_0 define the problem ***POM$_N$ (μ_0)*** as follows:

(i) Find for each initial observable state $s \in S$ the maximal N-stage expected reward with respect to μ_0 within the set of history-dependent policies, i.e.

$$v_N(\mu_0, s) := \sup\{v_{N\pi}(\mu_0, s) : \pi \in \Delta_N\}.$$

(ii) Find a policy $\pi^* \in \Delta_N$ which is optimal for $POM_N(\mu_0)$ in the sense that it maximizes $\pi \mapsto v_{N\pi}(\mu_0, s)$ on Δ_N for all $s \in S$.

Now we turn to the simplified setting in which the conditional distributions of the random variables with realizations ϑ_t are sequentially determined by a sufficient statistic. For fixed (μ, s, a) consider the probability space $(\Theta \times Z, \mathfrak{T} \otimes \mathfrak{Z}, P^{\mu})$ with

$$P^{\mu}(s, a, d(\vartheta', z)) := \int \mu(d\vartheta)\, P(\vartheta, s, a, d(\vartheta', z)).$$

Let τ_1 and η_1 be the coordinate variables on this probability space. We assume that the POM has a version of the so-called ***updating operator*** $\boldsymbol{\Phi}$ generated by P, defined by

$$(\mu, s, a, z) \mapsto \Phi(\mu, s, a, z, d\vartheta') := P^{\mu}_{\tau_1|\eta_1}(s, a, z, d\vartheta').$$

Note that a mapping Φ from $\mathbb{P}(\mathfrak{T}) \times D_S \times Z$ into $\mathbb{P}(\mathfrak{T})$ such that $\Phi(\mu, s, a, z, B)$, $B \in \mathfrak{T}$, is measurable in (s, a, z), is a version of the updating operator generated by P if and only if for $\mu \in \mathbb{P}(\mathfrak{T})$, $(s, a) \in D$ and all measurable $v \geq 0$ on $Z \times \Theta$ we have

$$\int P^{\mu}_{\eta_1,\tau_1}(s, a, d(\vartheta', z))\, v(\vartheta', z) = \int P^{\mu}_{\eta_1}(s, a, dz) \int \Phi(\mu, s, a, z, d\vartheta'')\, v(\vartheta'', z),$$

i.e. if and only if with Q from (26.2)

$$\begin{aligned} &\int \mu(d\vartheta) \int P(\vartheta, s, a, d(\vartheta', z))\, v(\vartheta', z) \qquad (26.3)\\ &= \int \mu(d\vartheta) \int Q(\vartheta, s, a, dz) \int \Phi(\mu, s, a, z, d\vartheta'')\, v(\vartheta'', z). \end{aligned}$$

Given a POM, let $(I, \mathfrak{I})$ be a measurable space and let $(i, B) \mapsto \hat{\mu}(i, B)$ be a transition probability from I into Θ. Then we speak of the model $(\boldsymbol{POM}, \hat{\boldsymbol{\mu}})$. It defines the family of problems $(\boldsymbol{POM_N}\ (\boldsymbol{i}) := \boldsymbol{POM_N}\ (\hat{\boldsymbol{\mu}}\,(\boldsymbol{i})),\ \boldsymbol{i} \in \boldsymbol{I})$.

A measurable mapping φ from $I \times D_S \times Z$ into I is called a ***sufficient statistic*** with respect to P for the family $\hat{\mu}$ of probability distributions $\hat{\mu}(i)$, $i \in I$, if for all $(i, s, a, z) \in I \times D_S \times Z$ we have

$$\Phi(\hat{\mu}(i), s, a, z, d\vartheta') = \hat{\mu}(\varphi(i, s, a, z), d\vartheta')$$

for some version of Φ.

If φ is independent of some of its arguments, these are dropped in the notation tacitly. Obviously for the existence of a sufficient statistic for $\hat{\mu}$ it is necessary that

(i) $\Phi(\hat{\mu}(i), s, a, z) \in \hat{\mu}(I)$ for all (i, s, a, z).

The notion of sufficiency becomes clearer by the following observation: Assume besides (i) that

(ii) $\hat{\mu}$ is injective and its inverse $\hat{\mu}^{-1}$ is measurable,
(iii) $(i, s, a, z) \mapsto \Phi(\hat{\mu}(i), s, a, z)$ is measurable.

Then $\varphi(i, s, a, z) := \hat{\mu}^{-1}(\Phi(\hat{\mu}(i), s, a, z))$ is sufficient for $\hat{\mu}$. In particular, if $I \subset \mathbb{P}(\mathfrak{T})$ and $\hat{\mu}(i) := i$, $i \in I$, then $\varphi := \Phi$ is sufficient for $\hat{\mu}$.

The next definition differs from the corresponding one in Chap. 25 only with respect to the definition of r'.

Definition 26.2.1 (The *associated MDPD'*) Consider some partially observable model $(POM, \hat{\mu}, \varphi) = (\Theta \times S, A, D_S, Z, P, K_S, \tilde{r}, V_0, \tilde{\beta}, \hat{\mu}, \varphi)$ which satisfies assumption (LUBF). We associate to it the following Markovian decision process with disturbances MDPD' $(I \times S, A, D', Z, K', r', V_0', \tilde{\beta})$:

- The states are $(i, s) \in I \times S$, with σ-algebra $\mathfrak{I} \otimes \mathfrak{S}$.
- $D'(i, s) := D_S(s)$ is independent of i, i.e. $D' = I \times D_S$.
- K' is determined as follows: for all measurable $v \geq 0$ on $Z \times I \times S$

$$\begin{aligned} K'v(i, s, a) &:= \int K'(i, s, a, d(z, i', s'))\, v(z, i', s') \\ &:= \int Q'(i, s, a, dz) \int K_S(s, a, z, ds')\, v(z, \varphi(i, s, a, z), s') \end{aligned} \quad (26.4)$$

exists. Here the disturbance transition law Q' in MDPD' is given by

$$Q'(i, s, a, dz) := \int \hat{\mu}(i, d\vartheta) \int Q(\vartheta, s, a, dz). \quad (26.5)$$

- The one-stage reward does not depend on (z, i', s') and equals

$$r'(i, s, a) := \int \hat{\mu}(i, d\vartheta)\, r(\vartheta, s, a), \tag{26.6}$$

which is assumed to be finite.
- The terminal reward equals $V_0'(i, s) := \int \hat{\mu}(i, d\vartheta)\, V_0(\vartheta, s)$, which is assumed to be finite.

The form of K' is intuitively clear: If we take action a in state (i, s) and if the disturbance z occurs, then the next index i' is $\varphi(i, s, a, z)$. The function V_0' exists as the MDPD defining the POM has an upper or lower bounding function.

In MDPD$'$ the sets $\mathbb{F}'$, $(\mathbb{F}')^N$ and Δ_N' denote the set of decision rules, the set of N-stage Markovian policies and the set of N-stage history-dependent policies, respectively. Note that $\Delta_1' = \mathbb{F}'$ and that $\pi_t' = \pi_t'(h_t')$, where $h_t' = (i_0, s_0, (z_1, i_1, s_1), \ldots, (z_t, i_t, s_t))$. Moreover, $(i, s) \mapsto b(s)$ is an upper or lower bounding function for MDPD$'$ so that $V_{N\pi'}'$, defined as in (21.6), exists for $\pi' \in \Delta_N' \supset (\mathbb{F}')^N$. By $V_N'(i, s)$ we denote the maximal expected N-stage reward within the set $(\mathbb{F}')^N$ of Markovian policies in $MDPD_N'$. For $f' \in \mathbb{F}'$ put $\varphi_{f'}(i, s, z) := \varphi(i, s, f'(i, s), z)$ for all (i, s, z).

Part (b) of our basic result Theorem 26.2.3 below shows that in maximizing $\pi \mapsto v_{N\pi}(i_0, s)$ on Δ_N one can in general restrict oneself to those history-dependent policies $\pi = (\pi_t)_0^{N-1}$ where π_t depends on the history h_t only via the momentary state s_t and the momentary posterior $\hat{\mu}(i_t)$, or equivalently, via (i_t, s_t). Policies π with this property are generated by Markovian policies π' in MDPD$'$ as in part (ii) of the next definition, which in both parts is literally the same as in Chap. 25.

Definition 26.2.2 (Posterior index functions and generation of history-dependent policies in $POM_N(\hat{\mu}(i))$ from Markovian policies in $MDPD_N'$)

(i) The ***posterior index functions*** $i_{t\pi'}(i, \cdot, \cdot): H_{t-1} \times Z \to I$, $1 \le t \le N$, generated by $i \in I$ and $\pi' \in (\mathbb{F}')^N$, are defined recursively by $i_{1,\pi'}(i, s_0, z_1) := \varphi_{\pi_0'}(i, s_0, z_1)$ and for $1 \le t \le N-1$ by

$$i_{t+1,\pi'}(i, h_t, z_{t+1}) := \varphi_{\pi_t'}(i_{t\pi'}(i, h_{t-1}, z_t), s_t, z_{t+1}).$$

(ii) The ***history-dependent policy*** $\pi^i = (\pi_t^i)_0^{N-1} \in \Delta_N$, generated by $i \in I$ and a Markovian policy $\pi' = (\pi_t')_0^{N-1} \in (\mathbb{F}')^N$ is defined by $\pi_0^i(s_0) := \pi_0'(i, s_0)$, $s_0 \in S$, and for $1 \le t \le N-1$ by

$$\pi_t^i(h_t) := \pi_t'(i_{t\pi'}(i, h_{t-1}, z_t), s_t), \quad h_t \in H_t. \tag{26.7}$$

From now on we write $v_{N\pi}(i, s)$ and $v_N(i, s)$ instead of $v_{N\pi}(\hat{\mu}(i), s)$ and $v_N(\hat{\mu}(i), s)$, respectively. We call $v_N: I \times S \to \overline{\mathbb{R}}$ the ***N-stage value function*** of the POM. The notation $v_{N\pi}(i, s)$ and $v_N(i, s)$ will also be used for cost minimization problems, while one-stage costs and terminal costs will be denoted by $\tilde{c}$ and C_0, respectively.

The proof of the following Basic Theorem is literally the same as the proof of the Basic Theorem 25.1.10. Note that in Theorem 26.2.3 the explicit formulae for $K'(\tilde{\beta} \cdot v_{n-1})$ and for r' follow from (26.4) and from (26.6), respectively.

Theorem 26.2.3 (Basic Theorem for a partially observable model) *If the VI holds for the associated MDPD′ then we have for $N \geq 1$:*

(a) The value function v_N equals the value function V'_N in MDPD′, hence can be found recursively by value iteration in MDPD′:

$$v_n(i,s) = \sup_{a \in D_S(s)} [r'(i,s,a) + K'(\tilde{\beta} \cdot v_{n-1})(i,s,a)], \tag{26.8}$$

$$1 \leq n \leq N, \ (i,s) \in I \times S.$$

Here

$$v_0(i,s) := \int \hat{\mu}(i, d\vartheta)\, V_0(\vartheta, s),$$

$$K'(\tilde{\beta} \cdot v_{n-1})(i,s,a)$$
$$= \int Q'(i,s,a,dz)\, \tilde{\beta}(z) \cdot \int K_S(s,a,z,ds')\, v_{n-1}(\varphi(i,s,a,z), s'), \tag{26.9}$$

and $r'(i,s,a) = \int \hat{\mu}(i, d\vartheta)\, r(\vartheta, s, a)$.

(b) Assume that at each stage $1 \leq n \leq N$ a maximizer $f'_n \in \mathbb{F}'$ exists in MDPD′. Then for each $i \in I$ the history-dependent policy $\pi^i \in \Delta_N$, generated by i and by $\pi' = (f'_n)_N^1 \in (\mathbb{F}')^N$, is optimal for problem $POM_N(\hat{\mu}(i))$.

For solving a problem $POM_N(\mu_0)$ for given μ_0 by means of Theorem 26.2.3, one must find a family $\hat{\mu}$ containing μ_0 and having a sufficient statistic φ. Note that while the VI in MDPD′ gives the value functions for all problems $POM(\hat{\mu}(i))$, $i \in I$, in general no policy $\pi \in \Delta_N$ is optimal for all these problems.

Of course, there exists a minimization version, an infinite-stage version and a non-stationary version of Theorem 26.2.3.

Proposition 26.2.4 (The associated MDPDs for special POMs)
The following hold:

(a) If the model $(POM, \hat{\mu}, \varphi)$ has a constant discount factor β the associated MDPD′ is equivalent (cf. Chap. 12) to the MDP′ $(I \times S, A, D', P', r', V'_0, \beta)$ where P' is determined by the property that for all measurable $v \geq 0$ on $I \times S$, using

Q' from (26.5),

$$P'v(i,s,a) = \int Q'(i,s,a,dz) \int K_S(s,a,z,ds')\, v(\varphi(i,s,a,z),s').$$

We call the MDP′ the associated MDP′.

(b) *In the MDP′ associated to a POMCM we have for all measurable $v \geq 0$ on $I \times S$*

$$P'v(i,s,a) = \int Q'(i,s,a,dz)\, v(\varphi(i,s,a,z),T(s,a,z)).$$

(c) *In the MDP′ associated to a simple POMCM we have for all measurable $v \geq 0$ on I*

$$P'v(i,a) = \int Q'(i,a,dz)\, v(\varphi(i,a,z)),$$

$$Q'(i,a,dz) = \int \hat{\mu}(i,d\vartheta)\, P(\vartheta,a,\Theta \times dz).$$

(d) *In the MDP′ associated to a POMDP the following holds for all measurable $v \geq 0$ on $I \times S$*

$$P'v(i,s,a) = \int Q'(i,s,a,ds')\, v(\varphi(i,s,a,s'),s'),$$

$$Q'(i,s,a,ds') = \int \hat{\mu}(i,d\vartheta)\, P(\vartheta,s,a,\Theta \times ds').$$

The proof is omitted since it is very similar to the proof of Lemma 25.2.2 due to the fact that the definitions of MDPD′ in Chap. 25 and here differ only in r'.

For our next results we generalize the notions of a transition probability and of a σ-finite measure, as follows. Let $(X,\mathfrak{X})$ and $(Y,\mathfrak{Y})$ be measurable spaces. A function $(x,B) \mapsto \rho(x,B)$ from $X \times \mathfrak{Y}$ into $\mathbb{R}_+$ is called a ***transition measure*** from X into Y if $\rho(x,B)$ is measurable in x and a measure in B. A transition measure ρ from X into Y is called ***uniformly σ-finite*** if there exists an increasing sequence of sets $B_n \in \mathfrak{Y}$, $n \geq 1$, converging towards Y such that $\sup_{x \in X} \rho(x,B_n) < \infty$ for all n. We do not distinguish between a σ-finite measure $\overline{K}$ on $\mathfrak{Y}$ and the σ-finite transition measure $(x,B) \mapsto K(x,b) := \overline{K}(B)$ from X into Y.

Most examples fulfill the subsequent assumption, which allows us to compute in Lemma 26.2.5 the updating operator Φ and then in Proposition 26.2.6 a sufficient statistic.

Assumption (A1) for a POM: P has a transition density function $(\vartheta,s,a,\vartheta',z) \mapsto p(\vartheta,s,a,\vartheta',z)$ with respect to $\kappa(s,a,d\vartheta') \times \nu(s,a,dz)$, where κ and ν are uniformly σ-finite transition measures from D_S into Θ and into Z, respectively.

If both Θ and Z are countable, (A1) holds with κ and ν (independent of (s, a)) as the counting measures on Θ and Z, respectively, and this is assumed from now on. In the classical POMDP the assumption (A1) holds, for example, if P_ϑ and $P_{S|\Theta}$ have transition density functions with respect to some σ-finite measure κ and ν, respectively. This follows from the definition of a transition density function and from Fubinis theorem for σ-finite measures (cf. Appendix B.2.3).

Under (A1) we put for $\mu \in \mathbb{P}(\mathfrak{T})$ and $(s, a, \vartheta', z) \in D_S \times \Theta \times Z$

$$p^\mu(s, a, \vartheta', z) := \int \mu(d\vartheta)\, p(\vartheta, s, a, \vartheta', z),$$

$$p^{\mu\kappa}(s, a, z) := \int \kappa(s, a, d\vartheta')\, p^\mu(s, a, \vartheta', z).$$

Lemma 26.2.5 (Computation of the updating operator for POMs) *Assume (A1), and for* $(\mu, s, a, z) \in \mathbb{P}(\mathfrak{T}) \times D_S \times Z$ *put*

$$\Phi(\mu, s, a, z, d\vartheta') :=$$
$$\begin{cases} p^\mu(s, a, \vartheta', z) \cdot \kappa(s, a, d\vartheta')/p^{\mu\kappa}(s, a, z), & \textit{if } 0 < p^{\mu\kappa}(s, a, z) < \infty, \\ \mu(d\vartheta'), & \textit{otherwise}. \end{cases}$$

Then Φ *is a version of the updating operator generated by P.*

Proof Several times we tacitly use Fubini's theorem for uniformly σ-finite transition measures (cf. Appendix B.2.3). Fix μ.

(a) We verify that $(s, a, z, B) \mapsto \Phi(\mu, s, a, z, B)$ is a transition probability from $D_S \times Z$ into Θ. Firstly, it is clear that $B \mapsto \Phi(\mu, s, a, z, B)$ is a probability distribution for all (s, a, z). Next, measurability of p^μ and of $p^{\mu\kappa}$ holds since measurability of functions is preserved under integration by a uniformly σ-finite transition measure (cf. Appendix B.2.2). It follows that $\{(s, a, z) \in D_S \times Z : 0 < p^{\mu\kappa}(s, a, z) < \infty\}$ is measurable, and that $(s, a, z) \mapsto \Phi(\mu, s, a, z, B)$ is measurable for all B.

(b) We now prove that Φ is a version of the updating operator. Obviously Q from (26.2) has the transition density function $(\vartheta, s, a, z) \mapsto \int \kappa(s, a, d\vartheta')\, p(\vartheta, s, a, \vartheta', z)$ with respect to $\nu(s, a)$. Fix (s, a) and suppress it in the sequel. Put $B := [0 < p^{\mu\kappa} < \infty]$, $B_0 := [p^{\mu\kappa} = 0]$ and $B_1 := [p^{\mu\kappa} = \infty]$. According to (26.3) it suffices to verify for all measurable $v \geq 0$ on $Z \times \Theta$ that

$$\int \mu(d\vartheta) \int \nu(dz) \int \kappa(d\vartheta')\, p(\vartheta, \vartheta', z) \cdot v(\vartheta', z) \tag{26.10}$$
$$= \int \mu(d\vartheta) \int \nu(dz) \int \kappa(d\vartheta')\, p(\vartheta, \vartheta', z) \cdot g(z),$$

where

$$g(z) := 1_B(z) \cdot \int \kappa(d\vartheta')\, p^{\mu}(\vartheta', z) \cdot v(\vartheta', z)/p^{\mu\kappa}(z)$$
$$+\, 1_{B_0+B_1}(z) \cdot \int \mu(d\vartheta'')\, v(\vartheta'', z).$$

Put $H_v(z) := \int \kappa(d\vartheta')\, p^{\mu}(\vartheta', z) \cdot v(\vartheta', z)$ and $G_v(z) := p^{\mu\kappa}(z) \cdot \int \mu(d\vartheta'')\, v(\vartheta'', z)$. It follows that (26.10) holds if

$$\int H_v \, d\nu = \int_B H_v \, d\nu + \sum_{i=0}^{1} \int_{B_i} G_v \, d\nu,$$

i.e. if

$$\sum_{i=0}^{1} \int_{B_i} H_v \, d\nu = \sum_{i=0}^{1} \int_{B_i} G_v \, d\nu. \tag{26.11}$$

For $z \in B_0$ we have $p(\vartheta, \vartheta', z) = 0$ for $\mu \times \kappa$-almost all (ϑ, ϑ'), hence $H_v(z) = 0$, thus $\int_{B_0} H_v \, d\nu = 0$. Next, $G_v = 0$ on B_0, hence $\int_{B_0} G_v \, d\nu = 0$. Finally, $\nu(B_1) = 0$ since $\int \nu(dz)\, p^{\mu\kappa}(z) = \int \mu(d\vartheta) \int \nu \times \kappa(d(\vartheta', z))\, p(\vartheta, \vartheta', z) = 1 < \infty$. This shows that $\int_{B_1} H_v \, d\nu = \int_{B_1} G_v \, d\nu = 0$. Now the proof of (26.11) is complete. □

Consider a model $(POM, \hat{\mu})$ where POM satisfies (A1). Then we put $p_i := p^{\hat{\mu}(i)}$, $p_{i\kappa} := p^{\hat{\mu}(i),\kappa}$, i.e.

$$p_i(s, a, \vartheta', z) := \int \hat{\mu}(i, d\vartheta)\, p(\vartheta, s, a, \vartheta', z),$$
$$p_{i\kappa}(s, a, z) := \int \kappa(s, a, d\vartheta')\, p_i(s, a, \vartheta', z).$$

If both Θ and Z are countable then

$$p_i(s, a, \vartheta', z) = \sum_{\vartheta} \hat{\mu}(i, \{\vartheta\}) \cdot p(\vartheta, s, a, \vartheta', z), \quad p_{i\kappa}(s, a, z) = \sum_{\vartheta'} p_i(s, a, \vartheta', z).$$

Note that equation (26.5) yields for $(i, s, a, z) \in I \times D_S \times Z$

$$Q'(i, s, a, dz) = p_{i\kappa}(s, a, z) \cdot \nu(s, a, dz), \tag{26.12}$$

which together with (26.9) implies that the term $K'(\tilde{\beta} \cdot v_{n-1})(i, s, a)$ in the VI (26.8) equals

$$\int \nu(s, a, dz)\, p_{i\kappa}(s, a, z) \cdot \tilde{\beta}(z) \cdot \int K_S(s, a, z, ds')\, v_{n-1}(\varphi(i, s, a, z), s').$$

In part (a) of the next result, where $I \subset \mathbb{P}(\mathfrak{T})$, we need a natural ***σ-algebra*** $\mathfrak{J}$ on I such that $i \mapsto \hat{\mu}(i) := i$ becomes a transition probability from I into Θ, i.e. such that all functions $i \mapsto i(B)$, $B \in \mathfrak{T}$, are $\mathfrak{J}$-measurable. This leads to a natural choice for $\mathfrak{J}$ according to the next definition, and this choice also ensures the measurability of the sufficient statistic $\varphi: I \times D_S \times Z \to I$ in part (a) of the next result.

Proposition 26.2.6 (Sufficient statistics for POMs) *Consider a model* $(POM, \hat{\mu})$ *where POM satisfies (A1).*

(a) Assume $\hat{\mu}(i) := i$, $i \in I \subset \mathbb{P}(\mathfrak{T})$, *and use as σ-algebra* $\mathfrak{J}$ *on I the ∗-σ-algebra. Define the mapping* $\varphi = \varphi_\kappa$ *from* $I \times D_S \times Z$ *into* $\mathbb{P}(\mathfrak{T})$ *by*

$$\varphi(i, s, a, z) := \text{the probability distribution on } \mathfrak{T} \text{ with } \kappa(s, a)\text{-density} \quad \vartheta' \mapsto p_i(s, a, \vartheta', z)/p_{i\kappa}(s, a, z), \tag{26.13}$$

if $0 < p_{i\kappa}(s, a, z) < \infty$, *and* $\varphi(i, s, a, z) := i$, *otherwise. Then* φ *is sufficient for* $\hat{\mu}$, *provided* $\varphi(i, s, a, z) \in I$ *for all* $(i, s, a, z) \in I \times D_S \times Z$, *in particular, if* $I = \mathbb{P}(\mathfrak{T})$.

(b) Assume $\Theta := \mathbb{N}_d$ *for some* $d \geq 2$, *let* $\kappa(s, a, d\vartheta')$ *be the counting measure on the power set* $\mathfrak{T}$ *of* Θ. *Let I be the set of discrete densities of probability distributions on* $\mathfrak{T}$. *Consider I as a subset of* $\mathbb{R}^d$ *and endow it with the σ-algebra* $\mathfrak{J} := I \cap \mathfrak{B}_d$. *Let* $\hat{\mu}(i)$ *be the probability distribution on* $\mathfrak{T}$ *with discrete density* $i \in I$. *Then* $\hat{\mu}$ *is a transition probability from I into* Θ. *Define* $\varphi: I \times D_S \times Z \to I$ *by*

$$\varphi(i, s, a, z) := p_i(s, a, \cdot, z)/p_{i\kappa}(s, a, z), \tag{26.14}$$

if $0 < p_{i\kappa}(s, a, z) < \infty$, *and* $\varphi(i, s, a, z) := i$, *otherwise. Then* φ *is sufficient for* $\hat{\mu}$.

Proof The proof of (a) and (b) are nearly the same. One easily checks that $\Phi(\hat{\mu}(i), s, a, z) = \hat{\mu}(\varphi(i, s, a, z))$ with Φ from Lemma 26.2.5. The proof that φ is measurable is left to the reader. □

We call a POM with sufficient statistic φ for the prior $\hat{\mu}$ (**(*POMCM*,** $\hat{\mu}, \varphi$**)** for short) ***simple*** if the POMCM is simple. Since then $|S| = 1$, we can drop s everywhere and the value functions v_n and the maximizers f'_n do not depend on s. Thus parts (a) and (b) of the following result are obvious from Theorem 26.2.3 and (26.12). Note that by Theorem 16.1.16 the VI holds in the associated MDP′ from Proposition 26.2.4 if (MA1) holds and if A is finite.

Proposition 26.2.7 (The solution of simple POMCMs) *Consider a simple $(POMCM, \hat{\mu}, \varphi)$. Assume that the VI holds in the associated MDP′.*

(a) The value functions v_n, $n \geq 1$, satisfy

$$v_n(i) = \sup_{a \in A}[r'(i,a) + \beta \cdot \int Q'(i,a,dz)\, v_{n-1}(\varphi(i,a,z))]$$
$$=: \sup_{a \in A} w_n(i,a),\ n \geq 1,\ i \in I. \tag{26.15}$$

Here $Q'(i,a,dz) = \int \hat{\mu}(i,d\vartheta) \int P(\vartheta,a,\Theta \times dz)$. Moreover, if (A1) holds and if Z is finite, then we have for $n \geq 1$ and $(i,a) \in I \times A$

$$w_n(i,a) = r'(i,a) + \beta \cdot \sum_{z \in Z} p_{i\kappa}(a,z) \cdot v_{n-1}(\varphi(i,a,z)). \tag{26.16}$$

(b) Fix $N \geq 1$. Assume that at each stage $1 \leq n \leq N$ a maximizer $i \mapsto f'_n(i)$ of w_n exists. Then for each $i \in I$ the history-dependent policy $\pi^i \in \Delta_N$ from (26.7), generated by i and by $\pi' = (g'_n)_N^1 \in (\mathbb{F}')^N$—and which is optimal for problem $POM_N(\hat{\mu}(i))$ according to Theorem 26.2.3(b)—has the property that $\pi^i_t(s_0, z_1, s_1, \ldots, z_t, s_t)$ is independent of $(s_0, s_1, \ldots, s_t)$, $1 \leq t \leq N-1$.

Consider the model $(POM, \hat{\mu}, \varphi)$ under (A1). Our goal consists in showing that under certain assumptions the value functions $(i,s) \mapsto v'_n(i,s)$ are increasing in i with respect to an appropriate relation on I. As an example we consider the $\leq_{\text{tp}}$-relation.

Assumption (A2) for a (POM, $\hat{\mu}$, φ):

(i) Θ and Z are Cartesian products $\times_1^k \Theta_l$ and $\times_1^d Z_l$, respectively, with Borel sets $\Theta_l \subset \mathbb{R}$ and $Z_l \subset \mathbb{R}$. Thus Θ, Z and products of these spaces are lattices and the relation $\leq_{\text{tp}}$ is defined on the set of non-negative functions on these spaces (cf. Sect. 18.3). Moreover, $\mathfrak{T}$ and $\mathfrak{Z}$ are the σ-algebras of Borel sets in Θ and Z, respectively,
(ii) (A1) holds and the measures $\kappa(s,a)$ and $\nu(s,a)$ are product measures for all (s,a),
(iii) $\hat{\mu}(i) := i$, $i \in I \subset \mathbb{P}(\mathfrak{T})$, and $\mathfrak{J}$ is the $*$-σ-algebra on I,
(iv) $0 < p_{i\kappa} < \infty$.

In order to prove isotonicity of the value functions in POMs we apply

Lemma 26.2.8 *Assume (A2) for the model $(POM, \hat{\mu}, \varphi)$ with φ from (26.13). If $p(\vartheta, s, a, \vartheta', z)$ is TP_2 in $(\vartheta, \vartheta', z)$, then for $i, j \in I$ with $i \leq_{\text{tp}} j$ and for all $(s,a) \in D_S$:*

(a) $\varphi(i,s,a,z) \leq_{\text{tp}} \varphi(j,s,a,z')$ whenever $z \leq z'$.
(b) $p_{i\kappa}(s,a,\cdot) \leq_{\text{tp}} p_{j\kappa}(s,a,\cdot)$ on Z.
(c) $Q'(i,a,dz) \leq_{\text{st}} Q'(j,a,dz)$.

Proof Fix s, a, and $z \le z'$, and drop (s, a) in the sequel.

(a) By (26.13) it suffices to show that

$$p_i(\cdot, z)/p_{i\kappa}(z) \le_{\mathrm{tp}} p_j(\cdot, z')/p_{j\kappa}(z') \text{ on } \Theta,$$

or, using Lemma 18.3.8 with $X := \Theta$, that

$$p_i(\cdot, z) \le_{\mathrm{tp}} p_j(\cdot, z'). \tag{26.17}$$

This means, using $g_1(\vartheta, \vartheta') := p(\vartheta, \vartheta', z)$ and $g_2(\vartheta, \vartheta') := p(\vartheta, \vartheta', z')$, that

$$\int i(d\vartheta)\, g_1(\vartheta, \cdot) \le_{\mathrm{tp}} \int j(d\vartheta)\, g_2(\vartheta, \cdot) \text{ on } \Theta. \tag{26.18}$$

Now Lemma 18.3.8(e) with $X := \Theta^2$ and $Y := Z$ implies, since p is TP_2 in $(\vartheta, \vartheta', z)$ and since $z \le z'$, that $g_1 \le_{\mathrm{tp}} g_2$ on Θ^2. Finally (26.18) follows from Lemma 18.3.8(f2) with $X := Y := \Theta$, $P_1 := i$ and $P_2 := j$.

(b) This follows from (26.17) with $z' = z$ and from Lemma 18.3.8(f1) with $X := \Theta$, $Y := Z$, $g_1 := p_i$ and $g_2 := p_j$.

(c) From (26.12) we know that $Q'(i, dz) = p_{i\kappa}(z) \cdot \nu(dz)$. It follows from (b) that $Q'(i, dz) \le_{\mathrm{tp}} Q'(j, dz)$, which implies the assertion by Proposition 18.3.6. □

Proposition 26.2.9 (Isotonicity of the value functions in POMs) *Assume (A2) for the model $(POM, \hat{\mu}, \varphi)$, and assume in addition:*

(i) $p(\vartheta, s, a, \vartheta', z)$ is TP_2 in $(\vartheta, \vartheta', z)$,
(ii) $\tilde{r}(\vartheta, s, a, z, \vartheta', s')$ does not depend on (z, ϑ'),
(iii) $\tilde{r}(\vartheta, s, a, s')$ and $V_0(\vartheta, s)$ are increasing in ϑ,
(iv) $\tilde{\beta}$ is constant,
(v) the value iteration holds in the associated MDPD′.

Then for all $n \ge 0$:
If $i, j \in I$ with $i \le_{\mathrm{tp}} j$ and $(i \le_{\mathrm{tp}} i) \vee (j \le_{\mathrm{tp}} j)$, then $v_n(i, s) \le v_n(j, s)$, $s \in S$.

Proof

(a) Isotonicity of $\tilde{r}$ in ϑ implies isotonicity of the function $\vartheta \mapsto \int K_S(s, a, z, ds')\, \tilde{r}(\vartheta, s, a, s')$. From Proposition 18.3.6 we know that $i \le_{\mathrm{tp}} j$ entails $i \le_{\mathrm{st}} j$. Now it follows from (26.6) that $r'(i, s, a) \le r'(j, s, a)$. In a similar way one shows that $v_0(i, s) \le v_0(j, s)$.

(b) We prove the assertion (I_n) by induction on $n \ge 0$. From now on fix (s, a) and drop it. Firstly, (I_0) is contained in (a). Now assume (I_n), and select $i, j \in I$ such that $i \le_{\mathrm{tp}} j$. Assume also $i \le_{\mathrm{tp}} i$. (The case where $j \le_{\mathrm{tp}} j$ is treated in the same way.) Put

$$H(i, z) := \int K_S(z, ds')\, v_n(\varphi(i, z), s'), \quad (i, z) \in I \times Z.$$

Since $D'_S(i,s) = D_S(s)$ is independent of i, it suffices by Theorem 6.3.5, (26.8) and (26.9) to show that $r'(i) + \beta \cdot \int Q'(i,dz)\,H(i,z) \le r'(j) + \beta \cdot \int Q'(j,dz)\,H(j,z)$, or, using (a), that

$$\int Q'(i,dz)\,H(i,z) \le \int Q'(j,dz)\,H(j,z). \tag{26.19}$$

(c) Since $i \le_{\text{tp}} j$ and $i \le_{\text{tp}} i$ we get from Lemma 26.2.8(a)

$$\varphi(i,z) \le_{\text{tp}} \varphi(j,z), \quad \varphi(i,z) \le_{\text{tp}} \varphi(i,z'), \quad z \le z'.$$

Now (I_n) implies for $z \le z'$

$$v_n(\varphi(i,z),s') \le v_n(\varphi(j,z),s'), \quad v_n(\varphi(i,z),s') \le v_n(\varphi(i,z'),s').$$

This entails

$$H(i,z) \le H(j,z), \quad H(i,z) \le H(i,z'),\ z \le z'. \tag{26.20}$$

Next, Lemma 26.2.8(c) shows, since $H(i,\cdot)$ is increasing by (26.20), that

$$\int Q'(i,dz)\,H(i,z) \le \int Q'(j,dz)\,H(i,z).$$

Moreover, from (26.20) we get

$$\int Q'(j,dz)\,H(i,z) \le \int Q'(j,dz)\,H(j,z).$$

Combining these two inequalities yields (26.19), i.e. (I_{n+1}). □

Remark 26.2.10

(a) In the classical POMDP the assumption (i) is ensured if, for example, P_Θ has a transition density $(\vartheta,a,\vartheta') \mapsto p_\Theta(\vartheta,a,\vartheta')$ with respect to a σ-finite measure $\kappa(d\vartheta')$, which is TP_2 in (ϑ,ϑ') and if $P_{S|\Theta}$ has a transition density $(\vartheta,a,\vartheta',s') \mapsto p_S(\vartheta,a,\vartheta',s')$ with respect to a σ-finite measure $\nu(ds')$, which is TP_2 in $(\vartheta,\vartheta',s')$. This follows from Lemma 18.3.8(a) and (c).

(b) A slight modification of that part of the proof below which leads to (26.20) shows that Lemma 18.3.8 remains true when assumption (iv) is replaced by the requirement that $\tilde{\beta}$ is decreasing and that $\tilde{r} \le 0$ and $V_0 \le 0$. This comprises the cost minimization problem in partially observed Markov renewal programs where $\tilde{\beta}(z) := e^{-\alpha z}$, $\alpha \in \mathbb{R}_+$, $z \in Z := \mathbb{R}_+$. ◇

Consider a simple model $(POMCM, \hat{\mu}, \varphi)$ with the following ***property (P)***: Θ, A and Z are finite, I is the uncountable set of discrete densities of probability distributions on Θ, $\hat{\mu}(i)$ is the probability distribution on Θ with discrete density $i \in I$ and φ is taken from (26.14). Note that then p, p_i and $p_{i\kappa}$ are finite. The value functions v_n are defined on I, considered as a subset of $\mathbb{R}^{|\Theta|}$. The validity of the VI and of the OC is ensured by finiteness of A, due to Corollary 12.1.8, since the associated MDP′ has finite transition law. This property also allows, despite the uncountability of I, the exact numerical computation of $v_N(i_0)$ for a single i_0 (which suffices for applications) by means of the method of reachable states from Chap. 12.

Before dealing with the method of reachable states we now present a method for computing $v_n(i)$ jointly for all i which exploits that the value functions $i \mapsto v_n(i)$ are convex and piecewise linear and that such functions are representable by finitely many reals. (When only $v_n(i_0)$ for a single i_0 is required, the method of reachable states is greatly superior.) We call a function v on the (not necessarily convex) set $X \subset \mathbb{R}^d$ **convex and piecewise linear** ($v \in$ ***CVPL*** for short) if v is the maximum of finitely many linear functions. This means that there exists a finite non-empty set $\Gamma \subset \mathbb{R}^d$ of vectors γ, such that

$$v(x) = \max_{\gamma \in \Gamma} \langle \gamma, x \rangle, \quad x \in X. \tag{26.21}$$

Each such set Γ determines v and is called a *representative* of v, and the system of all representatives of v is denoted by $\mathfrak{G}(v)$. As an example, $x \mapsto v(x) := \alpha_1 \cdot x^+ + \alpha_2 \cdot x^-$, $x \in X := \mathbb{R}$, and $\alpha_1, \alpha_2 \in \mathbb{R}_+$, belongs to *CVPL* with $\Gamma = \{\alpha_1, -\alpha_2\}$, since $v(x) = \max\{\alpha_1 \cdot x, -\alpha_2 \cdot x\}$. The name *CVPL* stems from the following two properties of functions $v \in$ *CVPL*: (i) v is convex, provided its domain X is convex. This holds since the maximum of finitely many convex functions on a convex set is convex (cf. Appendix D.2.3 and Appendix D.2.13). (ii) v from (26.21) is linear on each of the sets $\{x \in X : v(x) = \langle \gamma, x \rangle\}$, $\gamma \in \Gamma$. The union of these sets equals X. Note that $v \in$ *CVPL* may have several representatives: If, for example, $\Gamma \in \mathfrak{G}(v)$, $X = \mathbb{R}_+$, $\gamma_0 < \gamma$ for all $\gamma \in \Gamma$, then also $\Gamma + \{\gamma_0\} \in \mathfrak{G}(v)$. Finally, note that $v \in$ *CVPL* if and only if there exists a finite set $K \neq \emptyset$ and a family of vectors $y_k \in \mathbb{R}^d$, $k \in K$, such that

$$v(x) = \max_{k \in K} \langle y_k, x \rangle, \quad x \in X,$$

and then $\cup_{k \in K} \{y_k\} \in \mathfrak{G}(v)$.

The simple proof of the next result is omitted.

Lemma 26.2.11 (Closure properties of *CVPL***)** *The set CVPL of convex and piecewise linear functions on a set $X \subset \mathbb{R}^d$ is a convex cone and closed under forming the maximum of finitely many functions. Moreover, for a non-empty finite* *set J, $v_j \in CVPL$ and $\Gamma_j \in \mathfrak{G}(v_j)$, $j \in J$:*

$$\left\{ \sum_{j \in J} \gamma_j : (\gamma_j)_{j \in J} \in \times_{j \in J} \Gamma_j \right\} \in \mathfrak{G}\left(\sum_{j \in J} v_j \right),$$

$$\bigcup_{j \in J} \Gamma_j \in \mathfrak{G}\left(\max_{j \in J} v_j \right). \tag{26.22}$$

In the sequel the ϑ-th component of a vector $\gamma \in \mathbb{R}^d$ is written as $\gamma(\vartheta)$ rather than γ_ϑ.

Theorem 26.2.12 (Solution of simple POMCMs via convex and piecewise linear value functions) *Consider a simple model (POMCM, $\hat{\mu}, \varphi$) with φ from (26.14). Assume:*

(i) $\Theta = \mathbb{N}_d$, $Z = \mathbb{N}_m$ for some $d \geq 2$, $m \geq 2$, and A is finite,
(ii) $\hat{\mu}(i)$ is the probability distribution on Θ with discrete density $i \in I :=$ the set of all discrete densities of probability distributions on Θ, considered as subset of $\mathbb{R}^d$.

Then the associated MDP′ can be solved as follows:

(a) The value function $i \mapsto v_n(i)$ is convex and piecewise linear for $n \geq 0$, i.e. there exist a finite set Γ_n of vectors $\gamma \in \mathbb{R}^d$ such that $v_n(i) = \max_{\gamma \in \Gamma_n} \langle \gamma, i \rangle$, $i \in I$.
(b) For $\lambda = (\lambda_z)_{z=1}^m \in (\mathbb{R}^d)^m$ and $a \in A$ put

$$u_{\lambda a}(\vartheta) := \sum_{z=1}^{m} \sum_{\vartheta'=1}^{d} \lambda_z(\vartheta') p(\vartheta, a, \vartheta', z), \quad \vartheta \in \Theta.$$

One can choose $\Gamma_0 := \{V_0\}$, and recursively for $n \geq 1$

$$\Gamma_n := \cup \left\{ r(\cdot, a) + \beta \cdot u_{\lambda a}(\cdot) : (\lambda, a) \in \Gamma_{n-1}^m \times A \right\}.$$

(c) For $n \geq 1$ and $i \in I$ let $f_n'(i)$ be a maximum point of

$$a \mapsto r'(i, a) + \beta \cdot \max_{\lambda \in \Gamma_{n-1}^m} \langle u_{\lambda a}, i \rangle.$$

Then f_n' is a maximizer at stage n.

Proof

(a) We use induction on $n \geq 0$ of the assertion (I_n). Since $i \mapsto \langle \gamma, i \rangle$ belongs to *CVPL* for each vector $\gamma \in \mathbb{R}^d$, (I_0) holds with $\Gamma_0 := \{V_0\}$, and $i \mapsto r'(i, a) = \langle \gamma, i \rangle$ with $\gamma := r(\cdot, a)$ belongs to *CVPL*. Assume (I_{n-1}) for some $n \geq 1$. Put $Z(i, a) := \{z \in Z : p_{i\kappa}(a, z) > 0\}$. We use the VI (26.15) (which holds since

A is finite and r' and v_0 are bounded) with φ from (26.14) and observe that K_S can be dropped and that $\kappa(s, a, d\vartheta')$ and $\nu(s, a, dz)$ are counting measures. Now (26.12) yields

$$v_n(i) = \max_{a \in A}[r'(i, a) + \beta \cdot H_n(i, a)], \text{ where}$$

$$H_n(i, a) := \sum_{z \in Z(i,a)} p_{i\kappa}(a, z) \cdot v_{n-1}(p_i(a, \cdot, z)/p_{i\kappa}(a, z)).$$

By the closure properties of *CVPL* from Lemma 26.2.11, it suffices for (I_n) to show that for each $a \in A$ the function $i \mapsto H_n(i, a)$ belongs to *CVPL*. Observing that $p_{i\kappa}(a, z) = 0$ implies $p_i(a, \vartheta', z) = 0$ for all ϑ', and using (I_{n-1}) we obtain

$$\begin{aligned} H_n(i, a) &= \sum_{z \in Z(i,a)} p_{i\kappa}(a, z) \cdot \max_{\gamma \in \Gamma_{n-1}} \langle \gamma, p_i(a, \cdot, z)/p_{i\kappa}(a, z) \rangle \\ &= \sum_{z \in Z(i,a)} \max_{\gamma} \langle \gamma, p_i(a, \cdot, z) \rangle = \sum_{z=1}^{m} \max_{\gamma} \langle \gamma, p_i(a, \cdot, z) \rangle \\ &= \sum_{z=1}^{m} \max_{\gamma \in \Gamma_{n-1}} \langle u_{\gamma,a,z}, i \rangle, \end{aligned} \tag{26.23}$$

where $u_{\gamma,a,z}(\vartheta) := \sum_{\vartheta' \in \Theta} \gamma(\vartheta') \cdot p(\vartheta, a, \vartheta', z)$. Since *CVPL* is closed under summation, $i \mapsto H_n(i, a)$, $a \in A$, belongs to *CVPL*, which proves (I_n).

(b) We prove the assertion (I_n) by induction on $n \geq 0$. Firstly, (I_0) has been shown in (a). Assume (I_{n-1}) for some $n \geq 1$. Put $\Gamma := \Gamma_{n-1}$ and, using $u_{\gamma,a,z}$ from (a),

$$\Gamma_{n,a,z} := \{u_{\gamma,a,z} : \gamma \in \Gamma\}, \quad \Gamma_{n,a} := \{u_{\lambda a} : \lambda \in \Gamma^m\}.$$

Now (26.23) implies

$$H_n(i, a) = \sum_{z=1}^{m} \max_{\gamma \in \Gamma_{n,a,z}} \langle \gamma, i \rangle = \max_{\gamma \in \Gamma_{na}} \langle \gamma, i \rangle = \max_{\lambda \in \Gamma^m} \langle u_{\lambda a}, i \rangle.$$

Next, $r'(i, a) = \langle r(\cdot, a), i \rangle$ yields

$$\begin{aligned} r'(i, a) + \beta \cdot H_n(i, a) &= \langle r(\cdot, a), i \rangle + \max_{\lambda \in \Gamma^m} \langle u_{\lambda a}, i \rangle \\ = \max_{\lambda \in \Gamma^m} \langle r(\cdot, a) + \beta \cdot u_{\lambda a}, i \rangle &= \max_{\gamma \in \Gamma'_{na}} \langle \gamma, i \rangle, \end{aligned} \tag{26.24}$$

where $\Gamma'_{na} := \cup_{\lambda \in \Gamma^m} \{r(\cdot, a) + \beta \cdot u_{\lambda a}\}$. Now (I_n) follows from (26.22) since $\Gamma_n = \cup_{a \in A} \Gamma'_{na}$.

(c) This is obvious from (26.24). □

Example 26.2.13 (Maintenance problem of a processing system) A plant produces products having $k \geq 2$ different attributes (e.g. dimensions), each of which may be defective (e.g. beyond a given limit). A product is defective if and only if at least one of its attributes is defective. There is a processing system consisting of k tools, each of which is designed for the repair of one and only one attribute. Tools may be faulty. Processing a product means that each defective attribute is repaired by the corresponding tool, provided the latter is faultless. At the beginning of each period either the system processes one of the products (actions $a = 1, 2$ below) or the system is overhauled (actions $a = 3, 4$ below). Let $\vartheta_t \in \Theta := \{0, 1, \ldots, k\}$ denote the unknown number of faultless tools at the beginning of period $t \geq 0$. We assume:

(i) The $k - \vartheta_t$ faulty tools stay faulty until they are replaced sometime in the future, if at all. During period t each of the ϑ_t faultless tools becomes faulty, independently of each other, with probability $\alpha \in (0, 1)$.

(ii) Each of the k attributes of each processed product is defective before entering the system, independently of each other, with probability $1-\gamma \in (0, 1)$. Thus a product, after having been processed by the system at time t, is non-defective with probability $\gamma^{k-\vartheta_t}$. This follows from the fact that the product is non-defective if and only if the $k - \vartheta_t$ attributes corresponding to the $k - \vartheta_t$ faulty tools are non-defective at time t.

(iii) In each period the following actions are available:

  $a = 1$: process a product and examine afterwards its status: defective ($z = 0$) or non-defective ($z = 1$),
  $a = 2$: process a product without examining its status,
  $a = 3$: do not process a product, but inspect all tools and replace the faulty ones,
  $a = 4$: do not process a product, but replace all tools without inspection.

  The facility which inspects tools under $a = 3$ does not reveal to the decision maker which of the tools are faultless; therefore no information about ϑ_t is gained under $a = 3$.

(iv) Replacement of tools takes one period.

(v) There is a positive reward R for each product which is non-defective after its processing of the system, positive costs c_1 for the examination of the status of a processed product, c_2 for the inspection and c_3 for the replacement of a tool. Rewards and costs are discounted back to time $t = 0$ with the discount factor $\beta \in [0, 1]$.

(vi) Under action $a = 2$, where the status is not observed and under actions 3, 4, where no product is processed, we use the dummy disturbance $z := 2$ which carries no information about ϑ.

(vii) After N periods the remaining faulty tools are worthless, while the scrap value of each faultless tool equals the replacement cost c_3.

We assume $\hat{\mu}(i) := i \in I$, where I is the set of all discrete densities on Θ and want to maximize the N-stage expected reward, given an arbitrary prior $i \in I$ for the initial number ϑ_0 of faultless tools.

We model the problem by the following simple POMCM: Θ as above, $Z := \{0, 1, 2\}$ and $A := \{1, 2, 3, 4\}$.

Since Θ and Z are finite, $P(\vartheta, a, d(\vartheta', z))$ has a discrete density function $(\vartheta', z) \mapsto p(\vartheta, a, \vartheta', z)$. We can write it in the form $p(\vartheta, a, \vartheta', z) = p_\Theta(\vartheta, a, \vartheta') \cdot p_Z(\vartheta, a, \vartheta', z)$, where the discrete transition density functions p_Θ and p_Z are obtained as follows:

$a = 1$: In the current period the number of faultless tools cannot increase, hence $p_\Theta(\vartheta, 1, \vartheta') = 0$ for $\vartheta' > \vartheta$. Otherwise $p_\Theta(\vartheta, 1, \vartheta')$ equals the probability that from the ϑ faultless tools at the beginning of the period, ϑ' remain faultless. Thus $p_\Theta(\vartheta, 1, \vartheta')$ equals the binomial distribution $\mathrm{Bi}(\vartheta, 1 - \alpha)$, i.e.

$$p_\Theta(\vartheta, 1, \vartheta') = \binom{\vartheta}{\vartheta'} \cdot (1 - \alpha)^{\vartheta'} \cdot \alpha^{\vartheta - \vartheta'}.$$

Due to the definition of the binomial coefficients this also covers the case where $\vartheta' > \vartheta$. Next, by (ii) above we have

$$p_Z(\vartheta, 1, \vartheta', z) = (1 - \gamma^{k-\vartheta}) \cdot \delta_{z,0} + \gamma^{k-\vartheta} \cdot \delta_{z,1}.$$

$a = 2$: Obviously $p_\Theta(\vartheta, 2, \vartheta') = p_\Theta(\vartheta, 1, \vartheta')$ and $p_Z(\vartheta, 2, \vartheta', z) = \delta_{z,2}$.

$a = 3, 4$: At the end of the current period all tools are faultless and no product is processed. Thus

$$p_\Theta(\vartheta, a, \vartheta') = \delta_{\vartheta',k}, \quad p_Z(\vartheta, a, \vartheta', z) = \delta_{z,2}.$$

Next, from (ii) above we obtain $\tilde{r}(\vartheta, a, \vartheta', z)$ for $a = 1, 2$ below, while the other one-stage rewards are clear.

$$\tilde{r}(\vartheta, a, \vartheta', z) = \begin{cases} R \cdot \delta_{z,1} - c_1, & \text{if } a = 1, \\ \gamma^{k-\vartheta} \cdot R \cdot \delta_{z,2}, & \text{if } a = 2, \\ -kc_2 - (k - \vartheta)c_3, & \text{if } a = 3, \\ -kc_3, & \text{if } a = 4. \end{cases}$$

Finally $V_0(\vartheta) = \vartheta \cdot c_3$.

Although $\vartheta_0, \vartheta_1, \ldots, \vartheta_N$ remain hidden for the decision maker, he knows after each realization of the stochastic decision process the total N-stage reward.

Obviously (A1) is fulfilled, taking for κ and ν the counting measure on Θ and on $S := \{1\}$, respectively. Since S and A are finite, the VI holds in the associated MDPD$'$ by the Basic Theorem 12.1.5. By (26.8), (26.9), (26.12) and

Proposition 26.2.6 it has the following form, using $v_0(i) := c_3 \cdot \sum_{\vartheta=0}^{k} i(\vartheta) \cdot \vartheta$:

$$v_n(i) = \max_{1 \le a \le 4} [r'(i,a) + \beta \cdot H_n(i,a)], \quad n \ge 1, i \in I, \tag{26.25}$$

where

$$H_n(i,a) := \sum_z p_{i\kappa}(a,z) \cdot v_{n-1}(p_i(a,\cdot,z)/p_{i\kappa}(a,z)). \tag{26.26}$$

Here the sum extends over those $z \in \{0, 1, 2\}$ for which $p_{i\kappa}(a,z) > 0$.

We now collect in (26.27) and (26.28) all expressions needed for the computation of the value functions v_n via (26.25) and (26.26). Firstly note that $p_{i\kappa} < \infty$ since Θ is finite. Let τ be a discrete random variable on Θ with discrete density i. Denote expectation with respect to i by $\mathbb{E}_i$. Put

$$g(\vartheta, \vartheta') := \alpha^{\vartheta} \cdot \left(\frac{1-\alpha}{\alpha}\right)^{\vartheta'} \cdot \binom{\vartheta}{\vartheta'}. \tag{26.27}$$

Then we have:

$$\begin{aligned}
r'(i,1) &= R \cdot \mathbb{E}_i\, \gamma^{k-\tau} - c_1, \quad r'(i,2) = R \cdot \mathbb{E}_i\, \gamma^{k-\tau}, \\
r'(i,3) &= -kc_2 - (k - \mathbb{E}_i\, \tau) \cdot c_3, \quad r'(i,4) = -kc_3, \\
p_i(1, \vartheta', 1) &= \gamma^k \cdot \mathbb{E}_i(g(\tau, \vartheta')/\gamma^{\tau}), \\
p_i(1, \vartheta', 0) &= \mathbb{E}_i\, g(\tau, \vartheta') - p_i(1, \vartheta', 1), \\
p_{i\kappa}(1,1) &= \mathbb{E}_i\, \gamma^{k-\tau} = 1 - p_{i\kappa}(1,0),\ p_{i\kappa}(1,2) = 0, \\
p_i(2, \vartheta', z) &= \delta_{z,2} \cdot \mathbb{E}_i\, g(\tau, \vartheta'), \\
p_{i\kappa}(2,z) &= \delta_{z,2}, \\
p_i(3, \vartheta', z) &= p_i(4, \vartheta', z) = \delta_{\vartheta',k} \cdot \delta_{z,2}, \\
p_{i\kappa}(3,z) &= p_{i\kappa}(4,z) = \delta_{z,2}.
\end{aligned} \tag{26.28}$$

Obviously $p_{i\kappa}(1,1) > 0$, and $p_{i\kappa}(1,0) = 0$ if and only if $i = \delta_{k\cdot}$. Moreover, $p_{i\kappa}(a,z) > 0$ if $z = 2$, $a = 2, 3, 4$. Finally (26.26), (26.27), and (26.28) yield $H_n(i,a)$ for all (i,a). In particular we get: If $i = \delta_{k,\cdot}$ then $H_n(i,1) = v_{n-1}(g(k,\cdot))$, $H_n(i,2) = v_{n-1}(\mathbb{E}_i\, g(\tau,\cdot))$ and $H_n(i,3) = H_n(i,4) = v_{n-1}(\delta_{\cdot,k})$.

Now our collection of expressions is complete.

In Figs. 26.1 and 26.2 we add the graph of v_n for several data of the simple POMCM starting with no defective tools, i.e. $i_0 = \{0, 0, 1\}$. ♦

Consider as in Theorem 26.2.12 a simple model $(POMCM, \hat{\mu}, \varphi)$ with property (P). We now turn to the method of reachable states from Sect. 3.2 for computing

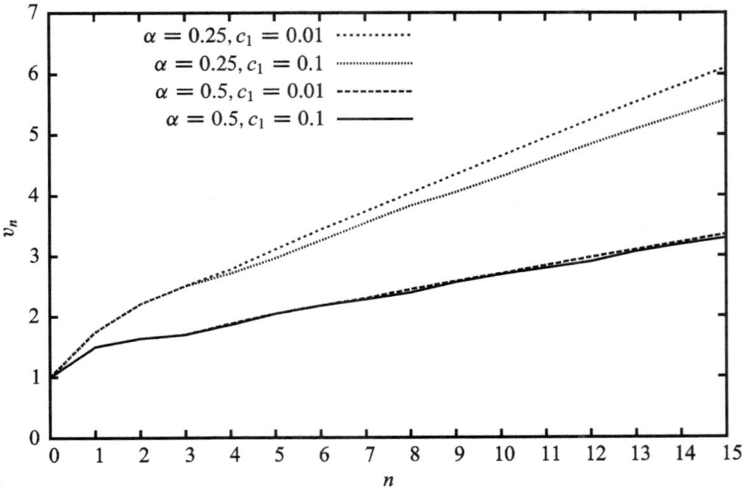

Fig. 26.1 v_n of Example 26.2.13 for $k = 2, R = 1, \beta = 0.99, \gamma = 0.25, c_2 = 0.25, c_3 = 0.5$

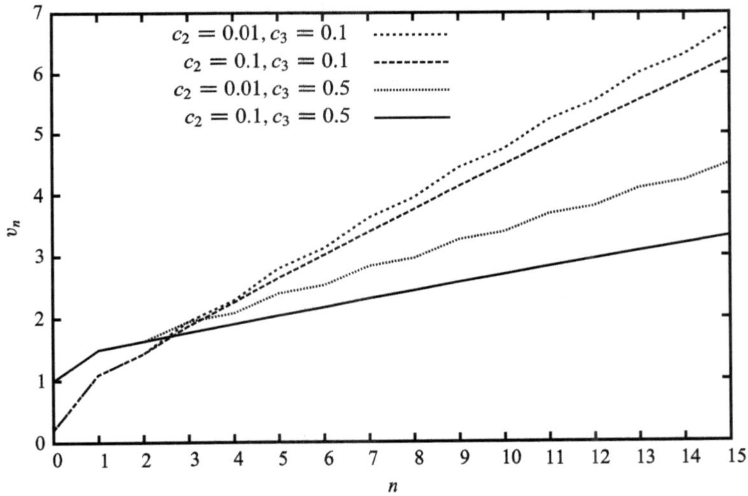

Fig. 26.2 v_n of Example 26.2.13 for $k = 2, R = 1, \beta = 0.99, \gamma = 0.2, \alpha = 0.5, c_1 = 0.1$

$v_n(i_0)$ for a single i_0. This is no restriction in practice, since there a fixed i_0 is given. Moreover, if one is interested in the dependence of the solution on i_0, the new algorithm still computes the solution for large finite sets of initial indexes much faster than the algorithm in Theorem 26.2.12.

Put $Z(i,a) := \{z \in Z : p_{i\kappa}(a,z) > 0\}$, which is nonempty for all $i \in I$, $a \in A$, since $\sum_{z\in Z} p_{i\kappa}(a,z) = 1$. The associated MDP′ has a finite transition law since Proposition 26.2.4(c) and (26.12) imply finiteness of the sets

$$R(i,a) := \{i' \in I : p'(i,a,i') > 0\} = \varphi(i,a,Z(i,a)), \quad (i,a) \in I \times Z.$$

Now we apply Proposition 12.1.9 to MDP′ by replacing S, P, r and V_n in Proposition 12.1.9 with I, P', r' and by v_n, respectively. Let $R_t(i_0) \subset I$, $t \geq 0$, $i \in I$, be defined as in Chap. 13 as the set of states in MDP′, reachable at time t from the initial state i_0. This means that $R_0(i_0) := \{i_0\}$ and

$$R_1(i) := \cup_{a\in A} R(i,a) = \cup_{a\in A} \varphi(i,a,Z(i,a)),\ i \in I,$$
$$R_{t+1}(i_0) := \cup_{i\in R_t(i_0)} R_1(i), \quad t \geq 0.$$

Proposition 26.2.14 (Solution of simple POMCMs via reachable states) *Consider a simple model (POMCM, $\hat{\mu}$, φ) with property (P)*

(a) We have, using $v_0(i) := \sum_{\vartheta\in\Theta} i(\vartheta) \cdot V_0(\vartheta)$,

$$\begin{aligned} v_n(i) &= \max_{a\in A} \left[r'(i,a) + \beta \cdot \sum_{z\in Z(i,a)} p_{i\kappa}(a,z) v_{n-1}(\varphi(i,a,z)) \right] \\ &=: \max_{a\in A} w_n(i,a), \quad 1 \leq n \leq N, i \in I. \end{aligned} \tag{26.29}$$

The recursion (26.29) also holds when i is restricted to $R_{N-n}(i_0)$. Thus one can compute $v_N(i_0)$ in finitely many steps.

(b) Each policy $(f'_n)_N^1 \in (\mathbb{F}')^N$ such that $f'_n(i)$ is a maximum point of $a \mapsto w_n(i,a)$ for $i \in R_{N-n}(i_0)$, $1 \leq n \leq N$, is optimal for problem $MDP'_N(i_0)$.

Proof The proof is a straightforward application of Proposition 12.1.9, combined with (26.16), which tells us that for all $n \geq 1$ and $i \in I$

$$v_n(i) = \max_{a\in A} \left[r'(i,a) + \beta \cdot \sum_{i'\in R(i,a)} p'(i,a,i') v_{n-1}(i') \right].$$

The recursion (26.29) with i restricted to $R_{N-n}(i_0)$ also holds since $i \in R_{N-n}(i_0)$ implies $R(i,a) \subset R_{N-(n-1)}(i_0)$. □

As far as the computation of the value functions in (26.29) is concerned one computes in a first step the sets $R_t(i_0)$, which is done by using the flowchart after Proposition 12.1.9 for MDP′. This requires the computation of $p_i(a,\cdot,z)$, $p_{i\kappa}(a,z)$, $Z(i,a)$, $\varphi(i,a,z)$ and $R_1(i)$ for $i \in R_t(i_0)$, $a \in A$, $z \in Z$. In a second step one computes by the partial VI (26.29) the values $v_n(i)$ and $f'_n(i)$ for $i \in R_{N-n}(i_0)$, $1 \leq n \leq N$.

26.3 Supplements

Supplement 26.3.1 (Embedding the prior into a family of distribution functions)
In applications $i \mapsto \hat{\mu}(i)$ is usually injective, in which case a given prior μ_0 can be embedded into $\hat{\mu}$ in at most one way. Otherwise, it may happen that μ_0 can be embedded into $\hat{\mu}$ in several ways. However, as the problem is defined independently of $\hat{\mu}$, the solutions for different embeddings must be the same.

Supplement 26.3.2 (If the disturbances coincide with the states) In POMDPs we have $z_\nu = s_\nu$, $\nu \geq 1$. Then the disturbances z_t could be dropped in h_t, i.e. h_t could be replaced by $s_0^t := (s_0, s_1, \ldots, s_t)$, $t \geq 1$, and Δ_N could be replaced by Δ_{NS}, the set of sequences $\pi = (\pi_t)_0^{N-1}$ where π_t is a measurable mapping from S^{t+1} into A such that $\pi_t(s_0^t) \in D_S(s_t)$ for all t and s_0^t. Moreover, $\omega_\nu = (z_\nu, \vartheta_\nu, s_\nu)$ could be replaced by (z_ν, s_ν).

For simplicity we treat POMDPs within the more general framework for POMs. It then turns out automatically that the posterior index functions and the history-dependent policies do not depend on disturbances.

Supplement 26.3.3 (Information for the decisionmaker) The framework for the problem $POM_N(\mu_0)$ implies the following situation: The decisionmaker which uses policy $\pi \in \Delta_N$ and starts with the observable state s_0 uses as information about the unknown $(\vartheta_\nu)_0^t$ for his decision at time $0 \leq t \leq N-1$ only the history h_t, but not possible further information which may be contained in the one-stage rewards or in the terminal reward. This can be imagined as if the latter become known to the decision maker only at time N when all decisions have been made.

Appendix A
Elementary Results on Optimization

A.1 Real and Extended Real Numbers

A.1.1 (**$\mathbb{N}$, $\mathbb{Z}$, $\mathbb{R}$**) $\mathbb{N}$, $\mathbb{Z}$ and $\mathbb{R}$ denote the set of positive integers, of integers and reals, respectively. $\mathbb{N}_0 := \{0\} + \mathbb{N}$; $\mathbb{N}_k := \{1, 2, \ldots, k\}$; $\mathbb{R}_+ := [0, \infty)$, $\mathbb{R}^+ := (0, \infty)$, $\overline{\mathbb{R}} := [-\infty, \infty]$. The phrase $n \geq 1$ means $n \in \mathbb{N}$, and similarly for $n \geq k$ for $k \in \mathbb{Z}$. $\mathbb{R}^n$ is the n-dimensional Euclidean space with elements $x = (x_i)_1^n := (x_1, x_2, \ldots, x_n)$; $\mathbb{R}^n_+ := [0, \infty)^n$, $(\mathbb{R}^+)^n := (0, \infty)^n$. The sets $\mathbb{R}^n$, $n \geq 1$, are called Euclidean spaces. The Kronecker symbol is

$$\delta_{ij} := \begin{cases} 1, & \text{if } i = j, \\ 0, & \text{else.} \end{cases}$$

The numbers $x^+ := \max\{x, 0\}$ and $x^- := \max\{-x, 0\} = (-x)^+$ are called the *positive* and the *negative part* of $x \in \overline{\mathbb{R}}$. Then $x = x^+ - x^-$ and $|x| = x^+ + x^- = \max\{x, -x\}$.

$\lfloor x \rfloor$ is the largest integer $\leq x \in \mathbb{R}$, and $\lceil x \rceil = -\lfloor -x \rfloor$ is the smallest integer $\geq x$. In case $a, b \in \mathbb{Z}$ with $b < a$ we define $\sum_{i=a}^b x_i := 0$ and $\prod_{i=a}^b x_i := 1$; moreover, $0^0 := 1$.

A.1.2 ($\overline{\mathbb{R}}$)

(a) The usual ordering on the set $\mathbb{R}$ of reals is extended to an ordering on the set $\overline{\mathbb{R}} := \mathbb{R} + \{-\infty, \infty\}$ of extended reals by defining $-\infty \leq a \leq \infty$ for $a \in \overline{\mathbb{R}}$.

(b) We now extend the four arithmetical operations on $\mathbb{R}$ to $\overline{\mathbb{R}}$ in such a way that many of the properties of the four operations remain true. The addition on $\mathbb{R}^2$ is extended to $M := \overline{\mathbb{R}}^2 - \{(\infty, -\infty), (-\infty, \infty)\}$ by

$$x + (\pm\infty) := (\pm\infty) + x := (\pm\infty) + (\pm\infty) := \pm\infty, \ x \in \mathbb{R}.$$

K. Hinderer et al., *Dynamic Optimization*, Universitext,
DOI 10.1007/978-3-319-48814-1

The terms $(\pm\infty) + (\mp\infty)$ are not defined. Not all properties of the addition on $\mathbb{R}^2$ carry over to M; as an example, the equation $x + \infty = 0$ does not have a solution. On the other hand, addition on M is associative and commutative. As a consequence, finite sums of extended reals are defined unless both ∞ and $-\infty$ occur as terms. Subtraction is defined on $\overline{\mathbb{R}}^2 - \{(\infty,\infty), (-\infty,-\infty)\}$ by $a - b := a + (-b)$, using $-(\pm\infty) := \mp\infty$.

Multiplication is extended to $\overline{\mathbb{R}}^2$ by defining for $a \in \overline{\mathbb{R}}$

$$a(\pm\infty) := (\pm\infty)a := \begin{cases} \pm\infty, & \text{if } a > 0, \\ 0, & \text{if } a = 0, \\ \mp\infty, & \text{if } a < 0. \end{cases}$$

In particular, $0(\pm\infty) = (\pm\infty)0 = 0$; this turns out to be very useful in measure-theoretic probability. Not all properties of the multiplication on $\mathbb{R}^2$ carry over to $\overline{\mathbb{R}}^2$. However, it is associative and commutative. Division is defined on $\overline{\mathbb{R}}^2$ by $a/b := a \cdot (1/b)$, using $1/\pm\infty := 0$ and $1/0 := \infty$.

We also use $\infty^+ := (-\infty)^- := \infty$, $\infty^- := (-\infty)^+ := 0$.

A.1.3 ($x^-, x^+, \min\{x,y\}, \max\{x,y\}$) For real x, y, z:

(a) Due to $x^- = (-x)^+$ and $x^+ = (-x)^-$ results for positive parts yield analogous results for negative parts, and vice versa; $x = x^+ - x^-$; $|x| = x^+ + x^- = \max\{x, -x\}$; $|x| \le y$ if and only if $\pm x \le y$; $x \mapsto x^+$ is increasing on $\mathbb{R}$ and $x \mapsto x^-$ is decreasing on $\mathbb{R}$. The same statements hold for $x, y \in \overline{\mathbb{R}}$.

(b) Due to $\min\{x,y\} = -\max\{-x,-y\}$ results for the maximum yield analogous results for the minimum, and vice versa; $(x+y)^+ \le x^+ + y^+$; $2\max\{x,y\} = x+y+|x-y|$ and $2\min\{x,y\} = x+y-|x-y|$, hence $\max\{x,y\}+\min\{x,y\} = x+y$ and $\max\{x,y\} - \min\{x,y\} = |x-y|$; $\max\{x,y\} \le x + y$ if $x \ge 0$, $y \ge 0$.

A.1.4 ($\max_i x_i$) For reals x_i, y_i, $1 \le i \le k$:

(a) $\max_i(x_i + y_i) \le \max_i x_i + \max_i y_i$.

(b) $|\max_i x_i - \max_i y_i| \le \max_i |x_i - y_i|$.

(c) If g is an increasing function on an interval I, then $g(\max_i x_i) = \max_i g(x_i)$ for $x_i \in I$.

A.1.5 ($\lceil x \rceil$, $\lfloor x \rfloor$) For real x:

(a) Due to $\lceil x \rceil = -\lfloor -x \rfloor$ properties of $\lfloor \cdot \rfloor$ yield corresponding properties of $\lceil \cdot \rceil$, and vice versa.

(b) For $n \in \mathbb{Z}$ we have $\lfloor x \rfloor = n$ if and only if $n \le x < n + 1$.

A.1.6 (Convergence) By $x_n \to x$ we mean $x_n \to x$ for $n \to \infty$. *Convergence* of a sequence of functions to a function is understood pointwise, unless stated otherwise.

A.2 Sets

A.2.1 (Basic algebraic notations) $A \subset B$ denotes $A \subseteq B$. $A - B$ denotes the *difference* between the set A and the set B; B need not be contained in A. $A^c := \Omega - A$ is the *complement* of the set A with respect to a given basic set Ω. The union of disjoint sets is denoted by $\sum$ (and by $+$ in case of two sets) rather than by $\cup$. A *decomposition* of a set M is a family of disjoint subsets of M whose sum is M. $|A|$ is the *cardinality* of the set A. The set A is called *countable* if it is either finite or countably infinite. Sets of sets are called *systems of sets*. The *power set* of a set A is the system of all subsets of A; it is denoted by $\mathfrak{P}(A)$.

A.2.2 (Basic topological notations) For a subset M of a metric space the set ∂M denotes the *boundary* and M° the *interior* of M. If the statement $R(x)$ holds for no $x \in M \subset \mathbb{R}$, we put

$$\sup\{x \in M : R(x)\} := \inf M, \text{ and } \inf\{x \in M : R(x)\} := \sup M.$$

A.2.3 (Intervals) ***Intervals in*** $\mathbb{R}$ may be closed, open, half-open, bounded or unbounded, i.e. an interval is a subset of $\mathbb{R}$ of the form $[a,b]$, $[a,b)$, $(a,b]$, (a,b), $(-\infty,b]$, $(-\infty,b)$, $[a,\infty)$, (a,∞) or $(-\infty,\infty)$ for some reals $a < b$ or $\{a\}$ for some real a; note that we do not exclude ***degenerate intervals*** $[a,a]$; a ***d-dimensional interval***, $d \geq 2$, is a Cartesian product of d intervals.

A set M of integers is called an ***interval*** in $\mathbb{Z}$ if it has one of the following forms: (i) $\{x, x+1, \ldots, y\}$ with $x \leq y$, (ii) $M = \{x, x+1, \ldots\}$ for some $x \in \mathbb{Z}$, (iii) $M = \{x, x-1, x-2, \ldots\}$ for some $x \in \mathbb{Z}$ or (iv) $M = \mathbb{Z}$. In particular, the singletons in $\mathbb{Z}$ are intervals in $\mathbb{Z}$. An interval in $\mathbb{Z}$ is called a ***discrete interval*** if it has at least two elements.

A.3 Mappings

A.3.1 (Functions) Mappings from a set into $\overline{\mathbb{R}}$ are called ***functions***. A mapping v from M into B is denoted by $v\colon M \to B$, by $x \mapsto v(x)$, $x \in M$ or simply by $v(\cdot)$. For $v_1, v_2\colon M \to \overline{\mathbb{R}}$ we write:

(a) $v_1 = (\leq, <) v_2 :\iff v_1(x) = (\leq, <) v_2(x)$ for all $x \in M$,
(b) $\max\{v_1, v_2\}: \iff M \ni x \mapsto \max\{v_1(x), v_2(x)\}$,
(c) v ***increasing*** or ***isotone*** (***decreasing*** or ***antitone***) $:\iff$ v non-decreasing (non-increasing).

A.3.2 (Section of a set and a function) For $M \subset M_1 \times M_2$ the set $M(x) := \{y \in M_2 : (x,y) \in M\}$, $x \in M_1$, is called the *x-section* of M; moreover, if $M(x)$ is non-empty, $v(x,\cdot)$ is the x-section of the mapping v on M, i.e. the function $y \mapsto v(x,y)$, $y \in M(x)$.

A.3.3 (Cartesian product) The *Cartesian product* $M_1 \times M_2 \times \cdots \times M_m$ (also denoted by $\times_{k=1}^m M_k$) of non-empty sets M_k is the set of all m-tuples $(x_k)_1^m$ with $x_k \in M_k$ for all k. This notion of a Cartesian product of finitely many set is generalized as follows. Let $(M_k, k \in K)$ be a family of non-empty subsets of a basic set M. The set of all mappings v from K into M such that $v(k) \in M_k$ for all k is also called the *Cartesian product* of the sets M_k, denoted by $\times_{k \in K} M_k$, or simply by $\times_k M_k$. Thus $M^K := \times_k M$ is the set of all mappings from K into M.

The projection pr_j from $M = \times_i M_i$ into M_j, $j \in I$, is defined by $\mathrm{pr}_j((x_i)) := x_j$. The set $\mathrm{pr}_j(G) := \{\mathrm{pr}_j(x) : x \in G\}$ is called the *projection* of the non-empty set $G \subset \times_i M_i$ into M, $j \in I$. Finally, the set $\bigcup_{j \in I} \mathrm{pr}_j(G)$ is called the *projection* $\mathrm{pr}(G \to M)$ of G into M.

A.3.4 (Argument and restriction of a function) Let $m \geq 2$ and let M_i, $1 \leq i \leq m$, be non-empty sets. If $(x_i)_1^m \mapsto v((x_i)_1^m)$ is a mapping from a non-empty subset G of $\times_1^m M_i$ into a set B, then we call x_i the *i-th argument of* v. More generally, if I is a non-empty subset of $\mathbb{N}_m$, we put $M_I := \times_{i \in I} M_i$ and call $x_I := (x_i, i \in I)$ the I-argument of v. We set $\mathrm{proj}(G \to M_I) := \{(\mathrm{pr}_j(x))_{j \in I}, x \in G\}$. For each $z \in \mathrm{proj}(G \to M_I)$ the variable x_I varies in the non-empty z-section G_z of G.

$v|B$ is the *restriction* to $B \subset M$ of a mapping v on M. The set $v(M) \subset C$ is the *range* of $v: M \to C$. $\sup f := \sup\{f(a) : a \in A\}$ for a function f defined on A, and similarly for $\inf f$.

A.3.5 (Indicator function) The ***indicator function*** of a subset A of a basic set Ω is defined on Ω by

$$1_A(\omega) := \begin{cases} 1, & \text{if } \omega \in A, \\ 0, & \text{if } \omega \notin A. \end{cases}$$

A.3.6 (Family of elements) A *family* $(x_i, i \in I)$ of elements $x_i \in M$ with index set I is just another notation for the mapping $i \mapsto x_i$ from I into M. In contrast to the elements of sets the elements of families need not be different.

A.3.7 (Direct and inverse image of a set under a function) For each mapping $v: M \to M'$ we call $v(B) := \{v(b) : b \in B\}$ the *direct image* of the set $B \subset M$ and $v^-(B') := \{x \in M : v(x) \in B'\}$ the *inverse image* of the set $B' \subset M'$. Note that $v(\emptyset) = v^-(\emptyset) = \emptyset$.

A.3.8 (Section of a set and a function) For $M \subset M_1 \times M_2$ the set $M(x) := \{y \in M_2 : (x, y) \in M\}$, $x \in M_1$, is called the *x-section* of M; moreover, if $M(x)$ is non-empty, $v(x, \cdot)$ is the x-section of the mapping v on M, i.e. the function $y \mapsto v(x, y)$, $y \in M(x)$.

A.4 Extrema of Functions

A.4.1 (Literature) There exists an extensive literature on numerical optimization methods. An introduction to the area is given in Peressini et al. (1988, Chapter 3). Standard references include Dennis and Schnabel (1983), Fletcher (1987) and Rao (1984).

An encyclopedia on optimization is edited by Göpfert et al. (1986). Introductions to general optimization include Foulds (1981) with an extensive bibliography, Horst and Tuy (1990) and Peressini et al. (1988). Practically all methods are iterative in the sense that one constructs a sequence of points $x_n \in M$ which converges *rapidly* towards a maximum point of f (or at least a local maximum point or a stationary point). Classical methods are Newton's method for solving $\operatorname{grad} v(x) = 0$ for the search for stationary points, and the so-called *method of steepest descent*. However, both methods are superseded by newer approaches in the literature mentioned above.

In this section we collect tools from classical optimization which are useful for solving a few of the one-stage problems to which N-stage Dynamic Programs are reduced. In contrast to this our presentation differs from traditional ones as follows:

- We deal exclusively with global rather than local extrema.
- Second derivatives and the Hessian are nearly always avoided.
- The importance of convexity, concavity, symmetry, transformations and substitutions is stressed.

A.4.1 *Extrema of Functions on Arbitrary Domains*

A.4.2 (sup v) Let v be a function on M. Then $\sup v := \sup v(M)$ is the unique number $\gamma \in \overline{\mathbb{R}}$ with the following two properties: (i) $v \leq \gamma$, (ii) for each $y < \gamma$ there exists an $x \in M$ such that $v(x) \geq y$. It follows that there exist $x_n \in M$, $n \geq 1$, such that $v(x_n) \uparrow \sup v$. Due to $\inf v = -\sup(-v)$ and $\sup v = -\inf(-v)$ results for the supremum of functions yield analogous results for the infimum, and vice versa.

For functions v and w on a set M:

(a) If $\alpha \in \mathbb{R}_+$ then
$\sup(\alpha v) = \alpha \sup v$, $\inf(\alpha v) = \alpha \inf v$,
$\sup(-\alpha v) = -\alpha \inf v$, $\inf(-\alpha v) = -\alpha \sup v$.
(b) $\sup(c + v) = c + \sup v$, $c \in \mathbb{R}$.
(c) If $u \leq v$ then $\sup u \leq \sup v$, $\inf u \leq \inf v$.
(d) $(\sup v)^+ = \sup v^+$, $(\sup v)^- = \inf v^-$, $(\inf v)^+ = \inf v^+$, $(\inf v)^- = \sup v^-$.

A.4.3 (inf v) For each function v on a set M:

$$\inf |v| \leq |\inf v|, \quad |\sup v| \leq \sup |v|.$$

Here the first inequality is easily proven if $v \geq 0$ or $v \leq 0$. If $v(x) < 0 < v(y)$ for some x, y, then $\inf |v| \leq -v(x) \leq \sup(-v) = -\inf v = |\inf v|$. The second inequality holds since $\pm v \leq |v|$ implies $\sup v \leq \sup |v|$ and $-\sup v = \inf(-v) \leq \inf |v| \leq \sup |v|$.

A.4.4 (inf($u \pm v$), sup($u \pm v$)) For functions u and v on a set M:

(a) $\inf u + \inf v \leq \inf(u + v)$, $\sup(u + v) \leq \sup u + \sup v$.
(b) If u and v are upper bounded then
$$\sup u - \sup v \leq \sup(u - v), \ |\sup u - \sup v| \leq \sup |u - v|.$$
(c) If u and v are lower bounded then
$$\inf(u - v) \leq \inf u - \inf v, \ |\inf u - \inf v| \leq \sup |u - v|.$$
(d) If u has a maximum [minimum] point, it is bounded above [below].

A.4.5 (sup $w(x, y)$)) For a function w on $M_X \times M_Y$ we have

$$\sup_{(x,y)} w(x, y) = \sup_x \sup_y w(x, y) = \sup_y \sup_x w(x, y).$$

A.4.6 (Existence of maximum points) Let v be continuous on an arbitrary set M. Assume that there is some $x_0 \in M$ with the following property: For each sequence (x_n) in M for which either $\|x_n\| \to \infty$ or for which (x_n) has a limit not belonging to M (which cannot happen if M is closed) we have

$$\inf_n v(x_n) \leq v(x_0).$$

Then the set of maximum points of v is nonempty and compact.

A.4.7 (sp v) For a function v on a set M the ***span*** (also called *total variation*) is defined by the number $\operatorname{sp} v := \sup v - \inf v \in [0, \infty]$.

(a) The following holds:

$$\operatorname{sp} v = \sup_{x,y \in M} [v(x) - v(y)] = \sup_{x,y \in M} |v(x) - v(y)| \leq 2 \sup |v|.$$

(b) v is bounded if and only if $\operatorname{sp} v$ is finite, and then $2 \sup |v| = \operatorname{sp} v + |\inf v + \sup v|$.
(c) The functional $v \mapsto \operatorname{sp} v$ on the vector space of bounded functions on a set M is a *semi-norm*.

A.4.8 (maximum point, minimum point) By a ***maximum point*** [***minimum point***] x^* of a function v on the set M we understand a ***global maximum point*** [***global minimum point***], i.e. a point where v assumes its supremum [infimum]. The (possibly empty) set of maximum points of v will be denoted by $M^* = M^*(v)$. Solving the problem *maximize* v means to find $\sup v$ and, if $M^*(v)$ is not empty, at least one maximum point of v or the whole set $M^*(v)$. If v assumes the value ∞, then $M^* = \{x \in M : v(x) = \infty\}$. As x^* is a minimum point of v if and only if it is a maximum point of $-v$, we usually state results either only for maximum points or only for minimum points. If M is a metric space, then $x^* \in M$ is called a ***local maximum point*** of v, if x^* is a maximum point of $v|B$ for some neighborhood B of x^*. For us there is rarely a need for using local extrema.

The search for maximum points of functions v often consists of the following two steps:

(i) Make sure (usually by a compactness and upper [lower] semicontinuity argument) that the set M^* is non-empty. Note that M^* is empty if v is upper unbounded.
(ii) Find a necessary condition, say (C), for maximum points (such as vanishing of partial derivatives, if these exist). Then M^* is contained in the (hopefully small) set M_C of points $x \in M$ which satisfy (C). In particular, if M_C turns out to be finite, a maximum point may be found by comparing the values of v on M_C.

A.4.9 (Maximizing a function by maximizing an auxiliary function) The task of maximizing a function $x \mapsto v(x)$ on a set X can sometimes be dealt with by maximizing an auxiliary function having one of the following two forms:

(i) $x \mapsto g(v(x))$, $x \in X$, where g is a function on $v(X)$. We say that $g \circ v$ is a ***transformation*** of v.
(ii) $b \mapsto v(h(b))$ for b in a set B, where h is a mapping from B into X. We say that $v \circ h$ is obtained from v by the ***substitution*** $x = h(b)$.

A.4.10 (Maximizing by transformation) Let X and $v: X \to \overline{\mathbb{R}}$ be arbitrary and let g be a strictly increasing function g on $v(X)$. Then v and $g \circ v$ have the same maximum points and the same minimum points.

A.4.11 (Maximizing by substitution) Let X and $v: X \to \overline{\mathbb{R}}$ be arbitrary and let h be a surjective mapping from a set B onto X. Then

(a) $\sup v = \sup v \circ h$.
(b) $M^*(v) = h(M^*(v \circ h))$, i.e. b^* is a maximum point of $v \circ h$ if and only if $h(b^*)$ is a maximum point of v.
(c) x^* is a maximum point of v if and only if some (and hence each) point b^* in the set $h^-(\{x^*\})$ is a maximum point of $v \circ h$. (Recall that h^- denotes the inverse image mapping of h.)
(d) Note that if h is not surjective, i.e. if X contains points which do not lie in $h(B)$, it may happen that $\sup v > \sup v \circ v$.

A.4.2 Extrema of Functions on Intervals

Abbreviations and Conventions

In this subsection v denotes, unless stated otherwise, a function on a non-degenerate interval $I \subset \mathbb{R}$ with endpoints a and b, $-\infty \leq a < b \leq \infty$. Finite endpoints of I

may or may not belong to I. Then $I \cap \delta I$ denotes the set of those endpoints of I which belong to I. Derivatives are used only in the interior $I^\circ = (a, b)$ of I. If v' exists at $x \in (a, b)$ and if $v'(x) = 0$, then x is called a ***stationary point*** of v.

Often the maximization of functions v on a compact interval $[a, b]$ is based on the following facts: If v is continuous then it has a (smallest and a largest) maximum point. Moreover, if v' exists on (a, b) and if v is evaluated at the stationary points and at a and b, the points with largest values are the maximum points. In calculus textbooks one sometimes uses the second derivative of v in order to exclude in this procedure stationary points which cannot be maximum points. According to our experience this rarely pays.

A.4.12 (Monotonicity)

(a) Let I be a non-degenerate interval. Let v be defined and continuous on I and differentiable on I°. Then v is increasing on I if and only if $v' \geq 0$ on I°. — Continuity of v on I may be weakened to continuity on $I \cap \delta I$ since differentiability of v on I° implies continuity of v on I°.

(b) Let v be defined on $[a, b]$, $-\infty < a < b < \infty$, let v be continuous at a and on b, and let v have a decreasing derivative v' on (a, b). (Hence v is concave, see Appendix D.2.5 and Appendix D.2.8(f).) Then a stationary point x^* of v is a maximum point of v. Functions with this property (in its weak form) are called ***unimodal with mode*** x^*.

A.4.13 (Unimodality criterion) Let v be defined on an interval I with endpoints $-\infty \leq a < b \leq \infty$. Assume that v is continuous on I, that v' exists on (a, b) and that $v'(x) = c(x)d(x)$, $x \in (a, b)$, for some non-negative [positive] function c and some decreasing function d with a [unique] zero x^*. Then x^* is a [unique] maximum point of v.

A.4.3 *Extrema of Functions of Several Real Variables*

Abbreviations and Conventions

Unless stated otherwise, $x = (x_i)_1^k \mapsto v(x)$ is a function on a set $M \subset \mathbb{R}^k$. $|\cdot|$ is any norm on $\mathbb{R}^k$. M° denotes the interior of M. The boundary of M is denoted by ∂M, hence $M \cap \partial M$ is the set of boundary points belonging to M.

The role of the derivative $v'(x)$ for functions of one real variable is now taken over by the vector of partial derivatives at the point x, i.e. by the *gradient* of v at x,

$$\operatorname{grad} v(x) := \left(\frac{\partial f}{\partial x_i}(x)\right)_1^k, \quad x \in M^\circ.$$

A point $x \in M^\circ$ is called a ***stationary point*** of v if grad v vanishes at x.

A.4.14 (The set of maximum points)

(a) If $M \subset \mathbb{R}^k$ is compact (i.e. bounded and closed) and if $v: M \to \mathbb{R}$ is continuous then the set of maximum points of v is nonempty and compact.
(b) If $v = \inf v$ on $M \cap \partial M$, and if v is not constant, then all maximum points are interior points.

A.4.15 (Existence of maximum points) Let $v: M \to \mathbb{R}$ be continuous and not constant. If $v(x_n) \to \inf v$ for each sequence (x_n) in M for which either $|x_n| \to \infty$ or for which (x_n) has a limit not belonging to M, then the set of maximum points of v is non-empty and compact.

A.4.16 (Fermat's criterion) If $M^\circ \neq \emptyset$ and if $\operatorname{grad} v$ exists on M° then each maximum point $x \in M^\circ$ of v is a stationary point.

A.4.17 (The set of stationary points) Assume that v is continuous on the compact set M, that $M^\circ \neq \emptyset$ and that $\operatorname{grad} v$ exists on M°. Let $M_{\text{st}} \subset M^\circ$ be the set of stationary points of v. Then the non-empty set of maximum points of v on $M_{\text{st}} + \partial M$ equals the set of maximum points of v. If in addition $v = \inf v$ on ∂M then M_{st} equals the set of maximum points of v.

A.5 Vector Spaces

Abbreviations and Conventions

Vectors in matrix products are column vectors. The *scalar product of the vectors x and y* in $\mathbb{R}^d$ is denoted by $\langle x, y\rangle$. The i-th unit vector in $\mathbb{R}^d$ is denoted by e_i, while $\mathbf{1}$ is the vector in $\mathbb{R}^d$ with all components equal to one.

A function v on $\mathbb{R}^d$ is called ***linear*** if $v(\alpha x) = \alpha \cdot v(x)$, $x \in \mathbb{R}^d$.

A function v on $\mathbb{R}^d$ is called ***affine*** if $v(x) = \langle p, x\rangle + b$, $x \in \mathbb{R}^d$ for some $p \in \mathbb{R}^d$ and $b \in \mathbb{R}$.

A subset K of a real vector space is called

- a *subspace*, if $x, y \in K, \alpha \in \mathbb{R} \Longrightarrow \alpha x \in K, x + y \in K$,
- a ***cone*** if $x \in K, \alpha \in \mathbb{R}_+ \Longrightarrow \alpha x \in K$,
- a ***convex set*** if $x, y \in K, \alpha \in (0, 1) \Longrightarrow \alpha x + (1 - \alpha) y \in K$.

A cone K is convex if $x, y \in K$ implies $x + y \in K$. This holds if and only if x, $y \in K, \alpha \in \mathbb{R}_+ \Longrightarrow x + y \in K, \alpha x \in K$, and then $\alpha x + \gamma y \in K$ for $x, y \in K$, α, $\gamma \in \mathbb{R}_+$. A set K of functions on a set M is called a (convex) cone if it is a (convex) cone in the vector space $\mathbb{R}^M$ of all functions on M.

A.6 Induction

Let (P) and (Q) be certain properties of V_n, $n \geq 0$, and of W_n, $n \geq 1$, respectively. (As an example consider Theorem 7.1.2, where (P) and (Q) mean that V_n and W_n are concave on S and on D, respectively.) Mostly we prove the assertions (A_n), $n \geq 0$, and (B_n), $n \geq 1$, that V_n and W_n have property (P) and (Q), respectively as follows: As a rule, (A_0) is ensured by assumption, and we show that (A_n) for some $n \geq 0$ implies (B_{n+1}) and that (B_{n+1}) implies (A_{n+1}).

Appendix B
Measure and Probability

B.1 Measurability

B.1.1 (σ-algebras) A *σ-algebra* $\mathfrak{G}$ on a non-empty set Ω is a system of subsets of Ω such that (i) $\Omega \in \mathfrak{G}$, (ii) $G \in \mathfrak{G}$ implies $G^c := \Omega - G \in \mathfrak{G}$, and (iii) the union of each sequence of sets in $\mathfrak{G}$ belongs to $\mathfrak{G}$. The pair $(\Omega, \mathfrak{G})$ is called a *measurable space* and the sets in $\mathfrak{G}$ are called *$\mathfrak{G}$-measurable (measurable for short)*. The σ-algebras are the natural domains of measures and transition probabilities.

B.1.2 (A generator of a σ-algebra) For each non-empty system $\mathcal{C}$ of subsets of a non-empty set Ω there exists a smallest σ-algebra $\sigma(\mathcal{C})$ which contains $\mathcal{C}$. We have $\sigma(\sigma(\mathcal{C})) = \sigma(\mathcal{C})$. If $\mathcal{C} \subset \mathcal{C}'$, then $\sigma(\mathcal{C}) \subset \sigma(\mathcal{C}')$. If $\sigma(\mathcal{C})$ equals a given σ-algebra $\mathfrak{G}$ on Ω, then $\mathcal{C}$ is called a *generator* of $\mathfrak{G}$. Generators play the following role in applications: often probabilities are only known on a generator of $\mathfrak{G}$, but these determine in a consistent way all probabilities on $\mathfrak{G} = \sigma(\mathcal{C})$.

B.1.3 (A trace of a σ-algebra) Let $(\Omega, \mathfrak{G})$ be a measurable space and let $\emptyset \neq B \in \mathfrak{G}$. Then

$$B \cap \mathfrak{G} := \{B \cap G : G \in \mathfrak{G}\} = \{G \in \mathfrak{G} : G \subset B\}$$

is a σ-algebra on B, the *trace* σ-algebra of $\mathfrak{G}$ on B.

B.1.4 (Borel σ-algebras) On a metric space (M, ρ) the most natural σ-algebra is $\mathfrak{B}(M) := \sigma(\mathfrak{T})$, i.e. the σ-algebra generated by the system of open sets. It is called the ***Borel σ-algebra*** on M, and the elements of $\mathfrak{B}(M)$ are called *Borel sets* in M. In particular, on $\mathbb{R}^d$ we use $\mathfrak{B}_d := \mathfrak{B}(\mathbb{R}^d)$, and on $(\overline{\mathbb{R}}, \overline{\rho})$ we use $\overline{\mathfrak{B}}_1 := \mathfrak{B}(\overline{\mathbb{R}})$; for $\overline{\rho}$ (cf. Appendix C.2). The σ-algebra $\overline{\mathfrak{B}}_1$ equals $\{B \cup C : B \in \mathfrak{B}_1, C \subset \{-\infty, \infty\}\}$, and it has the generator $\{[-\infty, \alpha], \alpha \in \mathbb{R}\}$.

B.1.5 (A metric with a countable base) If the metric space (M, ρ) has the countable base $\mathcal{C}$, then $\mathfrak{B}(M) = \sigma(\mathcal{C})$.

K. Hinderer et al., *Dynamic Optimization*, Universitext,
DOI 10.1007/978-3-319-48814-1

B.1.6 (Borel space) If X is a Borel set in a complete separable metric space (M, ρ) and if $\mathfrak{X}$ is the trace of $\mathfrak{B}(M)$ on X, then $(X, \mathfrak{X})$ is called a *Borel space.*

B.1.7 (Product σ-algebra) The Cartesian product $\times_1^d C_i$ of systems C_i of subsets of a set Ω_i is defined as $\{\times_1^d B_i : B_i \in C_i, 1 \leq i \leq d\}$. Let $(\Omega_i, \mathfrak{G}_i)$, $1 \leq i \leq d$, be measurable spaces. The σ-algebra $\otimes_1^d \mathfrak{G}_i := \sigma(\times_1^d \mathfrak{G}_i)$ on $\Omega := \times_1^d \Omega_i$ is called the *product-σ-algebra* of the σ-algebras $\mathfrak{G}_i$. Unless stated otherwise, products of measurable spaces are endowed with the product σ-algebra.

B.1.8 (Measurable mappings)

(a) If v is a mapping from a set Ω into a measurable space $(\Omega', \mathfrak{G}')$ then the system $v^{-1}(\mathfrak{G}') := \{v^{-1}(G') : G' \in \mathfrak{G}'\}$ is a σ-algebra in Ω. If in addition $\mathfrak{G}$ is a σ-algebra on Ω then v is called $\mathfrak{G}$-$\mathfrak{G}'$*-measurable* (or *measurable* for short) if $v^{-1}(\mathfrak{G}') \subset \mathfrak{G}$. Measurability of a function v on $(\Omega, \mathfrak{G})$ is simply called $\mathfrak{G}$-measurability. This property is required when the integral of v with respect to a measure on $\mathfrak{G}$ is defined. In a context where a probability distribution on $\mathfrak{G}$ is considered, the $\mathfrak{G}$-$\mathfrak{G}'$-measurable mappings are also called *random elements.*
(b) For a mapping v from Ω into Ω' and a system C' of subsets of Ω' we have $v^{-1}(\sigma(C')) = \sigma(v^{-1}(C'))$. Therefore v is $\mathfrak{G}$-$\mathfrak{G}'$-measurable if and only if $v^{-1}(C') \subset \mathfrak{G}$ for a generator C' of $\mathfrak{G}'$.
(c) A function v on $(\Omega, \mathfrak{G})$ is measurable if and only if $\{\omega \in \Omega : v(\omega) \leq \alpha\}$ belongs to $\mathfrak{G}$ for all $\alpha \in \mathbb{R}$.
(d) An upper semicontinuous function v on a metric space (M, ρ) is $\mathfrak{B}(M)$-measurable.
(e) The sum, the product, the quotient and the difference (if defined) of measurable functions is measurable.
(f) The restriction of a measurable mapping to a measurable set is measurable.
(g) Let $(\Omega, \mathfrak{G})$ and $(\Omega', \mathfrak{G}')$ be measurable spaces, and let $B \in \mathfrak{G}$. Measurability of $v_1: B \to \Omega'$ and of $v_2: B^c \to \Omega'$ implies measurability of

$$\omega \mapsto \begin{cases} v_1(\omega), & \text{if } \omega \in B, \\ v_2(\omega), & \text{if } \omega \in B^c. \end{cases}$$

(h) Let $v = (v_i)_1^d$ be a mapping from a set Ω into a measurable space $(\times_1^d \Omega_i, \otimes_1^d \mathfrak{G}_i)$. Then

$$v^{-1}(\otimes_1^d \mathfrak{G}_i) = \sigma(\cup_1^d v_i^{-1}(\mathfrak{G}_i)).$$

If $\mathfrak{G}$ is a σ-algebra on Ω, then v is $\mathfrak{G}$-$\otimes_1^d \mathfrak{G}_i$-measurable if and only if each mapping v_i is $\mathfrak{G}$-$\mathcal{G}_i$-measurable.
(i) Let C_i be a system of subsets of Ω_i such that $\Omega_i \in C_i$, $1 \leq i \leq d$. Put $\Omega := \times_1^d \Omega_i$. Then $\otimes_1^d \sigma(C_i) = \sigma(\times_1^d C_i)$.

B.1.9 (The Borel σ-algebra on the product of metric spaces) Let (M_i, ρ_i) be separable metric spaces, and endow $M := \times_1^d M_i$ with a product metric ρ. Then:

(a) The Borel σ-algebra $\mathfrak{B}(M)$ on M equals the product σ-algebra $\otimes_1^d \mathfrak{B}(M_i)$.
(b) If w on M is usc, then it is measurable with respect to the product σ-algebra.

B.2 Measures

B.2.1 (Uniformly σ-finite transition probabilities) A function $(x, B) \mapsto \rho(x, B)$ from $X \times \mathfrak{X}$ into $\mathbb{R}_+$ is called a ***transition probability*** from X into Y if $\rho(x, B)$ is measurable in x and a measure in B. A transition measure ρ from X into Y is called ***uniformly σ-finite*** if there exists an increasing sequence of sets $B_n \in \mathfrak{X}$, $n \geq 1$, converging to Y such that $\sup_{x \in X} \rho(x, B_n) < \infty$ for all n.

B.2.2 (Preservation of measurability of functions under integration by a uniformly σ-finite transition measure) Let v be a measurable function on $X \times Y$ and let ρ be a uniformly σ-finite transition measure from X into Y (in particular, a transition probability or a σ-finite measure). If there exists $g(x) := \int \rho(x, dy)\, v(x, y)$, $x \in X$, then g is measurable.

B.2.3 (Fubini's Theorem for uniformly σ-finite transition measures) Let ν be a σ-finite measure on X, let ρ be a uniformly σ-finite transition measure from X into Y and put $\mu := \nu \otimes \rho$. For each measurable function v on $X \times Y$:

(a) If $v \geq 0$, then

$$\int v\, d\mu = \int \nu(dx) \int \rho(x, dy)\, v(x, y). \tag{B.1}$$

(b) If $\int v\, d\mu$ exist, then $\int \rho(x, dy)\, v(x, y)$ exists for μ-almost all x and defines a measurable function if it is taken as zero on the exceptional set. And then (B.1) is true.

B.3 Probability

B.3.1 Let ν be a σ-finite measure on $\mathfrak{Y}$.

(i) A measurable function $g \geq 0$ on Y is called a ***ν-density function*** of a probability distribution P on $\mathfrak{Y}$ if $P(B) = \int_B \nu(dy)\, g(y)$, $B \in \mathfrak{Y}$.
(ii) A measurable function $h \geq 0$ on $X \times Y$ is called a ***transition density function*** with respect to ν of a transition probability Q from X to Y if $Q(x, B) = \int_B \nu(dy)\, h(x, y)$, $x \in X$, $B \in \mathfrak{Y}$.

B.3.2 (The total variation metric) Recall from Appendix A.4.7 the definition of the sp v of a bounded function v on a set Ω: $\operatorname{sp} v := \sup v - \inf v$. Let $(\Omega, \mathfrak{M})$ be a measurable space and $\mathbb{P}(\mathfrak{M})$ be the set of probability distributions on $\mathfrak{M}$. It is easily seen that

$$(P, Q) \mapsto \|P - Q\| := \sup_{B \in \mathfrak{M}} |P(B) - Q(B)|$$

is a metric on $\mathbb{P}(\mathfrak{M})$, the so-called ***total variation metric.***

B.3.3 (Properties of the total variation metric) The following properties hold:

(a) Let p and q be densities of P and Q, respectively, with respect to a σ-finite measure μ on $\mathfrak{M}$. Then

$$2 \cdot \|P - Q\| = \int |p - q| \, d\mu.$$

(b) Let $\mathcal{V}$ be the set of measurable functions on Ω. Then

$$\begin{aligned} \|P - Q\| &= \sup \left\{ \left| \int v \, dP - \int v \, dQ \right| : v \in \mathcal{V}, \operatorname{sp} v \leq 1 \right\} \\ &= \sup \left\{ \int v \, dP - \int v \, dQ : v \in \mathcal{V}, \operatorname{sp} v \leq 1 \right\}. \end{aligned} \tag{B.2}$$

(c) If v is a measurable bounded function on Ω with $\operatorname{sp} v > 0$ (i.e. if v is not constant), then replacing v in (B.2) by $v/(\operatorname{sp} v)$ yields the important inequality

$$\left| \int v \, dP - \int v \, dQ \right| \leq \|P - Q\| \cdot \operatorname{sp} v. \tag{B.3}$$

(d) The supremum in (B.3) and (B.2) is attained. As a consequence, $\|P - Q\| = 1$ if and only if there is a set $B \in \mathfrak{M}$ such that $P(B) = 1$ and $Q(B) = 0$, in which case P and Q are called *orthogonal.*

(e) Of course, a sequence (P_n) in the set $\mathbb{P}(\mathfrak{M})$ of probability distributions on $\mathfrak{M}$ is said to *converge in total variation* towards $P \in \mathbb{P}(\mathfrak{M})$ if $\|P_n - P\| \to 0$ for $n \to \infty$. This means that $P_n(B)$ converges to $P(B)$ uniformly in $B \in \mathfrak{M}$. If (Ω, ρ) is a metric space such that $\mathfrak{M}$ contains all Borel sets generated by ρ, then it follows from (B.2) that convergence in total variation implies weak convergence.

B.3.4 (Upper [lower] semicontinuous dependence of integrals on a parameter) For the next result we need the following definition: Let $M \neq \emptyset$ be a set, $(\Omega, \mathfrak{S}, \mu)$ a measure space and w a function on $M \times \Omega$. An integrable function h on Ω is called a *μ-integrable majorant of* w if $w(t, \cdot) \leq h$ for all $t \in M$.

Now let (M, ρ) be a metric space, $(\Omega, \mathfrak{G}, \mu)$ a measure space and w a function on $M \times \Omega$ such that $w(t, \cdot)$ is measurable for all t in M. Put

$$v(t) := \int w(t, \cdot)\, d\mu,\ t \in M,$$

in case the integrals exists. Then:

(a) If $w(\cdot, \omega)$ is usc for all ω and if w has a μ-integrable majorant, then v is usc.
(b) If $w(\cdot, \omega)$ is continuous for all ω and if $|w|$ has a μ-integrable majorant, then v is continuous.
(c) Let $(\Omega, \mathfrak{G}) := (\mathbb{R}, \mathfrak{B})$, let (M', ρ') be a metric space and endow $M := M' \times \mathbb{R}$ with the product topology. Put

$$v(x, y) := \int_{(-\infty, y]} w(x, y, \cdot)\, d\mu.$$

Then:

(c1) Assume that $w(x, y, \omega)$ is usc in (x, y) for all ω and that w has a μ-integrable majorant. If $w \geq 0$ or if (w is lower bounded and μ is continuous), then v is usc.
(c2) If $w(x, y, \omega)$ is continuous in (x, y) for all ω, if $|w|$ has a μ-integrable majorant and if μ is continuous, then v is continuous.

B.3.5 (Theorem of Ionescu Tulcea) Let $(\Omega_n, \mathfrak{G}_n)_{n \in \mathbb{N}}$ be a sequence of measurable spaces. Further let Q_0 be a measure on $\mathfrak{G}_1$ and Q_n transition probabilities from $\prod_1^n \Omega_i$ to Ω_{n+1}, $n \in \mathbb{N}$. Then there exists a unique probability measure Q on $(\prod_1^\infty \Omega_n, \bigotimes_1^\infty \mathfrak{G}_n)$ such that

$$Q(\prod_1^\infty B_n) = \int_{B_1} Q_0(d\omega_1) \int_{B_2} Q_1(\omega_1, d\omega_2) \cdots \int_{B_k} Q_{k-1}(\omega_1, \ldots, \omega_{k-1}, d\omega_k)$$

if $B_n = \Omega_n$ for $n > k$.

Appendix C
Metric Spaces

Elementary Facts on Metric Spaces

In the sequel, (M, ρ) always denotes an arbitrary metric space. No knowledge of general topology is required. Topologies induced by metrics are introduced below. However, we assume familiarity with the following basic notions for *metric spaces*: open sets, closed sets (as complements of open sets), Cauchy sequence, (uniform) continuity of mappings from (M, ρ) into a metric space (M', ρ'), vector space and normed space. Sequences $(x_n)_1^\infty$ of points in M are often denoted just by (x_n), and instead of $x_n \to x$ *for* $n \to \infty$ we often only write $x_n \to x$. The open ball with center $x \in M$ and radius $\varepsilon \in \mathbb{R}^+$ is the set $B(x, \varepsilon) := B_{M,\rho}(x, \varepsilon) := \{y \in M : \rho(x, y) < \varepsilon\}$.

We summarize elementary notions and facts, which we mostly use without mention. In particular we indicate the relations between a metric ρ and its induced topology. On $\mathbb{R}$ we use, unless stated otherwise, the *Euclidean metric* ρ_E, defined by $\rho_E(x, y) := |x - y|, x, y \in \mathbb{R}$.

(a) Recall that a set $B \subset M$ is said to be open if for each $x \in B$ there exists some $\varepsilon = \varepsilon(x) \in \mathbb{R}^+$ such that $B(x, \varepsilon) \subset B$. Recall also that a sequence $(x_n) \subset M$ is said to be converging to $x \in M$ if for each open ball $B(x, \varepsilon)$, $\varepsilon \in \mathbb{R}^+$, there exists an $n_0 = n_0(\varepsilon)$ such that $x_n \in B(x, \varepsilon)$ for $n \geq n_0$.
(b) A set $B \subset M$ is closed if and only if it contains the limits of all sequences in B which converge in M.
(c) Convergent sequences are Cauchy sequences. If the converse holds, the space is called *complete.*
(d) A subset B of M is called *bounded* if its *diameter*, the number $\sup\{\rho(b, b') : b, b' \in B\}$, is finite.
(e) A point $x \in M$ is called a *cluster value* of a sequence (x_n) in M, if x is the limit of some subsequence of (x_n), i.e. $\liminf_n \rho(x, x_n) = 0$. A set $B \subset M$ is closed if and only if it contains the cluster values of all sequences in B. A set $B \subset M$ is called *compact* if each sequence in B has a cluster value belonging to B. The

K. Hinderer et al., *Dynamic Optimization*, Universitext,
DOI 10.1007/978-3-319-48814-1

empty set is compact. Each compact subset of M is closed and bounded. Each closed and bounded subset of $\mathbb{R}^n$ is compact. Closed subsets of a compact space M are compact. The intersection of each decreasing sequence of non-empty compact sets is non-empty and compact (cf. Ash 1972, p. 381).

(f) M is called *separable* if there exists a countable set $B \subset M$ which is *dense* in M, i.e. for which each point in M is the limit of a sequence in B. M is separable if and only if it has a *countable base,* i.e. a countable system $\mathfrak{C}$ of open sets such that each open set in M is a union of sets in $\mathfrak{C}$. If B is dense in M then B has the countable base $\{B(x, 1/n) : x \in B, n \geq 1\}$. Compact spaces are complete and separable.

(g) The system $\mathcal{T}$ of open subsets of M is a *topology,* i.e. it has the following properties: M and $\emptyset$ are both open, the union of any number of open sets is open, the intersection of any pair of open sets is open. $\mathcal{T}$ is called the *topology on M induced by the metric ρ*. Note also that M and $\emptyset$ are both closed, the intersection of any number of closed sets is closed, the union of finitely many closed sets is closed.

Two metrics ρ and ρ' on M may induce the same topology, in which case they are said to be *topologically equivalent*. This holds if and only if for each sequence (x_n) in M and each $x \in M$ we have: $\rho(x_n, x) \to 0$ if and only if $\rho'(x_n, x) \to 0$. The bounded metric $1 \wedge \rho$ is topologically equivalent to ρ (cf. Dieudonné 1960, p. 53).

Notions which can be characterized solely via open sets (such as closed set, cluster value or limit of a sequence, compact set, continuity of a mapping) are the same for topologically equivalent metrics, and are called *topological* properties. Completeness is not a topological property (cf. Bertsekas and Shreve 1978, p. 109).

(h) If $B \subset M$ is non-empty, the restriction $\rho|B$ of ρ to $B \times B$ is a metric on B. $(B, \rho|B)$ is called a *subspace* of (M, ρ), and the open (closed) sets in $(B, \rho|B)$ are called *open (closed) in B*. Then $\{B \cap G : G \in \mathcal{T}\}$ is the topology on B induced by $\rho|B$ on B. It is called the *relative topology* of $\mathcal{T}$ on B. Note that, for example, the interval $B := (0, 1] \subset M := \mathbb{R}$ is open and closed in $(B, \rho_E|B)$ but not in $(\mathbb{R}, \rho_E)$. All subsets of $B \subset M$ which are closed [open] in $(B|\rho|B)$ are closed [open] in M if and only if B is closed [open] in (M, ρ). A subset B of M is compact if and only if it is compact as subspace of M. Each subspace of a separable space is separable. A subspace B of a complete metric space M is complete if and only if B is closed in M (cf. Wilansky 1970, p. 171).

(i) A sequence of mappings $(v_n)_1^\infty$ from a set K (not necessarily endowed with a metric) into (M, ρ) is called *uniformly convergent* to $v: K \to M$ if $\sup\{\rho(v_n(x), v(x)) : x \in K\}$ converges to zero for $n \to \infty$.

(j) Two *metrics* ρ and ρ' on M are called *equivalent* if $\alpha \cdot \rho \leq \rho' \leq \beta \cdot \rho$ for some positive constants α and β. Then ρ and ρ' are also topologically equivalent. The following notions depend on the metric, but are still the same for equivalent metrics: boundedness of a set, uniform convergence of a sequence of functions, completeness, Lipschitz continuity.

(k) If $(M, \|\cdot\|)$ is a normed linear space, then $(x, y) \mapsto \|x - y\|$ is a metric on each non-empty subset of M. It is called the *metric induced by the norm* $\|\cdot\|$. Two norms $\|\cdot\|_1$ and $\|\cdot\|_2$ on the same vector space M are called *equivalent* if $\alpha \cdot \|\cdot\|_1 \le \|\cdot\|_2 \le \beta \cdot \|\cdot\|_1$ for some $\alpha, \beta \in \mathbb{R}^+$. The metrics induced by equivalent norms are equivalent. Thus, since all norms on $\mathbb{R}^d$ are equivalent, all metrics on $\mathbb{R}^d$ induced by norms (such as the taxicab metric, the Euclidean metric and the maximum-metric, cf. below) are equivalent.
(l) We have $|\rho(x, z) - \rho(y, z)| \le \rho(x, y)$ for $x, y, z \in M$.
(m) Each metric ρ on M is uniformly continuous on $M \times M$.

C.1 (Product Metrics) Let (M_1, ρ_1) and (M_2, ρ_2) be metric spaces. We deal with the product of the topologies induced by ρ_1 and ρ_2 in the following metric space setting. We call a metric ρ on a set $M \subset M_1 \times M_2$ a *product metric* with respect to (ρ_1, ρ_2) if it has the following property: Each sequence of points $(x_k, y_k) \in M$, $k \ge 1$, converges with respect to ρ towards $(x, y) \in M$ if and only if $x_k \to x$ and $y_k \to y$.

For arbitrary (M_1, ρ_1) and (M_2, ρ_2) there exist several different product metrics. The most important ones are:

(a) The *maximum metric* $d_1 := \max\{\rho_1, \rho_2\}$, defined by

$$d_1((x, y), (x', y')) := \max\{\rho_1(x, x'), \rho_2(y, y')\}.$$

(b) The l_2*-metric* (which generalizes the Euclidean metric)

$$d_2((x, y), (x', y')) := \left[\rho_1(x, x')^2 + \rho_2(y, y')^2\right]^{1/2}.$$

(c) The l^1*-metric* (also called *taxicab metric*)

$$d_3((x, y), (x', y')) := \rho_1(x, x') + \rho_2(y, y').$$

The extension of the preceding definition of a product metric to $m \ge 2$ metric spaces (M_i, ρ_i) is obvious. This applies in particular to d_1, d_2 and d_3. As an example, the *maximum metric* is given by $d_1(x, y) := \max_{1 \le i \le m} \rho_i(x_i, y_i)$ for $x = (x_i)$, $y = (y_i) \in \times_1^m M_i$. Since $d_1 \le d_2 \le d_3 \le m \cdot d_1$, all three metrics are equivalent. The maximum metric has the advantage that the ε-ball $B((x_i)_1^m, \varepsilon)$ in $\times_1^m M_i$ is the Cartesian product of the ε-balls $B(x_i, \varepsilon)$ in M_i, $1 \le i \le m$. Therefore we use, unless stated otherwise, on $\mathbb{R}^m$, $m \ge 2$, the maximum metric.

All product metrics ρ are topologically equivalent, and the topology they induce on $M_1 \times M_2$ is called the *product topology* generated by (ρ_1, ρ_2). If both (M_1, ρ_1) and (M_2, ρ_2) are separable, then $M_1 \times M_2$ is separable under each product metric.

The use of a product metric on $M_1 \times M_2$ implies the following facts:

(a) For $x \in M_1$ and $B \subset M$ with non-empty x-section B_x the x-section $w(x, \cdot)$ of a continuous function w on B is continuous on B_x.

(b) For each continuous function v on M_2 the function $(x, y) \mapsto v(y)$ is continuous on M.
(c) If $B \subset M$ is closed then B_x is closed in M_2 for all $x \in M_1$.

C.2 (A Metric on the Extended Real Line) When discussing continuity of the value functions the usual assumption that these are finite is often cumbersome. We avoid this by using an appropriate notion of convergence in $\overline{\mathbb{R}}$. Intuitively most appealing is the following one: A sequence of numbers $x_n \in \overline{\mathbb{R}}$ is called *convergent* to $x \in \overline{\mathbb{R}}$ if for each $\varepsilon > 0$ there exists an $n = n_0(\varepsilon)$ such that for $n \geq n_0$ we have

$$|x_n - x| \leq \varepsilon, \text{ if } x \in \mathbb{R}, \quad \pm x_n \geq \pm\varepsilon, \text{ if } x = \pm\infty.$$

This notion of convergence is equivalent to convergence with respect to the following metric $\overline{\rho}_E$ which will be used on $\overline{\mathbb{R}}$ unless stated otherwise:

$$\overline{\rho}_E(x, y) := |g(x) - g(y)|, \text{ where } g(z) := \begin{cases} z/(1 + |z|), & \text{if } z \in \mathbb{R}, \\ \pm 1, & \text{if } z = \pm\infty. \end{cases}$$

Thus $\overline{\rho}_E = \rho_E \circ g$. As a consequence, a function on (M, ρ) is continuous in $x \in M$ with respect to $\overline{\rho}$ if and only if for each sequence $(x_n) \subset M$ converging to x the sequence of points $v(x_n) \in \overline{\mathbb{R}}$ is convergent to $v(x)$. The space $(\overline{\mathbb{R}}, \overline{\rho})$ is compact. The restriction of $\overline{\rho}$ to $\mathbb{R} \times \mathbb{R}$ is topologically equivalent but not equivalent to the Euclidean metric on $\mathbb{R}$.

For sequences (x_n) in $\overline{\mathbb{R}}$ we define (as for sequences in $\mathbb{R}$)

$$\liminf_{n\to\infty} x_n := \sup_{n\in\mathbb{N}} \inf_{m\geq n} x_m, \quad \limsup_{n\to\infty} x_n := \inf_{n\in\mathbb{N}} \sup_{m\geq n} x_m.$$

This definition does not require a metric on $\overline{\mathbb{R}}$. However, $\liminf_n x_n$ [$\limsup_n x_n$] equals the smallest [largest] cluster value with respect to $\overline{\rho}$.

Appendix D
Convexity

Proofs for many of the subsequent results may be found in textbooks. Our main reference is Roberts and Varberg (1973).

D.1 Convex Sets

D.1.1 (Convex sets in $\mathbb{R}^k$)

(a) The set $M \subset \mathbb{R}^k$ is called ***convex*** if $\alpha x + (1-\alpha)y \in M$ whenever $x, y \in M$ and $\alpha \in (0,1)$. If M is convex then $\sum_1^m \alpha_i x_i$ belongs to M for all $m \geq 1$, $x_i \in M$ and $\alpha_i \in [0,1]$, $1 \leq i \leq m$, such that $\sum_1^m \alpha_i = 1$. Also the empty set is convex. Unless stated otherwise, the convex sets considered below are assumed to be non-empty. We adhere to the *Emergency Axiom* of Valentine (1964, p. 198): "If a theorem is false because a set A is empty, we tacitly assume that A is not empty."

(b) A non-empty subset of $\mathbb{R}$ is convex if and only if it is an interval.

(c) The following are convex subsets of $\mathbb{R}^k$: the whole space $\mathbb{R}^k$; the singletons; the linear subspaces, in particular the closed half-spaces; the ***affine sets*** M, defined by the property that $\gamma x + (1-\gamma)y \in M$ whenever $x, y \in M$ and $\gamma \in \mathbb{R}$. The hyperplanes are affine convex sets which do not contain interior points.

(d) The intersection of any family of convex sets is convex. The intersection of a finite family of closed half-spaces is called a ***polyhedral set***; it is closed and convex. The strip $[0,1] \times \mathbb{R}_+$ and the cone $\{(x,y) \in \mathbb{R}_+^2 : y \leq x\}$ are unbounded polyhedral sets.

(e) The point $x \in \mathbb{R}^k$ is called a ***convex combination*** of points in the non-empty set $B \subset \mathbb{R}^k$ if $x := \sum_{i=1}^m \alpha_i x_i$ for some $m \geq 1$, some $\alpha_1, \alpha_2, \cdots, \alpha_m \in (0,1)$ such that $\sum_{i=1}^m \alpha_i = 1$ and some $x_1, x_2, \ldots, x_m \in B$. A subset B of $\mathbb{R}^k$ is convex if and only if it contains all convex combinations of points in B.

K. Hinderer et al., *Dynamic Optimization*, Universitext,
DOI 10.1007/978-3-319-48814-1

(f) The set of convex combinations of points in a non-empty set $B \subset \mathbb{R}^k$ is called the ***convex hull*** co(B) of B. It is the smallest convex set containing B and hence also the intersection of all convex sets containing B. If B is finite, co(B) is called a ***polytope***. Each polytope is compact. The polytopes are the bounded polyhedral sets. The polytopes in $\mathbb{R}^2$ are the convex polygons, the line segments and the singletons.

(g) If $B = \{b_0, b_1, \cdots, b_n\} \subset \mathbb{R}^n$ and if $b_1 - b_0, b_2 - b_0, \ldots, b_n - b_0$ are linearly independent then co(B) is called a ***simplex.*** A set $M \subset \mathbb{R}^n$ is a simplex for $n = 2$ if and only if M is a triangle and for $n = 3$ if and only if M is a tetrahedron.

(h) The Cartesian product of finitely many convex sets is convex. In particular the Cartesian product of k intervals in $\mathbb{R}$ is a convex set in $\mathbb{R}^k$.

(i) If $M \subset \mathbb{R}^d \times \mathbb{R}^k$ is convex and $x \in \mathbb{R}^d$, then $M(x) := \{y \in \mathbb{R}^k : (x, y) \in M\}$, the *x-section* of M, is convex (possibly empty).

(j) The projection of a convex set $M \subset \mathbb{R}^d \times \mathbb{R}^k$ into $\mathbb{R}^d$ is convex.

(k) The interior and the closure of a convex set are convex.

(l) Convex sets $M \subset \mathbb{R}^d$, $d \geq 2$, need not be Borel measurable. As a consequence, there exist non-measurable convex functions (see Appendix D.2.2(b) below), and we must assume in statements like Appendix D.2.14 below that both M and v are not only convex but also Borel measurable.

D.2 Convex Functions

D.2.1 Convex Functions on Subsets of $\mathbb{R}^k$

D.2.1 (Convex functions)

(a) The function v on the convex set $M \subset \mathbb{R}^k$ is called ***convex*** if

$$v(\alpha x + (1 - \alpha)y) \leq \alpha v(x) + (1 - \alpha)v(y) \tag{D.1}$$

for $x, y \in M$, $x \neq y$ and $\alpha \in (0, 1)$. If (D.1) holds even with strict inequality then v is called ***strictly convex***. Moreover, v is called ***[strictly] concave*** if $-v$ is [strictly] convex. Thus most results on convex functions are easily transformed into results on concave functions and vice versa; we shall mostly state results only for convex functions.

(b) For convex v on M one obtains by induction on $m \geq 1$ ***Jensen's inequality***

$$v\left(\sum_{i=1}^{m} \alpha_i x_i\right) \leq \sum_{i=1}^{m} \alpha_i v(x_i),\ m \geq 1,\ x_i \in M,\ \alpha_i \in \mathbb{R}_+ \text{ such that } \sum_{i=1}^{m} \alpha_i = 1.$$

(c) A [strictly] convex function on M is [strictly] convex on each convex subset of M.

D.2.2 (Properties of convex functions)

(a) Let v be convex on M. Then v is continuous on the interior M° of M, if M° is non-empty; Roberts and Varberg (1973, p. 93). In particular, convex functions on open convex sets are continuous. The function v on M is [strictly] convex and continuous if it is the restriction of some [strictly] convex function on an open superset of M. Convex functions on intervals are at least upper semicontinuous. However, upper semicontinuity need not be true for convex functions on $M \subset \mathbb{R}^2$.
(b) Convex functions on intervals are Borel measurable, which need not be true for convex functions on Borelian convex sets $M_1 \subset \mathbb{R}^2$.
(c) If the interior M° of M is non-empty, then the function v is [strictly] convex on M if it is [strictly] convex on M° and continuous in each point of $M \cap \partial M$.
(d) If v is convex then for each $\gamma \in \mathbb{R}$ the so-called *level sets* $\{x \in M : v(x) \leq \gamma\}$ and $\{x \in M : v(x) < \gamma\}$ are convex. The converse does not hold.

D.2.3 (Affine functions)

(a) Let $M \subset \mathbb{R}^k$ be convex. The function v on M is called ***affine*** if it is both convex and concave, i.e. if $x, y \in M$ and $\alpha \in (0, 1)$ implies

$$v(\alpha x + (1-\alpha)y) = \alpha v(x) + (1-\alpha)v(y).$$

The function v on M is affine if for some $p \in \mathbb{R}^k$ and some $b \in \mathbb{R}$

$$v(x) = \langle p, x\rangle + b, \quad x \in M.$$

The converse holds if M is an interval (see Roberts and Varberg (1973, p. 7), Problem A) or if $M = \mathbb{R}^k$ (see Rockafellar 1970, p. 27).
(b) A convex function v on an interval I is *not* strictly convex if and only if v is affine on some subinterval of I (see Roberts and Varberg (1973, p. 7), Problem A).

D.2.4 (Further properties of convex functions) Let v be convex on a convex set $M \subset \mathbb{R}^k$.

(a) v is ***affinely bounded below***, i.e. there exists an affine function H on $\mathbb{R}^k$ such that $H \leq v$ on M.
(b) Each convex function v on a polytope, in particular on a compact interval, is bounded. If M is bounded, then v is lower bounded by (a), but v need not be upper bounded.

D.2.2 Convex Functions on Intervals

D.2.5 (Derivatives) If v' exists, then v is [strictly] convex if and only if v' is [strictly] increasing. If v'' exists, then v is convex if and only if $v'' \geq 0$; if $v'' > 0$ then v is strictly convex, but the converse need not hold (see Roberts and Varberg 1973, p. 11).

D.2.6 (Slope inequalities and the law of decreasing marginal returns)

(a) Let v be a function on a non-degenerate interval I. For each pair x, y of different points in I put

$$\Delta v(x, y) := \frac{v(x) - v(y)}{x - y} = \Delta v(y, x).$$

This is the slope of the line segment joining the two points $(x, v(x))$ and $(y, v(y))$ on the graph of v.

(b) The study of the one-sided differentiability of convex functions can be based on the following result:

If v is a convex function on a non-degenerate interval I, then

$$\Delta v(x, y) \leq \Delta v(x, z) \leq \Delta v(y, z) \tag{D.2}$$

for each triple x, y, z of points in I with $x < y < z$. If (D.2) holds then v is convex.

(c) Le v be a convex function on an interval $I = I(a, b)$. Then for each $h \in (0, b-a)$ the function $x \mapsto v(x + h) - v(x)$ is increasing on $\{x \in I : x + h \in I\}$.

(d) In economic processes one often studies the influence of some *input* x (e.g. the amount of energy used for the production of some asset) on the *output* (e.g. the amount $v(x)$ of asset produced by the input x) when the latter varies in an interval $I = I(a, b)$. Then v is called an *output function*. These functions, besides being increasing, often satisfy the so-called ***law of decreasing marginal returns***: for each $h \in (0, b - a)$ the function $x \mapsto u_h(x) := v(x + h) - v(x)$ is decreasing on $\{x \in I : x + h \in I\}$. In fact, this holds if v is concave.

(e) If v is continuous and satisfies the law of decreasing marginal returns, then v is concave.

D.2.7 (One-sided derivatives)

(a) The result of Appendix D.2.5 is useful for modeling in DPs the functions r and V_0. However, in most DPs with convex value functions these are not everywhere differentiable. Then the subsequent properties are useful. If v is convex on (a, b) then:

(i) The left derivative v'_- and the right derivative v'_+ exist on (a, b).
(ii) We have $v'_- \leq v'_+$.
(iii) The functions v'_- and v'_+ are increasing.

(b) The functions v'_+ and v'_- are right continuous and left continuous, respectively, and $v'_+(x) = v'_-(x+)$, $v'_-(x) = v'_+(x-)$ for $x \in (a, b)$ (see Roberts and Varberg 1973, p. 7). In particular, if v' exists on (a, b), then v' is continuous.
(c) Let v be convex on a non-degenerate interval $I = I(a, b)$. If $a \in I$ we do not define $v'_+(a)$, and if $b \in I$ we do not define $v'_-(b)$. However, there exist the limits $v'_+(a+) = v'_-(a+) \in [-\infty, \infty)$ and $v'_-(b-) = v'_+(b-) \in (-\infty, \infty]$.

D.2.8 (Monotonicity of convex functions, convexity on intervals) Let v be a convex function on an interval $I = I(a, b)$.

(a) Let $x \in (a, b)$. Then v is [strictly] increasing on $[x, b)$ if and only if $v'_+(x) \geq 0$ [if $v'_+(x) > 0$], and v is decreasing on $[x, b)$ if and only if $v'_+ \leq 0$ on $[x, b)$. A similar result holds for monotonicity of v on $(a, x]$.
(b) The function v is monotone on some right neighborhood of a and monotone on some left neighborhood of b.
(c) v is increasing [decreasing] on (a, b) if and only if $v'_+(a+) \geq 0$ [$v'_+(b-) \leq 0$] if and only if $v'_+ \geq 0$ [$v'_+ \leq 0$]. In addition, v is strictly increasing [strictly decreasing] on (a, b) if $v'_+(a+) > 0$ [$v'_+(b-) < 0$].
(d) If v is convex on (a, b) then there exist

$$v(a+) := \lim_{x \downarrow a} v(x) \in \overline{\mathbb{R}}, \quad v(b-) := \lim_{x \uparrow b} v(x) \in \overline{\mathbb{R}}.$$

(e) We have

$$v(a+) \leq v(a) \text{ if } a \in I, \quad v(b-) \leq v(b) \text{ if } b \in I. \tag{D.3}$$

(f) The function v is convex on I if and only if (v is convex on (a, b)) $\wedge$ ((D.3) holds). In particular, a continuous function on a compact interval $[a, b]$ is convex if and only if it is convex on (a, b).
(g) The function v is strictly convex on I if and only if (v is strictly convex on (a, b)) $\wedge$ (v is convex on I).
(h) Note that (f) and (g) imply that v is strictly convex on I if and only if (v is strictly convex on (a, b)) $\wedge$ ((D.3) holds).

D.2.3 Convex Functions of Several Variables

D.2.9 (Closure properties of the set of convex functions) Checking convexity of complicated functions of one variable by means of the first or second derivative or of a function v of several variables using grad v or the Hessian of v is often cumbersome. Usually it is easier to combine knowledge about elementary convex functions with *closure properties* of the set of convex functions, such as the following ones.

(a) A sum of finitely many convex functions is convex. It is strictly convex if in addition one of the functions is strictly convex.
(b) If v is [strictly] convex and $\alpha \in \mathbb{R}_+$ [$\alpha \in \mathbb{R}^+$], then αv is [strictly] convex.
(c) The set of convex functions on a convex set is a convex cone $\mathcal{C}$, i.e. if $v_1, v_2 \in \mathcal{C}$ and $\alpha \in \mathbb{R}_+$ then $v_1 + v_2$ and αv_1 belong to $\mathcal{C}$.
(d) The (pointwise) maximum of finitely many convex functions is convex. The same holds for the supremum of arbitrarily many convex functions, provided the supremum is finite.
(e) The (pointwise) maximum of finitely many strictly convex functions is strictly convex.

D.2.10 (The composition of convex/concave functions) (Cf. also Appendix D.6.2.) Consider the composition $v \circ g$ of a function v on an interval I and a function g from a convex set M into I. Then the following holds:

(a) If v is convex then $v \circ g$ is convex if (g is convex and v is increasing) $\vee$ (g is concave and v is decreasing) $\vee$ (g is affine).
(b) If v is concave then $v \circ g$ is concave if (g is concave and v is increasing) $\vee$ (g is convex and v is decreasing) $\vee$ (g is affine).
(c) Assume that v is increasing and convex and that g is convex. Then $v \circ g$ is strictly convex under each of the following conditions.

  (c1) v is strictly increasing and g is strictly convex.
  (c2) v is strictly convex and g is injective.

D.2.11 (Convexity of sections of functions) If v is [strictly] convex on $M \subset \mathbb{R}^d \times \mathbb{R}^k$, if $x \in \mathbb{R}^d$ and if the x-section $M(x)$ is non-empty, then the x-section $v(x, \cdot)$ of v is [strictly] convex. If all one-dimensional sections of v are convex, v need not be convex.

D.2.12 (Convex functions of two variables)

(a) Each convex function v of one real variable can be considered as a convex function w of two real variables. More precisely: Let D be convex (hence S and A_0 are convex) and let v_1 and v_2 be convex functions on S and on A_0, respectively. Then $(s, a) \mapsto v_1(s)$ [$(s, a) \mapsto v_2(a)$] is convex.
(b) If both v_1 and v_2 are [strictly] convex on $M \subset \mathbb{R}^d$ and $K \subset \mathbb{R}^k$, respectively, then $(x, y) \mapsto v_1(x) + v_2(y)$ is [strictly] convex on $M \times K$.

D.2.13 (Affine transformations) Let $M \subset \mathbb{R}^k$ be convex. We call $v: M \to \mathbb{R}^d$ an ***affine transformation*** if $v(x) = Bx + b$, $x \in M$, for some $d \times k$-matrix B and some $b \in \mathbb{R}^d$. Then $v(M)$ is convex. If $g: v(M) \to \mathbb{R}$ is convex, then $g \circ v$ is convex.

D.2.14 (Preservation of convexity by integration) Let $(\Omega, \mathfrak{F}, \mu)$ be a measure space and let $M \subset \mathbb{R}^d$ be convex and measurable. Let g be a function on $M \times \Omega$ such that

(i) $g(x, \cdot)$ is μ-integrable for $x \in M$,

(ii) $g(\cdot, \omega)$ is convex and measurable for μ-almost all $\omega \in \Omega$ [and strictly convex for all ω in a set of positive μ-measure].

Then the following holds:

(a) The function

$$x \mapsto G(x) := \int g(x, \omega)\, \mu(d\omega)$$

is [strictly] convex.

(b) Moreover, let M be an open interval and let $g'_{\pm}(x, \omega)$ denote the one-sided derivatives of $g(\cdot, \omega)$ at $x \in M$. Then $g'_{\pm}(x, \cdot)$ is μ-integrable for all $x \in M$ and the one-sided derivatives of the above convex function G may be taken under the integral sign, i.e.

$$G'_{\pm}(x) = \int g'_{\pm}(x, \omega)\, \mu(d\omega), \quad x \in M.$$

D.3 Minimization of Convex Functions

D.3.1 (Minimum points of convex functions) Assume that v is convex on $M \subset \mathbb{R}^k$. Then:

(a) The set M^* of minimum points of v is convex, hence an interval in case $k = 1$ and if M^* is non-empty. Moreover, M^* is closed if v is continuous.
(b) If v is strictly convex, then v has at most one minimum point.
(c) If grad v exists on the interior of M, an interior point x of M is a minimum point if and only if it is stationary (see Roberts and Varberg 1973, p. 124).

D.3.2 (Minimization of symmetric convex functions) The minimization of a convex function v on a convex set $M \subset \mathbb{R}^n$ becomes particularly simple if both M and v are *symmetric*, i.e. if $x = (x_i)_1^n \in M$ implies that $\pi x := (x_{\pi(i)})_1^n$ belongs to M and that $v(\pi x) = v(x)$ for all permutations π on $\{1, 2, \ldots, n\}$. Then by the following result the minimization can be reduced to the minimization of a convex function of *one* variable.

Let v be a symmetric and [strictly] convex function on the symmetric and convex set $M \subset \mathbb{R}^n$. Then:

(a) $I := \{z \in \mathbb{R} : (z)_1^n \in M\}$ is an interval, and

$$z \mapsto w(z) := v((z)_1^n), z \in I,$$

is [strictly] convex. Moreover, v has a minimum point if and only if w has a minimum point z^*. And then $(z^*)_1^n$ is a [the unique] minimum point of v.

(b) If $K \in \mathbb{R}$, the constraint $\sum_1^n x_i = K$ is fulfilled for at least one point x in M if and only if $(K/n)_1^n \in M$. And then v has under the constraint the [unique] minimum point $(K/n)_1^n$.

D.3.3 (Minimum points of a convex function v on an interval $I = I(a,b)$) The following assertions hold:

(a) There need not exist a minimum point of v, even if I is compact. The point $x^* \in (a,b)$ is a minimum point if and only if $v'_-(x^*) \leq 0 \leq v'_+(x^*)$. The minimum point is unique if both inequalities are strict, but the converse is not true. In addition, v is decreasing left of x^* and increasing right of x^*. Moreover, if v' exists then a point in (a,b) is a minimum point if and only if it is stationary.
(b) If $v'_+(a+) < 0 \leq v'_+(y)$ for some $y \in (a,b)$ then $x^* := \min\{x \in (a,b) : v'_+(x) \geq 0\}$ is the smallest minimum point of v.
(c) If v is strictly convex and continuous on $M := [a,b]$ and differentiable on (a,b), then v has a unique minimum point x^*. Moreover, $x^* = a$ if $v'(a+) \geq 0$, $x^* = b$ if $v'(b-) \leq 0$, else x^* is the unique zero of v' on (a,b).
(d) If $a \in I$, then a is a minimum point of v if and only if $(v'_+(a+) \geq 0) \wedge$ (v is continuous in the point a) if and only if v is increasing on $[a,b)$. If in addition $v'_+(a+) > 0$ then a is the unique minimum point of v.
(e) If $I = \mathbb{R}$ then v has both a smallest and a largest minimum point if and only if $v'_+(-\infty) < 0 < v'_+(\infty)$.

D.4 Maximization of Convex Functions

D.4.1 (Extreme points of a set) A point z of a convex set $M \subset \mathbb{R}^n$ is called an ***extreme point*** of M if z is not representable as $z = \alpha x + (1-\alpha)y$ for some points $x \neq y$ in M and some $\alpha \in (0,1)$. Only boundary points can be extreme points. Half-spaces do not have extreme points, even if they are closed. The extreme points of a polytope $M = \text{co}(B)$ belong to B and are called the *vertices* of M. The extreme points of a simplex $M = \text{co}(B)$ are the points in B. Thus a simplex in $\mathbb{R}^n$ is a polytope with $n+1$ vertices.

D.4.2 (The convex hull of extreme points) Each non-empty compact convex set, in particular each polytope, is the convex hull of its set E of extreme points (see Roberts and Varberg 1973, p. 84).

Examples of polytopes M and their set E of extreme points are:

(a) M = the simplex $\Delta_n(K) := \{x \in \mathbb{R}^n_+ : \sum_{i=1}^n x_i \leq K\}$, $K \in \mathbb{R}^+$; $E = \{\mathbf{0}, Ke_1, Ke_2, \ldots, Ke_n\}$, where e_i is the i-th unit vector in $\mathbb{R}^n$.
(b) $M = \{x \in \mathbb{R}^n_+ : \sum_{i=1}^n x_i = K\}$, $E = \{Ke_1, Ke_2, \ldots, Ke_n\}$.
(c) M = the n-dimensional interval $\times_1^n [a_i, b_i] \subset \mathbb{R}^n$, $E = \{(\varepsilon_i a_i + (1-\varepsilon_i)b_i)_1^n : \varepsilon_i \in \{0,1\}$ for $1 \leq i \leq n\}$.

D.4.3 (Maximum points of convex functions) Let v be a convex function on a convex set $M \subset \mathbb{R}^n$.

(a) If v is not constant then each maximum point is a boundary point. Thus non-constant convex functions on open convex sets do not have maximum points.
(b) If v is strictly convex then each maximum point is an extreme point. Thus strictly convex functions on convex sets without extreme points do not have maximum points.
(c) If M is the convex hull of an arbitrary set B, then each maximum point b of $v|B$ is also a maximum point of v.
(d) If either M is a polytope or if (M is compact and $v|\partial M$ is continuous), then $v|E$ has a maximum point and each such point is a maximum point of v. Thus $\max v = \max_{x \in E} v(x)$. In particular, if $M = [a, b]$, then a or b is a maximum point, and $\max v = \max\{v(a), v(b)\}$.
(e) The set M^* of maximum points of a strictly convex function v on a compact interval $[x, y]$ equals

$$M^* = \begin{cases} \{x\}, & \text{if } v(x) > v(y), \\ \{y\}, & \text{if } v(x) < v(y), \\ \{x, y\}, & \text{otherwise.} \end{cases}$$

D.5 Convex Functions on Discrete Intervals

D.5.1 (Discrete convexity)

(a) A set M of integers is called a ***discrete interval*** if it has one of the following forms: (i) $M = \{x, x+1, \cdots, y\}$ with $x \leq y$, in particular, singletons in $\mathbb{Z}$; (ii) $M = \{x, x+1, \cdots\}$ for some $x \in \mathbb{Z}$; (iii) $M = \{x, x-1, x-2, \cdots\}$ for some $x \in \mathbb{Z}$; (iv) $M = \mathbb{Z}$. If $|M| \geq 2$ we call M a discrete ***non-degenerate interval.***
(b) A function v on a discrete non-degenerate interval M is called ***discretely convex*** if

$$x \mapsto v(x+1) - v(x) \text{ is increasing on } \{x \in M : x < \sup M\}. \tag{D.4}$$

If $|M| \geq 3$ this is equivalent to

$$v(x) \leq [v(x+1) + v(x-1)]/2 \text{ for } \inf M < x < \sup M. \tag{D.5}$$

Moreover v is called ***strictly discretely convex*** if isotonicity in (D.4) is strict. Note that $x \pm 1 \in M$ if $\inf M < x < \sup M$ and that (D.5) may be written as $v(x+1) - 2v(x) + v(x-1) \geq 0$. Of course, v is called ***[strictly] discretely concave*** if $-v$ is [strictly] discretely convex. If v is both discretely convex and discretely concave, it is called ***discretely affine.*** Each function v on M is discretely affine if $|M| = 2$. The restriction of a convex function on co(M) to M is discretely convex.

(c) Each function v on a discrete non-degenerate interval M has a piecewise ***affine extension*** to a function $\tilde{v}$, defined on the interval co(M) by $\tilde{v}|M := v$ and for x, $x+1 \in M$ by

$$\tilde{v}(\alpha x + (1-\alpha)(x+1)) := \alpha v(x) + (1-\alpha)v(x+1), \quad \alpha \in (0,1).$$

D.5.2 (Discretely convex functions) Let v be a function on a discrete non-degenerate interval M. Then:

(a) v is discretely convex if and only if its affine extension on co(M) is convex if and only if

$$v(\alpha x + (1-\alpha)y) \leq \alpha v(x) + (1-\alpha)v(y)$$

whenever $\alpha \in (0,1)$ and x, y, $\alpha x + (1-\alpha)y \in M$.

(b) v is discretely concave if and only if the function $x \mapsto v(x+k) - v(x)$ is decreasing on its domain $\{y \in M : y+k \in M\}$ for all $k \in \mathbb{N}$ with $k \leq |M| - 2$. This property is called the ***discrete law of decreasing marginal returns.***

(c) The set of discretely convex functions on M is a convex cone.

(d) If v is discretely convex and has a minimum point x^*, then $x \mapsto w(x) := v(x \vee x^*)$ is discretely convex and increasing on M. Moreover for each $x \in M$ we have $w(x) = \min\{v(y) : y \in M, y \geq x\}$, and the point $x \vee x^*$ is a minimum point of v on the set $\{y \in M : y \geq x\}$.

(e) The supremum w^* of a family of discretely convex functions on M is discretely convex if $w^* < \infty$.

(f) Let M be finite and let v be discretely convex on M. Then v trivially has a minimum point, and at most two minimum points if v is strictly discretely convex. Moreover, v assumes its maximum at one of the endpoints of M.

(g) Let v be discretely convex on M. The set of minimum points of v either is empty or a discrete interval. If $|M| \geq 3$ the point x^* with $\inf M < x^* < \sup M$ is a minimum point of v if and only if

$$v(x^*) - v(x^* - 1) \leq 0 \leq v(x^* + 1) - v(x^*).$$

This holds if and only if v is decreasing left of x^* and increasing right of x^*. If M has the left endpoint $a \in \mathbb{Z}$, then a is a minimum point of v if and only if $v(a) \leq v(a+1)$ if and only if v is increasing. A similar fact holds for a right endpoint of M.

(h) Let v be discretely convex on each of two discrete non-degenerate intervals I and J. Then:

If $|I \cap J| \geq 2$, then v is discretely convex on $I \cup J$.

If $|I \cap J| = \{b\}$ for some integer b, then v is discretely convex on $I \cup J$ if and only if $v(b+1) - v(b) \geq v(b) - v(b-1)$.

(i) Part (a) and (b) of Appendix D.2.14 about the preservation of convexity by integration remain true when there M is a discrete non-degenerate interval, when convex is replaced by discrete convexity and when using $g'_+(x, \omega) := g(x+1, \omega) - g(x, \omega)$ and $g'_-(x, \omega) := g(x, \omega) - g(x-1, \omega)$.

D.5.3 (The composition of discretely convex/concave functions) Consider the composition $v \circ g$ of a function v on a discrete non-degenerate interval I and a function g from a discrete non-degenerate interval M into I. The following hold:

(a) If v is discretely convex then $v \circ g$ is discretely convex if (g is discretely convex and v is increasing) $\vee$ (g is discretely concave and v is decreasing) $\vee$ (g is discretely affine).
(b) If v is discretely concave then $v \circ g$ is discretely concave if (g is discretely concave and v is increasing) $\vee$ (g is discretely convex and v is decreasing) $\vee$ (g is discretely affine).

D.6 Vector-Valued Convex Mappings

D.6.1 (Concavity with respect to a relation) We generalize Appendix D.2.10(a) in Appendix D.6.2, as follows. Let the composition $v \circ g$ of a function v on and g be a mapping from a convex set $M \subset \mathbb{R}^m$ into a convex set $K \subset \mathbb{R}^k$. Let $\preceq_K$ be a relation on K; M need not be structured. The mapping g is called ***[strictly] $\preceq_K$-concave*** if $x \neq y \in M$ and $\alpha \in (0, 1)$ implies

$$\alpha g(x) + (1-\alpha) g(y) \; [\prec_K] \; \preceq_K g(\alpha x + (1-\alpha) y). \tag{D.6}$$

(Recall that $u \prec_K z := (u \preceq_K z) \wedge (u \neq z)$ for $u, z \in K$.) Moreover, we say g ***[strictly] $\preceq_K$-convex*** if (D.6) holds with $\preceq_K$ replaced by $\succeq_K$.

Special cases are:

(a) g is $\leq_{\text{diag}}$-convex if and only if g is affine if and only if g is $\leq_{\text{diag}}$-concave.
(b) Let $K = \times_1^k K_i$ and let $\preceq_K$ be the product of relations $\preceq_i$ on K_i, $1 \leq i \leq k$. Then:

  (b1) $g = (g_i)_1^k$ is $\preceq_K$-concave if and only if $g_i: M \to \mathbb{R}$ is $\preceq_i$-concave for all i.
  (b2) If $k = 1$ and if $\preceq_K$ equals $\leq^1$ or the inverse of $\leq^1$ or the diagonal ordering, then g is $\preceq_K$-concave if and only if it is concave or convex or affine, respectively.
  (b3) If $I, J \subset \mathbb{N}_k$ and if $\preceq_i := \leq^1$ for $i \in I$ and $\preceq_i :=$ the inverse of $\leq^1$ for $i \in J$, then v is $\preceq_K$-concave if and only if g_i is concave for all $i \in I$ and convex for all $i \in J$ (hence affine for $i \in I \cap J$).

D.6.2 (Composition of vector-valued functions) Consider the composition $v \circ g$ of a function v on a convex set $K \subset \mathbb{R}^k$ and a mapping g from a convex set $M \subset \mathbb{R}^m$ into K. Let $\preceq_K$ be a relation on K. Let g be $\preceq_K$-concave [$\preceq_K$-convex] and let v be concave [convex] and $\preceq_K$-increasing. Then:

(a) $v \circ g \colon M \to \mathbb{R}$ is concave [convex].
(b) $v \circ g$ is even strictly concave [strictly convex] under each of the following additional conditions:

(b1) v is strictly $\preceq_K$-increasing and g is strictly $\preceq_K$-concave [strictly $\preceq_K$-convex].
(b2) v is strictly concave [strictly convex] and g is injective.

Index of Appendix

K. Hinderer et al., *Dynamic Optimization*, Universitext,
DOI 10.1007/978-3-319-48814-1

References

Adomian, G., & Lee, E. S. (1986). The research contributions of Richard Bellman. *Computers & Mathematics with Applications Series A, 12*, 633–651.

Ahlswede, R., & Wegener, I. (1979). *Suchprobleme*. Stuttgart: B. G. Teubner.

Almudevar, A. (2014). *Approximate iterative algorithms*. Leiden: CRC Press/Balkema.

Altman, E. (1999). *Constrained Markov decision processes* (Stochastic Modeling). Boca Raton: Chapman & Hall/CRC.

Aoki, N. (Ed.) (1989). *The study of dynamical systems* (World Scientific Advanced Series in Dynamical Systems, Vol. 7). Teaneck: World Scientific Publishing Co., Inc.

Ash, R. B. (1972). *Real analysis and probability*. New York/London: Academic Press.

Bäuerle, N., & Rieder, U. (2011). *Markov decision processes with applications to finance* (Universitext). Heidelberg: Springer.

Beckmann, M. J. (1968). *Dynamic programming of economic decisions*. New York: Springer.

Bellman, R. (1957). *Dynamic programming*. Princeton: Princeton University Press.

Berry, D. A., & Fristedt, B. (1985). *Bandit problems*. London: Chapman & Hall.

Bertsekas, D. P. (1975). Convergence of discretization procedures in dynamic programming. *IEEE Transactions Automatic Control, AC-20*, 415–419.

Bertsekas, D. P. (1976). *Dynamic programming and stochastic control*. New York: Academic Press.

Bertsekas, D. P. (1987). *Dynamic programming*. Englewood Cliffs: Prentice Hall Inc.

Bertsekas, D. P. (1995). *Dynamic programming and optimal control*. Belmont: Athena Scientific.

Bertsekas, D. P. (2001). *Dynamic programming and optimal control* (Vol. II, 2nd ed.). Belmont: Athena Scientific.

Bertsekas, D. P. (2005). *Dynamic programming and optimal control* (Vol. I, 3rd ed.). Belmont: Athena Scientific.

Bertsekas, D. P., & Shreve, S. E. (1978). *Stochastic optimal control*. New York: Academic Press.

Bertsekas, D. P., & Shreve, S. E. (1979). Existence of optimal stationary policies in deterministic optimal control. *Journal of Mathematical Analysis and Applications, 69*, 607–620.

Bertsekas, D. P., & Tsitsiklis, J. N. (1996). *Neuro-dynamic programming*. Belmont: Athena Scientific.

Blackwell, D. (1962). Discrete dynamic programming. *The Annals of Mathematical Statistics, 33*, 719–726.

Blackwell, D. (1965). Discounted dynamic programming. *The Annals of Mathematical Statistics, 36*, 226–235.

Cairns, S. S. (1963). Balance scale sorting. *The American Mathematical Monthly, 70*, 136–148.

Chang, H. S., Fu, M. C., Hu, J., & Marcus, S. I. (2007). *Simulation-based algorithms for Markov decision processes*. London: Springer.

K. Hinderer et al., *Dynamic Optimization*, Universitext,
DOI 10.1007/978-3-319-48814-1

Coolidge, J. L. (1909). The gambler's ruin. *Annals of Mathematics (2), 10*, 181–192.

Davis, M. H. A., & Vinter, R. B. (1985). *Stochastic modelling and control*. London: Chapman & Hall.

DeGroot, M. H. (1970). *Optimal statistical decisions*. New York: McGraw-Hill.

Denardo, E. V. (2003). *Dynamic programming*. Mineola: Dover Publications Inc.

Dennis, Jr., J. E., & Schnabel, R. B. (1983). *Numerical methods for unconstrained optimization and nonlinear equations*. Englewood Cliffs: Prentice Hall Inc.

Dieudonné, J. (1960). *Foundations of modern analysis*. New York: Academic Press.

Dijkstra, E. W. (1959). A note on two problems in connexion with graphs. *Numerische Mathematik, 1*, 269–271.

Dreyfus, S. E., & Law, A. M. (1977). *The art and theory of dynamic programming*. New York: Academic Press.

Dubins, L. E., & Savage, L. J. (1965). *How to gamble if you must. Inequalities for stochastic processes*. New York: McGraw-Hill.

Dubins, L. E., & Savage, L. J. (1976). *Inequalities for stochastic processes (how to gamble if you must)*. New York: Dover Publications Inc. Corrected republication of the 1965 edition.

Dudley, R. M. (1989). *Real analysis and probability*. Pacific Grove: Wadsworth & Brooks/Cole Advanced Books & Software.

Dvoretzky, A., Kiefer, J., & Wolfowitz, J. (1952a). The inventory problem. I. Case of known distributions of demand. *Econometrica, 20*, 187–222.

Dvoretzky, A., Kiefer, J., & Wolfowitz, J. (1952b). The inventory problem. II. Case of unknown distributions of demand. *Econometrica, 20*, 450–466.

Dynkin, E. B. (1965). *Markov processes* (Vols. I, II). New York: Academic Press.

Dynkin, E. B., & Yushkevich, A. A. (1979). *Controlled Markov processes*. Berlin: Springer.

Filar, J., & Vrieze, K. (1997). *Competitive Markov decision processes*. New York: Springer.

Fletcher, R. (1987). *Practical methods of optimization* (2nd ed.). Chichester: Wiley.

Foulds, L. R. (1981). *Optimization techniques*. New York: Springer.

Gani, J. (1984). Obituary: Richard Bellman. *Journal of Applied Probability, 21*, 935–936.

Girlich, H.-J., Köchel, P., & Küenle, H.-U. (1990). *Steuerung dynamischer Systeme*. Leipzig: Fachbuchverlag.

Gondran, M., & Minoux, M. (1984). *Graphs and algorithms*. Chichester: Wiley.

Göpfert et al. (Eds.). (1986). *Lexikon der Optimierung*. Optimierung und Optimale Steuerung. Berlin: Akademie-Verlag.

Hakansson, N. H. (1970/1971). Optimal entrepreneurial decisions in a completely stochastic environment. *Management Science, 17*, 427–449.

van Hee, K. M. (1978). *Bayesian control of Markov chains* (Mathematical Centre Tracts, Vol. 95). Amsterdam: Mathematisch Centrum.

Hernández-Lerma, O. (1989). *Adaptive Markov control processes*. New York: Springer.

Hernández-Lerma, O., & Lasserre, J. B. (1996). *Discrete-time Markov control processes*. New York: Springer.

Hernández-Lerma, O., & Lasserre, J. B. (1999). *Further topics on discrete-time Markov control processes*. New York: Springer.

Heyman, D., & Sobel, M. (1984). *Stochastic models in operations research: Stochastic optimization*. New York: McGraw-Hill.

Hinderer, K. (1970). *Foundations of non-stationary dynamic programming with discrete time parameter* (Lecture Notes in Operations Research and Mathematical Systems, Vol. 33). Berlin: Springer.

Hinderer, K. (1971). Instationäre dynamische Optimierung bei schwachen Voraussetzungen über die Gewinnfunktionen. *Abhandlungen aus dem Mathematischen Seminar University of Hamburg, 36*, 208–223.

Hinderer, K. (1976). Estimates for finite-stage dynamic programs. *Journal of Mathematical Analysis and Applications, 55*, 207–238.

Hinderer, K. (1977). Does the value iteration for finite-stage discrete dynamic programs hold "always"? In *Dynamische optimierung* (pp. 41–56). Bonner Math. Schriften, No. 98.

Hinderer, K. (1985). On the structure of solutions of stochastic dynamic programs. In *Proceedings of the seventh conference on probability theory (Braşov, 1982)* (pp. 173–182). Utrecht: VNU Science Press.

Hinderer, K., & Hübner, G. (1977). An improvement of J. F. Shapiro's turnpike theorem for the horizon of finite stage discrete dynamic programs. In *Transactions of the Seventh Prague Conference on Information Theory, Statistical Decision Functions, Random Processes* (Vol. A, pp. 245–255). Dordrecht: Reidel.

Hinderer, K., & Stieglitz, M. (1996). Increasing and Lipschitz continuous minimizers in one-dimensional linear-convex systems without constraints: The continuous and the discrete case. *Mathematical Methods of Operations Research, 44*, 189–204.

Hordijk, A. (1974). *Dynamic programming and Markov potential theory*. Amsterdam: Mathematisch Centrum.

Horst, R., & Tuy, H. (1990). *Global optimization*. Berlin: Springer.

Hotelling, H. (1931). Recent publications: Reviews: Mathematical introduction to economics. *American Mathematical Monthly, 38*, 101–103.

Howard, G. T., & Nemhauser, G. L. (1968). Optimal capacity expansion. *Naval Research Logistics Quarterly, 15*, 535–550.

Howard, R. A. (1960). *Dynamic programming and Markov processes*. Cambridge: Technology Press of Massachusetts Institute of Technology.

Hübner, G. (1980). Bounds and good policies in stationary finite-stage Markovian decision problems. *Advances in Applied Probability, 12*, 154–173.

Ibaraki, T., & Katoh, N. (1988). *Resource allocation problems*. Cambridge: MIT Press.

Iyengar, G. N. (2005). Robust dynamic programming. *Mathematics of Operations Research, 30*, 257–280.

Jungnickel, D. (1987). On a theorem of Ganley. *Graphs and Combinatorics, 3*, 141–143.

Karlin, S. (1955). The structure of dynamic programming models. *Naval Research Logistics Quarterly, 2*, 285–294 (1956).

Karlin, S., & Rinott, Y. (1980). Classes of orderings of measures and related correlation inequalities. II. Multivariate reverse rule distributions. *Journal of Multivariate Analysis, 10*, 499–516.

Kuratowski, K., & Ryll-Nardzewski, C. (1965). A general theorem on selectors. *Bulletin de L'Académie Polonaise Des Sciences: Série des Sciences Mathématiques, Astronomiques et Physiques, 13*, 397–403.

Lew, A. (1986). Richard Bellman's contributions to computer science. *Journal of Mathematical Analysis and Applications, 119*, 90–96.

Lippman, S. A. (1974/1975). On dynamic programming with unbounded rewards. *Management Science, 21*, 1225–1233.

Maitra, A. P., & Sudderth, W. D. (1996). *Discrete gambling and stochastic games*. New York: Springer.

Martello, S., & Toth, P. (1990). *Knapsack problems*. Chichester: Wiley.

Morin, T. L. (1978). Computational advances in dynamic programming. In *Dynamic programming and its applications* (pp. 53–90). New York: Academic Press.

Nemhauser, G. L., Rinnooy Kan, A. H. G., & Todd, M. J. (Eds.). (1989). *Optimization* (Handbooks in Operations Research and Management Science, Vol. 1). Amsterdam: North-Holland Publishing Co.

Nemhauser, G. L., & Wolsey, L. A. (1988). *Integer and combinatorial optimization*. New York: Wiley.

Neumann, K., & Morlock, M. (1993). *Operations research*. Munich: Carl Hanser Verlag.

Nilim, A., & El Ghaoui, L. (2005). Robust control of Markov decision processes with uncertain transition matrices. *Operations Research, 53*, 780–798.

Peressini, A. L., Sullivan, F. E., & Uhl, Jr., J. J. (1988). *The mathematics of nonlinear programming*. New York: Springer.

Porteus, E. L. (1975). An informal look at the principle of optimality. *Management Science, 21*, 1346–1348.

Porteus, E. L. (1971/1972). Some bounds for discounted sequential decision processes. *Management Science, 18*, 7–11.

Powell, W. B. (2007). *Approximate dynamic programming*. New York: Wiley.

Presman, È. L., & Sonin, I. N. (1990). *Sequential control with incomplete information*. San Diego: Academic Press.

Puterman, M. L. (1994). *Markov decision processes: Discrete stochastic dynamic programming*. New York: Wiley.

Rao, S. B. (1984). Characterizations of harmonious marked graphs and consistent nets. *Journal of Combinatorics, Information & System Sciences, 9*, 97–112.

Rieder, U. (1975a). Bayesian dynamic programming. *Advances in Application Probability, 7*, 330–348.

Rieder, U. (1975b). On stopped decision processes with discrete time parameter. *Stochastic Processes Application, 3*, 365–383.

Rieder, U. (1978). Measurable selection theorems for optimization problems. *Manuscripta Mathematica, 24*, 115–131.

Rieder, U. (1988). *Bayessche Kontrollmodelle: Vorlesung von Prof. Dr. Ulrich Rieder im WS 1987/88 an der Universität Ulm*.

Rieder, U. (1991). Structural results for partially observed control models. *Zeitschrift für Operations Research, 35*, 473–490.

Roberts, A. W., & Varberg, D. E. (1973). *Convex functions*. New York/London: Academic Press.

Rockafellar, R. T. (1970). *Convex analysis*. Princeton: Princeton University Press.

Ross, S. M. (1970). *Applied probability models with optimization applications*. San Francisco: Holden-Day.

Ross, S. M. (1983). *Introduction to stochastic dynamic programming* (Probability and Mathematical Statistics). New York: Academic Press.

van Roy, B. (2002). Neuro-dynamic programming: Overview and recent trends. In E. Feinberg & Shwartz, A. (Eds.) *Handbook of Markov decision processes: Methods and applications*. Boston: Springer.

Schäl, M. (1975). Conditions for optimality in dynamic programming and for the limit of n-stage optimal policies to be optimal. *Z. Wahrscheinlichkeitstheorie und Verw. Gebiete, 32*, 179–196.

Schäl, M. (1990). *Markoffsche Entscheidungsprozesse*. Stuttgart: B. G. Teubner.

Schneeweiss, H. (1974). *Ökonometrie*. Würzburg: Physica-Verlag.

Sennott, L. I. (1999). *Stochastic dynamic programming and the control of queueing systems*. New York: Wiley.

Sniedovich, M. (1992). *Dynamic programming*. New York: Marcel Dekker Inc.

Stoer, J., & Witzgall, C. (1970). *Convexity and optimization in finite dimensions. I*. New York/Berlin: Springer.

Sutton, R. S., & Barto, A. G. (1998). *Reinforcement learning*. Cambridge: The MIT Press.

Tsitsiklis, J. N. (1986). A lemma on the multiarmed bandit problem. *IEEE Transactions on Automatic Control, 31*, 576–577.

Valentine, F. A. (1964). *Convex sets*. New York: McGraw-Hill.

Veinott, Jr., A. F. (1969). Discrete dynamic programming with sensitive discount optimality criteria. *The Annals of Mathematical Statistics, 40*, 1635–1660.

Wald, A. (1947). Foundations of a general theory of sequential decision functions. *Econometrica, 15*, 279–313.

Weissinger, J. (1952). Zur Theorie und Anwendung des Iterationsverfahrens. *Mathematische Nachrichten, 8*, 193–212.

Wessels, J. (1977). Stopping times and Markov programming. In *Transactions of the seventh Prague conference on information theory, statistical decision functions, random P* (pp. 575–585). Dordrecht: Reidel.

Whittle, P. (1982). *Optimization over time* (Vol. I). Chichester: Wiley.
Wilansky, A. (1970). *Topology for analysis*. Toronto/London: Waltham.
Yakowitz, S. J. (1969). *Mathematics of adaptive control processes*. New York: American Elsevier Publishing Co., Inc.

List of the Most Important Examples

K. Hinderer et al., *Dynamic Optimization*, Universitext,
DOI 10.1007/978-3-319-48814-1

Index

K. Hinderer et al., *Dynamic Optimization*, Universitext,
DOI 10.1007/978-3-319-48814-1

MIX
Papier aus verantwortungsvollen Quellen
Paper from responsible sources
FSC® C105338

If you have any concerns about our products,
you can contact us on
ProductSafety@springernature.com

In case Publisher is established outside the EU,
the EU authorized representative is:
Springer Nature Customer Service Center GmbH
Europaplatz 3, 69115 Heidelberg, Germany

Printed by Libri Plureos GmbH
in Hamburg, Germany